W9-CRX-678

Ices in the Solar System

NATO ASI Series

Advanced Science Institutes Series

A series presenting the results of activities sponsored by the NATO Science Committee, which aims at the dissemination of advanced scientific and technological knowledge, with a view to strengthening links between scientific communities.

The series is published by an international board of publishers in conjunction with the NATO Scientific Affairs Division

A	Life Sciences	Plenum Publishing Corporation
B	Physics	London and New York
C	Mathematical and Physical Sciences	D. Reidel Publishing Company Dordrecht, Boston and Lancaster
D	Behavioural and Social Sciences	Martinus Nijhoff Publishers
E	Engineering and Materials Sciences	The Hague, Boston and Lancaster
F	Computer and Systems Sciences	Springer-Verlag
G	Ecological Sciences	Berlin, Heidelberg, New York and Tokyo

Series C: Mathematical and Physical Sciences Vol. 156

Ices in the Solar System

edited by

Jürgen Klinger
Laboratoire de Glaciologie du CNRS,
Saint-Martin d'Heres, France

Daniel Benest
Observatoire de Nice, Nice, France

Audouin Dollfus
Observatoire de Meudon, Meudon, France

Roman Smoluchowski
Astronomy Department, University of Texas at Austin,
Austin, Texas, U.S.A.

D. Reidel Publishing Company

Dordrecht / Boston / Lancaster

Published in cooperation with NATO Scientific Affairs Division

Proceedings of the NATO Advanced Research Workshop on
Ices in the Solar System
Nice, France
January 16-19, 1984

Library of Congress Cataloging in Publication Data

NATO Advanced Research Workshop on Ices in the Solar System (1984 : Nice, France)
    Ices in the solar system.

    (NATO ASI series, Series C, Mathematical and physical sciences ; v. 156)
    Proceedings of the NATO Advanced Research Workshop on Ices in the Solar System,
Nice, France, January 16—19, 1984.
    Published in cooperation with NATO Scientific Affairs Division.
    Includes indexes.
        1. Ice—Congresses. 2. Solar system—Congresses. 3. Astrophysics—Congresses.
4. Cosmochemistry—Congresses. I. Benest, Daniel. II. North Atlantic Treaty
Organization. Scientific Affairs Division. III. Title. IV. Series.
QB505.N38   1984   523.2   85-10777
ISBN 90-277-2062-2

Published by D. Reidel Publishing Company
P.O. Box 17, 3300 AA Dordrecht, Holland

Sold and distributed in the U.S.A. and Canada
by Kluwer Academic Publishers,
190 Old Derby Street, Hingham, MA 02043, U.S.A.

In all other countries, sold and distributed
by Kluwer Academic Publishers Group,
P.O. Box 322, 3300 AH Dordrecht, Holland

D. Reidel Publishing Company is a member of the Kluwer Academic Publishers Group

TABLE OF CONTENTS

FOREWORD

Audouin Dollfus

Observatoire de Paris,
Section de Meudon, 92195 Meudon, FRANCE

The North Atlantic Treaty Organization (NATO) and, in particular, its Department of Scientific Affairs headed by Dr. C. Sinclair, actively supports new fields of science. The recent exploration of the outer parts of the Solar System by spacecraft focused the attention of a large community of scientists on the problem of ices, which play a major role in the accretionary processes in space except for the close neighborhood of the Sun and of other stars. NATO responded to this new interest by agreeing to sponsor an Advanced Research Workshop "Ices in the Solar System", provided a proper organizing body could be set up.

It was a pleasure to organize such a workshop jointly with Professor Roman Smoluchowski who had earlier organized similar conferences. I knew from the experience of others who managed such meetings in the past that there would be much work, but the opportunity of cooperating with Smoluchowski was very attractive and convinced me to agree. If well organized, the whole project promised to be more than rewarding for a large community of scientists, both in the short run and in the long run, by clarifying certain outstanding questions in astrophysics. It became clear that a well-organized international conference would attract top scientists and help unravel many fundamental problems.

The undertaking became particularly attractive when Jürgen Klinger, of the Laboratoire de Glaciologie du CNRS in Grenoble, agreed to join us in the project and offered to devote much of his time. In addition, his participation made it possible to hold the conference in France in accord with NATO's wishes. Klinger added Daniel Benest of the Observatoire de Nice, which was an excellent choice in view of our plans to organize the conference in that part of the country.

Finally, Mrs. Jacqueline Advielle, from Observatoire de Meudon, joined the group, in order to share with Benest and Klinger the task of organizing all the details of the conference which was held in Nice, from 16 to 19 January 1984, with 60 speakers and many other participants from 12 countries.

The subject matter of the conference was divided into nine sessions, each organized by an expert in the particular area:

Session I, Remote Sensing of Ices, a fast moving and difficult field was headed by Thomas McCord. Session II, Physics of Ice, and Session III, Impacts of Ice, were in the hands of E. Whalley and of T. Ahrens respectively. Cosmochemistry under T. Yamamoto and Interplanetary Particles with Mayo Greenberg formed interrelated and complementary Sessions IV and V. Comets, Session VI, a basic and expanding field with many speakers, was put together by Armand Delsemme. For Session VII, Ice on Mars, the review papers were selected by Philippe Masson. Large Icy Satellites, Session VIII, was headed by David Morrison, and finally the last Session IX, Small Satellites and Rings, was expertly organized by Joseph Burns.

The classical Conclusions of the Conference, which so often result in a classical failure, were most successfully presented by Smoluchowski. I, for one, benefitted greatly from this synthesis of the results which emerged from the conference.

After the conference, the attention of the organizers turned to the book because the real impact of a workshop depends on the quality of the ensuing publication. It was felt that a simple collection of submitted papers in the form of Proceedings was not the right solution and that, in fact, the papers should be considered just a means for the preparation of a book. Thus, the nine sessions were reorganized and combined into five chapters, each topic including relevant review papers, usually but not always derived from the invited papers of the conference. Other specially requested papers written by experts were added.

Each paper of the book was carefully reviewed by referees chosen by the chairmen of the conference sessions. Very useful to the book were the discussions recorded during the conference; the reader should not miss the Introduction by Klinger and the Summary of the Highlights of the Conference by Smoluchowski.

The onerous editorial tasks were shared by Benest and Klinger who acted as Editor-in-Chief. The generous and efficient support by NATO is gratefully acknowledged.

LIST OF PARTICIPANTS

Dr. A'HEARN Michael F.
  Astronomy Program
  University of Maryland
  College Park, Md 20742   U. S. A.

Dr. AHRENS Thomas J.
  Seismological Laboratory 252-21
  California Institute of Technology
  Pasadena, CA 91125   U. S. A.

Dr. ANDERSON Duwayne
  Associate Provost for Research
  Texas A & M University
  College station - Texas 77843   U. S. A.

Dr. BAR-NUN Akiva
  Department of Geophysics and Planetary
  Sciences - Tel Aviv University
  Ramat Aviv - 69978   ISRAEL

Dr. BATTISTINI René
  86, rue G. Flaubert
  45100 - Orléans   FRANCE

Dr. BENEST Daniel
  Observatoire de Nice
  B. P. n° 139
  06003 - Nice Cedex   FRANCE

Dr. BIANCHI Remo
  I. A. S. Reparto di Planetologia
  Viale Dell'Universita 11
  00185 - Roma   ITALY

Dr. BORDERIES Nicole
  CNES/BGI
  18, avenue Edouard Bélin
  31055 - Toulouse Cedex   FRANCE

Dr. BROWN Robert H.
  Planetary Geosciences Div.
  Hawaii Institute of Geophysics
  2525 Correa Road
  Honolulu - Hawai 96822   U. S. A.

Dr. BURNS Joseph A.
      Thurston Hall
      Cornell University
      Ithaca NY 14853   U. S. A.

Dr. CAILLEUX André
      9, avenue de la Trémouille
      94100 - Saint-Maur   FRANCE

Dr. CAMPINS Humberto
      Astronomy Program
      University of Maryland
      College Park, Maryland 20742   U. S. A.

Dr. CASACCHIA Ruggero
      I. A. S. Reparto di Planetologia
      Viale Dell'Universita 11
      00185 - Roma   ITALY

Dr. COSMOVICI C. B.
      Istituto Fisica Spazio Interplanetario
      C. N. R.
      00044 - Frascati   ITALY

Dr. CRIFO J. F.
      L. P. S. P. - C. N. R. S.
      B. P. n° 10
      91370 - Verrières-le-Buisson   FRANCE

Dr. CRUIKSHANK Dale P.
      Institute for Astronomy
      2680 Woodlawn Dr.
      Honolulu. HI 96822   U. S. A.

Dr. DELSEMME Armand H.
      Department of Physics and Astronomy
      The University of Toledo
      Toledo, Ohio 43606   U. S. A.

Dr. DESCHAMPS Marc
      Laboratoire de Physique du Système solaire
      Observatoire de Meudon
      92190 - Meudon   FRANCE

Dr. DOLLFUS Audouin
      Laboratoire de Physique du Système solaire
      Observatoire de Meudon
      92190 - Meudon   FRANCE

Dr. EMERICH Claude
         L. P. S. P. - C. N. R. S.
         B. P. n° 10
         91370 - Verrière-le-Buisson   FRANCE

Dr. FANALE Frazer P.
         Hawaii Institute of Geophysics
         University of Hawaii
         Honolulu, HI 96822   U. S. A.

Dr. FARINELLA Paolo
         Dipartimento di Matematica
         Universita di Pisa
         Piazza Dei Cavalieri 2
         56100 - Pisa   ITALY

Dr. FECHTIG Hugo
         Max-Planck Institut für Kernphysik
         Postfach 10 30 80
         6900 Heidelberg 1   F. R. G.

Dr. FELDMAN Paul D.
         Physics Department
         John Hopkins University
         Baltimore, Md. 21218   U. S. A.

Dr. FORNI Olivier
         Laboratoire de géologie dynamique interne
         Université Paris IX
         91405 ORSAY Cedex   FRANCE

Dr. FOTI Gaetano
         Istituto di Fisica
         Corso Italia 57
         95125 - Catania   ITALY

Dr. FROESCHLE Claude
         Observatoire de Nice
         B. P. n° 139
         06003 - Nice Cedex   FRANCE

Dr. FULCHIGNONI Marcello
         Istituto Astrofisica Spaziale
         CNR - CP 67
         00044 - Frascati   ITALY

Dr. GAFFNEY Edward S.
         Los Alamos National Lab. MS F665
         Los Alamos, NM 87545   U. S. A.

Dr. GREENBERG J. Mayo
        Leiden University
        Huygens Laboratory
        Wassenaarseweg 78
        2300 RA - Leiden   NEDERLAND

Dr. GREGORI Giovanni P;
        Istituto di Fisica dell'Atmosfera
        Sturzo 31
        0144 - Roma   ITALY

Dr. GRUN Eberhard
        Max-Planck Institut für Kernphysik
        6900 Heidelberg 1   F. R. G.

Dr. HARTMANN William K.
        Planetary Science Institute
        2030 E. Speedway
        Tucson, Arizona 85719   U. S. A.

Dr. JOHNSON Robert E.
        University of Virginia
        Charlottesville - VA 22901   U. S. A.

Dr. KIRBY Stephen H.
        U. S. Geological Survey MS/77
        345 Middlefield Road
        Menlo Park, California 94025   U. S. A.

Dr. KLINGER Jurgen
        Laboratoire de Glaciologie
        B. P. n° 96
        38402 - Saint-Martin d'Hères Cedex   FRANCE

Dr. KRÄTSCHMER Wolfgang
        Max Planck Institut für Kernphysik
        p. o. Box 10 39 80
        6900 Heidelberg 1   F. R. G.

Dr. LAMY Philippe
        Laboratoire d'Astronomie Spatiale
        Traverse du Siphon - Les Trois Lucs
        13012 - Marseille   FRANCE

Dr. LANZEROTTI L.
        AT & T Bell Laboratories
        Murray Hill, N. J. 07977 - 74   U. S. A.

Dr. LEGER Alain
        GPS tour 23
        2, place Jussieu
        75251 - Paris Cedex 05   FRANCE

Dr. LEVASSEUR-REGOURD A. CH.
        Service d'Aéronomie du C. N. R. S.
        B. P. n° 3
        91370 - Verrière-le-Buisson   FRANCE

Dr. LUCCHITTA Baerbel
        U. S. Geological Survey
        2255 North Gemini Drive
        Flagstaff, Arizona 86001   U. S. A.

Dr. LUNINE Jonathan
        California Institute of Technology
        170-25 Caltech
        Pasadena - CA 91125   FRANCE

Dr. LUCEY
        Hawaii Inst. of Geophysics
        Planetary Geosciences Div.
        2525 Correa Rd
        Honolulu HI 96822   U. S. A.

Dr. MACKINNON William
        Department Earth and Planetary Sciences
        Washington University
        Saint-Louis, MO 63130   U. S. A.

Dr. MASSON Philippe
        Université Paris-Sud - Bat 509
        Laboratoire de Géologie Dynamique interne
        91405 - Orsay Cedex   FRANCE

Dr. MAYER Erwin
        Institut für Anorganische und Analytische
        Chemie, Innrain 52 a, Universitat Innsbruck
        6020 - Innsbruck   AUSTRIA

Dr. MENDIS Asoka
        Dept. of Applied Physics Information
        Science
        University of California
        San Diego - La Jolla CA 92093   U. S. A.

Dr. MICHARD Raymond
        Observatoire de Nice
        B. P. n° 139
        06003 - Nice Cedex   FRANCE

Dr. MIGNARD François
        C. E. R. G. A.
        Avenue   Copernic
        06130 - Grasse   FRANCE

Dr. MILLER Stanley
        Department of Chemistry B-017
        University of California, San Diego
        La Jolla, Ca. 92093   U. S. A.

Dr. MORRISON David
        Institute for Astronomy
        2680 Woodlawn Dr.
        Honolulu, Hawaii 96822   U. S. A.

Dr. NICOL Malcolm
        Department of Chemistry and Biochemistry
        University of California
        Los Angeles, CA 40024   U. S. A.

Dr. OWEN Tobias
        ESS/SUNY
        State University of New-York
        Stony Brook - N. Y. 11794   U. S. A.

Dr. PICCIOTTO Edgard
        Faculté des Sciences
        Université Libre de Bruxelles
        50, avenue F. Roosevelt
        1050 - Bruxelles   BELGIUM

Dr. PILCHER Carl
        Institut für Astronomie
        Türkenschanzstr. 17
        1180 Wien   AUSTRIA

Dr. PIRRONELLO Valerio
        Istituto di Fisica Facoltadi
        Ingegneria - Corso delle Provincie 47
        95129 - Catania   ITALY

Dr. PLESCIA Jeff
        Mail Stop 183/501 - Jet Propulsion
        Laboratory
        Pasadena, CA 91109   U. S. A.

Dr. PLETZER Rudolf
        Institut für Anorganische und Analytische
        Chemie, Innrain 52 a, Universität Innsbruck
        6020 Innsbruck   AUSTRIA

Dr. POIRIER Jean-Paul
        Institut de Physique du Globe de Paris
        4, place Jussieu - Tour 14
        75230 - Paris Cedex 05   FRANCE

Dr. REISSE Jacques
        Université Libre de Bruxelles
        Chimie Organique C. P. 165
        Avenue F. Roosevelt, 50
        1050 - Bruxelles   BELGIUM

Dr. REMY Frédérique
        C. E. R. G. A.
        Avenue   Copernic
        06130 - Grasse   FRANCE

Dr. RICKMAN Hans
        Astronomiska Observatoriet
        Box 515
        75120 - Uppsala   SWEDEN

Dr. SCHWEHM Gerhard H.
        European Space Operations Centre
        OAD
        Robert-Bosch.Str. 5
        6100 - Darmstadt   F. R. G.

Dr. SMOLUCHOWSKI Roman
        Astronomy Dep. RLM 15.308
        University of Texas
        Austin, TX 78712 - 1083   U. S. A.

Dr. SMYTHE William D.
        Jet Propulsion Laboratory 183/301
        4800 Oak Grove Drive
        Pasadena, Ca 91109   U. S. A.

Dr. SOTIN Christophe
        Laboratoire Physique et Mécanique
        des Matériaux Terrestres
        Institut de Physique du Globe
        Tour 14-15 - 3ème étage
        4, place Jussieu
        75230 - Paris Cedex 05   FRANCE

Dr. SQUYRES
  MS 245/3
  NASA Ames Research Center
  Moffett Field CA 94040   U. S. A.

Dr. STRAZZULLA Giovanni
  Osservatorio Astrofisico
  Citta Universitaria
  95125 - Catania   ITALY

Dr. THOMAS Pierre
  Laboratoire de Géologie Dynamique interne
  Université Paris XI - Bat 509
  91405 - Orsay - Cedex   FRANCE

Dr. TRAPPENIERS Nestor
  Van der Waals Laboratory, University of
  Amsterdam, Valckenierstraat 67
  XE Amsterdam - 1018   NEDERLAND

Dr. WEISSMAN Paul
  Jet Propulsion Laboratory
  183-301
  4800 Oak Grove Drive
  Pasadena CA 91109   U. S. A.

Dr. WHALLEY Edward
  Division of Chemistry National Research
  Council
  Ottawa - KIA OR9   CANADA

Dr. WHIPPLE Fred L.
  Smithsonian Astrophysical Observatory
  60 Garden Street
  Cambridge, Massachusetts 02138   U. S. A.

Dr. WRIGHT Ian
  Dept. of Earth Sciences
  The open University
  Walton Hall
  Milton Keynes MK7 6AA   U. K.

INTRODUCTION : Scientific Background

J. KLINGER

Laboratoire de Glaciologie et Géophysique
de l'Environnement
B.P.96, 38402 - St Martin d'Hères Cedex - France

According to a definition given by Whalley (1), what astrophysicists usually call "ices" are moderately volatile substances such as N_2, CH_4, CO, CO_2, NH_3, SO_2 and H_2O. One can add to this list solid argon, krypton and xenon and the hydrates of a number of the above mentioned compounds and elements.

Since Laplace several authors suggested that solid moderately volatile substances are present in the solar system. Laplace himself explained the cometary phenomena by the evaporation of some volatiles (2). At the beginning of this century the austrian engineer Hanns Hörbiger tried to explain the formation of the solar system by agglomeration of pieces of ice (3) and among others, he believed that comets are made of ice. Unfortunately, he made other statements such as, for instance, that the moon is ice covered and that hailstones are of cosmic origine, so that the few good ideas he had have never been seriously discussed.

In the early fifties, Whipple (4) proposed his, now generally accepted, icy conglomerate model for comets. A short time later Delsemme and Swings (5) suggested that in comets the volatiles other than H_2O are entrapped in clathrate hydrates.

The problem at that time was the stability of ices as a function of heliocentric distance at a time scale as long as the age of the solar system. This problem has been solved by Watson et al. in 1962 (6). In model calculations they considered the lifetimes of ice spheres of one km in size orbiting at a given heliocentric distance. This study revealed that such a sphere of water ice would survive for time span as long as the age of the

1

J. Klinger et al. (eds.), Ices in the Solar System, 1–5.
© 1985 by D. Reidel Publishing Company.

solar system at a heliocentric distance slighly smaller than Jupiter's orbit. Similar calculations can be done for ices more volatile than H_2O.

It must be said however that the study of Watson et al. did take into account evaporation as the only erosion mechanism. If we include sputtering by energetic particles the previous result is not altered significantly if we consider massive bodies. For instance, massive icy bodies exposed to the particles of the magnetosphere of Saturn will lose only a few meters of ice in $4.5 \ 10^9$ years (7). The situation is quite different if we consider small icy grains. In this case sputtering leads to a severe restriction to the lifetime of these grains. Other loss mechanisms like the Poynting – Robertson effect are, in comparison, negligible (6).

Having shown that ices are able to survive in the outer solar system, the question arises: are we sure that ices really occur there ? The first hint in this sense is due to Hepburn in the early twenties (8). Hepburn tried to determine simultaneously the albedo, the radius and the density of the satellites of Saturn. At that time, only the radius of Titan had been measured directly. Hepburn concluded that the albedo of these satellites must be high and that the density should be close to that of water. Spectroscopic identification of ices in the solar system was only possible after the second world war when great progress in infrared spectroscopy had been accomplished. The first spectroscopic identification of ices in the solar system is due to Kuiper (9). Since the late sixties, water-ice as well as solid methane have been detected.For detailed references see chapter V.

After having answered the question whether ices are stable and whether they exist in the solar system, let us ask where they come from. The cosmic abundance of elements (10) indicates that, if we disregard He and Ne the most abundant elements are H, O, C and N. Thus, it seems natural that compounds of these elements such as H_2O, CH_4, and NH_3, as well as their hydrates, or even more complicated organic molecules should be present in substantial quantity in the solar system. The solar system was formed from the presolar nebula which in turn formed by fractionation of an interstellar cloud. Presumably, the molecules we are interested in, have been synthetized during the interstellar cloud stage. Based on considerations of elemental abundance and of the heat of formation, van de Hulst (11) suggested in 1949 that interstellar particles or grains are mainly composed of H_2O. Similar suggestions have been made by Eddington twelve years earlier (12).In the meantime a great number of moderately volatile molecules have been detected in the interstellar medium (see for example (13,14,15,16)).

Theoretical studies (for review see (17)) and laboratory simulations (18,19) give us a fair idea about interstellar chemistry. We are quite sure that molecular hydrogene forms on interstellar grains. Molecules like H_2O, CH_4 and NH_3 may form in the same way. These "first generation" molecules are in exchange equilibrium with the gas phase where they are subject to photodissociation. Diatomic molecules as CO, CN and CS are formed by exchange reactions with C^+ ions. If these molecules collide with grains they can form more complicated substances some of them being moderatly volatile, others produce an organic refractory coating on the grains (19). It must be said however that this is only one of the possible scenarios of formation of interstellar molecules. In fact several of the above mentioned molecules may be formed by other mechanisms, ionic reactions in the gas phase for example. At present we are far from a complete knowledge of interstellar chemistry.

When the interstellar cloud fraction densifies, the opacity for ultraviolet radiation increases and photoprocessing stops. It is essential that ices condense before they accrete to form the bodies of the outer solar system. Condensation of particular molecular species depends of course on pressure and temperature and whether homogenous or heterogenous nucleation takes place. Let us give as an example the condensation of H_2O : it is well known that water vapor condenses as an amorphous solid at temperatures lower than 100 K (20). At higher temperatures, we obtain cubic or hexagonal ice (see (21)). If we adopt for the condensation of volatiles pressure conditions like those in the presolar nebula as it was done by Lewis (22) we find a condensation temperature of H_2O of about 170 K. On the other hand, if we condense H_2O under pressures like those in molecular clouds, we get condensation temperatures close to 100 K. But H_2O can condense, of course, at much lower temperatures if it is deposited on cold grains. Thus, we will obtain crystalline water ice, or amorphous water ice. It is useful to mention that the high frequency wing of the 3.1 µm absorption line found in some interstellar clouds is most likely due to amorphous H_2O. Thus, the presence of various forms of ice on bodies in the solar system gives valuable clues about the condensation and accretion mechanisms of these bodies. A great amount of laboratory work must still be done in order to improve the phase diagrams of ices and to make possible spectroscopic idenfications of ice phases.

When the ices are incorporated in larger bodies, various reprocessing changes may occur. The reprocessing evidently depends on the size and density of the body as well as on heliocentric distance and particular local conditions such as tidal heating, variation in solar heating etc..

Mars evidently is the most earth like body among the ice bearing bodies mentioned in this book. As on the Earth the volatiles on the surface and in the atmosphere probably have been accumulated in the post-accretion phase.Just as on the Earth, recycling of ices on Mars is subject to periodic climatic changes most likely due to a coupling between orbital variations and variations in the tilt of the rotation axis. Thus the variability of the ice   coverage on Mars could perhaps help us to understand the mechanism of climatic changes on our own planet.

Beginning with the orbit of Jupiter we meet a great variety of bodies that are either ice covered or, in most cases, dominated by ices. The giant planets themselves are excluded from our considerations because they are dominated by highly volatile substances such hydrogen and helium.

At the outer fringe of the solar system, we finally find the comets that are generally considered to be the least reprocessed objects that are accessible to investigation. Thus, the study of comets is particularly important if we want to acquire a better knowledge of the early history of our planetary system.

ACKNOWLEDGEMENTS

Numerous suggestions for the improovement of this text by Dr Benest, Prof. Dollfus, and Prof. Smoluchowski are gratefully acknowledged.

REFERENCES

(1)   Whalley, E. 1985, this book.
(2)   Laplace, P.S. 1808, Exposition du système du monde, 3rd ed. Paris, p.130.
(3)   Fauth, P. 1913, Hörbigers Glacial-Kosmogonie, Hermann Kaysers Verlag, Kaiserslautern
(4)   Whipple, F.L. 1950, Astrophys. J.111, pp.375-394.
(5)   Delsemme, A. and Swings, P. 1952, Annales Astrophys. 15, pp.1-6.
(6)   Watson, K., Murray, B.C. and Brown, H. 1963, Icarus 1, pp.317-327.
(7)   Johnson, R.E., Brown, W.L. and Lanzerotti, L.J., 1983, J. of Physical Chemistry 87, pp.4218-4220.
(8)   Hepburn, P.H. 1923, J. of the British Astron. Soc. 33, pp.244-250.
(9)   Kuiper, G.P. 1957, the Astronomical Journal 62, p.245.

(10) Cameron,A.G.W.1981,   Harward-Smithonian   Center   for
     Astrophysics preprint n°1357.
(11) Van der Hulst, H.C. 1949, Rech. Astron. Obs., Utrecht, 11,
     pp. 2-18.
(12) Eddington, A.S. 1937, Observatory 60, pp.99-103.
(13) Weinreb, S. , Barrett, A.H., Meeks, M.L. and Henry, J.C.
     1963, Nature 200,pp .829-831.
(14) Cheung, A.C., Rank, D.M., Townes, C.H., Thornton, D.D. and
     Welch, W.J. 1968, Phys. Rev. Lett. 21, pp. 1701-1705.
(15) Snyder, L.E., Buhl, D., Zuckerman, B. and Palmer, P. 1969,
     Phys. Rev. Lett.22, pp. 679-681.
(16) Cheung, A.C., Rank, D.M., Townes, C.H., Thornton, D.D. and
     Welch, W.J. 1969, Nature 221, pp. 626-628.
(17) Watson, W.D. 1976, Rev.of Mod. Phys. 48, pp.513-552.
(18) Tielens, A.G.G.M., Hagen, W. and Greenberg, J.M. 1983, J.
     Phys. Chem. 87, pp.4220-4229.
(19) Greenbeg, J.M. and d'Hendecourt, L.B. 1985, this book.
(20) Ghormley, J.A. 1968, J. Chem. Phys., 48, pp. 503-508
(21) Eisenberg, D. and Kauzmann, W. 1969, The structure and
     properties of water, Oxford University Press.
(22) Lewis, J.S. 1972, Icarus 16, pp. 272-252.

Part I

Physics and Remote Sensing of Ices

THE PHYSICS OF ICE: SOME FUNDAMENTALS OF PLANETARY GLACIOLOGY*

Edward Whalley

Division of Chemistry, National Research Council,
Ottawa, K1A 0R9, Canada

Abstract. Selected properties of ice, particularly those that
may be useful for remote sensing in the planetary system or
understanding the behavior of ice there, or that will help to
predict properties that planetary scientists need, are reviewed.
Among them are the phase diagram, including a new easy
transformation of ice Ih at 77 K near the extrapolated melting
line, the microwave spectrum of ice Ih as determined from an
extrapolation of the far-infrared spectrum and used to determine
the thickness of ice in Saturn's rings, and the use of halos to
detect crystals of hexagonal and cubic ice. Many properties of
ice that are needed for planetary studies may need to be
calculated from molecular potential functions. These can be
tested by predicting the energies of the phases of ice at zero
temperature, which can be evaluated from experimental
measurements.

1. INTRODUCTION

        Material in the planetary system is often classified
into three groups, the most volatile, i.e., hydrogen and helium;
moderately volatile, i.e., nitrogen, methane, carbon monoxide,
carbon dioxide, ammonia, hydrogen sulphide, sulphur dioxide,
water, sulphuric acid, sulphur, etc; and non-volatile, i.e.,
carbonate and silicate rocks, iron, etc. Free sulphur
occurs, for example on Jupiter's moon Io, and as its boiling
point at atmospheric pressure is 718 K it seems to be
intermediate between the moderately volatile and non-volatile

*N.R.C. No. 23718

9

J. Klinger et al. (eds.), Ices in the Solar System, 9–37.
© 1985 by D. Reidel Publishing Company.

materials. The first two groups and sulphur occur as solids,
liquids, and gases, and the third group as solids and viscous
liquids.

This workshop was called to discuss the role of the
moderately volatile solids in our planetary system. These
phases are often called by astronomers, although by no one else,
ices. It is an unfortunate use of a word that is already well
established as denoting the solid phase of water, and the solid
phase of water is so important to us in so many ways,
particularly those of us who live in northern countries, that it
deserves a name all to itself. We have only to imagine being
offered ice in our whiskey by a planetary astronomer and being
given solid hydrogen sulphide. The precedent is, of course, the
use of the term "Dry Ice" as a proprietary name for solid carbon
dioxide. As long as there was little overlap between physicists
and chemists seriously interested in ice and astronomers
seriously interested in other volatile solids as well, these
uses of the word "ice" caused little confusion. Now that there
is a significant and increasing overlap, we should consider a
new word for these moderately volatile solids.

The term "ices" can, of course, include with equal
legitimacy many other substances such as solid argon, krypton,
and xenon; the halogens, interhalogen compounds, hydrogen
halides, and pseudo halogens like cyanogen, cyanogen chloride,
etc; many metallic hydrides like boron, aluminum, and antimony
hydrides; boron and nitrogen trifluorides; hydrazine, hydrazine
hydrate, hydrazoic acid, hydrogen peroxide, and many others.
But it would be absurd to extend the definition so far. The
names "ammonia ice", etc. seem to have no advantages over "solid
ammonia", etc., and do we really want names like "ammonia ice
I", instead of "ammonia I"?

We need, therefore, a special word to encompass those
volatile solids that are or may be common in the planetary
system, namely nitrogen; methane, water, hydrogen sulphide, and
ammonia; carbon monoxide, carbon dioxide, and sulphur dioxide;
clathrate hydrates of all of these; and the ammonia hydrates.
Of these substances only ice and probably the clathrate hydrates
of methane and perhaps some other hydrocarbons occur naturally
on earth. The glaciologists are already interested in the
permafrost on Mars, and when naturally occurring clathrates are
better studied, perhaps the earth-bound glaciologists will
include them in their field of study. It may be appropriate
therefore, to consider designating the field of study concerned
with naturally occurring volatile solids in the planetary system
as "planetary glaciology". The field has much in common with
ordinary earth-bound glaciology and is in fact an extension of
it, and the term has the advantage that it emphasizes the unity
of the subject of naturally occurring volatile solids in the
planetary system. There is adequate precedent for this. Robert
Boyle in 1681 used "glacial" to mean frozen (1). "Glacial

acetic acid" is still a common term and "glacial acrylic acid" and "glacial phosphoric acid" are recognized by Webster (2). Furthermore, "glacial ammonia", "glacial carbon dioxide", etc. are euphonius terms. "Glace" is an obsolete English word for ice that came from the Latin and French, and could be used as a generic term for low-melting solids.

The purpose of this half-day session is to discuss some aspects of the physics of volatile solids that may contribute to the study of "planetary glaciology", and this, the first paper, is concerned specifically with the physics of ice. The physics of ice is a large subject, and ten years ago a book of almost a thousand pages was written about it (3). Much has been learned since then, and as there is too much to review in detail in a short time, I shall concentrate on those aspects that I am most familiar with and that may be of some interest in planetary physics. Later papers will discuss the physics of other volatile solids that may occur in the solar system.

2. THE PHASES OF ICE

2.1 The Crystalline Phases

Much has been learned about the phases of ice in the 85 or so years since Tammann (4) discovered the first high-pressure phases. It all needs to be carefully reviewed to obtain the most reliable values of their properties, but that is too big a job for a short talk. It is the properties at moderately low temperatures that are of greatest interest in planetary glaciology, and this is the region where there is greatest uncertainty because equilibrium is reached only very slowly. There are indeed many qualitative gaps in our knowledge, which occur partly because the time scales of laboratory experiments and planetary changes are enormously different, and I shall try to point out the major gaps.

The phase diagram of ice is drawn in Fig. 1. Phases II and III have been known since 1900 (4) and phases V and VI were added by Bridgman (5), who also cleared up some uncertainties in Tammann's measurements, and provided more accurate phase boundaries of II and III. Bridgman's pressures appear to be low by about 1% (6,7) and have been corrected in the figure. Phase IV is a wholly metastable phase that was detected by Bridgman in 1911, but was not fully confirmed until work on D_2O was done in 1935 (8). It has never been kept for more than a few minutes near its melting line. Its melting curve has been studied by Evans (9), Engelhardt and Whalley (10), and Nishibata (11), who also measured its boundary with VI. It is wholly metastable, but has been stored in our laboratory for at least several years at zero pressure and 77 K. Ice VII was added by Bridgman (12) when he increased his

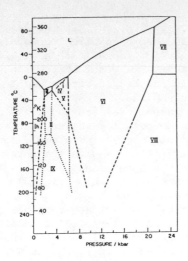

Fig. 1.   Phase diagram of ice.

pressure range to 45 kbar in 1937. One phase missing from Fig.
1 is cubic ice or ice Ic (13), which differs from ice Ih only in
the way the puckered hexagonal planes of water molecules are
stacked. Under all known conditions, however, ice Ic is less
stable than Ih, but the difference in energy between them is
unknown, but is probably <23 J mole^{-1} (14).

The phase diagram is complicated by the
configurational entropy that some phases have. It arises
because each oxygen atom is joined to its four nearest neighbors
by hydrogen bonds. Hydrogen bonds, at least in the ice phases,
are asymmetric, and the hydrogen can be on either side of the
center. Each water molecule can then have six orientations in
its cage of four near neighbors. Because each water molecule
can have two, and only two, hydrogen atoms near it (rare
imperfections in the crystal excepted), the orientations of
neighboring molecules are strongly correlated, and there is a
maximum of only approximately 1.5 orientations (more accurately
1.50685 (15,16)) per molecule. There can, therefore, be a
configurational entropy as high as approximately Rln3/2 (more
accurately 0.4100R) where R is the gas constant. Because the
vibrational entropy seems to differ little between the various
phases, the configurational entropy has a profound effect on the
phase diagram.

Ice Ih and Ic appear to have fully disordered
molecular orientations, within the ice rules, even when quenched
to zero temperature. Numerous reports have been published
purporting to show that some ordering occurs in pure ice, but
none is convincing. A recent calorimetric measurement on a
sample made by freezing water containing 0.1 mol L^{-1} of

potassium hydroxide and keeping at 60-65 K for a week (17) finds
a phase transition at 72 K with an entropy increase of about 0.7
of the expected value for a transition from fully ordered to
fully disordered ice Ih. An ordering transition in undoped ice
Ih or Ic could perhaps occur on the time scale of the planetary
system.
        Ice III, IV, V, and VI are essentially fully
disordered at high temperature by a variety of evidence (18).
Ice II, on the other hand is fully ordered at ~95 K according to
a neutron diffraction study of D_2O (19) and both its low
dielectric constant at high temperature (20) and its low entropy
relative to ice I, II, and V (21,22) are consistent with its
being fully ordered at high temperature. The variation of
temperature T and pressure p along an equilibrium line is
described by the Clausius-Clapeyron equation

$$\left(\frac{\partial p}{\partial T}\right)_e = \frac{\Delta S}{\Delta V} \; ,$$

where subscript e indicates equilibrium and ΔS and ΔV are
respectively the entropy and volume change at the transition.
If the entropy change is zero and the volume change is not, the
equilibrium line in Fig. 1 is vertical, and if the volume change
is zero and the entropy change is not it is horizontal. The
configurational entropy of ice II, as estimated from the volume
changes at and slopes of, the I-II, II-III, and III-I
boundaries, is zero within experimental uncertainty (21) and so
is also consistent with full order. Its low configurational
entropy causes its field of equilibrium to broaden as the
temperature decreases. The best extrapolated I-II line
intercepts the pressure axis at zero temperature at 0.14 ± ~0.2
kbar (23). It is possible that ice II is the stable phase at
zero temperature and pressure.
        The great importance of the configurational entropy of
ice I to our terrestrial glaciology can be understood by the
following fairy tale (24). Suppose that ice II were the stable
phase at zero temperature and pressure and that ice I and II had
the same configurational and vibrational entropy. The
equilibrium line would then be vertical, and so ice II would be
the stable phase at zero pressure. As it is significantly
denser than water, any ice that formed on the polar seas would
sink to the bottom and so would any glaciers that flowed into
the oceans from polar icecaps. The polar seas would then be
frozen from the bottom upwards to near the surface. As salt
does not dissolve in ice, the remaining water would be very
salty. Lakes and rivers would, of course, freeze from the
bottom upwards in the winter in temperate latitudes and would
take a long time to melt in the summer, and the deep lakes may
not melt at all. The consequences of all this on our world
would be very profound, and perhaps life as we know it may never
have developed. Perhaps we are here today discussing planetary

glaciology only because ice II does not have a configurational
entropy of Rln3/2.

Ice VIII is the only other fully ordered phase that is
stable at high temperatures. It is composed (see Ref. 25) of
two interpenetrating sub-structures each like ice Ic and having
a diamond-like arrangement of the oxygen atoms. The molecular
dipoles in each sub-structure are parallel, but the two
substructures have antiparallel orientations and the structure
is non-polar. Ice VIII and VII have the same volume to within
the experimental precision of ~ ±0.5 mm^3 mol^{-1} (26) as the
equilibrium line is independent of the pressure up to at least
100 kbar. Ice VII has a configurational entropy of about 2.8 J
K^{-1} mol^{-1} (27) compared with the maximum of 3.4 J K^{-1} mol^{-1} for
complete disorder consistent with the ice rules, and so it is
incompletely disordered. This conclusion is supported by the
dielectric properties (28). At higher temperatures, the
orientations should become more disordered, but no measurements
have been made.

Ice III becomes ordered on cooling, as was shown by
its dielectric permittivity (29). In 1967, the entropy of
transformation ice III-IX was measured in the first experiment
using high-pressure differential scanning calorimetry (30) as
-1.34 J K^{-1} mol^{-1}, which is little more than one third of the
value for the formation of completely ordered ice IX. This high
configurational entropy is, however, obtained by only a small
departure from complete order of the occupancy of the hydrogen
positions. There are six possible independent hydrogen
positions, all having the same multiplicity, which occur in
pairs I, I'; II, II'; III, III' at opposite ends of the same
kind of O--O bond. They are illustrated in Fig. 2. A sample
that had been quenched from 250 to 77 K in ~ 10 s had, according
to neutron diffraction, an occupancy of positions I' and II' of
0.051 ± 0.009 and of position III' 0.034 ± 0.007 (31). The
configurational entropy calculated from these values (30) agrees
with the measured entropy difference between ice III and IX.

No doubt, when ice IX is cooled slowly enough it will
become fully ordered. At the lowest temperature of the
dielectric measurements, 167.3 K, the amplitude of the
permittivity was 0.02 of the value for complete disorder
consistent to the ice rules. Unfortunately, the relation
between the degree of disorder and the permittivity is not
known. Attempts to form fully ordered ice IX by cooling slowly
in the region of 170 K would be very useful. If ice IX were
formed by the spontaneous cooling of ice III in the interior of
a planet or satellite, it may cool slowly enough to become fully
ordered.

The low configurational entropy of ice IX relative to
its neighbors Ih, V, and VI, causes its field of stability to
increase greatly as the temperature is lowered, as with ice II.
As ice IX is a wholly metastable phase, it may be less important

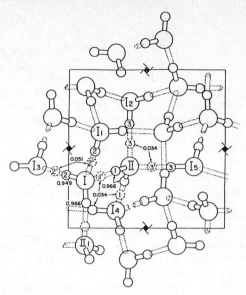

Fig. 2. Structure of ice IX.

in planetary glaciology than the stable phase in this region
i.e. II.

The boundary line of ice III and V is very nearly
vertical, showing (22) that both have almost the same entropy
and, as ice III is disordered, so also is ice V. The dielectric
properties of ice V have been studied only down to 227 K (20),
and down to this temperature, the amplitude of the permittivity
is consistent with complete disorder. The phase boundary has
been measured down to 228 K (32), and agrees with the
extrapolation in Fig. 1.

The boundary line of ice V with ice VI is nearly
vertical, showing that both have nearly the same entropy (22),
and so ice VI appears to be fully disordered within the
restrictions of the ice rules. Dielectric measurements on ice
VI (33) over the range 301 - 128 K give a linear Curie-Weiss
plot with an intercept on the temperature axis of 47 K, implying
that they fit the equation

$$\varepsilon_0 - \varepsilon_\infty = A/(T-47 \text{ K}),\tag{1}$$

where ε_0 and ε_∞ are the limiting low- and high-frequency
permittivities and A is a constant. Equation (1) implies that
the amplitude of the permittivity extrapolates to infinity at 47
K, and so that ice VI would become ferroelectric there.
However, single crystals of ice VI quenched inside a pressure
vessel to 77 K, removed from the vessel, and examined by neutron

diffraction (34) were partly ordered in an antiferroelectric
manner. It seems that an as yet unobserved transition must
occur between 130 K and ~ 10 kbar and 100 K and zero pressure.
There is even another complication in the phase diagram of ice
VI. Whalley and Heath in unpublished work referred to in Ref.
24, first showed that ice VI undergoes a slow transformation at
~ 120 K as detected by a drift in the permittivity at 5 kHz, and
Johari and Whalley (35) followed it in both H_2O and D_2O ice VI
at 117 K for 252 days, at the end of which it was still
drifting. The high-frequency permittivity of H_2O and D_2O ice VI
decreased by 0.057 and 0.012 respectively. This is too slow a
transformation to be easily studied in the laboratory, but could
be very important if ice VI occurs in the planetary system, and
deserves more study.

The greatest uncertainties about the phase diagram
that concern planetary physicists may be the disorder-to-order
transformations. Little is known about the rate of ordering of
ice I, and the transformation III-IX is not quite complete under
laboratory conditions. Ordering of ice IV is not known,
although it may be too unstable for our present concerns, only
partial ordering of uncertain origin occurs in ice V and VI, and
complete disordering of ice VII has not been reported.

The clathrate hydrates are really phases of ice having
other molecules dissolved in them. They are reviewed in another
paper in this volume (36) and a more detailed account has been
written by Davidson (37), so they will not be discussed here in
detail.

2.2   The Amorphous Phases

The crystalline phases of ice occur, of course, in
discrete structures because of their repetitive nature, or
alternatively of their symmetry, and although each phase can
have a range of properties depending on the physical and
chemical imperfections in them, they are clearly distinguished
from one another.

Amorphous phases also occur, but because the structure
does not repeat itself, and has no microscopic symmetry, they
form in principle a continuous distribution of structures. The
earliest amorphous phase of ice was made by Burton and Oliver
(38) by condensing water vapor onto a plate cooled below 158 K,
a technique invented by Tammann and Starinkewitsch (39), and
many investigations of ice made by this technique have been
published. The exact nature of the phase no doubt depends on
the temperature of its surface during the condensation, and that
depends on the rate of condensation, the thickness of the
amorphous film, and the strength of the coupling of the surface
of the substrate the film is condensed on to the coolant. Most
work has been done on phases condensed at nominally 77 K,
although the condensing surface was probably usually

significantly warmer. By judicious choice of conditions, samples of amorphous ice with a significant range of properties could no doubt be made, as Olander and Rice (40) verified when they first condensed it at temperatures around that of liquid helium.

The most common way of making amorphous phases is by cooling the liquid, usually more or less rapidly depending on whether the viscosity is low or high. The method has, of course, been used to make glass since before recorded history. Although aqueous solutions have been vitrified many times, vitrified pure water was not made until Brüggeller and Mayer (41) succeeded by injecting an emulsion of micrometer-sized droplets of water in n-pentane into a cooled cryogenic fluid like ethane, and later by injecting a thin jet of liquid water (42). Again, no doubt significantly different cooling rates would give significantly different samples of vitreous ice.

A third way of making amorphous ice has recently been discovered in our laboratory. It is based on the observation, as shown in Fig. 1, that the melting point of ice falls from 273.16 K at the solid-liquid-vapor triple point to 250.91 ± 0.1 K at the liquid-I-III triple point at 2090.5 ± 2 bar (7). The melting line does not, of course, end at the triple point, but can be extrapolated beyond it, and, indeed, the liquid-I equilibrium has been followed to 181 K and 2 kbar by using an emulsion of minute droplets of water to reduce the rate of nucleation (43). In principle, of course, the melting line can be extrapolated to zero temperature, and there is no reason to think that it must end at a finite temperature. The extrapolated melting pressure at 77 K is ~10 kbar. If ice I is compressed at 77 K to over 10 kbar, it is unlikely to transform to a crystalline phase because ice II, IX, V, VI, and VIII, when recovered at 77 K and zero pressure, do not transform to ice I in times of a few minutes until temperatures of 125-170 K depending on the phase are reached (44,45), and there is no reason to expect that ice I will transform to a high-pressure phase with very much greater ease. Therefore, if ice I is compressed to >10 kbar, it must either "melt" to form a liquid, which will in fact be an amorphous solid because the temperature is well below the glass transition of the liquid, or the solid will become superheated into the region of stability of the liquid (or glass). The results of a compression experiment on ice I at 77 K are plotted in Fig. 3 (46). There is a surprisingly sharp, for 77 K, transition at $10 \pm$ ~1 kbar, and it does not reverse on removing the pressure, as Fig. 3 shows. The density of the recovered phase was measured as 1.17 g cm^{-3} by weighing in liquid nitrogen. Its x-ray diffraction pattern is composed of a few broad lines, the lowest of which is at a distance corresponding to 3.0 Å at ~95 K. On heating for 10 min at each of several temperatures up to ~135 K and cooling, the

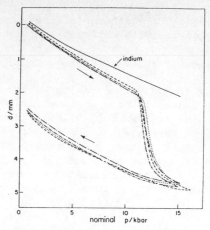

Fig. 3.   Compression of ice I contained in an indium
           cup, as a function of the nominal pressure on
           the sample for four independent runs at 77 K,
           and on a volume of indium equal to the
           volume of the indium cup and ice.

first peak increased smoothly in distance to 3.65 Å and after
heating to 155 K it became 3.67 Å. Several different kinds of
amorphous ice can therefore be made by annealing the original
forms. Other variations could no doubt be made by compressing
ice I at both lower and higher temperatures. As the experiments
were only begun in 1983 mid-November, a full account cannot yet
be given. If existing particles of ice I were to agglomerate to
form a planet whose interior pressure reached >10 kbar, this
high-density amorphous ice would undoubtedly be formed.

        A low-density form of amorphous ice can be made in
bulk quantities by heating the recovered high-density amorphous
ice to ~140 K (46). When this phase is compressed at 77 K, a
relatively sharp and irreversible transition occurs at ~6 kbar
with a decrease of volume of 21% (47). The dense phase can be
recovered at zero pressure and has a density of 1.17 g cm^{-3}.
Its x-ray diffraction pattern is characteristic of an amorphous
phase and is similar to that of the amorphous phase made by
compressing ice I at 77 K. The transition by which it is formed
appears to be another new kind of transition - a relatively
sharp transition from a low-density to a high-density amorphous
phase induced by pressure. This transition may be important in
the planetary system if particles of amorphous ice agglomerate
to form a planet or satellite.

        All these amorphous phases may have a part to play in
the planetary system. It is quite possible and even likely that
some of the water vapor in the early nebula condensed to

amorphous ice, as Klinger (48) has pointed out, and it is quite
possible that amorphous forms of the clathrate hydrates were
also produced. Amorphous ethylene oxide hydrate has already
been made by condensing a stoichiometric mixture of water vapor
and ethylene oxide at ~ 80 K (49). Vitreous water may be made
if a jet of liquid issues from the interior of a planet or
satellite into an atmosphere at very low pressure and is rapidly
cooled below the glass transition by evaporation of water.
Indeed, Torchet, Schwartz, Farges, de Feraudy, and Raoult (50)
have shown that small water droplets of < ~200 molecules formed
in a free jet expansion always freeze to an amorphous solid. If
low-density amorphous ice formed in this way agglomerates to
form a planet or satellite it will probably transform to the
high-density amorphous ice if the planet or satellite is large
enough.

          Amorphous phases of ice, as well as high-pressure
phases recovered at low pressure have hitherto always
transformed first to ice Ic, and only at higher temperature has
ice Ih formed. All the samples of the high-density amorphous
ice that we have made have contained small amounts of
untransformed ice Ih, and in one sequence of heatings
of a sample, all the amorphous ice transformed directly to ice
Ih by growth of already existing small single crystals without
any ice Ic forming (46). This process may also occur in the
planets and satellites, and we may need to be cautious about
firmly inferring the existence of Ic on the basis of history.

3. REMOTE SENSING OF THE PHASES OF ICE

3.1. Infrared Spectrum

          Ice I, both hexagonal and cubic, has a strong
absorption and reflection spectrum at about 3400 cm^{-1} due to the
O-H stretching vibrations, at about 850 cm^{-1} due to the
rotational vibration of the water molecules, and at about 230
cm^{-1} due to the translational vibrations of the water molecules,
and all three have been often used to recognize ice. No
noticeable difference in the spectra of ice Ih and Ic has been
reported above 350 cm^{-1} (51) and there are only small
differences below 350 cm^{-1} (52), where the translational
vibrations (53,54) may be more sensitive to the differences
between the two forms. Amorphous ice made by condensing the
vapor at low temperatures has similar bands, but they are
noticeably broader, and have been used to identify amorphous ice
in the planetary system (55).
          Astronomers have studied the rings of Saturn in the
microwave region for many years. By measuring the microwave
intensity emitted by Saturn and its rings at a particular
frequency when the rings are at various angles to the line of

sight, the brightness temperature, which is the temperature a
black body would have if it emitted the same amount of radiation
at the given frequency, can be measured. The brightness
temperature of a body differs significantly from the actual
temperature if the body transmits significant amounts of light.
If the absorptivity of ice at the same frequency is known, the
thickness of the rings can in principle then be determined. The
absorptivity of ice at 3 cm^{-1} at 85 K is only ~0.3 m^{-1}, and so
would be difficult to measure directly.

We have recently, therefore, measured the absorptivity
in the range 8–25 cm^{-1} and 80–202 K (56). In this region the
fundamental excitations are sound waves, and as ice I is a
disordered crystal it has no symmetry, and so all the vibrations
are active as fundamentals. The absorption intensity of the
disorder-induced fundamental vibrations is proportional to the
square of the frequency (53,57), and the density of vibrational
states in the sound waves is also proportional to the square of
the frequency. The sound waves contribute, therefore, a term to
the absorptivity that is proportional to the fourth power of the
frequency.

The dominant absorption particularly at low
frequencies, is due to difference bands, but only a small
fraction of all the possible difference bands can contribute
because wave vector is conserved in the interaction with light.
Furthermore, difference bands can arise only from excited
vibrations, so that only the low-frequency molecular
translational vibrations can contribute significantly. The
maximum in the density of states of these bands is at ~230 cm^{-1}
(54), and is no doubt due largely to vibrations near the center
of the Brillouin zone. Then Szigeti's (58) theory of the
absorptivity due to difference bands when specialized to this
region reduces to (56)

$$K(\nu,T) = \frac{A_o}{T} \frac{e^{a'\nu_o}}{(e^{a'\nu_o}-1)^2} \frac{1}{\nu_o^2} \nu^2 , \qquad\qquad (2)$$

where $K(\nu,T)$ is the absorptivity as a function of both the wave
number ν and the temperature T, A_o is a constant, $a' = hc/kT$,
where h is Planck's constant, c the speed of light, k
Boltzmann's constant, T the temperature, and ν_o a weighted mean
frequency of the transverse optic branches at the boundary of
the Brillouin zone. The best fit of the experimental
absorptivity to Eq. (2) with an additional term $B\nu^4$, where B is
a constant, to allow for the fundamentals, was obtained by
varying A_o, ν_o, and B. It yielded A_o= (1.188 ± 0.01) × 10^5 cm^{-1}
K, ν_o = 233 cm^{-1}, and B = (1.11 ± 0.03) × 10^{-6} cm^3. The
experimental measurements are compared with the fitted equation
in Fig. 4. The fit is excellent at low temperatures but
systematic deviations occur at high temperatures, perhaps caused
by thermal expansion, which was ignored in the theory.

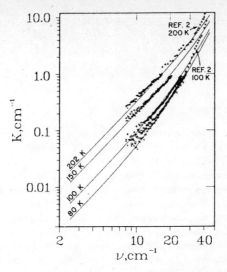

Fig. 4. Absorptivity of ice Ih in the range 2-40 cm^{-1}. The points are experimental measurements at the indicated temperatures and the lines are the fitted curves described by Eq. (2) with the values of the constants given in the text.

The absorptivity at the temperature of Saturn's rings, which is 85 K, extrapolated to 3.0 cm^{-1} is 0.48 ± ~0.1 m^{-1}. If Saturn's rings are represented by a single layer of particles with spaces between them, the measured absorptivity and brightness temperature (59,60) are consistent with an average thickness of 30 ± ~10 cm of ice. A more detailed discussion of the implication for rings is given in Ref. 60.

3.2 Absorption by the Sound Waves: a Disorder-Induced Piezo-
     Electricity

As has already been mentioned, the sound waves in ice are active as fundamentals (57). This occurs because the molecules are orientationally disordered and so ice has, on a molecular scale, no symmetry. All molecular motions, therefore, cause a change of dipole moment and so all absorb infrared radiation at the proper frequency. The absorptivity due to the sound waves was initially interpreted on a molecular model (57) in which the motion of neighboring molecules causes changes of the dipole moment. The absorption can also be interpreted in terms of a new macroscopic concept, namely disorder-induced piezoelectricity (61).

The ordinary piezoelectricity of crystalline solids is described by piezoelectric moduli, which can be represented by third-rank tensors $\underset{\sim}{d}$ that connect the induced polarization $\underset{\sim}{P}$ and the stress $\underset{\sim}{\sigma}$ by the equation

$$\underset{\sim}{P} = \underset{\sim}{d} \cdot \underset{\sim}{\sigma} \, ,$$

or

$$d_{\alpha\beta\gamma} = \partial P_\alpha / \partial \sigma_{\beta\gamma} ,$$

where the tensor summation convention is used for repeated Greek suffixes. Some crystals are symmetric enough that their piezoelectric moduli are zero.

Isotropic disordered solids, such as glasses and other amorphous solids, in which there is no long-range correlation of atomic positions, have by symmetry no ordinary piezoelectric properties. They do, however, have piezoelectric properties of another kind. Although the ensemble-averaged dipole moment of isotropic disordered solids is zero, the ensemble-averaged squared dipole moment is not. A specimen of an isotropic glass or amorphous solid having a volume V can be considered as made of n subspecimens i, each having a volume v_i so that

$$V = nv_i .$$

The dipole moment $\underset{\sim}{\mu}$ of the specimen is the sum of the dipole moments of the subspecimens i

$$\underset{\sim}{\mu} = \sum_i \underset{\sim}{\mu}_i ,$$

and the ensemble average of its square is

$$\langle \mu^2 \rangle = n\sum_i \langle \mu_i^2 \rangle , \tag{3}$$

if the subspecimens are large enough that the dipole moments of neighbors are uncorrelated. To ensure this, the dimensions of the specimens must be much greater than atomic correlation distances. When Eq. (3) is divided by the volume it becomes

$$\langle \mu^2 \rangle / v = n\sum_i \langle \mu_i^2 \rangle / V ,$$

$$= \sum_i \langle \mu_i^2 \rangle / v_i , \tag{4}$$

The mean square dipole moment per unit volume is, from Eq. (4), a property of the material.

The dipole moment may be induced by a stress. Then a disorder-induced piezoelectric coefficient d''^2 may be defined. For example

$$d''^2_{\alpha 23} = \frac{1}{V} \left(\frac{\partial \mu''(\sigma)}{\partial \sigma_{23}} \right)^2 ,$$

where σ_{23} is the 23 cartesian component of the stress. The spontaneous fluctuations of the thermal sound waves will induce dipole moments, which can be characterized by separate coefficients $d_\ell''^2$ and $d_t''^2$ for the longitudinal and transverse waves respectively, where

$$d_\ell''^2 = d_{xxx}''^2 + d_{yxx}''^2 + d_{zxx}''^2$$

$$= d_{xyy}''^2 + d_{yyy}''^2 + d_{zyy}''^2, \text{ etc.}$$

and

$$2d_t''^2 = d_{xxy}''^2 + d_{yxy}''^2 + d_{zxy}''^2 + d_{xyx}''^2 + d_{yyx}''^2 + d_{zyx}''^2, \text{ etc.}$$

If the measured squared dipole moments are generated by the natural fluctuations of the sample, then only the sum

$$d''^2 = d_\ell''^2 + 2d_t''^2 ,$$

or a weighted sum can, in general, be measured.

The fluctuating dipole moments cause the sound waves to absorb light according to the equation (61)

$$K(\nu) = \frac{16\pi^4}{3} c^3 \rho \frac{(n^2+2)^2}{9n} \left(\frac{c_{11}d_\ell''^2}{\nu_{D\ell}^3} + \frac{2c_{12}d_t''^2}{\nu_{Dt}^3} \right) , \qquad (5)$$

where $K(\nu)$ is the absorptivity of the sample at wave number ν as defined by the equation

$$I = I_o e^{-K\ell},$$

where I_o and I are the intensities of light before and after traversing the distance ℓ in the sample, c the speed of light, ρ and n the refractive index.

The fundamental absorptivity by the sound waves is proportional to the fourth power of the frequency with the proportionality constant $B = (1.11 \pm 0.03) \ \mp \ 10^{-6} \ cm^3$, as reported in Section 3.1. The value of $c_{11}d_\ell''^2/\nu_{D\ell}^3 + 2c_{12}d_t''^2/\nu_{Dt}^3$ is $5.69 \times 10^{10} \ D^2 \ bar^{-1}$. These sound waves may, of course, be useful for detecting ice and determining its thickness from the infrared absorbance.

The shear waves of ice are much weaker in the infrared than the longitudinal waves, as is shown by the small effect of the shear waves, whose maximum frequency is about $60 \ cm^{-1}$, on the absorptivity (62). To a first approximation, therefore, they can be neglected. The effective maximum frequency of the longitudinal waves is $\sim 250 \ cm^{-1}$ and the longitudinal elastic constant of ice is $c_{11} = 139.3 \ kbar$ (63), so that

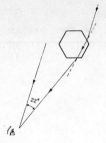

Fig. 5.   Origin of the 22° halo by refraction in
hexagonal crystals of ice Ih.

$$d_{\ell}^{"2} = 6.37 \; D^2 \; \mu m^{-3} \; bar^{-2}$$

$$= 6.37 \; D^2 \; Å^{-3} \; Mbar^{-2}$$

3.3. Halos

Halos around the sun or moon caused by ice particles
in the upper atmosphere occur about 100 times a year in
temperate latitudes. They are caused by refraction of sunlight
or moonlight at minimum deviation by prisms of ice (64). The
most common halo by far is the 22° halo, which occurs about 100
times a year in temperate latitudes. It is caused by refraction
between alternate side faces of hexagonal plates or columns of
ice Ih in the upper atmosphere, usually in altostratus and
similar clouds, as is illustrated in Fig. 5. The angle of
minimum deviation D is determined by the so-called halo
equation

$$D = 2 \; arcsin \; (n \sin \tfrac{1}{2}P) - P, \tag{6}$$

where n is the refractive index of the prism and P the prism
angle. It follows that a halo can occur only when P < 2 arcsin
n^{-1}. Red light is deviated the least, and so the refractive
index of ice for red light at −40°C, which is the temperature of
the upper atmosphere where ice crystals that cause the halos
occur, is needed. It is 1.309 without significant anisotropy
for our purpose (65), and it will predict the inner edge of the
halos. It also predicts that the maximum prism angle for the
formation of a halo is 99.6°. Crystals of ice Ih having faces
whose Miller indices are either 0 or 1, which are illustrated in
Fig. 6, can form the halos listed in Table 1, where the axial
ratio c/a of ice Ih at −40° of 1.6284 (66,67) has been assumed.
All of them have been observed, but all of them that require
refraction through a pyramidal face are very rare on earth. The
radii of the halos in Table 1 are for a point source of very

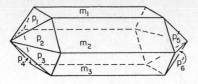

Fig. 6. Crystal of ice Ih with faces having Miller
indices of either 0 or 1 only.

Table 1 . Halos produced by crystals of ice Ih having faces
with Miller indices only 0 or 1.

Refracting faces	cos P^a	P^b	Halo radius^c	Name of halo
$p_1 m_4$ or $m_1 p_4$	$(3/4r^2+1)^{-\frac{1}{2}}$	28.01°	8.87°	Van Buijsen
$p_1 p_2'$	$\dfrac{2r^2/3+1}{4r^2/3+1}$	52.39°	18.08°	Rankin
$p_1 p_4$	$\dfrac{4r^2/3-1}{4r^2/3+1}$	56.01°	19.71°	Burney
$m_1 m_3$	$\frac{1}{2}$	60°	21.61°	22° halo
cp	$(4r^2/3+1)^{-\frac{1}{2}}$	62.00°	22.62°	22° halo
$p_1 m_3$	$\frac{1}{2}(3/4r^2+1)^{-\frac{1}{2}}$	63.80°	23.57°	Dutheil
$p_1 p_3$	$\dfrac{2r^2/3-1}{4r^2/3+1}$	80.25°	34.52°	Feuillee
cm	0	90°	45.09°	45° halo or great ring

^{a}P is the refracting angle and r the ratio c/a of the crystal
axes.
^{b}The ratio r is taken as 1.6284 (66,67).
^{c}The refractive index is taken as 1.309 (65).

distant light, and to determine the actual inner radius, half
the diameter of the source must be subtracted, or 0.25° for
either the sun or the moon as seen from the earth.
        Crystals of ice Ic may also occur in the upper
atmosphere. If the crystals are randomly oriented, they cause
halos, and all the halos expected for crystals that have faces
with Miller indices either 0 or 1 are listed in Table 2. The
halo at 19.12°, which is due to refraction in the angle between
faces o_1 and o_5 in Fig. 7, almost coincides with Burney's halo

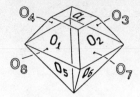

Fig. 7. Crystal of ice Ic with faces having Miller
indices of either 0 or 1 only.

Table 2. Halos produced by crystals of ice Ic and methane.

Refracting faces	Refracting angle	Halo radius in degrees		Name of halo from ice Ic
		Ice Ic[a]	CH₄[b]	
O_1O_3	70.53°	27.45°	29°	Scheiner
a_1O_5	54.74°	19.12°	21°	–
a_1a_2	90°	45.09°	50°	–

[a]The refractive index is taken as 1.309.
[b]The refractive index is taken as 1.33 (65).

at 19.71°, and the halo at 45.09°, which is caused by refraction
through adjacent cube faces, coincides exactly with the 45° halo
of ice Ih. Only the halo at 27.45° is characteristic of ice Ic,
and it shows the presence of octahedral faces. It was first
reported by Scheiner in 1629, and has been reported around the
sun or moon seven times in all (see Ref. 68 for a review). The
formation of Scheiner's halo is illustrated in Fig. 8. Crystals
of ice Ih that cause the 45° halo must also have 60° refracting
edges, and so should also cause the 22° halo. If the 45° halo
occurs without the 22° halo it may, therefore, constitute
presumptive evidence of ice Ic crystals with cube faces.
        Ice particles in the atmosphere usually form by
freezing water droplets. When the suggestion that Scheiner's
halo is produced by crystals of ice Ic in the upper atmosphere
was submitted for publication (69), the direct freezing of water
to ice Ic had not been reported. Since then, droplets of water
of a few μm diameter in emulsion in n-heptane have been frozen
directly to ice Ic by firing a jet of droplets of the emulsion
into liquid ethane at ~95 K (41), and Torchet, Schwarz, Farges,
de Feraudy, and Raoult (51) have shown that particles of ice of
diameter about 60 Å formed by a free jet expansion of water

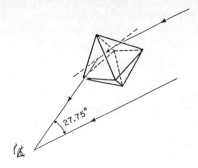

Fig. 8. Origin of the 28° halo by refraction in
         octahedral crystals of ice Ic.

vapor are always ice Ic. Other evidence for the common
occurence of ice Ic in the upper atmosphere is the frequent
occurrence of twinned crystals of ice Ih with their c axes at
~70.5° to one another. They are probably formed from small
octahedral crystals of ice Ic by the growth of Ih crystals with
their basal planes parallel to the octahedral faces of ice Ic
[Takahashi and Kobayashi (1983) and earlier papers].

       These halos may be formed by ice particles surrounding
the planets and their satellites, for example the particles in
Saturn's rings, and would be worth looking for from spacecraft
in suitable positions. Attempts to see from earth the 22° halo
in the atmosphere of Venus, which should occur near inferior
conjunction, have so far been unsuccessful (71-73).

       The halos are formed by randomly oriented crystals.
If the crystals are partly or completely oriented, for example
by aerodynamic forces when falling through an atmosphere, other
halos, arcs, and parhelia occur, and there are in addition a
number of colorless halos caused by reflections from the crystal
faces or by refraction by crystals without dispersion (74,75).

       Clathrate hydrates may form in the atmospheres of the
planets and satellites if the composition is correct. Methane,
nitrogen, and carbon dioxide clathrates are prime candidates.
There are two principal kinds of clathrate hydrate, both having
cubic strucutres, structures I and II having unit-cell edges of
~12.0 and ~17.0 Å respectively. They appear to form octahedral
crystals according to some century-old work (76,77) that D.W.
Davidson kindly told us about. They would, therefore, form
halos like those produced by ice Ic, but slightly shifted
because of the differences in refractive index, and may be
detectable by means of the halos.

       Methane crystals may form in the atmosphere of the
outer planets. Their structure is cubic, and so they may be
octahedra or cubes. The predicted halos are listed in Table 2.

Since this paper was presented, the diameters of halos produced
by substances that may form cubic crystals in the planetary
atmospheres have been calculated (78), and may be useful in
identifying them.

4. ENERGIES OF THE PHASES OF ICE

To understand the behavior of ice in the planets and
satellites, we will undoubtedly need to know many properties
that have not been measured. Perhaps some properties that we
need may not be measurable because the planetary system can
afford to wait much longer for equilibrium than the
experimenters can. If great accuracy is not needed, or if the
properties cannot be accurately measured because equilibrium is
reached too slowly or because there are other difficulties, the
properties that are needed can be calculated from suitably
chosen intermolecular potentials. Furthermore, in the for-
seeable future, molecular theory will often allow us to
calculate properties that we now measure, and so free the
experimenter for more interesting experiments. The potentials
will need testing on suitable known properties, and the most
elementary properties are the structures, densities, and heats
of sublimation at zero temperature. These tests were started
several years ago (79) on a model resembling ice Ih, and Morse
and Rice (80,81) have studied ordered forms of ice Ih, II, IX,
and VIII.

The densities and structures have been well determined
by x-ray and neutron diffraction [see Whalley (18) for a
review]. The heat of sublimation of ice Ih is 47.341 ± 0.015 kJ
mol^{-1} (82). Estimates of the difference of the energies of two
phases can be obtained from two principal sources, the
equilibrium pressure extrapolated to zero temperature and the
heat released when a high-pressure phase is transformed to ice
Ic on heating (44,45).

• At zero temperature and the equilibrium pressure p_e,
the difference of internal energy U of two phases is

$$\Delta U(p_e, T=0) = -\Delta V(p_e, T=0)p_e \ ,$$

where ΔV is the difference of the volumes of the two phases.
The difference of energy at zero pressure can then be obtained
from the equation

$$\Delta U(p=0) = \Delta U(p=p_e) + \int_0^{p_e} \Delta(V\kappa)p\,dp,$$

where κ is the isothermal compressibility. When allowance is
made for the effect of pressure on the volume differences, this
equation becomes (Whalley, 1984)

$$\Delta U(p=0, T=0) = -\Delta V(p=0, T=0)p_e + \tfrac{1}{2}\Delta(V\kappa)p_e^2.$$

The difference of volumes at zero temperature has been estimated
from the measured difference of volumes on the equilibrium lines
and from the difference of volumes determined by x-ray or
neutron diffraction at ~100 K and zero pressure. The
compressibilities of all the phases except VII and VIII may be
estimated from a model that assumes only first-neighbor
interactions and neglects changes in O--O--O angles. These
assumptions lead to the equation

$$\kappa = 9/Nkr^2 ,$$

where N is the number density of molecules, k the intermolecular
hydrogen-bonded O--O force constant, and r the mean O--O
distance. The force constant was assumed to be the same for all
phases of ice except VII and VIII and the compressibilities were
estimated from the measured compressibility of ice I by the
equation

$$\kappa = \kappa_I N_I r_I^2 /Nr^2 ,$$

where subscript I refers to ice I. The compressibility of ice
VIII was assumed to be the same as that of ice VII (83). From
the differences of energy between the various phases,
extrapolated to zero temperature and pressure, the energy of
each phase relative to ice Ih can be estimated with reasonable
accuracy (23) and the results are summarized in Table 3.
     The energies of ice Ih, II, IX, and VIII are compared
with the energies calculated by Morse and Rice (81) in Table 4.
The best potential is the ST2, which predicts an error in the
density of -3.7% and in the energy of +9.5%. It also has the
energies of the three phases in their right order, while the MCY
and RSL2 do not. However, the ST2 overestimates the difference
in energy between ice I and VIII by a factor of ~2.7.

Table 3. Energies of the crystalline phases of ice at zero
          temperature and pressure (23) relative to the value
          for ice Ih of 47.341 ± 0.025 kJ mol^{-1} (82)

Phase change	1-II	I-III	I-IX	I-IV	I-V	1-VI	1-VII	1-VIII
$\Delta U(T=0, p=0)$ kJ mol^{-1}	0.057	0.90	0.34	1.45	0.95	1.62	4.25	3.16

5. DISCUSSION

     The physics of ice is, of course, a large subject, and
as we learn more about the solar system, more of its properties
become important to astronomers. This review has touched on

Table 4. Comparison of the experimental and theoretical
         lattice energies of some phases of ice.

Phase	Experimental energy/kJ mol⁻¹a	Calculated energy/kJ mol⁻¹b		
		MCY	ST2	RSL2
I[c]	−63.86	−49.61	−60.97	−52.04
II	−63.80	−46.58	−57.77	−48.97
IX	−63.52	−47.87	−57.01	−51.50
VIII	−62.4	−40.56	−57.00	−44.69

[a]The effect of the zero-point energy of the intermolecular vibrations, but not the intramolecular vibrations,
has been allowed for.
[b]Ref. 81
[c]The calculations were for a hypothetical ordered form.

several topics that have been worked on recently in our
laboratory, but a large amount of work by others that is
also important for astronomy has had to be omitted. Some of it is covered in following papers, and some of it
has been referred to in other papers in this Workshop.
        It was quite clear throughout the Workshop that
the astronomers have discovered much about the planetary system, and in many areas a lack of knowledge of the
physics and chemistry of ice hinders interpretation of
astronomical observations. I hope that the publication
of this volume will encourage all of us to contribute
what we can.

REFERENCES

1. Murray, James A. H., ed., Henry Bradley, Author,
    1901, A new English dictionary. Clarendon, Oxford,
    Vol. IV, p. 192.

2. Gove, Philip Babcock, 1971, Webster's Third New
    International Dictionary. Merriam, Springfield,
    p. 961.

3. Hobbs, Peter V.. Ice Physics, Clarendon, Oxford,
    1974.

4. Tammann, G., 1900, Uber die Grenzen des Festens
    Zustandes IV. Ann. Physik 4te Series 2, pp. 1-31.

5. Bridgman, P. W., 1911, Water, in the liquid and fi-
    ve solid forms, under pressure. Proc. Am. Acad.
    Arts Sci., 47, pp. 441-558.

6.   Babb, Stanley E., 1963, Some notes concerning Brid-
     gman's manganin pressure scale. In "High pressure
     measurement", ed. A. A. Giardini and Edward C.
     Lloyd. Butterworths, Washington, pp. 115-123.

7.   Kell, G. S. and Whalley, E., 1968, Equilibrium line
     between ice I and III. J. Chem. Phys., 48, pp.
     2359-2361.

8.   Bridgman, P. W., 1935, The pressure-volume-tempera-
     ture relations of the liquid, and the phase dia-
     gram of heavy water. J. Chem. Phys., 3, pp. 597-
     605.

9.   Evans, L. F., 1967, Selective nucleation of the
     high-pressure ices. J. Appli. Phys., 38, pp. 4930-
     4932.

10. Engelhardt, H. and Whalley, E., 1972, Ice IV. J.
     Chem. Phys., 56, pp. 2678-2684.

11. Nishibata, K., 1972, Growth of ice IV and equili-
     brium curves between liquid water, ice IV, ice V
     and ice VI. Japan. J. Appl. Phys., 11, pp. 1701-
     1708.

12. Bridgman, P. W., 1937, The phase diagram of water
     to 45.000 kg/cm^2. J. Chem. Phys., 5, pp. 964-966.

13. König, Hans, 1943, Eine kubische eismodifikation.
     Z. Kristallogr., 105, pp. 279-286.

14. Gormley, J. A., 1968, Enthalpy changes and heat ca-
     pacity changes in the transformations from high-
     surface-area amorphous ice to stable hexagonal ice.
     J. Chem. Phys., 48, pp. 503-508.

15. Di Marzio, E. A. and Stillinger, F. H., 1964, Resi-
     dual entropy of ice. J. Chem. Phys., 40, pp. 1577-
     1581.

16. Nagle, J. F., 1966, Lattice statistics of hydrogen-
     bonded crystals. I. The residual entropy of ice.
     J. Math. Phys., 7, pp. 1484-1491.

17. Tajima, Y., Matsuo, T. and Suga, H., 1982, Phase
     transition in KOH-doped hexagonal ice. Nature,
     299, pp. 810-812.

18. Whalley, E., 1976, The hydrogen bond in ice. In
     "The hydrogen bond. III. Dynamics, thermodynamics
     and special systems". Ed. P. Shuster, G. Zundel
     and C. Sandorfy. North Holland, Amsterdam, pp.
     1425-1470.

19. Hamilton, W. C., Kamb, B., LaPlaca, S. J. and Pra-
     kash, A., 1971, Ordered proton configuration in
     ice II from single-crystal neutron diffraction. J.
     Chem. Phys., 55, pp. 1934-1945.

20. Wilson, G. J., Chan, R. K., Davidson, D. W. and
     Whalley, E., 1965, Dielectric properties of ices
     II, III, V and VI, J. Chem. Phys., 43, pp. 2384-
     2391.

21. Kamb, B., 1964, Ice II : A proton-ordered form of
     ice. Acta Cryst., 17, pp. 1437-1449.

22. Whalley, E. and Davidson, D. W., 1965, Entropy chan-
     ges at the phase transitions in ice. J. Chem. Phys.
     43, pp. 2148-2149.

23. Whalley, Edward. Relative energies of the phases
     of ice at zero temperature and pressure. J. Chem.
     Phys. in press.

24. Whalley, E., 1969, Structure problems of ice. In
     "Physics of Ice". Ed. N. Riehl, B. Bullemer and
     H. Engelhardt. Plenum, New York, pp. 19-43.

25. Wong, P. T. T. and Whalley, E., 1976, Raman spec-
     trum of ice VIII. J. Chem. Phys., 64, pp. 2359-
     2366.

26. Whalley, E., Davidson, D. W. and Heath, J. B. R.,
     1966, Dielectric properties of ice VII. Ice VIII :
     a new phase of ice. J. Chem. Phys., 46, pp. 3976-
     3982.

27. Brown, A. J. and Whalley, E., 1966, A preliminary
     investigation of the phase boundaries between ice
     VI and VII and ice VI and VIII. J. Chem. Phys.,
     45, pp. 4360-4361.

28. Johari, G. P., Lavergne, A. and Whalley, E., 1974,
     The dielectric properties of ice VII and VIII and
     the phase boundary between ice VI and VII. J.
     Chem. Phys., 61, pp. 4202-4300.

29,   Whalley, E., Heath, J. B. R. and Davidson, D. W.,
      1968, Ice IX : An antiferroelectric phase related
      to ice III. J. Chem. Phys., 48, pp. 2362-2370.

30.   Nishibata, K. and Whalley, E., 1974, Thermal effects
      of the transformation ice III-IX. J. Chem. Phys.,
      60, pp. 3189-3194.

31.   LaPlaca, S. J., Hamilton, W. C., Kamb, B. and Pra-
      kash, A., 1974, A nearly proton-ordered structure
      for ice IX. J. Chem. Phys., 58, pp. 567-580.

32.   Mishima, O. and Endo, S., 1980, Phase relations of
      ice under pressure. J. Chem. Phys., 73, pp. 2454-
      2456.

33.   Johari, G. P. and Whalley, E., 1976, Dielectric
      properties of ice VI at low temperatures. J. Chem.
      Phys., 64, pp. 4484-4489.

34.   Kamb, B., 1973, Crystallography of ice. In "Physics
      and chemistry of ice", ed. E. Whalley, S. J. Jones
      and L. W. Gold. Royal Society of Canada, Ottawa,
      pp. 28-41.

35.   Johari, G. P. and Whalley, E., 1979, Evidence for a
      very slow transformation in ice VI at low tempera-
      tures. J. Chem. Phys., 70, pp. 2094-2097.

36.   Miller, Stanley L.. Clathrate hydrates in the solar
      system. This volume.

37.   Davidson, D. W., 1973, Clathrate hydrates. In "Wa-
      ter, a comprehensive treatise", ed. F. Franks, Ple-
      num, New York, pp. 115-234.

38.   Burton, E. F. and Oliver, W. F., 1938, The crystal
      structure of ice at low temperatures. Proc. Roy.
      Soc. A 153, pp. 166-172.

39.   Tammann, G. and Starinkewitsch, J., 1913, Über die
      Bildung von Glas auf Dampf. Z. Physik Chem., 85,
      pp. 573-578.

40.   Olander, David S. and Rice, Stuart A., 1972, Prepa-
      ration of amorphous solid water. Proc. Nat. Acad.
      Sci. U. S., 69, pp. 98-100.

41.   Brüggeller, Peter and Mayer, Erwin, 1980, Complete
      vitrification in pure liquid water and dilute
      aqueous solutions. Nature, 288, pp. 569-571.

42. Mayer, Erwin and Brüggeller, Peter, 1982, Vitrifi-
     cation of pure liquid water by high pressure jet
     freezing. Nature, 298, pp. 715-717.

43. Kanno, H., Speedy, R. J. and Angell, C. A., 1975,
     Science 189, pp. 880-881.

44. Bertie, J. E., Calvert, L. D. and Whalley, E.,
     1963, Transformations of ice II, ice III, and ice
     V at atmospheric pressure. J. Chem. Phys., 38,
     pp. 840-846.

45. Bertie, J. E., Calvert, L. D. and Whalley, E.,
     1964, Transformation of ice VI and ice VII at
     atmospheric pressure. Can. J. Chem., 42, pp. 1373-
     1378.

46. Mishima, O., Calvert, L. D. and Whalley, E., 1984,
     "Melting" ice I at 77 K. A new method of making
     amorphous solids. Nature, 310, pp. 393-395.

47. Mishima, O., Calvert, L. D. and Whalley, E., In
     preparation.

48. Klinger, J., 1983, Extraterrestial ice. A review.
     J. Phys. Chem., 87, pp. 4209-4214.

49. Bertie, John E. and Devlin, J. Paul, 1983, Infrared
     spectroscopic proof of the formation of the struc-
     ture I hydrate of oxirane from annealed low-tempe-
     rature condensate. J. Chem. Phys., 78, pp. 6340-
     6341.

50. Torchet, G., Schwartz, P., Farges, J., de Feraudy,
     M. F. and Raoult, B., 1983, Structure of solid wa-
     ter clusters formed in a free jet expansion. J.
     Chem. Phys., 79, pp. 6196-6202.

51. Bertie, J. E. and Whalley, E., 1964, Infrared spec-
     tra of ices Ih and Ic in the range 4000 to 350
     cm^{-1}. J. Chem. Phys., 40, pp. 1637-1645.

52. Bertie, John E. and Jacobs, Stephen M., 1977, Far
     infrared absorption in ices Ih and Ic at 4.3 K and
     the powder diffraction pattern of ice Ic. J. Chem.
     Phys., 67, pp. 2445-2558.

53. Whalley, E. and Bertie, J. E., 1967, Optical spec-
     tra of orientationally disordered crystals. I.
     Theory for translational lattice vibrations. J.
     Chem. Phys., 46, pp. 1264-1270.

54. Bertie, J. E. and Whalley, E., 1967, Optical spec-
     tra of orientationally disordered crystals. II.
     Infrared spectrum of ice Ih and Ic from 360 to 50
     cm^{-1}. J. Chem. Phys., 46, pp. 1271-1284.

55. Greenberg, J. Mayo. Chemical evolution of inters-
     tellar dust ice. This volume.

56. Mishima, O., Klug, D. D. and Whalley, E., 1983, The
     far-infrared spectrum of ice Ih in the range 8-25
     cm^{-1}. Sound waves and difference bands, with appli-
     cation to Saturn's rings. J. Chem. Phys., 78, pp.
     6399-6404.

57. Whalley, E. and Labbé, H. J., 1969, Optical spectra
     of orientationally disordered crystals. III. Infra-
     red spectra of the sound waves. J. Chem. Phys., 51,
     pp. 3120-3127.

58. Szigeti, B., 1960, The infrared spectra of crystals.
     Proc. Roy. Soc. A 258, pp. 377-401.

59. Epstein, Eugene E., Janssen, Michael A., Cuzzi,
     Jeffrey N., Fogarty, William G. and Mattmann,
     John, 1980, Saturn's rings : 3 -mm observations
     and derived properties. Icarus, 41, pp. 103-118.

60. Epstein, Eugene E., Janssen, Michael A. and Cuzzi,
     Jeffrey N. Saturn's rings : 3 -mm low-inclination
     observations and derived properties. Icarus, in
     press.

61. Whalley, Edward and Klug, Dennis D. Disorder-indu-
     ced piezoelectricity. I. The theory and applica-
     tion to the sound waves in ice. J. Chem. Phys., in
     press.

62. Bertie, J. E., Labbé, H. J. and Whalley, E., 1969,
     Absorptivity of ice I in the range 4000-30 cm^{-1}.
     J. Chem. Phys., 50, pp. 4501-4520.

63. Gammon, P. H., Kiefte, H. and Clouter, M. J., 1983,
     Elastic constants of ice samples by Brillouin spec-
     troscopy. J. Phys. Chem., 87, pp. 4025-4029.

64. Greeler, Robert, 1980, Rainbows, halos and glories.
     Cambridge University Press.

65. Merwin, H. E., 1930, Internat. Crit. Tables, Vol.
     7, p. 17.

66. LaPlaca, S. and Post, B., 1960, Thermal expansion
     of ice. Acta Cryst., 13, pp. 503-505.

67. Brill, R. and Tippe, A., 1967, Gitterparameter von
     Eis I bei tiefen temperaturen. Acta Cryst., 23,
     pp. 343-345.

68. Whalley, E., 1983, Cubic ice in nature. J. Phys.
     Chem., 87, pp. 4174-4179.

69. Whalley, E., 1981, Scheiner's halo : evidence for
     ice Ic in the atmosphere. Science 211, pp. 389-
     390.

70. Takahaski, T. and Kobayashi, T., 1983, The role of
     the cubic structure in freezing of a supercooled
     water droplet on an ice substrate. J. Cryst. Growth
     64, pp. 593-603.

71. O'Leary, Brian T., 1966, The presence of ice in the
     Venus atmosphere as inferred from a halo effect.
     Astrophys. J. 146, pp. 754-766.

72. Veverka, J., 1971, A polarimetric search for a Ve-
     nus halo during the 1969 inferior conjunction.
     Icarus, 14, pp. 282-283.

73. Können, G. P., 1983, Polarization and intensity dis-
     tribution of refraction halos. J. Opt. Soc. Am.,
     73, pp. 1629-1640.

74. Tricker, R. A. R., 1979, Arcs associated with halos
     of unusual radii. J. Opt. Soc. Am., 69, pp. 1093-
     1100.

75. Tricker, R. A. R., 1979, Ice crystal halos. Optical
     Society of America, Washington.

76. de Forcrand, R., 1883, Recherches sur les hydrates
     sulphydrés. Ann. Chim. Phys., 28, pp. 5-66.

77. Villard, P., 1897, Etude expérimentale des hydrates
     de gaz. Ann. Chim. Phys. Ser. 7, 11, pp. 289-394.

78.   Whalley, Edward and McLaurin, Graham E.. Halos in
      the solar system. I. Halos from cubic crystals
      that mays occur in the solar system. J. Opt. Soc.
      Am. in press.

79.   Huler, E. and Zunger, A., 1976, Calculation of the
      equilibrium configuration and intermolecular fre-
      quencies of water dimers and hexagonal ice. Chem.
      Phys., 13, pp. 433-440.

80.   Morse, M. D. and Rice, S. A., 1981, A test of the
      accuracy of an effective pair potential for liquid
      water. J. Chem. Phys., 74, pp. 6514-6516.

81.   Morse, M. D. and Rice, S. A., 1982, Tests of effec-
      tive pair potentials for water : predicted ice
      structures. J. Chem. Phys., 76, pp. 650-660.

82.   Whalley, E., 1957, The difference in the intermole-
      cular forces of H_2O and D_2O. Trans. Faraday Soc.
      53, pp. 1578-1585.

83.   Walrafen, G. E., Abebe, M., Mauer, F. A., Block,
      S., Piermarini, G. J. and Munro, R. G., 1982, J.
      Chem. Phys., 77, pp. 2166-2174.

PARTIAL PHASE DIAGRAM FOR THE SYSTEM NH_3-H_2O:
THE WATER-RICH REGION

Mary L. Johnson, Andrea Schwake, and Malcolm Nicol

Department of Chemistry and Biochemistry
University of California
Los Angeles, CA 90024 USA

ABSTRACT. This paper describes the status of our diamond-anvil cell experiments of the NH_3-H_2O P-T-X phase diagram. Two parts of the diagram are discussed: the isothermal P-X diagram of $(NH_3)_X(H_2O)_{1-X}$ for $0 \leq X \leq 0.3$ and the melting line of $(NH_3)_{0.5}(H_2O)_{0.5}$. Five phases were observed, including: the liquid; ice VI and ice VII (which resembled their descriptions in the literature); a high-relief apparently isotropic ammonia monohydrate phase; and a low-relief, strongly anisotropic ammonia dihydrate phase. Our interpretation of the isothermal P-X diagram has been modified by recognition that difficulty in nucleating ice VI had led us to underestimate the importance of the ice VI plus liquid and ice VI plus ammonia dihydrate fields. Some unresolved questions and directions for future studies are discussed.

1.   INTRODUCTION

Ammonia and water have been proposed as major constituents of such bodies as Neptune [1], Uranus [1, 2, 3], Titan [4], and Comet Bowell [5]. Hunter et al. estimate the cosmic abundance of ammonia to be about 18% that of water [4]. Hence, the study of the system NH_3-H_2O at various temperatures and pressures has great relevance to the understanding of the geology of bodies in the solar system. The P-T diagram of water has been examined by a number of investigators, most recently by Mishima and Endo [6]; and Mills et al. [7] have determined the corresponding diagram for NH_3.

    The ammonia-water system at atmospheric pressure has been

J. Klinger et al. (eds.), Ices in the Solar System, 39–47.

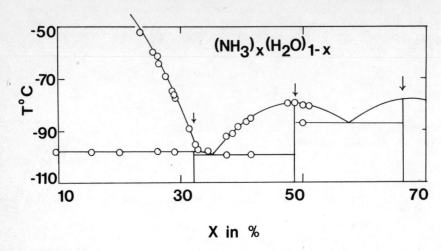

Figure 1. Part of the T-X phase diagram for the ammonia-water system at atmospheric pressure, from [8]. Arrows indicate the solid phases: ammonia dihydrate, ammonia hydrate, and ammonia hemihydrate.

explored by Rollet and Vuillard [8]. Their results are shown in Figure 1. They found that $NH_3 \cdot 2H_2O$ (ammonia dihydrate, D) melts incongruently and that $NH_3 \cdot H_2O$ (ammonia monohydrate, M), which is nonstoichiometric, melts at a significantly higher temperature than the dihydrate.

This paper is a progress report that describes our work on the phase diagram of the NH_3-H_2O system, our current interpretation of the available data, and additional studies that we intend to pursue. We describe experimental studies of phase boundaries in the water-rich region of the isothermal P-X diagram at room temperature and an approximate melting curve for the monohydrate composition. A tentative interpretation of these observations has been reported elsewhere [9,10]; however, recent experiments by Boehler [11] demonstrate that the earlier interpretation must be modified because liquid and ice VII and, possibly, glassy solids may occur metastably in the ice VI field.

2.   MATERIALS AND METHODS

Two sets of experiments have been performed. In the first, the phases present were determined as a function of pressure in mixtures of several compositions up to 30 mole percent NH_3 at room temperature, $21 \pm 2^{\circ}$ C; in the second, the solid-liquid boundary for the composition, $(NH_3)_{0.5}(H_2O)_{0.5}$, was monitored at several temperatures and pressures. For both sets of experiments, the samples were enclosed in a diamond-anvil high pressure cell of Holzapfel's design [12]; and phase transitions

were observed visually while the pressure and/or temperature was
cycled. The ruby-luminescence method [13] was used to determine
the pressure.

For the first set of experiments, each sample was prepared
by mixing distilled water with reagent ammonium hydroxide
(Mallinkrodt Analytical Reagent) and was loaded as a liquid into
the diamond cell. Concentrations were measured with a hydrometer,
which had been calibrated against titration with standard acid.
All phases were identified optically under polarized light.

In the second set of experiments, samples were prepared by
partially crystallizing approximately equimolar mixtures of
distilled water and liquid ammonia (Air Products Co., 99.99%
pure) at low temperatures and decanting the supernatant liquids.
Titration of these samples with standard acid indicated that
their NH$_3$ concentrations varied from 46 to 50 mole percent;
effects of these variations on the observed phase boundaries are
discussed below. The diamond cell was enclosed in a cryostat with
which the temperature of the cell was varied from above room
temperature to 120 K with a combination of resistance heating
and cryogenic cooling.

3.   RESULTS AND DISCUSSION

3.1. Room Temperature Studies

Our observations at room temperature are shown in Figure 2.
Filled circles represent ice VII; striped circles, ice VI;
hatched circles, ammonia dihydrate; stippled circles, ammonia
hydrate; open circles, liquid; and halved circles, two-phase
fields. Arrows adjacent to the circles indicate the direction in
which the pressure had been changed immediately before the
observation was made. In all cases, the phases were identified
visually, usually both with and without crossed polarizers.

The ice phases resembled their descriptions in the
literature [14,15]. The dihydrate phase had less relief than the
liquid, was anisotropic under crossed nicols, and tended to form
two networks of cracks at about a 70° angle to each other at low
pressures. When squeezed above 3.3 GPa, dihydrate reconstituted
into myrmekitic intergrowths of Ice VII with a higher-relief
phase which appeared to be isotropic. From calculation of the
area of the diamond cell occupied by the various phases, the
low-pressure phase was shown to have a composition near
NH$_3 \cdot$2H$_2$O, and the high-pressure phase to have composition near
NH$_3 \cdot$H$_2$O.

Nucleation problems complicated this work considerably, with
ice VI being especially difficult to nucleate: in 24 separate
measurements between 0 and 30 mole percent NH$_3$ where ice VI
should have occured, it was detected only 8 times, only twice at
compositions other than 5 mole percent NH$_3$, and never observed

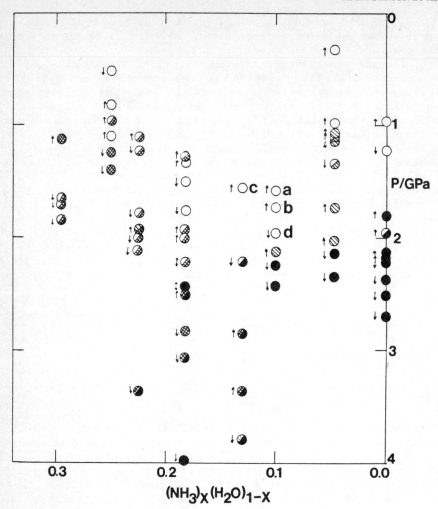

Figure 2. Our observations plotted on the isothermal P-X cut of the ammonia-water system for 21°C and mole percentages of ammonia to 0.3. The points are coded to indicate the observation of liquid (open), Ice VI (diagonal shading), Ice VII (solid), ammonia dihydrate (cross-hatching), or ammonia monohydrate (dots). Two-phase observations are indicated by bisected circles, with codes for both phases indicated. The arrow next to each point indicates the direction in which the pressure had been changed just before that observation was made. Points labeled (a) through (d) are discussed in the text.

at the pure water composition. The points in Figure 2 labeled (a) and (b) at 10 mole percent NH$_3$, and (c) at 13 mole percent NH$_3$ resulted from the melting of ice VII, not ice VI. A glaze of ice VI was observed during one decompression cycle as the pressure was lowered from point (d) at 10 mole percent NH$_3$ and 1.97 GPa; however, the pressure at which this glaze disappeared was not determined. This nucleation problem is not simply limited to this laboratory; both Yamamoto [14] and Piermarini et al. [16]

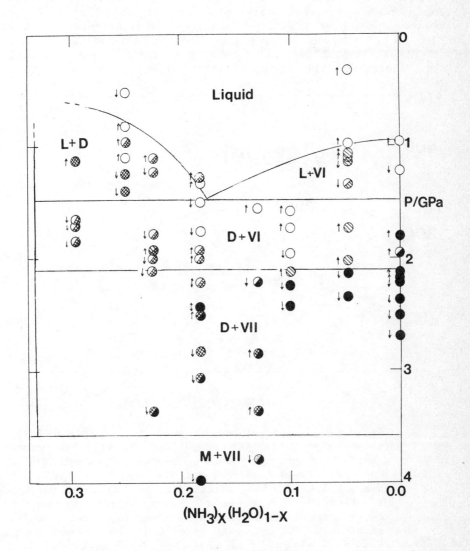

Figure 3. The revised isothermal cut of the phase diagram for the observations shown in Figure 2.

have described the metastable liquid-ice VII equilibrium in the
ice VI stability field.

The problems with ice VI are confirmed by Boehler's recent
adiabatic compression measurements of NH_3–H_2O mixtures in a
piston-cylinder apparatus [11]   which indicate that the
ice VI-liquid field extends further into the liquid region than
we found at several of the compositions described in Figure 2. In
passing, we note that adiabatic compression work in a
piston-cylinder apparatus is subject to other complications. The
sample cannot be seen   so   phases must be identified by
inference. Also, at the compression/decompression rates of the
order to 10 MPa/sec [11], glassy ammonia-bearing phases [17]
rather than crystalline solids may form; compare, for instance,
[18]. Therefore, the liquid-to-liquid plus ice VI transition

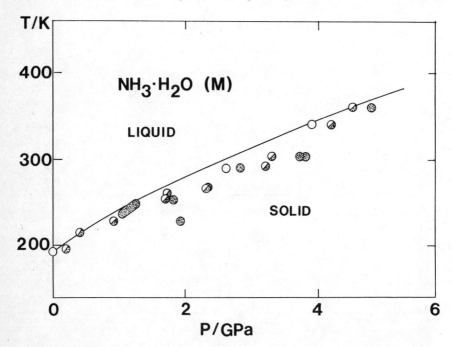

Figure 4. Melting observations for $(NH_3)_X(H_2O)_{1-X}$
samples for X near 0.5 from [10]. The stippled points
indicate where crystals were observed, open points show
where fluid was observed, and the coexistence of fluid
and solid phases is indicated by half-stippled points.
If $NH_3 \cdot H_2O$ is assumed to melt congruently at all
temperatures and pressures, its melting curve must lie
at or above the temperatures of these two-phase points.
Lower   melting temperatures may occur if the
monohydrate melts incongruently, as suggested by Lunine
[19].

pressures determined by Boehler should be upper limits. A combination of methods is more likely than any one technique to reveal the equilibrium phase diagram.

A phase diagram that is consistent with both sets of results is given in Figure 3. At compositions accessible with reagent grade ammonium hydroxide, the following phases were observed:

> Liquid
> Liquid and dihydrate (at or below 1.0 GPa)
> Liquid and Ice VI (at 0.98 GPa)
> Dihydrate and Ice VI (at 1.5 $\pm$ 0.3 GPa)
> Dihydrate and Ice VII (at 2.1 GPa)
> Monohydrate and Ice VII (at 3.6 $\pm$ 0.2 GPa)

An issue that must still be examined is whether the low-temperature and high-pressure hydrates are equivalent. Both NH$_3$ and H$_2$O have many solid phases. At low temperatures, NH$_3$·H$_2$O has an orthorhombic structure [18], while the structure of NH$_3$·2H$_2$O has not been determined. Although there is no evidence that the mono- and dihydrate phases observed at room temperature differ from their low-temperature counterparts, further studies of vibrational spectra or x-ray diffraction patterns are needed to identify the phases.

3.2. (NH$_3$)$_{0.5}$(H$_2$O)$_{0.5}$

Our observations near this composition are shown in Figure 4. The stippled and open points indicate where the samples were solid or fluid, respectively; half-stippled points represent conditions where solid and liquid coexisted. The variable (48 $\pm$ 2 mole percent) compositions of these samples complicate the interpretation of these data. If we assume that the monohydrate melts congruently at all temperatures and pressures, the melting curve must lie at or above (higher temperatures) the two-phase points shown in this figure; however, this picture may be too simple. Lunine [19] has suggested an alternative interpretation based on the data in Figure 2: that the monohydrate melts incongruently to dihydrate plus liquid at room temperature. Further work must be done by visual methods with polarized optics, vibrational spectroscopy, or x-ray diffraction to identify the phases present.

4.   SUMMARY

Although additional experimental work needs to be done, parts of the P-T-X phase diagram for the NH$_3$-H$_2$O system that are consistent with all available observations have tentatively been determined in the water-rich region at room temperature and near equimolar composition at high pressures. We intend to continue our studies of this system to include:

   a.   extension of the low-ammonia phase diagram to other interesting temperatures;

   b.   investigation of the structures and stoichiometries of the high-pressure hydrates; and

   c.   determination of the densities (and, possibly, viscosities) of the solids and fluids in the NH_3-H_2O system.

ACKNOWLEDGEMENTS

We are grateful to Dr. R. Boehler for sharing his unpublished results, to Dr. J. Lunine for a copy of his calculations, and to Drs. R. Boehler, J. Lunine and D.J. Stevenson for their enlightening comments. This work was supported by NASA Grant NAGW-104.

REFERENCES

1.   Smoluchowski, R. 1983, The Moon and Planets 28, pp. 137-154.
2.   Brown, R.H., Cruikshank, D.P., and Morrison, D. 1982, Nature 300, pp. 423-425.
3.   Gulkis, S., Olsen, E.T., Klein, M.J., and Thompson, T.J. 1983, Science 221, pp 453-455.
4.   Hunter, D.M., Tomasko, M.G., Flasar, F.M., Samuelson, R.E., Strobel, D.F., and Stevenson, D.J. 1983, ′Interior and Surface of Titan,′ in Gehrels, T., ed., Saturn (University of Arizona Press, Tucson) pp. 671-759.
5.   Campins, H., Rieke, G.H., and Lebofsky, M.J. 1983, Nature 301, pp. 405-406.
6.   Mishima, O., and Endo, S.J. 1980, J. Chem. Phys. 73, pp. 2454-2456.
7.   Mills, R.L., Liebenberg, D.H., and Pruzan, Ph. 1982, J. Phys. Chem. 86, pp. 5219-5222.
8.   Rollet, A.-P., and Vuillard, G. 1956, Comptes. rendus Acad. Sci. Paris 243, pp. 383-386.
9.   Johnson, M.L., Nicol, M., and Schwake, A. 1983, Trans. Am. Geophys. Union 64, p. 875.
10.   Johnson, M.L., Nicol, M., and Schwake, A. 1984, Lun. Planet. Sci. XV Abstracts, pp. 405-406.
11.   Boehler, R. 1984, [personal communication].
12.   Hirsch, R., and Holzapfel, W.B. 1981, Rev. Sci. Instrum. 52, pp. 52-55.
13.   Piermarini, G.J., Block, S., Barnett, J.D., and Forman, R.A. 1975, J. Appl. Phys. 46, pp. 2774-2780.
14.   Yamamoto, K. 1980, Jpn. J. Appl. Phys. 19, pp. 1841-1845.
15.   Whatley, L.S., and vanValkenburg, A. 1967, in Bradley, R.S. ed., Advances in High Pressure Research (Academic Press, London) v. 1, pp. 327-337.

16. Piermarini, G.J., Munro, R.G., and Block, S. 1983, Abstracts of the IX AIRAPT Conf. High Pressure Res., AF2-1.
17. vanKasteren, P.H.G 1973, Bull. Inst. Int. Froid Annexe 4, pp. 81.
18. Wada, T., Kimura, F., and Y. Matsuo, Y. 1983, Bull. Chem. Soc. Jpn. 56, pp. 3827-3829.
19. Lunine, J. 1984 [personal communication].

PHASE TRANSITIONS IN SOLID METHANE AT HIGH PRESSURE

N.J. Trappeniers

Van der Waals Laboratory, University of Amsterdam, The
Netherlands

ABSTRACT

   Methane, in both the solid and the non-condensed phases,
plays an important role in the structure of our planets, where extreme conditions of low temperature and high pressure are prevalent. This paper reports an investigation of the phase diagram of
solid methane up to 60 kbar and down to 2.1 K, using the method
of nuclear magnetic resonance and of a diamond anvil cell.

1. INTRODUCTION

   Methane is a major component of the outer planets and it is
present also in many of their moons. There is good evidence that
in these celestial bodies methane is subjected to extreme conditions of low temperature and high pressure. It is of importance,
therefore, to ascertain which phase changes solid methane may undergo in these circumstances and to establish its phase diagram
over a wide range of temperature and pressure. Such a diagram
could fulfil a useful purpose similar to that of the well-known
phase diagram of ice, when models for the structure of outer
planets and their moons are set up.
   The present paper will review briefly the literature data of
the subject of solid methane, and report on the present work at
the Van der Waals Laboratory regarding the behaviour of solid
methane at pressures up to 60 kbar.

49

J. Klinger et al. (eds.), Ices in the Solar System, 49–58.
© *1985 by D. Reidel Publishing Company.*

2. HISTORICAL

Liquid methane solidifies at 90.7 K into an f.c.c.-phase α or I (recent nomenclature), which has all the characteristics of a plastic crystal: low melting entropy, little resistance to defor- mation, high rotational freedom of the molecules. In fact the me- thane molecules, which have a spherical shape with slight tetra- hedral anisotropy, behave in phase I like almost free rotators.

In 1929 Clusius (1) discovered by a calorimetric experiment, that phase I undergoes a transition at 20.4 K to the β or II phase. The transition is of the well-known order-disorder type, showing a λ-type anomaly of the specific heat, indicating that phase II shows more order, i.e. less rotational freedom of the molecules.

An intriguing fact, found by Clusius and co-workers (2) in 1937, is that the heavy methane, CD_4, exhibits two phase transi- tions, respectively at 27.2 K (I → II) and at 22.4 K (II → III or γ). In a beautiful experiment Eucken et al. (3), in 1938, were able to prove from a study of mixtures of CH_4 and CD_4, with grad- ually diminishing concentration of CD_4, that the transition II → III disappears at concentrations below 20% CD_4. This was seen as definite proof that it is indeed phase III which is missing in the solid CD_4 at low temperature.

The problem of the missing phase III in CD_4 has stimulated much high pressure research on the subject of solid methane as it was surmized that the absence of an ordered phase III is due to the disturbing effect of the zero-point energy of CH_4, which is large compared to that of the heavy methane CD_4. It is possible then that an increase of density may overcome this disorder effect and thus also lead to a phase III in the light methane CH_4, stable under conditions of high pressure. A number of investigators (Trapeznikowa (4), 1939; Stevenson (5), 1957; Stewart (6), 1959, and others) have produced good evidence for such an occurrence but, due to the crude experimental methods, the phase diagram of solid methane remained rather confused as can be seen from Fig. 1, in which the data taken from literature are summarized.

3. NMR PULSE TECHNIQUE

The protons and deuterons in light and heavy methanes carry a nuclear spin respectively with $I = \frac{1}{2}$ and $I = 1$, which can be used to obtain nuclear magnetic resonance signals. In particular, the so-called NMR pulse technique, based on a saturation of the nuclear Zeeman levels by intense r.f. pulses, may be employed to obtain the spin lattice relaxation time T_1. This quantity, defined by the well-known rate equation $\frac{dM_z}{dt} = \frac{M_0 - M_z}{T_1}$, provides a measure for the time it takes for the nuclear magnetization M_z to return to its equilibrium value M_0 after an r.f. pulse. The relaxation

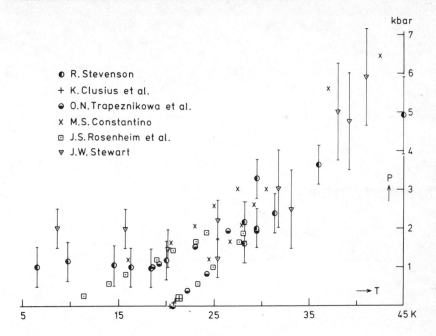

Figure 1. Phase diagram of CH_4 from literature data.

is the result of a coupling between nuclear spins and lattice vibrations.

Whenever the lattice undergoes a transformation as the result of a phase transition, the lattice vibrations will change and thus cause a discontinuity in the value of the spin lattice relaxation time T_1, plotted as a function of pressure or temperature. The NMR method of determining a phase transition by the measurement of T_1 has the great advantage of not disturbing the equilibrium of the sample, as the energy involved in saturating the nuclear Zeeman level is of the order of mJ only. Moreover, it is a very sensitive method, capable of detecting subtle transitions which would otherwise remain invisible in the normal thermodynamic methods.

As an example, in Fig. 2 the T_1 relaxation curve of CH_4 at 9.30 K is shown, featuring clearly the II → III transition and the accompanying hysteresis. In Fig. 3, the phase diagram of CH_4 (7) is depicted as it is obtained from NMR measurements. The remarkable feature of this diagram is that it provides a definitive proof for the existence of phase III at pressures above 400 bar together with the inversion of the II → III transition curve around 8 K. Phase III cannot, therefore, be obtained at atmospheric pressure. One can understand this feature from the knowledge of the structure of phase II, shown in Fig. 4 (deduced from the James-Keenan theory and confirmed by neutron diffraction), in which 1/4 of the methane molecules behave as free rotators. Although the precise

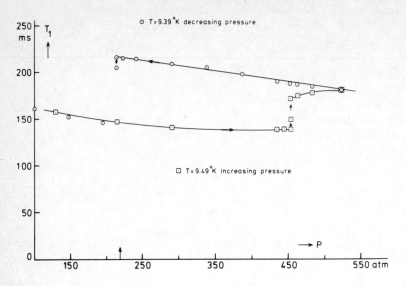

Figure 2. T_1 relaxation curve of CH_4 at 9.30 K showing II → III
transition.

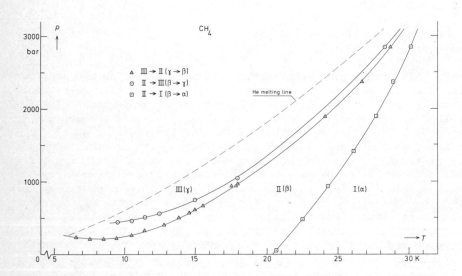

Figure 3. Phase diagram of CH_4 to 3 kbar (by NMR).

structure of phase III is, as yet, unknown, it is generally accepted that all it's molecules are in ordered positions.

If one now refers to the energy level diagram shown in Fig. 5,

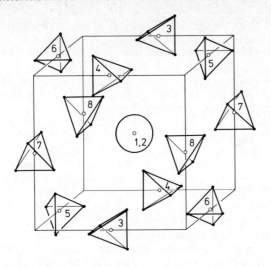

Figure 4. Structure of phase II.

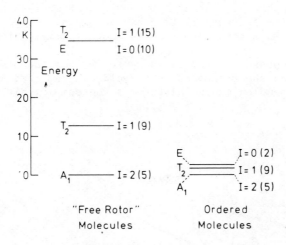

Figure 5. Energy level diagram in CH_4.

respectively of free rotators and of ordered molecules, it is clear that the entropy change of the II → III transition is given by $\Delta S = R[\log\Omega_{ordered(III)} - \log\Omega_{rotor(II)}]$, in which Ω designates the number of available states. At high temperatures, where the rotational contribution is dominant, this will result in $\Delta S < 0$, but at the lowest temperature, where only the ground states contribute, it gives $\Delta S = R(\log 16 - \log 5) > 0$. Coupled

with the knowledge that $\Delta V = V_{III} - V_{II} < 0$ this means that, according to the Clausius–Clapeyron relation $\frac{dP}{dT} = \frac{\Delta S}{\Delta V}$, the slope of the
II → III transition is <u>reversed</u> at an intermediate temperature.

A more extensive study by NMR of the solid CH_4 phases up to
10 kbar (8) has revealed the following features:

i.   Referring to the T_1 relaxation curve of Fig. 6 (open symbols)
     it is noticed that at the two lower pressures, two transitions
     occur, I → III and II → III, but at the two higher pressures
     only transition I → III survives. This establishes the exis-
     tence of a triple point, which was accurately fixed at <u>3.9
     kbar and 32.1 K</u>. The new feature is shown in Fig. 7 (dash–dot-
     ted lines) which, of course, incorporates at pressures below
     3 kbar the earlier phase diagram of Fig. 2.
ii. A rather new and interesting observation was made at low tem-
     perature and high pressure above 6 kbar, when it appeared that
     in the course of several hours phase III spontaneously trans-
     forms to the <u>new phase IV</u>, which shows an altogether different
     kind of relaxation curve, as indicated by the full symbols in
     Fig. 6. This new phase IV can be made to transform to the high
     temperature phase I, but at a much higher temperature than the
     previous III → I transformation and accompanied by a sizeable
     hysteresis. This transformation is shown in Fig. 8 for the
     6004 isobar.

In the complete phase diagram valid up to 10 kbar the new I → IV
transformation is super-imposed on the old diagram (see Fig. 7,
full lines). Depending on the history of the sample, either the

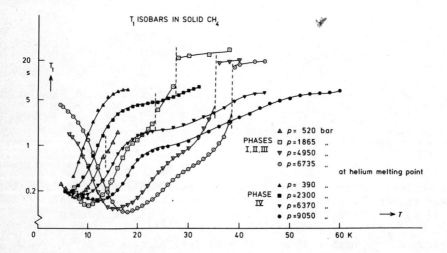

Figure 6. Four T_1 isobars of CH_4 showing the occurrence of a
          triple point and the III → IV transformation.

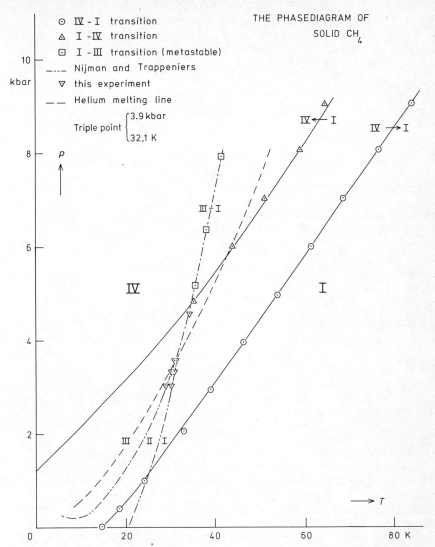

Figure 7. Phase diagram of CH to 10 kbar (by NMR) showing the
I → IV transition.

"old" or the "new" diagram is obtained. Thus, on the one hand,
when the solid is reached from the liquid phase at pressures below
6 kbar, the old phase diagram (Figs. 2 and 7) with phases I, II
and III comes into play; on the other hand, when the sample is
maintained at temperatures below 30 K and pressures above 6 kbar,
the new phase IV is produced at the expense of phase III, and it
can then be made to transform reversibly into phase I even down to
atmospheric pressure. However, heating of the solid methane sample

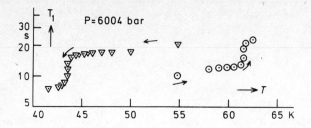

Figure 8. The new transition IV → I on the 6004 isobar
from T_1 relaxation.

to temperatures near the melting line eliminates all traces of
phase IV and the old diagram is recovered.

4. STUDY OF SOLID METHANE IN A DIAMOND ANVIL CELL (DAC)

At this stage the question presented itself as to which phases
would be stabilized at pressures in excess of 10 kbar, and for
this purpose a study was undertaken of solid methane in a DAC (9).
Fig. 9 shows a DAC specially adapted for filling with liquid methane. The cell can be used up to pressures of 100 kbar, the pressure being monitored on the ruby scale by means of laser spectroscopy. Phase transitions may be observed in two different ways:
i. by observing a discontinuity in the pressure as a result of
changing the temperature of the sample at constant volume; ii. by
observing the discontinuity in the Raman spectrum as function of
temperature or pressure.

In Fig. 10, the phase diagram up to 60 kbar is shown as it
is obtained from the study in the DAC. Between 10 and 60 kbar,
only one transition curve has been found with a sizeable hysteresis of nearly 20 K. This curve fits in perfectly well with the
I → IV transition obtained below 10 kbar by NMR technique. The
conclusion is thus that at pressures in excess of 10 kbar, the
only stable phases are the disordered plastic phase I and the new
phase IV.

Against the background of the vast pT-plane the "old" diagram,
showing phases I, II and III, must be seen as a local complication
which is confined to a small corner of the pT-diagram.

The exact structure of phase IV is as yet unknown, but it
seems fairly certain, both from Raman and NMR measurements, that
the methane molecules are in ordered positions on a lattice with
symmetry lower than that of cubic phase I and that there may be
several non-equivalent sites available to the CH_4 molecules.

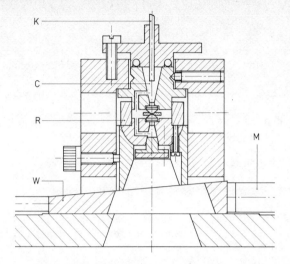

Figure 9. Diamond Anvil Cell for solid methane
measurements.

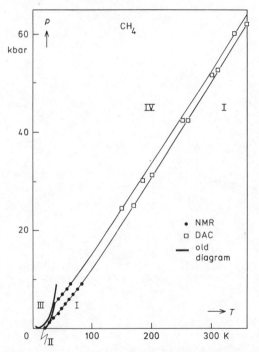

Figure 10. Phase diagram of methane to 60 kbar. The
lines in the left-hand lower corner represent the "old" diagram from Figs. 3 and 7.

REFERENCES

1. Clusius, K. (1929) Z. Phys. Chem. B3, pp.41-79.
2. Kruis, A., Popp, L. and Clusius, K. (1937), Z. Elektrochem.
   43, pp.664.
3. Bartolomé, A., Drikos G. and Eucken, A. (1938), Z. Phys.
   Chem. B39, pp.371-384.
4. Trapeznikowa, G. and Miljutin, G. (1939) Nature 144, p.632.
5. Stevenson, R. (1957) J. Chem. Phys. 27, pp. 656-658.
6. Stewart, J.W. (1959) J. Phys. Chem. Solids 12, pp.122-129.
7. Nijman, A.J. and Trappeniers, N.J. (1977) Chem. Phys. Lett.
   47, pp.188-193.
   Nijman, A.J. and Trappeniers, N.J. (1978) Physica 95B, pp.
   147-162
   Nijman, A.J. (1977), Thesis, University of Amsterdam.
8. Van der Putten, D., Prins, K.O. and Trappeniers, N.J. (1982)
   Physica 114B, pp.281-286.
9. Wieldraaijer, H., Schouten, J.A. and Trappeniers, N.J. (1983)
   High Temp.-High Press. 15, pp.87-92.

CLATHRATE HYDRATES IN THE SOLAR SYSTEM

Stanley L. Miller
Department of Chemistry, B-017
University of California at San Diego
La Jolla, California 92093

ABSTRACT

Clathrate hydrates are crystalline compounds in which an ex-
panded ice lattice forms cages that contain gas molecules. There
are two principal hydrate structures. Structure I, with a 12 Å
cubic unit cell, contains 46 water molecules and 8 cages of two
types, giving an ideal formula (for CH_4) of $CH_4 \cdot 5.75H_2O$. The
actual formula contains somewhat more water as the cages are not
completely filled. Other examples that form Structure I hydrates
are C_2H_6, C_2H_4, C_2H_2, CO_2, SO_2, OCS, Xe, H_2S. Structure II, with
a 17 Å cubic unit cell, contains 136 water molecules, and 8 large
and 16 small cages. The ideal formula for $CHCl_3$ is $CHCl_3 \cdot 17H_2O$.
Other examples of Structure II hydrates include C_3H_8, C_2H_5Cl, ace-
tone, and tetrahydrofuran. Small molecules such as Ar, Kr and
probably N_2 and O_2 also form a Structure II hydrate. The small
molecules occupy both the large and small cages, giving an ideal
formula of $Ar \cdot 5.67H_2O$. The conditions of pressure and temperature
for hydrate formation are discussed.

Methane hydrate should be a major constituent of the planets
Uranus and Neptune as well as several of the icy satellites of
Jupiter and Saturn. The hydrates of ammonia ($2NH_3 \cdot H_2O$, $NH_3 \cdot H_2O$,
and $NH_3 \cdot 2H_2O$), which are not clathrates, should also occur on
these planets. Comets are likely to contain clathrate hydrates.
The ice cap of Mars, which is mostly solid CO_2, should contain CO_2
hydrate. A hydrate of air [$(N_2, O_2) \cdot 6H_2O$] occurs in the Antarctic
ice cap. Very large quantities of methane hydrate occur in the
permafrost of Siberia, Alaska and Northern Canada as well as in
some methane rich oceanic sediments.

59

J. Klinger et al. (eds.), Ices in the Solar System, 59–79.
© *1985 by D. Reidel Publishing Company.*

INTRODUCTION

A clathrate compound is one in which a crystal lattice contains cages (or voids) that can incorporate guest molecules (i.e., CH_4). The clathrate hydrates, sometimes referred to as gas hydrates, are a special case of clathrate compounds in which the framework consists of water molecules [for reviews see ref. 1-4]. The clathrate hydrates can be considered as low pressure forms of ice, which are only stable when a gas molecule is present in the cages.

Molecules such as urea, thiourea, phenol, hydroquinone (p-dihydroxybenzene, sometimes called quinol) form crystal lattices with cages in them that can incorporate guest molecules [for reviews see ref. 5-8]. Silica can form clathrate structures, not only in the case of the zeolites (molecular sieves are an example of this), but also a structure analogous to one of the clathrate hydrates (9). This discussion will be confined largely to the clathrate hydrates, with particular emphasis on the gas hydrates.

The first clathrate hydrate was discovered by Sir Humphry Davy in 1810, who cooled an aqueous solution saturated with chlorine gas below $9^{\circ}C$ (10). He observed that crystals of an ice-like material formed which contained about one Cl_2 for each ten water molecules. The field lay largely dormant until the period 1880-1910 when Villard and de Forcrand, as well as others, investigated a large number of gases that formed hydrates (i.e., CH_4, CO_2, C_3H_8, etc.). The field again remained largely dormant until 1935-1945 when the clathrates of natural gas were found to precipitate in natural gas pipelines, thereby clogging them (11,12). This problem was solved by the simple expedient of drying the natural gas before putting it into the pipeline.

The modern investigations of clathrates begin with the work of Stackelberg and co-workers between 1949 and 1954 (13-15), who showed by X-ray diffraction that there were two types of gas hydrates (Structure I and Structure II). Stackelberg showed that the gas hydrates were clathrate compounds, the nature of which had been demonstrated by Palin and Powell (16,17) for the hydroquinone clathrates.

Some progress was made by Stackelberg in determining the crystal structure of the hydrates, but the first determination of the Structure I hydrate was done by Pauling and Marsh (18). The correct structure for the Structure II hydrates was guessed by model building (19) and subsequently determined by X-ray diffraction (14,20).

Calculations of clathrate hydrate properties, based on statistical mechanical treatment were first made by Barrer and Stuart (21) and by van der Waals and Platteeuw (1).

In 1960, Jeffrey and co-workers began a program of determining the structure of a large number of amine hydrates (reviewed in 2) including uncharged amines (i.e., t-butyl amine) and cationic amines (i.e., $(C_5H_{11})_4N^+F^-$). A wide variety of hydrate structures, in addition to the Structure I and Structure II hydrates, were determined. This discussion will not consider these neutral amine and cationic amine hydrates.

Recent work on clathrate hydrates include their use in desalination of sea water (22), discussions of their occurrence in the solar system (23-25), numerous studies of dielectric relaxation and nuclear magnetic resonance (3), and the use of gas hydrates as a basis for a theory of general anesthesia (26,27). Large amounts of CH_4 hydrate have been found in the permafrost of Siberia, Alaska and Northern Canada, as well as in some methane containing deep-sea sediments (28-30).

Clathrate hydrates have been used to stabilize free radicals at low temperatures and to study their electron paramagnetic resonance (31). The hydrates are frequently used as model compounds in discussions of non-polar gases dissolved in liquid water (32-34), and also in discussions of the hydrophobic bond that is responsible in part for the tertiary structure of proteins (35-37).

Water alone can form two different self-clathrate structures. Ice VII consists of two interpenetrating cubic ice lattices (38). There are no hydrogen bonds between the two different lattices that are held in place by van der Waals forces. Ice VI also consists of two interpenetrating lattices that are not hydrogen bonded to each other (39).

HYDRATE STRUCTURES

Structure I Hydrate

Gases that form this hydrate include CO_2, SO_2, OCS, Xe, CH_4, C_2H_6, C_2H_4, CH_3Cl, Cl_2 and cyclopropane. Molecules such as $CHCl_3$, C_2H_5Cl and propane are too large to fit in the cavities of the Structure I hydrate and so form Structure II hydrates.

The framework of the Structure I hydrate is shown in Fig. 1. There are 46 molecules in the unit cell that form 2 pentagonal dodecahedra (12 sided polyhedron) and 6 tetrakaidecahedra (14 sided polyhedron). Molecules 5.1 Å or less in diameter will fit into the 12-hedra (CO_2 and CH_4, but not CH_3Br), and molecules 5.8 Å or less in diameter will fit into the 14-hedra (CH_4 and

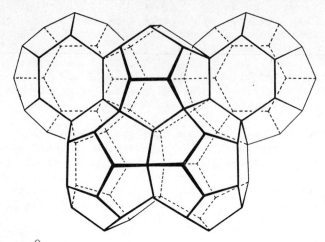

Fig. 1   The 12 Å hydrate lattice.   The pentagonal dodecahedron (upper center) is formed by 20 water molecules.   The tetrakai-decahedra formed by 24 water molecules have 2 opposite hexagonal faces and 12 pentagonal faces.

CH_3CH_3, but not $CH_3CH_2CH_3$).   The ideal formula for a hydrate with both types of cavities filled is 46/8 = 5.75 (i.e., $CH_4 \cdot 5.75H_2O$). If only the 14-hedra are occupied, the ideal formula is 46/6 = 7.67 (i.e., cyclopropane$\cdot 7.67H_2O$).

The actual formulas differ from the ideal formulas because the cavities are not 100% occupied by gas molecules.   For example, direct analysis (12) gives a formula of $CH_4 \cdot 7.1H_2O$ instead of the ideal $CH_4 \cdot 5.75H_2O$; however, the hydrate formula derived from a thermodynamic treatment of the dissociation pressures gives a formula close to the ideal formula (40).   The determination of the hydrate formulas is a difficult task and there is consider-able uncertainty as to correct formulas.   It is clear from the statistical mechanical treatment of hydrate stabilities that there is never 100% occupancy of the cavities.   However, it ap-pears that for most hydrates the actual formula is not far from the ideal formula.

It should be noted that the water molecules forming the cav-ities do not interact specifically with the encaged molecule.   In other words, the structure of methane hydrate is the same as car-bon dioxide hydrate.   Thus, mixed hydrates of methane and carbon dioxide can be prepared by using a mixture of gaseous methane and carbon dioxide.   The ideal hydrate formula is then $(CH_4, CO_2) \cdot 5.75H_2O$, rather than a mixture of $CH_4 \cdot 5.75H_2O$ and $CO_2 \cdot 5.75H_2O$.   A mixed hydrate is thus a solid solution of gases in the hydrate framework.

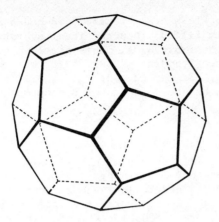

Fig. 2. The hexakaidecahedron formed by 28 water molecules in the 17 Å lattice. There are 4 hexagonal faces and 12 pentagonal faces.

Structure II Hydrate

This structure contains 136 molecules of water in the unit cell of 17 Å. There are 16 12-hedra very similar to those of the Structure I hydrate and 8 hexakaidecahedra (16 sided polyhedron). Molecules with diameters of 6.7 Å or less will fit into the 16-hedra ($CH_3CH_2CH_3$, but not $CH_3CH_2CH_2CH_3$). The unit cell cannot be visualized with less than a three dimensional model, but the 16-hedron is shown in Fig. 2. The ideal formula for a Structure II hydrate is 136/8 = 17 (i.e., $CHCl_3 \cdot 17H_2O$). The actual formula appears to be close to the ideal formula (41). Chloroform is much too large to fit into the 12-hedra but if a smaller molecule such as H_2S is mixed with the chloroform, a mixed hydrate, usually termed a double hydrate, is formed (i.e., $CHCl_3 \cdot 2H_2S \cdot 17H_2O$). In a double hydrate, the large molecule ($CHCl_3$) is confined to the large cavity and the small molecule (H_2S) is largely, but not entirely, confined to the small cavities. This differs from a mixed hydrate, i.e., $[(CH_4, CO_2) \cdot 5.75H_2O]$, where the methane and carbon dioxide are present in both the large and small cavities. The mole fraction of carbon dioxide is expected to be larger in the large cavities than in the small cavities because carbon dioxide is larger than methane. Double hydrates are only known for Structure II hydrates, although a Structure I double hydrate is a possibility (i.e., cyclopropane $\cdot 0.33H_2S \cdot 7.67H_2O$).

It has recently been shown that Ar, Kr, and probably N_2 and O_2 form a Structure II hydrate rather than Structure I (42). This came as a surprise since it was thought that small molecules form Structure I and large molecules Structure II. But the smallest-size gases go over to Structure II since they can fit into both the 16-hedra and the 12-hedra. If the 8 16-hedra and the 16 12-hedra are completely filled, then the ideal hydrate formula is 136/24 = 5.67 (e.g., $N_2 \cdot 5.67 H_2O$). This is close to the composition of the Structure I hydrate ($N_2 \cdot 5.75 H_2O$) but the structures are quite different. It may be that the stable structures change with temperature as has been found for cyclopropane and dimethyl-ether hydrates (43-45).

Tetragonal Hydrate

This hydrate was discovered with Br_2 (46) and the only other known example is dimethyl ether hydrate (45). The unit cell contains 172 H_2O with 4 15-hedra, 16 14-hedra, and 10 12-hedra. If Br_2 fills both the 14- and 15-hedra, then the formula would be 172/20 = 8.60 (i.e., $Br_2 \cdot 8.60 H_2O$). Although the only molecules that form this hydrate are intermediate in size between the largest that form Structure I and the smallest that fit into the 16-hedra of Structure II, it is possible that the tetragonal hydrate could be stable at some temperatures with small molecules as is the case with Structure II. Thus the ideal formula for the N_2 tetragonal hydrate would be $N_2 \cdot 5.73 H_2O$.

DISSOCIATION PRESSURES AND PHASE DIAGRAMS

· The dissociation pressures of gas in a hydrate, which can be considered as the vapor pressure of the hydrate, is a definite value when the hydrate is in equilibrium with ice and the gas (or with water and the gas). The dissociation pressures of a number of hydrates at 0^oC are given in Table 1. Just as the vapor pressure of a pure substance varies with temperature, that of a hydrate does also, but in a somewhat more complicated way. Figure 3 shows the phase diagram of methane hydrate. At -20^oC, the phase diagram shows that for pressures less than 13.5 atm of CH_4, ice + $CH_4(g)$ are the only stable phases. If the pressure is raised to 13.5 atm, ice + CH_4 hydrate + $CH_4(g)$ are in equilibrium. If the CH_4 pressure is raised above 13.5 atm, all the ice will be converted to hydrate. Above 0^oC, the equilibria involve liquid water. For example, at $+16^oC$, the dissociation pressure is 136 atm. Below this pressure CH_4 hydrate is unstable; if the pressure of CH_4 is raised above 136 atm, all the water will be converted to methane hydrate. The quadruple point, where ice + CH_4 hydrate + water + $CH_4(g)$ are in equilibrium is slightly below 0^oC (-0.19^oC) and 26 atm. Because the critical point of CH_4 is at -82.5^oC, the pressure (and fugacity) of CH_4 can be increased indefinitely without forming liquid CH_4. Thus methane hydrate has been prepared (47) at temperatures as high as 47° (3900 atm) and the only limit in temperature would be where the VI and VII become stable relative to liquid water or the ΔV for the reaction becomes unfavorable. Because the slope of the line describing

Table 1. Examples of Gas Hydrate Dissociation Pressures. Data
mostly from ref. 3.

Gas	Diameter (Å)	P_{diss}^{atm} ($0^{\circ}C$)	Hydrate Structure
N_2	4.1	160	II?
CO^*	4.1	135(estim.)	II?
O_2	4.0	120	II?
Ar	3.8	95.5	II
Kr	4.0	14.5	II
Xe	4.4	1.50	I
CF_4	5.1	41.5	I
CH_4	4.1	26.0	I
C_2H_6	5.3	5.2	I
C_2H_4	5.4	5.5	I
C_2H_2	5.5	5.7	I
CO_2	4.7	12.5	I
H_2S	4.1	0.92	I
SO_2	5.0	0.39	I
Cl_2	5.2	0.32	I
Br_2	5.7	0.056	Tetragonal
C_3H_8	6.3	1.74	II
$CHCl_3$	6.4	0.065	II
C_2H_5Cl	6.2	0.26	II
SF_6	5.8	0.77	II

*CO hydrate has not yet been prepared, but the dissociation pressure is estimated from its boiling point relative to N_2 and O_2 as
well as the relative dissociation pressures of the hydroquinone
clathrates.

the vapor pressure of liquid and solid CH_4 (very close to the line
hydrate + $CH_4(1)$ + $CH_4(g)$ is less than the slope of the dissociation pressure curve (Fig. 3), methane hydrate is stable to $0^{\circ}K$.

Figure 4 shows a somewhat different case of the CO_2 hydrate.
In this case the critical point is at $31.0^{\circ}C$, so that the dissociation pressure curve meets the vapor pressure curve at $10.20^{\circ}C$
and 44.50 atm (Q_2). The hydrate is not stable above this temperature along the line CO_2 hydrate + water + $CO_2(9)$, but it is
stable at somewhat higher temperatures along the line CO_2 hydrate
+ water + $CO_2(1)$. Thus, CO_2 hydrate has been prepared at $19.5^{\circ}C$
and 1840 atm (48).

Figure 5 shows the low temperature region of the CO_2 hydrate.
In this case the slope of the vapor pressure curve of CO_2 (dry

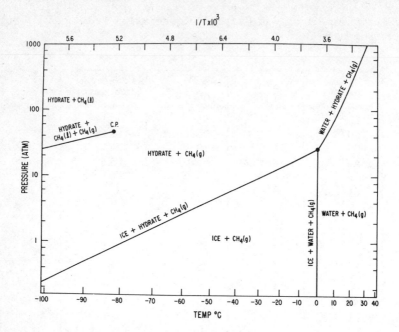

Fig. 3.   The phase diagram of methane hydrate.   The line hydrate
+ $CH_4(l)$ + $CH_4(G)$ ends very close to the critical point of methane
($191°K$).   Q is the quadruple point [ice + hydrate + water + $CH_4(g)$].

ice) is greater than the slope of the hydrate dissociation pressure curve, and thus the two curves meet at approximately $121°K$.
Below this temperature CO_2 hydrate is unstable relative to ice
and solid CO_2 (49).

Cyclopropane hydrate is a more complicated case, where both
Structure I and Structure II hydrates are formed (43).   Below
-16°C, Structure I hydrate is stable; between -16° and +4° Structure II hydrate is the stable form; above 4° Structure I again
becomes stable until 16° where the critical decomposition point
is reached.   The hydrates of cyclopropane were prepared with D_2O
(deuteriohydrates) (44).   Below 0°C the deuteriohydrate has a
higher dissociation pressure than the hydrate, but above 4°C the
deuteriohydrate dissociation pressure is lower.

STATISTICAL MECHANICS OF CLATHRATE FORMATION

The statistical mechanical treatment of clathrate formation
(1,21,50), of which only the basic results will be presented here,
treats hydrate formation as the sum of two reactions

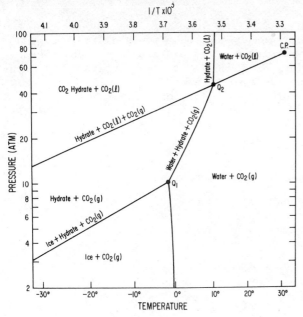

Fig. 4. The phase diagram of carbon dioxide hydrate near $0°$. C.P. is the critical point of carbon dioxide ($31.0°C$ and 72.8 atm). Q_1 is the quadrupole point [ice + water + hydrate + $CO_2(g)$] at $-1.77°C$ and 10.20 atm, Q_2 is the quadruple point [water + hydrate + $CO_2(1)$ + $CO_2(g)$] at $10.20°C$ and 44.50 atm.

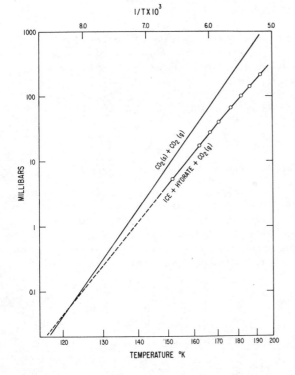

Fig. 5. The dissociation pressure of CO_2 hydrate and the vapor pressure of solid CO_2. The circles are the experimental dissociation-pressure measurements. Above about $121°K$, CO_2 hydrate is stable at pressures equal to or greater than the dissociation pressure. Below $121°K$, CO_2 hydrate is unstable with respect to decomposition to ice and $CO_2(s)$.

Mixed Hydrates

It is important to remember that when two hydrate-forming gases are present, only a single mixed hydrate is formed. Thus if the gas phase contains CH_4 and C_2H_6, then the hydrate will contain CH_4 and C_2H_6 and not be a mixture of CH_4 and C_2H_6 hydrates. An approximate formula for the stability of a mixed hydrate at a given temperature is (23)

$$\frac{P_a}{P_a^o} + \frac{P_b}{P_b^o} = 1$$

where P_a and P_b are the partial pressures of gas a and b and P_a^o and P_b^o are the respective dissociation pressures of the pure hydrates at this temperature. If

$$\frac{P_a}{P_a^o} + \frac{P_b}{P_b^o} < 1$$

then no hydrate can form. If the sum is greater than 1 then all the ice (or water) will be converted to hydrate.

There are no hydrate data sufficiently accurate to test this formula, but it cannot be far wrong provided both gases fit into both types of cages of the hydrate. It does not apply to Structure II hydrates with one large and one small molecule (e.g., $C_3H_8 + H_2S$). A more accurate equation can be obtained from the statistical mechanical analysis. Along the line ice + hydrate + gas we have

$$\frac{2}{46} \ln (1-y_{1a}-y_{1b}) + \frac{6}{46} \ln(1-y_{2a}-y_{2b}) = -\frac{\Delta G}{RT} .$$

The fractional occupancies can be expressed in terms of the partial pressures and the Langmuir binding constants giving $y_{1a} = C_{1a}P_a/(1 + C_{1a}P_a + C_{1b}P_b)$, $Y_{2a} = C_{2a}P_a/(1 + C_{2a}P_a + C_{2b}P_b)$, etc., where C_{1a} is the Langmuir binding constant for molecule a in the 12-hedra and C_{2a} is the binding constant of a in the 14-hedra. This equation then becomes

$$\frac{2}{46} \ln (1 + C_{1a}P_a + C_{1b}P_b) + \frac{6}{46} \ln (1 + C_{2a}P_a + C_{2b}P_b) = +\frac{\Delta G}{RT}$$

The approximate relative composition of the gases in the hydrate phase is proportional to their partial pressures and

ice $=$ empty hydrate lattice
$$\Delta G = 0.2 \text{ kcal}, \Delta H \sim 0$$
per mole H_2O \hfill (1)

empty hydrate lattice + gas $=$ hydrate
$$\Delta H = -4 \text{ to } -8 \text{ kcal}$$
per mole of gas \hfill (2)

---

ice + gas $=$ hydrate
$$\Delta G = 0 \text{ at equilibrium}$$
$$\Delta H = -4 \text{ to } -8 \text{ kcal}$$
per mole of gas \hfill (3)

The free energy of formation of the empty hydrate lattice from Ice I_h is unfavorable by ≈ 0.2 kcal per mole of water (1.15 kcal per mole of gas in the ideal Structure I hydrate). This unfavorable reaction together with the unfavorable entropy of compression of the gas is overcome by the enthalpy gained in putting the gas molecule into the cavity. The statistical mechanical treatment uses various methods to calculate the enthalpy (and entropy) of the second reaction, using various potential interactions (e.g., the Lennard-Jones and Devonshire 12,6 potential). When this is evaluated theoretically, and two arbitrary constants are assigned (ΔG for reaction (1) and a normalization factor based on the argon dissociation pressure), the calculation of the dissociation pressure for all Structure I hydrates can be carried out. The theoretical dissociation pressures agree surprisingly well (within about 20%) for the noble gases and spherical molecules (except for CF_4), and within a factor of 5 or better for molecules such as ethane, ethylene and acetylene.

The statistical mechanical treatment shows that the condition for stability of a Structure I hydrate in equilibrium with ice and the gas is

$$\frac{2}{46} \ln (1-y_1) + \frac{6}{46} \ln (1-y_2) = - \Delta G/RT$$

where ΔG is the free energy for reaction (1), y_1 is the fraction of the 12-hedra occupied by the gas molecules, and y_2 is the fraction of the 14-hedra occupied. For the Structure II hydrates, the 2/46 and 6/46 become 16/136 and 8/136, respectively. This equation shows that the cages can never be completely occupied (unless ΔG were infinite). Thus the clathrate hydrates are non-stoichiometric compounds, although many hydrates have occupancy numbers close to 100%, and so have compositions close to the ideal formulas.

inversely proportional to their dissociation pressures. Thus

$$\frac{n_a}{n_b} = \frac{P_a/P_a^o}{P_b/P_b^o}$$

where n_a and n_b are the number of molecules of a and b in the unit cell. This is an approximate formula (the statistical mechanical formula is complex) which shows that gases with low hydrate dissociation pressures concentrate in the hydrate phase. Thus if the partial pressures of CH_4 and H_2S are 13 atm and 0.46 atm respectively at 0^oC, then the CH_4-H_2S hydrate will be stable since $P_{CH_4}^o = 26.0$ atm and $P_{H_2S}^o = 0.92$ atm. There will be equal amounts of CH_4 and H_2S in the hydrate, so the H_2S in the hydrate has been concentrated in the hydrate by a factor of 28.

The hydrates of hydrogen, helium and neon have never been prepared and are thought not to be stable (1) at least near 0^oC. Stabilities may be different at very low temperatures. Even though the pure hydrates do not form, these gases are incorporated into hydrates of other gases (e.g., CH_4) and will stabilize this hydrate (21,51). This effect is very marked with Structure II hydrates (e.g., with $CHCl_3$) where the H_2 and He can occupy the 12-hedra but the $CHCl_3$ cannot.

NATURAL OCCURRENCE OF GAS HYDRATES

The occurrence of gas hydrates in natural gas pipelines was mentioned previously. These hydrates are mixed Structure I hydrates of CH_4 and C_2H_6, along with Structure II double hydrates of propane and methane (ideally $C_3H_8 \cdot 2CH_4 \cdot 17H_2O$).

In the solar system, the most abundant occurrence of hydrates must be on the planets Uranus and Neptune (23). Their densities are 1.56 and 2.22, respectively. These figures indicate, when allowance is made for the compression of material in the interior of the planet, that these planets must be made up largely of CH_4, NH_3 and H_2O. Methane has been detected in the atmospheres of Uranus and Neptune. Ammonia is believed to be frozen out below the cloud layer on these planets. The pressure at the cloud layer of Uranus is about 0.1 atm and the temperature is about 90^oK. By extrapolating the dissociation pressure curve of methane hydrate measured at higher temperatures, the dissociation pressure of the hydrate is calculated to be 2×10^{-6} atm at 90^oK. Thus, methane hydrate is very stable relative to solid or liquid methane (the triple point of methane is 90.8^oK and 0.12 atm). In fact, all the water on the planet except that deep in the interior should be converted to methane hydrate if there is an excess of water (the relative proportions of CH_4, H_2O and NH_3 are not known accurately). It has been calculated (56) that at pressures of 10 to 14 kilobars, methane hydrate decomposes to water or ice and methane gas because the ΔV of hydrate formation becomes unfavorable at these high pressures. Therefore CH_4 hydrate is apparently not stable to the center of these planets.

This picture is complicated by the reaction of NH_3 with water to form the ammonia hydrates, the phase diagram for which is shown in Fig. 6. The ammonia hydrates, $2NH_3 \cdot H_2O$, $NH_3 \cdot H_2O$ and $NH_3 \cdot 2H_2O$, are stoichiometric compounds rather than clathrates. The crystal structures of $2NH_3 \cdot H_2O$ and $NH_3 \cdot H_2O$ show that these are hydrogen bonded solids (52,53). Except for their thermal properties (54,55), which show ordered hydrogen bonds at $0^\circ K$, these 'ices' have received little attention, and they may prove to be very interesting compounds in their own right. Even with this complication, it seems very likely that Uranus and Neptune must be composed of methane hydrate, the ammonia hydrates, and perhaps water ice.

Jupiter and Saturn are believed to be composed largely of hydrogen and helium, but methane and ammonia have been detected in their atmospheres. It seems possible that methane hydrate is present on these planets along with the ammonia hydrates (23,57), but not enough is known about the structure of their atmospheres to make an accurate prediction.

The largest satellite of Saturn, Titan, is a very favorable case for clathrate hydrates. The atmosphere contains mainly N_2, Ar and CH_4. It has been proposed (58) that the ocean on Titan is mainly ethane with dissolved N_2, Ar, CH_4 and other minor species from the atmosphere. A mixed hydrate of these gases should form provided there has been adequate mixing with the ice, although the rate of this reaction could be very slow. Another possibility is that Titan was formed from hydrates that were stable in the solar nebula.

Mercury has no atmosphere and almost certainly no water. There is a great deal of CO_2 on Venus and small amounts of water, and thus the formation of CO_2 hydrate is possible, but the temperature and pressure conditions as presently accepted are not favorable for this.

Comets seem another likely place for the occurrence of hydrates (23,59-61). This was first proposed by Delsemme and Swings in 1952. The most popular model of comets is that of a mixture of 'ices' (a mixture of CH_4, NH_3 and H_2O ices), along with some silicate particles (62). It is expected that CH_4, CO and CO_2 hydrates and the ammonia hydrates would be present. Little is known about the chemical composition of interstellar

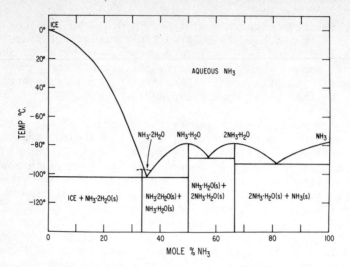

Fig. 6.   The phase diagram of NH_3-H_2O system at low temperature.
The melting points of NH_3, $2NH_3 \cdot H_2O$ and $NH_3 \cdot H_2O$ are -77.80^O,
-78.83^O and -79.00^O, respectively.   The hydrate $NH_3 \cdot 2H_2O$ melts
(incongruently) at -97.06^O.   Its composition-melting curve is
shown in the figure as the dotted line.

dust, but on the basis of the cosmic abundances of the elements,
it would be expected to contain solid methane, ammonia, and water.
The temperature of the interstellar dust is usually given as 3^OK,
and thus methane hydrate and the ammonia hydrates are thermodynamically stable.   However, the situation here is complicated by the
kinetics of hydrate formation and the presence of a strong ultraviolet radiation field.

     It appears very likely that the CO_2 hydrate is present in the
ice cap of Mars (49).   The Mariner 6 and 7 missions in 1967 determined a temperature of 148^OK for the ice cap and a partial pressure of CO_2 of 6.5 mbars (63-65).   This is the sublimation pressure of CO_2 at 148^OK.   The dissociation pressures of CO_2 hydrate
were measured at low temperature (see Fig. 5), and the P_{disc} =
3.48 mbars at 148^OK (49).   Therefore, the phases at equilibrium
would be CO_2 hydrate + CO_2(solid) + CO_2(gas) rather than CO_2
hydrate + ice + CO_2(gas).   It is not certain that equilibrium
would be attained as the formation of CO_2 hydrate at 148^OK is very
slow unless the ice is finely divided.   However, it would be expected that ice precipitating from the Martian atmosphere would
be very finely divided.

     The first reported natural occurrence of clathrate hydrates
on the earth (aside from the natural gas pipelines) is a hydrate

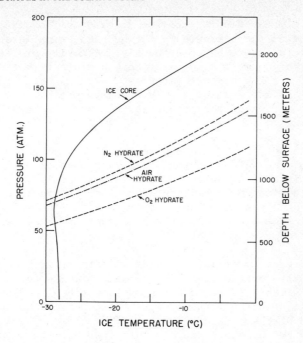

Fig. 7. The relation between temperature, pressure and depth in the Antarctic ice core. The dissociation pressures of N₂, O₂ and air hydrates are also shown.

of air in an antarctic ice core (66), which was drilled 2164 meters to bedrock (67). It was observed that bubbles of air were present in the ice core near the top, and they became smaller as the depth increased by the compression of the bubble due to the overlying weight of ice. However, the bubbles became smaller than predicted at 900 meters assuming hydrostatic equilibrium, and by 1200 meters the bubbles entirely disappeared and remained absent to the bottom of the ice core.

The dissociation pressures of nitrogen, oxygen, and air hydrates are shown in Fig. 7, together with the temperature and pressures in the ice core. The data show that the air hydrate should begin to form at 800 meters depth and the bubbles should completely disappear at 850 meters. Thus there is a discrepancy of about 200 meters.

Although the discrepancy between the predicted depth for the disappearance of bubbles and the observed depth is not great, it might be accounted for by errors in the hydrate data, the ice temperature, density, or depth. Some of the hydrate might also have decomposed to form bubbles in the 800 and 1200 meters part of the core when the pressure was released. The discrepancy could also be due to the pressure in the bubbles being less than the hydrostatic pressure because the high viscosity of ice makes the pressure readjustment slow after some hydrate formation, but it was shown that the hydrostatic pressure and bubble pressure are the same. It may well be that hydrate formation is very slow from the reaction of the compressed air and the large ice crystals in the core.

The amount of hydrate can be calculated on the assumption that 10% of the volume of the ice is gas bubbles when the snow is compacted into ice. On the basis of the formula $(N_2,O_2) \cdot 6H_2O$ 0.06% of the ice would be in hydrate form. This is a small amount, but Shoji and Langway (68) have been able to find the cubic hydrate crystals in the presence of the hexagonal ice in a core from the Greenland ice sheet.

The discovery of large fields of natural gas (CH_4) hydrate in Siberia (69) occurred at about the same time as the discovery in oceanic sediments (70). Makagon has described the Russian work (28). Similar natural gas hydrate fields have been found in Alaska and Northern Canada (29,30). The oceanic sediments were discovered by anamolous sound absorbtion in certain anaerobic sediments of about 4000 meter depth. This is only indirect evidence, but subsequently cores were obtained which were full of gas from decomposed hydrate (29). More recently cores containing intact hydrate have been obtained (71).

Acknowledgement: This work was supported by NASA Grant NAGW20.

REFERENCES

1. van der Waals, J. H., and Platteeuw, J. C. 1959, Advan.
    Chem. Phys. 2, pp. 1-57.

2. Jeffrey, G. A. and McMullan, R. K. 1967, Prog. Inorg. Chem.
    8, pp. 43-108.

3. Davidson, D. W. 1973, in *Water--A Comprehensive Treatise*,
    Franks, F. Ed. (Plenum Press, New York) Vol. 2, pp. 115-234.

4. Berecz, E. and Balla-Achs, M. 1983, *Gas Hydrates* (Elsevier,
    Amsterdam).

5. Mandelcorn, L. 1959, Chem. Rev. 59, pp. 827-839.

6. Hagan, M. 1962, *Clathrate Inclusion Compounds* (Reinhold,
    New York).

7. Powell, H. M. 1964, in *Non-Stoichiometric Compounds*,
    Mandelcorn, L. Ed. (Academic Press, New York) p. 438.

8. Bhatnagar, V. M. 1970, *Clathrate Compounds* (Chemical Pub. Co.,
    New York).

9. Kamb, B. 1965, Science 148, pp. 232-234.

10. Davy, H. 1811, Phil. Trans. Roy. Soc., London, 101, p. 1-35.

11. Hammerschmidt, E. G. 1934, Ind. Eng. Chem. 26, pp. 851-855.

12. Deaton, W. M. and Frost, E. M., Jr. 1946, *U. S. Bureau of
    Mines Monograph, No. 8.*

13. Stackelberg, M. von 1949, Naturwissenschaften 36, pp. 327-
    333, 359-362.

14. Stackelberg, M. von and Müller, H. R. 1951, J. Chem. Phys.
    19, pp. 1319-1320.

15. Stackelberg, M. von and Jahns, W. 1954, Z. Elektrochem. 58,
    pp. 162-164.

16. Palin, D. E. and Powell, H. M. 1945, Nature 156, pp. 334-335.

17. Palin, D. E. and Powell, H. M. 1948, J. Chem. Soc. pp. 815-
    821.

18. Pauling, L. and Marsh, R. E. 1952, Proc. Nat. Acad. Sci.
    U.S. 38, pp. 112-118.

19. Claussen, W. F. 1951, J. Chem. Phys. 19, pp. 259-260, 662.

20. Mak, T. C. and McMullan, R. K. 1965, J. Chem. Phys. 42, pp. 2732-2737.

21. Barrer, R. M. and Stuart, W. I. 1957, Proc. Roy. Soc., London, A243, pp. 172-189.

22. Barduhn, A. J. 1967, Chem. Eng. Prog. 63, pp. 98-103.

23. Miller, S. L. 1961, Proc. Nat. Acad. Sci. U.S. 47, pp. 1798-1808.

24. Miller, S. L. 1973, in *Physics and Chemistry of Ice*, Walley, E., et al. Eds. (Royal Soc. Canada, Ottawa) pp. 42-50.

25. Miller, S. L. 1974, in *Natural Gas in Marine Sediments*, Kaplan, I. R. Ed. (Plenum Press, New York) pp. 151-177.

26. Pauling L. 1961, Science 134, pp. 15-21.

27. Miller, S. L. 1961, Proc. Nat. Acad. Sci. U.S. 47, pp. 1515-1524.

28. Makogon, Y. F. 1981, *Hydrates of Natural Gas* (PennWell Pub. Co., Tulsa, Oklahoma).

29. Kaplan, I. R. Ed 1974, *Natural Gases in Marine Sediments* (Plenum, New York).

30. Holder, G. D., Kamath, V. A. and Godbole, S. P. 1984, Ann. Rev. Energy 9, pp. 427-445.

31. Goldberg, P. 1963, Science 142, pp. 378-379.

32. Frank, H. S. and Evans, M. W. 1945, J. Chem. Phys. 13, pp. 507-532.

33. Frank, H. S. 1970, Science 169, pp. 635-641.

34. Claussen, W. F. and Polglase, M. F. 1952, J. Amer. Chem. Soc. 74, pp. 4817-4819.

35. Kauzmann, W. 1959, Advan. Protein Chem. 14, pp. 1-63.

36. Ben-Naim, A. 1980, *Hydrophobic Interactions* (Plenum, New York.

37. Tanford, C. 1980, *The Hydrophobic Effect* (Wiley, New York).

38. Kamb, B. and Davis, B. L. 1964, Nat. Acad. Sci. U.S. 52, pp. 1433-1439.

39. Kamb, B. 1965, Science 150, pp. 205-209.

40. Glew, D. N. 1962, J. Phys. Chem. 66, pp. 605-609.

41. Glew, D. N. 1960, Can. J. Chem. 38, pp. 208-221.

42. Davidson, D. W., Handa, Y. P., Ratcliffe, C. I., Tse, J. S. and Powell, B. M. 1984, Nature 311, pp. 142-143.

43. Hafemann, D. R. and Miller, S. L. 1969, J. Phys. Chem. 73, pp. 1392-1397.

44. Hafemann, D. R. and Miller, S. L. 1969, J. Phys. Chem. 73, pp. 1398-1401.

45. Miller, S. L., Gough, S. R. and Davidson, D. W. 1977, J. Phys. Chem. 81, pp. 2154-2157.

46. Allen, K. W. and Jeffrey, G. A. 1963, J. Chem. Phys. 38, pp. 2304-2305.

47. Marshall, D. R., Saito, S. and Kobayashi, R. 1964, AIChE J. 10, pp. 202-205.

48. Takenouchi, S. and Kennedy, G. C. 1965, J. Geol. 73, pp. 383-390.

49. Miller, S. L. and Smythe, W. D. 1970, Science 170, pp. 531-533.

50. McKoy, V. and Sinanoglu, O. 1963, J. Chem. Phys. 38, pp. 2946-2956.

51. Stackelberg, M. von and Meinhold, W. 1954, Z. Elektrochem. 58, pp. 40-45.

52. Siemons, W. J. and Templeton, D. H. 1954, Acta Crystallogr. 7, pp. 194-198.

53. Olovsson, I. and Templeton, D. H. 1959, Acta Crystallogr. 12, pp. 827-832.

54. Hildenbrand, D. L. and Giauque, W. F. 1953, J. Amer. Chem. Soc. /5, pp. 2811-2818.

55. Chan, J. P. and Giauque, W. F. 1964, J. Phys. Chem. 68, pp. 3053-3057.

56. Lunine, J. I. and Stevenson, D. J. 1985, this volume.

57. Lewis, J. S. 1969, Icarus 10, pp. 365-378.

58. Lunine, J. I. Stevenson, D. J., and Yung, Y. L. 1983, Science 222, pp. 1229-1230.

59. Delsemme, A. H. and Swings, P. 1952, Ann. Astrophys. 15, pp. 1-6.

60. Delsemme, A. H. and Wegner, A. 1970, Planet. Space Sci. 18, pp. 709-715.

61. Delsemme, A. H. and Miller, D. C. 1970, Planet. Space Sci. 18, pp. 717-730.

62. Whipple, F. L. 1951, Astrophys. J. 113, pp. 464-474.

63. Neugebauer, G., Münch, G., Chase, S. C., Jr., Hatzenbeler, H., Miner, E. and Schofield, D. 1969, Science 166 pp. 98-99.

64. Kliore, A., Fjeldbo, G., Seidel, B. L. and Rasool, S. I., Science 166, pp. 1393-1397.

65. Neugebauer, g., Münch, G., Kieffer, H., Chase, S. C., Jr. and Miner, E. 1971, Astron. J. 76, pp. 719-728.

66. Miller, S. L. 1969, Science 165, pp. 489-490.

67. Gow, A. J., Ueda, H. T. and Garfield, D. E. 1969, Science 161, 1011-1013.

68. Shoji, H. and Langway, C. C., Jr. 1983, Nature 298, pp. 548-550.

69. Makogon, Yu. F., et al 1971, Dokl. Akad. Nauk SSSR (Earth Sci.), English Transl. 196, pp. 197-200.

70. Stoll, R. D., Ewing, J. and Bryan, G. M. 1971, J. Geophys. Res. 76, pp. 2090-2094.

71. Claypool, C. E. and Kvenvolden, K. A. 1983, Ann. Rev. Earth Planet. Sci. 11, pp. 299-327.

DISCUSSION

E. Gaffney :

        In response to a question by Klinger, Miller
commented that temperatures in the outer solar system
might be so low that kinetics would prohibit the formation of clathrate hydrates. I would caution that with
millions of years available for reactions, even reactions which cannot be observed in the laboratory in
days or even months may indeed occur. This is based on
the experience of experimental petrology in which it is
often difficult or impossible to synthesise known phases (i. e., naturally occurring ones) on laboratory time scale. Furthermore, phase equilibrium is considerably facilitated by the presence of a reactive liquid.
In much of the outer solar system, N_2 or CH_4 may play
such a role. Given millions (or even billions) of years,
nature has many ways of reaching equilibrium.

POLYMORPHISM IN VAPOR DEPOSITED AMORPHOUS SOLID WATER

Erwin Mayer and Rudolf Pletzer

Institut für Anorganische und Analytische Chemie, Universität Innsbruck, A-6020 Innsbruck, Austria

The physical properties of vapor deposited amorphous solid water, $(H_2O)_{as}$, are known to depend strongly on the details of preparation. Various parameters were found to have an influence, for example the rate of deposition, the substrate temperature and so on [1,2]. We observed a further parameter to be important for the preparation of $(H_2O)_{as}$ which has not been discussed yet: supersonic flow of the water vapor and condensation of the vapor to clusters already in the gas phase [3]. The usual procedure for preparing $(H_2O)_{as}$ is to admit water vapor through a capillary into a high vacuum system and condense the vapor on a cryoplate. The exit of the capillary can act as a nozzle and from water vapor pressure and nozzle diameter it can be calculated that for the usual deposition conditions supersonic flow and possibly formation of clusters in the gas phase will occur. In contrast, a baffle between nozzle exit and cryoplate has the effect to break up supersonic flow and give water vapor consisting only of monomers. The amorphous deposits prepared either way look very different: supersonic flow gives a white deposit with a very rugged surface, whereas baffled flow gives transparent deposits with a smooth surface. We have further characterized the two types of deposits by differential thermal analysis (DTA) and by nitrogen adsorption isotherms. In short we observed first that amorphous deposits from supersonic flows devitrify in the presence of

81

J. Klinger et al. (eds.), Ices in the Solar System, 81–88.
© *1985 by D. Reidel Publishing Company.*

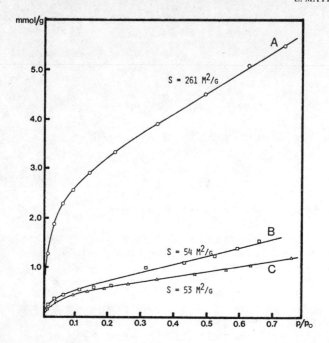

Figure 1. Isotherms for adsorption of N_2 on $(H_2O)_{as}$
           at 77 K: (A) $(H_2O)_{as}$ deposited without
           baffle at 77 K, (B) $(H_2O)_{as}$ prepared as
           above and heated after deposition for
           30 min to 115 K, (C) $(H_2O)_{as}$ deposited
           with baffle 77 K. (Deposition conditions:
           K_n = 0.07, P_o = 0.04 mbar, x/d = 4.3,
           d_n = 13 mm, 6 h deposition time).

adsorbed gas in several distinct steps, whereas
condensates from baffled flow devitrify giving
only one exothermic peak as expected (4), and
second that supersonic flow gives condensates with
a "high" surface area, whereas baffled flow
appears to give deposits with a lower surface area.
These results are shown in the following Figures.

    Figure 1 shows nitrogen adsorption isotherms
of $(H_2O)_{as}$ prepared from supersonic flow without
baffle by condensation at 77 K (curve A), prepared
in the same way but heated in addition for 30 min
to 115 K (curve B), and prepared with a baffle
(curve C). The samples for the isotherms A and B

have been prepared and investigated in the same
apparatus. The sample from baffled flow for isotherm
C had to be transferred from a high vacuum
apparatus to the adsorption apparatus and we
consider this curve only to be preliminary. The
surface area of sample A was calculated by BET
to be 261 m^2/g. This value is very similar to the
value reported by Ghormley (5,6), but is much higher
than the values reported by two other groups (7,8).
Both heating of a sample to 115 K and baffled flow
have the same effect to reduce the surface area
considerably. The effect observable in curve B -
- reduction of surface area due to heating of
$(H_2O)_{as}$ to 115 K - has been predicted already by
Ghormley. The reduction of surface area from
approximately 260 m^2/g (Fig. 1A) to about 50 m^2/g
(Fig. 1B and 1C) is not caused by the formation
of crystalline ice: X-ray diffraction patterns of
all three samples contained very little crystalline
material.

Figure 2 shows DTA warm-up curves of $(H_2O)_{as}$
prepared on top with baffle, and below without
baffle, demonstrating the effect of supersonic flow
on the DTA patterns. (The small peak at 199 K has
been characterized by X-ray diffraction to be due
to the ice Ic $\longrightarrow$ Ih transition). These samples
have been transferred from a high-vacuum apparatus
to a precooled DTA apparatus, and thus have been
exposed to air and liquid nitrogen. Figure 3
shows DTA warm-up curves of $(H_2O)_{as}$ samples which
have been prepared and devitrified in situ, using
identical conditions with only one exception:
curve a was obtained by warming up a sample in
vacuum and shows only one exothermic peak, and the
other curves by warming up the samples at 1 atm
of N_2, He and O_2. This Figure demonstrates two
things: first, that in addition to supersonic flow
the presence of adsorbed gas is necessary during
devitrification to obtained several peaks, and
second, that O_2 seems to have an additional effect
by causing the peak at 179 K.

We have investigated the influence of various
parameters in the DTA curves and want to show in
Figure 4 only the effect of substrate temperature
on the DTA patterns. $(H_2O)_{as}$ for curve (a) was
deposited at 77 K, for curve (b) again at 77 K but
warmed in addition afterwards to 110 K for 30 min.
Due to this treatment the peak at 177 K disappears

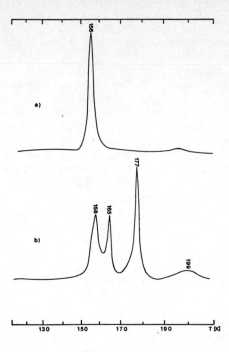

Figure 2.   DTA warm-up curves of $(H_2O)_{as}$ in air,
             (a) prepared with a baffle between nozzle
             exit and cryoplate, (b) without baffle.
             (Deposition conditions: K_n = 0.06,
             x/d = 6.3, d = 13 mm, p_o = 0.05 mbar,
             6 h desposition time).

completely during heating in the DTA curve. Curve (c)
was obtained with a $(H_2O)_{as}$ sample deposited at
110 K and shows only one exothermic peak. These
results obviously have to be combined with the
adsorption isotherm measurements, where thermal
treatment comparable to Figure 4b caused a reduction
of surface area from 260 to about 50 m^2/g.

     An interpretation for these effects is given
in the following. We suggest that clusters formed
in our experimental setup from supersonic flows
retain something of their identity during condensation
on a cooled substrate. By this we don't want to imply
that the gas phase microclusters are not influenced
by the condensation process on the cryoplate,
either breaking up into smaller units or

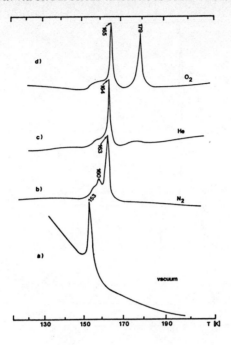

Figure 3.   DTA warm-up curves of $(H_2O)_{as}$, prepared
and devitrified in situ, (a) in vacuo,
(b), (c) and (d) at normal pressures
of N_2, He and O_2 respectively.
(Deposition conditions: K_n = 0.004,
x/d = 3.4, d = 5 mm, p_o = 0.5 mbar,
2 h deposition time).

coaleszing to larger aggregates. We suggest further
that freezing out of these gas phase clusters on a
cryoplate generates a network of voids or pores of
various sizes, and that $(H_2O)_{as}$ formed in this way
differs in this respect from $(H_2O)_{as}$ formed by
condensation of monomers from baffled flow. The
effect of adsorbed gas in Figure 3 can be explained
if we assume nucleation to be surface induced: the
delaying effect of adsorbed gas can be understood
by occupation of active sites on the surface or
in pores. As a working hypothesis we suggest that
in vapor deposited amorphous solid water polymorphism
exists. To be more specific we suggest that possibly
the structure of $(H_2O)_{as}$ prepared by condensation
of gas phase clusters is better described by a

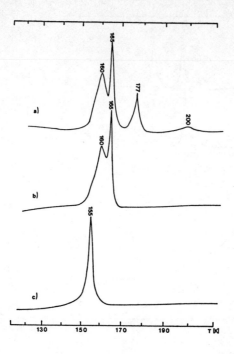

Figure 4.   DTA warm-up curves of $(H_2O)_{as}$ in air,
             demonstrating the influence of substrate
             temperature, (a) $(H_2O)_{as}$ prepared by
             condensation of water vapor at 77 K, (b)
             $(H_2O)_{as}$ warmed in vacuo to 113 K for
             30 min immediately after deposition at
             77 K, (c) $(H_2O)_{as}$ deposited at 110 K.
             (Deposition conditions: K_n = 0.06,
             x/d = 6.3, d = 13 mm, p_o = 0.05 mbar,
             6 h deposition time).

discontinuous model such as the amorphous cluster
model, whereas the structure of $(H_2O)_{as}$ prepared by
condensation of monomers is as usual described best
by a continuous model such as the continuous random
network model (9,10). These two descriptions
would represent only the extremes, and any real
sample might contain various parts with continuous
and discontinuous regions. A more detailed account
of the experimental work and the interpretation
is being published (3).

    It is tempting to speculate what influence

these results and interpretations might have on the
present ideas about extraterrestrial amorphous ice.
We see two areas where amorphous ice in form of a
high-surface area solid with a network of pores or
micropores might have an influence on the existing
concepts: the first is concerned with the role
amorphous ice might play in interstellar dust and
the hydrogen recombination rate on this solid,
and the second concerns the speculation that
amorphous ice might be present in the cometary
nucleus (11,12). Especially consideration of the
micropore effect, i.e. the enhancement of
interaction potential in pores of molecular
dimensions and the resultant increased adsorption
at low relative pressures (13), could well lead
to different conclusions than consideration only
of coarser pores in amorphous ice, because adsorption
in micropores would be strongly favored in space
due to the low pressures.

ACKNOWLEDGEMENT

     This work was supported by the "Fonds zur
Förderung der wissenschaftlichen Forschung" of
Austria.

Question Dr. F. L. Whipple
If the vapor is at rest and the cold plate meets
it supersonically will "gas clusters" form and a
porous ice be formed?

Answer
To our knowledge for this situation porous ice
with a high surface area has not been observed
experimentally. However, it seems to be accepted
that low-density wind tunnel experiments are able
to simulate the response of aerodynamic bodies
under conditions of supersonic flight at high
altitudes in planetary atmospheres (14). And
for supersonic flow in low-density wind tunnels
condensation of water vapor to gas clusters has
been observed many times, this effect having been
the starting-point for the pioneering work of
Oswatitsch (15) on condensation in wind tunnels.
Since in principle there is no difference between
the nozzle of a wind tunnel and of a cluster beam source
(16) ist seems to be conceivable that in the
encounter of a cold plate at supersonic speed with
vapor at rest gas clusters and high-surface-area
porous ice will be formed.

REFERENCES

(1) Sivakumar, T.C., Rice, S.A. and Sceats,
     M.G. 1978, J.Chem.Phys.69, pp. 3468-3476
(2) Hobbs, P.V. 1974, Ice Physics (Clarendon
     Press, Oxford)
(3) Mayer, E. and Pletzer, R. 1984, J.Chem.Phys.
     80, pp. 2939-2952
(4) McMillan, J.A. and Los, S.C. 1965, Nature 206,
     pp. 806-807
(5) Ghormley, J.A. 1967, J.Chem.Phys.46, pp.
     1321-1325
(6) Ghormley, J.A. 1968, J.Chem.Phys.48,
     pp. 503-508
(7) Adamson, A.W., Dormant, L.M. and Orem, M.
     1967, J.Colloid Interface Sci. 25, pp. 206-217
(8) Ocampo, J. and Klinger, J. 1982, J. Colloid
     Interface Sci.86, pp. 377-383
(9) Alben, R. and Boutron, P. 1975, Science 187,
     pp. 430-432
(10) Narten, A.H., Venkatesh, C.G. and Rice, S.A.
     1976, J.Chem.Phys.64, pp. 1106-1121
(11) Smoluchowski, R. 1981, Astrophys.J.244,
     pp. L31-L34
(12) Smoluchowski, R. 1983, J.Phys.Chem.87,
     pp. 4229-4233
(13) Gregg, S.J. and Sing, K.S.1982. Adsorption,
     Surface Area and Porosity (Academic Press)
(14) Potter, J.L. 1974. Molecular Beams and Low-
     -Density Gasdynamics, Ed. P.P. Wegener,
     (Marcel Dekker), Chapter 3, p. 198
(15) Oswatitsch, K. 1942, Z.angew.Math. mech. 22,
     p. 1

RHEOLOGIES OF H_2O ICES I_h, II, AND III AT HIGH PRESSURES:
A PROGRESS REPORT

S. H. Kirby[1], W. B. Durham[2], and H. C. Heard[2]

[1]U.S. Geological Survey, Menlo Park, CA 94025, U.S.A.
[2]Lawrence Livermore National Laboratory, Livermore, CA
94550, U.S.A.

ABSTRACT

Experimental studies of the rheology of ordinary hexagonal
ice have, until recently, been restricted to temperatures
greater than -30°C (243 K) and pressures less than 30 MPa. We
have greatly extended this pressure-temperature field to pressures as high as 350 MPa and temperatures as low as 77 K and
have established the basic features of ice I_h rheology. At the
lowest temperatures and highest strain rates, ice I_h fails by a
very unusual form of fracture. Above a confining pressure of
50 MPa, fracture strength is independent of pressure and varies
from about 160 MPa at 77 K to 120 MPa at 158 K. This fracture
process appears to be a shear instability and may be related to
localized transformation to a high-pressure phase of ice. At
higher temperatures, ice displays a steady-state flow behavior
at strains greater than about 10%. No fewer than three separate rheological laws are needed to adequately characterize
this steady-state ice I_h rheology. Although our rheological
data compare well with glaciological studies at high temperatures, the latter are inadequate to describe the flow of ice at
temperatures below about 240 K. Plastic relaxation times t_r
have been computed from our flow laws at differential stresses
that correspond to a range of topographic reliefs on Ganymede
and Callisto. For a relief $\underline{h}$ = 100 m, t_r > 3 Ga at temperatures below 110 K and for $\underline{h}$ > 1 km, t_r < 10 Ma at temperatures
greater than 100 K. Marked crater profile relaxation is therefore implied for large craters with $\underline{h}$ > 1 km if pure ice I_h
rheology applies to these Jovian satellites.

J. Klinger et al. (eds.), Ices in the Solar System, 89–107.
© 1985 by D. Reidel Publishing Company.

We have also performed preliminary experiments on the high-pressure ice polymorphs II and III at a fixed strain rate of about 4×10^{-4} s^{-1}. Ice III is markedly weaker than ice I$_h$ in its pressure-temperature stability field and shows a much greater weakening with increasing temperature than ordinary ice. Ice II, on the other hand, is significantly stronger than ice I$_h$ and has a very similar temperature effect on strength as ice I$_h$. Lastly, both ice II and III are strengthened by increasing confining pressure, in contrast with softening with increasing pressure for ice I$_h$. These pressure effects on strength parallel the freezing point depression with pressure for ice I$_h$ and its elevation with increasing pressure for ice III, suggesting that all of the high pressure forms of ice will exhibit pressure strengthening.

INTRODUCTION

Ordinary hexagonal ice (ice I$_h$) is the stable crystalline form of H_2O on the Earth's surface. It exists on Earth over a fairly narrow range of temperatures (~240 to 273 K) and pressures (0.1 to 40 MPa) (see Figure 1). Serious study of the rheological properties of ice started in the 1950s and it is now arguably the best characterized rock-forming mineral as far as its flow properties under these conditions (see reviews by Weertman, 1983; Mellor, 1980; Duval et al., 1983; Glen, 1975). Ice is known to exist elsewhere in the solar system over far wider ranges of temperature and pressure. Several of the moons of Saturn and Jupiter are composed predominately of H_2O and their surface temperatures are about 75 and 100 K (Consolmagno, 1983). The masses and sizes of the larger of these moons also imply that pressures may be as high as 3 GPa in their interiors. Not only is this field of temperature and pressure far beyond the conditions over which the rheological laws for ice I$_h$ can be confidently extrapolated, these conditions also place H_2O in the stability fields of several of the high-pressure crystalline phases of ice (Figure 1). Since 1981, we have conducted over 100 triaxial compression tests over a wide range of temperatures (77 to 258 K) and pressures (0.1 to 350 MPa). This paper is a progress report on the results of those experiments and follows a more detailed report on ice I$_h$ flow and fracture published earlier (Durham et al., 1983).

EXPERIMENTAL DETAILS

All of the experiments were performed in a high-pressure deformation apparatus designed and built expressly for this ice

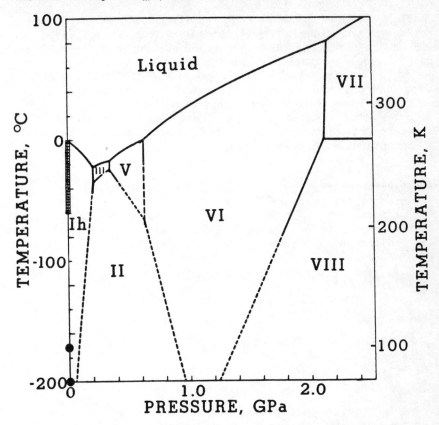

Figure 1.   Phase diagram for H$_2$O, modified after Hobbs
(1974).   The pressure-temperature region over which
previous rheological studies have been done is shown as a
ruled area on the vertical axis. The filled circles indicate the surface temperatures of the icy moons of Jupiter
(~100 K) and Saturn (~75 K).   Interior pressures in the
larger moons, such as Ganymede, Callisto and Titan, are
likely to reach several GPa, indicating the likelihood that
some of the high pressure phases of ice may exist in them.

deformation research program. A detailed description of the apparatus and technique may be found in Durham et al. (1983). Briefly, a jacketed 25.4 mm diameter, 62 mm-long right-circular cylinder of polycrystalline ice (grain size about 1 mm) is cooled to a constant temperature, pressurized to the desired pressure, and then shortened at a constant rate while the axial force, piston displacement, and temperature are recorded. The raw data are then reduced to true stress-strain curves and the steady-state flow stress σ is picked from the curve. Experimental uncertainties are as follows:

temperature	T	±2 K
confining pressure	P	±0.2 MPa
strain	ε	±0.001
strain rate	$\dot{\varepsilon}$	±10%
differential stress	σ	±0.2 MPa

EXPERIMENTAL RESULTS

Ice I_h

The steady-rate flow data may be represented by a thermally-activated power law:

$$\dot{\varepsilon} = A\sigma^n \exp(-Q^*/RT) \qquad (1)$$

that relates the imposed strain rate $\dot{\varepsilon}$, the observed steady-state stress σ, and the test temperature T in degrees absolute. $\underline{A}$, $\underline{n}$, and $\underline{Q^*}$ are material constants and $Q^* = H^* + PV^*$ where H^* is the activation enthalpy and V^* is the activation volume. The PV^* term accounts for a small pressure effect on steady-state flow. Taking common logarithms and rearranging:

$$\log \sigma = (\log \dot{\varepsilon})/n - (\log A)/n + Q^*/2.303nRT .$$

If we plot, therefore, the common logarithm of the steady-state flow stress at a fixed strain rate, it should be linear in $(1/T)$ with a slope proportional to Q^*/n. The $\underline{n}$ parameter is equal to the slope of the $\log \dot{\varepsilon}$ versus $\log \sigma$ relation. Inspection of Figure 2 immediately indicates that at least three different regimes of ductile flow are recognizable, each presumably representing different ductile flow mechanisms, each with different values of $\underline{Q^*}$, $\underline{n}$, and $\underline{A}$ and each therefore displaying different slopes $\overline{Q^*/n}$. At intermediate temperatures (~195 to 240 K), the value of $H^* = 61$ kJ mol^{-1} is fully consistent with determinations of H^* for single crystals at $273 < T < 223$ K and for polycrystals at $263 < T < 233$ K and $\overline{P} = 1$

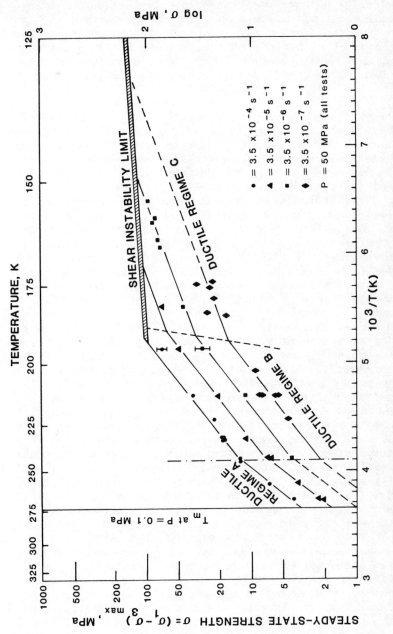

Figure 2. Ductile flow of ice I_h under 50 MPa confining pressure at four strain rates. The common logarithm of the steady-state strength σ is plotted against the inverse of the absolute temperature T. Three different flow regimes are identified by differences in slope. The ruled line represents the upper limit of ductile strength set by a shear instability though to be associated with the ice $I_h \rightarrow$ ice II phase transformation.

atmosphere (see reviews by Weertman, 1973, 1983). Moreover, $H^* = 61$ kJ mol^{-1} matches well the values for activation energy for molecular diffusion in ice (52 to 67 kJ mol^{-1}), as predicted from theories of high temperature flow controlled by diffusion-controlled healing of crystal-defect damage produced by deformation (see Weertman, 1983, for a review of this topic). Samples deformed in this regime are partially recrystallized.

At temperatures higher than about 240 K in our experiments, the activation energy is 91 kJ mol^{-1}, about 50% higher than at lower temperatures. A similar increase in H^* and creep rates is also observed in polycrystalline ice tested at atmospheric pressure, but the transition temperature is about 263 K (Mellor and Testa, 1969), about 17 K higher than indicated in Figure 2. This increase in H^* with increasing temperature is not observed in single crystal tests (Jones and Brunet, 1978; Homer and Glen, 1978) and electrical conductivity and NMR measurements indicate that molecular water exists in polycrystalline ice at $273 > T > 263$ K (Mellor and Testa, 1969; Barnes et al., 1971; Baker and Gerberich, 1979; Ohtomo and Wakahama, 1983). A number of workers have concluded that water at grain boundaries and triple junctions may facilitate deformation by grain boundary sliding or by mechanisms related to recrystallization (Barns et al., 1971; Duval, 1977; Ohtomo and Wakahama, 1983). The existence of molecular water at temperatures below the melting point T_m may be due to solute depression of T_m or due to the overall reduction of free energy of a polycrystalline aggregate by grain boundary wetting. We do not understand why our critical transition temperature is 17 K lower than in tests conducted at atmospheric pressure; T_m at a confining pressure of 50 MPa and $\sigma = 5$ MPa should have been reduced by only about 7 K. In any event, the strength values we obtained at high temperatures are consistent with previous measurements of steady-state strength at atmospheric pressure by Mellor and Cole (1983) who also obtained an identical $n = 4.0 \pm 0.1$. The flow regime at temperatures below about 195 K is the least understood. It is truncated abruptly by fracture at $\sigma =$ 130-160 MPa and so we have only a narrow range of temperature and strain rate over which to explore this flow behavior. Also, not much is known about the diffusion and transport behavior of ice at these temperatures. Thin sections of samples deformed in this regime clearly show that the deformation occurs by ductile mechanisms and that there is much more deformation in grain interiors and less recrystallization, suggesting reduced grain boundary mobility or reduced diffusion rates may be responsible for the transition to this flow behavior with reduced temperature. Obviously, additional tests are needed to better characterize this rheological regime.

Our work on ice I$_h$ has shown that the high temperature flow law that is so well determined and extensively applied to glaciological problems is a wholly inadequate description of the steady-state strength of ice at the low temperatures appropriate to the icy bodies of the outer solar system; ice is much weaker than expected from extrapolation of this high temperature flow law to lower temperatures.

Ice II and ice III

We have performed eleven deformation tests of high-pressure polymorphs of ice (Figure 3 shows the pressure-temperature conditions of these runs as well as the conditions of tests on ice I$_h$ at the same strain rate). The high-pressure phases were produced by pressurizing at a fixed temperature until a pressure drop was observed due to the volume reduction associated with the transformation (Figure 4). Apparent changes in sample length, detected by recording the piston displacement at the point the sample was "touched" by the loading system, also indicated that the transformation had taken place (Figure 5). The linear strains associated with the transformation corresponded well to the volumetric strain $\Delta V/V_o$ of the ice I $\rightarrow$ II transformation measured by Bridgman (1912) at 195 K, but at higher temperatures, the observed linear strains differed from those predicted from the appropriate $\Delta V/V_o$ values by as much as 3%. Apparently, the transformational strains are not isotropic. Corrections of the initial sample lengths ℓ_o and cross-sectional area A_o for the observed transformational strains were applied to the computation of the sample shortening strain $\varepsilon = \Delta \ell / \ell_o$ and differential stress $\sigma = F(1-\varepsilon)/A_o$, where $\Delta \ell$ is the sample shortening and F is the force applied by the piston. All of the experiments were performed at a fixed strain rate of about 4×10^{-4} s^{-1}. Typical stress-strain curves for ice II are shown in Figure 6 and are similar in form to those obtained for ice I$_h$ except that latter usually display more pronounced stress drops following yield. All of the stress-strain curves of the high-pressure phases developed a constant steady-state strength at strains greater than 0.1. After the deformation was completed, the piston was withdrawn 8 mm and the high-pressure phases were depressurized at the test temperature, during which time a pressure rise caused by the reversion to ice I was observed (Figure 4). Alternately, the sample was cooled to temperatures less than 160 K, depressurized, and the high-pressure phase was retained metastably. Sample volumes of the latter specimens were always less than the volume of the ice I$_h$ starting material, as required by their higher room-pressure densities.

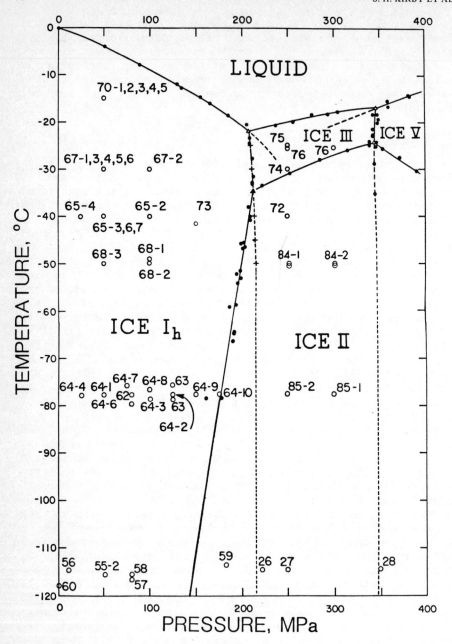

Figure 3.   Pressure and temperature conditions of the experiments performed at a strain rate of 3.5 to 4.5x10^{-4} s^{-1}.

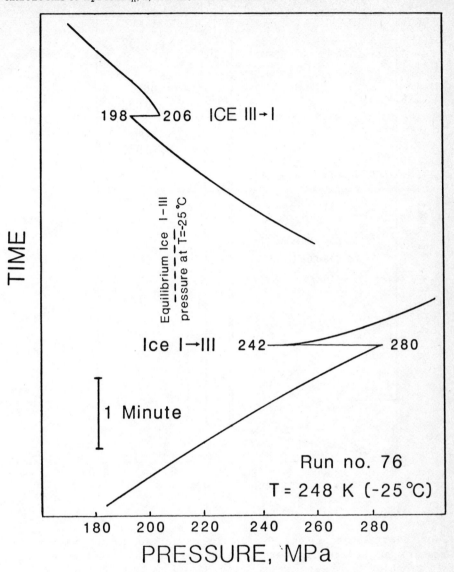

Figure 4.   Pressure-time recorder tracings of an experiment in which ice I$_h$ is pressurized at constant pumping rate. The discontinuity in the pressurization curve reflects the volume reduction of the sample upon transformation to ice III.   An analogous discontinuity occurs during depressurization.   The dotted line is the equilibrium transformation pressure at 248 K.

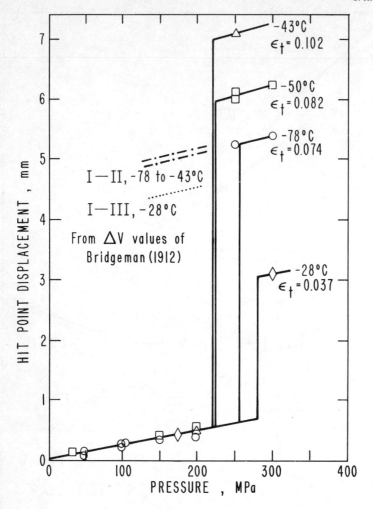

Figure 5. Apparent changes in ice specimen length with pressure at several different temperatures, as determined by "touching" the specimens with the load column. The apparent specimen shortening at P < 200 MPa is due both to volumetric compression upon pressurization as well as elastic stretching of the pressure vessel and elastic compression of the piston. At pressures above 200 MPa, discontinuous shortening of the samples occurs which is caused by the volumetric transformation strains. The linear transformation strains implied by this figure deviate from those expected from isotropic transformation by as much as 3%, indicating anisotropic transformation. The reasons for this anisotropy are unknown.

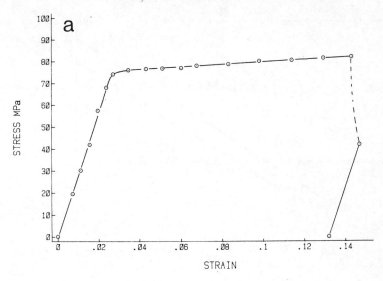

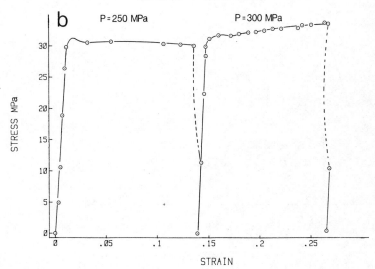

Figure 6. Stress-strain curves for ice II at a fixed strain rate of $\dot{\varepsilon} \simeq 4 \times 10^{-4}$ s^{-1}. Stress levels reach plateau values at $\varepsilon > 0.1$ which we interpret as representing the steady-state stress σ. a. T = $-78°$C (195 K), P = 300 MPa. b. T = $-50°$C (223 K), P = 250 and 300 MPa. Note slight strengthening with increasing pressure.

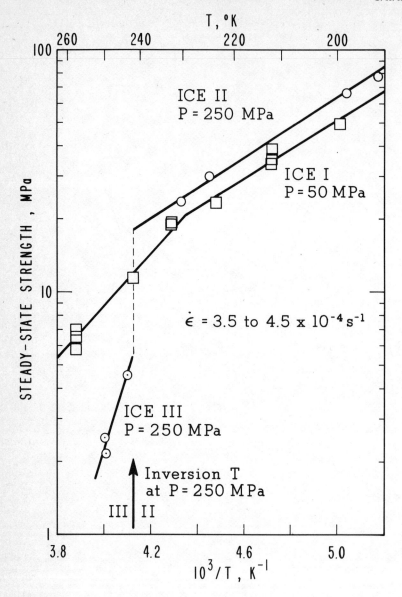

Figure 7. Comparison of $\log_{10}$ σ versus 1/T for ice I_h at P = 250 MPa and for ice II and III at P = 250 MPa. The strain rate is about 4×10^{-4} s^{-1} for these tests. Note that temperature has roughly the same effect on σ for ice II as for ice I_h but that ice II is significantly stronger than ice I_h. In contrast, ice III is markedly weaker than ice I_h and shows a much higher temperature sensitivity.

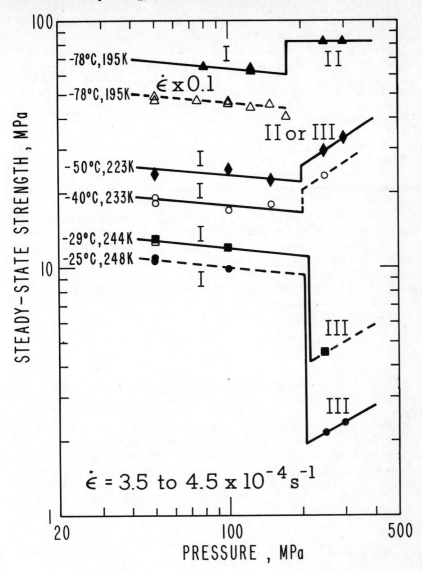

Figure 8.   Pressure effect on the strengths of ice I$_h$, II, and III.   Note pressure softening of ice I$_h$ and the pressure strengthening of ice III.   The pressure effect in ice II is either positive or zero.   The discontinuities in the curves are drawn at the transformation pressures at which ice I$_h$ is in equilibrium with the high pressure phase.

In Figure 7 ·we compare the steady-state strength values of ice I_h at P = 50 MPa with those for ice II and III at P = 250 MPa, all plotted on a semi-logarithmic basis to reveal differences in Q*/n values. Two major features of this plot are very apparent. First, ice II is significantly stronger than ice I_h at the same temperature and has roughly the same Q*/n. Second, ice III is markedly weaker than ice I_h at the same temperature and exhibits a much higher slope and thus higher Q*/n value. The data also show significant differences in the pressure sensitivities of steady-state strength. In Figure 8 we plot $\log\sigma$ vs log P. Ice I_h shows pressure soften-ing at all test temperatures (V* $\simeq$ -10×10^{-6} m^3 mol^{-1}). The strength data extrapolated to the phase transition pressures confirm the above conclusions drawn from Figure 8: ice II is stronger than ice I_h and ice III is markedly weaker than ice I_h. The reversal of the pressure effect on melting temperature (Figure 1), lends support to empirical correlations of creep properties with melting temperature (Sherby and Simnad, 1971; Weertman, 1970; Brown and Ashby, 1980; Sammis et al., 1981). It also suggests that the other higher pressure forms of ice that display positive pressure effects on T_m (ice V, VI, and VII) should also exhibit pressure strengthening.

DISCUSSION

A thorough discussion of the implications of this work in interpreting the surface features of the icy moons of Jupiter is beyond the scope of this paper. Some simple viscosity and strain-rate profiles have been presented in our earlier paper (Durham et al., 1983). We can make some additional general comments based on simple extension of the rheological data. Caldwell and Turcotte (1979) and Turcotte and Schubert (1982) have shown that the time t_r necessary for a given initial stress σ_o to relax to half its values by a non-linear plastic deformation is given by

$$t_r = \frac{(2^{n-1}-1)}{(n-1)EA\ \sigma^{n-1}} \ \exp(Q^*/RT) \qquad (2)$$

where $\underline{A}$, $\underline{n}$, and $\underline{Q^*}$ are defined as in eq. (1) and $\underline{E}$ is Young's modulus. The above authors actually computed the case for n = 3 but it is trivial to extend their solution to the case of arbitrary $\underline{n}$. Computing t_r (T) has been useful in terrestrial applications where the critical temperature T_c below which relaxation times are large relative to geologic time can be used to define the depth H to the base of the mechanically strong lithosphere and hence is useful in evaluating the response to surface topographic loads. We have computed the t_r versus T relations for the two lowest temperature rheological laws for ice I_h summarized in Table 1. The results are shown in Figure 9 for various initial stresses. The independent mechanism that produces the shortest relaxation time at a given temperature will obviously dominate the relaxation response. The low-temperature flow law obviously dominates relaxation for these loads except at T >> 150 K where the intermediate temperature flow law begins to dominate.

Table 1. Flow law parameters for ice at a confining pressure of 50 MPa.

Regime	$\log_{10} A$ $[MPa^{-n}s^{-1}]$	n	$Q*$ $[kJ\ mol^{-1}]$
Ductile C Low T (< 195 K)	-3.4	3.7 ± 0.4	27.4
Ductile B Intermediate T (195-240 K)	5.10 ± 0.03	4.0 ± 0.1	61 ± 2
Ductile A High T (240-258 K)	11.8 ± 0.4	4.0 ± 0.6	91 ± 2

One can use the results of Figure 9 to roughly predict the longevity of topographic loads on the icy moons of Jupiter. To our knowledge, the problem of gravitational relaxation of topography on a non-linear substrate with temperature-dependent viscosity has never been fully analyzed. Jeffreys (1976, p. 267-269) summarizes the elastic and plastic solutions for the maximum stress differences associated with a topographic load of uniform relief h and shows that S is given by

$$S = A\bar{\rho}gh \qquad (3)$$

where A = 0.39 to 0.64 depends on the rheological assumptions. Arbitrarily taking A = 0.5 and a surface g of 1.42 m s^{-2} and $\bar{\rho}$ = 0.92 for Ganymede

$$S(MPa) = 6.5 \times 10^{-4}\ h(m)$$

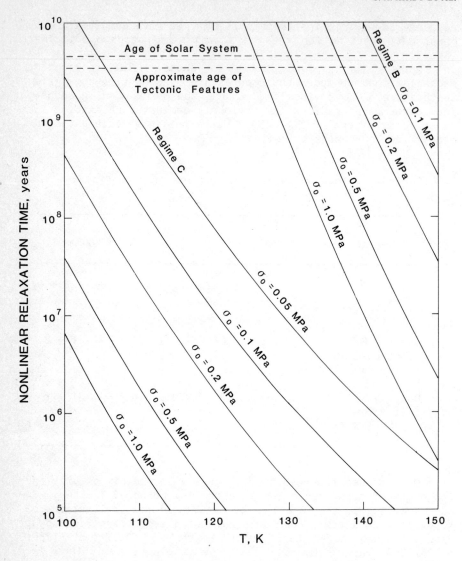

Figure 9.   Relaxation time versus temperature for various
initial stresses using the flow law parameters of Table 1
for ductile regimes B and C.   See text and eq. (2) for
details of calculation.

For h = 100 m, S = 0.065 MPa and for h = 1 km, S = 0.65 MPa. For a surface temperature of 100-110 K, the corresponding relaxation times from Figure 9 are >10^9 years and 10^7 years, respectively. These results indicate that topographic features with relief of a few hundred meters should be very long lived, whereas features with relief of kilometers should be largely relaxed in 10 to 100x10^6 years. This marked non-linear effect of the initial stress difference on relaxation time, also pointed out by Parmentier and Head (1979) and by Poirier (1982), is in addition to the effect of size (aerial extent) on the depth and temperature at which ductile relaxation takes place (Parmentier and Head, 1981); large, deep craters, for example, should relax much faster than small shallow craters because of the larger initial stress differences as well as the stress field of the topographic load penetrating into deeper, hotter ice. The predictions of t$_r$ based on Figure 9 are in overall agreement with the pattern of topographic features on the larger icy moons of Jupiter (see reviews in Parmentier and Head, 1981; Squires, 1980; Poirier, 1982). Large (>> 10 km) craters generally are under-represented relative to small ones and usually have low topographic relief relative to their diameters. Many of the larger craters also display central domes probably caused by backflow driven by isostatic relaxation (Squires, 1980; Parmentier and Head, 1981). Small craters with relief of less than a few hundred meters appear less modified by such non-linear viscous flow associated with isostatic relaxation, in agreement with predictions based on Figure 9.

We really have not provided full rheological laws for ice II and III, so comparison with ice I$_h$ at natural strain rates cannot be made at this time. The greatly lower strength of ice III compared to ice I$_h$ and the much higher temperature effect on strength are intriguing and invite speculation on possible runaway flows in large scale processes in the icy moon interiors where temperatures may approach the ice III stability field, and where thermal convection driven by density differences may be important or may have been important during the thermal evolution since accretion.

ACKNOWLEDGMENTS

We thank Pegi Daley and Carl Boro for technical assistance and acknowledge NASA support under order number W-15,070 that made our participation in this workshop possible. This work was performed under the auspices of the U.S. Department of Energy by the Lawrence Livermore National Laboratory under contract W-7405-ENG-48. Lastly, we give our appreciation to the organizers of the workshop, Jurgen Klinger and Daniel Benest, who made possible a most stimulating meeting.

REFERENCES

Baker, R.W. and Gerberich, W.W. 1979, The effect of crystal size and dispersed-solid inclusions on the activation energy for creep of ice, J. Glaciology 24, pp. 179-192.

Barnes, P., Tabor, D. and Walker, J.C.F. 1971, The friction and creep of polycrystalline ice, Proc. Roy. Soc. London A 324, pp. 127-155.

Bridgman, P.W. 1912, Water, in the liquid and five solid forms, under pressure, Proc. Am. Acad. Arts Sci. 47, pp. 347-438.

Brown, A.M. and Ashby, M.F. 1980, Correlations for diffusion constants, Acta Met. 28, pp. 1081-1101.

Consolmagno, G.J. 1983, Ice-rich planets and the physical properties of ice, J. Phys. Chem. 87, pp. 4204-4208.

Consolmagno, G.J. and Lewis, J.S. 1976, Structural and thermal models of icy galilean satellites, in "Jupiter," T. Gehrels, ed., Univ. of Arizona Press, pp. 1035-1051.

Durham, W.B, Heard, H.C. and Kirby, S.H. 1983, Experimental deformation of polycrystalline H_2O ice at high pressure and low temperature: preliminary results, J. Geophys. Res. 88 (Supplement), pp. B377-B392.

Duval, Paul 1977, The role of water content on the creep rate of polycrystalline ice, in "Isotopes and Impurities in Snow and Ice," Proc. of the Grenoble Conf., Aug. 1975, Internat. Assoc. of Hydrological Sciences publication 118, pp. 29-33.

Duval, Paul, Ashby, M.F. and Anderman, I. 1983, Rate-controlling processes in the creep of polycrystalline ice, J. Phys. Chem. 87, pp. 4066-4074.

Duval, Paul and LeGac, H. 1982, Recovery of ice, Annals of Glaciology 3, pp. 92.

Glen, J.W. 1975, The mechanics of ice, Cold Regions Science and Engineering Monograph 11-c2b, U.S. Army Cold Regions Research and Engineering Laboratory, 43 p.

Hobbs, P.V. 1974, "Ice Physics," Clarendon Press, Oxford, 837 p.

Homer, D.R. and Glen, J.W. 1978, The creep activation energies of ice, J. Glaciology 21, pp. 429-444.

Jones, S.J. and Brunet, J.-G. 1978, Deformation of ice single crystals close to the melting point, J. Glaciology 21, pp. 445-456.

Mellor, M. 1980, Mechanical properties of polycrystalline ice, in "Physics and Mechanics of Ice," P. Tryde, ed., Springer-Verlag, pp. 217-245.

Mellor, M. and Cole, D.M. 1982, Deformation and failure of ice under constant stress or constant strain rate, Cold Regions Science and Technology 5, pp. 201-219.

Mellor, M. and Cole, D.M. 1983, Stress/strain/time relations for ice under uniaxial compression, Cold Regions Science and Technology 6, pp. 207-230.

Mellor, M. and Testa, R. 1969, Effect of temperature on the creep of ice, J. Glaciology 8, pp. 131-145.

Ohtomo, M. and Wakahama, G. 1983, Growth mechanism of recrystallization in ice, J. Phys. Chem. 87, pp. 4139-4142.

Parmentier, E.M. and Head, J.W. 1979, Internal processes affecting surfaces of low-density satellites: Ganymede and Callisto, J. Geophys. Res. 84, pp. 6263-6276.

Parmentier, E.M., and Head, J.W. 1981, Viscous relaxation of impact craters on icy planetary surfaces: determination of viscosity variation with depth, Icarus 47, pp. 100-111.

Poirier, J.P. 1982, Rheology of ices: a key to the tectonics of the ice moons of Jupiter and Saturn, Nature 299, pp. 683-688.

Sammis, C.G., Smith, J.C. and Schubert, G. 1981, A critical assessment of estimation methods for activation volume, J. Geophys. Res. 86, pp. 10707-10718.

Squires, S.W. 1980, Topographic domes on Ganymede: ice vulcanism or isostatic upwarping, Icarus 44, pp. 472-480.

Weertman, J. 1970, The creep strength of the earth's mantle, Rev. Geophys. Space Phys. 8, pp. 145-168.

Weertman, J. 1973, Creep of ice in "Physics and Chemistry of Ice," E. Whalley, S.J. Jones and L.W. Gold, eds., Roy. Soc. Canada, Ottawa, pp. 320-337.

Weertman, J. 1983, Creep deformation of ice, in "Annual Review of Earth and Planetary Sciences," pp. 215-240.

CREEP OF HIGH-PRESSURE ICE VI

C. SOTIN, P. GILLET* & J.P. POIRIER

Institut de Physique du Globe, Université Paris VI,
4 Place Jussieu, 75230 Paris Cedex 05
* Now at Institut de Géologie, Université de Rennes,
35042 Rennes Cedex.

The creep of high pressure ice VI has been investigated at temperature close to the melting point (253 < T < 293 K) and pressures in the range 1.0 to 1.7 GPa. It is found to obey the equation :

$$\dot{\varepsilon} = 2.8 \times 10^{-11} \, \sigma^{1.9} \, \exp-\left[\frac{28500 + 8\times10^{-6} \, \bar{P}}{RT} \right]$$

($\dot{\varepsilon}$ in s^{-1}, σ and $\bar{P}$ in Pa)

The viscosity is in most cases lower than 3×10^{14} poise and is compatible with the possibility of convection inside the large icy satellites.

INTRODUCTION

High-pressure polymorphs of water ice may be present, or may have been present, at great depths within the large icy satellites of Jupiter and Saturn (Ganymede, Callisto, Titan).
Dynamical models of the interiors of these planets therefore cannot be set up without some .knowledge of the viscosity of the high pressure forms of ice and its temperature, pressure and stress dependence (e.g. 1). Following a preliminary report (2), we present here an experimental investigation of the creep parameters of ice VI at temperatures close to its melting point (253 < T < 293 K) and pressures in the range 1.0 to 1.67 GPa.

J. Klinger et al. (eds.), Ices in the Solar System, 109–118.
© *1985 by D. Reidel Publishing Company.*

EXPERIMENTAL TECHNIQUE AND PROCEDURES

The ice VI is prepared in a sapphire cell operating on the same principle as the diamond anvil cell. The unavoidable hydrostatic pressure gradient is used to drive the creep of ice, which is monitored by observing the displacement of markers. The sample, here initially a drop of tap water, is enclosed in a hole (0.5 to 1.5 mm diameter) drilled in an annealed copper gasket (0.3 mm thick). The gasket is squeezed between two sapphire anvils (single crystalline rods of Al_2O_3) until the water transforms into ice VI and the desired pressure is reached (see phase diagram of ice, Whalley, this volume). The uniaxial force is applied by an INSTRON testing machine through steel levers allowing observation of the sample through the sapphire anvils while it is under stress (fig. 1). The uniaxial force is kept constant by servo-control. Small (2 μm) particles of carborundum polishing grit are used as displacement markers, they are added to the water and get embedded in ice when it forms. Photographs of the sample with the markers are taken at successive time intervals through a microscope ; the experiment can also be monitored by a video camera. Runs have been performed at room temperature and at lower temperatures (0 °C and - 20 °C) obtained by immersing the jig in a bath of ethyl alcohol cooled with dry ice.

The hydrostatic pressure is higher at the center of the hole than at the edges and the ice creeps radially outwards along the pressure gradient ; markers accordingly move radially with a velocity u determined by measuring the distance covered as a function of time, on the photographs. The velocity u depends on the position of the markers, defined by the radius r (see fig. 2 for the definition of symbols). The local creep rate is then :

$$\dot{\varepsilon}\ (r) = \frac{u(r)}{h} \tag{1}$$

where h is the height of the ice disk, measured optically by focusing the microscope on the surfaces of the anvils. The rheological equation can in principle be found if the shear stress and hydrostatic pressure are known for a given value of r (the temperature is reasonably assumed to be constant over the whole ice disk). The main difficulty consists in determining the applied shear stress $\bar{\sigma}(r)$ and hydrostatic pressure $P(r)$ in the conditions of the experiment. These quantities can be calculated as functions of the geometrical quantities a, b, h (see fig. 2) and of the applied force F, using an approach devised for the Bridgman anvils (3). The following assumptions are made (σ_r, σ_θ and σ_z are the principal components of stress in cylindrical coordinates) :

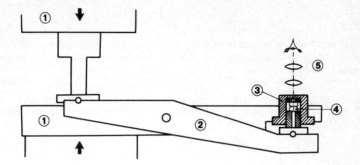

Fig. 1 : Experimental set up.

1. Compression platens of the Instron machine
2. Steel lever
3. Cell
4. Sapphire anvils
5. Viewing system

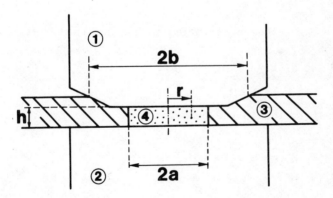

Fig. 2 : Detail of the sapphire anvil set up.

1. Top sapphire anvil.
2. Bottom sapphire anvil.
3. Gasket.
4. Hole with sample and markers.

i) $\sigma_r = \sigma_\theta$ $\qquad\qquad\qquad\qquad\qquad\qquad\qquad\qquad$ (2)

ii) There is adherence between ice and sapphire (no-slip condition). This assumption is found to be realistic : markers close to the anvils do not move.

iii) The ice and the copper gasket are considered as plastic solids, with a plastic limit given by Tresca's criterion.

$$\sigma_z - \sigma_r = \sigma_o^{ice} \qquad\qquad \text{for } 0 < r < a \qquad\qquad (3)$$

$$\sigma_z - \sigma_r = \sigma^{cu} \qquad\qquad \text{for } a < r < b \qquad\qquad (4)$$

iv) At the boundary between ice and gasket the normal stress is continuous :

$$\sigma_r^{ice}(a) = \sigma_r^{cu}(a) \qquad\qquad\qquad\qquad\qquad (5)$$

v) For steady-state conditions (u = const) we assume that the shear stress is independent of r and much larger than the plastic limit :

$$\sigma = \sigma_z - \sigma_r \gg \sigma_o^{ice} \qquad\qquad\qquad\qquad (6)$$

The local stress equilibrium in ice can then be written :

$$\frac{d\sigma_r}{dr} = - \frac{\sigma_z - \sigma_r}{h} = - \frac{\sigma}{h} \qquad\qquad\qquad (7)$$

Stress equilibrium in the copper gasket is written :

$$\frac{d\sigma_r}{dr} = - \frac{\sigma_o^{cu}}{h} \qquad\qquad\qquad\qquad\qquad (8)$$

Integration of (8) with the boundary conditions $\sigma_r(r=b) = 0$ gives :

$$\sigma_r = \sigma_o^{cu}\left(\frac{b-r}{h}\right) \qquad\qquad \text{for } a < r < b \qquad (9)$$

with (5) and (9), integration of (7) gives :

$$\sigma_r = \sigma \left(\frac{a-r}{h} \right) + \sigma_o^{cu} \left(\frac{b-a}{h} \right) \quad \text{for } 0 \leqslant r \leqslant a \tag{10}$$

The vertical stress σ_z is calculated from (3) (9) (6) and (10) :

$$\sigma_z = \sigma_o^{cu} \left(1 + \frac{b-r}{h} \right) \qquad \text{for } a \leqslant r \leqslant b \tag{11}$$

$$\sigma_z = \sigma \left(1 + \frac{a-r}{h} \right) + \sigma_o^{cu} \left(\frac{b-a}{h} \right) \qquad \text{for } 0 \leqslant r \leqslant a \tag{12}$$

It follows that the hydrostatic pressure P in the ice can be written :

$$P = \frac{1}{3} (2\sigma_r + \sigma_z) = \sigma \left(\frac{1}{3} + \frac{a-r}{h} \right) + \sigma_o^{cu} \left(\frac{b-a}{h} \right) \tag{13}$$

The uniaxial force F, applied by the Instron machine is :

$$F = \int_o^{2\pi} \int_o^b \sigma_z (r) r dr d\theta$$

or :

$$F = \pi a^2 \sigma \left(1 + \frac{a}{3h} \right) + \pi \sigma_o^{cu} \left(\frac{b^3 - a^3}{3h} + b^2 - a^2 \right) \tag{14}$$

The validity of equation (13) has been verified, using alkali halides (RbCl, NH$_4$F and KBr) with phase transitions at known pressures up to 20 kbar. Experimental and calculated values of P agree for σ_o^{cu} = 0.8 kbar, a value consistent with metallurgical data.

The hydrostatic pressure is found to decrease linearly as r increases, in agreement with the assumption that σ is independent of r.

We consider that the motion of a marker at a distance r from the center is representative of the rate of the strain of a disk of radius r under an average applied shear stress $\bar{\sigma}$ (r) :

$$\bar{\sigma} (r) = \frac{1}{\pi r^2} \int_o^{2\pi} \int_o^r \sigma_z(r) r dr d\theta - \sigma_r(r) = \sigma \left(1 + \frac{r}{3h} \right) \tag{15}$$

and an average hydrostatic pressure :

$$\bar{P}(r) = [P(0) + P(r)]/2$$

INVERSION OF THE DATA

The constitutive equation for creep of ice is assumed to have the usual power law form :

$$\dot{\varepsilon} = A \, \bar{\sigma}^{\,n} \, \exp\left(- \frac{Q + \bar{P}\Delta V}{RT} \right) \qquad\qquad (16)$$

where $\dot{\varepsilon} = u(r)/h$ is the strain rate measured from the motion of a marker at radius r, $\bar{\sigma}$ is the applied shear stress given by (15), A is a constant, R is the gas constant and P and T are the hydrostatic pressure (at r) and the temperature respectively. The rheological parameters n, Q and ΔV are the stress exponent, the activation energy and the activation volume, respectively.

For each experiment, corresponding to given (measured) values of F, T, a, b and h, we measured the maximum velocity of markers (for markers close to the mid-plane of the ice disk) at r = a/4, a/2, 3a/4, a. Using (1) (13) and (16), we can relate each measured value 4 of u(r) to the other data and parameters by :

$$\ln u(r) - \ln h - \ln A - n\ln\left[\sigma\left(1+\frac{r}{3h}\right)\right] + \frac{Q}{RT} + \frac{\Delta V}{RT}\left[\sigma\left(\frac{1}{3} + \frac{a-r/2}{h}\right) + \sigma_0^{cu}\left(\frac{b-a}{h}\right)\right] = 0$$

$$(17)$$

The best values of the parameters A, n, Q, ΔV, σ_0^{cu} and of the unknown quantity σ are simultaneously determined by a generalized non linear least squares inversion procedure (total inversion) taking experimental errors into account (4). For each experiment the measured data : F, T, a, b, h, u(a/4), u(a/2), u(3a/4), u(a), the unknown quantity σ and the values of the parameters A, n, Q, ΔV, σ_0^{cu} are linked by 4 equations of type (17) and 1 equation of type (14). We used 4 experiments which therefore yielded 20 equations ; the unknown and measured data (45 altogether) are treated similarly in the total inversion procedure which considers unknown as data with infinite variance.

The inversion is performed by iterations, starting with initial values of the data and reasonable a priori values for the unknown, it converges rapidly and yields best values for the data (fig. 3) as well as for the unknown parameters.

The variance of a parameter is found by recalculating this parameter using the 20 equations successively, with values of the other parameters found by the inversion and the experimental values of the data corresponding to each equation. The average of

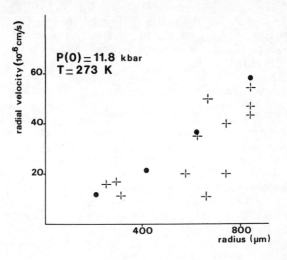

Plot of the radial velocity of markers against their distance to the center. For T = 273 K and P(O) = 11.8 kbar. Crosses correspond to individual measurements. Full circles represent the best values of maximum velocity given by the total inversion.

the 20 values found for the parameter is equal to the values found by inversion and the variance is adopted as the variance of the parameter.

Due to our lack of knowledge of the relevant physical properties of ice VI (like diffusion coefficients or dislocation structure), it is only possible to give a sketchy interpretation of the results.

RESULTS AND DISCUSSIONS

Creep of ice VI between 253 and 293 K at pressures ranging from 1 to 1.6 GPa is found to obey the usual creep law :

$$\dot{\varepsilon} = A\sigma^n \exp - \left(\frac{Q + \bar{P}\Delta V}{RT} \right)$$

with

$\ln A = - 24.3 \pm 0.4$ (σ in Pa)
$n = 1.93 \pm 0.02$
$Q = 28.5 \pm 1$ kJ/mole
$\Delta V = 8.1 \pm 0.8$ cm^3 mole

i) We may first notice that the value of the activation energy is rather low, it can be compared to the value Q = 31 kJ/mole found for ice I at low temperatures (158 < T < 195 K) by Durham et al (1983).
The structure of ice VI is tetragonal with chains of $(H_2O)_4$ tetrahedra along the edges of the cell, the chains are linked by hydrogen bonds, weaker than those of the tetrahedra. The structure is that of a self clathrate with 2 systems of chains enmeshed without bonding. We can only surmise (1) that deformation could be rather easy on the slip system {110} [001] which cuts only the weak bonds between C chains.

ii) The unit cell of ice VI is simple tetragonal with a motif of 2 tetrahedra. Each cell therefore contains 10 water molecules and its volume is $V = 2.276 \times 10^{-22} cm^3$ at P = 0 (5). Using the value for the bulk modulus of ice VI between 10 and 20 kbar, K = 210 kbar (6), we find that the activation volume corresponds to 0.8 volume of water molecule. Considering the inaccuracies of our measurements, we can only say that the activation volume for creep of ice VI is of the order of magnitude of the volume of a water molecule.
The pressure dependence of the creep rate of solids is often expressed in terms of the variation of the melting point with pressure

$$\dot{\varepsilon} = A\sigma^n \exp\left[-g\,T_m(P)/T\right] \qquad\qquad (18)$$

where g is an empirical constant, usually taken equal to 30 for silicates and to 18 for metals (7). In the case of ice VI, it has been taken equal to 25 (8). Comparing (18) to (16) we see that :

$$\Delta V = g\,R\,\frac{dT_m}{dP} \qquad\qquad (19)$$

From our value of ΔV and the value of dT_m/dP at 15 kbar from (9) we find :

$$g \simeq 20$$

iii) The stress exponent is quite low (n ≈ 2), making the viscosity almost newtonian, which is very unusual, not to say unheard of, for such high shear stresses.
The shear stress $\bar{\sigma}$ in our experiments has values in the range 0.5 to 1 kbar. The value of the shear modulus of ice VI can be calculated from the measurements of Polian and Grimsditch (1983) of the sound velocity in ice VI at P = 10 kbar by Brillouin scattering : $\mu \simeq 52.5$ kbar

We have therefore :

$$10^{-2} \leqslant \frac{\bar{\sigma}}{\mu} \leqslant 2 \times 10^{-2}$$

At this high stress levels, all solids we know of exhibit a highly non-newtonian viscosity, with high values of n, even above the usual values of 3 to 5. We have, so far, no explanation for this behaviour.
Finally it is interesting to calculate the extreme values of the viscosity found in our experiments.

For T = 253 K, $\bar{P}$ = 16 kbar, $\bar{\sigma}$ = 0.5 kbar, we find
$\dot{\varepsilon}$ = 2.3 x 10^{-3} s^{-1} hence $\eta \simeq 2 \times 10^{11}$ poise.

For T = 293 K, P = 10 kbar, σ = 1 kbar, we find
$\dot{\varepsilon}$ = 2.5 x 10^{-2} s^{-1} hence $\eta \simeq 4 \times 10^{10}$ poise

Using the values of the parameters found in our experiments we can extrapolate the values of the viscosity of ice VI to conditions more relevant to the deep interior of planets.

For $\bar{P}$ = 20 kbar, T = 273 K (still close to melting), $\bar{\sigma}$ = 0.01 kbar, we find $\eta \simeq 3 \times 10^{14}$ poise.

Isoviscosity contours can be drawn in the phase diagram of ice (fig.4), they are parallel to the melting curve of ice VI, which demonstrates that Weertman's equation using homologous temperature is valid for ice VI.

The low values found for the viscosity are in agreement with our preliminary results (obtained by more primitive methods) (2) and with our previous conclusions that solid state convection may have taken place (or may still take place) in the deep interior of the large ice moons.

ACKNOWLEDGMENTS

This work has been supported by ATP Planétologie 1983.
This is IPG contribution 790.

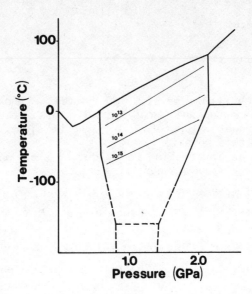

Fig. 4

Isoviscosity contours in the field of ice VI for σ = 10 bars. The viscosities are given in poises.

REFERENCES

(1) Poirier, J.P. 1982, Nature, 299, pp. 638-640.
(2) Poirier, J.P., Sotin, C., and J. Peyronneau. 1981, Nature, 292 pp. 225-227.
(3) Jackson, J.W. and Waxman, M. 1963, Giardini, A.M. Lloyd E.C. Eds. Butterworths, pp. 39-49.
(4) Tarantola, A. and Valette, B. 1982, Geophys. & Space Phys., 20, pp. 219-232.
(5) Kamb, B. 1965, Science, 150, pp. 205-209.
(6) Polian, A. and Grimsditch, M. 1983,. Phys. Rev. 8, 27, pp. 6409-6412.
(7) Weertman, J. 1970, Rev.Geophys. Space Phys., 8, pp. 145-168.
(8) Schubert, G., Stevenson, D.J. and Ellsworth, K. 1981, Icarus, 47, pp. 46-59.
(9) Mishima, O.and Endo, S. 1980, J. Chem. Phys., 73, pp. 2454-2456.

HUGONIOT OF WATER ICE

E. S. Gaffney

Earth and Space Sciences Division
Los Alamos National Laboratory
Los Alamos, NM 87544, USA

ABSTRACT. Hugoniot data for water ice are available for pressures ranging from about 150 MPa to about 50 GPa from initial states near 260 K. Limited data on porous ice (snow) at the same initial temperatures are available from 3.5 to 38 GPa and initial densities of 600 and 350 kg/m^3. At low stresses the shock velocity is a very complicated function of particle velocity due to elastic propagation, yielding and several possible phase changes. The Hugoniot elastic limit (HEL) of ice at these temperatures ranges from 150 to 300 MPa with the elastic waves travelling at about 3700 m/s. The mean stress at the HEL ranges from about 100 MPa to almost 200 MPa. Comparison with strength measurements at lower strain rate indicates that failure at the HEL probably involves fracture and is almost independent of both temperature and strain rate. Shocks with amplitudes between the HEL and about 500 MPa produce partial phase transformation either to melt or another solid phase. Above 500 MPa, the ice compression curve bends over very sharply indicating transition of large fractions of the material to the solid high pressure phases. By 690 MPa the transition to ice VI is virtually complete. Ice VI has been reported between 690 MPa and 2 GPa, and possibly as high as 3.7 GPa. Melting on the Hugoniot is definitely complete below 10 GPa. Above that level, the ice data are fairly well-fit by a linear relation between shock and particle velocity: D(km/s) = 1.79 + 1.42u. However, a quadratic form fits the data better: D(km/s) = 1.32 + 1.68u - 0.035u^2.

INTRODUCTION

Water ice is perhaps the most abundant solid constituent of

119

J. Klinger et al. (eds.), Ices in the Solar System, 119–148.
© 1985 by D. Reidel Publishing Company.

objects in the outer solar system, and it is the dominant constituent on the surface of most of the satellites observed by the Voyager missions (1-4). These surfaces generally show evidence of extensive impact cratering. Therefore, knowledge of the response of water ice to shock waves will be necessary for proper interpretation of the surficial geology of such bodies. For example, some crater scaling formulations require a knowledge of the shock properties of either or both of the target and the impactor (5). Detailed calculations of impact events will also require such a knowledge. Understanding of the Hugoniot of water ice will be essential in assessment of the likelihood of finding shock metamorphic products in icy regoliths (6).

The purpose of this report is to review the experimental data that are available on the response of water ice to dynamic loads. Most of the data are available only in technical reports which are not readily accessible to many researchers, especially those in other countries. In addition to presenting the data, some attempt is made to relate the shock wave data to other physical data on water ice, such as strength and phase composition, where that is possible and appropriate. Inasmuch as they come from many of the same data sources, some data for shock waves in frozen soils are also cited but not discussed in detail.

BACKGROUND

Shock Physics

The techniques for shock wave experiments and interpretation of the resultant data are based largely on application of the principles of conservation of mass, momentum and energy. In order that the reader may follow the discussion in the later part of this report more easily, a brief description of the salient features is presented here. For more detail, the reader is referred to several excellent review articles (7-10).

For our purposes we may consider a shock wave to be a discontinuity moving through a continuum at a velocity D (see Figure 1). In front of this discontinuity is unshocked material in some initial state, while behind it is shocked material in another state. Consider the properties of the states to be described by a pressure P, a density ρ, a particle (or material) velocity u, and a specific internal energy E; and define an operator [ ] such that

$$[x] = x(\text{shocked}) - x(\text{unshocked}),$$

where x is any property. Conservation of mass at the interface requires that

$$[\rho(D-u)] = 0.$$

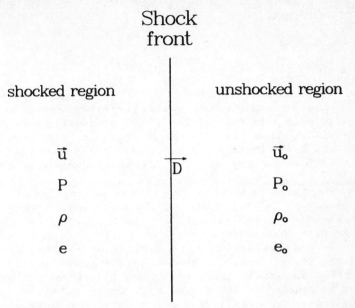

Fig. 1.   Schematic representation of a shock moving at a velocity
          D through a medium having an initial particle velocity
          u_o.

Conservation of momentum at the interface requires that

$$[P] = [\rho Du].$$

Finally, conservation of energy requires that changes in kinetic
energy be balanced by corresponding changes in internal energy:

$$[E] = [u^2]/2.$$

These three equations are adequate to describe the state of the
material behind the shock in terms of the unshocked properties.
(Strictly speaking, the conservation relation for momentum deals
with the component of stress normal to the shock front. However,
for much of the data discussed, especially for the high amplitude
shocks, the strength of the medium is insignificant and the
normal stress is indistinguishable from the mean stress or
pressure. Where strength is important, the term "stress" and the
symbol Σ will be used.)
    We will adopt the convention that unshocked properties will
be designated with subscript 0, whereas shocked properties will
be unsubscripted. It is usual to define the velocity u_o, pressure P_o and energy E_o in the unshocked material to be equal to
zero. In that case the equations above reduce to their more
familiar forms:

$$\rho/\rho_o = D/(D-u),$$

$$P = \rho_o Du \text{ and}$$

$$E = \frac{P}{2}\left(\frac{1}{\rho_o} - \frac{1}{\rho}\right) \ .$$

These three equations are collectively referred to as the Hugoniot relations in recognition of the pioneering work on shock physics of Hugoniot (11-13). Since there are five unknowns (ρ, D, u, P, and E) in these three equations, measurement of two of them is both necessary and sufficient to determine the other three.

The Hugoniot equations relate the properties of material behind the shock to those ahead of the shock and are based solely on conservation principles. They tell us little of the thermodynamic path that material takes when going through the shock front. We will return to this point shortly, but first must discuss the state reached by shocks. For a given initial state and a given material, there are only a restricted set of states that can be reached by shock loading. There exists a relation between pressure, density and energy for every material (termed an equation of state), and that relation provides a fourth constraint on our system. Thus, for any real material, there is only one degree of freedom. Given a specification of one variable, all the rest will follow. In the five-dimensional variable space (P, ρ, e, u, D) there is a locus of states which can be reached by a single shock and that locus is referred to as the Hugoniot. In the Soviet literature the term "shock adiabat" is synonymous with the Hugoniot. The most common projections of the Hugoniot are pressure-density (P-ρ), pressure-particle velocity (P-u), and shock velocity-particle velocity (D-u). P-ρ projections are particularly useful for comparing the results of shock experiments to other types of experiments, such as static experiments at high temperatures and/or pressures. Pressure-volume (P-V) projections are equivalent to P-ρ projections. The other two projections are peculiar to shock analysis because they involve the particle velocity which has no meaning (or at least, no relevance) in other types of experiments. P-u projections are particularly useful in the design and analysis of shock experiments because the initial and final states can be connected by a straight line of slope $\rho_o D$ in this space. D-u projections are useful because for most materials, in the absence of phase changes, the Hugoniot in this space can be represented as a straight line. (As we shall see below, this is not quite true of ice.) Furthermore, the shock velocity is usually very sensitive to phase changes, frequently showing large offsets over very small changes in particle velocity. Hugoniots in other spaces tend to be much smoother. Figure 2 is a representation of the relationships between these various Hugoniot curves subject, of

course, to the limitations of representing a five-dimensional
space on a two-dimensional surface.

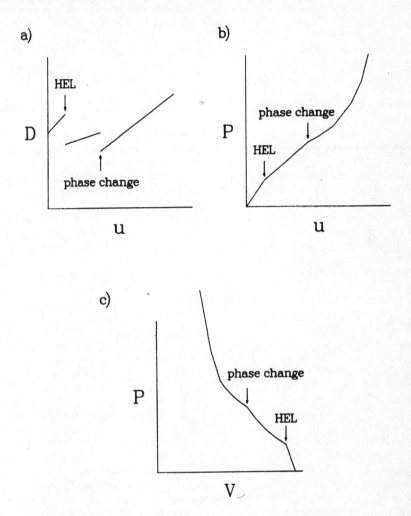

Fig. 2. Typical projections for the Hugoniot. a) Shock
         velocity-particle velocity. b) Pressure-particle
         velocity. c) Pressure-volume.

     As mentioned in the paragraph above, the Hugoniot provides
us with no description of the thermodynamic path followed by real
materials going through shocks. If the shock is truly a discontinuity, the path is merely a straight line between the initial
state and the Hugoniot. In fact, all shocks have finite widths
ranging from much less than a micrometer for shocks in gases,
liquids and in many homogeneous solids such as metals and

single crystals; millimeters for laboratory scale shocks in ice
and heterogeous materials like snow, soils and rocks; to meters
or even hundreds of meters for "shocks" produced by nuclear
explosions or large-scale impact events. The Hugoniot equations
still describe the relation of the final state to the initial
state if the wave is a steady wave, that is if the entire wave
can be characterized by a single value of the wave speed D,
because they are derived directly from conservation principles.
The steady wave assumption is a good one for many shocks in the
laboratory, especially in those materials which support narrow
shocks.

    Impedence matching, a technique commonly used in analysis of
shock wave data, assumes that the shocks are steady waves. In
this technique, the shock properties of a known material are used
as a constraint on the response of the material being investigated so that only a single property of the shock will be
sufficient to determine the state behind the shock. This is most
easily illustrated in the P-u plane (Figure 3). A shock is first
driven into a plate of the standard material, aluminum in this
illustration. Because the equation of state of aluminum is
known, a measurement of the shock velocity in the aluminum determines the pressure and particle velocity behind this shock
(point A). When the shock arrives at the aluminum/ice interface,
the wave is partially transmitted as a weaker shock moving into
the ice and partially reflected as a rarefaction moving back into
the shocked aluminum. The boundary conditions at the interface
require that both the pressure and the particle velocity be
continuous. The only way to accomplish this is for the rarefaction to accelerate the aluminum while the pressure drops; the
result is that both the aluminum and the ice will be a lower
pressure and higher particle velocity than the initial shocked
state in the aluminum (point B). Since the behavior of aluminum
is known (to a good approximation the curve AC is merely the
Hugoniot of aluminum reflected about the point A), the point B
must lie somewhere on AC. From the Hugoniot equation for pressure we see that the slope of a line connecting shocked and
unshocked states in the P-u plane is just $\rho_o D$. If the shock
velocity in the ice is also measured, we can draw the line with
that slope through the origin and determine B. Another method is
to measure the pressure of either the reflected wave in aluminum
or the transmitted wave in ice.

    Many important shock waves are not steady. In some cases
the final pressure declines as the shock propagates. This may
occur if the wave is diverging as from a point or line source, or
in planar flow if a release wave is following the shock and
travelling faster. For divergent flow, the Hugoniot equations
can be modified readily to account for the divergence. Since
none of the ice data discussed below involve divergent flow, that
case will not be discussed further.

a)

b)

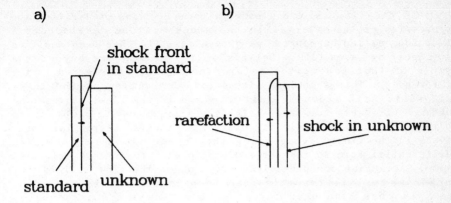

c)

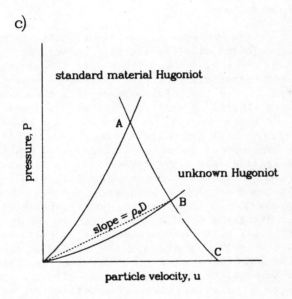

3. The impedence match technique. a) Incident shock loads
aluminum to state A. b) Reflected rarefaction and transmitted
shock bring both materials to state B. c) Impedence matching in
the P-u plane.

If the flow is not steady, alternate means of deriving the
properties of the material in the shock must be developed. A
technique called Lagrangian gauge analysis (14) has been modified
for decaying waves (15). Details of the method are beyond the
scope of this paper; the interested reader is referred to the
references. This technique also allows derivation of the thermodynamic path from the initial state to the final state. The
data requirements for this analysis are measurements of pressure
history or particle velocity history at several Lagrangian positions in the flow. By applying the conservation relations in
differential form to the three dimensional (pressure or velocity,
time, Lagrangian coordinate) description of the flow, all of the
other material properties can be obtained. This method of
analysis has been used for some of the data discussed below
(16,17).

Experimental Techniques

A brief review of some of the experimental techniques used
in generating Hugoniot data for ice will be helpful to the reader
unfamiliar with the methods. The review is not intended to be
exhaustive, but rather will describe only those techniques which
have been applied to ice. For more detail, there are excellent
review articles (18,19) as well as original references.

There are two principal ways to produce shock waves for
scientific study: high velocity impacts and explosions. The
former have the advantage of providing neater experiments, but
they require considerable capital expenditures for the sophisti-
.cated guns that are used and can only deal with rather small
samples (<50 mm) for higher pressure (>10 GPa) shocks. The
latter are relatively inexpensive, but require access to sites
where several kilograms of high explosive can be detonated. A
further difficulty with explosive sources is that when placed in
direct contact with the specimen, they do not produce a shock
pressure-time profile with a flat top; however, this desirable
property can be attained by using the explosive to accelerate a
metal plate to high velocities. The destructive nature of the
explosive test is, nevertheless, unavoidable.

One of the simplest methods for making measurements of shock
wave properties is the use of electrical pin switches to detect
the arrival time of the shock wave at the surface of the sample
(20). This is usually done by monitoring the current from a
capacitor discharging through a circuit which is initially open,
but which is closed by the motion of the surface of a conducting
sample. If the sample is an insulator, either of two approaches
may be taken. The surface of the sample may be covered with a
conducting foil which moves with the sample surface, or a coaxial
pin which is "self shorting" may be employed. In either event,
the surge of current gives a precise indication of the time of
arrival of the shock at the end of the pin. By placing pins at

several depths in a sample, the shock velocity can be measured. Typical accuracies for this technique at the time that ice was being studied were a few percent in the velocity.

Another method for determining the time of arrival of a shock wave at a surface is to monitor the reflectivity of a mirrored surface placed on the surface where arrival information is desired (21). Using a streak (or smear) camera to monitor the light is the usual technique. By placing mirrors at different levels of the sample, the shock velocity can be determined from the separation of the several mirrors and the respective relative arrival times derived from the separation of the events on the film and the writing rate of the camera. This technique depends on the shock wave to extinguish the mirror surface which occurs reliably for large amplitude shocks. However, for the relatively weak shocks that are of interest for many solid-solid phase changes in ice, extinction of the reflectivity may not occur. This was a particular difficulty in the experiments reported by Gaffney and Ahrens (22) and perhaps for the lowest pressure experiments of Anderson (23).

A straightforward extension of the streak camera technique has been used to measure the velocity of motion of a free surface subjected to shock. By lifting one end of the mirror off of the surface, the extinction of reflectivity will proceed from one edge of the mirror to the other as the surface moves up and intersects the mirror (24). This technique is referred to as the inclined mirror technique. A particularly good graphic representation of this method is presented in Figure 1 of Ahrens et al. (25). In order to obtain the particle velocity from the free surface velocity, it is assumed that the free surface moves at twice the particle velocity behind the incident shock. This assumption is exact for elastic waves and is very good for shocks that are not accompanied by considerable heating or that do not involve irreversible volume changes.

All of the above methods employ sensors at the surface of the sample. In order to obtain a direct measurement of the pressure or particle velocity behind the shock, it is necessary to place a gauge inside the sample. Such a gauge will move with the flow and is thus a Lagrangian gauge. As mentioned in the preceding section, such gauges make possible the application of powerful data analytic techniques. The first report of such measurements was by Zaitsev et al. who used large magnetic coils to produce a homogeneous magnetic field in the sample (26). Wires are placed in the sample at various depths and oriented so that the shock flow takes them across the field lines. This produces a current proportional to the velocity which can be recorded by standard methods. Because of the requirement of a homogeneous magnetic field, rapidly moving conductors such as metal flyers must be kept removed from the experiment. This restriction does limit the generality of the method, but it still remains probably the most powerful Lagrangian measurement technique.

Another Lagrangian technique is to employ gauges whose sensitive elements are piezoresistive. Three such materials have been in common use in shock experiments: manganin (27), ytterbium (28) and carbon (29). Gaffney used manganin gauges in his investigation of ice (30) so that specific type of gauge will be described here. Manganin is an alloy with a very small temperature coefficient of resistivity and a very large pressure coefficient of resistivity. The sensitive element is a foil grid of the piezoresistive metal encapsulated between two insulating foils, typically Kapton. A constant current is put through the grid during the duration of the shock experiment, and the resultant voltage across the sensitive element is monitored to determine the change in resistance of the manganin. The principle difficulty with Lagrangian stress measurements is the sensitivity of most piezoresistive components to nonhydrostatic stress and to details of their encapsulation. These two effects result in a typical accuracy of about ±5 percent.

DATA SOURCES

The first reported study of the Hugoniot of water ice was by Anderson (23). He used explosively driven flyer plates to load samples of ice initially at 263 K to shock stresses of 3.55 GPa to 30.1 GPa. The extinction of mirrors by the shock wave emerging from the surface of the samples was recorded with a streak camera to determine shock and particle velocities. The samples were about 25 mm in diameter and 5 mm thick, and the shock transit time through the specimens was on the order of a microsecond. Anderson also measured the particle velocities associated with release to about half of peak stress and to zero stress from initial shock states of 10.1, 20.6 and 26.6 GPa. Anderson investigated both single crystals (of unspecified orientation) and polycrystalline samples.

An extensive study of the response of ice to low amplitude shocks (<1 GPa) was reported by Larson et al. (16). A gas gun was used to launch flyers which impacted ice targets about 100 mm in diameter and 20 mm thick. They used the magnetic loop technique to measure particle velocity histories at several depths in the sample and Lagrangian analysis to derive both loading and unloading paths. Peak stresses in their experiments ranged from 258 to 695 MPa, and a precursor of 157 to 200 MPa was observed in all of their experiments. The shock transit times in their experiments were several microseconds. The initial temperature was 263 K. Most of these data were obtained on single crystals, but they included one polycrystalline sample.

In the same year Gaffney reported several experiments with peak stress below 1 GPa (30). These data were obtained from gas gun experiments on polycrystalline ice at an initial temperature of 263 K. Impact velocity and stress were measured, the latter with manganin gauges. The samples were about 100 mm in diameter and from 10 to 20 mm thick. In some cases, the stress gauges were embedded in the ice and in others they were in a buffer material. The shock durations were on the order of four microseconds.

Bakanova et al. reported data on ice and two artificial snows (ρ_0 = 600 and 350 kg/m^3) (31). The samples were loaded using explosively driven flyers to stresses of 3.43 to 50.3 GPa in ice and 6.8 to 35.4 GPa and 3.8 to 22.2 GPa in the snows, respectively. They measured the flyer plate velocity and the shock velocity using electrical shorting pins. The shock pressure and particle velocity were determined by impedence match. No information on the scale or duration of their experiments is presented. The initial temperature for these data was about 258 K.

Gaffney and Ahrens reported a single point at 1.91 GPa (22). As with all the other American studies, the initial temperature was 263 K. They used a powder-driven gun to launch a flyer at a target that was about 20 mm in diameter and about 2 mm thick. The shock transit time was about one microsecond. They measured the shock velocity with the streak camera technique and determined the particle velocity and shock pressure by impedence match with the measured flyer velocity.

Just this year, Larson published a much larger set of data including a reanalysis of his earlier work and results of seven new experiments with polycrystalline ice (17). Experimental particulars are similar to those of his earlier report (16), but peak stresses up to 3.56 GPa are included.

The American work on ice prior to 1979 was summarized by Gaffney, who also described the data available on frozen soils (32). Since this paper deals with ice and many occurrences of ice in the solar system will be as mixtures with silicates, these data will be briefly described here. However, a thorough treatment of the shock response of frozen soils is not the intent. All three of the early reports cited above contain data on frozen soils as well as on ice. Anderson (23) and Larson et al. (16) studied natural sands and silts, and Gaffney (30) studied mixtures of ice and pure quartz sand. Nakano and Froula also reported data on saturated and partially saturated artificial soils (33). Their experimental techniques and conditions were identical to those used by Gaffney (30). Louie reported data for the Hugoniot of a frozen silty clay initially at 263 to 271 K (34). His experimental technique was similar to that of Anderson.

RESULTS AND DISCUSSION

    The Hugoniot data for ice as taken from the sources outlined
above are summarized in Tables 1 and 2 and illustrated in
Figures 4 through 6. For the purposes of discussion the Hugoniot
will be divided into three regions: an elastic region, a low
pressure region and a high pressure region. In the elastic
region shocks propagate at approximately the velocity of p-waves
in ice and the material retains its shear strength. In the low
pressure region shock loading results in multiple wave structures
with an elastic precursor followed by one or more waves of lower
propagation velocity but higher stress amplitude. In the high
pressure regime the elastic precursor is not present because the
main shock travels at a greater speed than would the elastic wave
were it present.

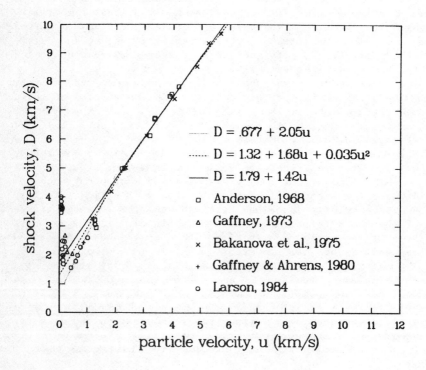

Fig. 4. Hugoniot of ice in the D-u plane.

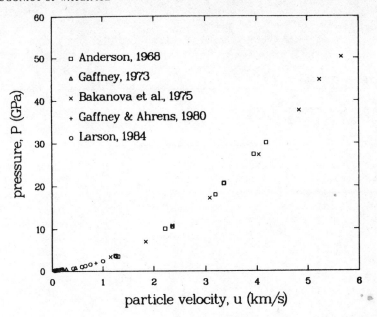

Fig. 5. Hugoniot of ice in the P-u plane.

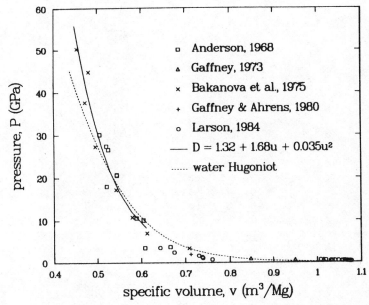

Fig. 6. Hugoniot of ice in the P-V plane. Hognoiot of water
from 293 K (48 is shown for comparison as the dashed
line.

Table 1.
SUMMARY OF HUGONIOT DATA FOR ICE

Initial State		Hugoniot State[*]				
Density (kg/m^3)	Temperature (K)	Stress (GPa)	Density (kg/m^3)	Shock Velocity (m/s)	Particle Velocity (m/s)	Ref
915	258	50.3	2205	9690	5670	31
915	258	44.8	2081	9350	5240	"
915	258	37.7	2118	8520	4840	"
917	263	30.1	1978	7810	4190	23
917	263	27.4	1920	7550	3940	"
915	258	27.3	2018	7390	4040	31
917	263	20.7	1837	6720	3360	23
917	263	20.6	1836	6690	3350	"
917	263	18.0	1919	6120	3190	"
915	258	17.2	1842	6120	3080	31
915	258	10.8	1726	5020	2360	"
917	263	10.6	1698	5000	2300	23
917	263	10.1	1654	4980	2220	"
915	258	7.05	1631	4190	1840	31
917	263	3.75	1495	3230	1250	23
917	263	3.75	1502	3210	1250	"
941	a	3.56	1555	3051	1260	17
917	263	3.55	1644	2950	1300	23
915	263	3.43	1411	3270	1150	31
935	b	2.43	1481	2600	1000	17
917	263	1.91	1416	2433	858	22
930	c	1.62	1370	2275	750	17
935	d	1.27	1355	1988	650	"
935	e	1.06	1353	1790	580	"
917	263	0.854	1179	2050	455	30
917	263	0.716	1590	1358	575	"
931	f	0.69	1316	1563	410	17
917	263	0.570	1316	1432	434	30
917	263	0.525	1053	2100	272	"
933	g	0.48	992	2283	195	17
917	263	0.457	986	2680	186	30
929	h	0.42	980	2467	165	17
931	i	0.305	961	1986	120	"
917	263[a]	0.30	941	3620	90	"
928	j	0.295	953	2302	105	"
929	k	0.29	969	1693	132	"
930	l	0.285	966	1818	125	"
930	m	0.26	945	2202	90	"

Table 1.
SUMMARY OF HUGONIOT DATA FOR ICE (Continued)

Initial State		Hugoniot State[*]				
Density (kg/m^3)	Temperature (K)	Stress (GPa)	Density (kg/m^3)	Shock Velocity (m/s)	Particle Velocity (m/s)	Ref
917	263[e]	0.24	935	3720	70	"
917	263[d]	0.23	935	3630	70	"
917	263[b]	0.23	935	3590	70	"
917	263[g]	0.20	933	3570	60	"
917	263[i]	0.19	931	3680	56	"
917	263[f]	0.18	934	3840	50	"
917	263[m]	0.18	930	3680	51	"
917	263[c]	0.17	930	3630	50	"
917	263[l]	0.16	930	3460	49	"
917	263[k]	0.16	929	3700	47	"
917	263[j]	0.16	928	4000	45	"
917	263[h]	0.15	929	3600	45	"

[*] Quantity underlined is that most directly measured.

[a,b,c,d,e,f,g,h] Sample polycrystalline, precursor has same superscript.

[i,m] Single crystal, shock propagating in $<11\bar{2}0>$ direction.

[j] Single crystal, shock propagating in $<10\bar{1}0>$ direction.

[k,l] Single crystal, shock propagating in $<0001>$ direction.

Table 2.
SUMMARY OF HUGONIOT DATA FOR SNOW (REF. 31)

Initial State		Hugoniot State[*]			
Density Temperature (kg/m^3) (K)		Stress (GPa)	Density (kg/m^3)	Shock Velocity (m/s)	Particle Velocity (m/s)
600	258	35.4	1720	9520	6200
600	258	30.7	1720	8860	5770
600	258	26.4	1626	8350	5270
600	258	18.6	1641	6970	4420
600	258	6.8	1424	4440	2570
350	258	22.2	1245	9390	6750
350	258	19.5	1165	8920	6240
350	258	16.2	1184	8120	5720
350	258	11.2	1217	6700	4770
350	258	3.8	1176	3930	2760

[*]
Quantity underlined is that most directly measured.

Elastic Region

The experiments of Larson et al. define and characterize shock propagation in the elastic region of ice. The amplitude of the Hugoniot elastic limit (HEL) varies with the direction of propagation in single crystal specimens. For shocks propagating along a crystallographic a-axis (<1120>) the maximum elastic amplitude is about 185 MPa, whereas shocks propagating between the a-axes (<1010>) or along the c-axis (<0001>) only reach about 160 MPa. Although the amplitudes of these stresses are quite large and the strain rates are high, so that detailed comparisons are not warranted, the orientational dependence of the deformation properties of single crystal ice are well documented (35-37). The shock velocity is slightly higher perpendicular to the c-axis (<1010> and <1120>) than parallel to it. This variation is consistent with the single crystal elastic constants measured by Dantl (38). The third order elastic constants c_{111} and c_{333} can be estimated using these wave speeds to be 78 GPa and 84 GPa, respectively. Using the revised data of Larson (17) the anisotropy is less apparent and no estimate of c_{iii} is possible.

The polycrystalline samples reported by Larson (17) had HEL's ranging from 150 to 300 MPa, depending to some extent on the amplitude of the main wave. This data is represented in the stress-volume plane in Figure 7 along with some of the data from the low pressure region. If we assume Poisson's ratio to be independent of stress and equal to the zero pressure value of 0.32 (39), then the lateral stresses at the HEL range from about 70 to 140 MPa. The stress differences of 80 to 160 MPa at mean stresses of 97 to 195 MPa are in fairly good agreement with the observations of Kirby et al. who observed brittle failure at about the same stress difference and mean stress for their highest strain rates ($\sim 4 \times 10^{-4}$ sec^{-1}) and at lower temperatures (40). This brittle failure level is very weakly dependent on temperature and essentially independent of strain rate in their results. Extrapolating their lower strain rate results, failure at 263 K and 130 MPa mean stress should be brittle if the strain rate exceeds about 10 sec^{-1}. Since the strain rate in the present case is on the order of 10^4 sec^{-1} or more, the HEL in polycrystalline ice is inferred to be the result of brittle failure.

The maximum observed stress difference of 160 MPa is about 6 percent of the value of the c_{44} elastic constant. Thus ice in shock loading has a strength approaching the maximum theoretical strength of the lattice. Similar behavior has been observed in many other materials and is usually accompanied by nearly complete loss of shear strength above the HEL (41). In order to interpret the data above the HEL, we must know how the strength behaves after yielding has begun. Two models of such behavior are commonly used. The elastic-plastic solid retains a constant

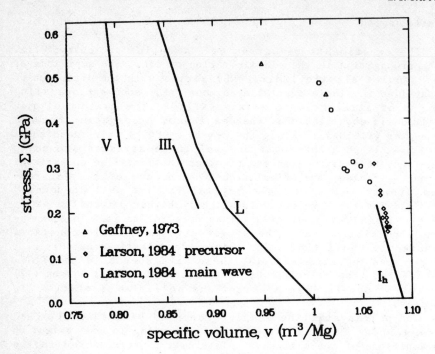

Fig. 7.   Hugoniot of ice in the low pressure regime (<625 MPa) in
          the Σ-V plane. The solid lines are the approximate P-V
          curves of the various phases of H_2O (indicated by Roman
          numerals, or L for the liquid) on the melting curve.

shear stress when loaded above its yield point in uniaxial strain
(42). Fowles first reported this type of behavior for aluminum
under shock loading (43), and it has since been found to be the
normal behavior for metals. Ahrens et al. have reported similar
behavior in polycrystalline alumina (25). For an elastic-plastic
solid, the proper value of P for the Hugoniot equations is the
normal stress Σ rather than the pressure or mean stress.
Wackerle reported that single crystal quartz loses its shear
strength completely above the HEL (44). Similar behavior has
been reported for several single crystal materials including
alumina (41). This type of behavior is termed elastic-isotropic.
For an elastic-isotropic solid, the proper value for P in thermodynamic models is the pressure because the mean stress and the
normal stress are identical. In Figure 8 the data from the
previous figure are replotted in terms of the mean stress or
pressure assuming complete loss of strength for the states above

the HEL. The dotted line is the linear fit to the initial state
and the HEL states extrapolated to 500 MPa. As can be seen from
this figure much the data in the range from 260 to 305 MPa do not
deviate significantly from the projected hydrostatic Hugoniot,
thus indicating elastic-isotropic behavior. (Two of the points
lie substantially to the left which indicates the likelihood of a
phase transition.) The deviation is +.004 m³/Mg for the point at
260 MPa and -.006 m³/Mg for the point at 305 MPa, a range only
slightly larger than that for HEL states at the same stress
(.004). Thus it appears that ice is like other materials which
approach their theoretical strength in that it is elastic-
isotropic in its shock response.

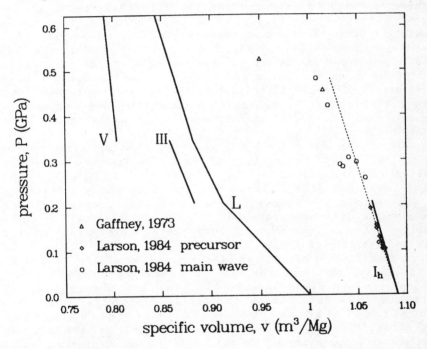

Fig. 8. Hugoniot of ice in the low pressure regime (<625 MPa) in
         the P-V plane.

Low pressure region (<8 GPa)

     Between the HEL and stresses of about 8 GPa, the Hugoniot of
ice is very complicated in both its shape and its interpretation.
There are at least two reasons for this complexity. First, a
shock wave with an amplitude above the elastic limit but below
about 8 GPa will not propagate as a single discontinuity. At
least two separate loading steps will be seen. The first will be
a wave travelling about 4 km/s with an amplitude equal to the
HEL, and the second will travel somewhat slower and bring the ice

to its final stress value. As can be seen from the shock velocity-particle velocity Hugoniot (Figure 4), this second wave may be as slow as 1.5 km/s. This two wave structure results in some ambiguity in the interpretation of shock data obtained in this region with either pins (31) or with mirrors (22,23). The elastic wave is weak enough that it has a negligible effect on the calculated Hugoniot pressure and volume provided that the shock velocity used in the Hugoniot relations is that of the main shock. The 7 GPa point of Bakanova et al. has a shock velocity almost exactly equal to the expected elastic velocity which indicates that caution should be exercised in the interpretation of the Hugoniot near this point. (However, the value is also about what would be expected for the main shock based on the higher pressure results.) In the 1-4 GPa region, there is no possibility that the observed shock velocities (2400-3300 m/s) are those of the elastic wave. However, it is possible that the elastic precursor led to a poor measurement of the true velocity of the main wave. Below 600 MPa, the time-resolved stress measurements of Larson indicate that the waves are not shocks but have significant rise times, even approaching one microsecond. The response of both pins and mirrors to such waves is not well determined.

All of the data below 1 GPa were obtained with Lagrangian gauges and so are not subject to the difficulty just described. However, the data of Gaffney were obtained with stress gauges which have some difficulties associated with proper calibration in non-hydrostatic stress fields (30). Thus, we will first discuss the data of Larson and then try to interpret Gaffney's results. Figures 7 through 10 show the low pressure Hugoniot data in a variety of planes.

A quantitative assessment of the response of ice to shocks just above the HEL can be obtained by study of the pressure-volume plane (Figure 8). In this plane, four of Larson's 1984 data lie substantially to lower volumes than projected from the precursor data corrected for nonhydrostaticity. The two data at about 290 MPa were obtained from shocks propagating in the <0001> direction in single crystal ice. The volume decrease observed can be accounted for by 7 to 10 percent transition to liquid or by slightly less conversion to solid phases (5 to 7 percent ice V for all four data or 8 to 9 percent ice III for the two lower points). Larson supports the former model and, indeed, groups the other 260 to 305 MPa data with the <0001> data and concludes that the occurrence of the precursor is due to the onset of melting (17).

However, Gaffney has argued that melting could not be involved because there was insufficient energy available to melt the amount of ice required for the observed volume change (45). This is illustrated in Figure 9 which shows the phase diagram of water (heavy lines) in the pressure-internal energy plane along with the projected Hugoniot for pure ice I$_h$ (dotted line) and the

data which clearly lie very close to the projected Hugoniot.
Production of 10 percent liquid would require about 30 kJ/kg more
energy than is available to the shock wave. Larson acknowledges
this problem but proposes that the energy could be provided by
spontaneous cooling of the ice fraction to the minimum in the
melting curve at 251 K (17). While this may be an adequate
explanation for the behavior near 290 MPa, it seems unlikely at
400 to 500 MPa.

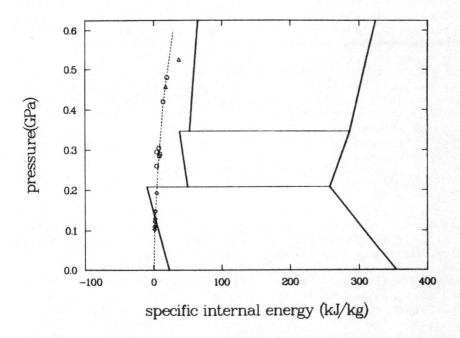

Fig. 9. Hugoniot of water ice below 600 MPa in the P-E plane.
Specific internal energies are from Burnham et al. for
the liquid (49) and Hobbs (50) for the solids. See text
for discussion.

Just above the HEL, Larson et al. report three records with
release wave speeds just below 3 km/s corresponding to a modulus
of about 8 GPa (16) (these speeds are not plotted in Figure 10).
Since these waves are just slightly stronger than the HEL, it is
reasonable to associate these waves with the bulk modulus of ice
which is of about the same value (39). Furthermore, the bulk
sound speed calculated from the HEL mean stress and volume is
2.8 km/s. This suggests that these waves represent hydrostatic
unloading of ice I_h with no shear strength from pressures up to
about 300 MPa. This interpretation, originally offered by

Gaffney (45), is contrary to that of Larson (16,17) who concluded that the HEL was caused by the onset of melting. However, presentation of the Hugoniot in other planes provides further support for the present interpretation (see previous paragraph).

The data that Gaffney obtained with stress gauges (30) does not seem to form a coherent picture when viewed in conjunction with the results of Larson. This is due in part to problems he encountered in the definition of the state of the standard material he used for impedence matching in two of the experiments (at peak stresses of 570 and 716 MPa), but there is a significant difference in the type of measurements for the other three experiments. Stresses in these experiments were recorded in a buffer material rather than in the ice. They therefore represent the state achieved at the interface without wave propagation in the ice, that is loading in a single step to some final state. One expects that such loading will lead to states that are less dense and more removed from equilibrium than would be produced by stepwise or gradual loading. This is indeed what is observed.

For shocks between about 600 MPa and 2.2 GPa, a straight line fits the data quite well in the shock velocity-particle velocity plane:

$$D = 677 \text{ m/s} + 2.05 \text{ u}.$$

This relationship is shown in Figure 10 as the dotted line. Although it was obtained as a least squares linear fit to the four data of Larson (17) and the datum of Gaffney and Ahrens (22), it passes almost exactly through the latter point and the stiffer two of the data of Anderson (23); hence, it is virtually identical to the curve shown in pressure-volume space by Gaffney and Ahrens. Ice VI is stable only up to 2.15 GPa and Larson's data above that level show slight softening relative to the projected ice VI line. This may be indicative of yet another phase change to ice VII or to liquid or may merely represent true scatter of the results in the absence of a phase change. If the latter were true, the higher pressure data should be included in the fit which would lower the slope to 1.79. The slope is related to dK/dP and Gaffney and Ahrens calculated that, based on hydrostatic compression data for ice VI, it should be 2.09. Therefore, it is concluded that another phase change commences in ice at about 2.2 GPa.

In the 3 to 4 GPa region, there is a wide discrepancy in the results of the three sets of investigators and even in Anderson's data alone. Anderson attributes this to poor data in one of his experiments (23), but all of his data in this region are distinct from that of Bakanova et al., and Larson reports a state between Anderson's points. In view of the general similarity of the experiments of the two earlier reports, this variability is surprising. However, Anderson used optical techniques to measure the shock velocity whereas Bakanova et al. used pins. There is no reason to expect that these two different techniques should respond to the low amplitude elastic precursor in the same manner.

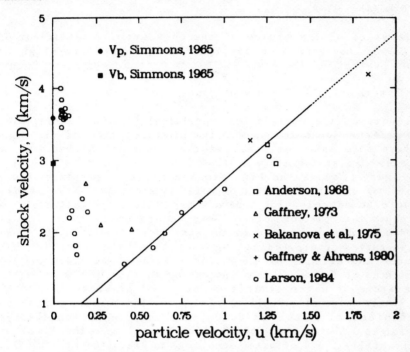

Fig. 10. Hugoniot of ice below 8 GPa in the D-u plane.

The transition between the simple high pressure region and the complex low pressure region has been taken here to be at about 8 GPa, excluding the datum of Bakanova et al. at 7 GPa, because that datum occurs almost exactly at the level where it is anticipated that the elastic precursor will be overdriven. Using the thermal model of Rice and Walsh (46), the volume measured at 7.03 GPa is consistent with water at a temperature of about 450 K. But the stable phase of water at 7 GPa and 450 K is ice VII (47). Hence either the datum is invalidated by the elastic precursor or melting is not complete on the Hugoniot at 7 GPa.

High pressure region (> 8 GPa)

Above 8 GPa the data indicate that the behavior of ice under shock loading is more straightforward. The data of two independent investigators (23,31) are consistent with each other and with an interpretation that the ice has melted behind the shock front. They can be fit to a straight line in shock

velocity-particle velocity space which is the usual relation
observed. A least squares fit to the 13 data above 8 GPa yields

$$D(km/s) = 1.79 + 1.42u$$

with a value of the square of the correlation coefficient of
$r^2 = 0.993$. However, a quadratic form fits the data slightly
better. Using

$$D(km/s) = 1.32 + 1.68u - 0.035u^2$$

the square of the correlation coefficient is 0.994. This quadratic relation lends credence to the interpretation of the high
pressure form as water, because the Hugoniot of liquid water
indicates similar curvature (48).

Figure 6 shows the high pressure data in pressure-volume
space along with the Hugoniot of water from Rice and Walsh (46).
As would be expected for a porous material, the Hugoniot of ice
is steeper than that of water. This occurs because the ice is
heated more when compressed to the same final density. At 21 GPa
the two curves cross, indicating that this excess heating has
overcome the effects of the lower initial temperature and the
latent heat of fusion of ice. According to Rice and Walsh, the
temperature of water at this density and pressure is 1644 K.
Using their value of 0.00013 $cm^3/g \cdot deg$ for the thermal expansion
coefficient of water at 8 GPa, the temperature on the ice
Hugoniot at 8 GPa is about 605 K which is very near the limit of
the stability field of water according to Pistorius et al. (47).
Thus, shocks above about 8 GPa result in liquid water and are
typical of shocks in many materials which result in melting.

Snow and Frozen Soils

Since the primary topic of this paper is the response of ice
to shocks, the following discussion of the behavior of snows and
frozen soils will be brief. The work of Bakanova et al. provides
our only experimental basis for describing the effect of porosity
on the shock response of snow (31). The porous ice data also
show evidence for slight curvature in the shock velocity-particle
velocity plane. For initial densities of 600 kg/m^3, the linear
relation is

$$D(km/s) = 0.830 + 1.40u \quad (r^2 = .9986)$$

and the quadratic relation is

$$D(km/s) = 0.767 + 1.44u - 0.004u^2 \quad (r^2 = .9986).$$

This difference is obviously not significant. For initial densities of 350 kg/m^3, the linear relation is

$$D(km/s) = 0.080 + 1.40u \quad (r^2 = .9982)$$

and the quadratic relation is

$$D(km/s) = -0.213 + 1.54u - 0.015u^2 \quad (r^2 = .9984).$$

Whether the nonlinearity of the lower density data is statistically significant is questionable. However, in practical terms there is very little difference in the values calculated with the two formulae, especially at 600 kg/m^3.

Because the initial volume of the sample is larger, snow shocked to a given pressure will undergo more compression than ice. The larger amount of work done results in higher temperatures on the Hugoniot of snow than at the same pressure on the Hugoniot of ice. For example, from their Table VI of the P-V-T relations of water, we can infer the following temperatures for ice shocked to about 20 GPa: ρ_o = 350 kg/m^3, P = 22.2 GPa, T = 5825 K; ρ_o = 600 kg/m^3, P = 18.6 GPa, T = 2025 K. We have just shown that the temperature on the Hugoniot of solid ice at 21 GPa is 1644 K, substantially lower than for snow.

The other main difference in the response of snows to shock relative to solid ice is that the snows should not have the complication of an elastic precursor because they have very little strength. Furthermore, snows at initial temperatures as high as 260 K will probably produce enough heat to melt at well below a gigapascal, and will not be complicated by solid-solid phase changes. This does not mean that high pressure phases can not be formed in regoliths in the outer solar system (6), however, because the much lower temperatures of those regoliths will require shock pressures in excess of a gigapascal which will be sufficient to produce high pressure phases up to ice VI.

The little work that has been reported on frozen soils indicates that there are many similarities between their response to shock and that of ice. Frozen sand and silt (16,30,32,33) both have elastic precursors of 100 to 200 MPa travelling at about 4 km/s followed by much slower waves presumably associated with either or both phase changes or yielding. The precursor is not reported in a silty clay where even waves as weak as about 100 MPa propagate at only about 1600 m/s (34). This lack of strength may be related to the unfrozen water content present in clays at temperatures as low as 235 K. At very high pressures (19 to 30 GPa) Anderson observed a softening in the Hugoniot of sand and silt which is associated with phase changes in the silicate components.

SUMMARY

There is a fairly extensive set of data on the response of water ice to shock waves. The interpretation is straightforward at high pressures and at very low stresses. In between is a region of very complicated response and some uncertainty in the validity of the reported results. Ice yields by brittle failure when the normal stress at the shock front exceeds 150 MPa to 300 MPa. There is disagreement as to whether compression above the yield point is accompanied by transition to the liquid phase or to ice III or ice V. Between 600 MPa and 2.2 GPa, ice VI is the phase of ice on the Hugoniot. Loading above 2.2 GPa is accompanied by further phase change. The multiplicity of "observed" specific volumes on the Hugoniot around 3.5 GPa is typical of the experimental problems in this region. Because of the uncertain effect of the elastic precursor on surface sensors such as pins and mirrors, Lagrangian gauges must be the preferred method for investigating the shock response in the region below 8 GPa.

Loading above 8 GPa in ice and at all measured stress levels in snow leads to melting. Because of the larger initial volume, the Hugoniot of ice is steeper than that of water in this region. The two cross at about 21 GPa where the temperature is about 1645 K. At 50 GPa Bakanova et al. estimate the temperature on the Hugoniot of ice to be about 4400 K.

The Hugoniots of snow are steeper than those of ice in the pressure volume plane because of the greater heating. At 22 GPa Bakanova et al. estimate a temperature for snow initially at 350 kg/m^3 on the order of 5000 K.

Data also exist and the sources have been listed for the shock response of several frozen soils. Frozen soils exhibit many phenomena seen in shocked ice. However, a detailed discussion of this data is beyond the scope of this paper.

REFERENCES

1.   Smith, B. A., L. A. Soderblom, T. V. Johnson, A. P. Ingersoll, S. A. Collins, E. M. Shoemaker, G. E. Hunt, H. Masursky, M. H. Carr, M. E. Davies, A. F. Cook, II, J. Boyce, G. E. Danielson, T. Owen, C. Sagan, R. F. Beebe, J. Veverka, R. G. Strom, J. F. McCauley, D. Morrison, G. A. Briggs and V. A. Soumi, 1979a, The Jupiter system through the eyes of Voyager 1, Science 204, pp. 951-971.

2.   Smith, B. A., L. A. Soderblom, R. F. Beebe, J. Boyce, G. A. Briggs, M. H. Carr, S. A. Collins, A. F. Cook, II, G. E. Danielson, M. E. Davies, G. E. Hunt, A. P. Ingersoll, T. V. Johnson, H. Masursky, J. F. McCauley, D. Morrison, T. Owen, C. Sagan, E. M. Shoemaker, V. A. Soumi, R. G. Strom and J. Veverka, 1979b, The Galilean satellites and Jupiter: Voyager 2 imaging science results, Science 206, pp. 927-950.

3. Smith, B. A., L. A. Soderblom, R. Beebe, J. Boyce, G. Briggs, A. Bunker, S. A. Collins, C. J. Hansen, T. V. Johnson, J. L. Mitchell, R. J. Terrill, M. Carr, A. F. Cook II, J. Cuzzi, J. B. Pollack, G. E. Danielson, A. Ingersoll, M. E. Davies, G. E. Hunt, H. Masursky, E. Shoemaker, D. Morrison, T. Owen, C. Sjogren, J. Veverka, R. Strom and V. E. Suomi, 1980, Encounter with Saturn: Voyager 1 imaging science results, Science 212, pp. 163-191.

4. Smith, B. A., L. A. Soderblom, R. Batson, P. Bridges, J. Inge, H. Masursky, E. Shoemaker, R. Beebe, J. Boyce, G. Briggs, A. Bunker, S. A. Collins, C. J. Hansen, T. V. Johnson, J. L. Mitchell, R. J. Terrill, A. F. Cook, II, J. Cuzzi, J. B. Pollack, G. E. Danielson, D. Morrison, T. Owen, C. Sagan, J. Veverka, R. Strom and V. E. Suomi, 1982, A new look at the Saturnian system: The Voyager 2 images, Science 215, pp. 504-537.

5. Kawakami, S., H. Mizutani, Y. Takagi, M. Kato and M. Kumazawa, Impact experiments on ice, J. Geophys. Res. 88, pp. 5806-5814.

6. Gaffney, E. S., and D. L. Matson, 1980, Water ice polymorphs and their significance on planetary surfaces, Icarus 44, pp. 511-519.

7. Duvall, G. E., and G. R. Fowles, 1963, Shock Waves, Ch. 9 of R. S. Bradley, ed., High Pressure Physics and Chemistry 2, Academic, New York.

8. Al´tshuler, L. V., 1965, Use of shock waves in high-pressure physics, Sov. Phys. Usp. 8, pp. 52-91, (originally published Usp. Fiz. Nauk 85, pp. 197-258, February, 1965).

9. Duvall, G. E., 1968, Shock Waves in Solids, pp. 19-29 in, B. M. French and N. M. Short, eds., Shock Metamorphism of Natural Materials, Mono Book Corp., Baltimore.

10. Duvall, G. E., and R. A. Graham, 1977, Phase transitions under shock wave loading, Rev. Mod. Phys. 49, pp. 523-579.

11. Hugoniot, H., 1885, Sur la propagation du mouvement dans les corps, et specialement dans les gaz parfaits, Paris Acad. Sci., C. R. 101, pp. 794-796.

12. Hugoniot, H., 1887, Sur la propagation du mouvement dans les corps, et specialement dans les gaz parfaits, J. Ec. Polyt. Paris 57, pp. 3-97.

13. Hugoniot, H., 1889, Sur la propagation du mouvement dans les corps, et specialement dans les gaz parfaits, J. Ec. Polyt. Paris 58, pp. 1-125.

14. Fowles, G. R., and R. F. Williams, 1970, Plane stress wave
     propagation in solids, J. Appl. Phys. 41, pp. 360-363.

15. Seaman, L., 1974, Lagrangian analysis for multiple stress or
     velocity gages in attenuating waves, J. Appl. Phys. 45,
     pp. 4303-4314.

16. Larson, D. B., G. D. Bearson and J. R. Taylor, 1973, Shock
     wave studies of ice and two frozen soils, pp. 318-325 in
     Permafrost: The North American Contribution to the Second
     International Conference, Nat. Acad. Sci., Washington.

17. Larson, D. B., 1984, Shock wave studies of ice under
     uniaxial strain conditions. J. Glaciol. (in press).

18. Fowles, G. R., 1973, Experimental technique and instrumenta-
     tion, Ch. 8, pp. 405-480, P. C. Chou and A. K. Hopkins,
     Dynamic Response of Materials to Intense Impulsive Loading,
     Air Force Materials Laboratory, Wright-Patterson AFB, Ohio.

19. Graham, R. A., and J. R. Asay, 1978, Measurement of wave
     profiles in shock loaded solids, High Temp.-High Press. 10,
     pp. 355-390.

20. Minshall, S., 1955, Properties of elastic and plastic waves
     determined by pin contactors and crystals, J. Appl. Phys.
     26, pp. 463-469.

21. Coleburn, N. L., 1964, Compressibility of pyrolytic
     graphite, J. Chem. Phys. 40, pp. 71-77.

22. Gaffney, E. S., and T. J. Ahrens, 1980, Identification of
     ice VI on the Hugoniot of ice I_h, Geophys. Res. Lett. 7,
     pp. 407-409.

23. Anderson, G. D., 1968, The Equation of State of Ice and
     Composite Frozen Material, US Army Cold Reg. Res. & Eng.
     Lab. Res. Rept. RR-257, Hanover, NH, (June 1968).

24. Doran, D. G., 1963, in High Pressure Measurement, ed. by
     A. A. Giardini and E. C. Lloyd, Butterworths, Washington,
     p. 59.

25. Ahrens, T. J., W. H. Gust and E. B. Royce, 1968, Material
     strength effect in the shock compression of alumina,
     J. Appl. Phys. 39, pp. 4610-4616.

26. Zaitsev, V. M., P. F. Pokhil and K. K. Shvedov, 1960, An
     electromagnetic method for measuring the velocity of detona-
     tion products, Doklady Akad.Nauk SSSR 132(6), pp. 529-530
     (originally published as Doklady Akad.Nauk SSSR 132(6) pp.
     1339-1340).

27. Keough, D. D., and J. Y. Wong, 1970, Variation of the
     piezoresistance coefficient of manganin as a function of
     deformation, J. Appl. Phys. 41, pp. 3508-3515.

28. Grady, D. E., and M. J. Ginsberg, 1978, Piezoresistive effects in ytterbium stress transducers, J. Appl. Phys. 48, pp. 2179-2181.

29. Krehl, P., 1978, Measurement of low shock pressures with piezoresistive carbon gauges, Rev. Sci. Instr. 49, pp. 1477-1484.

30. Gaffney, E. S., 1973, Study of the Nature of Shock Waves in Frozen Earth Materials, Systems Science and Software Report SSS-R-73-1557, La Jolla.

31. Bakanova, A. A., V. N. Zubarev, Yu. N. Sutulov and R. F. Trunin, 1975, Thermodynamic properties of water at high temperatures and pressures, Sov. Phys.-JETP 41, pp. 544-548 (originally published Zh. Eksp. Teor. Fiz. 68, pp. 1099-1107).

32. Gaffney, E. S., 1979, Equation of state of ice and frozen soils, Lunar Plan. Sci. 10, pp. 416-418.

33. Nakano, Y., and N. H. Froula, 1973, Sound and shock transmission in frozen soils, pp. 359-369 in Permafrost: The North American Contribution to the Second International Conference, Nat. Acad. Sci., Washington.

34. Louie, N. A., 1968, Equation of State of Frozen Material, Shock Hydrodynamics Report SH2155-08, Sherman Oaks, CA (Feb 1968).

35. Higashi, A., and N. Sakai, 1961, Movement of small angle boundary of ice crystal, J. Phys. Soc. Japan 16, pp. 2359-2360.

36. Bartlett, J. T., and C. J. Readings, 1968, Some optical effects in deformed single crystals of ice, IAHS Publ. 79, pp. 316-325.

37. Higashi, A., S. Mae, and A. Fukuda, 1968, Strength of ice single crystals in relation to the dislocation structure, Trans. Japan Inst. Metals 9, pp. 784-789.

38. Dantl, G., 1968, Die elastichen Moduln von Eis-Einkrystallen, Phys. Condens. Mater. 7, pp. 390-397.

39. Simmons, G., 1965, Single crystal elastic constants and calculated aggregate properties, J. Grad. Res. Center, So. Meth. U. 34, 1-269.

40. Kirby, S. H., W. B. Durham and H. C. Heard, 1985, Rheologies of H_2O ices I_h, II, and III at high pressures: A progress report.

41. Graham, R. A., and W. P. Brooks, 1971, Shock-wave compression of sapphire from 15 to 420 kbar. The effects of large anisotropic compressions, J. Phys. Chem. Solids 32, pp. 2311-2330.

42. Wood, D. S., 1952, On longitudinal plane waves of elastic-plastic strain in solids, J. Appl. Mech. 19, pp. 521-525.

43. Fowles, G. R., 1961, Shock wave compression of hardened and annealed 2024 aluminum, J. Appl. Phys. 32, pp. 1475-1487.

44. Wackerle, J., 1962, Shock-wave compression of quartz, J. Appl. Phys. 33, pp. 922-937.

45. Gaffney, E. S., 1975, Hugoniot elastic limit and phase changes in ice I, Bull. Amer. Phys. Soc. 20, p. 1514.

46. Rice, M. H., and J. M. Walsh, 1957, Equation of state of water to 250 kilobars, J. Chem. Phys. 26, pp. 824-830.

47. Pistorius, C. W. F. T., E. Rapoport and J. B. Clark, 1968, Phase diagrams of H_2O and D_2O at high pressures, J. Chem. Phys. 48, pp. 5509-5514.

48. Walsh, J. M., and M. H. Rice, 1957, Dynamic compression of liquids from measurements on strong shock waves, J. Chem. Phys. 26, pp. 815-823.

49. Burnham, C. W., J. R. Holoway and N. F. Davis, 1969, Thermodynamic properties of water to $^{\circ}C$ and 10,000 bars, Geol. Soc. Am., Spec. Paper 132, 96 p.

50. Hobbs, P. V., 1974, Ice Physics, Clarendon, Oxford, 837 p.

MEASUREMENT OF THE EXTINCTION OF WATER ICE PARTICLES

W. KRÄTSCHMER, N. SORG

Max-Planck-Institut für Kernphysik
Postfach 10 39 80
6900 Heidelberg
W. Germany

ABSTRACT

To study the extinction of a real small particle system in the infrared, water ice fog has been produced by blowing air through a pipe system which was held at liquid nitrogen temperature. The fog extinction between 2.5 and 15 μm wavelength was determined and compared with ice grain spectra calculated by Mie-theory. It was found that Mie-theory could reproduce the infrared extinction data provided the optical constants of partially amorphous water ice and grain radii of about 0.3 μm were used. Extinction measurements in the near UV confirmed the 0.3 μm grain radii deduced from the IR data. Mie-theory therefore seems to yield meaningful results when applied to sub-micron sized water-ice particles.

1.0 INTRODUCTION

Dust particles are important not only in our terrestrial environment but also in interplanetary and interstellar space. To obtain information on the chemical composition, size and shape of the dust grains, the technique usually applied is spectroscopy. For this purpose, the dust cloud under investigation has to be illuminated by some kind of light source such that the extinction or scattering of the light by the dust grains can be measured. A comprehensive treatment of this subject is given e.g. in (1).

In a series of laboratory experiments we have studied the extinction spectrum of a simple small particles system, namely water ice fog. The spectra measured from the UV to the IR were

149

J. Klinger et al. (eds.), Ices in the Solar System, 149–154.
© *1985 by D. Reidel Publishing Company.*

compared with Mie calculations using single-sized homogeneous
ice grains as model particles. The aim of this study was to in-
vestigate the extend to which dust particle spectra (e.g. of
interstellar grains) can be interpreted; we were especially
curious to learn whether both, namely (a) the structure of the
ice which forms the grains and (b) the grain radii could be deduced from the infrared extinction spectra.

2.0 EXPERIMENTAL PROCEDURE

Pressurized air was blown (flow rates between 5 and 10 1/min)
through a system of glass pipes which were floated in liquid
nitrogen (77 K). The fog produced by condensation of the humidity
of the air was carried by the flow into the sample compartment
of a double-beam spectrophotometer. The transmission (T) of the
dust cloud was recorded. To reduce the signal noise introduced
by fluctuations in the optical density of the cloud, repeated
scans with about 10 sec signal-integration time were performed.
For the range beyond 2.5 µm wavelength, a Beckman IR 11/12 was
used, and for the domain between 2.5 and 0.2 µm wavelength a
Perkin Elmer 330 was employed.

Before cooling, the humidity of the air usually was enhanced by blowing the flow through a bottle containing liquid
water. No special effort was made to keep the feedthrough of air
respectively fog constant in time. When we realised that fog production had ceased (which usually happened when the flow in the
cold tubes was blocked by snow-like ice crystals) the system was
warmed up and the measurement repeated. The extinctions at the
absorption peaks at 3.1 and 12 µm were repeatedly checked to
ensure that the fog feedthrough did not change within 10-20%.
The individual spectra obtained were normalized to unity at the
extinction maximum (usually at 3.1 µm wavelength), averaged, and
smoothed by hand. The absolute optical densities (-ln T) of the
fog ranged between 0.1-0.2 at 3.1 µm wavelength and between
0.4-0.8 in the near UV.

3.0 RESULTS AND DISCUSSION

The experimentally obtained IR spectrum of the water ice fog
is shown in Fig. 1 along with a number of calculated extinction
spectra. In each of the calculated spectra, a different structure
of the particle forming ice is assumed, with poly-crystalline and
amorphous ice as the limiting cases. The optical constants of the
ices are taken from the literature (2)-(5). The model particles
are assumed to be spherical in shape, homogeneous in composition,
and uniform in radius (here 0.3 µm). Obviously, only the partially
amorphous ice produces an extinction maximum at the proper

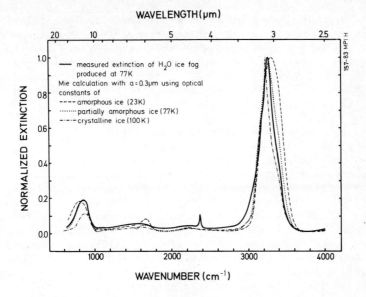

Figure 1. The infrared extinction of laboratory pro-
          duced water ice fog compared with the cal-
          culated spectra of ices of various structures
          (Mie-theory). In the fog extinction, the ab-
          sorption at 4.25 μm originates from CO_2 ice.

wavelength. In the case of the other structural modifications of
ice, the extinction maximum is either shortward (for amorphous
ice) or slightly longwards (for crystalline ice). One might
assume that an appropriate change of the particle size would result in a convincing fit of the extinction maximum for the other
ice modifications as well. But this turns out to be not true:
a smaller grain size leaves the positions of the extinction
maxima almost unchanged, a larger gain size significantly increases the amount of scattering in the extinction (as usual we
use the convention that extinction equals absorption plus scattering). The increased scattering can be noticed in the calculated
spectra by its additional extinction at the long-wavelength portion of the 3.1 μm absorption band; the observed spectrum however
seems to show only a relative weak scattering wing and thus an
increase of the grain size (e.g. to 0.4 μm) would significantly
degrade the fit. Using a grain size distribution (e.g. Gaussian
or log-normal) rather than single-sized particles in the Mie calculations not significantly improves the fit. It thus appears
that the fog particles exhibit a rather narrow size distribution

(relative width in the order of a few percent). According to
nucleation theory, this is not unexpected (6). From these observations we get the impression that our infrared spectra seem
to provide information on both, grain sizes as well as ice structure in such a way that the two parameters can be deduced independently from each other. It remains to be discussed how trust-
worthy these parameters really are.

By spectroscopy of water-ice particles of dense interstellar
molecular clouds it has been demonstrated that the structure of
the grain material can in fact be determined. The interstellar
particles mainly consist of structurally amorphous water ice (4),
(5). In the laboratory experiment discussed here, the structure
of the ice is distinctly different. The ice appears to be very
similar to that produced by water vapour condensed at 77 K, for
which the optical constants of (4) are valid. These optical constants lie between those of (poly)-crystalline ice and the amorphous ice obtained by condensation of water vapour at 10 K or
22 K (5), (7). We therefore call the 77 K ice "partially amorphous". Notice that the measured extinction structures at 6 and
12 µm wavelength are also much better reproduced by the partially
amorphous ice. We conclude from this fit, although it is far from
perfect that we have specified the structure of our laboratory
ice grains to sufficient accuracy and we obtain the impression
that our Mie-calculations produce reasonable results as far as
the ice structure is concerned.

The second parameter deduced from our model-calculations is
the grain radius of about 0.3 µm. In order to check the validity
of this value, it would be desirable to collect the particles and
to examine directly their diameters under a microscope. However,
we could find no way to make this feasible. Instead we investigated the scattering of our laboratory grains shortwards of 3.1 µm
wavelenght. In these experiments, the spectra recorded down to
0.2 µm wavelength showed always very similar shapes, independent
of the absolute extinction. We thus believe that under the conditions applied, distorting effects, like e.g. multiple scattering are unimportant. Fig. 2 shows the resulting spectrum covering
the 0.2 to 15 µm range. The calculated and measured spectra were
scaled such that they coincided at the extinction maxima at 3.1
and 0.2 µm wavelength. Thus the measured spectra do not perfectly
merge at 2.5 µm, the range limits of the IR instrument and the
NIR-UV spectrometer. What is important in this figure is the
shape of the calculated and measured extinction in the NIR-UV
range. Since the water ice has no major absorption in this range,
the shape of the extinction curve is entirely determined by the
size parameter X (= $2\pi a/\lambda$), i.e. entirely determined by the particle radius a. Therefore, our scattering measurement provides
an independent check of the particle size. An inspection of Fig. 2
shows that model particles of 0.3 µm radius produce an extinction

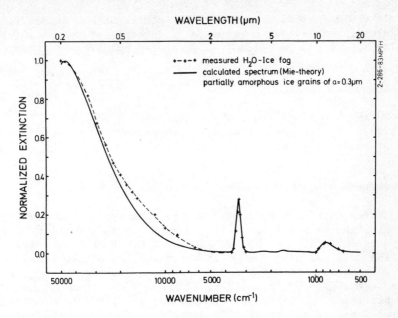

Figure 2. The extinction spectrum of ice fog extended
          into the UV region. The increased extinction
          at wavelengths shortwards of 3 μm originates
          entirely from scattering. A Mie-calculation
          indicates that the scattering ice particles
          seem to have a radius of about 0.3 μm.

spectrum rather similar in shape to that of the real particles.
We conclude that the Mie-calculations performed to evaluate the
IR extinction spectra yield meaningful results on the size of
the laboratory-produced ice particles as well.

The extinction spectrum of small particles depend not only
on the size but also on the shapes of the dust grains. Since we
applied Mie theory, we assumed that the particles are spherical
in shape. Cross-polarisation measurements of scattered light performed on fog particles produced by the same technique indicate
that the ice particles in fact are almost spherically (8). Despite this convenient feature of the fog particles, we in general
believe that shape effects in the case of water ice particles are
small. The reason for this lies in the fact that in the spectral
range considered the optical constants do not vary sufficiently
strongly (i.e. real part of the dielectric function in any ice
modification does not become negative). Thus the excitation of
shape depending modes at the particle surfaces will not
significantly contribute to the extinction (1).

It would therefore seem that IR spectra of sub-micron ice
particles in other environments (e.g. interplanetary and interstellar space) can be reliably interpreted by using Mie theory.

4.0 ACKNOWLEDGEMENTS

We thank D.R. Huffman, who encouraged us to perform this
experiment, for the many fruitful discussions we had during his
stay at our institute.

REFERENCES

(1) Bohren, C.F., Huffman, D.R., 1983: Absorption and scattering
     of light by small particles. John Wiley and Sons Inc.
(2) Bergren, M.S., Schuh, D., Sceats, M.G., Rice, S.A., 1978:
     The OH stretching region infrared spectra of low density
     amorphous solid water and polycrystalline ice Ih.
     J.Chem.Phys. 69, pp. 3477-3482.
(3) Bertie, J.E., Labbe, H.J., Whalley, E., 1969: Absorptivity
     of ice I in the range 4000-30 cm^{-1}. J.Chem.Phys. 50, pp. 4501.
(4) Léger, A., Gauthier, S., Defourneau, D., Rouan, D., 1983:
     Properties of amorphous H_2O ice and origin of the 3.1 μm
     absorption. Astron.Astrophys. 117, pp. 164-169.
(5) Kitta, K., Krätschmer, W., 1983: Status of laboratory ex-
     periments on ice mixtures and on the 12 μm H_2O ice feature.
     Astron.Astrophys. 122, pp. 105-110.
(6) Yamamoto, T., Hasegawa, H., 1977: Grain formation through
     nucleation process in astrophysical environment. Prog.
     Theor.Phys. 58, pp. 816.
(7) Hagen, W., Tielens, A.G.G.M., Greenberg, J.M., 1981: The
     infrared spectra of amorphous solid water and ice Ic bet-
     ween 10 and 140 K. Chemical Physics 56, pp. 367.
(8) Huffman, D.R., 1984: private communication.

SPECTRAL PROPERTIES OF WATER ICE AND CONTAMINANTS

Paul G. Lucey[1] - Roger N. Clark[2]

1. Planetary Geosciences Division - Hawai Institute of Geophysics - University of Hawai - Honolulu, Hawai 96822   USA
2. U.S. Geological Survey - Box 25046, Denver Federal Center, Mail Stop 964 - Denver, Colorado 80255

ABSTRACT

The spectral properties of mixtures of ice with particulates are reviewed. Water ice has absorption coefficients that range from $\sim$ 0.01 cm^{-1} in the visual region of the spectrum to 14000 cm^{-1} at 3.075 μm and its spectrum has prominent overtone absorptions at 2.0, 1.5, 1.25, and 1.04 μm. This wide variation in absorption coefficient allows different wavelengths to be used to probe to different depths below the surface and each absorption has a different sensitivity to mineral impurity concentrations. In the visual part of the spectrum, scattering conditions and contaminants are equally important for control of the color and reflectance level. Thus, the reflectance level and color of a surface is a poor indicator of ice content. A very clear ice can have a reflectance < 0.05 for particulate contaminations < 10^{-7} by weight for submicron dark particles. By using several ice absorptions, empirical and/or theoretical models might be used for deriving the characteristic grain size of ice or impurity minerals and for deriving abundance of the ice and rock or soil components. Quantitative analysis of remotely obtained reflectance spectra can only be performed by using the absorption features in the spectra of ice and soil and not by broadband response. This analysis might be done by special selection of several narrow band filters in the near-infrared that will adequately define the ice absorptions.

155

J. Klinger et al. (eds.), Ices in the Solar System, 155–168.
© 1985 by D. Reidel Publishing Company.

INTRODUCTION

        Understanding the optical properties of water
ice and mixtures of ice and particulate materials are
essential for remote sensing studies of ices in the solar system. The amount of sunlight reflected from a surface composed of ice and particulates depends on the
abundances and grain sizes of each mineral component
(ice is a mineral), the absorption coefficients of the
minerals and the physical state of the ice-particulate
association. The absorption coefficient is a description of how much light a material absorbs per unit distance, and is expressed as the inverse of the distance
the light travels in the material to the point that all
but e^{-1} (36.788 %) of the light has been absorbed. The
variation in the absorption coefficient as a function
of wavelength gives the material its spectral properties. The spectral properties of an ice-particulate intimate mixture (e. g., frozen mud) are a highly complex
(nonlinear) combination of the spectral properties of
the individual components. Intimate mixtures are one of
a range of possibilities that may be found. Another possibility is called an areal mixture, where the materials
are optically isolated patches of the pure end members
located on the surface. The spectrum of the latter is a
simple linear addition of the spectra of the optically
isolated patches of minerals weighted by the fractional
areal coverage of each patch in the field of view of
the detector. Another possibility is the vertically
stratified mixture such as a frost layer on soil. In
nature, there is a continuous range of combinations of
intimate, areal, and vertically stratified mixtures.
Inversion of a remotely obtained spectrum of a surface
to determine mineral abundance is a formidable problem,
if a quantitative and unique answer is desired. Fortunately, water ice has optical properties that are different from those of other minerals and this may allow
reasonable abundances to be derived from remotely obtained spectra. This paper reviews the current understanding of the spectral properties of ice and mixtures of
ice and particulates for the purpose of understanding
the remote sensing of ices.

ICE

        H_2O ice contains absorptions in the infrared
owing to O-H stretches (3 µm), H-O-H bend (6 µm), and
many translation and rotational modes at longer wavelengths. In the spectral region of reflected solar radiation, $\sim$ 0.4 to 3.0 µm, ice displays overtones of the
above modes at 2.02, 1.52, 1.25, 1.04, 0.90, and

0.81 μm. In reflectance, scattering controls the light
returned from the surface to the detector, and scattering can occur from ice-vacuum interfaces (grain boundaries or crystal imperfections) or from impurities mixed in the surface such as a particulate mineral. Thus,
even in a pure ice, the effective grain size greatly
controls the strength of absorptions, and the overall
shape of the reflectance spectrum (Fig. 1).

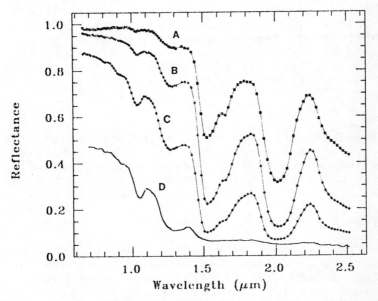

Fig. 1. Near-infrared spectral reflectance of (A) a
fine-grained (∿ 50 μm) water frost (grains of ice), (B)
medium-grained (∿ 200 μm) frost, (C) coarse-grained
(400-2000 μm) frost and (D) an ice block containing
abundant microbubbles. The larger the effective grain
size, the greater the mean path that photons travel in
the ice, and the deeper the absorptions become. Frost
data from Clark (1). Ice data from Clark and Lucey (2).

    The mean path length that photons travel in a
mineral is determined by scattering and by the absorption coefficient of the material. For instance at 3.075
μm the absorption coefficient of ice is 14008 cm^{-1} (3),
and a path length of 1 μm results in absorption of 25 %
of the light, whereas at the center of the 1.04 μm ice
absorption, the absorption coefficient is 0.34 cm^{-1} and
a path length of 4.1 cm is required for the same 25 %
absorption. At visual wavelengths, the ice absorption

coefficient is ~ 0.008 cm^{-1} (4) and a path length of 36 cm is required for 25 % absorption. Thus, different wavelengths probe to different depths in an icy surface. This large range in optical path length is an advantage for remote sensing studies because each absorption is sensitive to different amounts of impurities and to scattering due to different grain sizes.

Change in grain size causes changes in apparent band depth. The relative depth of an absorption is defined with respect to a continuum (a smooth line such as a cubic spline polynomial) fit to the reflectance peaks between absorption bands (1-5). If R_b is the reflectance at the absorption band center and R_c is the reflectance of the fitted continuum at the same wavelength as R_b, then the band depth, D, is (5):

$$D = (R_c - R_b)/R_c \tag{1}$$

The band depths for the ice absorptions increase as the mean optical path length increases until significant absorption occurs in the wings of the absorption bands, then the bands become saturated and the depth decreases. A saturated band may almost completely disappear as seen with the 1.5 and 2.0 μm bands in the pure ice spectrum in Fig. 2.

The behavior of a band during different phases of saturation has been quantified by Lucey and Clark (6). Fig. 3 shows curves describing the wavelength positions of ice band edges at 10, 25, 50, 75, and 90 percent of the maximum band depth of the 1.52 , 2.02 , and 3.07 μm bands of pure ice plotted against the log of the mean optical path through ice. The theoretical results in Fig. 3 were derived by Clark and Lucey (2) from ice absorption coefficients of Irvine and Pollack (3) using the methods of Clark and Roush (5). As the mean optical path through ice increases, growth and saturation of an ice band causes the wings of the absorption to move away from the band center. The overlapping of a mineral band by an ice band edge inhibits detection in three ways : the presence of the mineral absorption as a shoulder on the steep wing of the ice absorption makes detection and characterization more difficult ; the mineral band is completely suppressed if its position is at or near the bottom of a saturated ice band where no photons are returned except those from first surface reflection ; and close proximity of a mineral absorption to a strong ice band center reduces the signal to noise ratio which can be important in detecting weak features.

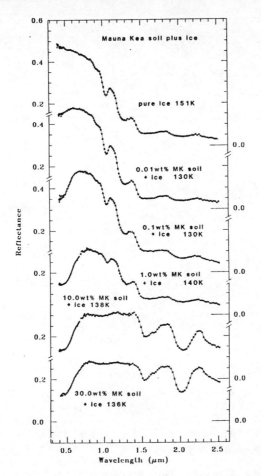

Fig. 2   Reflectance spectra of ice blocks containing
Mauna Kea soil of 15 μm mean grain size (2). Note how
the water-ice band depths at 2.0, 1.5, 1.25, 1.04, and
0.90 μm change as a function of the weight fraction of
the particulates in the ice. The particulate soil limits
the photons from penetrating into the ice, thus a greater particulate weight fraction results in less photon
path length in the ice and smaller absorption bands. The
particulate reflectance also limits the photon path ;
darker grains reduce the multiple scattering and also
the absorption band depths.

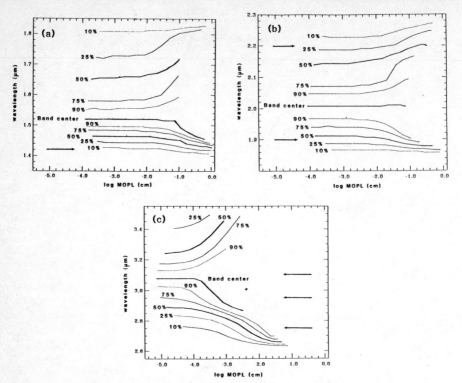

Fig. 3   Curves describing ice band edges at 10, 25, 50, 75, and 90 percent of the maximum band depths of (a) 1.52 , (b) 2.02 , and (c) 3.07 μm bands of pure ice plotted against the log of the mean optical path (MOPL) through ice. Values derived from spectra theoretically calculated by Clark and Lucey (2) from ice absorption coefficients of Irvine and Pollack (3) using the methods of Clark and Roush (5). Arrows indicate the center positions of major OH⁻ and H_2O bound water fundamentals and overtones.

SPECTRAL PROPERTIES OF ICE-PARTICULATE MIXTURES

Understanding of the optical properties of H_2O ice and the spectral properties of ice-particulate mixtures has advanced considerably during the last several years. Warren (4) reviewed the optical properties of snow, concentrating on how the visual reflectance is affected by small amounts of impurities. Clark (7) discussed the implications of using both visual and near infrared broadband detectors for remote sensing studies. The visual reflectance of pure frozen water can range from 1.0 for a fine grained frost (less than about 50 micrometers) to ∿ 0.4 for an ice with little internal scattering as is characteristic of some terrestrial ice in the Arctic and Antarctic (e. g. TB1 in Fig. 4). With small fractions of a weight percent soot or dust, the visual reflectance can range from unity to less than 0.1 (4-7), as is illustrated in Fig. 2. The visual re- flectance of ice varies so much with scattering condi- tions due to grain size and small amounts of impurities that using the visual albedo slone for any quantitative remote sensing study is unreliable.

     Clark (1-8) studied the spectral properties of ice blocks, frosts of various grain sizes, and ice-mi- neral mixtures. Impurities in an ice or frost layer can radically change the appearance of the reflectance spec- trum in the near infrared, as well as in the visual part of the spectrum. The addition of mineral grains dusted on the surface of a frost layer decreases the 1.5 and 2.0 μm absorptions roughly proportional to the fractional areal coverage. However, because the surface of the frost is not a flat layer but a "fairy castle" structure, the mineral grains become to a degree inti- mately mixed in the uppermost part of the surface, and the probability of a photon encountering a soil grain is higher. This higher probably results in reduced scattering and path length in ice. Also, the computed albedo of the surface may be drastically reduced many times more than expected for a fractional areal covera- ge using an additive model for areal mixtures because the mixture is not areal but is intermediate between areal and intimate. Thus, mineral grains on a frost layer are detectable in very small quantities. For example, a fractional areal coverage < 0.005 is detec- table if the mineral has suitable absorptions outside the major water ice features and the data are of suf- ficiently high quality, about 0.5 % precision (8).

     In a frost layer on a soil, a thick frost layer (> 1 mm) is required to mask the soil at wave- lengths smaller than 1.4 μm owing to the low absorption coefficient of ice. The thickness required also depends on the mineral reflectance features and the frost grain size. Frost on a very dark surface (e. g., reflectance 6 %) is easily discerned even when a layer of frost on- ly a few micrometers thick is present.

     In intimate mixtures, a dark material intima- tely mixed with water ice can completely mask the wa- ter absorptions at wavelengths shorter than 2.5 μm (8).

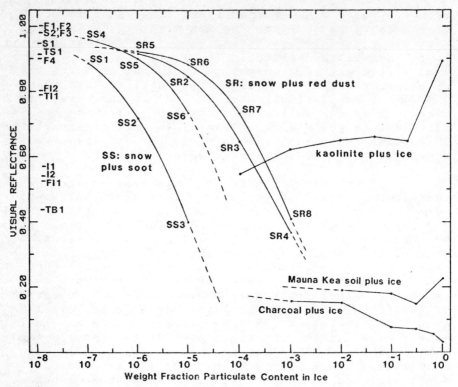

Fig. 4   The visual reflectance of selected spectra of
ice, frost and mixtures of ice and particulate materials (7). The dotted lines indicate the expected reflectance trends for unmeasured ice-particulate values.
Samples of charcoal plus ice, kaolinite plus ice, and
the Mauna Kea soil plus ice for which the visual reflectance values were measured were made from samples of water plus particulate which were was frozen rapidly from
room temperature and atmospheric pressure, so the ice
contained abundant exsolved microbubbles (data from
Clark an Lucey (2)). This process resulted in an increasing reflectance as the particulate concentration decreased, because the scattering by bubbles became more
dominant, except in the kaolinite. In the kaolinite plus
ice, the reflectance decreases as the kaolinite content
decreases because the pure kaolinite is higher in visual reflectance than the pure ice. Clear ice would contain fewer or no bubbles, thus the only scattering centers would be the particulate contaminants. The resulting reflectance levels (low for dark particulates)
could be easily maintained for concentrations below

parts per million or less contaminants. The codes used
are also given in Tables I and II of Clark (7). SS indicates snow plus soot ; SR indicates snow plus red
dust ; F are frosts ; S, snows ; I, laboratory ices I ;
TI, terrestrial ices ; and TB, terrestrial blue ice.
The visual color of ice with particulate contaminants
varies with grain size. For instance, SR3 and SR7 have
the same weight fraction of red dust but the finer
grain size (SR3) results in a more reddish color in the
snow plus dust. Grain-size differences of the frost
alone show similar results.

The higher overtones (0.8-1.25 µm) are more readily masked than the stronger 1.5 and 2.0 µm absorptions ; in
fact, these weaker absorptions become suppressed when
minerals are dusted on frost. Even relatively high reflectivity (e. g., 0.8) grains, intimately mixed in a
frost layer, can greatly reduce the scattering, limiting the path length and suppressing the higher overtone absorptions. These effects cause a decrease in the
ratio of higher to lower overtone absorption band depths (e. g., the depth of the 1.25 µm band/1.5 µm band
or 0.8 µm band/1.5 µm band). This decrease is in contrast to a thin frost layer on an ice surface whose
spectra show an increase in the corresponding apparent
band depth ratios (1).

More recently, Clark and Lucey (2) have studied
in more detail the spectral properties of mineral particulates intimately mixed in ice blocks and one of
their series of spectra are shown in Fig. 2. Note how
the ice bands at 2.0, 1.5, 1.25, 1.04, and 0.90 µm
change with the particulate content.

In order to help provide a calibration of ice
purity from reflectance spectra, Clark and Lucey (2)
derived a relation between absorption band depth and
optical path length, called a curve of growth, for several bands using the methods of Clark and Roush (5).
In Fig. 5 several theoretically derived curves of
growth are shown that are similar to curves of growth
produced from laboratory spectra such as those in Fig.
2. The albedo of the particulate impurity greatly affects the spectrum and the curves of growth. Darker
particulates cause less scattering in the surface, resulting in a lower optical path length in ice and a
smaller band depth. Clark and Lucey (2) found that the
band depth D, divided by the continuum reflectance R_c,
removed the effects of particulate albedo (Fig. 6).
Polynomials fit to such data for the 1.04 , 1.25 ,

1.52 , and 2.02 μm bands provide calibration curves to
mineral abundance.

Spectral studies of minerals with adsorbed water show that the physically and chemically bound water
absorptions, which occur at 1.4, 1.9, and 2.2 μm, are
distinguishable from the broader water ice absorptions
at 1.5 and 2.0 μm and do not shift appreciably in wavelength (< 0.01 μm) over the temperature range from 273
to 150K (1). However, the OH⁻ and boundwater fundamentals at 2.9 and 3.1 μm lie completely within the strong
ice fundamental causing difficulties in distinguishing
spectra within saturated bands. Any identification of
hydrated or hydroxylated minerals based on fundamentals
near 3 μm on a surface where there is evidence for ice
is hazardous.

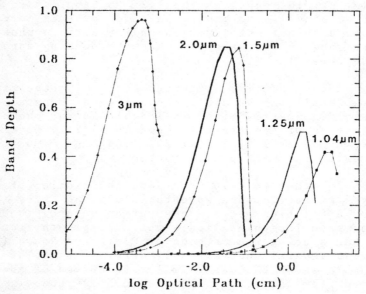

Fig. 5   Theoretically derived curves of growth (from
Clark and Lucey, (2)) for five ice absorptions are
shown. The optical path is the photon mean path length
in ice at the absorption band center. Because ice is
less absorbing at wavelengths outside the absorption
bands, the mean optical path lengths are as much as an
order of magnitude greater than at the band centers.

   The direct use of OH⁻ or H₂O bands for identification of hydrous minerals must be limited to the
overtones and combination tones if ice overtones are
present, especially those at 1.04 and 1.25 μm. For example, the identification of bound water on Callisto is
not justified on the basis of the 3.0 μm band. The
strength of the 1.52 and 2.02 μm ice overtone absorptions in the Callisto spectrum ensures that the ice
fundamental will be strong. An attempt to attribute
part of the 3.0 μm absorption to bound water must be
made carefully to avoid the ambiguity described above.

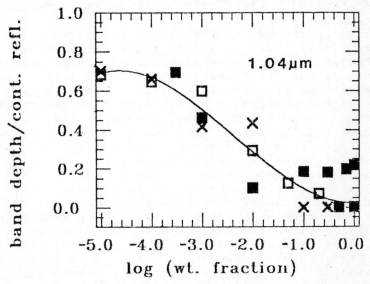

Fig. 6   The data from Fig. 2 (crosses) plus reflectance data for ice-charcoal mixtures (solid boxes) and ice-
kaolinite mixtures (open boxes) were used to derive the
1.04 μm ice absorption band depth divided by the continuum reflectance (2) as a function of the log weight
fraction of particulates in the ice. Although there is
a factor of about 20 between the reflectance of the
bright kaolinite grains and the dark charcoal grains,
division of the band depth by the continuum reflectance removes the effects of particulate contaminant reflectance. Thus, the curve can be used for crude abundance determinations from remotely obtained reflectance
data. Other ice absorption bands are sensitive to different amounts of impurities. Some scatter in the charcoal data is due to near-zero band depths and very low
continuum reflectance, thus approaching zero divided
by zero.

Hydrous bands can probably be detected at the
1-wt % level at 1.4 μm and the 5-wt % level at 1.9 and
2.2 μm if the ice contains few scattering defects and
the grains are large, but the detection limits will be
greater if the ice has a small grain size or the grains
have many defects. Decreased grain size of ice reduces
the spectral contrast of the OH⁻ signature.

QUANTITATIVE REMOTE SENSING

The previous discussion implies that some quantitative information can be obtained from remotely sen-
sed spectra of an icy surface, given adequate spectral
coverage and data precision. Recently, Hapke (9) presented a theory for computing bidirectional reflectance
of intimately mixed, multimineralic surfaces from the
complex indices of refraction, grain sizes, weight
fractions, and densities of the mineral components. Most
presentations of reflectance theories are in terms of
deriving the absorption coefficient from laboratory
spectra of a powdered sample, given the grain size of
the material. However, Clark and Roush (5) presented a
method for deriving abundance from reflectance spectra.
In a remote sensing study, if the materials present can
be identified from absorption features in the reflectance spectrum, then the absorption coefficients of the
materials are known (or can be measured). The reflectance theories can be used to derive the average-parti-
cle single scattering-albedo from observations of the
surface at many phase angles. Only the product of grain
size and absorption coefficient can be derived from the
theories if the absorption coefficient is unknown. If
the absorption coefficients are known, however, then a
non-linear least-squares algorithm can be used to solve
for the grain size and the weight fraction of each component.

The ice absorptions may be adequately defined
with as few as nine wavelengths centered near 0.95,
1.04, 1.12, 1.25, 1.40, 1.52, 1.83, 2.02, and 2.24 μm.
With these wavelengths, either the previously discussed
theoretical or empirical methods might be used to derive abundance from a reflectance spectrum. If a detector
system were to obtain images of a planetary surface at
wavelengths such as those listed above, in principle
the abundance of ice and other minerals can be mapped.

On a body with no atmosphere, an ideal data set
would involve ∿ 4 parameters : intensity at many wave-

lengths ($\sim$ 100 spectral channels from 0.5 to 3.0 μm), position (2 spatial coordinates), and viewing angle with the surface. The viewing angle data are required to determined the scattering properties (and thus the microstructure) within the optical surface. Many spectral channels are required for mineral identification from observed absorption bands. Current technology is adequate for obtaining such a data set. A NASA project plans to acquire such a data set on a planetary scale beginning in 1989/90 during the Galileo mission to Jupiter where three satellites, with abundant surface ices (ranging in size from Mercury to our moon), will be mapped at 204 wavelengths and at a spatial resolution as high as 1 km.

CONCLUSIONS

Reflectance spectra of ice-soil intimate mixtures are complex, nonlinear functions of the optical and physical properties of the components that comprise the surface. These spectra can be understood, and some quantitative information regarding mineral abundance can be derived. The mineral abundance from remotely obtained spectra can be crudely determined from empirically derived relationships of the optical properties of ice-particulate mixtures. A potentially more powerful technique is a nonlinear least-squares analysis of spectra obtained at several different viewing geometries. Any independent information about the physical properties of components is useful in the deconvolution.

ACKNOWLEDGMENTS

This is Planetary Geosciences publication number 423. This study was supported by NASA interagency agreement W15805.

REFERENCES

1. Clark, R. N., 1981, The spectral reflectance of water-mineral mixtures at low temperatures : Journal of Geophysical Research, 86, pp. 3074-3086.

2. Clark, R. N., and Lucey, P. G., 1984, Spectral properties of ice-particulate mixtures : Implications for remote sensing I : Intimate mixtures : Journal of Geophysical Research, 89, pp. 6341-6348.

3. Irvine, W. M., and Pollack, J. B., 1968, Infrared optical properties of water and ice spheres : Icarus, 8, pp. 324-360.

4. Warren, G. S., 1982, Optical properties of snow :
    Review of Geophysics and Space Physics, 20, pp.
    67-89.

5. Clark, R. N., and Roush, T., 1984, Reflectance
    spectroscopy : Quantitative analysis techniques
    for remote sensing applications : Journal of Geo-
    physical Research, 89, pp. 6329-6340.

6. Lucey, P. G., and Clark, R. N., 1983, Constraints
    on the detectability of minerals with physically
    or chemically absorbed water bands in the spectra
    of ice-mineral mixtures : Bull. of the American
    Astron. Society, 15, pp. 856.

7. Clark, R. N., 1982, Implications of using broad-
    band photometry for compositional remote sensing
    of icy objects : Icarus, 49, pp. 244-257.

8. Clark, R. N., 1981, The spectral reflectance of
    water-mineral mixtures at low temperatures : Jour-
    nal of Geophysical Research, 86, pp. 3074-3086.

9. Hapke, B., 1981, Bidirectional reflectance spec-
    troscopy 1. Theory : Journal of Geophysical Re-
    search, 86, pp. 3039-3054.

DISCUSSION

J. M. Greenberg :

Amorphous ice has been measured by Léger, Kråt-schmer, Hagen et al. at $\lambda > 2.5$ µm. The degree of amor-phocity modified the shape of the ice band. 10 K depo-sites is _more_ amorphous than 50-80 K deposites. Broad-ness of crystalline ice band occurs as a result of high optical depth when peak approaches saturation and wings are raised.

OUTER SOLAR SYSTEM MATERIALS: ICES AND COLOR SYSTEMATICS

William K. Hartmann, Dale P. Cruikshank, David J. Tholen

Planetary Science Institute, Tucson, AZ, and Institute
for Astronomy, University of Hawaii, Honolulu, HI

Our program of VJHK colorimetric observations and spectrophotometry is reviewed and updated. Attention is given to the probable
connections between VJHK colors and the ratios of bright ices and
dark soils on many outer solar system small bodies' surfaces.
VJHK color systematics may clarify ice/soil behaviors in bodies
too faint for high resolution spectra. Comets appear colored by
soils most closely resembling D-class Trojan asteroid soils. Our
recent model of comet surface properties, involving weakly-bonded
ice/soil regoliths, helps explain comet eruptive phenomena. The
work suggests the importance of considering ices in the context
of their intimate mixture with non-volatile soils.

BACKGROUND OF OBSERVING PROGRAM

This paper reviews results from a multi-year study of
spectrophometric relationships among small bodies in the outer
solar system (henceforth OSS), including new observational results from 1984. We emphasize the relation of ices and soils in
the OSS. The target objects include Trojan asteroids, outer-belt
asteroids, comets, and small satellites. (This paper ignores
objects such as Io and Titan, whose unique spectra do not involve
water ice.) The program began in 1977 when our simultaneous
visual and IR thermal photometry of Trojan asteroid 624 Hektor
confirmed Cruikshank's finding of Hektor's low albedo and unex-
pectedly large size, ~150 x 300 km.(1) Cruikshank and Hartmann
proved that the light variations stemmed from elongated shape,
not albedo spottiness (2),and proposed a possible origin for
Hektor as a compound asteroid, formed from two smaller Trojans.

169

J. Klinger et al. (eds.), Ices in the Solar System, 169–181.
© *1985 by D. Reidel Publishing Company.*

This in turn spurred further theoretical studies of Hektor's
shape and composition, and of Trojan dynamics (3,4).

  We then pursued relationships among Trojans and other small
bodies of the OSS. These had been less studied than most asteroids, due to their faintness. In a collaborative program,
Hartmann, Cruikshank, and Degewij obtained VJHK photometry of
selected brighter Trojan asteroids and outer satellites already
characterized by spectra (5,6,7). This allowed us to establish
correlations between VJHK colors and spectral classes among the
brighter objects. We found a range of VJHK colors among OSS small
bodies, from relatively bluer colors to relatively redder colors,
with some clumping toward the two extremes, as shown in Figs. 1
and 2. In all cases with measured albedo and known spectrum, the
bluest objects have high albedo and spectrally ice-dominated surfaces, while the reddest objects have low albedos of only a few
percent and surfaces classified in the C, D (sometimes called RD),

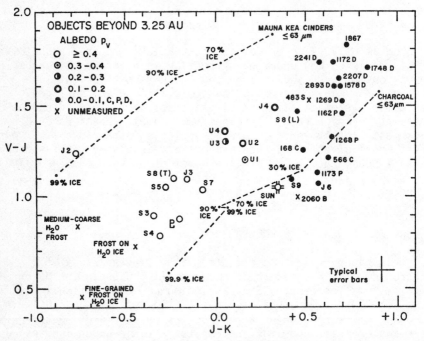

Fig 1: VJK color-color diagram for OSS objects (beyond the 2:1
Jupiter resonance at 3.25 A.U.). Albedos are indicated by symbols
showing the progression from bright icy objects to dark surfaces.
The fields of D-class asteroids (upper right) and C-class asteroids (right center) are well separated in JHK colors. Lab data
on various forms of pure H_2O ice are in lower left. Dashed lines
show "mixing tracks" defined by adding varying amounts of ice to
soil samples (see text and Fig. 2). "B" marks Chiron, designated
as class "B" by Tholen.

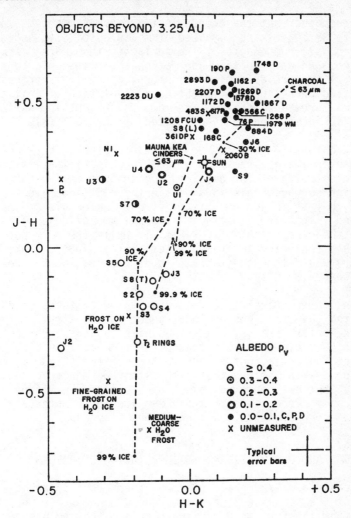

Fig. 2: JHK color-color diagram for OSS objects. Diagram format
is as in Fig. 1. The progression of albedos, increasing strength
of ice spectral bands toward lower left, and laboratory "mixing
tracks" are all consistent with the hypothesis that many OSS surfaces are regolith mixtures of dark soils and bright ices, with
mixing ratio a major factor in determining the spectrophotometric
appearance.

and P asteroidal spectral classes. Ice is absent in the spectra
of the latter, and their surfaces are spectrally dominated by
dark carbonaceous soil. In the color-color plots of Figs. 1 and
2, the correlation of color position and albedo is well shown by
the symbols.

We proposed that the trends in the figures might be interpreted if OSS bodies are composed of mixtures of ices and dark
(carbonaceous?) condensates. The surfaces, then, are mixtures of
bright ices and dark soils whose ratios determine color and albedos (6,7) and are determined, in turn, by geologic processes.
We discussed processes that might cause evolution of initial
surface mixtures into the two end members, bright icy surfaces
and dark surfaces spectrally dominated by the soil (6).

VJHK colors (as opposed to JHK colors alone, obtained by
most observers) are a remarkable tool for discriminating among
OSS objects too faint for detailed spectrophotometry. Not only
do ice-rich surfaces fall in a distinct field (lower left in both
figures), but the C and D subtypes of dark materials are markedly
segregated. The redder D-class (abundant among Trojans) is isolated in the upper right of VJK Fig. 1, while the more neutral
C's and P's (abundant in the outer belt) fall in the right center. Our 1984 data suggest that P's extend from among the C's
upward toward the D's.

Using these principles, we investigated surfaces of OSS
objects too faint for conventional spectrophotometry. For example, the outer dark satellites, J6 Himalia and S9 Phoebe, lie
not in the D field or on the trend line from ices to D's, where
most inner moons lie, but closer to the C field (Figs. 1 and 2).
This accords with the hypothesis that they were not formed among
the moons, but are captured objects. Similarly, we found that
2060 Chiron is inconsistent with clean ices or D materials, but
closer to the C-P fields, as seen in the figures.

THEORETICAL AND OBSERVATIONAL INTERPRETATION OF COLORS

As early as 1972, Lewis (9) and others made detailed calculations of the sequence of expected solid condensates in order
outward from the sun, in a cooling nebula. They pointed out that
the expected bulk compositions change as a function of solar distance, roughly matching the sequence of planet compositions. Bulk
compositions were rich in refractories near the sun. At increasing distances were added silicate rock materials, hydrated silicates, carbonaceous materials, and abundant ices. These results
are summarized in Fig. 3. The important point is that we can expect from this theory that dark soils (colored by opaque carbonaceous minerals and organic materials) and bright ices are the
two spectrally extreme materials in the OSS. This justifies
interpreting the colors in terms of their mixtures.

The elegance of this model increased with the finding by

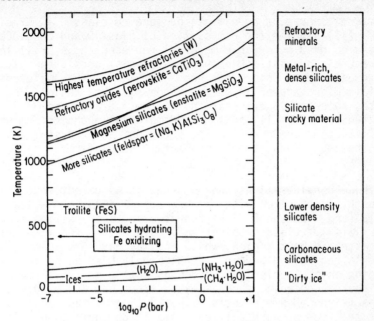

Fig. 3: The condensation sequence in the solar nebula. The left
box shows the sequence of appearance of different mineral suites
as the nebula cooled. The right box shows the dominant planet-
forming materials available in the various temperature regions,
consistent with the observed sequence from refractory rocky material through hydrated to black carbonaceous materials to mixtures of ice and carbonaceous soils, as we move from the inner
solar system through the asteroid belt into the OSS. The diagram
supports the concept that primordial OSS materials were mixtures
of dark carbonaceous soils and bright ices.

Gradie and Tedesco (10) that a zonal structure of compositional
classes exists among asteroids. Here we see the transition from
the metal-and-silicate rocks of the inner and central asteroid
belt (classes E, S, and M) to the carbonaceous asteroids in the
outer belt (classes C and P), and finally to the P's and redder
D's of the Trojan swarms. The outer belt C's are believed to
resemble carbonaceous meteorites, containing up to some 20%
water. These asteroids are known to be black objects with albedos around 5% and flat spectra. Lebofsky (11) discovered water
of hydration in C-class asteroid 1 Ceres. The Trojan D's have
similar low albedo but a more reddish-black color. (They were
thought to be C's due to their low albedo, until Degewij and van
Houten (12) discovered their red colors and defined them as class
"RD" for "reddish dark". As the single-letter classification

scheme grew in use, they came to be labeled class D.) Gradie and
Veverka (13) studied the composition of class D and concluded
they are colored by organic material representing a still lower
temperature condensate than the C materials, consistent with
their greater mean solar distance. In Fig. 3, the suite of C
materials might appear in the hydrated silicate region, and the
D suite, somewhat lower, near the bottom. Gradie and Veverka
predicted, as a result, that D-type material ought to dominate
the silicaceous component in the outermost solar system and ought
to be the dusty constituent in comets. We confirmed this (see
below).

These considerations suggest that as we go out into the OSS,
ices ought to be mixed with carbonaceous C-class or D-class materials (and with the P-class representing a transition from C to
D materials?). Class C objects may have originated more from the
outer asteroid belt region, while D's may have originated in
Jupiter's region and beyond. Perhaps only D's can be ice-rich.
Depending on the geological processes acting on the surface mixtures (both endogenic, such as water eruption, and exogenic, such
as impact-induced preferential vaporization of the ice component),
either the ices or dark soil can come to dominate spectrally,
thus explaining the end-member classes in Figs. 1 and 2.

Laboratory observations by Roger Clark (14, 15) greatly
clarified the interpretations of these data. Clark demonstrated
the important fact that in finely pulverized regoliths of ice/rock
mixtures, a tiny fraction (1% or less) of opaque black material
can reduce the albedo nearly to the albedo of the black component.
The black soil then masks ice absorption bands, and the soil's
spectral features dominate in spite of low soil abundance. The
reason is that light passes through microscopic ice grains until
it is absorbed by the black opaques. This mechanism works only
in intimately mixed, pulverized soils, but not, for example, in
coarse mixtures of cobbles. Jupiter's satellite, Callisto, provides an example, having albedo p_V only 0.17, but believed to
have an ice content of 30 to 90% by weight, according to Clark's
analysis. In principle, a regolith of 95% ice and 5% carbonaceous dust could have an albedo of only 0.05 and the spectral
appearance of a C- or D-class asteroid. On the other hand, the
same composition in a coarse texture could appear much brighter.
Thus, while the VJHK colors provide a powerful empirical dis-
criminator between bright surfaces of relatively clean ice (as in
Saturnian satellites) and dark, spectrally soil-dominated surfaces (such as Trojan asteroids), they do not give a rigorous indication of the composition -- a point emphasized by Clark (15).

As shown by the dashed lines in Figs. 1 and 2, Clark's own
samples show the same empirical discrimination of bright ices in

the lower left and dirty ice mixtures to the right and upper right of the ice field (7). Moreover, the dashed lines, from Clark's data, dramatically show the "mixing track" across the JHK and VJK diagrams as the ice/soil ratio changes in two-component systems. The samples of pure charcoal and reddish Mauna Kea cinders lie in the upper right, near the D field of Fig. 1 and C/D field of Fig. 2. Addition of H_2O ice causes displacement to the lower left, across the middle of the field, occupied by some OSS moons. Addition of 99% ice moves the samples into the ice field occupied by moons such as Europa and Tethys. Pure ices and frosts also lie in this corner of each diagram. The slope of the band defined by the dashed mixing tracks matches the band of OSS objects from icy surfaces, through Ganymede, Uranian satellites, and Callisto, to D asteroids.

In summary, we have developed a working hypothesis relevant to ices in the OSS: most OSS surfaces can be represented, to first order, as two-component "salt and pepper" mixtures or bright ices and dark soils, whose ratios determine the spectral and colorimetric appearance. The soils may have different subtypes, including C, D, and P classes. Going a step further, Fig. 1 suggests that surfaces such as seen on Rhea, Ganymede, Uranian satellites, Iapetus, and Callisto can be modeled by varying mixtures of D-type soils and ices. Cruikshank et al. (16) discuss a similar model that explains the dark side of Iapetus with a mixture of ices and reddish, organic, D-like material spectrally similar to organic residues from the Murchison carbonaceous chondrite. The C and P classes, together with outer (captured?) moons S9 Phoebe and J6 Himalia, may represent a suite of soils associated more closely with the outer belt than with the outermost solar system.

Because the JHK colors, or rather the positions in the VJHK color-color diagrams, appear to be such a useful indicator of surface properties, we have developed a shorthand indicator of the positions in these diagrams. This we call the alpha index; it can be thought of as a simple indicator of the position from lower left to upper right across both diagrams, and is defined in (7). We found empirically that this position correlated very strongly with known albedoes of OSS objects (7). We were thus able to define the alpha indices from both diagrams as numerically equal to albedoes. We suspect that when we observe VJHK and determine the alpha index of objects too faint for spectra or bolometry, we have an estimate of the albedo. In the case of the Uranian satellites, this procedure has accurate predicative value; albedoes for four moons determined from bolometry by Brown et al. were 0.18 to 0.30, matching the alpha indices of 0.2 to 0.3 we had assigned earlier (see data and plots in ref. 7).

RECENT OBSERVATIONS OF OSS ASTEROIDS: COLORS AND SPECTRA

    We have continued our observing program, emphasizing OSS
objects and a search for ices. By observing VJHK colors of additional objects with known spectra in 1983 and 1984, we have improved our mapping of the C, D, and P fields over that reported
in (7). Preliminary new data, insofar as reduced by us, are included in Figs. 1 and 2, and were used to define the C, P, and D
fields shown schematically in Figs. 4 and 5. We anticipate a
more complete data report in a separate paper.

    We have obtained spectra of a number of these objects with
circular variable filters (CVF's), ranging from 0.8 - 2.5 μm
wavelength. Neither the colors nor the spectra give evidence of
ice absorption bands in the surface materials, but as noted
earlier, ices are easily masked by black soils in regoliths. We
consider it possible, if not probable, that remote asteroids (e.g.,
D-class Trojans or Chiron) may have abundant ices in their surface
or subsurface materials.

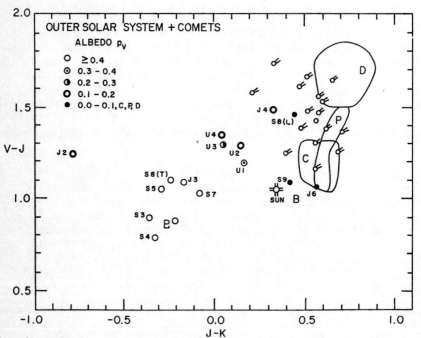

Fig. 4: VJK data on comets, presented in the format of Fig. 1.
Comet data are from our own observations plus those of other observers as referenced in (7) C, P, and D asteroid fields are
shown schematically based on our data for 6 C's, 5 P's, and 6 D's.
Comet colors would be consistent with dirty ice materials colored
by dark soils (esp. D-class soils) characteristic of OSS asteroids.

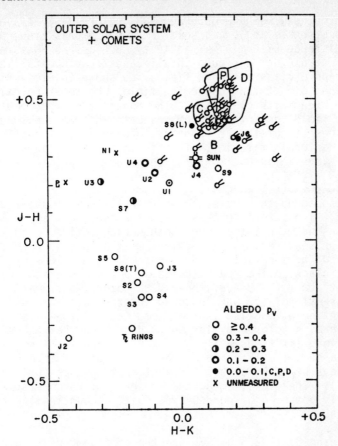

Fig. 5: As in Fig. 4, except using JHK data. In JHK colors, the
C, P, and D asteroid fields more nearly overlap. The fields were
defined from our data for 7 C's, 6 P's, and 7 D's. Comets again
are consistent with dirty ices colored by these soils, especially
D soils. Several comets are too red (toward upper right) to involve C soils, and fall .in the D field.

RECENT OBSERVATIONS OF COMETS: COLORS AND SPECTRA

     Having mapped the color fields of various OSS asteroids and
small satellites, we have paid particular attention to comparisons
with comets. Do comets (especially during low activity when far
from the sun and too faint for spectra) match the colors of icy
bodies, C asteroids, D asteroids, or Chiron? Or do they have
unique colors? Such questions are complicated by the fact that
we rarely, if ever, see the nuclear surfaces; observations are
affected by scattering and phase effects of coma particles (18,

19).   Nonetheless, we have observed VJHK colors of comets at a
wide range of phase angle and solar distance, emphasizing comets
at large distance with minimal activity such as P/Schwassmann-
Wachmann 1 at 6 A.U. We found in all cases that comets lie near
the D, P, and C class asteroid fields, as shown in Figs. 4 and 5,
in many cases offset toward the ice field. No comet revealed a
color consistent with clean ice. The majority of the comets lie
in a color field just offset from the D field. We concluded (7)
that the ices of comets are probably colored by dark dust similar
to D-class asteroid material. This is in excellent agreement with
the Gradie-Veverka prediction (13) that comets contain D-type
dust.

       We also found that comet colors correlate with solar dis-
tance (19). The comets furthest from the sun, such as P/
Schwassmann-Wachmann 1 (observed both during active outburst and
quiescent phases, ref. 7) have colors most displaced from the D
field toward the ice field. These resemble the colors of Clark's
samples of black soil mixed with ice. Comets closer to the sun,
however, at about 1 - 3 A.U., have VJHK colors most nearly matching D-class asteroids. Our proposed interpretation, consistent
with current comet models, is that comet comas at $\geq$4 A.U. contain
dirty ice grains colored by ice and D dust, and that these grains
have very long lifetimes at large solar distance. At smaller
solar distances, the lifetime of the ice component in a grain is
shorter, days or weeks at 2.3 A.U. but only an hour or less at
1 A.U. (18). Therefore, among the closer, active comets, we see
comas mostly composed of pure D-type dust, accounting for the D-
like colors.

       Obtaining 0.8 - 2.5 μm CVF spectra of a number of comets,
including several remote examples, we hoped to obtain evidence of
the water ice absorption bands near 2.0 and 2.4 μm. This would
be an important result, since direct spectral detections of ice
in comets have been marginal to date. Clean spectra of brighter
comets near the sun show no evidence of these ice bands. Our
results were limited by noise among faint distant comets, and
while we found marginal suggestions of the ice bands in P/
Schwassmann-Wachmann 1 and P/Bowell-Skiff, we are continuing the
work in hopes of obtaining better data.

       Our work on eruptive comet P/Schwassmann-Wachmann 1, together
with experimental work on gas-charged regoliths, led to a new
model of comet surfaces (7). We visualize remote inactive comets
as having a regolith of dark, dirty ice particles, colored
reddish-black by the D dust particles embedded in, or mixed among,
most ice grains. As the comet nears the sun, interstitial gas
pressure from subliming ice increases. An essential feature of
the model is that the regolith acquires a weak tensile strength
through bonding of the grains. This explains why an incoming

comet may have outbursts, rather than smoothly increasing dust
production. Gas pressure within the regolith must rise above the
tensile strength of the crust before sections of dusty crust and
debris are blown away in discrete eruptive episodes. This model
also qualitatively explains P/Schwassmann-Wachmann 1's outbursts.
A quiescent comet in a relatively circular orbit at a critical
solar distance would have enough sublimation to build gas pressure
slowly in the crusty regolith. As sketched in Fig. 6, this
pressure would rise until it overcame the crustal strength, blowing out material. The gas pressure is released. The temporary
coma would fade, with particles escaping or falling back. The
eruptive mechanism is thus reset. The process would then repeat
until the next eruptive event.

The behavior of ices in the outer solar system, in summary,
may best be analyzed not in isolation, but in the context of its
intimate mixtures with non-volatile soils.

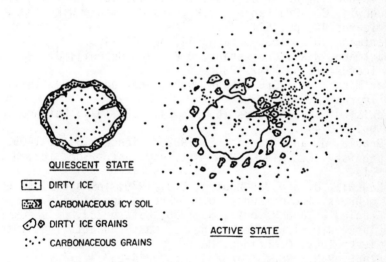

Fig. 6: Comet nucleus model, developed from observations, esp.
of P/Schwassmann-Wachmann 1. "Quiescent state" shows a dirty ice
core overlain by ice-depleted, weakly bonded dusty regolith. In
"active state", ice sublimation has increased gas pressure in
regolith pore spaces until it exceeds regolith tensile strength,
producing an outburst of dirty ice grains. At low-solar distance,
these sublime rapidly to yield a coma rich in dust grains.

ACKNOWLEDGMENTS

Thanks to C. Chapman, D. Davis, R. Greenberg, and S. Weiden-schilling for helpful comments and to K. Hankey and P. Watson-McBride for assistance in manuscript preparation. This paper was partially supported by NASA's Planetary Astronomy program. It is PSI Contribution No. 203. PSI is a division of Science Applications International Corporation.

REFERENCES

1. Cruikshank, D.P. and Hartmann, W.K. 1978, Icarus 36, pp. 353-366.
2. Hartmann, W.K. and Cruikshank, D.P. 1980, Science 207, pp. 976-977.
3. Weidenschilling, S.J. 1980, Icarus 44, pp. 807-809.
4. Weidenschilling, S.J. 1981, Icarus 46, pp. 124-126.
5. Degewij, J., Cruikshank, D.P., and Hartmann, W.K. 1980, Icarus 44, pp. 541-547.
6. Hartmann, W.K. 1980, Icarus 44, pp. 441-453.
7. Hartmann, W.K., Cruikshank, D.P., and Degewij, J. 1982, Icarus 52, pp. 377-408.
8. Hartmann, W.K., Cruikshank, D.P., and Degewij, J. 1981, Icarus 47, pp. 333-341.
9. Lewis, J. 1972, Icarus 16, pp. 241-252; 1974, Science 186, pp. 440-443.
10. Gradie, J. and Tedesco, E. 1982, Science 216, p. 1405.
11. Lebofsky, L.A. 1978, Mon. Not. Roy. Astr. Soc. 182, pp. 17p-21p.
12. Degewij, J. and Van Houten, C.J. 1979, in "Asteroids", ed. T. Gehrels, Tucson: Univ. of AZ Press, pp. 417-435.
13. Gradie, J. and Veverka, J. 1980, Nature 283, pp. 840-842.
14. Clark, R.N. 1980, Ph.D. Thesis, Massachusetts Institute of Technology, Cambridge, MA.
15. Clark, R.N. 1982, Icarus 49, pp. 244-257.
16. Cruikshank, D., Bell, J., Gaffey, M., Brown, R., Howell, R., Beerman, C., and Rognstad, M. 1983, Icarus 53, pp. 90-104.
17. Brown, R.H., Cruikshank, D.P., and Morrison, D. 1983, Icarus 55, pp. 83-92.
18. Hanner, M.S. 1981, Icarus 47, pp. 342-350.
19. Hartmann, W.K., and Cruikshank, D.P. 1984, Icarus 57, pp. 55-62.

COMMENT   by P. Weissman :

IRAS has seen comet P/Schwassmann–Wachmann 1 and it is very bright in the infrared, indicating a substantial dust coma even during quiescent periods. Thus, comments about possibly seeing a bare nucleus are likely incorrect, but the possibility of seeing dirty ice grains is good and a very interesting result. P/Schwassmann–Wachmann 1 seems to have a well behaved minimum coma at its quiescent stage, driven by some more volatile ice than H_2O, likely CO_2.

Part II

Cosmochemistry of Ices and Interplanetary Particles

EVOLUTION OF ICES FROM INTERSTELLAR SPACE TO THE SOLAR SYSTEM

J. Mayo Greenberg and L.B. d'Hendecourt
Laboratory Astrophysics
The Leiden University

ABSTRACT.   The chemical evolution of interstellar grains is
followed theoretically and observationally up to the time of
the prestellar nebula. The difference in the degree of
volatility of the various grain components is used to delineate
their survival within the solar system. It is shown that only
comets can preserve the interstellar dust in its primordial
form. If the parent bodies of meteorites were formed at temperatures less than 500 K they would contain not only primordial
silicate and carbon but also the complex organic molecules
which are the residue products of interstellar cosmochemistry.

1. INTRODUCTION

Interstellar grains are the small solid particles which atte-
nuate the light of distant stars. They are generally well
correlated with the clouds of gas which move about in the galaxy. These clouds, when sufficiently dense, are the material out
of which stars and planetary systems are born so that ultimately one can trace some of the constituents of our own solar
system back to their beginning some 4.5×10^9 years ago in a
dense cloud of gas and dust grains. Before a typical
interstellar grain is swallowed up in the process of star
formation, it has its own lifetime of $\sim 5 \times 10^9$ years during
which it goes through many stages of growth and evolution. It
is this history we wish to follow leading up to how a grain may

185

J. Klinger et al. (eds.), Ices in the Solar System, 185–204.

appear in the densest clouds just before star formation takes place. We shall focus our attention on those grains which are the carriers of ices. Whatever other kinds of particles are known to exist will be considered as part of the general environment along with the gas in the clouds. In section 2 we shall discuss the seedlings or nucleation cores which provide the initial surfaces for accretion of ice mantles.

In section 3 we shall describe how these core-mantle particles are cycled in and out of molecular clouds as well as how they evolve in the molecular cloud phase. The next section shows some observational (primarily infrared) aspects of grains in the diffuse cloud medium and in molecular clouds. Section 5 summarizes the projected composition of grains in the pre solar cloud. Throughout the discussion reference will be made to results from laboratory studies of interstellar grains which provide a basis for both the theory of grain evolution and of infrared spectra of grains.

In section 6 the degree of volatility of the various components of the dust in the nebula and its surroundings will be considered in relation to the chance of preservation of presolar dust during the early evolution of the solar system. A specific model is suggested for the composition of a comet based on the assumption that it is purely aggregated interstellar dust which has never been heated beyond its initial temperature characteristic of the cool dense cloud phase.

2. SOURCES AND SINKS OF GRAINS

The differences in volatility and physical resistance to destruction of grain materials in the wide variety of interstellar conditions has been taken up by a number of authors (1, 2, 3, 4, 5). One ultimate sink for all types of grains is incorporation into stars. At the very longest a grain can survive an interstellar material turnover time which is determined by the star formation rate (SFR). According to Kennicutt (6) the present SFR yields ISM (interstellar medium) consumption time scales of a few times 10^9 years. This is equivalent to the statement that 2% of the ISM is converted to stars each passage through a spiral arm (every 10^8 years at the solar radius) which gives 5×10^9 years as the ISM turnover time or an SFR $\simeq 2 \ M_0 yr^{-1}$.

According to Kwok (7) the mass loss from late type stars is $\sim 6 \times 10^{-10} \ M_0 \ pc^{-2} \ yr^{-1}$ which is equivalent to $\lesssim 1.0 \ M_0 \ yr^{-1}$. If late type stars are the principal sources of silicate grains they are just about the right order to maintain the present population if there is no major destruction at a rate higher than that by star formation. However, according to Draine and Salpeter (3,4), the maximum lifetime in the inter-

stellar medium of grains in the size range 0.01-0.1 μm is at
most only ≃ 5 x 10^8 yr! Since we certainly observe silicates as
an abundant interstellar constituent they must have acquired a
protective shield after ejection from the late type stars.
There is a beautifully simple way to build up such an ablation
coating which depends on accretion and photoprocessing of a
mantle of ices from the gas. This will be described later.

Carbon (or graphite) grains are formed either in carbon
stars or in novae. The rate of production of these particles is
not easily estimated. Their lifetime against destruction should
be about the same as that of silicates. If the only carbon
particles are $\lesssim$ 0.01 μm in radius (8) they are not generally
likely to accrete mantles as a shield (see Greenberg (9) for
reasons why very small particles do not accrete) and according
to the stability of the 220 nm hump, do not normally do so
(10). It is therefore difficult to explain their abundance unless the sources which produce them have about a 10 times
higher mass loss rate than that producing the silicates, or,
that the graphite producing objects produce carbon in much more
than cosmic abundance.

The mantles which form on the larger silicates - hereafter
characterized as cores with a mean radius of 0.05 μm - accrete
by the sticking of the condensable atoms and molcules
containing O, C and N from the gas at a rate (9),

$$\frac{da}{dt} \simeq 3.4 \times 10^{-22} \; n_H cm \; s^{-1} \qquad\qquad (1)$$

which in a molecular cloud of density $n_H = 10^3$ cm^{-3} adds a
thickness of 0.1 μm in 10^6 years. We shall show how this
accretion is kept in balance with the gas the next section.

3. CYCLING OF INTERSTELLAR GRAINS BETWEEN DIFFUSE CLOUDS AND MOLECULAR CLOUDS

There appears to be about equal amounts of material distributed
at low density $n_H \lesssim 10$ cm^{-3}) and at high density ($n_H \gtrsim 10^2$ cm^{-3})
in the interstellar medium. During the dense (molecular cloud)
phase the grains accrete mantles of molecular ices which are
slowly photoprocessed to produce a refractory combination of
the organics called organic refractory or yellow stuff (because
of its color). This yellow stuff as produced in laboratory
analog experiments (11, 12) provides the protective coating or
mantle which preserves the silicate cores in the diffuse cloud
phase from destruction during cloud - cloud collisions and by
sputtering from supernova shocks (13, 3). Reentering a
molecular cloud after one full cycle, which is estimated as
~ 2.10^8 yrs a silicate core plus yellow stuff mantle particle
again accretes a mantle of ices. The passage from the molecular
cloud to the diffuse cloud phase is generally actuated by star

formation, which ejects the matter energetically, leading to rapid evaporation of mantle ices – but still leaving observable quantities in the neighbourhood of protostars; e.g. H_2O and many protostellar sources (14) In the mean, therefore we expect to have volatile mantles on grains in molecular clouds and only yellow stuff mantles in diffuse clouds.

a) Diffuse Cloud Grains

It has been shown (8, 15a) that the shape of the wavelength dependence of extinction in the ultraviolet leads to the conclusion that the particles responsible for the full extinction constitute a trimodal distribution: 1) a ≃ 0.1 – 0.15 µm core-mantle particles which produce the visual extinction and polarization 2) a_g < 0.01 µm carbon (graphite ?) particles which produce the 2160 Å absorption 3) a_{si} < 0.01 µm silicate (magnetite ?) particles which are responsible for the rise in the ultraviolet extinction beyond $\lambda^{-1} \simeq 1700$ Å. The numbers of these core-mantle particles relative to hydrogen is $n_{C-M} \simeq 10^{-12}$ to 10^{-13} n_H. The number of bare carbon and silicate particles relative to the core-mantle particles are (depending on precise dimensions) $n_C \simeq n_{Si} \simeq 10^{-3} n_{C-M}$. The core-mantle particles are presumed to consist of elongated silicate cores of radius a_c = 0.05 µm with yellow stuff mantles of radius $a_m \simeq 0.12$ µm. The size and number of the core-mantle particles are described by the wavelength dependence and amount of visual extinction. The composition of the mantle is dictated by two conditions: 1) It must be made up of the cosmically abundant condensable organic constituents (O, C, N); 2) it must be approximately as resistant to sputtering as silicates. These conditions are statisfied by the complex organic nonvolatile molecular residues which are created as a result of ultraviolet photolysis. It is a material which, starting with simple oxygen rich molecules, evolves to carbon rich mantles.

b) Chemical Evolution within Molecular Clouds

The interplay between the atoms and molecules in the gas and in the grains is a complex one. Because the mean temperature of the grains is of the order of 10 K, all condensable species stick after collision at thermal speeds. If this were to continue unchecked and if there were no way to return the accreted atoms and molecules to the gas, the grains would freeze out the gas in a relatively short time. The depletion of the gas by accretion occurs at the rate

$$\frac{d\,n_G}{dt} = n_G\,v_G\,n_d A_d \tag{2}$$

where n_G is the number of gas phase condensable atoms per cm^3, v_G = gas velocity, n_d = number of core-mantle grains per cm^3,

A_d = area per grain. Using $v_G = 10^4$ cm s^{-1}, $n_d = 10^{-8}$cm^{-3} ($n_H = 10^4$ cm^{-3}), one derives an e-fold depletion time scale of

$$\tau_{dep} < 10^6 \text{ yr} \tag{3}$$

which is comparable with or shorter than the 10^6–10^7 yr characteristic of individual cloud life times, the value 10^6 being on the short side and 10^7 years being the mean time between cloud – cloud collisions (15b).

The most likely mechanism for returning accreted volatile mantles to the gas within molecular clouds seems to be the explosion of grain mantles resulting from the sudden release of stored energy in the form of free radicals produced by ultraviolet radiation of the grains (16). A scheme for including this in the total line dependent chemistry of clouds has been developed (17). The gas phase ion-molecule reactions are considered along with accretion and grain surface reactions leading to gradual formation of grain mantles which are sporadically triggered to explode by grain-grain collisions induced by cloud turbulence or by occasional impacts of heavy cosmic ray ions (18). The time dependence of the composition of both the gas and grain mantles is followed starting with various initial conditions of cloud density and ultraviolet flux spanning a wide range of possibilities.

We present here a small sample of the results. In Figs. 1a, 1b, 1c, we present the time dependence of the gas and mantle components for what we call the standard case defined by density $n_H = 2 \times 10^4$ cm^{-3}, and U.V. flux given by an attenuation of the mean diffuse cloud field by e^{-2A_V} where A_V is the visual extinction. We note several important results for the grain mantle:

1) At early times the grain mantle is dominated by $H_2O + CH_4 + NH_3$ more or less as given by Van de Hulst in his dirty ice grain model (19).

2) At intermediate times (10^6–10^7 yr) the mantle is dominated by $H_2O + CO_2$ or, eventually CO, (if we exclude the effect of photoprocessing).

3) At longer times $H_2O + CO_2$ remain the dominant constituents with no significant amount of CH_4 or NH_3.

For lower gas densities and extinctions ($n_H = 2 \times 10^3$ cm^{-3} $A_V = 2$) the results are similar but for greatly reduced U.V. and increased density ($n_H = 10^5$ cm^{-3}, $A_V = 8$) the grain mantle ultimately, after $t \gtrsim 5 \times 10^7$ yr loses almost all its H_2O, the oxygen becoming locked up in the O_2 molecule.

Although grain mantle photolysis is used to account for the storing of radicals and consequent explosions, it has not here been considered in terms of chemical changes within the mantle. This will produce some changes from the results reported here and will be taken up in the future.

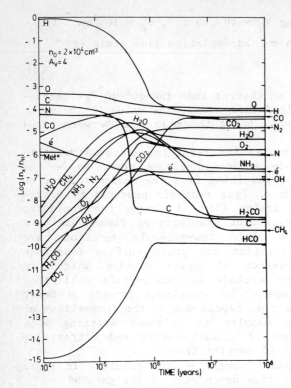

Fig. la. Time evolution of the abundances of the molecular components of the gas in a cloud with density $n_H = 2 \times 10^4$ cm^{-3} and extinction $A_V = 4$ magnitude.

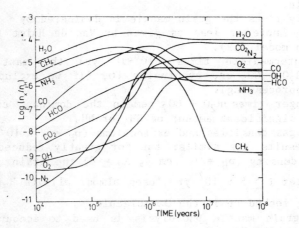

Fig. 1b. Same as Fig. la for the grain mantle.

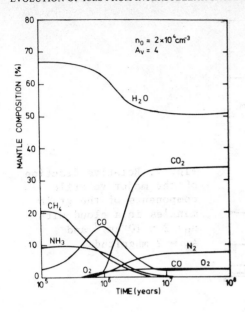

Fig. 1c. Relative fractions of the major volatile components of the grain mantles in a cloud with $n_H = 2 \times 10^4 cm^{-3}$ and extinction $A_V = 4$ magnitudes

One very general statement which can be made is that the gas and mantle compositions are <u>not</u> at all the same. For example the H_2O abundance in the solid phase is almost always considerably larger than in the gas phase, with ratios > 10 and as high as 10^2-10^4 being common; i.e., water does not appear on grains as a result of accretion of water molecules from the gas but is rather made on the grains.

Although the H_2O abundance in grain mantles has sometimes been stated to be small <u>even in molecular clouds</u> we see that throughout a substantial portion of a cloud's duration the H_2O fraction in the accreted mantle should be quite large. In very young clouds (t $\lesssim 10^5$ years) a characteristic fraction of H_2O is $> 60\%$. In Figure 1c, 2 and 3 are shown the time variation of the H_2O fraction for the three basic cases treated. Since a mean time scale for cloud-cloud collisions is $\sim 10^7$ years we take this to be a canonical mean cut-off time for unperturbed cloud chemistry. In no case is the predicted H_2O concentration $< 50\%$ for t $< 10^7$ years so that one <u>should</u> find large amounts of solid H_2O in molecular cloud grains. This is indeed confirmed by observations as shown in Section 4.

The theoretically predicted drop in the H_2O mantle fraction in dense clouds for t $> 10^7$ years is probably prevented from occurring because such clouds are likely to have much shorter lifetimes. The free-fall lifetime of a cloud with

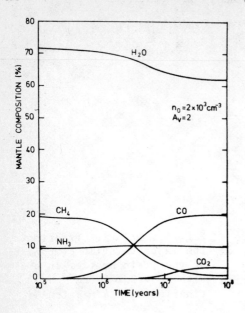

Fig. 2. Relative fraction of the major volatile components of the grain mantles in a cloud with $n_H = 2 \times 10^3 cm^{-3}$ and $A_V = 2$ magnitudes

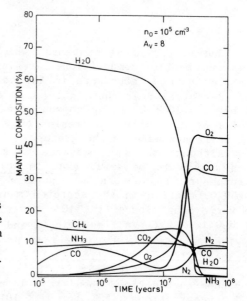

Fig. 3. Relative fractions of the major volatile components of the grain mantles in a cloud with $n_H = 10^5 cm^{-3}$ and extinction $A_V = 8$ magnitudes.

$n_H = 10^5 cm^{-3}$ is $\tau \approx \dfrac{4 \times 10^7}{n_H^{\frac{1}{2}}}$ yrs $\approx 10^5$ yrs which is a lower limit to the cloud lifetime, so it is not likely to remain static for 100 times this value.

4. INFRARED SPECTRA OF DIFFUSE GRAINS

a) Diffuse Clouds

No H_2O ice absorption at 3.07 μm is observed in the diffuse
cloud grains even when the 9.7 μm silicate feature is quite
strong (20). This has often been interpreted as implying that
there are no mantles on the silicate particles in diffuse
clouds. However, there are often infrared features which can be
interpreted as being caused by mantles of a material other then
such volatiles as H_2O. The most studied of these infrared absorptions is the one at 3.4 μm. It is well known that the
3.4 μm region encompasses the C-H stretch absorption in compounds containing $-CH_3$ and $=CH_2$. The organic nonvolatile organic residues created in our laboratory by ultraviolet photoprocessing of icy grain mantles exhibit such a feature. In
Figure 4a we show a comparison between the absorption spectrum
of the galactic center object IRS7 and that of one of our
yellow residues. The points of resemblance, not only in the
3.4 μm region but even in the 3 μm and 5.8 μm region
(particularly the features at 6 μm), are quite suggestive (see
Fig. 4b). We do not claim that any of these particular residues
is the grain mantle material on diffuse cloud grains but that
the evolution of interstellar grains in molecular clouds ultimately leads to the formation of such materials as components
of grain mantles which can survive the harsh environment of the
diffuse cloud space (13). It may be shown that the amount of
organic refractory material required to produce the observed
depth of the 3.4 μm feature is consistent with the silicate
core-organic refractory grain model (21).

b) Molecular Clouds

In molecular clouds the predominant absorption feature is due
to the O-H
stretch of H_2O ice at 3.08 μm. In fact the strength of this
feature generally implies that at least ~ 60 % of the grain
mantle composition is H_2O. (12, 10, 22). In addition, features
of H_2O at 6 μm can be detected but not with ground based
observations. Carbon monoxide in solid form has been detected
along with evidence of a CN feature in W33A which is a particularly suitable object for weaker features because it has an
enormous optical depth as well as a very strong infrared background (23). An analysis of the 6.8 μm band in the 5-8 μm
spectral region of this object indicates that ~ 30 % of all
available carbon is bound up in the mantles in the form of
molecules resembling our organic refractories (24). A comparison of laboratory results with astronomical spectra showing
the evidence for a number of these features is made by
d'Hendecourt (25).

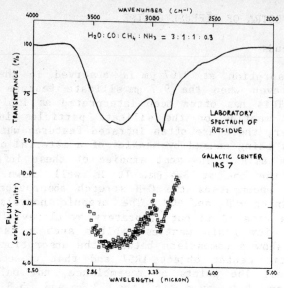

Fig. 4a. Upper curve: the infrared transmittance of
a laboratory organic refractory residue resulting
from photoprocessing of a starting mixture
$H_2O:CO:CH_4:NH_3 = 3:1:1:0.3$ (21).
Lower curve: the spectrum of the galactic center source
IRS7.

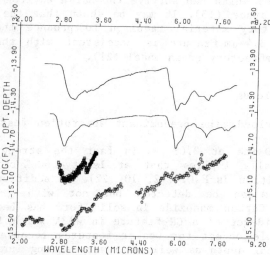

Fig. 4b. Infrared absorption spectra of two organic refractory
residues and the galactic center sources IRS7 and Sgr AW. From
top to bottom:
1) Residue of $CO:H_2O:CH_4:NH_3 = 2:2:1:2$, 2) Residue of $CO:H_2O:$
$NH_3 = 5:5:1$, 3) IRS7, 4) Sgr AW.

5. THE GRAIN MODEL

A working model for interstellar grains which provides the
essential observed features in diffuse, molecular and
protostellar clouds is shown in Figure 5. In addition to the
core mantle grains we require the co-existence of large numbers
of very small particles required for the 2200 Å hump which are
presumed to be some form of carbon, and further large numbers
of small particles, possibility silicate (or perhaps partly
magnetite), to produce the far ultraviolet extinction beyond
1700 Å. The relative numbers of these two populations of small
particles to the core-mantle particles is of the order
of ~ 500:1 or 1000:1 depending on their exact size which has
not yet been completely determined.

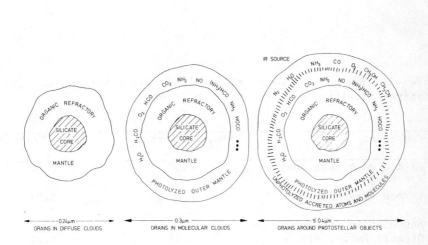

Fig. 5. Schematic configurations of the larger interstellar
grain in various regions of space.

A representative distribution of molecular constituents of
the grains assuming they have accreted all condensable matter
from the gas is shown in Table 1 (26). Sulfur containing
molecules have not been included in our theoretical molecular
evolution schemes (see A'Hearn and Feldman this volume).

6. INTERSTELLAR DUST IN THE EARLY SOLAR SYSTEM

The degree to which the prestellar dust grains remain intact in the early solar system determines the degree of relevance of the cosmochemistry of dust grains to ices in the solar system.

Bodies which preserve H_2O and all less volatile components are those which form where the temperature is T < 90 K. Among these would be comets and the outer planets; certainly beyond Uranus according to Cameron (27). Bodies which preserve the organic refractory yellow stuff are those which form where the temperature is not higher than T < 500 K. These would be the middle planets and the parent bodies of meteorites. Except for Mercury it could be that the silicates and graphite are preserved from primordial interstellar dust before aggregating into planets.

Table 1. Suggested mass and volume distribution of the principal chemical constituents of a comet.

Component	Mass Fraction	Volume Fraction
Silicates + Metal oxides	0.20	0.08
Carbon (Graphite ?)	0.06	0.03
Nonvolatile Complex Organic Refractory	0.19	0.21
H_2O	0.20	0.28
CO	0.03	0.04
CO_2	0.04	0.05
Other Molecules + Radicals (H_2CO, NH_3, HCN, CN, N_2, O_2 HCO, $HCONH_2$, ...)	0.23	0.31

We conclude that a major fraction of the initial interstellar dust may be retained in all but the inner region of the pre solar nebula. The chemical composition of such bodies as comets found at temperatures as low as 10-20 K (28) would contain a distribution of materials as itemized in Table 1. Note that, although this composition was derived from known properties of interstellar and protostellar dust their remain some uncertainties.

On a percentage basis, the silicates, carbon, organic refractory and H_2O are probably on the securest grounds. The relative amounts of CO and CO_2 are highly uncertain and the precise distribution within what is called "other molecules and radicals" is extremely difficult to predict. A very interesting and important feature is the possible survival of radicals in a comet since aggregation. It has been shown that the diffusion of HCO in matrices at temperatures T < 15 K is exceedingly slow (29) and that such free radicals could survive intact in an Oort cloud comet for longer than the age of the solar system. The survival of radicals and the presence of such highly volatile components as CO could play an important role in initiating bursts of activity.

REFERENCES

1.   Barlow, M.J. 1978 , MNRAS, **183**, pp. 367-395.

2.   Barlow, M.J. 1978 , MNRAS, **183**, pp. 397-415.

3.   Draine, B.T. and Salpeter, E.E., 1979 , Astrophys. J. **231**, pp. 77-94.

4.   Draine, B.T. and Salpeter, E.E., 1979 , Astrophys. J. **231**, pp. 438-455.

5.   Burke, J.R. and Silk, J. 1974, Astrophys. J. **190**, pp. 1-10.

6.   Kennicutt, R.C.C. 1983, Astrophys. J. **272**, pp. 54-67.

7.   Kwok, S., 1980, J. Roy. Astron. Soc. Can., **74**, no. 4, pp. 216-233.

8.   Greenberg, J.M. and Chlewicki, G., 1983, Astrophys. J. **272**, pp. 563-578.

9.   Greenberg, J.M. 1978, in "Cosmic Dust", ed. J.A.M. McDonnell (New York: Wiley-Interscience), pp. 187-294.

10.   Van de Bult, C.E.P.M. and Greenberg, J.M. 1984, MNRAS, **210**, pp. 803-811.

11.   Hagen, W., Allamandola, L.J. and Greenberg, J.M., 1979, Astrophys. & Sp. Sci., **65**, pp. 215-240.

12.   Greenberg, J.M., 1982a, in "Comets", ed. L. Wilkening, Un. of Arizona Press, Tucson AZ, pp. 131-163.

13.   Greenberg, J.M., 1982b, in "Submillimetre Wave Astronomy" ed. J.E. Beckman and J.P. Phillips Cambridge Univ. Press., pp. 261-306.

14.   Willner, S.P., Gillett, F.C., Herter, T.L., Jones, B., Krassner, J., Merrill, K.M., Pipher, J.L., Puetter, R.C., Rudy, R.J., Russell, R.W. and Soifer, B.T. 1982, Astrophys. J. **253**, pp. 174-187.

15a. Chlewicki, G., Greenberg, J.M., Aiello, S., Barsella, B., Patriarchi, P., Perinotto, M. 1984, Proc. 4th European IUE Conf. ESA SP-218, pp. 507-510.

15b. Greenberg, J.M. 1968, in "Stars and Stellar Systems" Vol. VII, ed. B.M. Middlehurst and L.H. Aller, U. of Chicago Press, pp. 221-364.

16.   Greenberg, J.M. and Yencha, A.J. 1973, in "Interstellar Dust and Related Topics", ed. J.M. Greenberg and H.C. van de Hulst, D. Reidel, Dordrecht, pp. 369-373.

17.   d'Hendecourt, L.B., Allamandola, L.J. and Greenberg J.M. 1984, Submitted to Astronomy and Astrophysics.

18.   Jura, M. and Léger, P. 1984, preprint.

19.   Van de Hulst, H.C. 1949, Rech. Astr. Obs., Utrecht, 11, part 2.

20.   Willner, S.P. and Pipher, J.C. 1982, Proc. Workshop on the Galactic Centre, Cal. Inst. Techn.

21.   Schutte, W. 1984, unpublished.

22.   Van de Bult, C.E.P.M., Greenberg, J.M. and Whittet, D.C.B. 1984, MNRAS in press.

23.   Lacy, L.H., Baas, F., Allamandola, L.J., Persson, S.E., McGregor, P.J., Lonsdale, Carol J. and Geballe, T.R., Van de Bult, C.E.P.M., 1984, Astrophys. J. 276, pp. 533-543.

24.   Tielens, A.G.M., Allamandola, L.J., Bregman, J., Goebel, J., d'Hendecourt, L.B. and Whitteborn, F.C. 1984, Astrophys. J., accepted for publication.

25.   d'Hendecourt, L.B. 1984, Ph. D. Thesis, University of Leiden.

26.   Greenberg, J.M., 1983, in "Asteroids, Comets, Meteors", eds. C.-I. Lagerkvist and H. Rickman, Uppsala Univ. Press, pp. 259-268.

27.   Cameron, A.G.W. 1977, The Moon and the Planets, **18**, 5.

28.   Greenberg, J.M., 1983, in "Cometary Exploration", ed. T.J. Gombosi, Proc. Conf. Cometary Exploration, II pp. 23-54

29.   Van IJzendoorn, L.J., unpublished results.

DISCUSSION

C. Pilcher :

There are a number of diffuse interstellar absorption bands in the visible. Have you examined the spectrum of the residue in this spectral region and, if so, does it show these bands?

J. M. Greenberg :

We see at least two of the bands consistently. It appears that one requires not just the residue, but a continually ultraviolet-irradiated residue.

T. Owen :

Laboratory experiments described by Dr. Greenberg lead to production of CO_2 in grain mantles. A good test of this model for interstellar grains is to search for the corresponding CO_2 absorption in the interstellar spectrum, since the absorption is very strong and diagnostic in the laboratory spectrum of the irradiated sample.

J. M. Greenberg :

I agree that one should try to observe CO_2 in grain mantles. This will have to wait for a space infrared telescope (ISO) because of the opacity of the atmosphere at this frequence.

M. F. A'Hearn :

        The C^{12}/C^{13} ratio is much more uncertain even than is normally quoted. The contamination problems may be solved by recent measurements by Lambert and Danks which utilize different emission bands and may therefore give better intensity ratios.

        A more serious problem, however, is that $C^{12}C^{12}$ is homonuclear, so that Vib - V_o transitions are strictly forbidden. Since $C^{12}C^{13}$ is not strictly homonuclear, there may be a large difference in the dipole moment and therefore in the fluorescence spectrum. This makes interpretation of intensity ratios totally uncertain.

J. M. Greenberg :

        I agree that the C^{12}/C^{13} ratio in comets is highly uncertain. However, I still suggest that the interstellar dust model predicts an enhanced C^{12}/C^{13} ratio in the volatiles and if the comet observations bear this out, they further confirm the interstellar dust model.

P. Weissman :

        Just a remark on Biermann's making comets in neighboring cloud fragments : Bert Donn suggested the same thing in 1974. If you assume typical cloud fragment random velocities, then it is difficult for the Solar System to capture enough comets if they are made somewhere else. Thus, to make comets, it is necessary to make them in the same cloud fragment as the Sun and planets.

J. M. Greenberg :

        All I want to show is that no matter where comets formed, they will be made out of interstellar dust so long as the prevailing temperature is low (say lower than $\sim$ 15 - 50 K).

G. Foti :

        In the picture that you give about grain evolution, why do not take into account the effects induced by ion bombardment ? For example, the outgoing molecules from grains occur without the hypothesis of grain-grain interactions ; our experimental data on proton bombardment of frozen mixture show that complex molecules come out from the target <u>during</u> irradiation at 10 K.

J. M. Greenberg :

   I do not take cosmic ray proton bombardment into account for consideration of grain evolution in the interstellar medium because I believe it becomes relevant <u>only</u> in the deepest interiors of very dark (and quiet) clouds. When I assume a flux of 2 MeV protons of 3 $cm^{-2}s^{-1}$, the energy impacted to a 0.1 mm radius grain is only of the order of 10^{-1} that due to the ultraviolet flux which is, in general, never lower than 10^{-4} of the diffuse cloud flux of 10^8 $h\nu$ $cm^{-2}s^{-1}$ for $E_{h\nu} > 6$ eV. I have chosen the factor 10^{-4} as a lower limit to the UV flux in clouds which may be derived as due to the Lyman α production by cosmic ray protons (see S. S. Prasad and S. P. Tarafdar, 1983, Ap J. 267, pp. 603-609). Furthermore the molecule ejection by protons is not sufficient to replenish the molecules in the gas even in the less dense regions of a molecular cloud.

M. Mac Kinnon :

   In a planetary environment, let us assume that a fresh optical surface of ammonia dihydrate is created. Under UV-photolysis similar to your experiments, what will happen chemically to this film and at what rate ?

J. M. Greenberg :

   I have not directly examined this problem.

A. Mendis :

   Aren't you just confirming Fred Hoyle's early idea that interstellar grains were made of complex organic polymers ?

J. M. Greenberg :

   Hoyle's idea of creating organic polymers was by accretion of H_2CO leading to polyformaldehyde, etc... This process certainly does not occur in the interstellar medium because H_2CO is a very <u>low</u> abundance molecule. However, the formation of complex organic molecules in grains by ultraviolet photoprocessing certainly oc-

curs. This was an idea which was investigated by Khare
and Sagan (using mercury lamps and LN_2 formaldehyde)
about the same time that I (with my colleagues) at State University of N. Y. at Albany did experiments with
vaccum ultraviolet on very low temperature ices (1970)
simulating the more appropriate interstellar conditions.

J. Lunine :

        Have you subjected your organic-coated particles
to heating in a hydrogen-rich environment ? I wonder
how much back reaction to CH_4 would occur ; hence how
relevant is the composition of your particles to the
composition of condensates in the Solar Nebula and in
outer planets ?

J. M. Greenberg :

        One needs a very warm environment indeed to pro-
duce a back reaction to CH_4. In fact one would not only
evaporate all the grain volatiles, but also the organic
refractories at the temperatures required for this reaction. I do not believe that the grains ever get to higher temparature than $\sim$ 50 K and probably much lower
temperatures prevail where grains aggregate into comets.

J. Klinger :

        Do you have any idea where the strong nonspheri-
city observed in interstellar grains does come from ?

J. M. Greenberg :

        No. This question has been raised over and over
again with many suggestions for possible explanations.
It is my guess that the silicate cores are formed non-
spherically and the mantles do not completely obscure
this substructure. However, I do not know exactly why
the silicates are nonspherical.

M. F. A'Hearn :

        Shul'man, Whipple and Donn at various times dis-
cussed of irradiation by cosmic rays as a means of explaining the anomalous behavior of dynamically new comets when they vaporize their first few meters of ice.
On the other hand, irradiation of ice mixtures by 1 MeV
protons, e. g. by M. Moore as well as others, leads to
all of the same qualitative effects observed in your ul-

traviolet experiments (CO $\longrightarrow$ CO_2, chemi-luminescence, explosions, refractory yellow residue, etc...). What is different about the case of irradiation of the outer layer of comets ?

J. M. Greenberg :

        The outer layer of a comet in the Oort cloud is indeed further modified by cosmic rays after aggregation of the previously UV irradiated dust. I suppose that the effects of this additional irradiation are not easily distinguished from what exists underneath. Perhaps it produces a "brown" gloop (higher carbon content). However, after the comet enters the inner Solar System just once this outer layer has disappeared and we are left with the original ultraviolet processed material.

A. Leger :

        1°) The residue spectrum you have shown has a 3 μm feature. Why is it not present in diffuse ISM spectra ?
        2°) Is there not a conflict between blowing out mantle by chem. explosions and remaining residues on core ?

J. M. Greenberg :

        1°) The broadness and weakness of the 3 μm absorption by organic residues makes it difficult to observe in the diffuse ISM except over very long optical paths. This is the case towards the galactic center where it is observed and compares well with laboratory spectra.
        2°) The chemical explosions will undoubtedly blow off some, or even a large fraction, of the complex organic molecules. However, annealing (low speed) collisions build up a sublayer of the organic residue which is not torn away by the outer explosive mixture.

P. Weissman :

        What are the typical dimensions of the silicate core grains you assume and do you consider grain-grain accretion or agglomeration processes in the clouds ?

J. M. Greenberg :

        I assume that the typical dimension of the silicate core grains are about thickness (twice radius) ∿ 0.1 μm and elongation about 3 : 1. I do not believe that agglomeration occurs in the clouds <u>except</u> in the latest stages of contraction leading to comets as well as planets. In all other molecular clouds, accretion is the dominant process. In my Summer Lectures at Les Houches (1983), I gave some preliminary attention to why agglomeration is not a likely process.

FORMATION HISTORY AND ENVIRONMENT OF COMETARY NUCLEI

Tetsuo Yamamoto

The Institute of Space and Astronautical Science
4-6-1 Komaba, Meguro-ku, Tokyo 153, Japan

ABSTRACT

We discuss the thermal history of the ice of a cometary nucleus and the formation environment of comets from the point of view of the chemical composition of the ice of the nucleus. For this basis, we propose a scheme for the formation of comets based on the theories on the evolution of the solar system. We describe the chemical composition of the ice of the nucleus inferred from the comparison of the abundances of cometary and interstellar molecules and from a condensation diagram of a gas having the interstellar molecule abundance. It is shown that compared with the interstellar abundance the ice of a cometary nucleus is depleted of very volatile species, which we assume to have lost by sublimation in the primordial solar nebula. We present a sequence of sublimation in the solar nebula of the ice having the interstellar molecule composition. From this sequence, we obtain the allowed range of the nebular temperature in which the composition of the ice of the nucleus is realized. Combining this result with models of the solar nebula, we discuss the formation region of cometary nuclei in the primordial solar nebula.

1. INTRODUCTION

A comet is a fossil of the primordial solar nebula in particular of its outer cold region, and will be an object which keeps a record of the evolution of the early solar system.

A scheme for the formation of cometary nuclei is presented in Fig. 1. We can divide the formation process of comets into

205

J. Klinger et al. (eds.), Ices in the Solar System, 205–219.

two stages from the viewpoint of the thermal history of the ice
of cometary nuclei.

The first is the stage of the parent interstellar molecular
cloud. At this stage, molecules in the cloud condensed on the
surface of grains to form ice mantle on it. The grains at this
stage may be characterized by those as proposed by Greenberg (1):
they were coated with ice mantles composed of the molecules up to
very volatile ones because of a low temperature in the cloud. We
can expect that the chemical composition of the ice mantle was
similar to that of interstellar molecules. It is to be noted,
however, that the size of these grains were of the order of 0.1
μm, much smaller than the size of cometary nuclei. This stage is
regarded to be a stage of the formation of original raw materials
of cometary nuclei.

The second is the stage of the primordial solar nebula,
which formed by gravitational contraction and fragmentation of
the parent interstellar cloud. In the inner region of the solar
nebula, the grains would have suffered strong heating and
sublimed completely. As a result the gaseous composition would
have become the solar abundance. Subsequent cooling led to
condensation of refractory grains. The condensation sequence in

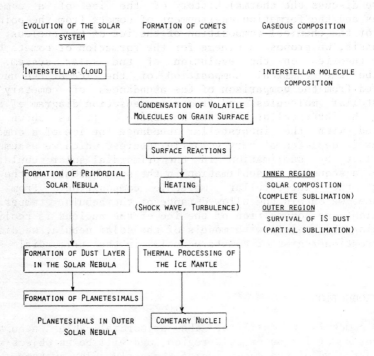

Fig. 1. A scheme for the formation of cometary nuclei based on
the evolution of the solar system.

a cooling gas having the solar composition has been discussed by many authors on the basis of the equilibrium calculation in connection with the chemical history of meteorites (2, 3, 4, 5, 6). In the outer region, on the other hand, the grains would not have been heated much strongly, and there must have been a region where the grains coated with the ice mantle survived. It is to be noted, however, that the original ice mantle would have suffered thermal processing to a certain degree, and lost volatile components, because the nebular temperature is considered to have been higher than the temperature of the parent interstellar cloud. The degree of the thermal processing depended upon the distance from the center of the solar nebula. It is the formation region of cometary nuclei where the ice mantle came to have the composition similar to that of the ice of a cometary nucleus. Cometary nuclei are regarded to be the planetesimals or their aggregates (7) formed by gravitational fragmentation of the dust layer (8, 9, 10) in this region.

2. THE CHEMICAL COMPOSITION OF THE ICE OF A COMETARY NUCLEUS

In the recent papers (11, 12), we discussed the composition of the ice of a cometary nucleus by comparing the abundances of cometary and interstellar molecules. We summarize the results in the following. See ref. (11) for details.

The interstellar molecule abundance used is listed in Table 1. We have taken into account abundant stable species for the possible components of the ice mantle in the parent interstellar cloud. Surface reactions are ignored for the first estimate. The abundance values are taken from the compilations of the observed results (13, 14, 15, 16). The results of the model computations of interstellar chemistry are referred (17, 18, 19) for the species that have not been detected but predicted to be abundant. These include N_2, CH_4, CO_2, C_2H_2, and C_2H_4.

Assuming that the nuclear ice (i.e. parent molecules) have the interstellar molecule composition, we evaluate the expected abundance of cometary molecules from the balance between the production and photodissociation rates in the whole coma. (For N_2^+, electron recombination is the main dissociation mechanism.) Namely, the abundance [x] of a cometary molecule x is estimated from

$$[x] \propto \text{(abundance of the parent X)} \times \text{(branching ratio for the production of x from X)} \times \text{(life time of x).} \quad (1)$$

For cometary CO, we have investigated two possibilities; (a) CO itself is the parent, and (b) CO is produced by photodissociation of CO_2. In the case (a), the second factor of the right hand side of the relation (1) is equal to unity. The abundance thus evaluated is compared with the observed one in Fig. 2.

Table 1. Interstellar molecule abundance

	$\log [X]/[H_2]$	adopted value
H_2	0	0
CO	$-4.5 \sim -3.5$	-4
N_2	$-4.5 \sim -3.5$	-4
H_2O	~ -4	-4
CH_4	$-5 \sim -4$	-4.5
CO_2	$-5.5 \sim -4.5$	-5
HCN	$(-8 \sim)-7 \sim -6$	-7
C_2H_2	$-7 \sim -6.5 (\sim -5)$	-7
C_2H_4	$-7 (\sim -8)$	-7
NH_3	$-8 \sim -5 \ (-7)$	-7
CH_3C_2H	$-9 \sim -7.5$	-8
CH_3CN	$-9 \sim -7.5$	-8

We can derive the following results from Fig. 2 and Table 1.

(i) Many of the possible parents, which are assumed to have the interstellar abundance, can reproduce the observed abundance of the cometary molecules within a factor of 10. Thus the ice of a cometary nucleus can be approximately regarded to be a condensate of interstellar molecules.

(ii) However, the candidate parents N_2 and CO are notable exceptions. As can be seen from Fig. 2, N_2 and CO produce overabundant cometary N_2^+ and CO with the amounts larger by an order of magnitude or more than the observed ones. Thus the nuclear ice is much depleted of N_2 and CO (and H_2, of course) compared with their interstellar abundance, although we have to keep it in mind that the CO abundance varies from comet to comet in contrast with other species as pointed out by Feldman (20). The relative abundance of cometary CO shown in Fig. 1 is derived from the observed result of Comet West (1976 VI) by Feldman and Brune (21), and from the compilation by Ip (22) and Delsemme (23).

(iii) The main parent of CO will be CO_2. CO_2 can account for the observed abundance of CO, and the observed upper limit of the cometary CO_2 abundance does not contradict the expected value. The parent of N_2^+ other than N_2 is not clear. Hydrazine N_2H_4 may listed as suggested by Delsemme (24). Another possible

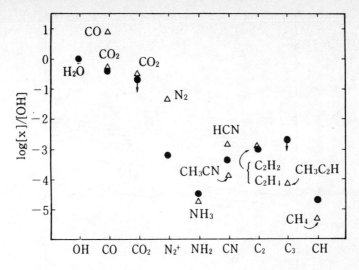

Fig. 2. Comparison of the expected abundance of cometary molecules produced from the ice assumed to have the interstellar molecule abundance with the observed abundance. The filled circles show the observed abundance, and the open triangles show the expected abundance produced from the parent molecules attached to them. The H_2O abundance is set to equal the OH abundance.

parents may be NH_3, from which N_2^+ is produced by photodissociation to form NH_2, collisional reaction of NH_2 with H^+ and NO to form N_2, and subsequent ionization of N_2 (25).

(iv) Photolysis of CH_4, whose interstellar abundance is estimated to be relatively large, can yield the observed abundance of CH. We have to point out, however, that CH, one of the radicals with minor abundance, may be produced as a trace product in the collisional reactions among CO, OH, H, C, and their ions, which are abundant species in a cometary coma.

(v) The main components of the nuclear ice will be H_2O and CO_2 with the abundance ratio of $[H_2O]/[CO_2] \sim 10$. In addition, CH_4 might be another main component.

Figure 3 shows the equilibrium temperatures for condensation of the molecular species listed in Table 1 with their abundance, as a function of the density n of the gas mainly of H_2. It is to be noted that both N_2 and CO have the lowest equilibrium temperatures following H_2, i.e., these are very volatile species. The equilibrium temperature does not depend much on the gas density. For $n = 10^5$ cm^{-3}, a typical gas density in molecular clouds, the equilibrium temperatures T_e of N_2 and CO are 15 K and

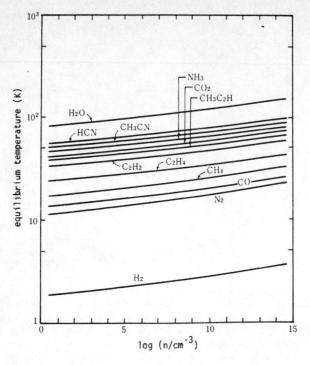

Fig. 3. The equilibrium temperature for condensation of a gas
with the interstellar composition as a function of the number
density n of the gas.

17 K, respectively, and T_e for H_2 is 2.2 K. Since the temperature in the parent interstellar cloud would have been as low as
∼ 10 K, both N_2 and CO (but probably not H_2) would have condensed
on the grain surface, and would have been the components of the
ice mantle (1), as mentioned in Sect 1. It is thus considered
that N_2 and CO are the species lost by thermal processing at the
primordial solar nebula stage.

3. THE TEMPERATURE IN THE FORMATION ENVIRONMENT

We assume sublimation of the ice mantle on the surface of
grains in the dust layer to be the main thermal processing in the
solar nebula. The grain temperature T_d in the dust layer, which
is assumed to be optically thick for visible and UV radiation but
thin for thermal infrared radiation, is determined by a balance
between collisional heating by gas molecules and radiative
cooling. The gaseous species contributing to the grain heating

are H_2 and He, which are the main composition of the nebular gas. The energy balance of a grain having a radius a and an emissivity ε in a gas of density n and temperature T is given (26) by

$$\xi n <v> \pi a^2 \; \alpha(T) 2k (T - T_d) = 4\pi a^2 \varepsilon \sigma (T_d^4 - T_b^4), \qquad (2)$$

where ξ is a constant determined by the abundance ratio of H_2 and He, $<v>$ is the mean thermal velocity of gas molecules, T_b is the background radiation temperature, $\alpha(T)$ is the accomodation coefficient (27) averaged over the impinging fluxes of H_2 and He, k is the Boltzmann constant, and σ is the Stefan-Boltzmann constant. Assuming that H_2 and He are of the solar abundance (28), we have $\xi = 1.02$, and we can use $\alpha(T)$ for H_2 molecules. In the following calculation, we put $\varepsilon = 0.01$ (29) and $T_b = 15$ K (30), and use $\alpha(T)$ for H_2 molecules impinging onto H_2O ice. The expression and relevant data for the $\alpha(T)$ are given by Hollenbach and Salpeter (27).

We assume for a first estimate that each kind of molecular species of the ice mantle sublimes independently, and that diffusion of molecules in it is rapid enough to supply molecules subliming from the surface. Then the sublimation rate r_X of molecular species X having molecular weight μ_X and vapor pressure $P_{v,X}$ is written as

$$r_X(T_d) = \frac{P_{v,X}(T_d)}{(2\pi\mu_X m_H k T_d)^{1/2}} \cdot 4\pi a^2, \qquad (3)$$

where m_H is the mass of the hydrogen atom. The vapor pressure data used are given by Yamamoto et al. (11). On the other hand, the impinging rate $p_X(T)$ is expressed as

$$p_X(T) = f_X n <v_X> \pi a^2, \qquad (4)$$

where f_X and $<v_X>$ are the relative abundance and mean thermal velocity of the molecules of species X, respectively. If $r_X(T_d) > p_X(T)$, the species X will be lost from the ice mantle due to net sublimation. The critical temperature at which the net sublimation rate is zero is determined by $r_X(T_d) = p_X(T)$, namely

$$f_X n k T / P_{v,X}(T_d) = (T/T_d)^{1/2}. \qquad (5)$$

We call the critical temperature for the gas the sublimation temperature from now on.

From eqs. (2) and (5), we can obtain the sublimation temperature as a function of gas density n. The result is shown in Fig. 4 for H_2O, CO_2, CH_4, CO, and N_2, which are expected to be abundant interstellar molecules, and will be the key species for the study of the formation environment of cometary nuclei.

In Table 2 we list the sublimation temperatures T_{subl}

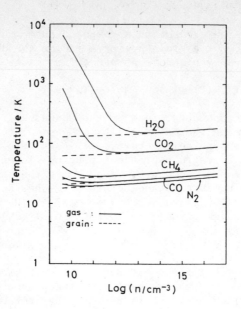

Fig. 4. The sublimation temperatures (solid curves) with the corresponding grain temperatures (dashed curves) versus gas density n for possible abundant species of the ice composition of a cometary nucleus.

including those of other less abundant possible candidate species of the nuclear ice with adopted values of f_X, according to the order of higher sublimation temperatures. The values of T_{subl} in this table are those for a typical nebular density of $n = 10^{13}$ cm^{-3}, at which density the temperature difference between gas and grains is small as can be seen in Fig. 4. The sequence of sublimation does not change for other gas density concerned. We have put $T_b = 0$ K here to see a full sequence of sublimation. For $T_b > 0$ K, those species having lower T_{subl} than T_b will vaporize in such an environment.

It is to be pointed out that in this sequence H_2O, HCN, CH_3CN, NH_3, and HC_3N have been reported to be detected in comets by radio observations (31, 32, 33, 34, 35). Furthermore CS_2 is a plausible parent of CS and S as suggested by Jackson et al. (36) from the analysis of the observed spatial profiles of CS and S. These results imply that at least the species whose sublimation temperatures are higher than that of HC_3N in Table 2 have been able to be retained as the components of the nuclear ice, and in consequence that the temperature of the formation region of cometary nuclei is at least lower than the sublimation temperature of HC_3N.

Table 2. Sublimation temperatures[*])

	T_{subl}	$\log [X]/[H_2]$
H_2O	152	-4
HCOOH	112	-7
CH_3OH	99	-8
HCN	95	-7
CH_3CN	91	-8
SO_2	83	-8
NH_3	78	-7
CS_2	78	-8
HC_3N	74	-8
CO_2	72	-5
CH_3C_2H	65	-8
H_2CO	64	-8
C_2H_2	57	-7
H_2S	57	-7
C_2H_4	42	-7
CH_4	31	-4.5
CO	25	-4
O_2	24	-5
N_2	22	-4
H_2	5	0

[*]) gas density $n = 10^{13}$ cm^{-3}

From the results (iii) and (ii) mentioned in Sect. 2, we can further restrict the temperature range. The degree of the thermal proceesing is conservatively estimated to be such that (1) CO_2 has been retained as one of the main components of the nuclear ice, and that (2) N_2 and CO have been lost from the original ice mantle. Thus it follows that the temperature range of the formation region is restricted to

$$T_{subl}(N_2) < T < T_{subl}(CO_2) \qquad (6)$$

in Fig. 4, where we have adopted $T_{sub1}(N_2)$ as the lower limit.

In respect to CO, we comment on the result of the UV observations (20) that CO abundance varies from comet to comet, whereas other species exhibit similar relative abundance. A possible interpretation of this result is that the formation region of cometary nuclei is just around the region of CO sublimation so that some comets retain substantial amount of CO and other comets are lacking in a large portion of CO. If this is actually the case, the temperature in the formation region nearly equals the sublimation temperature of CO, which is slightly higher that that of N_2.

4. FORMATION REGION OF COMETARY NUCLEI IN THE SOLAR NEBULA

On the basis of the condition (6), we discuss the region in the primordial solar nebula where cometary nuclei formed. For this purpose, we require a model for the solar nebula in which the density and temperature of the gas are given as a function of the distance from the center. The temperature distribution is more important. We shall discuss the formation region by adopting two types of temperature distributions: one is a radiative equilibrium distribution, and the other is an adiabatic one. The solar nebula models that provide these temperature distributions are Hayashi's model (37) and Cameron's model (38), respectively. In the Hayashi's model, the nebular disk is assumed to have a mass of ~ 0.01 $M_{\odot}$, and is in a dynamically eqilibrium state, whereas in the Cameron's model the disk, whose initial mass is assumed to be $\sim 1M_{\odot}$, never settles to a dynamically equilibrium state but evolves with time during the infall accretion and subsequent mass loss from the disk. For the Cameron's model, we used the temperatures and density distributions near the end of the infall accretion, at which phase both temperature and density reach near their maximum values in each of the planetary formation regions.

The distance from the center at which the abundant species of the original ice mantle attain their sublimation temperatures are shown for both models in Figs. 5(a) and 5(b), where we extrapolated their models to larger distances.

We can specify the distance r of the formation region of cometary nuclei in the solar nebula from the condition (6) in Sect. 3. For the radiative equilibrium distribution, we obtain $14 < r < 110$ AU as can be seen from Fig. 5(a). The inner limit is placed between the formation regions of Saturn and Uranus. If we take $T_{sub1}(CO)$ instead of $T_{sub1}(N_2)$ as the lower limit of the temperature range, the outer limit is replaced by 82 AU. For both cases, the outer limits lie beyond the formation regions of the planets. For the adiabatic distribution, the condition (6) yields $15 < r < 79$ AU as can be seen from Fig. 5(b): the inner limit corresponds to the formation region of Uranus. The outer

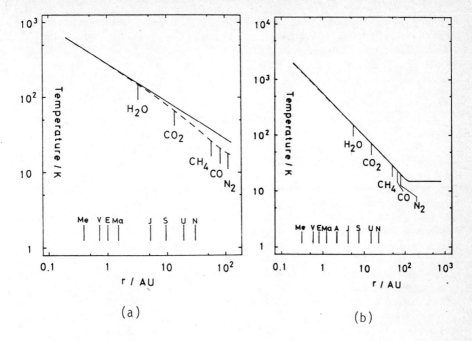

(a)   (b)

Fig. 5.   The distance r from a center of the solar nebula   inside
which the molecular species of the ice mantle sublimes, for two
types of the temperature distributions: (a) the radiative equilibrium one (Hayashi's model),   and (b) the adiabatic   one
(Cameron's model).   The solid curves show the gas temperature,
and the dashed curves, the grain temperature.   No substantial
temperature difference can be seen in the Cameron's model because
of large gas density.   The distances of the planetary formation
regions are also shown in the figures.

limit is replaced by 65 AU if we take $T_{subl}(CO)$. These upper
limits fall onto the region beyond the planetary formation
regions as in the radiative equilibrium case.
      Cameron himself and Biermann and Michel (39) proposed the
formation of cometary nuclei at the periphery of the solar
nebula.   Cameron considered the formation in the region at $r \sim$
10^3 AU.   At this distance, N_2 and CO are retained even if we
adopt his assumed background radiation temperature of 20 K,
though this is marginal since their sublimation temperatures are
only slightly higher than 20 K.   Biermann and Michel adopted, for
the formation environment, a gas density of $\sim 10^5$ cm^{-3} and a
grain temperature of 17 K, which they regarded to be equal to the
gas temperature.   In this case, species less volatile than CO are
retained, and CO is marginal because its sublimation temperature
is 17 K for the adopted condition.   Anyway it is required to make

the physical conditions, in particular the temperature, more definite in order to examine the theories of comet formation at the periphery of the solar nebula.

5. CONCLUSIONS

We have discussed the thermal history of the ice of a cometary nucleus and the formation environment of comets in the solar nebula on the basis of the scheme for the formation of cometary nuclei with the evolution of the solar system in the background. From the point of view of the chemical composition, the thermal history have been divided into two stages: (1) condensation of molecules onto the grain surface in the parent interstellar cloud; the ice mantle formed at this stage is regarded as the unprocessed raw materials of the nuclear ice; (2) sublimation of volatile species in the ice mantle in the promordial solar nebula.

By comparing the composition of the nuclear ice inferred from the abundance of the observed cometary molecules with that of interstellar molecules, it has been shown that the ice of the nucleus has the composition similar to that of interstellar molecules, but is lacking in the very volatile species such as N_2, CO, and H_2. In consequence, the main composition of the ice of the nucleus is H_2O and CO_2 with the abundance ratio of $[H_2O]/[CO_2] \sim 10$. It is an open question whether or not CH_4 is another main composition.

Using the results on the composition of the nuclear ice, we have investigated the allowed range of the temperature in the formation region of cometary nuclei in the solar nebula, by calculating the sublimation sequence of the molecular species composing the ice mantle.

The formation region of cometary nuclei derived from the temperature range has been discussed for the two types of the temperature distributions of the nebular gas: the radiative equilibrium and adiabatic distributions. It has been shown that the distance of the inner limit from the center of the nebula has to be, at least, larger than 14 or 15 AU, that is, outer than the region between the formation regions of Saturn and Uranus, in order for CO_2 ice to be retained as one of the main components of the nuclear ice. On the other hand, the outer limit is estimated, from the sublimation temperature of N_2, to be 80 to 110 AU from the center, depending on the temperature distribution. It is interesting to note the abundance difference of CO from comet to comet. If this is due to the abundance difference of CO in the ice of a cometary nucleus, we can expect that the formation region is around the region of CO sublimation. Its distance is 60 to 80 AU from the center.

It is important to clarify observationally to which volatile molecular species have been retained in the ice of a cometary nucleus in order to make the formation environment more definite. It is to be pointed out that several candidate molecules composing the ice have vibro-rotational bands in near to middle infrared wavelengths strong enough to be detected by recent infrared observation techniques (40, 41, 42, 43). Of these molecular species, CH_4 is the most important key species for better specification of the formation environment. If CH_4 is detected and found to be one of the main components, the formation temperature is restricted to a narrow range as $T_{subl}(N_2) < T < T_{subl}(CH_4)$ (see Fig. 4 and Table 2) instead of the condition (6). CH_4 has a strong emission band at 3.31 μm, from which we can expect the emission flux larger than 10^{-17} W cm^{-2} for a comet having the production rate of H_2O of 10^{29} s^{-1}, when observed with a diaphragm larger than the scale length at r = Δ = 1 AU, where r is the heliocentric distance of the comet, and Δ is the comet-observer distance. One should note that this band may be observed by ground-based observations since its wavelength is placed in the window of the earth atmosphere.

NOTE ADDED IN REVISION:

In Comet IRAS-Araki-Alcock (1983d), the detection of HCO and H_2S^+ has been reported (C.B. Cosmovici and S. Ortolani, Nature 310, 122, 1984). The authors suggest H_2CO and HCOOH as the parents of HCO. H_2S^+ is considered to be produced by ionization of H_2S. Of these possible parents, H_2S has the lowest sublimation temperature (see Table 2). If H_2S is actually the composition of the ice of the nucleus and is found to have substantial abundance, the retention of CO_2 is confirmed, and we can adopt $T_{subl}(H_2S)$ instead of $T_{subl}(CO_2)$ as the upper limit in the condition (6). Accordingly the inner limit of the distance of the formation region is calculated to be 21 AU for both of the temperature distributions. This distance corresponds to the formation regions of Uranus and Neptune.

REFERENCES

(1) Greenberg, J.M. 1983, in "Cometary Exploration II, Proc. Int. Conf. Cometary Exploration", ed. T.I. Gombosi, pp. 23-54.
(2) Lord III, H.C. 1965, Icarus 4, pp. 279-288.
(3) Larimer, J.W. 1967, Geochim. Cosmochim. Acta 31, pp. 1251-1238.
(4) Anders, E. 1968, Acc. Chem. Res. 1, pp. 289-298.
(5) Grossman, L. 1972, Geochim. Cosmochim. Acta 36, pp. 597-619.
(6) Grossman, L. and Larimer, J.W. 1974, Rev. Geophys. Space Phys. 12, pp. 71-101.

(7) Donn, B. 1981, in "Comets and the Origin of Life", ed.
     C. Ponnamperuma, D. Reidel Pub. Co., Dordrecht, pp. 21-29.
(8) Safronov, V.S. 1972, "Evolution of the Protoplanetary Cloud
     and the Formation of the Earth and the Planets", Israel
     Program for Scientific Translations, Jerusalem.
(9) Hayashi, C. 1972, in "Proc. Lunar and Planetary Symp.",
     ed. K. Takayanagi and M. Shimizu (Inst. Space Aeronautical
     Sci., Univ. Tokyo), pp. 13-18.
(10) Goldreich, P. and Ward, W.R. 1973, Astrophys. J. 183,
     pp. 1051-1061.
(11) Yamamoto, T. Nakagawa, N., and Fukui, Y. 1983, Astron.
     Astrophys. 122, pp. 171-176.
(12) Yamamoto, T. 1983, in "Cometary Exploration I, Proc. Int.
     Conf. Cometary Exploration", ed. T.I. Gombosi, pp. 85-91.
(13) Watson, W.D. 1977, Acc. Chem. Res. 10, pp. 221-226.
(14) Mann, A.P.C. and Williams, D.A. 1980, Nature 283,
     pp. 721-725.
(15) Fukui, Y. 1980, in "Proc. Workshop the Reactions of
     Organic Molecules in Space", ed. M. Shimizu (Inst. Space
     Aeronautical Sci., Univ. Tokyo), pp. 34-42.
(16) Irvine, W.M. and Hjalmarson, Å. 1983, in "Cosmochemistry
     and Origin of Life", ed. C. Ponnamperuma, pp. 113-142.
(17) Mitchell, G.F., Ginsburg, J.L., and Kuntz, P.J. 1978,
     Astrophys. J. Suppl. Ser. 38, pp. 39-68.
(18) Suzuki, H. 1979, Prog. Theor. Phys. 62, pp. 936-956.
(19) Prasad, S.S. and Huntress, W.T. Jr. 1980, Astrophys. J
     Suppl. 43, pp. 1-35.
(20) Feldman, P.D. 1983, Science 219, pp. 347-354.
(21) Feldman, P.D. and Brune, W.H. 1976, Astrophys. J. Lett.
     209, L45-L48.
(22) Ip, W.-H. 1979, in "Proc. Workshop Cometary Missions", ed.
     W.I. Axford, H. Fechtig, and J. Rahe, Bamberg, pp. 69-91.
(23) Delsemme, A.H. 1977, in "Comets, Asteroids, Meteorites -
     Interrelations, Evolution, and Origins", IAU Coll. 39,
     ed. A.H. Delsemme, Univ. Toledo, pp. 3-13.
(24) Delsemme, A.H. 1975, Icarus 24, pp. 95-110.
(25) Swift, M.B. and Mitchell, G.F. 1981, Icarus 47, pp. 412-430.
(26) Yamamoto, T. 1984, Astron. Astrophys. (in press).
(27) Hollenbach, D. and Salpeter, E.E. 1970, J. Chem. Phys.
     53, pp. 79-86.
(28) Cameron, A.G.W. 1973, Space Sci. Rev. 15, pp. 121-146.
(29) Greenberg, J.M. 1978, in "Cosmic Dust", ed. J.A.M. McDonnel
     John Wiley and Sons, pp. 187-294.
(30) Kusaka,T., Nakano, T., and Hayashi, C. 1970, Prog. Theor.
     Phys. 44, pp. 1580-1595.
(31) Jackson, W.M., Clark, T., and Donn, B. 1976, in "The Study
     of Comets, Part I", NASA SP-393, pp. 272-280.
(32) Huebner, W.F., Snyder, L.E., and Buhl, D. 1974, Icarus
     23, pp. 580-584.
(33) Ulich, B.L. and Conclin, E.K. 1974, Nature 248,
     pp. 121-122.

(34) Altenhoff, W. J., Batria, W., Kuchtmeier, W. K., Schmidt, J., Stumpff, P., and Walmsley, M. 1983, Astron. Astrophys. 125, L19 - L22.

(35) Hasegawa, T., Ohishi, M., Morimoto, M., Suzuki, H., and Kaifu, N. 1984, Icarus (to be published).

(36) Jackson, W.M., Halpern, J.B., Feldman, P.D., and Rahe, J. 1982, Astron. Astrophys. 107, pp. 385-389.

(37) Hayashi, C. 1981, Suppl. Prog. Theor. Phys. No. 70, pp. 35-53.

(38) Cameron, A.G.W. 1978, Moon and Planets 18, pp. 5-40.

(39) Biermann, L. and Michel, K.W. 1978, Moon and Planets 18, pp. 447-464.

(40) Yamamoto, T. 1982, Astron. Astrophys. 109, pp. 326-330.

(41) Encrenaz, Th., Crovisier, J., and Combes, M. 1982, Icarus 51, pp. 660-664.

(42) Crovisier, J. and Encrenaz, Th. 1983, Astron. Astrophys. 126, pp. 170-182.

(43) Weaver, H.A. and Mumma, M.J., 1984, Astrophys. J. 276, pp. 782-797.

STABLE ISOTOPIC COMPOSITIONS OF HYDROGEN, CARBON, NITROGEN, OXYGEN AND SULFUR IN METEORITIC LOW TEMPERATURE CONDENSATES

I P Wright

Planetary Sciences Unit, Department of Earth Sciences, The Open University, Walton Hall, Milton Keynes, MK7 6AA, England.

ABSTRACT

The study of light element stable isotopes (D/H, $^{13}C/^{12}C$, $^{15}N/^{14}N$, $^{18}O/^{17}O/^{16}O$, $^{36}S/^{34}S/^{33}S/^{32}S$) in meteorites gives an insight into processes which occurred during, and indeed before, the formation of the solar system. The more refractory light element-containing components preserve isotopic signatures recorded in pre-solar events or imparted in the high temperature regions of the solar nebula. The characteristics of the low temperature condensates in meteorites (organic matter etc.) are the result of processes which occurred in the cooler regions of the nebula. The origin of these materials is obviously relevant to the formation of the other low temperature solar system bodies such as the outer planets and comets.

The paper briefly reviews the light element isotopic compositions of hydrogen, carbon, nitrogen, oxygen and sulfur in meterorites and meteoritic components. Appraisal of similar isotopic ratios observed in stars, interstellar environments, planetary atmospheres and comets allows tentative genetic relationships to be invoked for certain materials found in meteorites.

1. INTRODUCTION

The light elements hydrogen, carbon, nitrogen, oxygen and sulfur found in meteorites are present as, amongst other things, a diverse set of minerals. Indeed, oxygen is the most abundant component of the silicates which comprise the chondritic and

J. Klinger et al. (eds.), Ices in the Solar System, 221–249.
© 1985 by D. Reidel Publishing Company.

achondritic meteorites. Elemental sulfur and troilite (FeS) are
prevelant in iron meteorites along with minor amounts of
carbides and nitrides. Small amounts of unusual minerals such
as titanium nitride and silicon oxynitride (sinoite, Si_2N_2O) are
found in the enstatite chrondites. The "low temperature
condensates" on the other hand are composed predominantly of the
light elements and engross the complex organic compounds which
are found in carbonaceous chrondites and the least
"metamorphosed" ordinary chondrites. This organic matter is
formed from relatively simple building blocks - veritable ices
in some cases - either in the interstellar medium or at some
time during the formation of the solar system.

The goal of research connected with low temperature
condensates is to devise consistent theories which can explain
the relationships of such material with the proposed precursors
and the subsequent evolutionary forms. Furthermore, it is
desirable to understand not only the relationship between the
light elements and the other rock-forming elements and minerals,
but also the mechanisms involved in segregating the light
elements between the refractory and volatile forms. A central
question concerns the relative timing and the formational
environment of the organic material present in meteorites.
Ultimately this problem will probably need the analysis
(structural, isotopic etc.) of the individual components of the
organic matter as inevitably there may be contributions from
primary sources (pre-solar matter or compounds formed in the
solar nebula) and secondary sources (subsequent contamination or
alteration in the nebula, space or indeed the laboratory).

A premise used when considering the origin of the solar
system is that at the outset of the nebula phase, practically
all of the elements which subsequently form the Sun, planets,
meteorites, comets, etc. are already synthesised. They have
been made in various nucleosynthetic environments; hydrogen,
deuterium, lithium, beryllium and boron predominantly from the
big bang, the light elements from nuclear fusion in the various
CNO-cycles and the heavier elements from slow or rapid neutron
capture processes in evolved stars and supernovae etc. The seed
materials were present as a pre-solar cloud of gas and dust
presumably similar to a presently observable interstellar cloud.
At the beginning of the solar system formation when the cloud
underwent transformation to the nebula environment the proto-sun
and proto-planets were formed. The exact action which caused
formation is still unsure but the presence in
calcium/aluminum-rich inclusions from certain carbonaceous
chondites of ^{26}Mg excesses, thought to arise from the decay of
^{26}Al (see 1 for a review) has prompted Cameron and Truran (2) to
suggest that a late-stage supernova injected the live
radioactive ^{26}Al and indeed the shock-wave caused the

interstellar   cloud   to   collapse.   However,   alternative
explanations exist for the existence of [26]Al (e.g. 3,4) and so a
supernova shock-wave may not have been causal. Regardless, for
the purpose of this paper, it is only necessary to consider that
gravitational instabilities and turbulent motions of some kind
produced the Sun and planets.

The nature of the conditions in the solar nebula remains an
unresolved   issue.   Before   the   detection   of   isotopic
heterogeneity within meteorites, it seemed necessary to
postulate that the gas and dust of an unknown number of
different sources was homogenised in the nebula. This was best
envisaged as taking place in a hot environment (>2000 K) where
all the components existed as an atomic gas (5).   From this
point the Sun was formed and as the gas began to condense, the
elements formed into minerals according to their volatility.
Thermodynamic calculations are able to predict the sequence of
minerals which should appear upon cooling, (e.g. 6, 7).   In
practice, it is thought that some of the minerals would become
removed from the system and preserved while others reacted with
the   cooling gas to form lower temperature species.   The
theoretical predictions seem to match the suite of minerals
observed in meteorites like Allende (8).   An important
consequence of condensation in this fashion is that at lower
temperatures (400 K) earlier formed phases such as magnetite
(Fe_3O_4),   hydrated   silicates   and   iron-nickel   metal   are
subsequently available for catalysis reactions involving the
light elements. Laboratory simulations have shown that simple
starting materials such as CO, H_2, NH_3, etc., can react in the
presence of these catalysts via Fischer-Tropsch-type syntheses
to form the complex organic materials similar to those found in
the meteorites (see 9 for historical review).   Further studies
have been successful in producing reduced and oxidised forms of
carbon which display a difference in isotopic composition of the
same sign and magnitude as the analogous forms in carbonaceous
chondrites (10).

After   the   identification   in meteorites of isotopic
anomalies (in particular oxygen, 11) it became necessary to
invoke a certain amount of heterogeneity in the primitive solar
nebula. At first, such heterogeneities were thought to result
from injections of small amounts of freshly synthesised,
isotopically distinct materials to the relatively homogeneous
pool constituting the solar nebula. Subsequent interpretations
have appealed to modifications of pre-existing material by
gas/dust fractionation or proton irradiations etc. In all cases
a heterogeneous environment is a necessity.

The presence in meteorites of noble gases with isotopic
compositions seemingly diagnostic of certain nucleosynthetic

sites has made it clear that some pre-solar material may have survived the nebula environment intact (e.g. 12). Further studies of light element isotopic compositions have confirmed this view (13, 14) and indeed studies of hydrogen isotopic composition (15, 17) suggest that organic molecules processed in interstellar clouds have survived to some extent. Thus the environment where the planets and meteorites formed is now thought to be much more complicated than was first envisaged. Indeed, the dogma of an initially hot gaseous solar nebula is no longer universally accepted. An alternative hypothesis (3) proposes the contraction of a cold dusty nebula with consequent preservation of isotopic structure that is a record of events which occurred in nucleosynthetic environments and interstellar clouds prior to the solar system formation. This scenario has important implications for the origin of "low temperature" material (i.e. organic compounds, carbonates, etc.): clearly the "cosmic chemical memory" may be preserved in the light elements.

Yet a further explanation (18, 19) for the origin of the solar system involves the collapse of a fully ionized rotating plasma where although molecules and dust are "cold", electron temperatures reach 10^5 K. This sort of environment could produce somewhat unexpected chemical effects (see 20) and may indeed provide trivial explanations for the existence of seemingly isotopically anomalous material. Certainly, photochemistry can produce isotopic effects in oxygen and nitrogen of the type which traditionally have been assigned a nucleosynthetic origin (e.g. 20, 21). Furthermore, photochemistry and reactions similar to the Miller-Urey synthesis may form complex organic molecules from starting materials such as CH_4, H_2O, NH_3, etc. (22).

While theoreticians continue to develop models of solar system formation and experimentalists attempt more realistic simulation studies it is desirable to persevere by extracing as much useful information as possible from meteorites and indeed cosmic dust particles. Carbon, nitrogen and hydrogen isotope ratios can be obtained by analysis in a mass spectrometer of gases released from solid samples by relatively simple stepped heating experiments under vacuum (pyrolysis), or an atmosphere of oxygen (combustion). The difficult part of the procedure is to ensure that the ubiquitous blank and contamination problems are reduced to a minimum. Clearly, analysis of the smallest possible samples allows the greatest amount of information to be extracted from an individual meteorite. Thus it has been necessary to develop suitable analytical techniques capable of providing the desired information (23 describes the historical development of these techniques). These include low blank extraction systems, appropriate sample handling procedures and even novel forms of mass spectrometry (24, 25). The static mass spectrometer used for nitrogen isotope ratio measurement (26, 27) is a variant of the type of instrument developed for noble gas determination (28); it allows samples of nitrogen about 1000 times smaller in size than employed conventionally to be measured isotopically to a precision of $\pm$ two parts in a thousand. This same type of instrument is now being used for carbon, oxygen and hydrogen isotopic measurements (29).

2. LIGHT ELEMENT STABLE ISOTOPE RATIOS IN OBSERVABLE ASTROPHYSICAL ENVIRONMENTS.

This section of the paper is not intended to be a comprehensive review, but merely attempts to summarize the pertinent isotopic data available from observations of various astrophysical environments (interstellar clouds, stars, the galactic disc, comets, planetary atmospheres, etc.). More detailed surveys can be found in 30-34 amongst others.

2.1 Hydrogen

Having the largest isotopic mass ratio this element is expected to show the largest fractionation effects during chemical or physical reactions. The D/H ratio of terrestrial sea water (used as reference material for laboratory isotope studies) is about 1.6×10^{-4} which is considerably enriched in deuterium relative to the best estimates for the interstellar gas (1.5×10^{-5}) or the protosolar gas (2×10^{-5}). In contrast to the interstellar gas, hydrogen-bearing interstellar molecules are in some cases dramatically enriched in deuterium with D/H ratios greater than unity (35, 36). Although cloud temperatures are low enough to permit large isotopic fractionations between neutral molecular species the timescales involved are far too long. Thus it is widely believed that the large difference in isotopic compositions between hydrogen in the interstellar gas and hydrogen in the interstellar molecules are caused by ion-molecule reaction of the type:

$$H_3^+ + HD \rightleftharpoons H_2D^+ + H_2 \ (\Delta E/k = 178 \ K)$$

and $$CH_3^+ + HD \rightleftharpoons CH_2D^+ + H_2 \ (\Delta E/k = 300 \ K)$$

The net result of such reactions is that deuterium from the relatively massive reservoir of interstellar hydrogen becomes concentrated in those species which subsequently combine to form ever more complex organic molecules.

A further reference point includes the D/H value of the solar wind, measured by release of implanted gases from lunar soils (37-39). Although these measurements were difficult, due to the extremely low abundance of the hydrogen and water in these samples, a D/H value of 3×10^{-6} is estimated which suggests that deuterium in the present day solar wind is somewhat depleted relative to the protosolar gas. This might suggest that the isotopic composition of the hydrogen expelled from the solar surface is modified in some way or that convection has cycled deuterium through the Sun's core subsequently destroying the isotope by nuclear burning.

The D/H values of planetary atmospheres are between 2×10^{-5} and 7×10^{-5} for the giant planets (Jupiter, Saturn and Uranus) which are similar albeit a little higher than the proposed value for the protosolar gas (see 32). It is thought that during solar nebula formation when temperatures had decreased to below 200 K large isotopic fractionations were induced between the protosolar hydrogen gas and any hydrogen-bearing molecules such as H_2O, CH_4, NH_3, etc., the latter becoming enriched in deuterium. If temperatures were cold enough to allow the hydrogen-bearing molecules to condense as ices then clearly substances composed of such species would have higher D/H ratios. In the case of the giant planets, which swept up relatively large amounts of protosolar hydrogen, the D/H ratios of the presently observable atmospheres reflect the protosolar gas diluted with small amounts of ices (i.e. making the D/H ratios slightly higher than that of the protosolar gas). On the other hand, a planet such as Earth would not have captured so much hydrogen but instead may have condensed a relatively higher proportion of ices. This could explain why the D/H ratio of Earth is somewhat higher than the value of 2×10^{-5}.

If a genetic relationship exists between the atmospheres of the giant planets and comets then isotopic determinations of the cometary ices becomes important and desirable. An upper limit for the D/H ratio in comets of 1×10^2 has been determined through measurements of OH (40). High precision isotope measurements of hydrogen in putative cometary dust particles have also been determined. Using an ion probe technique, Zinner et al. (41) have found that dust collected in the stratosphere, and identified as extraterrestrial, has a D/H value of 3.4×10^{-4} (i.e. enriched in deuterium relative to terrestrial ocean water). Thus it seems that the solid parts of comets may be formed in the solar nebula from condensed ices rather than the protosolar gas. Alternatively of course, cometary dust may contain interstellar molecule precursors.

The only other extraterrestrial hydrogen isotope measurements available, other than from meteorites, come from the atmosphere of Venus where the D/H value is estimated to be between 1 and 1.6×10^{-2} (42, 43). This is thought to be a secondary feature arising from isotopic fractionation during hydrogen escape out of the atmosphere.

2.2 Carbon

At temperatures of about 10^6 K and densities of 10^2 g cm^{-3} ^{4}He nuclei are formed in stars undergoing hydrogen burning (44). Carbon is subsequently produced at temperatures of 10^8 K and densities of 10^5 g cm^{-3} when three helium nuclei fuse together to form a stable ^{12}C nucleus. This primary seed undergoes further reactions with protons to form other species in the CNO cycles as follows:

$$^{12}C \ (p, \gamma) \ ^{13}N \ (e^+\nu) \ ^{13}C \ (p, \gamma) \ ^{14}N \ (p, \gamma) \ ^{15}O \ (e^+\nu) \ ^{15}N \ (p, \alpha) \ ^{12}C$$

$$^{15}N \ (p, \gamma) \ ^{16}O \ (p, \gamma) \ ^{17}F \ (e^+ \nu) \ ^{17}O \ (p, \alpha) \ ^{14}N$$

$$^{17}O \ (p, \gamma) \ ^{18}F \ (e^+ \nu) \ ^{18}O \ (P, \alpha) \ ^{15}N$$

and also $^{13}N \ (p, \gamma) \ ^{14}O \ (e^+\nu) \ ^{14}N$ in the "hot" CNO cycle (45).

Due to the difference in relative reaction rates of these competing processes most of the material involved in the CNO cycle under equilibrium conditions becomes transformed into ^{14}N; the $^{12}C/^{13}C$ ratio stabilises at a value of about 3/1. Modifications to the CNO cycle occur in stars at higher temperatures where CNO helium burning takes place. In this case the following reactions occur:

$$^{12}C \ (\alpha, \gamma) \ ^{16}O$$

$$^{14}N \ (\alpha, \gamma) \ ^{18}F \ (e^-\nu) \ ^{18}O$$

$$^{16}O \ (\alpha, \gamma) \ ^{20}Ne$$

A number of other reactions liberate a slow and steady flux of neutrons as the star evolves to the red giant phase. The most important of these reactions are:

$$^{13}C \ (\alpha, n) \ ^{16}O$$

$$^{22}Ne \ (\alpha, n) \ ^{25}Mg$$

At this stage in the development of the star elements heavier than iron are formed by s-process neutron captures (44).

The isotopic compositions and element distributions of the light elements become modified in explosive nucleosynthetic events. Red giant stars can experience partial runaways in the helium burning zone leading to a mixing with the hydrogen burning shell ("helium flash"). In this environment significant amounts of ^{13}C may be formed as well as the s-process .elements

(46). ^{13}C is also formed during nova outbursts (47), although
during explosive supernovae events ^{12}C may become concentrated
especially in zones where helium is the most abundant component
(44).

Observational measurements of red giant stars show $^{12}C/^{13}C$
ratios in the range 5/1 to 50/1 (48) which agrees well with the
calculated values of 20/1 to 30/1 (49).   The most extreme
examples of isotopic composition occur in the cool carbon stars
where $^{12}C/^{13}C$ values of between 2/1 and 20/1 have been detected
(50). $^{12}C/^{13}C$ ratios from molecules such as H_2CO and CO in the
galaxy range from 20/1 to 143/1 (30) with the most reasonable
estimate for the galactic plane being 67 $\pm$ 10/1 (31).   Henkel et
al. (51) more recently have applied additional corrections to
observational data and now propose a $^{12}C/^{13}C$ ratio of 80 $\pm$ 7/1
(based on H_2CO data) or 93 $\pm$ 8/1 (based on CO data) for the
solar neighborhood. Although there does not appear to be a
radial isotopic gradient in the galaxy the center is definitely
enriched in ^{13}C with a $^{12}C/^{13}C$ ratio of somewhere between 20/1
and 30/1 which is considerably lower than the terrestrial value
of 89/1.

A potential complication to interpretation of data from
interstellar molecules is that like hydrogen, carbon too may
become isotopically fractionated by ion-molecule reactions.
Watson (52) has suggested that this may occur in the reaction
involving CO:

$$^{13}C^+ + {}^{12}CO \rightleftharpoons {}^{12}C^+ + {}^{13}CO \quad \Delta E/k = 35K$$

In other words, the CO molecule becomes enriched in ^{13}C.
Because the vapor pressure of CO is higher than any other
carbon-bearing species, molecules in dense clouds that freeze
onto dust grains will be preferentially enriched in ^{12}C. Thus
the grains become enriched in ^{13}C while the gas phase
concentrates ^{13}C.   As only the interstellar molecules are
determined for their isotopic composition by observational
techniques, the $^{12}C/^{13}C$ value for the cloud as a whole may be
underestimated. However such effects are only thought to be
important at the outer edges of clouds and so determinations of
$^{12}C/^{13}C$ ratios made from cloud cores should be representative.
The important implication for the origin of the solar system is
that the pre-solar cloud may not be isotopically homogeneous
with respect to the carbon isotopic composition.

The carbon isotopic composition of the solar wind as
measured from lunar soils appears to be very similar to the
terrestrial value with just a slight enrichment of ^{13}C ($^{12}C/^{13}C$
= 87/1). On the other hand direct measurements of the sun have
yielded a $^{12}C/^{13}C$ of 84/1 (53) although the errors on this

measurement allows it to be similar to the present day solar wind. Measurement of other solar system materials such as planetary atmospheres yield a spread in $^{12}C/^{13}C$ of 55/1 to 160/1 but generally the data show an enrichment of ^{12}C relative to the solar neighborhood value of 80/1 (54). $^{12}C/^{13}C$ measurements of comets also show a large range but have an average value of about 100/1 (55). Thus in keeping with the model of galactic evolution proposed by Vigroux et al. (56), the present day solar neighborhood $^{12}C/^{13}C$ value is between 0.75 and 0.5 times the value for the putative primitive solar material preserved in comets and giant planetary atmospheres such as those on Jupiter.

$^{12}C/^{13}C$ ratios have also been measured from cosmic rays and yield a value of somewhere between 12/1 and 14/1 (57, 58). These data need to be corrected for the effects of propagation through interstellar space and the solar system (nuclear fragmentation, ionization energy loss, etc.), and so the galactic cosmic ray source regions appear to have $^{12}C/^{13}C$ ratios of about 50/1.

2.3 Nitrogen

The two stable isotopes of nitrogen are secondary products of nucleosynthesis; ^{14}N is formed during CNO hydrogen burning (the main reactions were summarized in the previous section) while the much less abundant ^{15}N is thought to be produced by explosive hydrogen burning in novae or supernovae. The $^{14}N/^{15}N$ ratio attained at equilibrium in the hydrostatic CNO cycle is 25,000/1. A value of about 10 times less than this is observed in IRC + 10216 (a highly evolved intermediate mass carbon star) and shows that for this object, either the products of CNO burning have become diluted, or there has been incomplete processing (33). Unfortunately, there are no other data from stars for comparison purposes.

Nitrogen from interstellar molecules (HCN and NH_3) have $^{14}N/^{15}N$ values between 250/1 and 635/1 (30, 59). In the galaxy there appears to be an isotopic gradient with $^{14}N/^{15}N$ at the centre being 500/1 and at the outskirts 300/1 (the terrestrial atmosphere has a value of 272/1. Wannier (30) claims that accompanying the decrease in $^{12}C/^{13}C$ ratio, which occurs as the galaxy evolves, there is an increase in the $^{14}N/^{15}N$ ratio.

Again the isotopic composition of nitrogen in interstellar molecules may be affected by ion-molecule reactions. The species $^{14}N_2H^+$ has been detected but as yet ^{14}N ^{15}N H^+ has not. Interpretation of the isotopic ratio of this compound might be hampered because of the reaction;

$$^{14}N_2H^+ + {}^{14}N\ {}^{15}N \rightleftharpoons {}^{14}N\ {}^{15}N\ H^+ + {}^{14}N_2 \quad \Delta E/k = 10K$$

As the ion-chemical pathways from N_2 to HCN are not well established it can not yet be assessed how the nitrogen isotope ratio observed in interstellar molecules is related to the isotopic composition of N_2 (60).

It is widely held that the rare isotope ^{15}N is formed during explosive nucleosynthesis. Indeed novae are believed to be the primary producers of ^{15}N. However, Clayton and Hoyle (61) considering the formation of grains in nova ejecta show that in the post-nova star large ^{14}C concentrations can be built up by the following reactions:

$$^{13}C (n,\gamma) {}^{14}C \text{ (highly } {}^{13}C\text{-rich environments)}$$

$$^{14}N (n,p) {}^{14}C \text{ (less } {}^{13}C\text{-rich environments)}$$

Thus combined enrichments of ^{13}C and ^{14}C can result and so condensing grains may be rich in ^{13}C and ^{14}C. The ^{14}C subsequently decays to ^{14}N with the possibility that these grains will eventually contain pure ^{14}N.

Nitrogen released from various components of lunar soils has quite intriguing isotopic compositions. One component, characterized by an enrichment of ^{15}N, is thought to arise from the spallation of oxygen in silicate minerals by solar and galactic cosmic rays. Of the other components it appears that one component is enriched in ^{15}N ($^{14}N/^{15}N = 230$) and the other is enriched in ^{14}N ($^{14}N/^{15}N = 350$). Due to the nature of the release of gases upon heating it has been postulated that the nitrogen isotope variation records a secular change of the $^{14}N/^{15}N$ ratio of the solar wind (62). A number of other interpretations have been offered however; for instance Geiss and Bochsler (34) consider that one of the components is primordial (^{14}N-enriched) while the other represents the present day solar wind (with a $^{14}N/^{15}N$ value of $250 \pm 20/1$).

Other examples of solar system materials include the planetary atmospheres. The $^{14}N/^{15}N$ value of Venus is similar to the terrestrial value, however, both Mars and Jupiter show enrichments in ^{15}N ($^{14}N/^{15}N = 160$). In the case of Mars it is proposed that this is due to loss of atomic ^{14}N with time from the atmosphere (63). Measurements of SNC meteorites which on the basis of a number of characteristics (e.g. young crystallization ages) are thought to be martian ejecta, show a trapped atmospheric gas component which is relatively enriched in ^{15}N and another component which has a $^{14}N/^{15}N$ ratio similar to the terrestrial value (64). If this latter component represents unfractionated primodial nitrogen then it seems that Mars has an isotopic composition similar to the Earth. The data for Jupiter are open to question and Tokunaga et al. (65) suggest that the $^{14}N/^{15}N$ ratio could in reality be similar to the terrestrial value.

2.4 Oxygen

Oxygen, unlike hydrogen, nitrogen and carbon, has three stable isotopes ^{16}O, ^{17}O and ^{18}O. The most abundant isotope ^{16}O can be regarded as a primary product of nucleosynthesis being formed by helium burning in the envelopes of massive stars. The secondary isotope ^{17}O is formed from hydrogen burning during the CNO cycles (especially in explosive events such as nova outbursts) while ^{18}O is produced by helium burning on ^{14}N (44).

On the basis of observational measurements of the CO and OH molecules it appears that the $^{18}O/^{17}O$ ratio of the galaxy is constant at about 3.2. The lack of source-to-source variation seems to rule out significant isotopic fractionation in the interstellar environment. The terrestrial $^{18}O/^{17}O$ ratio is about 5.5 while observational measurements of IRC + 10216 (the late-type star showing high ^{14}N enrichment) yield a value of <1. This has prompted Wannier (30) to propose an evolutionary decrease in the $^{18}O/^{17}O$ ratio of the galaxy, with the present day terrestrial (and thus solar) value of 5.5 reflecting conditions in the galaxy 5×10^9 years ago.

Apparently it is more difficult to measure the $^{16}O/^{18}O$ or $^{16}O/^{17}O$ ratios of interstellar molecules and thus there are more uncertainties about these data. However, it seems that the galactic plane has a $^{16}O/^{18}O$ ratio similar to the terrstrial value of 500 while the center is somewhat enriched in ^{18}O ($^{16}O/^{18}O$ = 300). These data contrast markedly with the $^{16}O/^{18}O$ value of 27,000 in IRC + 10216 (66); clearly nucleosynthetic processes have enriched the ^{16}O (and ^{17}O) with respect to ^{18}O in this carbon star environment. An outstanding problem with the galactic data is to explain why the $^{18}O/^{17}O$ ratio remains constant while the $^{16}O/^{18}O$ ratio exhibits a radial variation.

The oxygen isotopic composition of the martian atmosphere has been determined as $^{16}O/^{18}O$ = 499 (67) and is thus indistinguishable from the terrestrial value. Unfortunately, the oxygen isotopic compositions of other solar system materials (planetary atmospheres, comets, etc.) remain unknown and thus complete models of evolutionary processes cannot be devised at present.

2.5 Sulfur

At temperatures in excess of 10^9 K hydrostatic oxygen burning takes place in stellar envelopes and ^{32}S is formed by fusion of two ^{16}O nuclei (44):

$$^{16}O + ^{16}O \rightarrow ^{32}S + \gamma$$

The principal product of explosive oxygen burning is ^{28}Si:

$$^{16}O + ^{16}O \rightarrow ^{28}Si + ^{4}He$$

This species in turn may be involved in the production of ^{32}S as follows (silicon quasi-equilibrium):

$$\gamma + ^{28}Si \rightleftharpoons ^{24}Mg + ^{4}He$$

$$^{28}Si + ^{4}He \rightleftharpoons ^{32}S + \gamma$$

It is presumed that somewhere in this environment ^{33}S and ^{34}S are also formed although neither the exact mechanisms nor detailed calculations have yet been formulated. ^{36}S is probably formed in rapid neutron capture events (^{35}S is radioactive and decays to ^{35}Cl). ^{33}S and ^{34}S may also be formed by neutron capture processes.

Observational measurements of $^{32}S/^{34}S$ and $^{34}S/^{33}S$ in interstellar CS yield ratios which are similar to the terrestrial values of 22 and 5.5 respectively (30). The $^{32}S/^{34}S$ ratio appears to be constant everywhere in the galaxy, but with so few data and no appropriate theoretical models available it is difficult at present to understand the relationships of sulfur and its isotopes.

Sulfur in lunar rocks has been found to have an almost identical $^{32}S/^{34}S$ value to the terrestrial ratio while lunar soils are slightly enriched in ^{34}S (68-70). The increased ^{34}S content of the soils is thought to arise from processes operating at the lunar surface such as impact vaporization or solar wind sputtering. It is not sure what proportion of the sulfur in lunar samples is indigenous, meteoritic or of solar wind origin, but as the solar abundance of sulfur is so low, it is considered that the solar wind contribution is probably negligible. Thus no conclusive information concerning the solar sulfur isotope ratios has yet been derived from lunar soils.

3. LIGHT ELEMENT STABLE ISOTOPE RATIOS OF EXTRATERRESTRIAL MATERIALS MEASURED IN THE LABORATORY

The comparatively high levels of precision that can be obtained by laboratory analyses of light element stable isotope ratios arise because relative isotope enrichments, or depletions, are measured rather than the absolute isotopic compositions. The analytical technique involves comparison of the measured isotope ratio of the sample with that of a reference material thereby eliminating any experimental artefacts. The isotopic composition of the reference is known relative to that of an internationally accepted standard and thus results are expressed as deviations from the standard material according to the following:

$$\delta I = \left[\frac{R(SAMPLE) - R(STANDARD)}{R(STANDARD)} \right] \times 1000 \; (\permil)$$

Where δI (the 'delta' value) is the difference in isotopic composition in parts per mil, R(SAMPLE) is the measured ratio of the rare isotope to the most abundant isotope in the sample material and similarly R(STANDARD) is the ratio measured from the internationally accepted standard material. In the case of hydrogen isotope studies δD records the difference in isotopic composition (in parts per mil) of the sample with that of the international standard SMOW (which has a D/H value of 1.5576×10^{-4}, ref.71). The standards used for carbon, nitrogen, oxygen and sulfur are respectively PDB ($^{13}C/^{12}C = 0.0112372$, ref.72), AIR ($^{15}N/^{14}N = 3.676 \times 10^{-3}$, ref.73), SMOW ($^{17}O/^{16}O = 3.72 \times 10^{-4}$, ref.74, $^{18}O/^{16}O = 2.0052 \times 10^{-3}$, ref.75) and CDT ($^{33}S/^{32}S = 8.0998 \times 10^{-3}$, $^{34}S/^{32}S = 0.045$, $^{36}S/^{32}S = 1.79 \times 10^{-4}$, ref.76). Correspondingly, $\delta^{13}C$, $\delta^{15}N$, $\delta^{17}O$, $\delta^{18}O$, $\delta^{33}S$, $\delta^{34}S$ and $\delta^{36}S$ values are derived.

In order to establish a consistent theory for the origin of the low temperature condensates in meteorites it is necessary to account for their stable isotopic compositions. This in turn involves understanding the relationship of the isotope ratios observed in the various forms that the light elements assume.Some of the isotopic variation will result from the physical and chemical processes which occurred during the nebula phase but evidence of isotopic heterogeneities and preservation of·pre-solar components is always sought.

Historically, the study of light element isotopic compositions in meteorites yielded some anomalous results at the very first attempts, although often little importance was attached to the data at the time. A goal of the early research was to try and establish, for instance, that components such as organic matter in meteorites were indigenous and not the result of contamination. Thus the deuterium enrichments observed by Boato (77) and Briggs (78) in organic matter from carbonaceous chondrites were taken merely as indicators of an extraterrestrial origin; no explanations to account for the enrichment were offered (it must be remembered here that deuterium enrichments in interstellar molecules were not discovered until 1973). In 1963 Clayton (79) measured the carbon isotopic composition of meteoritic carbonates and found

them to be extremely enriched in ^{13}C, compared to the coexisting
indigenous organic material. Although he postulated a possible
nucleosynthetic origin for the carbon in the carbonate (with all
the implications of isotopic heterogeneity in the solar nebula)
Lancet and Anders (10) preferred to explain the isotopic
distribution, between carbonate and organic matter, as resulting
from carbon isotopic fractionation in the solar nebula (i.e.
during a Fischer-Tropsch-type reaction). However, Bunch and
Chang (80) propose that the carbonate minerals in carbonaceous
chondrites were formed by secondary aqueous alteration in a
regolith-type environment on a pre-accreted parent body. Closer
examination has revealed that carbonates have variable chemical
compositions (Ca, Fe, Mg and Mn) and are present not only in the
matrix, but also inside chondrules (81). Clearly therefore, the
term "carbonate" incorporates a number of components which
undoubtedly have different origins.

The 1960's and 1970's saw several isotopic surveys
undertaken in order to set limits on isotopic compositions and
attempt to produce meteorite classification schemes. The main
conclusion from carbon isotope work was that for bulk meteorites
the $\delta^{13}C$ values were similar to those from the terrestrial
environment (82-84). Acid-insoluble organic compounds of
carbonaceous chondrites (which constitute up to 95% of the
carbon in some meteorites) were found to be slightly enriched in
^{13}C relative to analagous terrestrial material ('kerogen').
However, the isotopic composition of the bulk earth (thought to
be somewhere between -5 and -25‰) remains controversial and so
comparisons with meteorite data are somewhat ambiguous.
Furthermore, terrestrial kerogens are composed of organic
materials which have of course undergone biological processes.
Ordinary chondrites were found to have $\delta^{13}C$ values between -5
and -30‰ (82,85) while graphite nodules from iron meteorites
are about -5‰ (86,87). Nitrogen isotope data from meteorites
show more variability (88) and have been used to devise a loose
classification scheme (89). Bulk $\delta^{15}N$ values in meteorites
range from about -40 to +170‰ which is in marked contrast to
the variability observed on earth ($\pm$ 20‰). The oxygen isotopic
composition of meteoritic organic matter has never been
determined, but data ($\delta^{17}O$, $\delta^{18}O$) from bulk meteorites (which is
mostly silicate oxygen) have been used to establish quite a
successful classification system (90). Furthermore, the oxygen
isotopic compositions of minerals separated from various
meteorites have been used to determine their temperaturs of
condensation (assuming a cooling solar gas model). On this
basis, the hydrous silicate matrix minerals and carbonates of
carbonaceous chondrites are thought to have condensed below 400
K (91). The sulfur isotopic composition of meteorites has been
studied by Hulston and Thode (92) and Kaplan and Hulston (93)
and shows very little variability between different groups.

More detailed studies of stable hydrogen, carbon and nitrogen isotopic compositions were begun in the late 1970's and early 1980's when a number of groups started to analyse various meteorite extracts (solvent-soluble fractions, acid-resistant fractions, etc.) using stepped heating extractions. These experiments not only allow terrestrial contaminants to be burned off at low temperatures, but also cause different light element-containing components to degrade over diagnostic temperature ranges.

Robert et al. (15) detected a component in the unequilibrated ordinary chondrite Chainpur which had a δD value of about + 4000‰ . Deuterium enrichments in a number of unequilibrated ordinary chondrites have since been uncovered by McNaughton et al., (16, 94. Fallick et al., ref.95, discuss in detail the implications of these results). Most other workers have tended to concentrate on the carbonaceous chondrites and large δD values have now been discovered in several meteorites (17, 96-99). Robert and Epstein (17) and Yang and Epstein (99), have analysed in detail a number of meteoritic fractions and conclude that the water in hydrated silicates (which have δD values very similar to terrestrial values) acquired its isotopic composition as a result of hydrogen/water equilibration at 200 K in the solar nebula, while some of the organic fraction represents interstellar material which survived nebular processes. Further work by Becker and Epstein (100) demonstrates that solvent-soluble organic materials in carbonaceous chondrites were probably introduced into the meteorite during the same event which formed the hydrated silicates and deposited carbonates.

The first detailed study of carbon isotopic variation in meteoritic extracts was undertaken by Chang et al. (101), who found that the amino acid extracts from the Murchison meteorite had approximately the same δ^{13}C value (+ 23 to 44‰) as the carbonate fraction. Other solvent soluble fractions were also found to be enriched in ^{13}C, a result more recently confirmed by Becker and Epstein (100). These data are in marked contrast to the isotopic composition of the coexisting acid-insoluble organic matter (with δ^{13}C values around -16‰). Advances in the study of low and high molecular weight organic material (i.e. that which can be released as gases or that which is soluble in solvents, water, etc.) are forthcoming with the analysis of individual compounds using gas chromatographic techniques (102, 103).

The study of carbon in macromolecular organic material in meteorites (which is acid-insoluble) is hampered to some extent because other forms of carbon (amorphous carbon, graphite, carbon in solid solution, etc.) are often isolated along with the organics during the extraction procedure (i.e. acid-dissolution). The use of elementary stepped heating analyses have been investigated but early studies tended to involve fairly large temperature increments and so only limited information concerning the organic fraction was retrieved (17, 99). Swart et al. (104) demonstrated that elemental forms of carbon oxidize at higher temperatures than organic material and so the use of stepped combustion appears to be much more suitable. High resolution stepped combustion studies have subsequently been attempted by Swart et al. (105) and Kerridge (98), the latter study being dedicated to the analysis of the acid insoluble organic material. Pushing the techniques even further, Kerridge and Chang (106) have measured the δD, $\delta^{13}C$ and $\delta^{15}N$ values in the organic fraction of Murchison by stepped combustion using 29 temperature increments. This study has demonstrated that different parts of the insoluble organic fraction (characterized by their different temperatures of oxidation) show variations in isotopic compositions. This apparently demonstrates that the various molecular and structural fractions of the macromolecular material have isotopic compositions which record events that occurred in the solar nebula. Further experiments of this kind will undoubtedly improve our understanding of the formational processes.

Detailed studies of the nitrogen isotopic composition of the organic fractions of meteorites are also underway (17, 98-100). At present it is known that the nitrogen in the solvent-soluble organics is generally enriched in ^{15}N relative to that in the insoluble organic material within the same meteorite. However, the most striking feature of the nitrogen isotope data for bulk organic material is that the variations observed between meteorites are extremely large (+30 to +170‰). Nitrogen from inorganic fractions of enstatite chondrites have $\delta^{15}N$ values of between -30 and -40‰ (89), although even lighter isotopic compositions have been identified (187). Furthermore, $\delta^{15}N$ values of between -50 and -90‰ are recorded for certain iron meteorites (108). Thus it appears that a difference in nitrogen isotopic composition of at least 60-80‰ exists between meteoritic organic and inorganic fractions. To date, nitrogen isotope fractionations of only 3‰ (for Fischer-Tropsch-type reactions) or about 11‰ (Miller-Urey rections) have been obtained during laboratory simulation experiments (109).

The oxygen isotopic composition of meteoritic organic material remains unknown and intriguing. However, oxygen isotope analyses of high temperature silicate fractions have been performed in abundance (e.g. 11, 110, 111) and indeed provided the first indisputable evidence that isotopic heterogeneities survived the solar nebula. Low temperature silicate minerals have also been studied (91) and in Murchison show evidence for isotopic fractionation during aqueous alteration (112). It is possible that the oxygen isotopic composition of calcite from Murchison also records evidence of isotopic fractionation during aqueous alteration although a detailed study by Swart et al. (113) has demonstrated the possibility of oxygen isotope anomalies in meteoritic carbonates. As only a few data have so far been acquired, coupled with the lack of isotope data from other low temperature oxygen-bearing components, it is not yet possible to fully understand the formation mechanisms.

Isotope studies of various sulfur-containing entities in the Orgueil meteorite were undertaken in 1965 by Monster et al. (114). They found that the sulfur in the sulfate mineral $MgSO_4.7H_2O$ was slightly depleted in ^{34}S ($\delta^{34}S$ =-1.3‰) relative to elemental and troilite sulfur ($\delta^{34}S$ = +1.5‰ and +2.6‰ respectively). Small ^{34}S enrichments have been detected in the solvent-soluble organic compounds of a number of carbonaceous chondrites (78). The small magnitude of isotope effects in meteoritic sulfur-bearing compounds has been confirmed by Hulston and Thode (92) although more recently Rees and Thode (115) have detected a sulfur isotope anomaly (a relative ^{33}S enrichment) in the Allende meteorite. It is perhaps the lack of large isotope variations in sulfur which has deterred more detailed and extensive studies but having four stable isotopes it has the potential for providing information concerning mass-dependent isotopic fractionation (i.e. resulting from physical and chemical reactions) as well as nuclear processes; the relatively low "natural" abundance of ^{36}S on the other hand should allow spallogenically produced sulfur to be detected.

3.1 Exotic components

Perhaps the most dramatic stable isotope effects in meteoritic materials have been observed in those acid-resistant fractions which have previously been shown to contain isotopically anomalous noble gases. It has long been thought that the noble gases record nucleosynthetic events which occurred some time before the solar nebula formed; the preservation of these components strongly argues the case for incomplete homogenization in the solar nebula. Swart et al. (13) analysed a suite of residues from the Allende and Murchison meteorites and found that a small percentage of the total carbon was extremely enriched in ^{13}C ($\delta^{13}C$ up to +1100‰, i.e. $^{12}C/^{13}C$ = 42/1). Similar enrichments have since been detected by Yang and Epstein (99) and Carr et al. (116). The major isotopically heavy carbon component is thought to be the host phase of xenon which shows an s-process isotopic signature and is thought to

have been formed in a highly evolved red giant star (13). A
second isotopically heavy component is possibly related to yet
another noble gas component called neon-E which is essentially
pure ^{22}Ne and though to be the decay product of ^{22}Na (116). A
further isotopically anomalous xenon component characterized by
an enrichment of the light and heavy isotopes (and referred to
in the literature as CCF-Xe, Xe-X, DME-Xe or HL-Xe) is found to
be associated with carbon whose isotopic composition is very
similar to the terrestrial value. However, the accompanying
nitrogen is enriched in ^{14}N ($\varsigma^{15}N$ = -326‰ , i.e. $^{14}N/^{15}N$ =
400/1) and it is now believed that this material probably formed
in a supernova (14).

The reason it has proved possible to analyse these
isotopically anomalous materials is because they are extremely
refractory and are resilient to the acids used to concentrate
them. Naturally, it is desirable to provide adequate
explanation for the existence of the material, but perhaps the
importance of the data is that they provide an insight into the
nature and mixture of materials which coalesced in the solar
nebula. Furthermore it is important to remember that an unknown
number of light element-bearing components are lost during the
acid-dissolution process; each of these has the potential to
provide information on any number of nucleosynthetic processes
or chemical/physical events.

4. CONCLUSIONS

Obvious isotopic anomalies in meteorites give us an insight
into some of the more bizarre components which were originally
input to the solar nebula. Other isotopically distinct
materials although possibly pre-solar remnants are perhaps more
easily explained as resulting from processes operating at the
time of solar system formation. The task of research connected
with the study of the low temperature condensates in meteorites
is to place constraints on the conditions which were prevalent
during the solar nebula phase. The rewards lie in the
understanding of the solar nebula formation and evolution as
well as the relationships between the Sun, meteorites,
asteroids, planets and comets.

Progress will be made when more observational isotope ratio
data become available for the giant planets and their
satellites, as well as comets. The smaller the error limits on
these measurements, the better the chance of being able to
relate the results with those from meteorites. The samples of
meteorites available for study are undoubtedly the remains of a
variety of processes which occurred before, or in, the solar
nebula. A number of meteoriticists have tried to account for

chemical and physical variations within the more refractory materials (i.e. the inner planets, the Earth, meteorites, etc.). In parallel, studies concerning interstellar molecules and grains, solar system ices, the outer planets and satellites, planetary atmospheres and the formation of comets have also been in progress. It is entirely possible that certain meteorites (in particular, the primitive carbonaceous chondrites) can shed light on the formational environment of comets or the outer planets. The study of the recently acquired interplanetary dust particles will help to bridge further the gaps between the disciplines of meteoritics and solar system ice studies and this in turn will ultimately allow better understanding of the origin of the solar system.

ACKNOWLEDGEMENTS

I gratefully acknowledge the receipt of funds from N.A.T.O to allow attendance of the "Ices in the Solar System" conference. Financial support to the Planetary Sciences Unit from The Science and Engineering Research Council is also acknowledged. I would like to thank M.M. Grady, P.K. Swart, A.E. Fallick and C.T. Pillinger for coaching and interaction through the tribulations of becoming a meteoriticist.

REFERENCES

1.   Lee,T. (1979). New isotopic clues to solar system formation. Rev. Geophys. Space Phys., 17, pp. 1591-1611.

2.   Cameron, A.G.W. and Truran, J.W. (1977). The supernova trigger for formation of the solar system. Icarus, 30, pp. 447-461.

3.   Clayton, D.D. (1982). Cosmic chemical memory: A new astronomy. Q. J. R. Astr. Soc., 23, pp. 174-212.

4.   Chen, J.H. and Wasserburg, G.J. (1983). The isotopic composition of silver and lead in two iron meteorites: Cape York and Grant. Geochim. Cosmochim. Acta., 47, pp. 1725-1737.

5.   Cameron, A.G.W. (1962). The formation of the Sun and Planets. Icarus, 1, pp. 13-69.

6.   Lord, H.C. (1965). Molecular equilibria and condensation in a solar nebula and cool stellar atmospheres. Icarus, 4, pp. 279-288.

7.   Grossman, L. (1972). Condensation in the primitive solar nebula. Geochim. Cosmochim. Acta., 36, pp. 597-619.

8.   Grossman, L. (1980). Refractory inclusions in the Allende meteorite. Ann. Rev. Earth Planet. Sci., 8, pp. 559-608.

9.   Hayatsu, R. and Anders, E. (1981). Organic compounds in meteorites and their origins. In: Topics in current chemistry, Cosmo- and Geochemistry, 99, pp. 1-37.

10. Lancet, M.S. and Anders, E. (1970). Carbon isotope fractionation in the Fischer-Tropsch synthesis and meteorites. Science, 170, pp. 980-982.

11. Clayton, R.N., Grossman, L. and Mayeda, T.K. (1973). A component of primitive nuclear composition in carbonaceous chondrites. Science, 182, pp. 485-488.

12. Anders, E. (1981). Noble gases in meteorites: Evidence for presolar matter and superheavy elements. Proc. R. Soc. Lond. A., 374, pp. 207-238.

13. Swart, P.K., Grady, M.M., Pillinger, C.T., Lewis, R.S. and Anders, E. (1983). Interstellar carbon in meteorites. Science, 220, pp. 406-410.

14. Lewis, R.S., Anders, E., Wright, I.P., Norris, S.J. and

Pillinger, C.T. (1983). Isotopically anomalous nitrogen in primitive meteorites. Nature, 305, pp. 767-771.

15.   Robert, F., Merlivat, L. and Javoy, M. (1979). Deuterium concentrations in the early solar system: Hydrogen and oxygen isotope study. Nature, 282, pp. 785-789.

16.   McNaughton, N.J., Borthwick, J., Fallick, A.E. and Pillinger, C.T. (1981). Deuterium/hydrogen ratios in unequilibrated ordinary chondrites. Nature, 294, pp. 639-641.

17.   Robert, F. and Epstein, S. (1982). The concentration and isotopic composition of hydrogen, carbon and nitrogen in carbonaceous chondrites. Geochim. Cosmochim. Acta., 46, pp. 81-95.

18.   Alfvén, H. and Arrhenius, G. (1970). Structure and evolutionary history of the solar system, I. Astrophys. Space Sci., 8, pp. 338-421.

19.   Arrhenius, G. (1978). Chemical aspects of the formation of the solar system. In: The origin of the solar system (Ed. S.F. Dermott). pp. 521-581.

20.   Thiemens, M.H. and Heidenreich, J.E. (1983). The mass-independent fractionation of oxygen: A novel isotope effect and its possible cosmochemical implications. Science, 219, pp. 1073-1075.

21.   Arrhenius, G., Fitzgerald, R., Markus, S. and Simpson, C. (1978). Isotope fractionation under simulated space conditions. Astrophys. Space Sci., 55, pp. 285-297.

22.   Miller, S.L., Urey, H.C. and Oró, J. (1976). Origin of organic compounds on the primitive Earth and in meteorites. J. Mol. Evol., 9, pp. 59-72.

23.   Pillinger, C.T. (1984). Light element stable isotopes in meteorites - from grams to picograms. Geochim. Cosmochim. Acta. (in press).

24.   Norris, S.J., Swart, P.K., Wright, I.P., Grady, M.M. and Pillinger, C.T. (1983). A search for correlatable, isotopically light carbon and nitrogen components in lunar soils and breccias. Proc. 14. Lunar Planet. Sci. Conf., Part 1, pp. B200-B210.

25.   Mattey, D.P., Carr, R.H., Wright, I.P. and Pillinger, C.T. (1984). Carbon isotopes in submarine basalts. Earth Planet. Sci. Lett. (in press).

26. Brown, P.W. and Pillinger, C.T. (1981). Nitrogen concentrations and isotopic ratios from separated lunar soils. Meteoritics, 16, pp. 298.

27. Frick, U. and Pepin, R.O. (1981). Microanalysis of nitrogen isotope abundances: Association of nitrogen with noble gas carriers in Allende. Earth Planet. Sci. Lett., 56, pp. 64-81.

28. Reynolds, J.H. (1956). High sensitivity mass spectrometer for noble gas analysis. Rev. Sci. Instrum., 27, pp. 928-934.

29. Wright, I.P. and Pillinger, C.T. (1984). Measurements of $\delta^{13}C$, $\delta^{15}N$ and δD at the nanogram level in extraterrestrial material using static mass spectrometry. Abstract, 27th International Geological Congress, 5, pp. 439-440.

30. Wannier, P.G. (1980). Nuclear abundances and evolution of the interstellar medium. Ann. Rev. Astron. Astrophys., 18, pp. 399-437.

31. Penzias, A.A. (1980). Nuclear processing and isotopes in the galaxy. Science, 208, pp. 663-669.

32. Geiss, J. and Reeves, H. (1981). Deuterium in the solar system. Astron. Astrophys., 93, pp. 189-199.

33. Wannier, P.G., Linke, R.A. and Penzias, A.A. (1981). Observations of $^{14}N/^{15}N$ in the galactic disk. Astrophys. J., 247, pp. 552-529.

34. Geiss, J. and Bochsler, P. (1982). Nitrogen isotopes in the solar system. Geochim. Cosmochim, Acta., 46, pp. 529-548.

35. Snell, R.L. and Wooten, H.A. (1979). Observations of interestellar HNC, DNC, and $HN^{13}C$: Temperature effects on deuterium fractionation. Astrophys. J. 228, pp. 748-754.

36. Brown, R.D. and Rice, E. (1981). Interstellar deuterium chemistry. Phil. Trans. R. Soc. Lond. A., 303, pp. 523-533.

37. Epstein, S. and Taylor, H.P. (1970). The concentration and isotopic composition of hydrogen, carbon and silicon in Apollo 11 lunar rocks and minerals. Proc. Apollo 11 Lunar Sci. Conf., 2, pp. 1085-1096.

38. Epstein, S. and Taylor, H.P. (1971). O^{18}/O^{16}, $Si\ 30/Si^{28}$, D/H and C^{13}/C^{12} ratios in lunar samples. Proc. 2nd Lunar Sci. Conf., 2, pp. 1421-1441.

39. Epstein, S. and Taylor, H.P. (1972). O^{18}/O^{16}, Si^{30}/Si^{28}, C^{13}/C^{12} and D/H studies of Apollo 14 and 15 samples. Proc. 3rd Lunar Sci. Conf., 2, pp. 1429-1454.

40. A'Hearn, M.F., Schleicher, D.G., Donn, B. and Jackson, W.M. (1980). Fluorescence equilibrium in the ultraviolet spectra of comets Seargent (1978m) and Bradfield (1979l). In: The Universe at ultraviolet wavelengths, the first two years of International Ultraviolet Explorer (NASA conference publication 2171). pp. 73-81.

41. Zinner, E., McKeegan, K.D. and Walker, R.M. (1983). Laboratory measurements of D/H ratios in interplanetary dust. Nature, 305, pp. 119-121.

42. Donahue, T.M., Hoffman, J.H., Hodges, R.R. and Watson, A.J. (1982). Venus was wet: a measurement of the ratio of deuterium to hydrogen. Science, 216, pp. 630-633.

43. McElroy, M.B., Prather, M.J. and Rodriguez, J.M. (1982). Escape of hydrogen from Venus. Science, 215, pp. 1614-1615.

44. Audouze, J. and Vauclair, S. (1980). An introduction to nuclear astrophysics. D. Reidel, Dordrecht, Holland. 167 pp.

45. Caughlan, G.R. (1977). The CNO cycles. In: CNO isotopes in astrophysics (Ed. J. Audouze). pp. 121-131.

46. Audouze, J. and Tinsley, B.M. (1976). Chemical evolution of galaxies. Ann. Rev. Astron. Astrophys., 14, pp. 43-79.

47. Starrfield, S., Truran, J.W., Sparks, W.M. and Kutter, G.S. (1972). CNO abundances and hydrodynamic models of the nova outburst. Astrophys. J., 176, pp. 169-176.

48. Dearborn, D.S.P. (1977). CNO isotopes and red giants. In: CNO isotopes in astrophysics. (Ed. J. Audouze). pp. 39-44.

49. Dearborn, D.S.P., Eggleton, P.P. and Schramm, D.N. (1976). $^{12}C/^{13}C$ ratios in stars ascending the giant branch the first time. Astrophys. J., 203, pp. 455-462.

50. Climenhaga, J.L., Harris, B.L., Holts, J.T. and Smolinski, J. (1977). The $^{12}C/^{13}C$ ratio in 18 cool carbon stars. Astrophys. J., 215, pp. 836-844.

51. Henkel, C., Wilson, T.L. and Bieging, J. (1982). Further ($^{12}C/^{13}C$) ratios from formaldehyde: A variation with distance from the galactic center. Astron. Astrophys., 109, pp. 344-351.

52. Watson, W.D. (1977). Isotope fractionation in interstellar molecules. In: CNO isotopes in astrophysics. (Ed. J Audouze). pp. 105-114.

53. Hall, D.N.B. (1973). Detection of the ^{13}C, ^{17}O and ^{18}O isotope bands of CO in the infrared solar spectrum. Astrophys. J., 182, pp. 977-982.

54. Courtin, R., Gautier, D., Marten, A. and Kunde, V. (1983). The $^{12}C/^{13}C$ ratio in Jupiter from the Voyager infrared investigation. Icarus, 53, pp. 121-132.

55. Vanysek, V. and Rahe, J. (1978). The $^{12}C/^{13}C$ isotope ratio in comets, stars and interestellar matter. Moon and Planets, 18, pp. 441-446.

56. Vigroux, L., Audouze, J. and Lequeux, J. (1976). Isotopes of C, N and O and chemical evolution of galaxies (II) Astron. Astrophys., 52, pp. 1-9.

57. Jonsson, G. Kristiansson, K., and Malmqvist, L. (1970). The relative abundance of the carbon isotopes ^{12}C and ^{13}C in primary cosmic radiation. Astrophys. Space Sci., 7, pp. 231-251.

58. Wiedenbeck, M.E. and Greiner, D.E. (1981). High-resolution observations of the isotopic composition of carbon and silicon in the galactic cosmic rays. Astrophys. J., 247, pp. L119-L122.

59. Wilson, T.L. and Pauls, T. (1979). The detection of interstellar $^{15}NH_3$. Astron. Astrophys., 73, pp. L10-L12.

60. Adams, N.G. and Smith, D. (1981). $^{14}N/^{15}N$ isotope fractionation in the reaction $N_2H^+ + N_2$: Interstellar significance. Astrophys. J., 247, pp. L123-L125.

61. Clayton, D.D. and Hoyle, F. (1976). Grains of anomalous isotopic composition from novae. Astrophys. J., 203, pp. 490-496.

62. Kerridge, J.F. (1975). Solar nitrogen: Evidence for a secular increase in the ratio of nitrogen-15 to nitrogen-14. Science, 188, pp. 162-164.

63. McElroy, M.B., Yung, Y.L. and Nier, A.O. (1976). Isotopic composition of nitrogen: Implications for the past history of Mars' atmosphere. Science, 194, pp. 70-72.

64. Becker, R.H. and Pepin, R.O. (1983). Nitrogen isotopic compositions in EETA 79001. Meteoritics, 18, pp. 264-265.

65. Tokunaga, A.T., Knacke, R.F., Ridgway, S.T. and Wallace, L.
(1979). High-resolution spectra of Jupiter in the 744-980
inverse centimeter spectral range. Astrophys. J., 232, pp.
603-615.

66. Wannier, P.G. and Linke, R.A. (1978). Isotope abundance
anomalies in IRC +10216. Astrophys. J., 225, pp. 130-137.

67. Owen, T., Biemann, K., Rushneck, D.R., Biller, J.E.,
Howarth, D.W. and Lafleur, A.L. (1977). The composition of the
atmosphere at the surface of Mars. J. Geophys. Res., 82, pp.
4635-4639.

68. Thode, H.G. and Rees, C.E. (1971). Measurement of sulphur
concentrations and the isotope ratios $^{33}S/^{32}S$, $^{34}S/^{32}S$ and
$^{36}S/^{32}S$ in Apollo 12 samples. Earth Planet. Sci. Lett., 12, pp.
434-438.

69. Rees, C.E. and Thode, H.G. (1974). Sulphur concentrations
and isotope ratios in Apollo 16 and 17 samples. Proc. 5th Lunar
Sci. Conf., pp. 1963-1973.

70. Kerridge, J.F., Kaplan, I.R. and Petrowski, C. (1975).
Evidence for meteoritic sulfur in the lunar regolith. Proc. 6th
Lunar Sci. Conf., pp. 2151-2162.

71. Gonfiantini, R. (1978). Standards for stable isotope
measurements in natural compounds. Nature, 271, pp. 534-536.

72. Craig, H. (1957). Isotopic standards for carbon and oxygen
and correction factors for mass-spectrometric analysis of carbon
dioxide. Geochim. Cosmochim. Acta., 12, pp. 133-149.

73. Junk, G. and Svec, H.J. (1958). The absolute abundance of
the nitrogen isotopes in the atmosphere and compressed gas from
various sources. Geochim. Cosmochim. Acta., 14, pp. 234-243.

74. Craig, H. (1961). Standard for reporting concentrations of
deuterium and oxygen-18 in natural waters. Science, 133, pp.
1833-1834.

75. Baertschi, P. (1976). Absolute ^{18}O content of standard
mean ocean water. Earth Planet. Sci. Lett., 31, pp. 341-344.

76. Nielsen, H. (1978). Sulfur. In: Handbook of geochemistry.
(Ed. K.H. Wedepohl).

77. Boato, G. (1954). The isotopic composition of hydrogen and
carbon in the carbonaceous chondrites. Geochim. Cosmochim.
Acta., 6, pp. 209-220

78. Briggs, M.H. (1963). Evidence for an extraterrestrial
origin for some organic constituents of meteorites. Nature,
197, pp. 1290.

79. Clayton, R.N. (1963). Carbon isotope abundances in
meteoritic carbonates. Science, 140, pp. 192-193.

80. Bunch, T.E. and Chang, S. (1980). Carbonaceous
chondrites-II. Carbonaceous chondrite phyllosilicates and light
element geochemistry as indicators of parent body processes and
surface conditions. Geochim. Cosmochim. Acta., 44, pp.
1543-1577.

81. Fredriksson, K., Mason, B., Beauchamp, R. and Kurat, G.
(1981). Carbonates and magnetites in the Renazzo chondrite.
Meteoritics, 16, pp. 316.

82. Belsky, T. and Kaplan, I.R. (1970). Light hydrocarbon
gases, C^{13}, and the rrigin of organic matter in carbonaceous
chondrites. Geochim. Cosmochim. Acta., 34, pp. 257-278.

83. Krouse, H.R. and Modzeleski, V.E. (1970). C^{13}/C^{12}
abundances in components of carbonaceous chondrites and
terrestrial samples. Geochim. Cosmochim. Acta., 34, pp.
459-474.

84. Smith, J.W. and Kaplan, I.R. (1970). Endogenous carbon in
carbonaceous meteorites. Science, 167, pp. 1367-1370.

85. Grady, M.M., Swart, P.K. and Pillinger, C.T. (1982). The
variable carbon isotopic composition of Type 3 ordinary
chondrites. Proc. 13th Lunar Planet Sci. Conf., Part 1, pp.
A289-A296.

86. Deines, P. and Wickman, F. (1973). The isotopic
composition of "graphitic" carbon from iron meteorites and some
remarks on the troilitic sulfur of iron meteorites. Geochim.
Cosmochim. Acta., 37, pp. 1295-1319.

87. Deines, P. and Wickman, F. (1975). A contribution to the
stable carbon isotope geochemistry of iron meteorites. Geochim.
Cosmochim. Acta., 39, pp. 547-557.

88. Injerd, W.G. and Kaplan, I.R. (1974). Nitrogen isotope
distribution in meteorites. Meteoritics, 9, pp. 352-353.

89. Kung, C-C. and Clayton, R.N. (1978). Nitrogen abundances
and isotopic compositions in stony meteorites. Earth Planet.
Sci. Lett., 38, pp. 421-435.

90. Clayton, R.N., Onuma, N. and Mayeda, T.K. (1976). A classification of meteorites based on oxygen isotopes. Earth Planet. Sci. Lett., 30, pp. 10-18.

91. Onuma, N., Clayton, R.N. and Mayeda, T.K. (1972). Oxygen isotope cosmo-thermometer. Geochim. Cosmochim. Acta., 36, pp. 169-188.

92. Hulston, J.R. and Thode, H.G. (1965). Variations in the S^{33}, S^{34} and S^{36} contents of meteorites and their relation to chemical and nuclear effects. J. Geophys. Res., 70, pp. 3475-3484.

93. Kaplan, I.R. and Hulston, J.R. (1966). The isotopic abundance and content of sulfur in meteorites. Geochim. Cosmochim. Acta., 30, pp. 479-496.

94. McNaughton, N.J., Fallick, A.E. and Pillinger, C.T. (1982). Deuterium enrichments in Type 3 ordinary chondrites. Proc. 13th Lunar Planet. Sci. Conf., Part 1, pp. A297-A302.

95. Fallick, A.E., Hinton, R.W., McNaughton, N.J. and Pillinger, C.T. (1983). D/H ratios in meteorites: Some results and implications. Annales Geophys., 1, pp. 129-133.

96. Kolodny, Y., Kerridge, J.F. and Kaplan, I.R. (1980). Deuterium in carbonaceous chondrites. Earth Planet. Sci. Lett., 46, pp. 149-158.

97. Smith, J.W. and Rigby, D. (1981). Comments on D/H ratios in chondritic matter. Earth Planet. Sci. Lett., 54, pp. 64-66.

98. Kerridge, J.F. (1983). Isotopic composition of carbonaceous-chondrite kerogen: Evidence for an interstellar origin of organic matter in meteorites. Earth Planet. Sci. Lett., 64, pp. 186-200.

99. Yang, J. and Epstein, S. (1983). Interstellar organic matter in meteorites. Geochim. Cosmochim. Acta., 47, pp. 2199-2216.

100. Becker, R.H. and Epstein, S. (1982). Carbon, hydrogen and nitrogen isotopes in solvent-extractable organic matter from carbonaceous chondrites. Geochim. Cosmochim. Acta., 46, pp. 97-103.

101. Chang, S., Mack, R. and Lennon, K. (1978). Carbon chemistry of separated phases of Murchison and Allende meteorites. Lunar Planet. Sci., IX, pp. 157-159.

102. Gilmour, I., Swart, P.K. and Pillinger, C.T. (1983). The carbon isotopic composition of individual high molecular weight alkanes in the Murchison carbonaceous chondrite. Meteoritics, 18, pp. 302.

103. Yuen, G., Blair, N., Des Marais, D.J. and Chang, S. (1984). Carbon isotope composition of low molecular weight hydrocarbons and monocarboxylic acids from Murchison meteorite. Nature, 307, pp. 252-254.

104. Swart, P.K., Grady, M.M. and Pillinger, C.T. (1982). Isotopically distinguishable carbon phases in the Allende meteorite. Nature, 297, pp. 381-382.

105. Swart, P.K., Grady, M.M., Wright, I.P. and Pillinger, C.T. (1982). Carbon components and their isotopic compositions in the Allende meteorite. Proc. 13th Lunar Planet. Sci. Conf. Part 1, pp. A283-A288.

106. Kerridge, J.F. and Chang, S. (1983). Isotopic characterization of carbonaceous matter in Murchison. Meteoritics, 18, pp. 323.

107. Carr, L.P., Grady, M.M., Wright, I.P., Pillinger, C.T. and Fallick, A.E. (1983). Carbon, nitrogen and hydrogen in enstatite chondrites. Meteoritics, 18, pp. 276.

108. Prombo, C.A. and Clayton, R.N. (1983). Nitrogen isotopes in iron meteorites. Lunar Planet. Sci., XIV, pp. 620-621.

109. Kung, C-C., Hayatsu, R., Studier, M.H. and Clayton, R.N. (1979). Nitrogen isotope fractionations in the Fischer-Tropsch synthesis and in the Miller-Urey reaction. Earth Planet. Sci. Lett., 46, pp. 141-146.

110. Clayton, R.N. and Mayeda, T.K. (1977). Correlated O + Mg isotope anomalies in Allende inclusions, 1: Oxygen. Geophys. Res. Lett., 4, pp. 295-298.

111. Clayton, R.N., Onuma, N., Grossman, L. and Mayeda, T.k. (1977). Distribution of the pre-solar component in Allende and other carbonaceous chondrites. Earth Planet. Sci. Lett., 34, pp. 209-224.

112. Clayton, R.N. and Mayeda, T.K. (1982). Oxygen isotopes in carbonaceous chondrites and in achondrites. Lunar Planet. Sci., XIII, pp. 117-118.

113. Swart, P.K., Grady, M.M., Wright, I.P. and Pillinger, C.T. (1984). The carbon and oxygen isotopic composition of meteoritic carbonates (in prep.).

114. Monster, J., Anders, E. and Thode, H.G. (1965). $^{34}S/^{32}S$ ratios for the different forms of sulphur in the Orgueil meteorite and their mode of formation. Geochim. Cosmochim. Acta., 29, pp. 773-779.

115. Rees, C.E. and Thode, H.G. (1977). A ^{33}S anomaly in the Allende meteorite. Geochim. Cosmochim. Acta., 41, pp. 1679-1682.

116. Carr, R.H., Wright, I.P., Pillinger, C.T., Lewis, R.S. and Anders, E. (1983). Interstellar carbon in meteorites: Isotopic analysis using static mass spectrometry. Meteoritics, 18, pp. 277.

DUST OF VARIABLE POROSITIES (DENSITIES) IN THE SOLAR SYSTEM

H. Fechtig[*] and T. Mukai[+]

[*]Max-Planck-Institut für Kernphysik, P.O. Box 10 39 80,
6900 Heidelberg, F.R.G.
[+]Kanazawa Institute of Technology, Nonoichi, Ishikawa
921, Japan

ABSTRACT. Recently released cometary particles are most likely
aggregates of core-mantle building blocks. Their structures are
therefore porous (fluffy particles). Lunar Microcrater studies
and in situ dust experiments have shown that the majority of the
interplanetary dust particles are compact grains with densities
between 3 and 8 g/cm³, although ∿ 20% of the interplanetary dust
is of low density (< 1 g/cm³). Under the assumption that cometary
grains slowly lose their mantle material, the grains become denser.
Simultaneously the orbits also become more circular as compared
to the original cometary orbit. This has been observed by the
Helios dust experiment.

1. INTRODUCTION

In the Solar System, there are potentially two main sources for
interplanetary dust: the comets and the asteroids.

Comets release considerable amounts of dust grains during
their perihelion passages. Depending on the sizes of the dust
grains, they are either directly pushed out of the solar system
by solar irradiation (submicronsized dust) or orbit the sun within a meteor stream slowly spiralling towards the sun due to the
Poynting-Robertson forces.

Asteroids are potentially and continuously producing dust by
hypervelocity collisions. The observations of the asteroidal source
by experiments on Pioneer 10/11 travelling through the asteroidal
belt (1,2) show no enhancement of dust grains there which leads
to the conclusion that the asteroids play at most only a minor
role as a dust source. This conclusion is also reached by theoretical considerations of Dohnanyi (3,4).

251

J. Klinger et al. (eds.), Ices in the Solar System, 251–259.
© 1985 by D. Reidel Publishing Company.

Therefore the cometary source is considered as the main
source for interplanetary dust. One of the generally accepted
properties of cometary dust is its porosity as suggested by the
appearence of the Brownlee particles (5). These are aggregates
of submicron-sized building blocks. Their overall densities
(= mass within the envelopes of the Brownlee particles) are
about 1 g/cm³. Much lower densities have been reported from the
meteor stream particles (6).

As discussed already in a review paper (7) there are direct
observations of the densities of interplanetary particles from
the studies of lunar microcraters and from in situ experiments.
As shown in Fig. 1 for example, the histogram of the diameter
over depth ratio D/T of lunar microcraters shows 2 major and 1
minor peaks. These peaks correspond to projectile densities of
8 (iron), 3 (silicates) and 1 g/cm³. The latter value agrees
with the density of the Brownlee particles. For details of these
analyses see publications 8,9,10,11,12.

The in situ dust experiments on the HEOS-2 satellite and
the Helios spaceprobe clearly indicate the same results. HEOS-2
has observed dust swarms within 10 earth radii distance from
Earth (13). These swarms are interpreted as recently fragmented
meteoroids by electrostatic charging. Ceplecha and McCrosky
(14) have reported a meteoroid class with densities below 1 g/cm³.
Both observations lead to the same conclusion as the lunar micro-

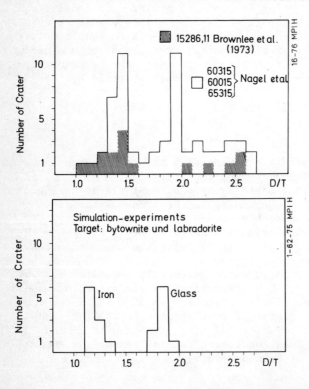

Fig. 1: Histogram of
D/T-values for lunar
micro-craters on
glass targets (upper
diagram: lunar micro-
craters; lower dia-
gram: simulated cra-
ters): the peak at
D/T = 1,4 corres-
ponds to iron pro-
jectiles, the peak
at D/T = 1,9 to glass
(silicates) and for
D/T = 2,5 to a lower
density projectile.

crater studies: only a minority (< 25%) of the meteoroids and micrometeoroids are of the fluffy type and hence of low density.
These observations seem to contradict a cometary nature with its
porous structure and low densities.

However, an interesting observation on the dust experiment
on the Helios spaceprobe has been made which leads to a possible
solution of this contradiction. The Helios dust experiment (15,
16) observed two principally different dust populations: dust
orbiting the sun on ellipses with low eccentricities (ε < 0.4)
of normal densities between 3 and 8 g/cm³ and dust grains orbiting the sun on ellipses with high eccentricities (ε > 0.4) of
low densities (ρ < 1 g/cm³). This observation suggests that the
low density particles on elliptical orbits are younger, i.e. more
recently released from a comet than the normal dense particles
which are in quasicircular orbits due to the action of the Poynting-Robertson effect (which continuously decreases the eccentricity of ellipses). But what mechanism could be responsible for
a time-dependent increasing density?

2. VARIATION OF MASS DENSITY OF COMETARY PARTICLES

Greenberg (17,18,19,20) has developed a model for the nature and
composition of cometary particles following his simulation experiments. Cometary dust grains are expected to be conglomerates of
small submicron-sized building blocks. These building blocks consist of elongated core-mantle grains. The cores are most likely
silicates of dimensions in the order of 0.1 μm. The mantles are
likely to consist of quite stable organic photoprocessed material
consisting of the elements H,C,N,O and quite volatile icy mantle
material (water and other frozen parent molecules).

Fig. 2 shows an idealized model of a cometary fluffy particle

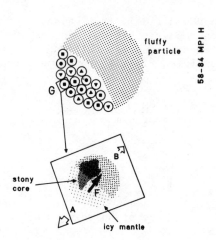

Fig. 2: Idealized model
of a cometary dust grain

consisting of spherical core/mantle building blocks. The uppermost layer of the grain is exposed to sunlight. Each of the building blocks of this uppermost layer is irradiated on one hemisphere
and shadowed on the other hemisphere.

2.1. Loss of mantle material by heating

According to Fig. 2 the mantle material is heated up and sublimates due to its temperature T_a at the sunlit side and T_b at the
shadowed side, with $T_a > T_b$. The escaping gas has velocities of

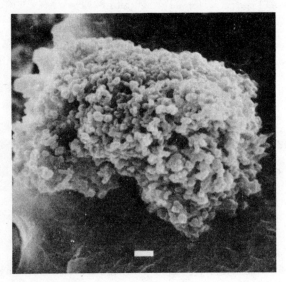

Fig. 3: Brownlee particle (bar = 1 μm)

v_a (sunlit side) > v_b (shadowed side). This causes a packing
force F by analogy with the nongravitational forces produced by
outgasing of volatiles from comet nuclei (21) and of the Yarkovsky-
effect acting on small particles (22). The packing force F can be
calculated from

$$F = 2\pi s_g^2 \frac{dM}{dt} (v_a - v_b) = 2\pi s_g^2 \frac{dM}{dt} \sqrt{\frac{2k}{m_m}} \cdot \Delta\tau$$

with
$$\Delta\tau = \sqrt{T_a} - \sqrt{T_b} \qquad v_b = \sqrt{\frac{2kT_b}{m_m}}$$

$$v_a = \sqrt{\frac{2kT_a}{m_m}} \qquad\qquad s_g = \text{size of building block}$$

$$\frac{dM}{dt} = \text{mass loss rate} \qquad k = \text{Boltzmann constant}$$

$$m_m = \text{average mass of mantle molecule}$$

After a certain time the uppermost building blocks will have lost all mantle material. The cores "fall" back on the surface of the second layer of the building blocks. This goes on until finally an aggregate of the building blocks, called a Brownlee particle, remains (Fig. 3). As shown in detail by Mukai and Fechtig (23) it is possible to calculate the time required for all mantle material to be lost. With continuous sublimation of mantle material the overall density of a fluffy cometary grain increases. The density varies with time as

$$\rho(t) = \frac{V_m(t) \cdot \rho_m + V_c(t)\,\rho_c}{V_m(t) + V_c(t) + V_s(t)} \qquad \text{with} \quad V_m(t) = \text{Volume of mantle material at time t}$$

$$V_c(t) = \text{Volume of core material at time t}$$

$$V_s(t) = \text{Volume of space between blocks at time t}$$

$$\rho_m = \text{density of mantle material}$$

$$\rho_c = \text{density of core material}$$

For Halley dust particle released at time $t = 0$ and $V_m(0)/V_c(0) = 9$ the value of $\rho(0)$ is $\rho(0) = 0.6$ gcm^{-3}. The diagram of Fig. 4 shows the density versus time for a sublimation rate of dM/dT = 5×10^{-13}; = 10^{-13} and 0.5×10^{-13} gcm^{-2}sec^{-1}. After about 10^5 years the mantle material will have completely evaporated and the density will have risen to about 3 gcm^{-3}.

If one calculates the time which is needed to change the original Halley-orbit of a Halley dust particle into a circular orbiting particle, this will happen after about 2×10^5 years, a time which is comparable to the above calculated sublimation time (22). This agrees exactly with the observation of the Helios dust experiment.

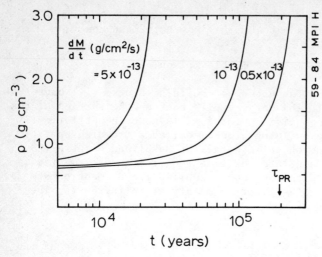

Fig. 4:

Variation of mass density of a fluffy particle with time.

2.2. Loss of mantle material by sputtering

Another process which removes material is sputtering by solar wind. Mukai and Schwehm (24) give sputter rates for water ice, obsidian and magnetite to be 10.6 Å/year for water, 0.3 for obsidian and 0.7 for magnetite, and McDonnell and Flavill (25) give $\leq$ 0.1 A/year for silicates, all valid at 1 AU sun distance (the latter value is derived for the moon which therefore is in agreement with Mukai and Schwehm).

If one calculates the mass loss for a fluffy cometary Halley particle due to sputtering, one ends up with a mass loss of only 1.3 μm per 10^5 years. This process can therefore be neglected.

3.   CONCLUSIONS

If one assumes that the organic mantles of cometary dust grains sublimate at a rate of $10^{-13} gcm^{-2}sec^{-1}$, a Halley dust grain needs about 10^5 years to completeley get rid of the mantle material of the building blocks. During this time the cores remain to form the Brownlee particles and even more compact grains approaching the core material density. During the same time the particle changes its orbit from the original Halley orbit into a quasicircular orbit around the sun due to the Poynting-Robertson effect.

This is exactly what has been observed in space by the Helios dust experiment. As a conclusion one cannot expect that all inter-planetary dust particles are of low density. Only the young particles are low in density, but the majority of the particles, that means the old particles, are of normal densities.

If this suggestion is correct one should possible expect some indications by the study of the Brownlee particles which might be young particles and which might have lost their mantles during entry into the Earth's atmosphere. Indeed the Brownlee particles look black and could therefore be coated by a thin dark layer (5). Could this coating be a refractory residue of the vaporized mantle material?

A further indication can be seen in the observation of Comet Bowell 1980b by A'Hearn et al. (26). These authors see a production of OH at 4-5 AU sun distance which could partly be interpreted as the first stage of vaporization of volatiles from the cometary grains.

REFERENCES

1. Humes D.H., Alvarez J.M., O'Neal M.L. and Kinard W.H., 1974, "The interplanetary and near-Jupiter meteoroid environments", J.Geophys.Res. 79, pp. 3677-3684.

2. Weinberg J.C., Hanner M.S., Mann H.M., Hutchinson P.B. and Fimmel R., 1973, "Observations of zodiacal light from the Pioneer 10 Asteroid-Jupiter probe: preliminary results", Space Research XIII, (M.J. Rycroft and S.K. Runcorn, eds.), Akademie-Verlag Berlin, pp. 1187-1192.

3. Dohnanyi J.S., 1972, "Interplanetary objects in review: statistics of their masses and dynamics", Icarus 17, pp. 1-48.

4. Dohnanyi J.S., 1976, "Sources of Interplanetary dust: Asteroids" in Lecture Notes in Physics (eds. H. Elsässer and H. Fechtig), Springer Verlag, Berlin-Heidelberg, Vol. 48, pp. 187-205.

5. Brownlee D.E., 1978, "Microparticle Studies by Sampling Techniques", in "Cosmic Dust" (ed. J.A.M. McDonnell), John Wiley % Sons, Chichester, pp. 295-336.

6. Verniani F., 1973, "An Analysis of the Physical Parameters of 5759 Faint Radio Meteors", J.Geophys.Res. 78, pp. 8429-8462.

7. Fechtig H., 1982, "Cometary Dust in the Solar System" in "Comets" (L.L. Wilkening, ed.), The University of Arizona Press, Tucson, Arizona, pp. 370-382.

8. Brownlee D.E., Hörz F., Vedder J.F., Gault D.E. and Hartung J.B., 1973, "Some physical parameters of micrometeoroids, Proc.Lunar Sci.Conf. 4, pp. 3197-3212.

9. Nagel K., Neukum G., Eichhorn G., Fechtig H., Müller O. and
    Schneider E., 1975, "Dependencies of microcrater formation
    on impact parameters", Proc.Lunar Sci.Conf. 6, pp. 3417-3432.

10. Nagel K., Neukum G., Dohnanyi J.S., Fechtig H. and Gentner
    W., 1976, "Density and Chemistry of interplanetary dust par-
    ticles derived from measurements of lunar microcraters", Proc.
    Lunar Sci.Conf. 7, pp. 1021-1029.

11. Nagel K., Neukum G., Fechtig H., Gentner W., 1976, "Density
    and composition of interplanetary dust particles", Earth
    Planet.Sci. Letters 30, pp. 234-240.

12. Fechtig H., Gentner W., Hartung J.B., Nagel K., Neukum G.,
    Schneider E. and Storzer D., 1975, "Microcraters on lunar
    samples", Proc. Soviet-American Conference on Cosmochemistry,
    Moscow, pp. 453-472 (Translation: NASA SP-370, pp. 585-603).

13. Fechtig H., Grün E. and Morfill G., 1979, "Micrometeoroids
    within ten Earth radii", Planet.Space Sci. 27, 511-531.

14. Ceplecha Z. and McCrosky R.E., 1976, "Fireball end heights:
    a diagnostic for the structure of meteoritic material",
    J.Geophys.Res. 81, pp. 6257-6275.

15. Grün E., Pailer N., Fechtig H. and Kissel J., 1980, "Orbital
    physical characteristics of micrometeoroids in the inner so-
    lar system as observed by Helios 1", Planet.Space Sci. 28,
    pp. 333-349.

16. Grün E., 1981, "Physikalische und chemische Eigenschaften
    des interplanetaren Staubes - Messungen des Mikrometeoritenex-
    perimentes auf Helios", Habilitationsschrift Universität
    Heidelberg.

17. Greenberg J.M., 1978, "Interstellar Dust", in "Cosmic Dust",
    (ed. J.A.M. McDonnell), John Wiley & Sons, Chichester,
    pp. 187-294.

18. Greenberg J.M., 1982, "What are comets made of? a model based
    on interstellar dust", in "Comets" (L.L. Wilkening, ed.),
    The University of Arizona Press, Tucson, Arizona, pp. 131-
    163.

19. Greenberg J.M., 1983, "Laboratory dust experiments - tracing
    the composition of cometary dust" in "Cometary Exploration
    II", (ed. T.I. Gombosi), Central Research Institute of Phy-
    sics, Hungarian Academa of Sciences, Budapest, pp. 23-53.

20. Greenberg J.M., 1985, this volume.

21. Whipple F.L., 1978, "Comets", in "Cosmic Dust", (ed. J.A.M. McDonnell), John Wiley & Sons, Chichester, pp. 1-74.

22. Burns J.A., Lamy Ph. and Soter S., 1979, "Radiation Forces on small particles in the Solar System", Icarus 40, pp. 1-48.

23. Mukai T. and Fechtig H., 1983, "Packing effect of fluffy particles", Planet. Space Sci. 31, pp. 655-658.

24. Mukai T. and Schwehm G., 1981, "Interaction of grains with the solar energetic particles", Astron.Astrophys. 95, pp. 373-382.

25. McDonnell J.A.M. and Flavill R.P., 1974, "Solar wind sputtering on the lunar surface: equilibrium crater densities related to past and present microparticle influx rates", Proc. Lunar Sci.Conf., 5th, pp. 2441-2449.

26. A'Hearn M.F., Schleicher D.G., Feldmann P.D., Millis R.L. and Thompson D.T., 1984, "Comet Bowell 1980b", in press.

DISCUSSION

M. A'Hearn:
Did I understand correctly that you just assumed a vapourization rate? It seems very low for vapourization of H_2O.

H. Fechtig:
The vapourization rate was chosen by fitting to the Helios data. It must refer to the "inner" mantle of Greenberg rather than the "outer" mantle of H_2O.

L. Lanzerotti:
Both Strazulla and I show that ion irradiations of organic ices (e.g. CH_4; CH_4-CO_2) can make "black stuff". This must be considered in addition to the U-V produced "yellow stuff".

MOLECULE FORMATION IN COMETARY ENVIRONMENTS

Valerio Pirronello

Istituto di Fisica, Facoltà di Ingegneria, Università
di Catania, Corso delle Provincie 47, I-95129 Catania
Italy
Osservatorio Astrofisico, V.le A.Doria Città Universitaria, I-95125 Catania, Italy

ABSTRACT

The formation of molecules induced in comets by the release
of energy of fast ions is presented. Results of laboratory simulations and the applications to cometary environments, both inside
and outside the solar system are discussed. Particular attention
is payed to the suggestion of an independent method for obtaining
estimates of or limits on the low energy cosmic ray flux, which
is poorly known in the interstellar medium.

1. INTRODUCTION

It is widely accepted that comets, which occasionally enter
the inner part of the solar system, come from a region called the
Oort cloud, extending to about 50,000 A.U. from the Sun (1,2).
Our understanding of the physics of comets is based on the Whipple model which describes the nucleus as "a matrix of meteoric
material with little structural strength, mixed together with
frozen gases"(3); a model that has been confirmed by many observational evidences. A strong controversy still exists about the
origin of comets. Whether they formed in the solar nebula and then
ejected in the Oort cloud or they formed in situ by aggregation
of interstellar dust is still debated. The model of Greenberg
is basically consistent with that Whipple but specifies more
clearly the various components (4).

261

J. Klinger et al. (eds.), Ices in the Solar System, 261–272.
© *1985 by D. Reidel Publishing Company.*

In order to have a better insight into this problem, it is mandatory to reveal the chemical composition of cometary nuclei. Such a task has to be carried out mainly through the study of atoms, molecules and ions observed in the comae of approaching comets. The chemistry of cometary comae has been investigated in several papers from the theoretical point of view (5-14) and the experimental one (15, 16).

It is also accepted that, the age of comets is about the same of that of the solar system (17, 18); their nuclei have then spent about 4.6×10^9 years in the Oort region suffering the continuous bombardment of cosmic rays and UV photons and experiencing solar wind and photon irradiation during close approaches to the Sun.

The interaction of UV photons with frozen gases has been simulated experimentally for years in Leiden by the Greenberg group and the consequences of astrophysical relevance have been carried out (4,19,20).

Also the interaction of energetic particles with frost plays a role and should be studied in order to assess its relevance in comets. Other contributions on this subject are given in this meeting, here I would like to focus my attention only on the production of molecules induced in cometary environments by particle bombardment . In the next section a brief review of the experimental technique and results is given; then applications to comets will follow. A final suggestion on a possible way to determine the flux of bombarding particles in the interstellar medium is presented.

2. LABORATORY RESULTS AND THEIR APPLICATIONS TO COMETS

The experimental simulation of the interaction of energetic ions (cosmic rays and wind particles) with comets has been performed using an electrostatic Van de Graaf accelerator or an ion implanter to induce erosion and chemical reactions in targets obtained by depositing pure gases or mixtures on cold substrates. Their temperature could be continuously varied between 9K and room temperature value. The vacuum in the scattering chamber was better than 2.0×10^{-9} torr during measurements performed at the Bell Laboratories and about two orders of magnitude higher on measurements carried out in Catania. A quadrupole mass spetrometer with an electron impact ionizer, detected chemical species produced and ejected during irradiation or after when molecules, which had been trapped, were released during the sublimation of the ice layer. The dose rate of projectiles was always in the range where

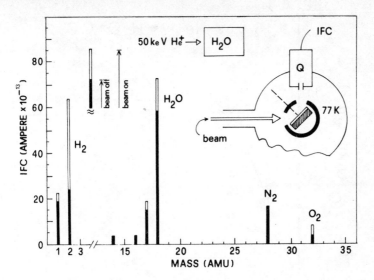

Fig. 1.   Mass spectra collected with the beam off (solid bars)
          and the beam on the ice target (solid plus light bars).
          A schematic of the apparatus is shown in the insert.

a linear relationship holds between the erosion yield and the
current, in order to be sure that only single particle effects
and not collective ones were induced (21). A typical mass spectrum for H_2O bombarded by keV ions is presented in fig.1. Solid
bars show the background, while solid plus light ones represent
the molecule release when the beam is on the target (22). In this
way together with a suitable external calibration, the temperature
dependence of release of H_2, O_2 and H_2O ejection by water ice
bombarded with 1.5 MeV helium ions has been obtained (23). The
yield "Y" (number of sputtered molecules per impinging particle)
of H_2O is equal to about 6 and is almost independent of the
substrate temperature below 130K, at which value sublimation becomes the dominant effect. The ejection of H_2, on the contrary,
occurs following a power law between 50K and 130K and O_2 is released with a less steep rate. Similar kind of results, even if more
difficult to assess quantitatively, have been obtained for CO_2,
CO from CO_2 ice and for SO, SO_2 from SO_2 ice.
     The same technique was adopted to measure the production rate
of "complex" molecules in ice mixtures. More precisely CO_2 and
H_2O mixed in gas phase and deposited at 9K were bombarded with
1.5 MeV He^+ at the same temperature. Typical thicknesses of accre-

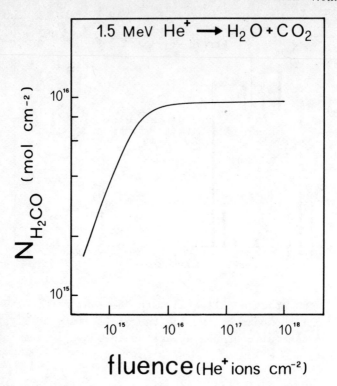

Fig. 2. Experimental production yield of formaldehyde is a H_2O-
         CO_2 mixture versus the fluence of helium ions. The
         thickness of the film is about 1.8×10^{18} mol cm^{-2}.

ted layers were about 1.8×10^{18} mol cm^{-2}. Formaldehyde molecules
were formed and remained trapped in the frozen film till and
released when the sample was warmed letting it sublime. No formic acid was detected. The H_2CO formation is an increasing function with the total dose of projectiles until saturation occurs
(see fig.2) corresponding to a fluence of about 8.0×10^{15} He$^+$ cm^{-2}
at the target (24).

     The formation of molecules not present in the original depo-
sit has to be ascribed to the recombination of fragments of parent
molecules cracked along the ion track. The saturation is due to
the increasing probability to break already formed molecules and
to the continuosly diminishing concentration of water and carbon
dioxide.

     Together with these experiments those carried out at the

Goddard Space Flight Center by Donn and coworkers also need to be
mentioned (25,26). In their work they analyze how radiation
synthesis changes the outer layers of comets. Various ice
mixtures including H_2O, NH_3, CH_4, N_2 and CO_2 were irradiated
by protons and synthesized molecules were identified, either
by their IR signature or by gas chromatographic analysis of the
volatile fraction obtained warming up the processed mixture itself.
Their results give evidence supporting the fact that when CH_4 is
present in the deposited ice mixture long chained volatile hydrocarbons are produced by the bombardment and about 1% of the ice
is converted into a "residue" stable at room temperature. Thermoluminescence, pressure bursts during warm up and color changes
of samples with irradiation were also reported. All this phenome-
nology, even if only on a qualitative ground, is very similar to
that obtained by Greenberg and his collaborators in Leiden; the
main difference being the penetration depth of particles and
photons.

The experimental results, we have described, show that the
most mobile among the new formed molecules in the high temperature
region around the ion track can escape from the bombarded ice layer,
while the most complex and heaviest ones remain trapped in it.
When applying these results to comets, one should envisage scenarios where the irradiation takes place: one in which the nucleus
resides in the Oort region, suffering the bombardment by cosmic
rays which alter the original chemical composition of its external layers; the other in which the comet is irradiated by the solar wind during a close approach.

During close approaches the comet releases, by sputtering,
atoms and molecules enriching the interplanetary medium (27). In
fig. 3 the production of H_2, O_2 and H_2O, per unit area and unit
time, versus the solar distance is shown for a water dominated
object. The calculations have been performed using the laboratory
data given by Brown et al. (23), scaled for 1 keV particles, and
using an average flux of the solar wind adopted by Mukai and
Schwehm (28). Wind particles can directly hit the nuclear surface
at distances greater than approximately 5 A.U. from the Sun (29).
In this case the total production of those molecules per unit
time will be just given by the results presented in fig.3 multiplied by the surface area of the nucleus exposed to the wind. At
smaller distances, the formation of an extended atmosphere prevents the solar wind from hitting the nucleus by collision de-
flection. The penetration depth in this case is a function of the

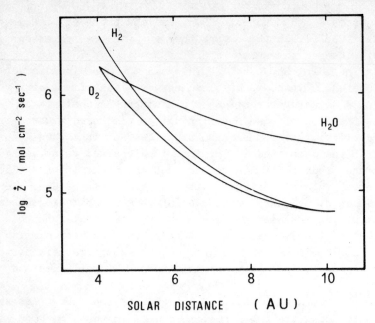

Fig. 3. H_2O, H_2 and O_2 production rate for a water dominated
comet bombarded by wind protons versus the solar distance.

comet distance from the Sun and of the cometary albedo. The possibility of forming molecules by this mechanism is then strictly
connected with the existence of icy grains in the coma, grains
which have been finally detected by Campins et al.(30). In order
to get the total production rate of molecules with the solar
distance, one must evaluate the total surface exposed by those
grains dispersed in the coma; an estimate with the distance from
the Sun is given by Pirronello et al. (31).

Particularly interesting results can be obtained for comets
entering the inner solar system for their first time. They maintain recorded in their external layers informations about their
past history in the Oort cloud. For such comets the release of
formaldehyde has been calculated, and is shown in fig.4, based
only on the hypothesis that H_2CO was formed by cosmic ray bombardment at saturation during its long life outside our planetary
system (32)and that H_2O and CO_2 are among the main constituents
of cometary nuclei, a composition which is not strongly different
from that which results from interstellar dust aggregation even
before cosmic ray bombardment (4). Sublimation results, shown in

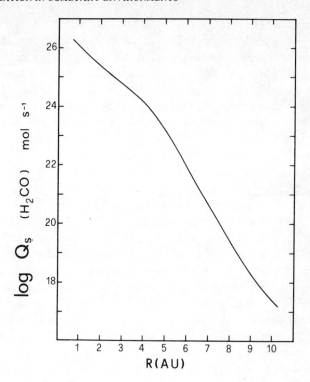

Fig. 4. Release of H_2CO versus the solar distance by an hypothe-
         tical comet bombarded at saturation by cosmic rays in
         the Oort cloud.

fig.4, were obtained by integrating the Clayperon equation. According to the experimental results (24) the process has been assumed
to be controlled by H_2O (i.e. the latent heat of sublimation of
the mixture has been chosen to be equal to that of water).

3. COSMIC RAYS IN THE OORT REGION

These calculations together with the laboratory data, described in the previous section, can be used to obtain evaluation
of the low energy cosmic ray flux in the interstellar medium. The
knowledge of such a flux is important for any consideration about
the chemical composition, the stability of clouds and hence the
formation of new stars, since cosmic rays can release considera-

ble amounts of energy throughout the whole cloud. They can
vary the chemical composition of molecular clouds either through
the direct formation and ejection of new species from frozen
mantles on grains or through the control of the kinetics of ion-
molecule reactions obtained by affecting the ionization degree
of clouds themselves (33-35). Also the interaction between mass
motions and magnetic fields are influenced by the cosmic ray
induced ionization of atoms and molecules and hence the internal
dynamics of clouds can be affected.

Unfortunately such a low energy cosmic ray flux is almost
unknown. This occurs since in the planetary region, where we can
measure it, the magnetic field carried by the solar wind gives
rise to a strong screening effect.

Our suggestion consists into using observations of appropriate molecules in the comae of comets entering for the first time
the inner solar system, in order to infer the total dose of projectiles which reached their nuclei when they were in the Oort
cloud. From the estimated total dose of particles suffered per
cm^{-2} together with the age of the comet one can then obtain an
average flux of cosmic rays, valid for a region where no screening
exists. The choice of molecules to observe is naturally restricted
to those which are synthesized in ice mixtures by particle irradiation and which remain trapped in the film until sublimation
occurs. Here the case of formaldehyde is presented.

Of course a quantitative estimate of the abundance of molecules produced by particle irradiation needs a preliminary subtraction of the abundance of those already present in the gas
that aggregated to form the cometary nucleus and those formed
by UV photon irradiation which enrich the very external layer
of the nucleus while the comet is still in the Oort cloud.

As a first possibility it is worthy to mention that of
getting either a lower or an upper limit to the flux. This comes
from the observed saturation in the formation of molecules during
bombardment; saturation which happens when about 8.0×10^{15} He^{+}
cm^{-2} have already reached the icy target, and which is shown in
fig.2. The detection of formaldehyde in a comet, whic is approaching for its first time, corresponding to what is shown in fig.4
would say that at least a fluence $f_s = 8.0 \times 10^{15}$ cm^{-2} of helium equivalent projectiles has reached the cometary nucleus. A release of
H_2CO not corresponding to figure 4 would imply that "f_s" should be
considered an upper limit. Such an "f_s" correspond to a flux of
cosmic rays equivalent to about .05 cm^{-2} s^{-1} helium ions or to
about .8-1.0 H^{+} cm^{-2} s^{-1}.

In the case in which observations would show that the nucleus of the entering comet has not been bombarded at saturation a more precise estimate of the cosmic ray flux can be obtained. The general solution of our problem would be reached having, for each kind of projectiles (H^+, He^+, C^+,...and so on) on an experimental ground, the production yield of the molecule we want to monitor (in this case H_2CO) as a function of energy "$Y(E)$". With the laboratory data we have to date on formaldehyde production by MeV helium ions, we can anyway estimate the cosmic ray flux if we restrict our analysis to a very thin layer of the nucleus, considering only the release of the molecule we want to monitor just when sublimation is getting efficient.

The production rate "$Q(mol\ s^{-1})$" of formaldehyde, measured in terms of that one relative to the saturation. "Q_s", versus the fluence of helium projectiles that can produce it, is given in fig.5 for an hypothetical comet at its first approach. Once from observations of a real comet the production rate of formaldehyde is measured, using the result of fig.5 it is possible to estimate the total fluence of particles which irradiated its nucleus when it still was in the Oort cloud and hence deduce the flux of cosmic rays pervading the Galaxy outside the solar system.

In this paper we have shown an example of how experimental results on ion irradiation of ices can be used in astrophysics; a more precise evaluation of the flux of cosmic rays can be obtained by monitoring the release of the chosen chemical species with the distance of the comet.

In this way, because sublimation will occur at deeper layers in the comet, while it is approaching the Sun, it will be possible to analyze the abundance of molecules, produced by the cosmic ray bombardment, as a function of the depth in the nucleus. It will be then possible to obtain informations also on the shape of the energy spectrum of cosmic rays even at E<100 MeV.

ACKNOWLEDGEMENTS

Very helpful discussion with J.M.Greenberg during the meeting are gratefully acknowledged. Thanks are due to Mr.Piparo and M.Carcaci for their help during experiments, to Mr.A.Calì for drawing the figures and to Mrs.M.G.Nastasi for typing the manuscript.

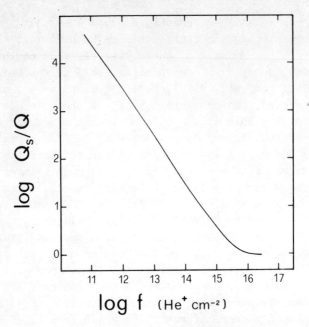

Fig. 5   Production rate of formaldehyde for an hypothetical comet entering for the first time the inner solar system as a function of the fluence of energetic particles which irradiated its nucleus.

REFERENCES

1.   Oort, J. H., 1950, Bull. Astr. Inst. Netherlands 11, pp. 91-110.

2.   Oort, J. H. and Schmidt, M., 1951, Bull. Astr. Inst. Netherlands 11, pp. 259-269.

3.   Whipple, F. L., 1950, Astrophys. J. 111, pp. 375-394.

4.   Greenberg, J. M., 1982, in "Comets", ed. L. L. Wilkening, Univ. of Arizona Press, pp. 131-163.

5.   Shimizu, M., 1975, Astrophys. Space Sci. 36, pp. 353-361.

6.   Oppenheimer, M., 1975, Astrophys. J. 196, pp. 251-259.

7.   Oppenheimer, M., 1978, Astrophys. J. 225, pp. 1083.

8. Ip, W-H and Mendis, D. A., 1976, Icarus 28, pp. 389-400.

9. Ip, W-H and Mendis, D. A., 1977, Icarus 30, pp. 377-384.

10. Ip. W-H, 1979, in "Proc. Workshop on Cometary Mission", W. I. Axford, H. Fechtig, J. Rahe, Bamberg, p. 69.

11. Huebner, W. F. and Giguere, P. T., 1980, Astrophys. J. 238, pp. 753-762.

12. Mitchell, G. F., Prasad, S. S. and Huntress, W. T., 1981, Astrophys. J. 244, pp. 1087-1093.

13. Swift, M. B. and Mitchell, G. F., 1981, Icarus 47, pp. 412-430.

14. Biermann, L., Giguere, P. T. and Huebner, W. F., 1982, Astron. Astrophys. 108, pp. 221-226.

15. Huntress, W. T., 1977, Astrophys. J. Suppl. 33, pp. 495-514.

16. Huntress, W. T., McEvan, M. J., Karpas, Z. and Anicich, V. G., 1980, Astrophys. J. Suppl. 44, pp. 481-488.

17. Delsemme, A. H., 1977, in "Comets, Asteroids and Meteorites", ed. A. H. Delsemme, Univ. of Toledo Press, pp. 3-13.

18. Donn, B. and Rahe, J., 1982, in "Comets", ed. L. L. Wilkening, Univ. of Arizona Press, pp. 203-226.

19. Greenberg, J. M., 1981, in "Comets and the Origin of Life", ed. C. Ponnamperuma, Reidel, p. 11.

20. Hagen, W., Allamandola, L. J. and Greenberg, J. M., 1979, Astrophys. Space Sci. 65, pp. 215-240.

21. Pirronello, V., Strazzulla, G., Foti, G. and Rimini, E., 1981, Astron. Astrophys. 96, pp. 267-270.

22. Ciavola, G., Foti, G., Torrisi, L., Pirronello, V., Strazzula, G., 1982, Radiation Effects 65, p. 167.

23. Brown, W. L., Augustyniak, W. M., Simmons, E., Marcantonio, K. J., Lanzerotti, L. J., Johnson, R. E.,

Boring, J. W., Reimann, C., Foti, G. and Pirronello, V, 1982, Nucl. Instr. and Methods 198, p.
1.

24. Pirronello, V., Brown, W. L., Lanzerotti, L. J.,
     Marcantonio, K. J. and Simmons, E., 1982, Astro-
     phys. J. 262, pp. 636-640.

25. Moore, M. H., 1981, Ph. D. Thesis, Univ. of Mary-
     land.

26. Moore, M. H., Donn, B., Khanna, R., A'Hearn, M. F.,
     1983, Icarus 54, pp. 388-405.

27. Pirronello, V., Strazzulla, G. and Foti, G., 1982,
     in "Sun and Planetary System", eds. W. Fricke and
     G. Teleki, Reidel, p. 253.

28. Mukai, T. and Schwehm, G., 1981, Astron. Astrophys.
     95, pp. 373-382.

29. Mendis, D. A., Hill, J. R., Houpis, H. L. F. and
     Whipple, Jr. E. C., 1981, Astrophys. J. 249, pp.
     787-797.

30. Campins, H., Rieke, G. H. and Lebofsky, L. A.,
     1983, Nature 301, pp. 405-406.

31. Pirronello, V., Strazzulla, G. and Foti, G., 1983,
     Astron. Astrophys. 118, pp. 341-344.

32. Pirronello, V., Strazzulla, G., Foti, G., Brown,
     W. L. and Lanzerotti, L. J., 1984, Astron. Astro-
     phys. 134, pp. 204-206.

33. Prasad, S. S. and Huntress, W. T., 1980, Astro-
     phys. J. Suppl. 43, p. 1.

34. Mitchell, G. F. and Deveau, T. J., 1983, Astro-
     phys. J. 266, pp. 646-661.

35. Watt, G. D., 1983, M. N. R. A. S. 205, pp. 321-335.

INTERACTION BETWEEN SOLAR ENERGETIC PARTICLES AND INTERPLANETARY GRAINS

G. Strazzulla[1], L. Calcagno[2], G. Foti[2], K.L. Sheng[2]

1) Osservatorio Astrofisico, Città Universitaria, I-95125 Catania, Italy
2) Istituto Dipartimentale di Fisica, Catania,
   Italy

ABSTRACT

Some laboratory-studied effects induced by the fluence of fast ions on frosts of astrophysical interest are summarized. The results are applied to the interaction between energetic solar ions and interplanetary dust grains assumed to be cometary debris which spends $\sim 10^6$ yr before being collected in the Earth's atmosphere or colliding on the Moon's surface.

The importance of erosion by particles to the stability of ice grains is confirmed. The build up of carbonaceous material by ion fluence on hydrocarbon containing grains is discussed. It is suggested that these new materials could be the glue which cements submicron silicate particles to form a complex agglomeration whose density increases with increasing proton fluence (packing effect).

The IR spectra of laboratory synthesized carbonaceous material is compared with that ones observed in some carbonaceous meteoritic extracts.

I. INTRODUCTION

The hypothesis that the bulk interplanetary grains, whose main observational evidence is the production of the zodiacal light, is cometary debris (1) is now widely accepted in the scientific community.

J. Klinger et al. (eds.), Ices in the Solar System, 273–285.
© 1985 by D. Reidel Publishing Company.

A wide disagreement however exists on the size distribution and physical properties of interplanetary dust grains (2). Greenberg and Gustafson (3) proposed a comet fragment model for zodiacal light particles. They found that the principal features of scattering of visible radiation are reasonably represented by the scattering by birdsnest-like agglomerations of particles characteristic of interstellar grains of core-mantle type.

However they also pointed out that a number of unanswered questions remain, in particular the physical modifications these aggregates suffer after they leave the comet. Among these other questions there are :

a) How does erosion (sputtering) by solar energetic particles (wind and flares) modify the surface of the aggregates.

b) How effective are the energetic particles along with the solar ultraviolet in changing the chemical and physical properties of the aggregate.

Although we believe it is not yet possible to fully answer to these questions, esperimental results on the modification of molecular solids (frozen gas) by energetic ions bombardments have been performed in some laboratories that should help to gain some insight into the problem.

II. EXPERIMENTAL

In recent years much experimental work has been performed to study the effects of energetic ion fluences on molecular solids as frozen gases, with a view to its astrophysical relevance (see refs 4-5-6-7 and references therein).

The experimental results refer mainly to :

i) The measurements of erosion yields ; i. e. the number of atoms ejected per impinging ion for various projectile-target combinations ; and the observations of new chemical species formed and/or eroded i. e. restored to the gas phase ; and the velocity distribution of the ejecta.

ii) Measurements of the cross section of the polymerization process observed when fast protons collide with hydrocarbon rich frozen layer. The collisions result in the production of a dark complex (polymer-like) material made up of long chain carbon rich molecules. The characterization of these new materials is mainly through their IR signatures.

Details on the experimental apparatus, techniques and results can be found in the bibliography given above.

Here we like to add something on (ii) that will be useful in the next sections. In a recent work (8) thick methane frozen layers (up to 4 x 10^{19} C-atoms cm^{-2} or $\sim$ 19μm) were bombarded with 1.5 MeV proton beams. With low fluences ($\lesssim$ 5 x 10^{16} protons cm^{-2}) we obtained dark polymer-like thick residues with dimensions 2 mm x 2 mm x 15 μm and a density of $\simeq$ 0.5 g cm^{-3}.

In the future we plan to bombard a mixture of μm sized dust imbedded in a methane frost matrix. In the light of the macroscopic residues already obtained we expect the dust will be trapped and extracted as a unique body with the polymer-like residue acting as a glue.

Another results potentially important in the context of this paper has been obtained by Calcagno and Foti (9). When bombarding commercial polymers containing H and C at fluences up to 10^{17} ions/cm^2, they observed a continuous increase of the density of the material ranging from the initial $\sim$ 0.5 g cm^{-3} to $\sim$ 1 g cm^{-3}. Accordingly we expect that our newly synthesized polymer-like materials also evolve toward higher densities if bombardment occurs at higher doses. This as a probable consequence of a progressive loss of the hydrogen content shown by IR analysis (10) and backscattering measurements (11).

The residues extracted after moderate ($\sim$ 10^{16} ions cm^{-2}) ion fluences indicate a ratio H / C = 2.5 / 1 to have evolved from the H / C = 4 / 1 ration in the intial frozen methane deposit.

III. ASSUMPTIONS

To apply in the next sections the above sketched experimental results some assumptions on the fluxes of energetic ions, and composition and shape of interstellar dust grains at their leaving the parent comet are needed. In order to derive the dust properties at its arrival at the upper atmosphere of the Earth where it can be collected (12), we assume, following Greenberg (13), that this event occurs after a time as long as 10^6 yrs since its release.

In the next we shall consider only the effect of protons although the relative helium ion abundance is high and since the erosion yields measured for helium are higher than for protons, helium could contribute as much as protons to the grain destruction (14). The bulk of solar wind protons has a velocity of $\sim$ 400 km s^{-1} and a density $\sim$ 10 p cm^{-3}. Assuming, as is customary, that the density is controlled by adiabatic expansion, it decreases as r^{-2}, r being the distance from the Sun.

The average proton flux from solar flares has been variously estimated. Following Lanzerotti et al. (14) we assume ϕ(E $\gtrsim$ 100 keV) $\sim$ 10^{11} p cm^{-2} yr^{-1} at 1 AU and varying as E^{-2}.

The composition and shape of interplanetary dust as it is released from the parent comet is not known. In the following we consider a spherical shape and two distinct grain populations : (1) pure ice grains and (2) bird's nest type grain i. e. constituted of submicron sized (silicate) dust ($\sim$ 50 %) imbedded in a volatile matrix ($\sim$ 50 %) in which important components (say $\sim$ 25 %) are carbon containing molecules (H$_2$CO, CH$_4$, HCN, HCO, HCONH$_2$ etc...).

The real grain population is, of course a more varied and complicated than assumed here. For a detailed description of the present thinking on the chemicophysical properties of comet dust see e. g. Greenberg (13).

IV. ICE GRAIN DESTRUCTION

Lanzerotti et al. (14) showed the importance of the high erosion yield for fast ions flowing on frozen H$_2$O on the stability of ice grains in the interplanetary medium.

To see if erosion by particles is the dominant mechanism of destruction we have to compare the production rates (g cm^{-2} sec^{-1}) with those expected by sublimation that are a very strong function of the temperature as well as of the grain dimension. Patashnick and Rupprecht (15) predicted, both by theoretical and experimental work, that the spectral absorption properties of the grains together with the spectral distribution of solar photons, leads to a quasi stationary size ($\sim$ 20 μm) of interplanetary ice particles independent of solar distances. Based on this finding Lanzerotti et al. (14) showed the dominant mechanism of erosion for the 20 μm sized ice grains to be the interaction with solar wind protons at distances as low as ∿ 1.5 AU.

The results by Patashnick and Rupprecht (15) have been seriously questioned by Parravano and Ferrin (16) in a detailed study on the sublimation rates of ice particles in circular orbits around the Sun. From their Fig. 4 we see however that at 2 AU the lifetime against sublimation of ice grains with radius $\gtrsim$ few μm is infinite. The lifetime of these grains against "sputtering" by solar wind protons is :

$$\tau = \frac{M}{YJ \ m_{H_2O} \ \pi a^2} \quad (sec) \tag{1}$$

where M = grain mass (g)
      Y = erosion yields $\simeq$ 0.2 for keV protons (see Lanzerotti et al. (14))
      J = proton (1 keV) flux $\simeq 10^8$ p cm^{-2} sec^{-1} at 2 AU
      m_{H_2O} = mass of the sputtered H_2O molecules
      a = grain radius

For a = 10 μm we find τ = 6 x 10^4 yr and this demonstrates once again the fundamental role played by ion erosion for the stability of interplanetary ice grains.

V. BUILD-UP OF NEW MATERIALS

In this section we consider the grains of type (2) described in sect. 3. This experimental results show that the effect of particle bombardment on frozen carbon-containing volatiles is not only destructive, as for pure ice, but is also the way to induce substantial changes in the physical properties as well as in the chemical composition of the bombarded targets ; i. e., to build up new materials.

The experiments have shown that the polymerization process occurs along the entire ion path i. e. the initial hydrocarbon layer is polymerized from the surface to a depth comparable with the range of the incoming ion. This is shown in Fig. 1 where the thickness (in C-atom cm^{-2}) of the solid residues obtained from frozen benzene bombarded by A_r^+ beams at various energies is plotted versus ion fluence.

The range of 100 keV protons in a matrix of the

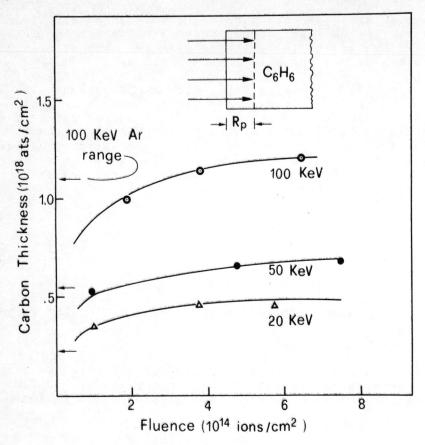

Fig. 1   The thickness (C-atoms cm⁻²) of solid residues obtained by bombarding benzene frosts with A⁺ᵣ ion
at three energies is plotted versus ion fluence. The ranges of ions are indicated by arrows.

type assumed for our grain is of the order of microns.
Thus the carbon atoms contained in the volatile portion
of the grain should evolve towards a long-chain polymeric material of the type observed in the lab when bombarded by the assumed flux of 100 keV solar flare protons. This as a consequence of the high energy deposition rate that produces along the ion track (a cylinder
of tens of A around the ion path) a high concentration
of radicals whose recombination gives rise to long chain
stable residue.

        Experiments show also that, if the ion range is
greater than the thickness of the deposited film, almost
all of the film is changed into a stable residue. This

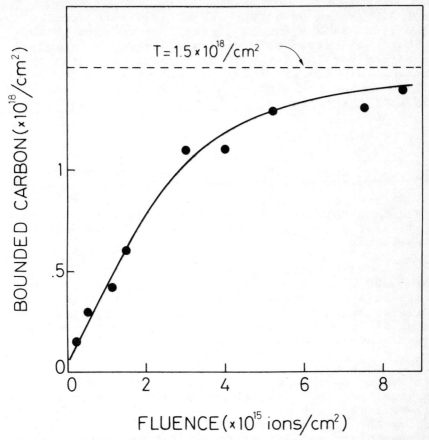

Fig. 2   The thickness (C-atoms cm^{-2}) of a residue obtained by bombarding a 1.5 x 10^{18} C-atoms cm^{-2} thick benzene layer with 100 keV protons is plotted versus ion fluence.

is shown in Fig. 2 where the thickness of solid residues obtained from a 1.5 x 10^{18} carbon atoms cm^{-2} thick benzene layer bombarded by 100 keV protons is plotted versus ion fluence. The result confirms that the process is a volume one ; i. e., it occurs along the entire ion path. During bombardment the film is also eroded but this is a surface phenomenon whose yield for benzene has not been measured. However assuming Y = 1 mol/ion we find that after a dose of ∿ 8.10^{15} ions cm^{-2} only 5 x 10^{16} carbon atoms cm^{-2} are sputtered. From Fig. 2 we see that at the same dose a solid residue as thick as ∿ 1.4 x 10^{18} C = atoms cm^{-2} has been synthesized.

The cross section for polymerization induced by

1.5 MeV proton on methane frost has been measured (8) and found be $\sigma = 4 \times 10^{-17}$ cm^2. For 100 keV protons σ should be greater but even considering the same value as obtained for 1.5 MeV and the formula (1) given by Strazzula et al. (7) we find a relation between the proton flux (J) and the time ($t_{50\%}$) needed to bond together the 50 % of the carbon atoms in a grain :

$$Jt_{50\%} = 1.7 \times 10^{16} \text{ protons cm}^{-2} \qquad (2)$$

Using the flux assumed in sect. 3 we find $t_{50\%} = 6.8 \times 10^5$ yr at 2 AU. This means that an important fraction of carbon atoms can be bonded together in $\sim 10^6$ yr spent by a grain in the interplanetary medium before reaching the Earth or the Moon where it can be collected or revealed, a posteriori, by its impact on the surface.

Many of the so called Brownlee particles consist of ~ 100 A silicate particles bonded together and mixed with carbonaceous material to form a μm sized aggregate. Here we suggest that the carbonaceous material could be similar to that ones synthesized in the lab and the glue that binds the silicate dust.

In the absence of particle bombardment the volatile component of grains would be completely volatilized by sublimation and single submicron silicate particles be revealed with the only organic polymeric material being the original silicate mantles which evolved from ultraviolet photoprocessing in the interstellar medium (17).

The total proton ($\gtrsim 100$ keV) fluence during 10^6 yr on interplanetary grains is $\sim 10^{17}$ p cm^{-2} i. e. of the same order used in the lab by Calcagno and Foti (9) to change the density of commercial polymers from 0.5 to 1 g cm^{-3}. Thus we expect that this "packing" effect should work also on interplanetary grains on which polymers are built up by the above described mechanism or are the relic of pre-cometary interstellar dust (13). This packing effect could contribute to understand the results by Le Sergeant d'Hedencourt and Lamy (2) who find, from analyses of lunar microcraters, that interplanetary dust has a density of the order of several g cm^{-3} while grains of low density (~ 0.3 g cm^{-3}) are probably not present to a large extent.

The IR spectra of solid residues extracted after the bombardment have been obtained (18-10). In Fig. 3

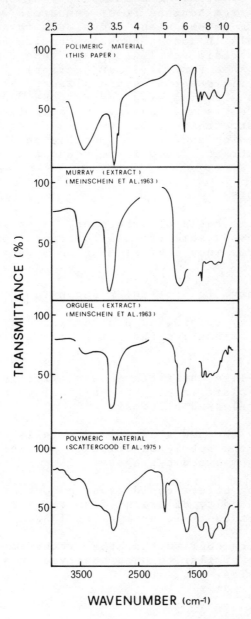

Fig. 3   The IR spectrum of a laboratory synthesized
polymer (upper section) is compared with that of organic materials extracted from meteorites and that of a
polymeric material obtained by Scattergood et al. (21)

the infrared spectrum of a laboratory synthesized residue obtained from benzene frost bombarded with 10^{16}p
cm^{-2} (100 keV) is shown and compared with those from organic materials in carbonaceous chondrite extracts as
found by Meinschein et al. (19) and reported by Knacke
(20). In the same Fig. 3 (lower section) the spectrum
from a polymeric material synthesized by Scattergood et
al. (21) by bombarding with MeV protons dense hydrocarbon atmospheres is also shown. The main features around
2.9 µm (Q-H stretching), 3.4 µm (C-H stretching) and
6 µm (O-H bending and/or C = C double bond) observed in
meteoritic extracts are well reproduced in the spectrum
of our residue.

VI. CONCLUSIONS

        On the basis of laboratory experimental results
we have discussed some effects that fast ions from solar
wind and flares induce by flowing on interplanetary
grains.

        We have in particular shown that :

i)   Erosion by solar wind protons is a very effective
      (often dominant) mechanism of ice grain destruction
      (14).

ii)   Polymer-like materials should effectively be produ-
      ced by solar flares protons flowing on complex
      grains released by comets.

iii) Possibly the new synthesized material is the glue
      bonding the submicron silicate particles to form a
      complex µm sized aggregate.

iv)   A packing effect could cause a continuous increase
      in the density of grains from the time they leave
      the comet to their arrival at 1 AU.

v)   IR spectra confirm that the synthesized materials
      could be similar to ones observed in some meteori-
      tic samples.

REFERENCES

 1. Whipple, F. L. : 1976, in "Interplanetary Dust and
     Zodiacal Light", H. Elsasser and H. Fechtig eds.,
     Springer-Verlag, pp. 403-414.

 2. Le Sergeant d'Hedencourt, L. B., and Lamy, Ph. L. :
     1980, Icarus 43, pp. 350-372.

3. Greenberg, J. M., and Gustafson, B. A. S. : 1981, Astron. Astrophys. 93, pp. 35-42.

4. Johnson, R. E., Lanzerotti, L. J., and Brown, W. L.: 1982, Nucl. Instr. Methods 198, pp. 147-156.

5. Tombrello, T. A. : 1982, Rad. Effects 65, pp. 389-398.

6. Strazzulla, G., Pirronello, V., and Foti, G. : 1983 , Astrophys. J. 271, pp. 255-258.

7. Strazzulla, G., Calcagno, L., and Foti, G. : 1983 , Mon. Not. Royal Astron. Soc. 204, pp. 59P - 62P.

8. Foti, G., Calcagno, L., Sheng, K. L., and Strazzulla, G. : 1984, Nature 310, pp. 126-128.

9. Calcagno, L., and Foti, F. : 1984 (work in progress).

10. Strazzulla, G., Cataliotti, R. S., Calcagno, L., and Foti, G. : 1984, Astron. Astrophys. 133, pp. 77-79.

11. Brown, W. L., and Lanzerotti, L. J. : 1983, (private communication).

12. Brownlee, D. E. : 1978, in "Cosmic Dust", J. A. M. Mc Donnell ed., John Wiley and Sons, pp. 295-336.

13. Greenberg, J. M. : 1983, in "Asteroids, Comets, Meteors", C. I. Lagerkvist and H. Rickman eds., Uppsala University, pp. 259-272.

14. Lanzerotti, L. J., Brown, W. L., Poate, J. M., and Augustyniak, W. M. : 1978, Nature 272, pp. 431-433.

15. Patashnick, H., and Rupprecht, G. : 1977, Icarus 30, pp. 402-412.

16. Parravano, A., and Ferrin, I. : 1981, Rev. Mexicana Astron. Astrophys. 6, pp. 347-350.

17. Greenberg, J. : 1982, in "Comets", ed. L. L. Wilkening, Univ. of Arizona Press, pp. 131-163.

18. Moore, M. H., and Donn, B. : 1982, Astrophys. J. Lett. 257, pp. L47-L50.

19. Meinschein, W. G., Nagy, B., and Henessy, D. J. :
     1963, Ann. N. Y. Acad. Sci. 108, pp. 553.

20. Knacke, R. F. : 1977, Nature 269, pp. 132-134.

21. Scattergood, T., Lesser, P., and Owen, T. : 1975,
     Icarus 24, pp. 465-474.

DISCUSSION

Johnson :

1) If you are irradiating Greenberg particles then the
   carbonaceous materials already exists. If not in
   Greenberg particles there is no CH_4.

2) You mentioned solar wind particles. These only penetrates a few hundred angstrom.

Strazzulla :

(1) We suggest that irradiation of the submicron core-
    mantle Greenberg particles mixed together in a fro-
    zen matrix containing whatever hydrocarbon (not on-
    ly CH_4) produces new organics that bind the submi-
    cron particles to form µm-sized stable aggregates.

(2) We mentioned solar wind particles ($\sim$ 1 keV) as res-
    ponsible of ice grain destruction and ions from fla-
    re ($\gtrsim$ 100 keV) as responsible of the polymerization
    process. These latter penetrate more than few µm.

Mc Kinnon :

(1) How refractory are your "carbonaceous" residues ?

(2) Is there any evidence for elemental carbon in your
    bombarded films ?

(3) Have you run experiments on films composed solely of
    water and ammonia ice ?

Strazzulla :

(1) We have tested the stability of our residues at
    T $\simeq$ 500 K ;

(2) No direct evidence ; we however observe an hydrogen

depauperament with ion fluence and the formation of C = C bonds from IR spectra ;

(3) No. Experiments on water and ammonia ice have been performed at FOM in Amsterdam (see e. g. Haring et al., 1983, Nucl. Instr. and Methods 211, p. 529).

SPUTTERING OF WATER ICE AT 30 - 140 K
BY 0.5 - 6.0 keV H+ AND Ne+ IONS

A. Bar-Nun[1], G. Herman[1], M.L. Rappaport[2] and Yu. Mekler[1]

1. Dept. of Geophysics and Planetary Sciences, Tel Aviv
   University, Tel Aviv, Israel.
2. Dept. of Physics and Astronomy, Tel Aviv University,
   Tel Aviv, Israel.

The sputtering from water ice at 30-140K by H^+ ions of 0.5-6 KeV
was studied experimentally. The sputtering of water molecules
was found to be temperature independent over the whole temperature range. Neon ions deposit their energy in the ice by
nuclear collisions, while for protons the energy deposition
mechanism shifts gradually from predominantly nuclear collisions
to predominantly electronic processes from 0.5 to 6 KeV. The
existing theory of nuclear sputtering predicts very well the
yield of ejected water molecules and the experimental results
in the region of electronic processes agree well with the
experimental results of Lanzerotti, Brown and Johnson. However,
the major mass loss of water bombarded by ions is via the
ejection of O_2 and H_2 molecules and of H atoms, which exceed
the ejection of water molecules. O_2 and H_2 production is
markedly enhanced at temperatures exceeding ~100K whereas H_2O
and H production are temperature independent, suggesting that
O_2 and H_2 are produced in the bulk of the ice whereas H_2O and
H atoms are ejected from the surface or near surface layers.
About 2% of the mass loss is due to the ejection of positive
ions and clusters.

INTRODUCTION

Most of the ices in the solar system are subject to bombardment
by either solar wind or magnetospheric particles. Comets,
interplanetary icy grains and the moons which lie outside the
planetary magnetospheres are swept by the solar wind, which

2.Present address: Elscint Cryotronics Ltd. 10 Tower Road,
                   Berinsfield, Oxford, England

J. Klinger et al. (eds.), Ices in the Solar System, 287–299.
© 1985 by D. Reidel Publishing Company.

consists of protons and electrons (with a few percent helium
ions) with energies of ~1 KeV per atomic mass unit and,
occasionally, after sporadic eruptions on the sun, by particles
with energies up to 10's of MeV. The moons of Jupiter and
Saturn's icy ring particles which are within their magnetospheres, are subject to bombardment by magnetospheric particles
whose population varies from planet to planet both in composition and in energy, as well as with place and time. For
example, the Jovian magnetospheric plasma, which is dominated
by ejecta from Io, consists of oxygen and sulfur ions mainly in
the KeV energy range, but contains also a considerable flux in
the MeV energy range.

The pioneering experimental and theoretical work of
Lanzerotti, Brown and Johnson (L.B.J.) (e.g. 1-8) on sputtering
of water ice and other ices (SO_2, CH_4, CO_2) by protons and
heavier ions (He, C, O, Ar) covered the energy range from
10 KeV to a few MeV, thus rendering important information on
sputtering by high energy ions, as well as providing valuable
information on the mechanism of interaction of these ions with
ices. We extended their study to the low energy range: 0.5-
6 KeV, with protons and neon ions, for direct application of
the sputtering data to ices in the solar system which are
bombarded by solar wind protons and by low energy, magnetospheric, heavier ions. In addition, we gained more insight
into the mechanism of interaction of KeV ions with water ice,
which differs from the interaction with the more energetic
particles. The coverage of a wide range of ion energies and
masses by L.B.J. and by us, makes it possible to distinguish
between the two mechanisms of energy deposition by ions in ice,
namely electronic and nuclear, and provides a means for calculating the results of bombardment of water ice in the solar
system by ions with varying masses and energies.

EXPERIMENTAL PROCEDURE AND RESULTS

The experimental setup is shown schematically in Figure 1. The
details of the experimental procedure were given elsewhere (9)
and will be described here only briefly. Water vapor was
frozen as amorphous ice on a plate cooled to 30K, in a chamber
whose pressure was maintained by cryopumping at ~10^{-8} Torr.
The temperature of the few microns thick ice layer was controlled
between 30 and 140K to within ±1K. Protons and neon ions with
energies between 0.5 and 6 KeV bombarded the ice with a flux
of ~10^{14} ions cm^{-2} sec^{-1}. The fluxes of the various species
which were ejected from the ice were measured by a precalibrated
quadrupole mass-filter. Similar results were obtained with
amorphous ice and with cubic ice, which was formed by heating
the amorphous ice to 150K and cooling down to 30-140K.

The sputtering yield for water by protons: water molecules

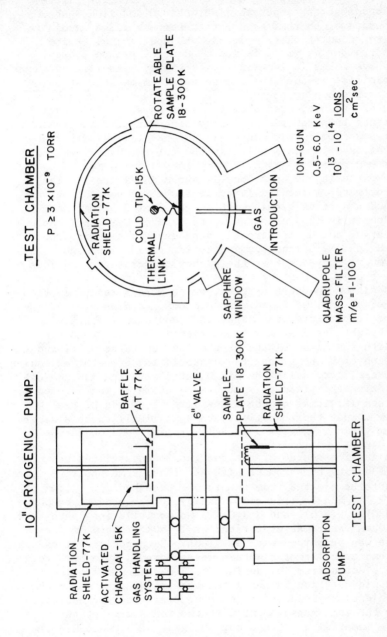

Figure 1: Schematic illustration of the experimental arrangement

ejected per proton impinging on the ice, is shown in Figure 2
to be independent of temperature between 30 and 140K. Above
this temperature, thermal evaporation becomes large enough to
mask the ejection of water molecules by sputtering. Below
2 KeV the sputtering yield is independent of proton energy,
whereas above 2 KeV it increases with energy. As will be shown
later, this rise is due to a changeover from nuclear stopping
to electronic stopping of the protons in the ice. Oxygen
molecules are also ejected from the ice, as shown in Figure 2,
with a strong temperature and energy dependence. Ejection of
molecular and atomic hydrogen could not be studied with proton
bombardment and will be studied in the future, using deuterated
water.

The sputtering by neon ions is shown in Figure 3. The
ejection of water molecules is temperature independent between
30 and 140K and rises slowly with energy from 0.5 to 2 KeV,
above which it remains constant up to 6 KeV. Here too, the
ejection of O_2 and H_2 is strongly dependent on temperature.
For both H^+ and Ne^+ ions the sputtering yield of H_2 and O_2 was
strongly energy dependent at the higher temperatures and less
so at lower ones.

In addition to these species, which were already reported
by Brown et al (3), we found that H atoms are also ejected, and
seem to be temperature and energy independent. In obtaining
the peak height for H atoms we subtracted the background of
H atoms with the high voltage of the ion gun turned off, and the
contribution of H atoms from the dissociation of water molecules
in the mass-filter, using the H/H_2O ratio with the ion gun off.
In calculating the yield of sputtered H atoms it was assumed
that the sensitivities of the mass-filter to H and H_2 are
identical. Since the sensitivity to H is likely to be smaller,
the H atom yield in Figure 3 is a lower limit and could be
somewhat higher.

When the ion source of the quadrupole mass-filter was
turned off during bombardment with Ne^+ ions, some positive ions
were seen to be ejected from the ice. These had m/e ratios of
17, 18, 19, 37 and 55 (Figures 4 and 5), corresponding to OH^+,
H_2O^+, $H(H_2O)^+$, $H(H_2O)_2^+$ and $H(H_2O)_3^+$. No positive ions were
observed when water was evaporated at T>150K or when an ice-
free plate was bombarded with Ne^+ ions. Such positive ions and
clusters and even larger clusters were observed by others (e.g.
10) to be ejected from ice during its bombardment with ions.
Yet, we could not find in the literature any reference to the
relative ejection yields of ions to neutral water molecules.
By making the same measurement with the ion source of the
quadrupole mass-filter turned on, the neutral H_2O and $H(H_2O)^+$
ion were measured simultaneously. Assuming a ~5% efficiency
for the ionization of neutral H_2O molecules in the ion source
of the quadrupole mass-filter, which is not off by more than a
factor of 2 (11), would imply, from Fig. 5, an approximate

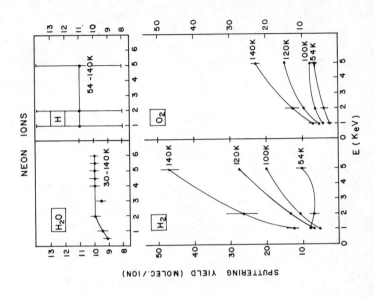

Figure 3: Sputtering yields for water, oxygen and hydrogen molecules and atomic hydrogen, from the bombardment of ice at 30–140K by 0.5–6 KeV neon ions.

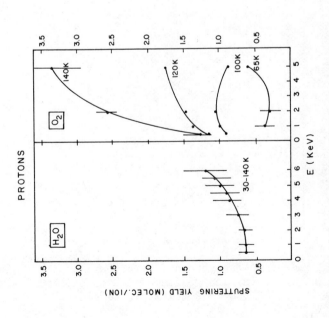

Figure 2: Sputtering yields for water and oxygen molecules from the bombardment of ice at 30–140K by 0.5–6 KeV protons.

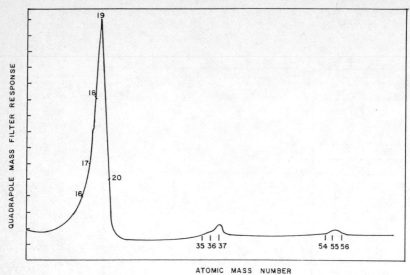

Figure 4: A mass–spectrum obtained by bombarding ice at 30K with 1.5 KeV Ne⁺ ions, with the ion source of the quadrupole mass–filter turned off.

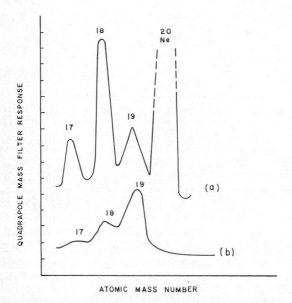

Figure 5: A mass–spectrum obtained by bombarding ice at 30K with 2 KeV Ne⁺ ions. (a) ion source of the mass–filter is turned on. (b) ion source of the mass–filter is turned off.

sputtering yield for $H(H_2O)^+$ of 0.2 $(^{+0.2}_{-0.1})$, as compared with 10 for neutral H_2O. This species being the major ion sputtered all other ions would be sputtered with even smaller efficiencies (Fig. 4). Thus, the ejection of neutral species exceeds the ejection of ions by a factor of about 50.

DISCUSSION

The Sputtering Mechanism

The impinging ions can deposit their energy in the ice by one of two mechanisms: elastic nuclear collisions or electronic processes, depending on the nature of the target, the impinging ion and the ion energy. L.B.J. found for H^+, He^+, C^+ and O^+ ions at energies from 10 KeV to several MeV that the sputtering of water molecules from ice is proportional to the square of the electronic stopping power - $(dE/dx)^2_{elect}$. Haff (12) found a similar dependence for MeV F^+ ions. Their results are represented by the straight line in Figure 6, taken from Brown et al (3). Our

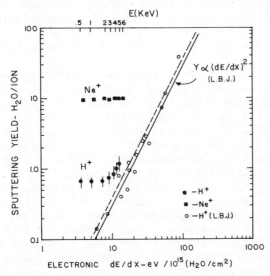

Figure 6: Sputtering yield for water molecules, from the bombardment of ice at 30-140K by 0.5-6 KeV H^+ and Ne^+ ions vs the electronic stopping power - $(dE/dx)_{elect}$. calculated from Anderson and Ziegler (13). The L.B.J. results for sputtering of ice by ~10 KeV - 1.5 MeV protons (open circles and solid line) are shown for comparison. The dashed line represents a 30% increase of the L.B.J. sputtering yield, as expected from the 60° angle of incidence of the ion beam in our experiments. The ion energies at the top scale apply only to our results.

results of sputtering yield vs electronic stopping, calculated
as $2(dE/dx)_H + (dE/dx)_0$ from the tables of Anderson and
Ziegler (13), are also plotted in Figure 6. The dashed line
in the figure is 30% higher than the (solid) L.B.J. line. This
is because the ion beam in their experiments was perpendicular
to the ice, whereas in our experiments the beam was 30° away
from perpendicular, which should increase the sputtering yield
by ~30% according to Sigmund (14, Fig. 11). Comparison of our
results with the dashed line in Figure 6 clearly shows that only
the sputtering yields for protons of 4-6 KeV approach the
electronic $(dE/dx)^2$ line, whereas with protons of lower energies
and with Ne^+ ions of all energies the sputtering yields are much
higher than those expected from electronic stopping. It seems
therefore that energy transfer through nuclear collisions con-
tributes considerably to the sputtering of water molecules by
ions of 0.5-6 KeV. This is demonstrated in Figure 7, where the
sputtering yields obtained in the experiments are compared with
the sputtering yields calculated for nuclear energy transfer,
from Sigmund (14) and the sputtering yields from electronic processes which were obtained by L.B.J. (the dashed line in Fig. 6).
As seen in Figure 7, Ne^+ ions at 0.5-6 KeV deposit their energy
exclusively by the nuclear process, whereas H^+ ions deposit
their energy by the nuclear process at low energies with a
gradual shift towards electronic energy transfer as their energy
increases.

In the 5-6 KeV energy range for protons, the sputtering is
divided between ~1/3 nuclear and ~2/3 electronic processes. If
our experimental points at these energies (Fig. 6) are lowered
to 2/3 of their values, the agreement between our results and
those of L.B.J. (dashed line in Fig. 6) is very good. This
agreement is encouraging, in view of the entirely different
methods used for obtaining the experimental sputtering yields.
Whereas L.B.J. used Rutherford back-scattering of 1.5 MeV He^+
ions to measure the thickness of the ice film and hence its
depletion due to sputtering, we measured directly the flux of
sputtered H_2O molecules by the quadrupole mass-filter.

In conclusion, the sputtering of water molecules from ice
is temperature independent between 30 and 140K. Ne^+ ions deposit
their energy exclusively by nuclear collisions over the energy
range 0.5 to 6 KeV, whereas with protons there is a gradual
shift from nuclear collisions to electronic processes. Thus, the
sputtering yield of water molecules from ice up to 140K can be
now calculated for a wide range of ion masses and energies.
The contribution to the sputtering from nuclear processes can be
calculated from the theory of Sigmund (14) and the contribution
of electronic processes from Anderson and Ziegler's (13) tables
of electronic stopping vs ion energy and mass and from L.B.J.'s
experimental curve of sputtering yield vs electronic stopping.

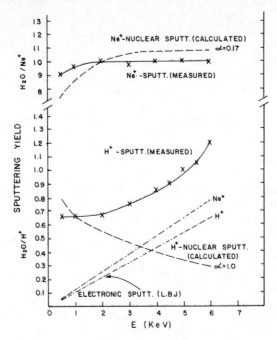

Figure 7: Comparison of the theoretical and experimental sputtering yields of water molecules from ice at 30–140K vs ion energy for H$^+$ and Ne$^+$ ions. The nuclear sputtering yield was calculated from Sigmund (14), with α = 0.17 for Ne$^+$ and α = 1.0 for H$^+$ and increased by 30% to account for the 60° angle of incidence of the ion beam. The electronic sputtering yield was obtained by calculating the electronic stopping power $(dE/dx)_{elect.}$ for each ion energy and using the L.B.J. correlation of sputtering yield with $(dE/dx)_{elect.}$ – dashed line in Fig. 6.

Sputtering of O_2, H_2, H and ions

The sputtering of O_2, H_2 and H by Ne$^+$ ions is shown in Figure 3 and sputtering of O_2^- by H$^+$ ions is shown in Figure 2. In the bombardment by protons, only O_2 ejection was measured, because of the interference of hydrogen from the ion gun. At all temperatures, the major mass loss from the ice is through H_2 and O_2 and much more so at higher temperatures, where the ejection of H_2 and O_2 increases markedly.

This strong temperature dependence is understandable if we assume that H_2 and O_2 are formed along the track of the penetrating ion, down to a considerable depth. The atoms thus formed have to migrate through the ice and find other atoms to recombine

with. Then, the newly formed H_2 and O_2 molecules have to
migrate outward, to the surface. The higher the temperature of
the ice lattice, the larger is the mobility of these species.
Yet, even at 54-64K, considerable amounts of H_2 and O_2 are
formed and find their way to the surface. Brown et al (3) found
that the rate of ejection of O_2 and H_2 decreases with the decrease in the ice thickness, which demonstrates that O_2 and H_2
are formed throughout the ice layer. The sputtering of H_2O
molecules and H atoms does not depend on temperature, because
both are probably ejected from the surface and do not have to
migrate through the lattice. H atoms, which tunnel easily
through the ice lattice, would recombine with each other when
formed deep in the lattice and only those formed at or near the
surface can escape as atoms.

No other species, up to m/e = 100 were detected and neither
was H_2O_2 seen when all the ice was evaporated, at 180K. The
amount of ejected hydrogen (H_2 + H) was twice the amount of
ejected O_2 within the experimental error. Thus, there is no
enrichment by either oxygen or hydrogen as a result of bombardment by Ne^+ ion and probably H^+ as well.

Yet another observation is the increase of the sputtering
of H_2 and O_2 with ion energy at high temperatures and the
diminishing effect at lower temperatures (Figs. 2,3). It seems
that the longer paths of the more energetic ions in the ice can
manifest themselves more easily when the matrix is softer.
Since H_2O and H production is almost energy independent, the
behavior of H_2 and O_2 again support the suggestion of bulk
production of the latter species as compared with the surface
production of the former ones. Since the formation of H_2 and O_2
takes place throughout the lattice, while the ejection of H_2O
molecules is a surface phenomenon, it is not surprising that the
ice is sputtered mainly through H_2 and O_2 and not as water
molecules. In addition, the ratio of $H/H_2O \approx 1$ in the ejecta,
which is indicative of the relative rates of H formation vs H_2O
ejection, suggests that the breaking of bonds of water molecules
at the surface is comparable to the ejection of whole water
molecules. The ejection of positive ions and clusters is
certainly a surface phenomenon (10). Its small efficiency
relative to that of neutral water ejection ($\sim$1/50) is a measure
of the relative importance of ion processes vs neutral processes
at the ice surface.

CONCLUSIONS

The major mass loss from water ice bombarded by ions in the
0.5-6 KeV energy range is through the ejection of O_2, H_2 and H.
At $\sim$50K this mass loss is comparable to that of H_2O ejection,
while at higher temperatures and energies ejection of O_2, H_2 and
H becomes dominant. Positive ions contribute about 2% to the
mass loss.

At low energies the mechanism of energy deposition in the ice matrix is via nuclear interactions and, as the energy increases, electronic processes take over. The transition from nuclear to electronic processes depends on the impinging ion. For protons this transition occurs at $\sim$ 3 KeV, whereas for Ne^+ ions it should occur at energies much higher than those reached in our experiment (Fig. 7).

The theory of sputtering by nuclear processes predicts very well the sputtering of water molecules by H^+ and Ne^+ ions. It fails, however, to predict the total mass loss from ice by ion bombardment, by not taking into account the ejection of considerable amounts of O_2, H_2 and H, which dominates the mass loss above $\sim$ 50 K.

The wide range of ion masses and energies studied by L. B. J. and by us permits an accurate determination of the rate of mass loss and the nature of ejected species for a variety of icy bodies in the solar system to be made.

REFERENCES

1.   Brown, W. L., Lanzerotti, L. J., Poate, J. M. and Augustyniak, W. M., 1978, Phys. Rev. Lett. 40, pp. 1027-1030.

2.   Brown, W. L., Augustyniak, W. M., Brody, E., Cooper, B., Lanzerotti, L. J., Ramirez, A., Evat, R. and Johnson, R. E., 1980, Nucl. Instr. and Meth. 170, pp. 321-325.

3.   Brown, W. L., Augustyniak, W. M., Simons, E., Marcantonio, K. J., Lanzerotti, L. J., Johnson, R. E., Boring, J. W., Reimann, C. T., Foti, G. and Pirronello, V., 1982, Nucl. Instr. and Meth., 198, pp. 1-8.

4.   Lanzerotti, L. J., Brown, W. L., Poate, J. M. and Augustyniak, W. M., 1978, Geophys. Res. Lett., 5, pp. 155-158.

5.   Lanzerotti, L. J., Brown, W. L., Augustyniak, W. M., Johnson, R. E. and Armstrong, T. P., 1982, Astrophys. J., 259, pp. 920-929.

6.   Johnson, R. E., Lanzerotti, L. J. and Brown, W. L., 1982, Nucl. Instr. and Meth., 198, pp. 147-157.

7.   Johnson, R. E., Brown, W. L. and Lanzerotti, L. J.,
      1983, J. Phys. Chem., 87, pp. 4218-4220.

8.   Johnson, R. E., Boring, J. W., Reimann, C. T., Bar-
      ton, L. A., Sieveka, E. M., Garret, J. W., Farmer,
      K. R., Brown, W. L. and Lanzerotti, L. J., 1983,
      Geophys. Res. Lett. 10, pp. 892-895.

9.   Bar-Nun, A., Herman, G., Rappaport, M. L. and Mek-
      ler, Yu, 1984, Surface Sciences (In press).

10.   Lancaster, G. M., Honda, F., Fukuda, Y. and Raba-
      lais, J. W., 1979, J. Am. Chem. Soc., 101, pp.
      1957-1958.

11.   Riber Manual No. 6.016, Quadrupole Analyzer Series
      QX.

12.   Haff, P. K., 1979, J. Geophys. Res., 84, pp. 8436-
      8442.

13.   Anderson, H. H. and Ziegler, J. F., 1977, Hydro-
      gen : Stopping powers and ranges in all elements,
      Pergamon, N. Y.

14.   Sigmund, P., 1969, Phys. Rev., 184, pp. 383-415.

This work was supported by a grant fron the Adler Fund
for Space Research received through the Israel Foundation Trustees.

DISCUSSION

V. Pirronello :

     What is the thickness of the ice layer you accre-
te on the substrate before ion bombardment ?

     I'm asking this because too thick sample could
be in a poor thermal coupling with the substrate during
irradiation.

A. Bar-Nun :

     The ice layer is several microns thick. Its ther-
mal conductivity is large enough so that its surface is
not heated during ion bombardment to temperatures where
thermal evaporation becomes significant. We are certain
of this point, since we observed with thick ice layers
a strong temperature dependance of the sputtering yield

for water molecules, which was actually a contribution
from skin heating and subsequent thermal evaporation.
As the bulk temperature of the ice was increased, its
surface temperature approached 150-160K, where thermal
evaporation is high.

        With thin ice layers, the sputtering yield for
water molecules did not depend on temperature, between
30 and 140K. Hence, in this case, heat conduction
through the ice is fast enough to prevent skin heating
to 150K even when the bulk of the ice is at 140K.

R. E. Johnson :

        The observation of H atoms which is temperature
independent is very interesting and would confirm the
difference between photon modification of the material
and proton modification. These differences need to be
explored.

A. Bar-Nun :

        The sputtering of H_2O molecules and H atoms is
temperature independent between 30 and 140K, whereas the
sputtering of H_2 and O_2 molecules depend strongly on ice
temperature. H and O atoms, which are fragmented from
the water, have to migrate in the ice matrix and find
other atoms to recombine with. The newly formed H_2 and
O_2 molecules then have to migrate to the surface. At
higher temperatures, the ice matrix is softer and these
migrations are facilitated. Hence, the increase in O_2
and H_2 production at higher temperatures.

        The H atoms seem to be sputtered from the ice
surface since, if they were to tunnel through the ice,
their high mobility would lead to recombination with
other atoms or radicals and their subsequent disappearance. Similarly, H_2O molecules ares sputtered from the
surface and not from deep layers. Hence the different
dependance on temperature of H_2 and O_2 vs H_2O and H
atoms.

CHARGED PARTICLE MODIFICATION OF ICES IN THE SATURNIAN AND JOVIAN SYSTEMS

R. E. Johnson,[+] L. A. Barton,[+] J. W. Boring[+],
W.A. Jesser,[*] W.L Brown,[□] and L.J. Lanzerotti[□]

[+]Department of Nuclear Engineering   [*]Department of Materials
    & Engineering Physics   Science
University of Virginia   University of Virginia
Charlottesville, VA 22901   Charlottesville, VA 22901

[□]AT&T Bell Laboratories
Murray Hill, NJ 07974

ABSTRACT

    We discuss the modification by ion bombardment of the surfaces of icy objects in the Saturian and Jovian systems. Chemical changes in ices are induced by breaking of bonds and by implantation of incident ions. Long-term irradiation by fast ions produces physical changes such as increasing the surface reflectivity and ability to scatter light. On large satellites molecules which are ejected by ion bombardment are redistributed across the surfaces of large satellites. For small satellites and ring particles bombarded by ions, such as those of Saturn, most or all of the sputtered material is lost to space, forming a neutral torus in the locale of the satellite orbits and rings and supplying ions to the magnetosphere. Noting the existence of such a torus the sputter erosion and possible stabilization of the E-ring of Saturn is discussed.

J. Klinger et al. (eds.), Ices in the Solar System, 301–315.
© 1985 by D. Reidel Publishing Company.

INTRODUCTION

Pioneer and Voyager spacecraft measurements showed that the magnetospheres of Saturn and Jupiter confine high fluxes of energetic ions (1,2). These ions continuously bombard the surfaces of the icy satellites and rings. In this paper we briefly explore some of the changes induced in the ices and the effect of the bombardment on the composition of the magnetospheric plasma.

The magnetospheric plasma ions are arbitrarily divided by kinetic energy into two components: (a) an energetic keV to 100's of keV component of light ions, mostly protons, which are assumed to bombard the satellites nearly isotropically, (b) a component of slow and/or heavy ions having kinetic energies much less than the energy associated with the rotation of the magnetic field and having gyroradii much less than the satellite radii. This component can be thought of as co-rotating with the magnetic field lines and preferentially bombards the trailing hemisphere of the satellites. Small ring particles can be thought of as being isotropically bombarded by all of the ions.

Because of this bombardment the icy surfaces observed in the Jovian and Saturnian systems are considerably modified from that of a pristine, freshly deposited ice. Ion bombardment results in the modification of the surface microscopic roughness (3), sputtering (or ejection) of material from the satellite (4) and redistribution across the surface (5), changes in the surface concentration of impurities (3), and modification of the composition by inducing chemical reactions (6-10) and by ion implantation (11,12). That the albedo of the surface is affected by such exogenic processes is indicated by a number of observations. Photographs of Ganymede and Europa show leading/trailing asymmetries and visual differences from equators to poles (13). Lane et al. (11) have found differences in the U.V. from the leading to the trailing hemisphere of Europa and Nelson et al. (14) have shown that systematic leading-trailing (as well as pole to equator) asymmetries in the ratio of U.V. to violet reflectance spectra exist for Europa and Ganymede. Initially it was uncertain whether, in fact, the ions could reach the satellite surfaces if these objects had small atmospheres or magnetospheres. The above observations, however, provide evidence that such bombardment does occur. In the following we review and discuss, briefly, a few of the effects produced by ion bombardment of the satellites of Jupiter and Saturn and the E-ring of Saturn. We first consider the chemical and physical changes induced in the surface layers and then the loss of surface (sputter ejection) and redistribution of this material.

CHEMICAL AND PHYSICAL CHANGES

Ions bombarding ices break chemical bonds producing free radicals. At very low temperatures and/or low density of excitation this produces an icy surface containing trapped radicals. As the temperature is increased such species may become mobile and react. In D_2O ice this results in the formation of D_2 and O_2 (15-18), in CO_2 ice, in the formation of CO, O_2 (8), and in SO_2 ice, in the formation of SO_3 (19). In addition to such new volatile species, it has been found that non-volatile residues are produced by ion irradiation of carbon and sulfur containing species (papers by Lanzerotti et al. and Strazzula et al. in these proceedings).

The measured sputter yield of ice was found to be temperature dependent (15,20). This was shown to be due to the production of new species (O_2 and D_2) as seen in Fig. (1) from the measurements of sputter yield for He^+ incident on D_2O ice. The decomposition of ice by radiation does not occur for incident U.V., x-rays or γ-rays at normal fluxes. It is a distinct effect of charged particle bombardment, a result of the high ionization densities produced by secondary electrons. Such differences may help distinguish between the importance of U.V. and proton irradiation of comets (8).

The yields of new molecular species, D_2 and O_2, which are seen in Fig. 1 to be ejected from the ice film, are found to have rather different dependences on the irradiation time (fluence) than the ejected D_2O molecules (18). From Fig. 2 it is seen that D_2O is ejected promptly on initiation of the ion bombardment and this signal is independent of temperature over a broad range of temperatures 10K to 140K (Fig 1). In contrast, the mass 32 (O_2) and mass 4 (D_2) yields are both temperature dependent (Fig. 1) and fluence dependent (Fig. 2) . Any off-on sequence in the ion beam results in an immediate return of the yield to a value equal to that before it was turned off. This indicates an irreversible change in the ice composition has occurred due to ion bombardment. For D_2 erosion it is seen in Fig. 2 that there is a prompt signal when the beam first strikes a fresh film followed by a delayed transient increase. This suggests that rapid formation of D_2 can occur on electronic recombination, in addition to diffusion and chemical reaction of fragments. At higher temperatures than shown in Fig (2) two transients are seen in the O_2 signal and both transients depend on temperature. The temperature dependence observed in these transients indicates that thermally activated processes are occurring having activation energies of the order of 0.05 to 0.07 eV (18).

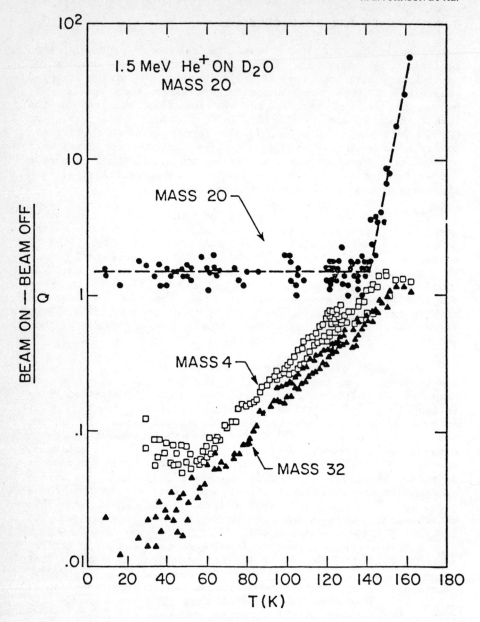

Fig. 1 Quadrupole mass spectrometer signal for neutral molecules ejected from D_2O ice by 1.5 MeV He^+ ions vs film temperature (15).

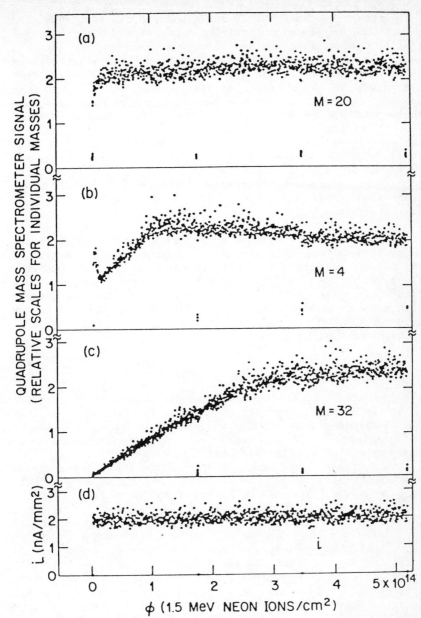

Fig. 2   Quadrupole mass spectrometer signal for yields of
neutral mass 20, 4 and 32 molecules ejected from a 10K ice layer
plotted vs the fluence for 1.5 MeV neon ions. Plot (d) is the
ion current. Beam was turned off and back on again at points
marked and signals returned to their original levels. (18).

The strength of observable spectroscopic features resulting from ion-induced chemical activity will depend on irradition time until the creation and destruction of new species attain a balance. For mass 80 (SO_3 or S_2O) produced by ions incident on SO_2 (19) this balance occurs when a fluence of ~10^{14}ions/cm^2 has been received by the surface, similar to the result for D_2O in Fig. (2). For larger fluences a new equilibrium composition of molecules may be produced in the ice. The resulting column density in an ice of a new molecular species depends primarily on the ion penetration depth if the ice layer is thick (i.e. not fully eroded). We have shown, for example, that an observable (from earth) SO_3 spectroscopic feature in an SO_2 ice layer on Io could only be produced by fast magnetospheric ions bombarding Io as the slow co-rotating ions at Io do not penetrate to sufficient depths (19).

The incident ion itself is a radical which can react after stopping in the ice. For sulfur ions bombarding the surface of Europa this is thought to be the source of an SO band observed on the trailing side (11). The strength of such a spectroscopic feature will depend on the ion penetration depth and the rate of erosion of the surface by the ions (12). The equilibrium column density (N_I) of the implanted feature in an ice layer is independent of bombarding flux

$$N_I \approx f_I \, [(\bar{Y}/n\bar{R}_p) + \bar{\sigma}_D]^{-1} \tag{1}$$

Here $\bar{R}_p$ is the projected range (the perpendicular penetration depth of the ions of interest), f_I is the fraction of such ions in the bombarding flux, and n the molecular number density of the ice (molecules per unit volume). Incident ions may remove the new molecular species produced by previous bombarding ions at a rate determined by some average destruction cross section, $\bar{\sigma}_D$ (an effective area for collisions), and by erosion of the surface at a rate determined by the average sputter yield, $\bar{Y}$ (molecules ejected per ion incident). For 1 keV S^+ in water ice, n$\bar{R}_p$~ 3×10^{15} mol/cm^2, $\bar{Y}$ ~ 10^2, assuming no destruction and f_I~ $1/3$, then N_I~ 10^{13} (S/cm^2) implanted. This rather low value for the implantation column density is a result of the small penetration depths and large yield. On an object with non-negligible gravity, however, $\bar{Y}$ is not simply the laboratory sputter yield, since all of the molecules ejected are not sufficiently energetic to escape or even travel very far from their ejection site; further, the surface may be very rough thereby restricting ejection. In such a case $\bar{Y}$ is the net removal (molecules/ion incident) of surface, locally, after transport across the satellite surface is accounted for. Sieveka and Johnson (5) have carried out sample calculations of this transport for the Jovian and Saturnian satellites. Recent laboratory measurements of the velocity distributions of the sputtered species have led to accurate escape fractions for use in such calculations (21,22). Although the composition of the incident magnetospheric plasma is required in order to determine the strength of such a spectroscopic feature, the observation by Lane et al.,(11) of an SO feature with column density of the order of 10^{15} SO/cm^2 on the trailing hemisphere of Europa requires incident sulfur ions much more energetic than the co-rotation energy in order to obtain reasonable projected ranges in Eq (1). Confirmation of this spectroscopic feature, therefore, would have an important bearing on the composition of the energetic plasma ion flux at the Europa orbit. This composition was not uniquely determined by Voyager in the vicinity of the Galilean satellites.

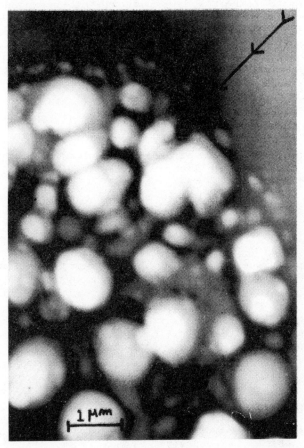

Fig. 3 High voltage electron micrograph of an initially uniform film of H_2O at about 95K after irradiation with 80 keV He$^+$ ions to a fluence of 10^{16} ions/cm^2. Ion beam was incident at 20 to the normal. The ion beam diameter was a few millimeters.

Long-term bombardment of icy surfaces has been thought of as a smoothing process. However, we have shown that clear deposits become highly reflecting and scatter light efficiently after bombardment by fast ions (23,24). In addition, in Fig. (3) we show a transmission-electron-microscope photograph of an initially uniform, thick ice layer which experienced 'long-term' ion and electron bombardment. The resulting layer is very non-uniform, with many areas having micron size pits. Such a surface would not scatter light according to Lambert's law, as would a snow-covered surface. Veverka and Gradie (25) have shown this for micron-size holes produced in an icy surface by other means. It is quite possible that such holes, produced by ion bombardment, can account for the fact that even the brightest ice covered satellites (e.g. Enceladus) do not scatter light according to Lambert's law, although other explanations have been given (25). Therefore, accurately measuring, in the laboratory, physical and chemical changes induced by ion bombardment is necessary for understanding the observed spectroscopic details of the reflectivity (albedos) of the icy satellites. We note in this regard that low energy ions which do not penetrate very far (e.g., hundreds of angstrom) do not seem to produce light-scattering surfaces.

LOSS OF SURFACE AND REDISTRIBUTION

Ions bombarding surfaces redistribute molecules by sputter ejection. On objects with little gravitational pull all the material is ejected to space, whereas on large objects (e.g. Ganymede) most of the material is redeposited on the surface. Sieveka and Johnson (5) have calculated such surface redistribution due to non-isotropic ion bombardment and sublimation. The non-isotropy in the ion bombardment comes from the measured ion pitch-angle distributions in the magnetosphere; the non-isotropy in the sublimation results from the latitudinal and longitudinal dependence of the temperature of the satellite's surface. Molecules redeposited on the surface by either process may form ice grains depending on the deposition conditions (26). These grains in turn can be destroyed by ion bombardment and by sublimation. Therefore, on Ganymede, grain growth might be expected to dominate in the polar regions where there is a net deposition of material, while grain destruction would dominate in the equatorial, trailing-hemisphere region (5). This, therefore, could result in latitudinal variations in the albedo much like that observed.

Material which does escape the gravitational field and is ejected into space can survive, temporarily, as neutral molecules. These molecules are eventually ionized by the magnetospheric plasma electrons or solar U.V. photons. Once ionized they are a source of ions for the magnetospheres of the planets. Using measurements of the sputter yield and velocity distributions we have calculated the azimuthally averaged source rates of oxygen ions for the magnetosphere of Saturn (27,28) and show the results in Fig. (4). Not surprisingly, production occurs, primarily, close to the orbital plane, particularly for the inner satellites. Because the ejected molecules have relatively low mean velocities, the ions produced will also have large pitch angles, on the average, when picked-up by the rotating magnetic field. Therefore, the observed vertical gradient in the production of heavy ions is found to be similar in size to the estimated plasma scale height of co-rotating plasma ions found in the Voyager spacecraft data (29). In fact, as the hydrogen atoms associated with H_2O are also eventually ionized, the net supply rate of ions to the magnetosphere is three times that depicted in Fig. (4). The total source strength used here for the four satellites is of the order of 10^{24} ions/sec.

In the above we have made a very conservative estimate of the source of neutrals considering only those ions (keV-MeV) measured by the LECP instrument on Voyager (22,30) and, further, assuming that they are all protons. The average yields for incident protons are of the order of one to two water molecules

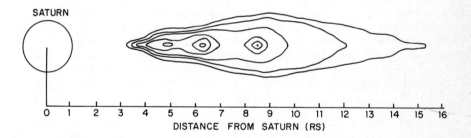

SATURN

DISTANCE FROM SATURN (RS)

Fig. 4 Contours for heavy ion production in the vicinity of Saturn's satellites. Outer contour 10^{-13} ions/cm^3/sec, contours increase by powers of ten. These result from ionization of water molecules sputtered from satellites by keV-MeV energy protons. Azymuthal symmetry about Saturn was assumed and the ejected neutral were assumed to travel in ballistic orbits until ionization. Multiplying by the life time against ionization at the Saturn orbit one obtains the number density of oxygen atoms in the neutral torus.

per ion incident and, in equilibrium, enough H or H_2 to account for the loss of the incident proton by implantation into the surface. Therefore, bombardment by protons is a <u>net source</u> of both new protons and heavy ions, O^+. The yields for keV to MeV O^+ bombardment are approximately two orders of magnitude larger (20). Including a large fraction of O^+ in this energetic plasma would, therefore, significantly change these estimates (30); the net source in this instance could exceed the estimated supply of nitrogen atoms from Titan. (Note: Our published values of ejection yields (Table III, reference (22)) are close to an order of magnitude too large; an erratum is published. The corrected data is used here and in reference (30).)

We have neglected in the above discussion bombardment by the low energy plasma ions measured by the plasma analyzer on Voyager. Co-rotating protons with impact energies of ~2 eV at Mimas to ~30 eV at Rhea should contribute negligibly to the yield (31). Co-rotating O^+ ions with impact energies 16 times larger could, in principle, contribute significantly (22,32) if one employs the standard expressions for the yield that is used when sputtering atomic solids (33). Evidence suggests, however, that the actual ejection yields from the icy surfaces are much smaller (32) because of either surface roughness and/or the molecular nature of the material (e.g. vibrational excitation). We also note that if the yield for incident low energy O^+ approaches unity then the net production of fresh heavy ions approaches zero as one oxygen atom is implanted for each water molecule lost. Therefore, the role of the incident O^+ remains to be determined. Results presented at this conference (31) show that the yields are of the order of 8 for 200 eV Ne^+ on ice. This suggests that there is a significant contribution from sputtering by co-rotating O^+ at the more distant icy satellites.

Since the neutral molecules ejected from the Saturnian satellites will survive for more than a few orbits, a toroidal atmosphere of neutral molecules will exist in the vicinity of the satellites and the E-ring. The lifetime time of these molecules is limited primarily by electron impact ionization and photoionization. Integrating the resulting neutral densities produces a line-of-sight column density of the toroidal atmosphere much too small to be observed from earth. It is possible, however, that these neutral molecules play a role in maintaining the E-ring.

Based on reasonable extrapolations of the plasma ion measurements of Voyager and Pioneer to the satellite orbital plane near Enceladus, which lies within the E-ring, it is found that the E-ring particles (~2 μm) would be eroded away by keV-MeV proton bombardment in about 10^5 years (34) and by keV O^+ in the order of 10^3 years (30). If indeed this is the case, then this ring is a relatively young feature of the Saturnian system. For the E-ring to be a stable feature a supply of water molecules is needed to balance the sputter loss. As the surface area of Enceladus is much larger than the total surface area of E-ring particles it was thought possible that neutrals ejected from Enceladus can stabilize the ring. Haff et al.(32) have shown that if only Enceladus is considered as a source, and if the sputter yields for the surfaces of Enceladus and the ring particles are equal, then stablization is unlikely based on present estimates of the lifetimes of the neutral molecules and their velocities of ejection. Using the numbers in Fig. (4) and the approximate ionization time to estimate the total number density of neutrals in the vicinity of the E-ring our calculations would confirm this conclusion. Taking the extreme case, however, in which there is much less plasma in the vicinity of the ring and/or essentially no sputtering, then neutrals ejected from the neighboring satellites could, in principle, be a source of water molecules for such a ring feature. At present, however, best estimates suggest this ring is a recent feature and its lifetime is limited by sputtering and micrometeoroid bombardment (35). This provides a high incentive for carefully describing the plasma characteristics and the sputter rate of water ice in this region of the Saturnian magnetosphere.

At Jupiter, a heavy particle plasma of sulfur and oxygen ions is known to be produced in the vicinity of Io. This was thought to be produced by sputter ejection from the surface of Io of SO_2 ice patches deposited by the volcanic activity on this satellite (2,36,37). Recent measurements of the sputter ejection rate and sputtered molecule energy spectra indicate that, unlike the water molecules ejected from the surfaces of the icy Saturnean satellites, only a small fraction of the sputtered SO_2 molecules are energetic enough to escape Io's gravitational field (19). Recent modelling of the processes occurring near Io strongly suggests a small atmosphere of sublimated SO_2 occurs (38) from which the plasma ions can eject sufficient sulfur and oxygen (36,39,40). A small atmosphere, sufficient for magnetospheric erosion, could possibly be maintained by ion sputtering (41). The existence on Io of such a sublimated or sputtered generated atmosphere above an SO_2 ice surface remains an interesting and open problem.

ACKNOWLEDGMENTS

Work supported at Virginia by the NSF Astronomy Division under grant AST-82-00477 and NASA Planetary Science Geophysics and Geochemistry Division under grant NAGW-186.

REFERENCES

1. Bridge, H. S., Belcher, J. W., Lazarus, A. J., Ol-
    bert, S., Sullivan, J. D., Bagenal, F., Gazis, P.
    R., Hartle, R. E., Ogilvie, K. W., Scudder, J. D.,
    Sittler, E. C., Eviatar, A., Siscoe, G. L., Goertz,
    C. K. and Vasyliumas, V. M., 1981, Sciences, 212,
    pp. 217-224.

2. Krimigis, S. M., Carbary, J. E., Keath, E. P., Bos-
    trom, C. O., Axford, W. I., Gloeckler, G., Lanze-
    rotti, L. J. and Armstrong, T. P., 1981, J. Geo-
    phys. Res., 86, pp. 8227-8257.

3. Haff, P. K., Watson, C. C. and Tombrello, T. A.,
    1979, Proc. 10th. Lunar Planet. Sci. Conf., pp.
    1685.

4. Matson, D. L., Johnson, T. V. and Fanale, F. P.,
    1974, Astrophys. J., 192, pp. L43-L46.

5. Sieveka, E. M. and Johnson, R. E., 1982, Icarus, 51,
    pp. 527-548.

6. Cheng, A. F. and Lanzerotti, L. J., 1978, J. Geo-
    phys. Res., 83, pp. 2597-2602.

7. Pirronello, V., Brown, W. L., Lanzerotti, L. J.,
    Marcantonio, K. T. and Simmons, E. H., 1982, Astro-
    phys. J., 262, pp. 636-640.

8. Johnson, R. E., Lanzerotti, L. J., Brown, W. L., Au-
    gustyniak, W. M. and Mussil, C., 1983, Astron. As-
    trophys., 123, pp. 343-347.

9. Moore, M. H., Donn, B., Khanna, R. and A'Hearn, M.
    F., 1983 , Icarus, 54, pp. 388-405.

10. Strazzula, G., Pirronello, V. and Foti, G., 1983,
    Astron. Astrophys., 123, pp. 93-97.

11. Lane, A. L., Nelson, R. M. and Matson, D. L., 1981,
    Nature, 292, pp. 38-39.

12. Johnson, R. E., Lanzerotti, L. J. and Brown, W. L.,
    1982, Nucl. Instrum. Methods, 198, pp. 147-158.

13. Smith, B. A., Soderblom, L. A., Johnson, T. V., In-
    gersoll, A. P., Collins, S. A., Shoemaker, E. M.,

Hunt, G. E., Masursky, H., Carr, M. H., Davies, M. E., Cook, A. F., Boyce, J., Danielson, G. E., Owen, T., Sagan, C., Beebe, R., Morrison, D., Briggs, G. A. and Suomi, V. E., 1979, Science, 204, pp. 951-971.

14. Nelson, M. L., McCord, T. B., Clark, R. N., Johnson, T. V. and Soderblom, L. A., 1984, Icarus (in press).

15. Brown, W. L., Augustyniak, W. H., Simmons, E., Marcantonio, K. J., Lanzerotti, L. J., Johnson, R. E., Boring, J. W., Reimann, C. T., Foti, G. and Pirronello, V., 1982, Nucl. Inst. & Meth., 198, pp. 1-8. Brown, W. L., Augustyniak, W. M., Marcantonio, K. J., Simmons, E. H., Boring, J. W., Johnson, R. E. and Reimann, C. T., 1984, Nucl. Inst. 1 Meth. (in press).

16. Ciavola, G., Foti, G., Torrisi, L., Pirronello, V. and Strazzula, G., 1982, Rad. Effects, 65, pp. 167-172.

17. Haring, R. A., Haring, A., Klein, F. S., Kummel, A. C. and deVries, A. E., 1983, Nucl. Inst. & Meth, 211, pp. 529-538.

18. Reimann, C. T., Boring, J. W., Johnson, R. E., Garret, J. W., Farmer, K. R. and Brown, W. L., 1984, Surface Science (in press).

19. Moore, M. H., 1984, Icarus, 59, pp. 114-128. Johnson, R. E., Garrett, J. W., Boring, J. W., Barton, L. A. and Brown, W. L., 1984, J. Geophys. Res., 89, Supp., pp. B711-B715.

20. Brown, W. L., Augustyniak, W. M., Brody, E., Cooper, B., Lanzerotti, L. J., Ramirez, A., Evatt, E. and Johnson, R. E., 1980, Phys. Rev. Lett., 45, pp. 1632-1635.

21. Johnson, R. E., Boring, J. W., Reimann, C. T., Barton, L. A., Sieveka, E. M., Garrett, J. W., Farmer, K. R., Brown, W. L. and Lanzerotti, L. J., 1983, Geophys. Res. Lett., 10, pp. 892-895.

22. Lanzerotti, L. J., MacLennan, C. G., Brown, W. L., Johnson, R. E., Barton, L. A., Reimann, C. T., Garrett, J. W. and Boring, J. W., 1983, J. of Geophys. Res., 88, pp. 8765-8771 ; erratum 89, p. 9157.

23. Brown, W. L., Lanzerotti, L. J., Poate, J. M. and Augustyniak, W. M., 1978, Phys. Rev. Lett., 40, pp. 1027-1030.

24. Smythe, W. D., 1984, presented at this workshop ; not published.

25. Veverka, J. and Gradie, J., 1983, Bull. Amer. As. Soc., 15, p. 853.

26. Mayer, E. and Pletzer, R., These proceedings.

27. Barton, L. A., 1983, "Magnetospheric ion erosion of the icy satellites of Saturn". Masters Thesis, University of Virginia, Charlottesville, VA 22901.

28. Johnson, R. E., Sieveka, E. M., Cheng, A. F. and Lanzerotti, L. J., 1984, J. Geophys. Res., to be submitted.

29. Lazarus, A. J. and McNutt, R. M. Jr., 1983, J. Geophys. Res., 88, pp. 8831-8846.

30. Johnson, R. E., Lanzerotti, L. J. and Brown, W. L., 1984, Proceedings of the XXV COSPAR Conference, Graz, in press.

31. Bar-Nun, A. These proceedings.

32. Haff, P. K., Eviatar, A. and Siscope, G. L., 1984, Icarus, in press. Sigmund, P., 1969, Phys. Rev., 184, pp. 383-416.

33. Sigmund, P., 1969, Phys. Rev., 184, pp. 383-416.

34. Cheng, A. F., Lanzerotti, L. J. and Pirronello, V., 1982, J. Geophys. Res., 87, 4567-4570.

35. Morfil, G. E., Fechtig, H., Grun, E. and Goertz, C. K., 1983, Icarus, 55, pp. 439-447.

36. Haff, P. K., Watson, C. C. and Yung, Y. L., 1981, J. Geophys. Res., 86, pp. 6933-6938.

37. Lanzerotti, L. J., Brown, W. L., Augustyniak, W. M., Johnson, R. E. and Armstrong, T. P., 1982, Astrophys. J., 259, pp. 920-929.

38. Kumar, S., 1982, J. Geophys. Res. 87, pp. 1677-1684.

39.   Cheng. A. F., 1984, Geophys. Res. Lett., in press.

40.   Sieveka, E, M. and Johnson, R. E., 1984, Astrophys. J., in press.

41.   Lanzerotti, L. J. and Brown, W. L., 1983, J. Geophys. Res., 88, pp. 989-990.

LABORATORY STUDIES OF ION IRRADIATIONS OF WATER, SULFUR DIOXIDE, AND METHANE ICES

L. J. Lanzerotti and W. L. Brown
AT&T Bell Laboratories
Murray Hill, New Jersey 07974, USA

R. E. Johnson
University of Virginia
Charlottesville, Virginia 22901, USA

ABSTRACT. We review the results obtained at AT&T Bell Laboratories for the erosion and modification of water, sulfur dioxide, and methane ices by energetic (~10 - 2000 keV) ions. Such ions are common constituents of the solar wind and planetary magnetospheres and can interact with the ices on the surfaces of the Jovian and Saturnian satellites, in comets, and in interplanetary grains. The particle irradiations produce high erosion rates and significant chemical modifications of the ices. Of particular importance in the outer solar system, at Uranus and beyond, is the alteration of methane ice -- or mixtures of methane ice and water -- in which polymer residues are produced by ion irradiation.

1. INTRODUCTION

The environment of the ices in the solar system consists of photons from the sun (primarily in the visible and UV wavelengths) and various plasma populations such as the solar wind and the magnetospheres of the planets. The plasmas are primarily hydrogen and electrons, with a wide range of energies and fluences. A few percent helium ions are typical in the solar wind, while the magnetosphere of Jupiter, for example, contains intense fluxes of keV-energy sulfur and oxygen ions.

The solar wind, continually flowing out from the sun with a density of a few particles per cubic centimeter at Earth, is composed of ions and electrons with energies of approximately 1 keV per atomic mass unit. Eruptions on the sun produce sporadic outbursts of solar cosmic rays with energies from tens of thousands to tens of millions of electron volts per atomic mass unit. The magnetosphere particle populations (fluences and compositions) vary from planet to planet and have spatial and temporal dependencies at each planet. Examples of energetic ($E \geq 20$ keV) ion spectra measured at Jupiter and Saturn by the Low Energy Charged Particle (LECP) experiment (1) on the Voyager spacecraft are shown in Fig. 1, in order to provide some idea of the general intensity

317

J. Klinger et al. (eds.), Ices in the Solar System, 317–335.
© *1985 by D. Reidel Publishing Company.*

levels. Significant particle intensities extend down to energies as low as a few eV and up to energies as high as several 100 of MeV per atomic mass unit.

The recognition that many solar system ices in their myriad of manifestations are inevitably bombarded by at least some of these ions led to the initiation of experiments to measure the consequences of such bombardment (2). Studied were the erosion and modification of ice layers and the formation of molecular fragments as well as more complex molecules. It is well-known that energetic ions incident on more conventional solids results in the ejection of atoms and molecules. Indeed, this physical process was investigated in considerable depth in the process of understanding the surface properties of material returned from the moon in the Apollo program (3-5).

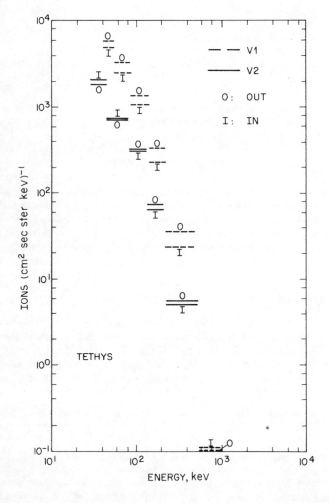

Figure 1a. Ion spectra measured by the Voyager Low Energy Charged Particle Experiment (LECP) in the vicinity of Tethys (Saturn).

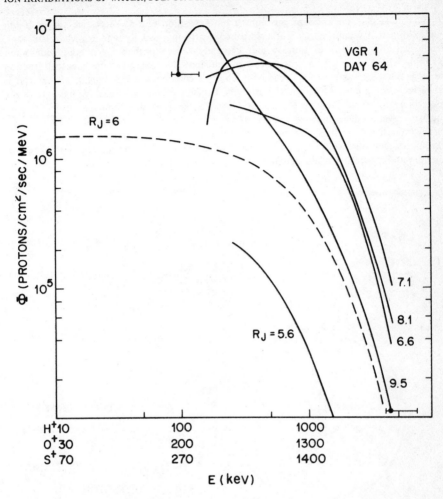

Figure 1b. Ion spectra measured by the Voyager Low Energy Charged Particle Experiment (LECP) in the vicinity of Io (Jupiter).

The studies of the ion bombardment of ices at AT&T Bell Laboratories have been carried out primarily using the experimental arrangement sketched in Fig. 2. Ion energies from ~10 keV to ~3 MeV per atomic mass unit are available. Ice films are condensed from the vapor on cold metallic substrates. The substrate is beryllium, covered with a 50 Å gold layer to serve as a marker in measurements of ice film thicknesses. Films from a few hundred to a few thousand Angstroms in thickness are grown at a typical rate of 1000 Å per minute. Substrate temperatures are chosen to be well below the effective sublimation temperature for the ice being studied and, for many experiments, are ~10 K. The base vacuum in the region of the substrate, $<10^{-8}$ torr, is maintained by a combination of turbomolecular and cryogenic pumping.

The molecular thickness of the films after deposition and at successive stages of ion bombardment is monitored by Rutherford backscattering, usually with 1.5-MeV He$^+$ ions. Such scattering is due to the low but precisely known probability of an incident ion colliding with a nucleus in a nearly head-on collision. The energy of the backscattered ion depends on the mass of the nucleus with which it collided and the amount of material it traversed (losing energy) before and after the collision. The most sensitive measure of ice film thickness is provided by the energy of the backscattered helium ions from the gold marker layer under the ice. This gives a thickness sensitivity of ∼10 Å. The backscattered helium ions also directly reveal heavy-atom constituents of a film, such as carbon, oxygen, and sulfur, giving an independent measure of the molecular thickness of the film and a measure of its stoichiometry at various stages of erosion (6).

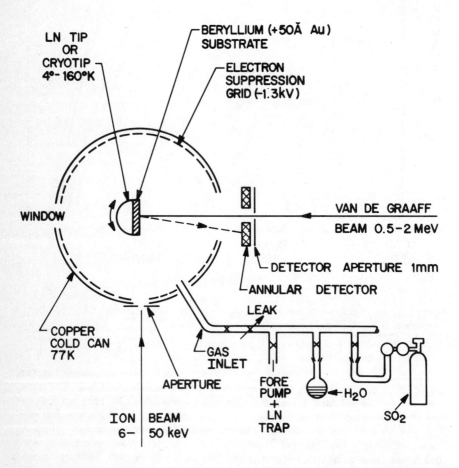

Figure 2. Schematic illustration of experiment arrangement, including cold finger, copper cold can, gas manifold, annular solid state detector, and ion beams from Van de Graaff and electrostatic accelerators.

Beams ~1-2 mm in diameter are defined by collimation. The erosion yield Y (molecules of film material lost per incident ion) is found to be independent of beam current. Thus, the erosion ocurrs as a result of individual ions of the beam, not through macroscopic heating of the films by the total beam (2).

A number of different ices have been studied under ion irradiation at AT&T Bell Laboratories. Three of these (H_2O, SO_2, CH_4) are of particular interest for several solar system objects, and they are reviewed and discussed here. Our laboratory results for ion bombardment of CO_2 ice and CO ice are discussed in (7) and (8), respectively. Physical mechanisms proposed for the erosion process have been reviewed in (9) and (10), and are not elaborated upon in this paper. Water ice is found in the satellites of the giant planets (for example, refs. 11-14), in the rings of Saturn (e.g., ref. 14), in the nuclei of comets (refs. 15, 16) and is likely to be a constituent of interplanetary grains (for example, refs. 15, 17-21). Volcanic Io, the first Galilean satellite of Jupiter, has sulfur dioxide frost deposits (for example, refs. 22-26). Pluto (and its companion moon) are likely to be covered with methane ice or frost (27-29).

2. CONSIDERATIONS FOR SOLAR SYSTEM OBJECTS

The gravitational attraction of a body is a major factor in determining the net loss of ejected material by ion bombardment of its surface. Atoms and molecules will be ejected from a surface with a distribution in energy that depends on the physical mechanisms of the erosion process. For collision cascade sputtering of metals this distribution is well characterized (30); it is beginning to be measured for electronically-induced erosion of condensed volatiles (31). If the energy required for gravitational escape from a body is much larger than the mean ejection energy, then sputtering will predominantly redistribute material across the surface and the escape fraction will be determined by the tail of the energy distribution (32). For small icy objects with negligible gravitational attraction, such as comets (a few kilometers in diameter), planetary ring particles (centimeters to meters), and interplanetary ice grains (tens to thousands of micrometers), most of the sputtered material will be lost to space. The relevant physical parameters for several solar system bodies of interest are given in Table 1.

In order to be an altering influence on a planetary satellite, the incident particle radiation must first reach the surface. If a body has an intrinsic magnetic field, or is immersed in the field of another body, the trajectory of incident ions will be altered so as to exclude bombardment of some regions of the surface for ions lower than a certain energy. Areas of preferential particle bombardment (e.g., ref. 33) can also result from the interaction of the planetary magnetic fields with the intrinsic field or surface conductivity of a moon. This appears to occur with the Jovian satellite Io. Electrons and ions flowing along Jupiter's magnetic field lines are directed into the polar regions of the moon (34).

An atmosphere around a satellite will prevent particles with energies lower than a certain value from reaching any part of the surface. In fact, charged particles can produce a self-limiting atmosphere (35); the erosion of surface ices by such ions can build up an atmosphere that will ultimately prevent the lower energy particles from striking the surface. A balance will result between exclusion of the low-energy particles and erosion of the surface by the higher energy particles at a sufficient rate to maintain an atmospheric equilibrium (36).

TABLE I. Escape energies and velocities from the surfaces of various solar system bodies.

Object	Radius (km)	Density (g/cm³)	g (cm/sec²)	Escape energy (eV/amu)	Escape velocity (m/sec)
Jovian satellites					
Io	1,820	3.5	178	0.035	2,580
Europa	1,500	3.5	147	0.023	2,080
Ganymede	2,640	2.0	147	0.040	2,790
Callisto	2,500	1.6	114	0.029	2,390
Saturnian satellites					
Mimas	195	1.2	6.6	0.00013	150
Enceladus	250	1	7	0.0002	190
Tethys	525	1.1	15	0.0008	400
Dione	560	1.4	22	0.0013	500
Rhea	765	1.3	28	0.0023	660
Titan	2,570	1.9	136	0.0036	2,640
A-ring object	~ 0.001	~ 1	~ 2.6×10^{-5}	~ 2.6×10^{-15}	~ 7.2×10^{-4}
Moon	1,738	3.34	164	0.029	2,400
Earth	6,378	5.5	980	0.64	11,200

A molecule or atom eroded from the surface of a moon by a charged particle can suffer a variety of fates, some of which are illustrated in Fig. 3. It can escape directly or can strike the surface again (Fig. 3a), depending on the gravitational attraction of the body and the sputtered particle energy. If a moon has sufficient atmosphere, the outgoing sputtered particle will collide with atmospheric constituents, lose energy, and ultimately be absorbed on the surface (Fig. 3b). A molecule or atom eroded from a body can be ionized by solar or stellar photons or by an external plasma (Fig. 3c). Such ionized species can escape from the body, especially if there is an externally imposed magnetic field (such as a planetary or interplanetary magnetic field). In the case of Jupiter, for example, the planetary rotation speed, and hence the rotation speed of the planetary magnetic field, at the orbital locations of the large Galilean satellites (Table 1) is greater than the orbital speeds of the moons. Ionized species that find themselves on the planetary field lines are thus quickly swept away from the moon. In the actual situation for solar system moons, a combination of the simplified processes illustrated in Figs. 3a,b,c are generally operative, in particular the processes sketched in Figs. 3a and 3c.

3. WATER ICE

3.1 Experiments

Figure 4 shows the erosion yield of H_2O ice bombarded with hydrogen ions (6). The measured erosion yield Y peaks at ~100 keV and falls rapidly at lower and higher energies. Also shown is a calculation of the expected loss of H_2O from conventional ion sputtering (i.e., sputtering due to the formation of collision cascades initiated in the ice by direct momentum-imparting collisions of the incident ions with the nuclei of the ice atoms; ref. 37). Such a process is well understood and is known to be responsible for the loss of material in ion bombardment of metallic and semiconducting materials (30). The ices studied are all electronic insulators and it is clear that the material erosion from ices has a different origin.

The peak of Y in Fig. 4 occurs at an energy close to the maximum in the electronic stopping power of protons in the ice (6). The electronic stopping power of an incident ion, $(dE/dx)_e$, is the energy given up by the ion to ionization and excitation of the electrons of the material per unit path length of the ion. The erosion yield data plotted as a function of the stopping power in Fig. 5 (liquid nitrogen temperature) show clearly that electronic processes, and not nuclear collisions, control the erosion of the ice films. The figure includes the data of Fig. 4, which are ordered very well to a single line in the plot. For erosion by hydrogen and helium, the dependence of Y on $(dE/dx)_e$ is approximately quadratic. The results for the heavier ions suggest a steeper than quadratic dependence at the larger stopping powers. This regime of high-energy heavy ions is not of appreciable significance in solar system problems.

3.2 Applications

Icy bodies in interplanetary space -- for example, grains, comets, and minor bodies such as the recently-discovered Chiron -- can be eroded by solar flare and solar wind particles.

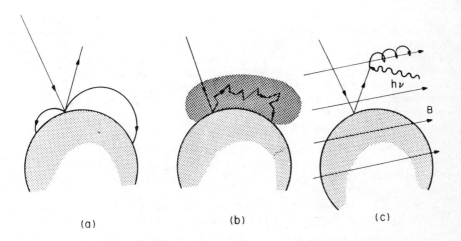

(a)   (b)   (c)

Figure 3. Schematic illustration of possible fates of material eroded from a surface, depending on the gravitational attraction and external environment of the object. (a) An eroded atom or molecule can escape or enter into a ballistic trajectory, depending on its ejection energy and the gravitational attraction of the object; (b) an atmosphere can cause scattering and reimpact of the ejected species on the surface; or (c) ionization by solar photons ($h\nu$) of an ejected species in the presence of an externally imposed magnetic field can cause a loss of the species. In practice, various combinations of these three conditions can also exist.

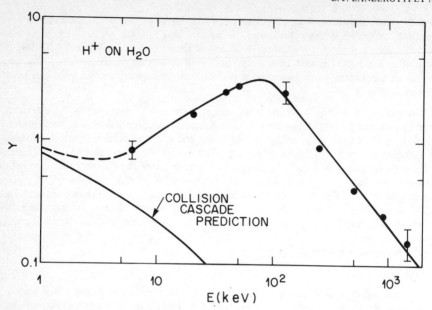

Figure 4. Erosion yield as a function of the energy of incident hydrogen ions.

The water ice erosion rate on grains expected as a function of heliosphere distance is shown in Fig. 6 (38), and is compared to that expected for sublimation of ~20 μm particles (particles with the longest lifetimes under solar photon irradiation; refs. 39, 40). Under the assumption that the solar wind is controlled by adiabatic expansion (ignoring enhanced densities in high-speed streams and interplanetary shocks), erosion by the wind is the dominant determinant of grain lifetimes beyond ~1.5 AU. Neglecting the solar wind, even the ocurrence of one typical solar flare particle event per year (41) will dominate the particle erosion rate beyond ~2 AU.

The ice particles comprising the faint E-ring of Saturn have sizes comparable to interplanetary grains. Under bombardment by Saturn magnetosphere ions, the lifetime of the E-ring, without replenishment, must be $\lesssim 10^5$ years (42,43), substantially less than the lifetime of the Saturn system (several $\times$ 10^9 years). Erosion of ring particles, as well as the surfaces of the Saturian satellites, could be a local source (42,44) of the heavy ion plasma in this region of Saturn's magnetosphere (45,46).

On the Galilean moons of Jupiter, particle erosion of the ices will result primarily in redistribution of the material on the surface (32). In the case of Saturn's moons, the eroded species will escape into the planet's magnetosphere (44,47; Johnson et al., this volume). On the Earth's moon, the redistribution of any frozen volatiles by sputtering by the solar wind and by magnetosphere particles would effectively eliminate the possibility of surface water ice in the polar regions of that body (48).

Sputtered water molecules could form the basis for a tenuous atmosphere around some of the Galilean moons (31,35,36,49,50). Calculations based on particle fluxes measured

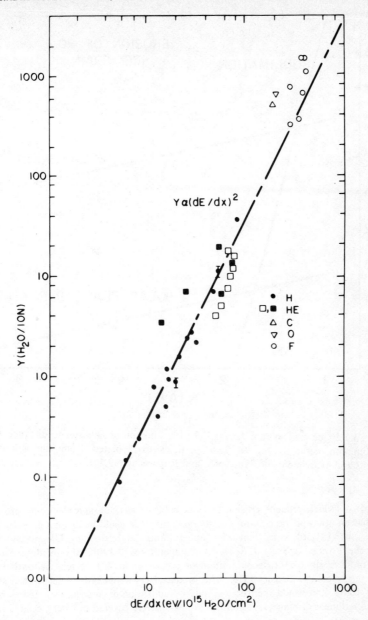

Figure 5. Erosion yield for water ice at liquid nitrogen temperature as a function of $(dE/dx)_e$, the electronic energy loss of hydrogen, helium, oxygen, carbon, and fluorine ions. ($\bullet$, $\blacksquare$, $\triangle$, and $\triangledown$) data from Brown et al., (6,67); ($\square$) data from Ollerhead et al., (68); ($\bigcirc$) data from Cooper and Tombrello (69) and Seiberling et al., (70).

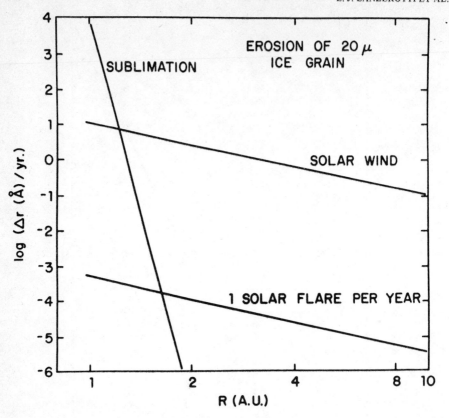

Figure 6. Water ice erosion rate ($\Delta r/yr$, Å) as a function of heliosphere distance (AU). The sublimation curve is for $\sim 20 \mu m$ particles extrapolated from the sublimation calculations and experiments of Patashnick and Rupprecht (39,40).

by Voyager 1 (which produce erosion by both collision cascade and electronic processes) suggest that an atmospheric column density on Ganymede could be as high as $\sim 10^{14}$ H_2O molecules cm^{-2} (31,36), consistent with upper limits established by UV measurements made by the Voyager spacecraft (13). At temperatures $\geq 120K$, the incident energetic ions can produce significant decomposition of the water ice (9). Such a situation could lead to a larger column density (31). The produced O_2 would not condense on a surface with $T \geq 100$ K and would then be lost only by ionization or dissociation. Issues related to ion bombardment of planetary satellites are extensively covered in Cheng et al. (51).

4. SULFUR DIOXIDE ICE

A set of experiments was conducted to study the erosion of sulfur dioxide ice in order to obtain quantitative data which could be used in evaluations of the influences of charged particles on the surface of Io. The erosion yield as a function of the incident ion energy for four ion species (hydrogen, helium, sulfur, oxygen) are shown in Fig. 7 (52). The experimental arrangement is the same as for water ice. The expected (calculated) contributions from collision-cascade sputtering, most significant at the lower energies, are also shown. Recent results suggest that these lower energy yields may be lower limits.

With the information of Fig. 7 and the measured particle fluxes from the Voyager spacecraft, the erosion yields can be calculated for Io for keV-MeV energy ion bombardment. The results are shown in Fig. 8 as a function of surface temperature (52,53). The erosion yields are shown to increase with higher temperature, reflecting the temperature dependence of the yields (9). If Io's surface temperature is less than ~90K, the erosion of SO_2 frost by the plasma ions and the energetic sulfur ions will dominate sublimation processes. This situation is likely to hold on the night side of the moon and in the polar regions. The higher energy ions may dominate the erosion process if Io has a small, intrinsic magnetic field that can shield the surface from the plasma ions. The eroded ions do not have sufficient energy to be ejected directly into the magnetosphere (31). Rather, the eroded ions can form the basis of an Io atmosphere,

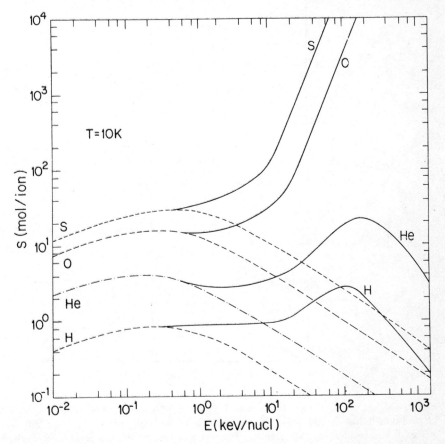

Figure 7. Sulfur dioxide erosion yield as a function of incident ion energy for four ion species. Dashed lines: contributions from collision cascade sputtering; solid lines: contributions from electronic sputtering based upon laboratory measurements.

which can also be sputtered away (49,50), and/or which can be ionized by solar photons and Jovian magnetosphere plasma and then become part of the magnetosphere population (e.g, Fig. 3c).

5. METHANE ICE

The major identification of methane in the solar system has been associated with Pluto and its "moon" Charon, where methane surface frost has been reported. There are also reports of possible methane atmospheres associated with Pluto and Neptune's moon Triton (54-57), although there appears to still be some uncertainty about the density of

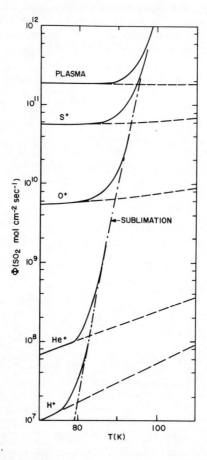

Figure 8. Flux of SO_2 from Io (assumed 100% SO_2 frost-covered) from sublimation (dashed-dotted line) and from sputtering plus sublimation (solid lines) for different species of ions incident on the satellite. The fluxes of ions were taken from data obtained during the Voyager 1 encounter with Jupiter and Io.

the atmospheres and the amount of detected signal that should be apportioned between surface and atmosphere.

It is well known that organic materials (such as methane) can be polymerized by irradiation. Indeed, irradiation is a step in the fabrication process of many polymers, including the polyethylene used in cable sheathing. A number of authors have discussed polymers in connection with the very dark portion of the surface of Iapetus (e.g., ref. 58). Mendis and Axford (59) discussed possible solar wind irradiation effects on satellites, including the dark features of Iapetus. Pirronello et al. (60) reported the formation of an organic compound (formaldehyde: H_2CO) in irradiating an ice mixture of equal parts of CO_2 and H_2O with 1.5 MeV helium ions. Changes in the surface layers of ice mixtures after irradiation by 1 MeV protons were reported by Moore and Donn (61).

We began experiments several years ago to determine quantitatively the effects of particle irradiation on methane ice, in order to assess the effects of the solar wind and solar cosmic rays on solar system objects that may contain such ices. Cheng and Lanzerotti (62) used these initial results to propose that darkening on the rings of Uranus could be produced by irradiation of initially methane-covered rings by the ions and electrons trapped in the planet's magnetosphere.

Recently, laboratory experimental results have been reported on the irradiation, by 600-1500 keV protons, of a methane atmosphere and on the polymerization of benzene ice (63,64). These results have been applied to a variety of astrophysical situations, including Pluto (65) and interstellar grains (66).

Presented here are our early results for the irradiation, by 1.5 MeV helium ions, of methane ice. Our results for lower energy helium and proton irradiations are contained in a paper in preparation. The results can most appropriately be discussed in the context of the rate, as a function of fluence, of the polymerization process. By the Rutherford backscatter measurement technique, we measured both the overall change in the film "thickness" (by observation of the change in position of the gold marker layer in the backscattered spectrum) and the overall decrease in the number of carbon atoms in the ice film, ΣC (by observation of the number of carbon atoms in the carbon peak in the backscattered spectrum). The ratio of the changes in these two quantities is shown in Fig. 9 for an initial methane ice layer of a few 1000 Å thickness.

It is immediately evident that the overall ice layer "thickness" decreases more under the irradiation than does the amount of carbon on the substrate ($\Delta C_{Au}/\Delta C_{Au_0} < \Sigma C/\Sigma C_o < 1$ after irradiation). The decrease in the number of carbon atoms indicates that there is some erosion of methane from the fresh ice film. However, the larger proportional decrease in the gold marker layer position shows that substantially more than four hydrogen atoms are released from the film for every carbon released. This results in a carbon-enriched film left on the cold substrate. The results of Fig. 9 suggest that after an incident fluence of $\sim 2\times 10^{15}$ He ions/cm^2, the carbon enrichment process nears "saturation", leaving a polymer residue that is a mixture of carbon and hydrogen.

The results of Fig. 9 can be used to determine the gross chemical composition of the residue material; i.e., the amount **n** of hydrogen in the "compound" CH_n. The results of this determination, shown in Fig. 10, indicate that the "equilibrium" value of **n** = H/C $\sim$ 2. The resultant polymer might be characterized as CH_2 or, more likely, C_2H_4, ethylene.

The results of Figs. 9 and 10 are different from those reported for 100 keV proton

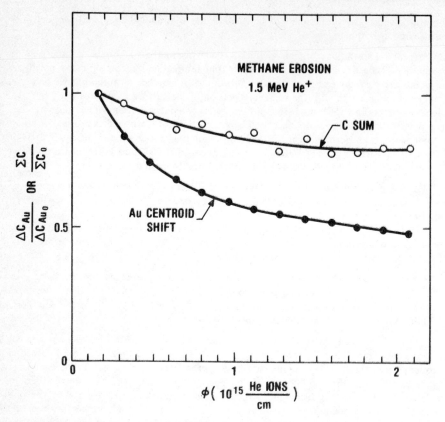

Figure 9. Fractional change in the carbon content of the methane ice (10K) film under irradiation by 1.5 MeV He$^+$ ions. The fractional change in the gold marker label position gives the change in the total ice film thickness.

irradiation of benzene (64). These authors report an exponential (asymptotic) approach, as a function of fluence, of the number of residual carbon atoms to a number equal to the initial carbon content of the film. However, for our 1.5 MeV helium irradiation of methane ice, we find a net loss of carbon (approximately 20% of the initial film (few thousand Angstroms) from Fig. 9) and a residual film with the composition ratio H/C ~ 2. Backscatter analysis of the residual film from the methane ice at room temperature indicates that the residual composition ratio remains essentially equal to two. Whereas the original methane film at 10K was essentially transparent, the residual material appears black.

6. SUMMARY

The particle radiation environment of the solar wind and the planetary magnetospheres can significantly affect the surfaces of icy bodies in the outer solar system. In this paper we have concentrated on the laboratory measurements of the ion erosion and modification of H_2O, SO_2 and CH_4 ices, which are important frozen volatile constituents in the outer

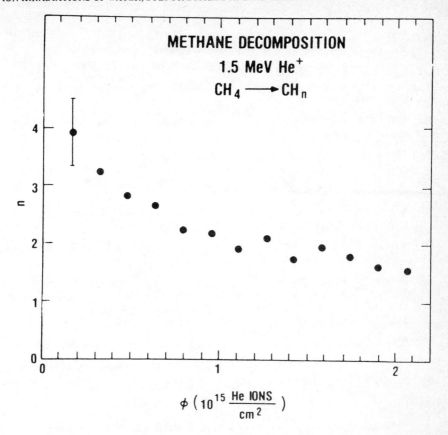

Figure 10. Production of residual polymer CH_n as a function of 1.5 MeV helium ion fluence.

solar system. A companion article in this volume (Johnson et al.) discusses other chemical and physical modifications of water ice surfaces and the production of magnetospheric plasmas from the ices. The solar system particle radiation environments can be important factors in producing "space weathering" effects on icy matter ranging in size from grains to planetary satellites and other icy bodies.

7. REFERENCES

(1) Krimigis, S. M., Armstrong, T. P., Axford, W. I., Bostrom, C. O., Fan, C. Y., Gloecker, G., and Lanzerotti, L. J.: 1977, *Space Sci. Rev., 21*, 329.

(2) Brown, W. L., Lanzerotti, L. J., Poate, J. M., and Augustyniak, W. M.: 1978, *Phys. Rev. Letters, 40*, 1027.

(3) Wehner, E. K., KenKnight, C. E., and Rosenberg, D. L.: 1963, *Planet. Space Sci., 11*, 1257.

(4) Maurette, M., and Price, P. B.: 1973, *Science, 187*, 121.

(5) Switkowski, Z. E., Haff, P. K., Tombrello, T. A., and Burnett, D. S.: 1977, *J. Geophys. Res., 82*, 3797.

(6) Brown, W. L., Augustyniak, W. M., Brody, E., Cooper, B., Lanzerotti, L. J., Ramirez, A., Evatt, R., and Johnson, R. E.: 1980, *Nucl. Instrum. Meth., 170*, 321.

(7) Johnson, R. E., Lanzerotti, L. J., Brown, W. L., Augustyniak, W. M., and Mussil, C.: 1983, *Astron. Astrophys., 123*, 343.

(8) Brown, W. L., Augustyniak, W. M., Lanzerotti,L. J., Marcantonio, K. J., and Johnson, R. E.: 1983, in *Desorption Induced by Electron Transitions I*, ed. N. H. Tolk, M. M. Traum, J. C. Tulley, and T. E. Madey, Springer-Verlag, New York, 234.

(9) Brown, W. L., Lanzerotti, L. J., and Johnson, R. E.: 1982, *Science, 218*, 525.

(10) Johnson, R. E., and Brown, W. L.: 1982, *Nucl. Instr. Methods, 198*, 103.

(11) Pilcher, C. B., Ridgway, S. T., McCord, T. B.: 1972, *Science, 178*, 1087.

(12) Lebofsky, L. A.: 1977, *Nature, 269*, 785.

(13) Broadfoot, A. L., Belton, M. J. S., Takacs, P. Z., Sande, B. R., Shemansky, D. E., Holberg, J. B., Ajello, J. M., Atreya, S. K., Donahue, T. M., Moos, H. W., Bertaux, J. L., Blamont, J. E., Strobel, D. F., McConnell, J. C., Dalgarno, A., Goody, R., and McElroy, M. B.: 1979, *Science, 212*, 163.

(14) Smith, B. A., Soderblom, L., Beebe, R., Boyce, J., Briggs, G., Bunker, A., Collins, S. A., Hansen, C. J., Johnson, T. V., Mitchell, J. L., Terrile, R. J., Carr, M., Cook, A. F., II, Cuzzi, J., Pollack, J. B., Danielson, G. E., Ingersoll, A., Davies, M. E., Hunt, G. E., Masursky, H., Shoemaker, E., Morrison, D., Owen, T., Sagan, C., Veverka, J., Strom, R., Suomi, V. E.: 1981, *Science, 212*, 163.

(15) Whipple, F. L.: 1950, *Ap. J., 11*, 375.

(16) Greenberg, J. M.: 1982, in *Comets* ed. L. L. Wilkening, U. Arizona Press, Tucson, 131.

(17) Bertaux, J. L., and Blamont, J. E.: 1976, *Nature, 262*, 263.

(18) Delsemme, A. H., and Wenger, A.: 1970, *Planet. Space Sci., 18*, 709.

(19) Greenberg, J. M. and Gustafson, B. A. S.: 1981, *Astron. Astrophys. 93*, 35.

(20) Greenberg, J. M.: 1983, in *Cometary Exploration*, ed. T. I. Gombosi, Hungarian Acad. Sci., Budapest, 23.

(21) Greenberg, J. M.: 1983, in *Asteroids, Comets, Meteors*, ed. C. I. Lagerkvist and H. Rickman, Uppsala Univ., 259.

(22) Bertaux, J. L., and Belton, M. J. S.: 1979, *Nature, 282*, 813.

(23) Smythe, W. D., Nelson, R. M., and Nash, D. B.: 1979, *Nature, 280*, 766.

(24) Fanale, F. P., Brown, R. H., Cruikshank, D. P., and Clark, R. N.: 1979, *Nature, 280*, 761.

(25) Nelson, R. M., Lane A. L., Matson, D. L., Fanale, F. P., Nash, D. B., and Johnson, T. V.: 1980, *Science, 210*, 784.

(26) Matson, D. and Nash, D.: 1983, *J. Geophys. Res., 88*, 4771.

(27) Cruikshank, D. P., Pilcher, C. B., and Morrison, D.: 1976, *Science, 194*, 835.

(28) Lebofsky, L. A., Rieke, G. H., and Lebofsky, M.: 1979, *Icarus, 37*, 554.

(29) Soifer, B. T., Neugebauer, G., and Matthews, K.: 1980, *Astron. J., 85*, 166.

(30) Behrisch, R.: 1981, ed. *Sputtering by Particle Bombardment*, Springer, Berlin.

(31) Johnson, R. E., Boring, J. W., Reimann, C., Barton, L. A., E. Sieveka, M., Garrett, J. W., Farmer, K., Brown, W. L., and Lanzerotti, L. J.: 1983, *Geophys. Res. Lett., 10*, 892.

(32) Sieveka, E. M., and Johnson, R. E.: 1982, *Icarus, 51*, 528.

(33) Ip, W. H.: 1981, *J. Geophys. Res., 86*, 1596.

(34) Hill, T. W., Dessler, A. J., and Fonale, F. P.: 1979, *Planet Space Sci., 27*, 419.

(35) Lanzerotti, L. J., Brown, W. L., Poate, J. M., and Augustyniak, W. M.: 1978, *Geophys. Res. Letters, 5*, 155.

(36) Johnson, R. E., Lanzerotti, L. J., Brown, W. L., and Armstrong, T. P.: 1981, *Science, 212*, 1027.

(37) Sigmund, P.: 1969, *Phys. Rev., 184*, 383.

(38) Lanzerotti, L. J., Brown, W. L., Poate, J. M., and Augustyniak, W. M.: 1978, *Nature, 272*, 431.

(39) Patashnick, H., and Rupprecht, G.: 1975, *Ap. J., 197*, L79.

(40) Patashnick, H., and Rupprecht, G.: 1977, *Icarus, 30*, 402.

(41) Lanzerotti, L. J., and Maclennan, C. G.: 1973, *J. Geophys. Res., 78*, 3935.

(42) Cheng, A. F., Lanzerotti, L. J., and Pirronello V.: 1982, *J. Geophys. Res., 87*, 4567.

(43) Haff, P. K., Eviatar, A., and Siscoe, G. L., *Icarus*, in press, 1983.

(44) Lanzerotti, L. J., Maclennan, C. G., Brown, W. L., Johnson, R. E., Barton, L. A., Reiman, C. T., Garrett, J. W., and Boring, J. W.: 1983, *J. Geophys. Res., 88*, 8765.

(45) Frank, L. A., Burek, B. G., Ackerson, K. L., Wolfe, J. H., and Mihalov, J. D.: 1980, *J. Geophys. Res., 85*, 5695.

(46) Bridge, H. S., Bagenal, F., Belcher, J. W., Lazarus, A. J., McNutt, R. L., Sullivan, J. D., Gazis, P. R., Hartle, R. E., Ogilvie, K. W., Scudder, J. D., Sittler, E. C., Eviatar, A., Siscoe, G. L., Goertz, C. K., and Vasylunias, V. M.: 1982, *Science*, *215*, 563.

(47) Barton, L. A.: 1983, *Magnetosphere Ion Erosion of the Icy Satellites of Saturn*, M.S. Thesis, Faculty of School of Engineering and Applied Science, Univ. Virginia.

(48) Lanzerotti, L. J., Brown, W. L., and Johnson, R. E.: 1981, *J. Geophys. Res.*, *86*, 3449.

(49) Haff, P. K., Watson, C. C.: 1979, *J. Geophys. Res.*, *84*, 8436.

(50) Watson, C. C.: 1981, *Proc. Lunar Planet. Sci. Conf.* 12th, 1569.

(51) Cheng, A. F., Haff, P. K., Johnson, R. E., and Lanzerotti, L. J.: 1985, in *Natural Satellites*, Univ. Arizona Press.

(52) Lanzerotti, L. J., Brown, W. L., Augustyniak, W., Johnson, R. E., and Armstrong, T. P.: 1982, *Ap. J.*, *259*, 920.

(53) Lanzerotti, L. J., and Brown, W. L.: 1983, *J. Geophys. Res.*, *88*, 989.

(54) Cruikshank, D. P., and Silvaggio, P. M.: 1980, *Icarus*, *41*, 96.

(55) Combes, M., Encrenaz, T., Lecacheux, J., and Perrier, C.: 1981, *Icarus*, *47*, 139.

(56) Johnson, J. R., Fink, U., Bradford, A., Smith, B. A., and Reitsema, H. J.: 1981, *Icarus*, *46*, 288.

(57) Trafton, L.: 1981, *Revs. Geophys. Space Phys.*, *19*, 43.

(58) Cruikshank, D. P., Bell, J. F., Gaffey, M. J., Brown, R. H., Howell, R., Beerman, C., and Rognstad, M.: 1983, *Icarus*, *53*, 90.

(59) Mendis, D. A., and Axford, W. I.: 1974, *Ann. Rev. Earth Planet. Sci.*, *2*, 419.

(60) Pirronello, V., Brown, W. L., Lanzerotti, L. J., Simmons, E., and Marcantonio, K. J.: 1982, *Ap. J.*, *262*, 636.

(61) Moore, M. H., and Donn, B.: 1982, *Astrophys. J.*, *257*, L47.

(62) Cheng, A. F., and Lanzerotti,L. J.: 1978. *J. Geophys. Res.*, *83*, 2596.

(63) Calcagno, L., Strazzulla, G., and Foti, G.: 1983, *Lett. Nuovo Cim.*, *37*, 303.

(64) Calcagno, L., Strazzulla, G., Fichera, M., and Foti, G.: 1983, *Rad. Eff. Lett.*, in press.

(65) Strazzulla, G., Calcagno, L., and Foti, G.: 1984, *Astron. Astrophys.*, submitted.

(66) Strazzulla, G., Calcagno, L., and Foti, G.: 1983, *Mon. Not. R. Soc.*, *204*, 59.

(67) Brown, W. L., Augustyniak, W. M., Lanzerotti, L. J., Johnson, R. E., and Evatt, R.: 1980, *Phys. Rev. Lett.*, *45*, 1632.

(68) Ollerhead, R., Bottiger, J., Davis, J. A., L'Ecuyer, J., Haugen, J. H., and Matsunami, M.: 1980, *Rad. Eff.*, *49*, 203.

(69) Cooper, B. H., and Tombrello, T. A.: 1983, *Rad. Effects*, in press.

(70) Seiberling, L. E., Meins, C. K., Cooper, B. H., Griffith, J. E., Mendenhall, M. H., and Tombrello, T. A.: 1982, *Nucl. Instr. Meth., 198*, 17.

(71) Johnson, R. E., Barton, L. A., Boring, J. W., Jesser, W. A., Brown, W. L. and Lanzerotti, L. J., in this volume.

COMMENT ON THE EVOLUTION OF INTERPLANETARY GRAINS

R. E. Johnson

Dept. of Nuclear Engineering & Engineering Physics
University of Virginia
Charlottesville, VA 22901

Energetic ions and electrons penetrating the interfaces between metallic, semiconducting and insulating solids have been shown to enhance the adhesion between these materials. This process is proposed as a means of fusing aggregates of silicate-iron core particles and/or core-mantle particles to form Brownlee particles. It is discussed in light of models for the evolution of cometary grains described at this conference.

BACKGROUND

Cometary grains (dust) have been described by Greenberg and co-workers (1) as bird-cage-like aggregates of processed micron to submicron sized core-mantle particles. These particles have silicate-iron cores, often a processed organic-refractory mantle and finally a mantle of frozen volatiles. The evolution of such particles in interplanetary space was discussed at this con-ference by Mukai and Fechtig and Strazzulla and Foti (these proceedings).

The evolutionary sequence proposed by Strazzulla and Foti was similar to that proposed by Cheng and Lanzerotti (2) for forming the darkened ring particles at Uranus. If the particles are covered with a layer containing a considerable amount of organic molecules (e.g. CH_4) then it has been shown that the energetic ions and electrons drive-off a large fraction of the hydrogen atoms (Lanzerotti et al., these proceedings) leaving behind a stable residue. The cohesion in this residue is provided by carbon-carbon bonds. In the case of methane and other organic molecules, ion bombardment has been shown to result in the formation of a blackened residue of non-volatile

337

material the I.R. spectra of which has been examined by
Strazzulla and Foti (these proceedings). This alteration of the
material occurs as ionizing radiations break chemical bonds
which subsequently reform creating more stable structures.
Since the outer icy mantle of the grains may contain only a
small fraction of organic molecules, Strazzulla and Foti propose
that most of the volatiles are driven off by sputtering (3) and
a residue is formed either on the silicate core or processed
with the organic refractory mantle primarily by solar flare
ions. Whether or not the aggregate of grains can remain intact
during the loss of volatiles (due to rapid sputtering and/or
charging) remains a question. However, the resultant
constituent particles are silicate-iron cores with carbon-based
mantles. These may in fact be indistinguishable from the
core-mantle particles of Greenberg (1) prior to their
accumulation on the icy outer mantle.

On the other hand, using observations from the Helios dust
experiments, Mukai and Fechtig (these proceedings) conclude that
the cometary grain aggregates can lose their volatiles slowly
while decaying from elliptical orbits to circular orbits and
remain intact. The resulting particles would be bird-cage-like
aggregates of core particles. This structure however is not a
Brownlee particle as the grains are only attached by the weak
contact forces (e.g., van-der-Waals forces) and the structure is
not stable. In some cases the constituent grains of the
particle may have organic mantles in which ion processing could
also occur, as described above, producing adhesion of the
grains. However, the Brownlee particles are often found to be
composed primarily of more dense materials such as the
silicate-iron core materials. The fusing of such materials is
discussed below.

DISCUSSION

The ability of charged particles, both ions and electrons,
to process materials is much more universal than indicated by
the experiments involving organic molecules. Tombrello and
co-workers (4-6) have shown that ions penetrating the interface
between two materials whether insulators, metals, semiconductors
or any combination can enhance the adhesion of these materials.
In particular, silicates have been attached to metals and even
more easily (lower doses) metals to metals. Mitchell et al. (7)
have shown this adhesion is also produced by 5 keV to 30 keV
electron bombardment and, therefore, must certainly occur for
protons, although this has not been tested. Following the
penetration of the interface by ions or electrons, enhanced
adhesion results when neighboring bonds are disrupted allowing
across-surface, covalent bonds to form as well as atoms to be
displaced between materials. This effect, requiring a high local

density of energy deposition, is produced by energetic ions and electrons directly and by the secondary electrons produced by these particles. Such adhesions will not be produced by U.V. irradiation at normal fluxes, although X-rays or γ-rays might, but with low efficiencies.

From the above it is clear that in conjunction with any process by which the volatile material in the cometary particle is lost, so that the aggregate remains intact, solar or cosmic ray ions and electrons crossing grain-grain surfaces can fuse the aggregates producing stable structures. This adhesion is very readily attained if an organic mantle covers the constituent grains as described above (Strazzulla and Foti, these proceedings). However, it is a quite general process in which core particles can be fused to core particles in the absence of organics and, hence, can play an important role in forming Brownlee particles. Additional laboratory studies are required to determine the amount of adhesion produced by interplanetary ions and electrons for the materials of interest. Also, the role of impurities on the surfaces need further study to see if a residual volatile layer helps enhance the adhesion, for example, by forming oxides on metallic surfaces.

This work is supported by NSF Astronomy Division under grant AST-82-00477.

REFERENCES

1. Greenberg, J. M., 1982, in "Comets", ed. L. L. Wilkening, University of Arizona Press, pp. 131-163.

2. Cheng, A. F. and Lanzerotti, L. J., 1978. J. Geophys. Res., 83, pp. 2597-2602.

3. Johnson, R. E., Lanzerotti, L. J. and Brown, W. L., 1983, Astron. Astrophys., 123, pp. 343-346.

4. Griffith, J. E., Qiu, Y. and Tombrello, T. A., 1982. Nucl. Inst. & Meth., 198, pp. 607-612.

5. Werner, B. T., Vreeland, T., Mendenhall, M. H., Qiu, Y. and Tombrello, T. A., 1983. Thin Solid Films, 104, pp. 163-166.

6. Tombrello, T. A., 1984. Materials Res. Soc. Proceedings, (to be published).

7. Mitchell, I. V., Williams, J. S., Smith, P. and Elliman, R. G., 1984. Appl. Phys. Lett., 44, pp. 193-195.

Part III

The Icy Nuclei of Comets

PRESENT STATUS OF THE ICY CONGLOMERATE MODEL

Fred L. Whipple

Smithsonian Astrophysical Observatory

A brief history is presented of the concept that the nucleus of a comet is a discreet body, an icy conglomerate of solar-type materials that would be solid at low temperatures, $\ll 100$ K. A summary describes briefly the observational success of the comet model, both quantitatively and qualitatively. The surprising aspect of the model is its usefulness in spite of its vagueness with regard not only to chemical composition but also to the physical structure of the nucleus, including such basic quantities as dimensions, density and albedo. Some emphasis will be placed on our increasing understanding of the morphology of comet nuclei.

Major attention will be centered on the interplay among the observations, the theory and the laboratory experiments, particularly with regard to the nature of the ices related to comet activity. The interest and progress in this field during the past decade has been most encouraging.

The icy conglomerate model (ICM) of a cometary nucleus (1) was introduced as a functional concept to make scientific sense of the already diverse and massive store of cometary observations. Laplace (2) and Hirn (3) had earlier suggested that comets were made of ice, but the physics and astronomy of Laplace's day were yet too primitive for a serious development to follow. Both visualized the recondensation of the vapor on the nucleus in the outer portions of the comet orbit, an unsupportable concept today. In the 1860's the clearcut identification of some comets as parents of meteor streams provided a cometary model, "the gravel bank," subjectively so appealing that its obvious faults were not considered damning for nearly a century. In 1948 Swings (4) concluded that solid ices must exist in comets.

J. Klinger et al. (eds.), Ices in the Solar System, 343–366.
© 1985 by D. Reidel Publishing Company.

The ICM involves a discrete cometary nucleus of radius up to tens of kilometers made of solid compounds primarily of C, N, O and H, conglomerated with a comparable amount of stony materials, expected to approximate a solar mix of the elements that would form solids at low temperatures, probably $\ll 100$ K. Note Table 1 for species observed to date in comets. The noble gases and solid hydrogen should not be present in appreciable quantities because of their high vapor pressures in near vacuum at temperatures even as low as those in interstellar clouds. Some noble gases as well as some atoms and molecules of stony material may be trapped in the ices while most of the stony or meteoroidal material appears as fine dust or "dust balls" (Öpik's term), typical of the fragile low density structures observed as meteors. The observed nature of the stony component of comets will be discussed in a following section.

Table 1. Species observed in comets.

Coma, Head	H, C, C_2, $^{12}C^{13}C$, C_3, CH, CN, CO, CS, HCN, CH_3CN, NH, NH_2, NH_3, HCO. O, OH, H_2O, S, S_2.
(Near Sun)	Na, K, Ca, Cr, Mn, Fe, Co, Ni, Cu, V, Ti.
Ions (tail)	C^+, CH^+, CN^+, CO^+, CO_2^+, N_2^+, OH^+, H_2O^+, Ca^+, H_2S^+.
Dust	Silicates, mostly dielectrics.

The basic concept of the ICM involves inactivity of the nucleus at great solar distances coupled with sublimation of the surface ices as the comet approaches the Sun towards perihelion. The nucleus is too small to be resolved optically. The outgoing gases carry with them fine dust and dust balls up to dimensions of even meters very near the Sun. The escaping material forms the dust coma of diameter several tens of thousands of kilometers. Solar radiation pressure drives away the fine dust in highly eccentric orbits to form the curved, generally short dust tails of many comets. The ionized solar wind engages the photo-ionized or charge-transfer ions of the gas to blow them away with acceleration many times the solar gravity to form the great ion tails, sometimes observable for more than 1 AU. All the escaping material is lost forever to the comet.

For a rotating nucleus an expected lag in sublimation on the morning side will concentrate the main thrust of the sublimating gases on the afternoon side. As a consequence, the jet reaction of the gas will develop a component of force forward or backwards with respect to the orbital motion about the Sun, depending on whether the rotation is prograde or retrograde, respectively, compared with the orbital motion. The orbital period will therefore increase or decrease with respect to Newtonian motion. Correspondingly, an oblate nucleus with its pole tilted to its orbital plane will receive a torque causing a precession of the polar axis.

The ICM predicts qualitatively or quantitatively a number of the basic observed characteristics of comets as follows:

a. The longevity of some comets such as P/Encke that has survived hundreds and probably thousands of revolutions about the Sun, losing $\sim 10^6$ tons per revolution. The brightest comets lose $\sim 10^6$ tons of gas per day near perihelion. The void of interplanetary space can provide no source of replenishment for the gases lost from a gravel-bank model (see (5), concerning the estimated lifetimes of comets).

b. The formation of dusty tails and meteor streams from the dust and dust balls ejected by the subliming gas.

c. Survival of several Sun-grazing comets.

d. The non-gravitation (NG) motions now measured for some 42 comets (6, 7). About half of the period changes are positive and half negative suggesting a random orientation of spin axes.

e. The large change in the NG motion of P/Encke caused by the precession of its polar axis (8).

f. Radar echoes from comets: P/Encke (9); comet IRAS-Araki-Alcock, 1983d (10-12); and comet Sugano-Saiggusa-Fujikawa, 1983e.

g. The lifetimes (short) of several components of split comets correlated with their radial solar NG motions (13) in the sense that the short-lived components have larger NG accelerations than long lived ones.

h. Dust jets and evidence for discrete active areas on specific spots correlating with spin periods and axes for P/Schwassmann-Wachmann I (14) and for P/Swift-Tuttle 1862III (15).

i. The occurrence of luminosity bursts of one to several magnitudes as the result of some property of cometary ices that is not yet clearly specified. The gravel-bank model does not seem to possess this potential.

j. The periodic recurrence of halos and parabolic envelopes in several comets, indicating a specific spin period for each such comet (16).

k. An acceptable mode of origin, such as the aggregation of ices and dust in the cold outer reaches of the forming planetary system or else within associated interstellar clouds.

l. A number of consistent results from theories based on the ICM, such as asymmetric halos indicating spin axes (17), halo expansion velocities (18), rates of production of sublimating water ice and relative abundances of elements (19).

COMPOSITION OF THE ICM ICES

In spite of many successful applications of the ICM in clarifying cometary behavior, the basic icy composition remains unhappily vague as well as the physical dimensions, masses, albedos and general morphology. The author's original ICM, following suggestions by Bobrovnikoff

(20) and Wurm (21), included specifically H_2O ice as the major ice with the though that CH_4, NH_3, CO_2 and radicals of C, N, O and H might be present. Clearly ices more volatile than H_2O are required to explain vigorous activity at solar distances greater than 5 AU, such as that of P/Schwassmann-Wachmann 1. But what materials?

Delsemme and Swings (22) pointed out the obvious temperature problem of maintaining solid CH_4 in comets and suggested that, if present, CH_4, NH_3 and possibly other highly volatile ices should appear as solid clathrates, or ionic hydrates, embedded in the H_2O ice or snow. Although the clathrates or ionic hydrates would solve the problem of containing the highly volatile molecules, they would not explain active sublimation at great solar distances. Unless, for example, the CH_4 clathrate should exceed some 15% of the H_2O content. Haser (23) suggested that the HO radical might solve this latter problem. Later (24) he developed a basic theory for the production and lifetimes of daughter species and parent molecules in the coma. To explain bursts in comets, Donn and Urey (25, 26) proposed exothermic chemical reactions in comet nuclei involving free radicals.

In the 1960's the theories multiplied to explain the life-times and distributions of various species observed in the comae of comets. The serious reader is referred to important summary articles if he wishes to review thoroughly the progress during the last 2 decades on problems of identifying parent molecules and coma structure from spectra and other observation: Swings (27), Arpigny (28), Herbig (29), Keller (30), Whipple and Huebner (31), Delsemme (18, 19), Huebner et al. (32) and Wyckoff (33, for a general review of comets).

An early problem concerned the rather short life times of species observed in comets as compared with the laboratory determined lifetimes against primarily photo-ionization and photo-dissociation by solar radiation. These vary with the observed species typically from a few hours to a few days at 1 AU from the Sun. Huebner and Weigert (34) suggested an icy grain halo in which the gases would be added more slowly to the coma by sublimation of the grains. Efforts to observe H_2O ice in the near infrared have generally been unsatisfactory although Crifo (35) finds evidence for H_2O icy grains in C/Kohoutek, 1973XII, and Hartmann and Cruikshank (36) report the probable detection of H_2O ice in P/Schwassmann-Wachmann 1.

For the short-period comets that may have made several dozen to even thousands of perihelion passages, H_2O ice appears to be sufficiently volatile to account for most of their average activity. Fig. 1 shows the fit of observed luminosity with model for P/Encke (Period = 3.3 yr) by Delsemme (37). His best fit for P/Encke suggests the unreasonable albedo of 0.7 in the visual and a low value of 0.1 in the infrared, probably explained by the surface of the nucleus being mostly covered with meteoroidal debris. His calculations for the idealized vaporation rates for H_2O, CO_2, CH_4, CO and N_2 are shown in Fig. 2. The inclusion of up to 15% clathrates has little effect on the H_2O curve.

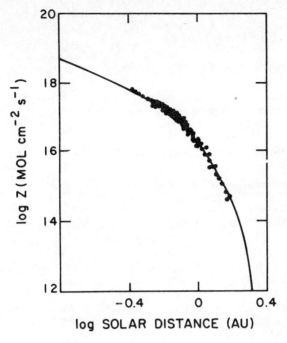

Figure 1. Production rate vs. solar distance for P/Encke (after Delsemme).

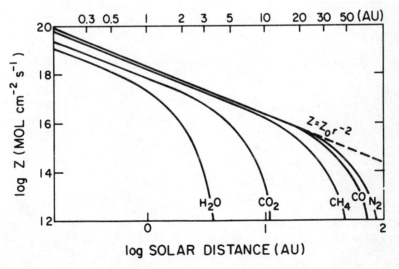

Figure 2. Production rates for H_2O, CO_2, CH_4, CO and N_2 vs. solar distance (after Delsemme).

The current identifications of species in comets are listed in Table 1. Recent additions have been made by C. B. Cosmovici and S. Ortolani (38). They find with the Asiago 1.82 m telescope for comet IRAS-Araki-Alcock (1983d) lines of the new molecules HCO and H_2S^+, identified by G. Herzberg. They strongly suspect the presence of H_2CO, DCO and NH_4.

The mean bulk composition in comets by atoms is now fairly well known. Delsemme's (18) evaluation is shown in Table 2, normalized to Si = 1.0. He bases the gaseous abundances (by number, col. 3) on the production rates determined in several recent bright comets by numerous observers. For the dust (col. 2) he deduces the mass ratio of dust to gas as 0.8, a quantity that is usually unmeasurable because the size distribution of the dust is rarely known (see Donn, (39), and Delsemme, this volume). For the abundance ratios in the dust Delsemme assumes Mason's (40) values measured for Cl carbonaceous chondrites. This assumption is founded on Millman's (41, 42) calculations of the abundance ratios of Na, Mg, Ca and Fe in cometary meteor spectra as similar to those in the Cl chondrites and upon the abundance ratios in the "Brownlee" particles. Brownlee et al. (43) find that these interplanetary micrometeorites, collected in the upper atmosphere, have elemental abundances also very similar to the Cl chondrites. For comparison with the cometary values, (col. 4) Table 2, col. 5 lists the "cosmic" abundances for solar-system elements as derived by Cameron (44).

Table 2. Atomic abundances[*] in bright comets.

Element	Dust no.	Gas no.	Total no.	Cosmic no.
H	2.00	22.30	24.30	26,600.00
C	0.70	3.03	3.73	11.70
N	0.05	1.46	1.51	2.31
O	7.50	14.80	22.30	18.40
S	0.50	0.05	0.55	0.50
Mg	1.06	—	1.06	1.06
Si	1.00	—	1.00	1.00
Fe	0.90	—	0.90	0.90
Ni + Cr	0.06	—	0.06	0.06
Total No.	13.77	41.64	55.41	—
Total Mass	254.5	318.3	572.8	—
Mass Percent	44.4	55.6	100.0	—

[*] By number, normalized to Si = 1.00 after Delsemme (18).

Our interest centers on the abundances ratios of C, N, O and H. The high depletion of H in comets compared to the Sun is, of course, to be expected if comets are the frozen residue from a similar mix of elements. The low H/O ratio rules out the presence of liquid or solid H and, indeed, greatly restricts the possible abundances of CH_4, NH_3, and the numerous hydrocarbons that appear in the interstellar medium. Consistent with this gross abundance ratio is the conclusion of Huebner et al. (32) from theirs and other studies of the gas-phase chemistry that "The comet models most consistent with observations indicate that only trace amounts (total amount 2%) of molecules bearing CN, C_2, C_3 and NH_2 can be present in the nucleus." Delsemme notes that C appears to be distinctly underabundant compared to O and N in a solar mix, or in the expected gross interstellar gas and dust.

Cochran (45) has applied non-equilibrium chemical modeling to a set of spatially and temporally resolved spectra of P/Stephen-Oterma. She concludes that near the nucleus the likely parent for CN is HCN; for OH, H_2O and CH_3OH; for CH, CH_4; for C_2, C_2H_2; and for C_3, model dependent, possibly involving grain photolysis. The vaporization rate appears to be controlled by H_2O.

The subject of parent molecules is too complicated for adequate review in this summary and so the reader is referred to other relevant papers in this colloquium for explanations and theories.

THE METEOROIDAL COMPONENT OF COMETS

Our knowledge of the dust and stony or meteoroidal material of very low vapor pressure in comets stems from surprisingly diverse sources:

    a. Cometary meteors and meteor streams.

    b. Brownlee particles captured in the high atmosphere.

    c. Impacts on artifical satellites, space probes and the Moon.

    d. Comae, dust tails, and near-nucleus jets, showing solar reflection spectra.

    e. Spectra of heavy elements in comets near the Sun.

    f. Solar directed tails of a few comets when the Earth passes through the orbit plane.

    g. Radar reflections from sizeable particles (centimeters) near the nucleus of the near-Earth Comet IRAS-Araki-Alcock, 1983d.

    h. The Zodiacal Light and Gegenschein.

Two important general references with many contributors to dust in the Solar System are Cosmic Dust, edited by J.A.M. McDonnell (46) and Solid Particles in the Solar System, edited by I. Halliday and B.A. McIntosh (47).

The observations of impacts in space or on the Moon and observations of the Zodiacal Light and Gegenschein provide information about the orbital distribution and mass content of tiny particles, mostly cometary, in orbits. About detailed composition, little has yet been learned. For relevant reviews see: McDonnell (46) for space studies; Ashworth (48) for lunar impact studies; Dohnanyi (49) for space dynamics of small particles; and Weinberg and Sparrow (50) for Zodiacal Light and Gegenschein results.

The studies of lunar microcraters summarized by Fechtig (51) show diameter/depth ratios between 1.0 and 2.7 with two conspicuous peaks at 1.4 and 1.9 and an extension to 2.5. From laboratory data he

attributes the highest values to fluffy cometary particles, comprising
≤ 30 percent of the total by number. This is in accord with the results
of Ceplecha and McCrosky (52) for large fireballs.

Even the polarization properties of the particles in the Zodiacal
Light appear as yet not to give clearcut evidence as to their character.
The difficulty lies in unknown morphology, probable irregular shapes
and probable fluffy structure, all highly variable from particle to particle
and not easily susceptible to theoretical or laboratory simulation. That
they are chiefly fluffy particles in the 1-100 μm range probably represents much of contemporary opinion.

The lifetimes of zodiacal particles against destruction or escape is
relatively short and the source generally thought to be comets, although
Delsemme (53), Röser (54), Kresák (55), and Mukai et al. (56) doubt
that the periodic comets can supply the few tens of tons per second
(Whipple (57)) necessary for continuous supply. Such estimates may
be increased significantly when the full implications of the large clouds
of sizeable particles observed to accompany comet IRAS-Iraki-Alcock
(10-12) and Comet Bowell, 1980b (58, 59) are fully interpreted.
Conceivably the zodiacal cloud is denser at present than in the past
or on the average over long periods of time. The impact rates on the
Moon, averaged over 10^4 to 10^5 years, appear to be smaller than
present rates on artificial satellites (see, e.g. Ashworth (48)).

From the Super-Schmidt photographic meteors of ~1 g mass,
Verniani (60) finds a logarithmic mean density of 0.28 g cm^{-3} with some
in the 0.01 g cm^{-3} range (specifically the Draconids from P/Giacobini-
Zinner).   These appear to be almost entirely cometary debris with the
interesting possible exception of the Geminids, now associated with the
asteroid 1983 TB (Whipple, (61), by comparison with Bardwell's elements (62)) discovered by the Infrared Astronomy Satellite and reported
by S. Green (63) of Leicester. Verniani finds that the mean density of
the Geminid meteoroids is ~1.0 g cm^{-3} or more than three times the
mean for cometary meteoroids. The Geminid stream, with a perihelion
distance of 0.14 AU and aphelion just beyond Mars' orbit, may be
asteroidal or, possibly, the asteroid may be a stony comet nucleus.
The fluffy or low-density nature of cometary non-volatiles is thus
firmly established. However, the analysis of fireball data of the Prairie
Network and in Central Europe show that a few stronger and denser
bodies appear among them, comparable to weak Cl chondrites (64).

The most extensive observational source of information about cometary dust lies in the coma and dust tails. The analysis of the kinetics
and dust tails by Finson and Probstein (65) has been continued by
Sekanina (66, review) and by Saito et al. (67). Their important results
will be summarized in the following pages.

Infrared photometry has developed into an extraordinary powerful
tool for analyzing comet dust and larger particles. Coupled with red
optical photometry the nearly blackbody radiation from the particles can
be separated from that of the scattered sunlight. Temperature, area,
some measure of particle size distribution and information about the

scattering nature of the particles can thus be derived with the addition
of polarization measures. Ney's (68) report is the major source of
the highly condensed summary that follows. Five bright dusty comets
all showed a silicate signature at 10 and 18 μm indicating the presence
of small refractory grains of radius < 5 μm in the comae and tails. The
anti-tail or sun-ward tail of C/Kohoutek (1973 XII) did not show the
silicate signature, thereby indicating nonsilicates or large silicate grains
> 30 μm. One (1975 IX) of two comets showing only ion tails gave thermal
emissions that could be interpreted as from large grains. The calculated
mass loss ratios for H_2O/solids for this comet was 1.2, the smallest
among five determinations. The other ratios of H_2O/solids, were 4.2
(1970 II); 9.6 (P/Encke); 9.2 (1973 XII); and 1.6 (1976 VI, C/West that
split), but all these calculated ratios are subject to great uncertainties.

The temperatures of the fine dust exceeded the blackbody temperatures at the observed solar distances by 8% for 1975 IX (ion tail), 26%
(1970 II) and 48% (1976 VI) for the fine grained dust, the latter two
comets showing fine grained silicate dust. Because the IR radiation is
almost exactly equal to the radiation absorbed from the sunlight, the
scattered component of which is measured at shorter wavelengths, the
albedo of the particles can be calculated from the photometry. For 1973
XII, C/Kohoutek, the albedo of the dust is 0.14 to 0.20 at scattering
angles in the range 80° to 135° (see also Crifo (35), for other results on
the grains of C/Kohoutek).

Forwarding scattering was remarkably strong for C/West indicative
of a dirty dielectric grain mixture with dominant size ~1 μm. The ratio
of reflected to absorbed energy for five comets is shown in Fig. 3 (68).
The albedos of the grains are surprisingly similar among these five
comets. During the splitting of C/West, the 10-μm feature was always in
evidence, indicating that the interior regions of the broken nucleus produced dust of the same character as the original surface and quite as
abundantly.

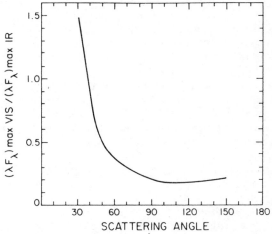

Figure 3.   Radio of reflected to absorbed flux vs. scattering angle
at coma, i. e. Earth–Sun angle (after Ney).

Campins and Hanner (69) and Hanner (70) are making rapid progress in interpreting the IR dust measures in terms of a mixture of hot absorbing grains (such as magnetite) and cold dielectric silicate grains, the latter providing the silicate feature. Such an interpretation seems needed in view of Sekanina and Farrell's (71) evidence for submicron-sized particles of strongly absorbing and others of essentially dielectric character in the striated tail of C/Mrkos, 1957 V. The absorbing particles with β lying between 1 and 2 (β = ratio of solar radiation acceleration to solar gravity) require a distribution peak just above $0.1 \mu m$, falling rapidly in number to $0.3 \mu m$. The nearly dielectric particles with $\beta \sim 0.6$ appear to be rare near $0.1 \mu m$ and increase rapidly in number to $0.3 \mu m$.

In his earlier study of the dust jets near the nucleus of P/Swift-Tuttle, 1862 III, the parent of the Perseid meteor stream, Sekanina (15) found clearcut evidence for jets having dust particles with $\beta < 0.5$, dielectric, and other jets that also include absorbing dust with $\beta > 1.0$. Saito et al. (67) point out that the lack of observed values of $\beta > 2.5$ excludes the presence in comets of significant numbers of graphite particles in the size range between 0.02 and $0.2 \mu m$.

The physical and chemical character of the refractory material in comets is intimately involved in theories of their mode of origin. The Brownlee (Fig. 4) particles with the electron microscope frequently look like clusters of grapes or of tiny fish eggs (0.1 to $1.0 \mu m$ in size) precisely as one might envision an aggregate of interstellar dust. But, of course, dust formed in the outer reaches of the proto-planetary nebula might give the same appearance. Fraundorf et al. (72) find that in 57 stratospheric micrometeorites the mean elemental abundances of Na, Mg, Al, S, Ti, Cr, Mn, Fe, and Ni (compared to Si = 1) all match the average Cl chondrite abundances well within a 30% range. The scatter is large, the order of a factor of 2, because the particles are so small, $\gtrsim 10 \mu m$, and their components variable enough to bias the measures for any single particle. The major deviant is Ca, underabundant by a fact of 3 or more with about twice the average spread. Fraundorf et al. observe that the Ca deficiency occurs in particles that are often smooth on a scale of microns while particles that are clearly aggregates with high porosity show normal Ca abundances. They suggest that the Ca depletion may be due to mobilization in the parent body.

Three Mg isotopes measured in several stratospheric micro-meteorites by Esat et al. (73) show normal solar-system ratios to an uncertainty of 1% with the exception of only 1 particle. The particles included chondritic aggregates, particles composed of single grains of olivine and pyroxene, and spherules depleted in Fe and enhanced in the more refractory Ca and Al.

The stratospheric micrometeorites are similar to soot in size and optical absorptivity. The aggregates vary greatly in porosity although the submicron components are compact, being mostly composed of amorphous and crystalline silicate materials with Ni bearing iron sulphides and carbonaceous material. Fraundorf et al. note that if the particles are cometary, the above sub-micron components were first assembled and aggregated along with lesser numbers of similar sized or larger

monomineralic grains, usually of olivine, enstatite or Ni-bearing iron
sulphides. Presumably the voids and surfaces once contained ices. Even
though proof of this cometary origin sequence is still elusive, the circumstantial evidence is stiking and probably our best foundation, today,
for more detailed theories of the formation of comets. Certainly the
primitive character of these particles is manifest, as we expect of cometary material.

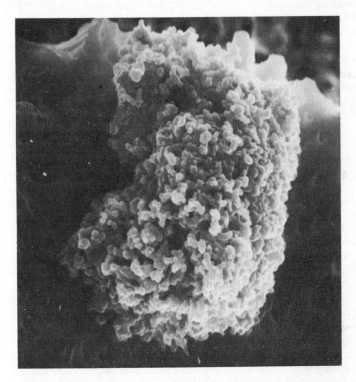

1μm

Figure 4. Extraterrestrial particle (courtesy D.E. Brownlee).

MORE EXOTIC ICES THAN H_2O IN COMETS

The conclusion that H_2O ice with or without clathrates is generally
satisfactory for even the short-period comets is not acceptable. Comet
bursts of 1-3 magnitudes occur in about 3/4 of the light curves of these
comets and splitting can also occur in what appear to be rather docile
old comets. The classical example is P/Biela, first observed in 1772
with a period of just under 7 years and perihelion distance of 0.72 AU.
It was observed again as 1806 I, as 1826 I and as 1832 III, without showing any unusual behavior or brightness changes. But as 1846 II it had
split into two components which lasted until 1852 and then disappeared
never to be seen again. Their remnants remain as a meteor shower,

spectacular in November 1872 (the Andromides), but active regularly and usually weak since 1772. (See Sekanina (13,58) and Hughes (74), for thorough accounts of comet splitting and flaring, respectively.)

A number of explanations for comet bursts have been proposed, including the statistically untenable idea of impacts by interplanetary boulders. Hughes finds that the burst distribution curve peaks near the solar distance of 1 AU with no evidence of a bump in the asteroid belt. The collapse of unstable structures on the nucleus, either vertical or cavity ceilings, remains a simple but untestable explanation. All evidence points to an extreme lack of homogeneity in cometary nuclei (75, 76). Under very low surface gravity, sublimation could produce bizarre formations. Wasting, slumping or collapse of such irregular features could expose large volumes of ices to sunlight, initiating an outburst. More detailed studies of these processes and their observable consequences are clearly needed.

The core-mantle processes of Mendis and Brin (77) and Brin and Mendis (78) deals effectively with the problem of meteoroidal blanketing in the post-perihelion fading of comets and the problem of the removal of the blanket near perihelion. Thus, a maximum of activity and intrinsic brightness, as observed statistically, occurs soon after perihelion. This concept has been elaborated by Horanji et al. (79) with an added process involving the breakup of friable sponge grains as they are blown away from the surface of the nucleus. Weissman and Kieffer (80) have included in their theory the effect of opacity and scattering by ejected dust on the thermal properties of the active nucleus.

The magnificent outbursts of P/Schwassmann-Wachmann I at more than 6 AU from the Sun require not only more volatile ices than H_2O but specific mechanisms and morphology both to start and stop them. Whitney (81) suggested pockets of highly volatile ices such as CH_4 and some storage mechanism for the solar heating between outbursts to build up gas pressure. The identification of CO^+ emission in the comet both during and between outbursts by Cochran et al. (82, 83) points strongly to CO_2 (or CO) as the active ice. Cowan and A'Hearn (84) base their theory on this premise. They suggest a cut-off mechanism involving accumulation of H_2O ice particles (and meteoroidal particles?). The concept of "pockets" with areas the order of square kilometers and thicknesses the order of centimeters strains the imagination. Whipple (14) suggested a fall-back mechanism of large particles from a small initial active area to break the crust and to expand the active area to an effective size. Cowan and A'Hearn propose that the diurnal temperature variations fracture the surface by differential expansion.

Slow warming to relevant depths over large areas and over intervals of weeks and months seems to be a requirement for preparing the subsurface material to produce the outbursts. The proposals by Pataschnik et al. (85, 86), Klinger (87-88) and Smoluchowski (89, 90) that amorphous ice may be present in comets and that its transition into the crystalline state may provide an internal source of activity, is most attractive. Discussion of the theory of cometary expectations from amorphous ice is presented by Klinger and Smoluchowski in these reports.

Shulman (91) doubts the occurrence of amorphous ice in comets on the basis of formation temperatures greater than the 155 K transition temperature, contrary to the assumptions of most investigators. He suggests instead the formation of nonequilibrium ion molecular clusters of H_2O^+ and H_2^+ induced by the solar wind at > 100 AU from the Sun in its T-Tauri stage.

The suggestion of "Platt" particles in comets will only be mentioned. They are (92), if they exist, the order of 10 Å in dimension and should scatter light of wavelength shorter than some limit, above which they are essentially transparent. See, for example, the discussion by Misconi and Whitlock (93), for background and references.

The formation and nature of interstellar grains as possible components of comets or as direct processes in comet formation are of prime importance to an understanding of the nature of comets. J.M. Greenberg and his collaborators have made remarkable progress in demonstrating by laboratory experiments, by reference to interstellar matter studies, and by theoretical studies that much of cometary material may well be of interstellar origin or else has been formed in a somewhat similar environment. His review article (94) is of vital importance to anyone interested in the nature of cometary nuclei, and provides references to all aspects of this research. Only a brief summary will be made here because the reader should refer to Greenberg's accompanying presentation. The laboratory experiments involve the deposition of H_2O, CO, NH_3, CH_4 etc. on a cold finger at T ~10 K while irradiated with ultraviolet photons. Extremely complex mixtures of molecules and radicals are so produced. They are explosive with accompanying luminescence when heated to ~27 K and to successively higher temperatures. Many of the puzzling aspects of cometary behavior can be explained on Greenberg's model including composition, activity at great solar distances, bursts etc. We have noted earlier the similarity of some of the Brownlee particles to an imagined collection of interstellar dust particles. Yamamoto (95) has discussed possible condensation processes in the interstellar environment as related to comet formation.

Shulman (96) proposed that deep space cosmic rays play a significant role in radiation synthesis to alter the chemical composition of the upper meter or so on cometary nuclei and that solar cosmic rays produce near-surface changes that may result in cometary outbursts. Comets that are "new," making their first approach to the Sun after a long residence in the Opik-Oort cloud, appear to be unusually bright at great solar distances, C/Kohoutek, 1973 XII, being a famous example. Marsden et al. (6) note an excessive number of "new" comets discovered at large perihelion distances, indicating that they are perhaps ~2 mag. brighter on the first approach than on later returns.

Whipple (97) supported the idea that cosmic rays in the Opik-Oort cloud are responsible for significant chemical effects in the outer meters of such comets to produce the unusual temporary activity of new comets. Donn (39) suggested that cosmic rays would produce complex hydrocarbons by polymerization to darken the outer layers. Laboratory experiments by Moore et al. (98) involved irradiation of thin ice films at T ~20 K by 1 ~Mev protons from a van de Graff accelerator. The ice mixtures included H_2O, NH_3, CH_4, N_2, C_3H_8, CO and CO_2. New molecules were synthesized in all the solid-phase mixtures. The irradiated mixtures contain reactive species exhibiting thermoluminescence and pressure enhancements during warming and also ~1% nonvolatile residue of complex carbon compounds appreciably darkened. Moore et al. expect exothermic activity to begin on "new" comets at solar distances > 100 AU when the surface temperature rises to 20-32 K. The strange particle cloud about C/Bowell, 1980b, suggests that "new" comets may indeed become active at large solar distances (58, 59). Other exothermal activity could occur on warming to $100 < T < 150$ K at $2 < r < 5$ AU. The laboratory irradiation matched that from cosmic rays in deep space over 4.6×10^9 years to a depth of more than 1 m in compact ices.

Evidence is strong, although not definitive, that faint P/Holmes 1892 III experienced two outbursts of 6^m or more separated by 72 days as the consequence of an encounter with a companion satellite followed by final impact (61, 75, 76). Peculiar to both these outbursts is further evidence for unusually rapidly expanding faint halos with about twice the velocity of the observed normal very bright halos, noted first by Orlov (99). If these impacts initiated actual explosions the second high-velocity halos could have resulted from gas (and dust?) ejected at about the velocity of sound for the prevalent molecules. The velocities involved are: for the first outburst, Nov. 5, 1892, 0.38 and 0.76 km s^{-1}, at $r = 2.3$ AU; and for the second on Jan. 16, 1893, 0.36 and 0.58 km s^{-1}, at $r = 2.6$ AU. The question is whether the impact of a satellite nucleus of ~100 meters in dimension impacting at a velocity of only a few meters per second could, by compression, initiate an explosion in a buried mixture of exotic ices?

CONCLUDING COMMENTS

It appears that the icy conglomerate model of comets is alive and well as a basic concept. Most of the details of cometary morphology and composition remain to be determined beyond the clear indications that H_2O ice is a major constituent and that comet nuclei contain a sizeable fraction of very fine particles much as one would expect to be aggregated either in the outer reaches of the protosolar nebula or in associated interstellar clouds. Exotic ices must certainly be present, some that probably provide exothermic reactions when warmed to rather low temperatures. Space missions to comets, specifically rendezvous missions, are clearly required to answer many of the basic questions. Laboratory studies of ices at very low temperatures and in various radiation environments are also vital to progress in understanding the nature of comets. Continued observations of comets with both classical and novel techniques remain, as ever, vital to this progress. The end result will be a great leap forward in our understanding of how the Solar System originated and, possibly, how life on Earth could arise.

The author is indebted to Zdenek Sekanina for discussions and a critical reading of the manuscript and also to Armand H. Delsemme. The study was supported by the Planetary Geology Division of the National Aeronautics and Space Administration under Grant NSG 7082.

REFERENCES

1. Whipple, F. L., 1950, A comet model. I. The accele-
    ration of Comet Encke. Astrophys. Journ. 111, pp.
    375-394 ; 1951, A comet model. II. Physical rela-
    tions for comets and meteors. ibid. 113, pp. 464-
    474 ; 1955, A comet model. III. The Zodiacal
    Light. ibid. 121, pp. 710-750 ; 1963, On the struc-
    ture of the cometary nucleus. In "The Solar Sys-
    tem ", vol. 4, ed. B. M. Middleburst and G. P.
    Kuiper, Univ. Chicago Press, pp. 639-664.

2. Laplace, P. S., 1813, "Exposition du système du
    Monde", 4th edition, Paris, p. 130.

3. Hirn, G. A., 1889, "Constitution de l'espace céles-
    te", Gauthier-Villars, Paris.

4. Swings, P., 1948, Le spectre de la comete d'Encke,
    1947c. Ann. Astrophys., 11, pp. 124-136.

5. Krezak, L., 1981, The lifetimes and disappearance
    of periodic comets. Bull. Astron. Inst. Czech.,
    32, pp. 321-339.

6. Marsden, B. G., Sekanina, Z. and Yeomans, D. K.,
    1973, Comets and nongravitational forces. V. Astron.
    Journ., 78, pp. 211-225.

7. Marsden, B. G., 1982, Catalogue of cometary orbits,
    4th edition, I. A. U. Smithsonian Astrophys. Obs.

8. Whipple, F. L.and Sekanina, Z., 1979, Comet Encke :
    Precession of the spin axis, nongravitational mo-
    tion, and sublimation. Astron. Journ., 84, pp.
    1894-1909.

9. Kamoun, P. G., Campbell, D. B., Ostro, S. J., Pet-
    tingill, G. H. and Shapiro, I. I., 1982, Comet
    Encke : Radar detection of nucleus. Science 216,
    p. 293.

10. Goldstein, R. M., Jurgens, R. F. and Sekanina, Z.,
    1983, Radar observations of Comet IRAS-Araki-Al-
    cock at two wavelengths. Bull. Amer. Astron. Soc.,
    15, p. 800 (abstract listed).

11. Campbell, D. B., Harmon, J. K., Hine, A. A., Sha-
    piro, I. I. and Marsden, B. G., 1983, Arecibo ra-

dar observations of Comets IRAS-Araki-Alcock and Sugano-Saigusa-Fugikawa. Bull. Amer. Astron. Soc., 15, p. 800 (abstract).

12. Shapiro, I. I., Marsden, B. G. and Whipple, F. L., 1983, Interpretations of radar observations of comets. Bull. Amer. Astron. Soc., 15, p. 800 (abstract listed).

13. Sekanina, Z., 1982, The problem of split comets in review. In "Comets", ed. L. L. Wilkening, Univ. Arizona, Tucson, p. 251-287.

14. Whipple, F. L., 1980, Rotation and outbursts of Comet P/Schwassmann-Wachmann 1, Astron. Jour., 85, pp. 305-313.

15. Sekanina, Z., 1981, Distribution and activity of discrete emission areas on the nucleus of periodic Comet Swift-Tuttle. Astron. Journ., 86, pp. 1741-1773.

16. Whipple, F. L., 1982, The rotation of comet nuclei. In "Comets", ed. L. L. Wilkening, Univ. Arizona, Tucson, pp. 227-250.

17. Sekanina, Z., 1981, Rotation and precession of cometary nuclei. Ann. Rev. Earth Planet. Sci., 9, pp. 113-148.

18. Delsemme, A. H., 1982, Chemical composition of Cometary nuclei. In "Comets" ed. L. L. Wilkening, Univ. Arizona, Tucson, pp. 85-130.

19. Delsemme, A. H., 1977, The pristine nature of comets. In "Comets, Asteroids, Meteorites", ed. A. H. Delsemme, Univ. Toledo, pp. 3-13.

20. Bobrovnikoff, N. T., 1942, Physical theory of comets in the light of spectroscopic data. Rev. Mod. Phys., 14, pp. 164-178.

21. Wurm, K., 1943, Die natur der kometen. Mitteilungen der Hamburger Sternwarte in Bergedorf 8, no. 51, pp. 57-92.

22. Delsemme, A. H. and Swings, P., 1952, Hydrates de gaz dans les noyaux cometaires et les grains interstellaires. Ann. d'Astrophys., 15, pp. 1- 6.

23. Haser, L., 1955, La conservation des radicaux li-
     bres à basse température et la structure des
     noyaux de comètes. Comptes Rendus 241, pp. 742-743.

24. Haser, L., 1957, Distribution d'intensité dans la
     tête d'une comète. Liège Inst. d'Astrophys., report
     no. 394.

25. Donn, B. and Urey, C. H., 1956, On the mechanism of
     comet outbursts and the chemical composition of
     comets. Astrophys. Journ. 123, pp. 339-342.

26. Donn, B. and Urey, C. H., 1957, Chemical heating
     processes in astronomical objects. Mem. Soc. Roy.
     Sc. Liège, ser. 4, 18, p. 124.

27. Swings, P., 1965, Cometary spectra. Quart. Jour.
     Roy. Astron. Soc., 6, pp. 28-69.

28. Arpigny, C., 1976, Interpretation of comet spectra.
     In "The Study of Comets", vol. II, ed. B. Donn et
     al., NASA SP-393, pp. 797-838.

29. Herbig, G., 1976, Review of cometary spectra. In
     "The Study of Comets" vol. I, ed. B. Donn et al.,
     NASA SP-393, pp. 136-154.

30. Keller, H. U., 1976, The interpretation of ultra-
     violet observations of comets. Space Sci. Rev.,
     18, pp. 641-684.

31. Whipple, F. L. and Huebner, W. F., 1976, Physical
     processes in comets. Ann. Rev. Astron. Astrophys.
     14, pp. 143-172.

32. Huebner, W. F., Giguere, P. T. and Slattery, W. L.,
     1982, Photochemical processes in the inner corona.
     In "Comets" ed. L. L. Wilkening, Univ. Arizona,
     Tucson, pp. 496-515.

33. Wyckoff, S., 1982, Overview of comet observations.
     In "Comets", ed. L. L. Wilkening, Univ. Arizona,
     Tucson, p. 3-55.

34. Huebner, W. F. and Weigert, A., 1966, Eiskörner in
     der koma von kometen, Z. Astrophys., 64, pp. 185-
     201.

35. Crifo, J. F., 1983, Visible and infrared emissions
     from volatile and refractory cometary dust. A new
     interpretation of Comet Kohoutek observations. In

"Cometary Exploration", ed. T. I. Gombosi, Hungarian Acad. Sci. II, pp. 167-176.

36. Hartmann, W. K. and Cruikshank, D. P., 1983, Systematics of ices among remote comets, asteroids and satellites. Bull. Amer. Astron. Soc., 15, p. 808 (abstract).

37. Delsemme, A. H., 1975, Physical interpretation of the brightness variation of Comet Kohoutek. In "Comet Kohoutek", ed. G. A. Gary, Washington, NASA SP-355, pp. 195-203.

38. Cosmovici, C. B. and Ortolani, S., 1984, I. A. U. Circ. 3915.

39. Donn, B., 1977, A comparison of the composition of new and evolved comets. In "Comets, Asteroids, Meteorites", ed. A. H. Delsemme, Univ. Toledo, pp. 15-23.

40. Mason, B., 1971, Handbook of Elemental Abundances in Meteorites, Gordon and Breach, New York, pp. 21-28, 81-107.

41. Millman, P. M., 1972, Cometary meteoroids. In "From Plasma to Planet", Proc. 21 st Nobel Symposium, Saltsjobaden, Sweden, 1971, eds. Almquist and Wiksell, Stockholm, pp. 157-168.

42. Millman, P. M., 1977, The chemical composition of cometary meteoroids. In "Comets, Asteroids, Meteorites", ed. A. H. Delsemme, Univ. Toledo, pp. 127-132.

43. Brownlee, D. E., Rajan, R. S. and Tomandl, D. A., 1977, Chondrites and dust. In "Comets, Asteroids, Meteorites", ed. A. H. Delsemme, Univ. Toledo, pp. 137-141.

44. Cameron, A. G. W., 1982, in "Essays in Nuclear Astrophysics", ed. C. A. Barnes, D. D. Clayton and D. N. Schramm, Cambridge Univ. Press, London, pp. 23-43.

45. Cochran, A. L., 1982, The chemical evolution of the coma of Comet P/Stephan-Oterma, Univ. Texas Pub. in Astronomy, no. 21.

46. McDonnell, J. A. M., 1978, Microparticle studies by

space instrumentation. In "Cosmic Dust", ed. J. A. M. McDonnell, John Wiley and Sons, pp. 337-426.

47.   Halliday, I. and McIntosh, B. A., eds., 1980, Solid Particles in the Solar System, Reidel.

48.   Ashworth, D. G., 1978, Lunar and planetary impact erosion. In "Cosmic Dust", ed. J. A. M. McDonnell, John Wiley and Sons, pp. 427-526.

49.   Dohnanyi, J. S., 1978, Particle dynamics. In "Cosmic Dust", ed. J. A. M. McDonnell, John Wiley and Sons, p. 527.

50.   Weinberg, J. L. and Sparrow, J. G., 1978, Zodiacal light as an indicator of interplanetary dust. In "Cosmic Dust", ed. J. A. M. McDonnell, John Wiley and Sons, pp. 75-122.

51.   Fechtig, H., 1982, Cometary dust. In "Comets", ed. L. L. Wilkening, Univ. Arizona, Tucson, pp. 370-382.

52.   Ceplecha, Z. and McCrosky, R. E., 1976, Fireball end heights : A diagnostic for the structure of meteoric material. Journ. Geophys. Res. 81, pp. 6257-6275.

53.   Delsemme, A. H., 1976, The production rate by comets. In Proc. IAU Colloq. No 31, "Interplanetary Dust and Zodiacal Light", ed. H. Elsasser and H. Fechtig, Springer-Verlag, New York 48, pp. 314-318.

54.   Röser, S., 1976, Can short-period comets maintain the Zodiacal cloud ? Proc. IAU Colloq. No 31 "Interplanetary Dust and Zodiacal Light", ed. H. Elsasser and H. Fechtig, Springer-Verlag 48, pp. 319-322.

55.   Kresak, L., 1980, Source of interplanetary dust. In "Solid Particles in the Solar System", ed. E. Halliday and B. A. McIntosh, Reidel, pp. 211-222.

56.   Mukai, T., Mukai, S., Schwehm, G. H. and Giese, R. H., 1983, Supply of interplanetary grains from comets. In "Cometary Exploration", ed. T. I. Gombosi, Hungarian Acad. Sci. II, pp. 135-141.

57.   Whipple, F. L., 1976, Sources of interplanetary

dust. In "Interplanetary Dust and Zodiacal Light", ed. by H. Elsasser and H. Fechtig, Springer-Verlag, Heidelberg, pp. 403-415.

58. Sekanina, Z., 1982, Comet Bowell (1980b) : An active-looking dormant object. Astron. Journ. 87, pp. 161-169.

59. A'Hearn, M. F., Dwek, E., Feldman, P. D., Millis, R. L., Schleicher, D. G., Thompson, D. T. and Tokunaga, A. T., 1983, The grains and gas in Comet Bowell. In "Cometary Exploration", ed. T. I. Gombosi, Hungarian Acad. Sci., Budapest, II, pp. 159-166.

60. Verniani, F., 1967, Meteor masses and luminosity. Contr. Astrophys., 10, pp. 181-195 ; 1969, Structure and fragmentation of meteoroids. Space Sci. Rev., 10, pp. 230-261.

61. Whipple, F. L., 1983, I. A. U. Circ. No. 3881.

62. Bardwell, C. M., 1983, I. A. U. Circ. 3879.

63. Green, S., 1983, I. A. U. Circ. No. 3878.

64. Wetherill, G. W. and ReVelle, D. O., 1982, Comets and large meteors. In "Comets", ed. L. L. Wilkening. Univ. Arizona, Tucson, pp. 297-319.

65. Finson, M. L. and Probstein, R. L., 1968 , A theory of dust comets. I. Model and equations. Astrophys. Journ., 154, pp. 327-352 ; A theory of dust comets. II. Results for Comet Arend-Roland, ibid. 154, pp. 353-380.

66. Sekanina, Z., 1980, Physical characteristics of cometary dust from dynamical studies : A review. In "Solid Particles in the Solar System", ed. E. Halliday and B. A. McIntosh, pp. 237-250.

67. Saito, K., Isobe, S., Nishioka, K. and Ishii, T., 1981, Substances of cometary grains estimated from evaporation and radiation pressure mechanisms. Icarus, 47, pp. 351-360.

68. Ney, E. P., 1982, Optical and infrared observations of bright comets in the range of 0.5 μm to 20 μm. In "Comets", ed. L. L. Wilkening, Univ. Arizona, Tucson, pp. 323-340.

69. Campins, H. and Hanner, M. S., 1982, Interpreting
     the thermal properties of cometary dust. In "Co-
     mets", ed. L. L. Wilkening, Univ. Arizona, Tucson,
     pp. 341-356.

70. Hanner, M. S., 1983, The nature of cometary dust
     from remote sensing. In "Cometary Exploration ",
     ed. T. I. Gombosi, Hungarian Acad. Sci., Budapest,
     vol. II, pp. 1-22.

71. Sekanina, Z. and Farrell, J. A., 1982, Two dust po-
     pulations of particle fragments in the striated
     tail of Comet Mrkos 1957V. Astron. Journ., 87,
     pp. 1836-1853.

72. Fraundorf, P., Brownlee, D. E. and Walker, R. M.,
     1982, Laboratory studies of interplanetary dust.
     In "Comets", ed. L. L. Wilkening, Univ. Arizona,
     Tucson, pp. 383-409.

73. Esat, T. M., Brownlee, D. E., Papanastassiou, D. A.
     and Wasserburg , G. J., 1979, Magnesium isotopic
     composition of interplanetary particles. Science
     206, pp. 190-197.

74. Hughes, D. W., 1975, Cometary outbursts, a brief
     survey. Quart. Journ. Roy. Astron. Soc., 16, pp.
     410-427.

75. Whipple, F. L., 1983, Cometary nucleus and active
     regions. In "Cometary Exploration", ed. T. I.
     Gombosi, Hungarian Acad. Sci., Budapest, I, p. 95-110.

76. Whipple, F. L., 1983, Comet P/Holmes, 1982 III. A
     case of duplicity ? Bull. Amer. Astron. Soc., 15,
     p. 805 (abstract).

77. Mendis, D. A. and Brin, G. D., 1977, Monochromatic
     brightness variations of comets. II. Core mantle
     model. The Moon 17, pp. 359-372.

78. Brin, G. D. and Mendis, D. A., 1979, Dust release
     and mantle development in comets. Astrophys. Journ.
     229, pp. 402-408.

79. Horanyi, N., Gombosi, T. I., Cravens, T. E., Kecske-
     mety, K., Nagy, A. F. and Szego, K., 1983, The
     friable sponge model of cometary nucleus. In "Co-
     metary Exploration", ed. T. I. Gombosi, Hungarian
     Acad. Sci., Budapest, I, pp. 59-73.

80.   Weissman, P. R. and Kieffer, H. H., 1981, Comet ther-
      mal modeling. Icarus, 47, pp. 302-331.

81.   Whitney, C. A., 1955, Comet outbursts. Astrophys.
      Journ. 122, pp. 190-195.

82.   Cochran, A. L., Barker, E. S. and Cochran, W. D.,
      1980, Spectrophotometric observations of P/Schwass-
      mann-Wachmann 1 during outburst. Astron. Journ.,
      85, pp. 474-477.

83.   Cochran, A. L., Cochran, W. D. and Barker, E. S.,
      1982, Spectrophotometry of Comet Schwassmann 1.
      II. Its color and CO$^+$ emission. Astrophys. Journ.
      254, pp. 816-822.

84.   Cowan, J. J. and A'Hearn, M. F., 1982, Vaporiza-
      tion in comets. Outbursts from Comet Schwassmann-
      Wachmann 1. Icarus, 50, pp. 53-62.

85.   Pataschnik, H., Rupprecht, G. and Schurman, D. W.,
      1974, Energy source for comet outbursts. Nature
      250, pp. 313-314.

86.   Pataschnik, H. and Rupprecht, G., 1977, Ice in spa-
      ce : An experimental and theoretical investigation.
      Final Report NAS 8-30566, Dudley Observatory, pp.
      1-185.

87.   Klinger, J., 1980, Influence of a phase changes of
      ice on the heat and mass balance of comets. Scien-
      ce 209, pp. 271-272.

88.   Klinger, J., 1981, Cometary water ice phase transi-
      tions. Icarus, 47, pp. 320-324.

89.   Smoluchowski, R., 1981, Amorphous ice and the beha-
      vior of cometary nuclei. Astrophys. Journ., 244,
      pp. L31-L34.

90.   Smoluchowski, R., 1981, Heat content and evolution
      of cometary nuclei. Icarus, 47, pp. 312-319.

91.   Shulman, L. M., 1983, A correction to the icy model
      of cometary nuclei, In "Cometary Exploration", ed.
      T. I. Gombosi, Hungarian Acad. Sci., Budapest, I,
      pp. 51-54 ; Have cometary nuclei any internal sour-
      ces of energy ? Ibid., pp. 55-58.

92. Platt, J. R., 1956, On the optical properties of interstellar dust. Astrophys. Journ., 123, pp. 486-490.

93. Misconi, N. Y. and Whitlock, L. A., 1983, On the unexplained outbursts of Comet P/ Tuttle-Giacobini-Kresak. In "Cometary Exploration", ed. T. I. Gombosi, Hungarian Acad. Sci., Budapest, II, pp. 185-189.

94. Greenberg, J. M., 1982, Interstellar dust model. In "Comets", ed. L. L. Wilkening, Univ. Arizona, Tucson, pp. 131-163.

95. Yamamoto, T., 1983, On the molecular composition of a cometary nucleus. In "Cometary Exploration", ed. T. I. Gombosi, Hungarian Acad. Sci. I, pp. 85-91.

96. Shulman, L. M., 1972, The chemical composition of cometary nuclei. In "The Motion, Orbital Evolution and Origin of Comets", eds., G. A. Chebotorev, E. I. Kazimirchak-Polonskaya, and B. G. Marsden, Reidel, Publ. Co., pp. 264-270.

97. Whipple, F. L., 1977, in "Comets-Asteroids-Meteorites", ed. A. H. Delsemme, Univ. Toledo, pp. 25-35.

98. Moore, M. H., Donn, B., Khanna, R.and A'Hearn, M. F., 1983, Studies of proton irradiated cometary type ice mixtures. Icarus, 54, pp. 388-405.

99. Orlov, S. V., 1940, Exceptional comets. I. Comet 1892 III. Astr. Journ. Sov. Union 17, no. 1, pp. 8-12.

DISCUSSION

F. Fanale :

If the image you showed of the Brownlee aggregate is at all typical (and I suspect it is), then published "grain" size distribution for cometary dust (which peak $\sim$ 0.7 µm) must refer almost entirely to <u>aggregate</u> size.

E. Gaffney :

Since Dr. Whipple ended his talk with a question, "Is there a way that a 100 m body impacting a comet at

$\sim$ 3 m/s can produce a chemical explosion ?", I will attempt to provide some answer to that.

Impacts of $\sim$ 3 m/s will produce very small temperature changes. For example, ice into ice will increase temperature of less than 1/30 K, an amount inadequate
to yield the desired result. Even with 90 % porosity in
the comet, we would still have less than 1/3 K increase
due to P x ΔV work alone. This is certainly too small
even to initiate transition of metastable phases. However, the energy dissipated in crushing a porous medium
is not evenly distributed. Although there is no good experimental data, one could expect that heating would be
concentrated at grain boundaries or other discontinuities sufficient to initiate stabilization transitions.
In addition to temperature increase, there will also be
substantial plastic deformation concentrated in these
same locations. These are the probable mechanisms of
initiation of normal explosives at impact velocities of
km/s. At m/s velocities, one certainly could not get a
detonation but initiation a slower exothermic reaction
which would still be fast enough to satisfy the observations of comet Holmes cited by Whipple.

THE SUBLIMATION TEMPERATURE OF THE COMETARY NUCLEUS: OBSERVATIONAL EVIDENCE FOR H₂O SNOWS

A. H. Delsemme

The University of Toledo, Toledo, OH U.S.A.

The sublimation temperature of the cometary nucleus essentially depends on the vapor pressure and the latent heat of the most volatile material available in abundance at the surface of the nucleus. This sublimation temperature sets in turn the characteristic distance r_0 separating the sublimation steady state (small dependence on distance) from the radiation steady state (steep dependence on distance). r_0 can therefore be used, not only as a measure of the surface temperature of the nucleus, but also as an identification of the nature of the prevailing snow. Three techniques to measure r_0 are described here. Two have given significant results because they can cover large ranges of distances: The dependence on distance of the Non Gravitational Forces suggests that the short period comets' sublimations are controlled by water snow only; the light curves of "new" comets concur to the same conclusion. Among all the comets studied so far, none seems to be controlled by anything more volatile than water snow or ice. The large production rates of H and OH observed in cometary atmospheres suggest that they come from the dissociation of H_2O in the vapor state. The theory of vaporization only is able to close the gap: it concludes that this water vapor was indeed released by the sublimation of water ice.

I. THE CHARACTERISTIC DISTANCE r_0

Circumstantial arguments have demonstrated that the cometary nucleus is an icy conglomerate, in which the mass of the volatile icy fraction is of the same general order of magnitude as that of the refractory, dusty fraction (Whipple 1950, 1951).

367

J. Klinger et al. (eds.), Ices in the Solar System, 367–387.
© *1985 by D. Reidel Publishing Company.*

Fluid dynamics (Finson and Probstein 1968) and the observation of collisional effects in the inner coma (Malaise 1970) have confirmed large production rates of gases, consistent with the sublimation of a few square kilometers of ices by the solar radiation (Delsemme & Rud 1973).

These large production rates imply that a large amount of heat is absorbed by the change of state of the sublimating ices; this latent heat effect makes the nuclear surface much colder at short heliocentric distances than implied by the black body radiation laws. However, the radiative steady state still prevails at large heliocentric distances. The characteristic distance r_0 at which the vaporization heat becomes much smaller than the radiated heat depends very much on the nature of the sublimating snows because of the differences, not only in the latent heat of the different snows, but mainly in their vapor pressures which control the sublimation rates and therefore the temperature.

A reasonable definition of r_0 has been adopted by Marsden et al. (1973); it will be used hereafter; r_0 is the distance at which the solar energy spent in reradiation is 40 times that spent in sublimation. On a theoretical log-log diagram of the production rates versus heliocentric distances, point r_0 is therefore about 4 astronomical magnitudes lower than the extrapolation of an inverse-square law of the distance, namely a straight line of slope -2 (see Fig. 1).

The elementary sublimation theory (Delsemme and Miller 1971, Delsemme 1972, Marsden et al. 1973) ignores geometrical or structural details and computes general averages for a rotating nucleus. In such a theory r_0 is rather well represented by the formula:

$$r_0 = \frac{1}{2T_0^2} \left(\frac{F}{\sigma}\right)^{\frac{1}{2}} \left(\frac{1 - A_0}{1 - A_1}\right)^{\frac{1}{2}} \left(1 + \frac{E_V}{E_R}\right)^{-\frac{1}{2}} \tag{1}$$

here, T_0 is the steady-state effective temperature of the rotating nucleus; F is the solar flux; σ the Stefan's constant; A_0 and A_1 the albedos of the nucleus in the visual and in the infrared (near 20 μm) respectively; E_V and E_R the energies used for the vaporization (sublimation) of the ices and for the radiation back to space.

Formula 1 shows that r_0 is strongly temperature-dependent; besides, the temperature itself is controlled by the nature of the snows (see Delsemme 1983 for details). In particular, it is easy to establish that for water:

$$r_0 \cong 2.8 \left(\frac{1 - A_0}{1 - A_1}\right)^{\frac{1}{2}} \text{ (in A.U.).} \tag{2}$$

Table I gives r_0, T_0 and T_1 as well as the vaporization rate Z_0

for different snow compositions, assuming $A_1 = 2A_0$ (a legitimate assumption, because at 20 micrometers, the albedo is usually much larger than in the visible).

TABLE I

Production Rates for Different Snows of the Cometary Nucleus

Snows Controlling the Vaporizations	Z_0 in 10^{18} mol cm^{-2}s^{-1}a	T_0b K	T_1c K	r_0d AU
Nitrogen	14.3	40	35	77.6
Carbon monoxide	13.0	44	39	62.5
Methane	10.6	55	50	38.0
Formaldehyde	5.0	90	82	14.1
Ammonia	3.7	112	99	9.7
Carbon dioxide	3.5	121	107	8.3
Hydrogen cyanide	2.3	160	140	4.8
Gases from ammonia water (aqueous ammonia)	2.7	213	193	2.6
Gases from methane clathrate	1.9	214	194	2.5
Water	1.7	215	195	2.5

$^a Z_0$ is the vaporization rate at subsolar point of a perfectly absorbing nucleus at 1 AU from the Sun.
$^b T_0$ is the steady-state temperature of Z_0 (subsolar point, nonrotating nucleus).
$^c T_1$ is the effective mean temperature of a rotating nucleus ($Z_1 = 1/4 \ Z_0$).
$^d r_0$ is the heliocentric distance beyond which the vaporization rate becomes negligible ($\leq 2.5\%$ of the solar flux is used for vaporization, $\geq 97.5\%$ for reradiation).

Now, how can r_0 be determined observationally? The dependence law of the production rate on heliocentric distance is difficult to establish. However, r_0 separates the weak and the steep dependences on distance: even crude measures should show where the switch occurs.

Three techniques can conceptually be used. Each can produce a reasonable approximation of the gaseous production rates. They are: the dependence on distance of the non-gravitational forces (N.G.F.), that of specific molecular production rates, and that of the brightness law (light curve in the visual).

II. THE NON-GRAVITATIONAL FORCE LAW

In their prevailing interpretation, the NGF come from the "jet" effect due to the anisotropic sublimation (mainly sunward) of the nuclear ices. The jet effect can be assumed to be more or

less in proportion to the total sublimation rate. (The larger
concentration of the vaporizations near the subsolar point at
larger heliocentric distances can be neglected in a first
approximation). I remarked (Delsemme 1972) that for comet
Schwassmann-Wachmann 2, the NGF dependence on distance that produced the least residuals in the computed orbit, was strangely
similar in shape with the prediction of the vaporization theory
applied to <u>water</u>, and to water only (see Fig. 1); namely, r_0 had
to be close to 2.8 AU suggesting a 200°K temperature and therefore a complete control of the sublimation by <u>water</u> ice. Later,
following my suggestion, Marsden et al. (1973) showed that this
remained indeed true for <u>all</u> of these short-period comets for
which enough good positions were available, so that orbital data
could be linked through several passages. Finally, Yeomans
(1977) showed that it was also true for comet Halley.

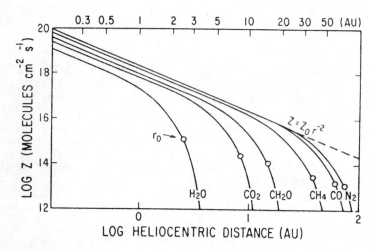

Fig. 1. *Vaporization rate, in molecules* $cm^{-2}\ s^{-1}$, *for
various snows as a function of the heliocentric distance
and assuming a steady-state rotating cometary nucleus,
with* $A_0 = 2A_1$ *and where* r_0 *is the distance defined in
Table I. Variation of the albedos' ratio shifts the
curves slightly (see Fig. 2).*

The results from the NGF can be summarized as follows: The
short-period comets show evidence that their production rates
are usually controlled by the sublimation of <u>water</u> ice, implying
nuclear surface temperatures close to 200°K for the usual observational distances (0.2 to 2.0 A.U.). The study of the NGF
dependence on distance cannot be done with any accuracy for
long-period comets, because they have been observed during one
single perihelion passage only.

III. THE MOLECULAR-EMISSION LAW

The second possible technique is the study of the dependence on distance of the molecular emissions. A crude qualitative assessment had already been proposed by Delsemme and Swings (1952) when they remarked that, in spite of the fourteen orders of magnitude difference between the vapor pressures of possible parent molecules, CH, NH, NH_2 and OH appear almost simultaneously, along with C_2 and CN, in the spectra of most comets when, in their approach to the sun, they reach distances of the order of 3 AU. Near $110°K$ (that they used as a typical surface temperature at 3 AU) H_2O is near 10^{-11} mm Hg, whereas NH_3 is near 10^{-4} mm Hg, CO_2 near 10^{-2} mm Hg and CH_4 near 10^3 mm Hg.

In modern terms, r_0 was close to 3 A.U., because of the simultaneous appearance of all molecular bands at that distance. They concluded that water ice was in control of the sublimations in most comets. In order to explain the simultaneous appearance of all other molecular species, they proposed that hydrates bind these molecules to water ice, namely ionic hydrates for ammonia and clathrate hydrates for the other parent molecules. The sublimation of the water lattice releases all other parent molecules in proportion to water.

In order to use this technique more quantitatively, we must use measured production rates at many heliocentric distances, so that we can establish where the slopes drastically change on the log-log diagram. The underlying assumption would be that the rates of those radicals observed would remain more or less in proportion to the total gas production rate. This is rather well suggested by the correlations between C_2, C_3, CN, CH, and OH studied in several comets by A'Hearn and Millis (1980), and A'Hearn et al. (1981, 1983). Their observations are particularly significant because OH is clearly a result of water photo-dissociation and its production rates show that it comes from a major, if not the major constituent.

The distance law established by A'Hearn et al. (1981) can be reasonably fitted to a straight line for comet Bradfield, on the log Q versus log r diagram; its slope is close to $r^{-3.2}$ in a first approximation; however, the range covered corresponds to less than 0.5 in log r (a factor of 3 on r) for C_2, C_3 and CN, and less than 0.25 in log r for OH. Weaver et al. (1981) also report production rates for the same comet; they give a slope of $r^{-3.7}$, but they also report an extra point for 1.55 AU at the extreme end of the range which shows a drop by a factor of 2 below the slope of $r^{-3.7}$, suggesting that the exponent is growing for large distances. The fact that the slope for comet Bradfield's OH is already much steeper than -2 for distances between 0.7 and 1.4 A.U. has been used by Wallis (1982) as an

argument that less volatile outgassing material than H_2O was
involved in comet Bradfield; however, no positive identification
is proposed (apart from the suggestion of "organic compounds").
Forthcoming Table II will show that many explanations may be
invoked for a short-range change or distortion of the slope,
therefore only very long ranges should be used to establish r_0.
Unfortunately, none of the studies of molecular emissions published so far cover a distance range much wider than that of
A'Hearn et al. (1981); this does not seem enough to observe the
two types of steady state in the same comet, and therefore the
transition from smaller to steeper slopes has never been observed
quantitatively from molecular emissions. One of the best early
data of this type were the two parallel slopes of -2.3 for OH
and H in comet Bennett. These slopes obtained by Keller and
Lillie (1974) correspond well to the sublimation steady state
of water ice, but they do not escape the previous criticism
since the range is only from -0.15 to +0.09 in log r.

IV. THE LIGHT CURVES OF COMETS

In the final analysis, it appears therefore that the third
technique, namely the study of the light curves of comets, in
spite of the intrinsic inaccuracy of visual observations, can be
used to cover a much longer range of distances and therefore
can be more successful than the observations of molecular production rates, at least to establish the transition from the
smaller to the steeper slopes expected from the sublimation
theory.

A'Hearn et al. (1980) have confirmed the existence of a rather
good correlation between intrinsic production rates and visual
magnitudes reduced to Δ = 1 A.U. (Δ: geocentric distance). The
interpretation of this correlation comes from the fact that dust
(which is often responsible for a good fraction of the comet's
luminosity) is dragged away by gas, more or less in proportion
to the gas itself; of course, for very dusty or less dusty
comets, the proportion coefficient would not be the same; however, we are not interested in absolute measures, but only in
the shape of the dependence law on distance for the same comet.
For this reason, we can use reduced magnitudes (or rather,
luminosities) as a crude measure proportional to the production
rate for a given comet. However, when actual light curves of
comets are considered (and properly reduced to Δ = 1 A.U.), only
about 25% of these reduced light curves can be fitted easily to
those predicted by the sublimation theory (Delsemme 1983). The
other 75% show outbursts and very often wide brightness fluctuations; although these phenomena have not, by far, been explained
in quantitative terms, the ideas that have been explored are
mostly based on the intrinsic complexity of the geometry and

dynamics of the nuclear region in comets. An incomplete list of
these ideas is given in Table II. Useless to say that they are
also applicable to the dependence laws of the molecular emissions.
It is interesting to note that, by and large, the observed
fluctuations (and for that matter, the interpretations given in
Table II) have characteristic times shorter, and very often much

TABLE II

Origin of Outbursts and of Brightness Fluctuations in Comets

1. *Heterogeneity of Nucleus*
   *Chemical differences in minerals (pockets, inclusions,
   etc.) differences in volatility, albedo, mineralogy, etc.*
2. *Intrinsic Sources of Heat*
   *Chemical: Reaction from free radicals when onset of
   heating.*
   Physical: Changes of state (like amorphous ice into cubic).
3. *Geometry and Dynamics*
   *Irregular nuclear shape combined with spin and precession,
   changing polar irradiation; changing torque due to
   asymmetrical vaporization; growing spin rate brings
   splitting of nucleus or of smaller chunks; behavior of
   double or multiple nucleus.*
4. *Dust Separations from Gas*
   *Fine dust dragged away by gas, coarse dust grains accumu-
   late in outgassed mantle; zonal mantle migrate toward
   less active regions; periodic blowaway of mantle by
   steadily growing sublimations; crust formation through
   grain soldering by contact in vacuum; loss of large
   cometary meteoroids; change in size distribution of dust
   grains with distance.*
5. *Interaction of Nuclear Region with Solar Radiation*
   *Optical thickness of dust halo; variable apparent cross
   section of nuclear zone to solar light; radiative transfer
   in dust, feedback on sublimation; hunting of feedback
   mechanism and oscillations in sunward "fountain" of dust;
   infrared radiation of dust halo; variable Swings and
   Greenstein's effects with radial velocity of comet; variable pumping of fluorescence by the vacuum ultraviolet
   emission lines of the sun.*
6. *Interaction of Nuclear Region with Solar Wind*
   *Variations in solar wind velocity, intensity and magnetic
   field. Variable position of contact surface (cometary ions
   with solar wind). Variations in comet's molecules photo-
   ionization from variable far solar UV. Alfvén waves in
   onset of tail; variable focussing of ions; periodic
   penetration of ions close to nucleus.*

shorter than the 2 to 3 years' transit time of comets from the radiative steady state of the nucleus (say 5 A.U.) down to the sublimation staedy state (say 1 A.U.). For this reason, if the light curve is established for a long enough range, the respectively <u>steep</u> and <u>weak</u> dependences on distance of the two steady states will be recognized as the general trend underlying the shorter-period phenomena.

Many light curve data exist in the cometary literature. Unfortunately, most authors reduce magnitudes by the formula:

$$m = H_0 + 5 \log \Delta + 2.5 \, n \log r$$

(H_0 absolute magnitude, Δ geocentric and r heliocentric distance) and they force-fit the data to a <u>constant</u> n, whatever the range, whereas r_0 is established from the observed distance where n changes from 2 (sublimation steady state) to the very large (steep) values of the <u>radiation</u> steady state. For this reason, in most of the cases we must go back to the unreduced data and plot them on a log-log diagram. As an example which is well documented, Morris and Green (1982) have collected most of the data corresponding to comet Halley's 1910 passage. However, a large gap exists from 1.0 to 0.6 A.U. on its preperihelion light curve, and another gap from 1.1 to 3.4 A.U. on its post perihelion light curve. Because a larger magnitude means a fainter brightness, it is worth mentioning that the <u>negative</u> slope on the brightness diagram corresponds to a <u>positive</u> n in the magnitude formula. Ernst (1911) had found an average n = 5.4; Bobrovnikoff (1941), 1942) finds 5.5 before perihelion and 5.4 after perihelion, but he complains twice that the observations are very discordant (implying that the two values for n are very crude averages). Yeomans (1981) finds 5.2 before perihelion; finally, in their exhaustive study, Morris and Green (1982) reduce n to 4.4 (preperihelion) and 3.1 (post perihelion).

If the condition that n be kept constant on a large range of r is waved, it becomes possible to see that r_0 was of the order of 3.6 A.U. before perihelion, but about 5.4 A.U. after perihelion, with a large uncertainty which makes it consistent with the results from the NGF (Yeomans 1977). However, comet Halley (as many bright comets) is a difficult case because it had many outbursts, including a major one just at and after perihelion. If we include the two major gaps mentioned earlier, plus the practical difficulty of reducing discordant 1910 data without enough information to find the source of the discordances, the 1910 light curve of comet Halley <u>after</u> perihelion, in spite of the interest brought about by its forthcoming 1986 passage, certainly is <u>not</u> a good case study for the theory of sublimation. A reasonable interpretation of what occurs in bright dusty comets for short heliocentric distances involves the feedback due to multiple scattering of the solar radiation by the dust

halo present close to the nucleus; in accordance with item 5 in
Table II, the multiple scattering makes the effective cross
section of the nuclear region grow with diminishing distance,
and for bright dusty comets it is not difficult to develop models
where, for shorter distances than r_o, n goes asymptotically not
towards 2 but rather towards 3 to 3.5 (Keller 1979, Hellmich
1981). It is not unlikely that these models could also explain
the slope mentioned before for comet Bradfield, as well as the
fraction of the light curve of comet Halley, for its shortest
distances to the sun.

The well-behaved preperihelion light curve of old periodic comet
Encke (Fig. 2) represents the other extreme: I have shown
(Delsemme 1975) that its preperihelion light curves, from Beyer's

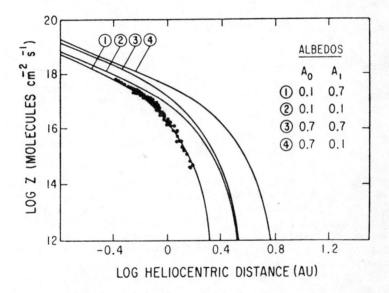

Fig. 2. *Lightcurve of P/Encke, fitted to the vaporization
curve predicted for a rotating nucleus vaporizing water,
assuming the total brightness is in proportion to the production rate of water molecules. The best fit is for A_o =
0.7 and A_1 = 0.1, shifting the average curve of Fig. 1
sunwards to r_o = 1.5 AU. None of the other molecules
represented in Fig. 1 could be shifted enough without using
completely unrealistic values of the albedos. Typically,
the albedos for CO_2 that would fit the lightcurve of
P/Encke would be $A_o \geq 0.99$ and $A_1 \leq 0.1$.*

observations (1950a, b, 1955, 1962) for the two passages of 1947
and 1951, superpose rather well with a slight vertical shift of
less than 0.2 magnitudes on that of 1961. The 0.2 magnitude
shift in 1961 may come from an artifact, namely the deterioration
of the observing conditions in Hamburg in 1961 (brighter night
sky). The mean light curve is extremely close to that defined
by the vaporization theory, with r_0 = 1.6 A.U. This very small
value suggests that, although the sublimation of the icy conglomerate still is controlled by water snow, the nucleus is not
homogeneously covered by snow but must show a large cross
sectional area covered by a material much less volatile than
snow, like a mantle of silicate dust (Delsemme and Rud 1973).
The post-perihelion light curve of comet Encke, as reported by
several authors, is somewhat different and should probably be
explained by one of the phenomena discussed in Table II.

V.   THE LIGHT CURVES OF NEW COMETS

The literature on the light curves of comets consists of a vast
amount of heterogeneous data that are difficult to reduce and
homogenize. The two best sources of predigested data that do
not hide or modify individual observations probably are those of
Beyer and of Bobrovnikoff. I have previously used Beyer's (1950,
1955, 1962) data for the perihelion light curves of comet Encke,
reproduced in Fig. 2; I will use here Bobrovnikoff's (1941, 1942)
data, for all available "new" comets.

New comets were selected to compare with short-period comets,
because they represent the other extreme possibility, namely
pristine comets that might have never lost their volatiles.
Since the sublimation of short-period comets has been shown hereabove to be controlled by water ice (from the NGF laws), it is of
some interest to see whether pristine comets, newly entering the
inner solar system, show a sublimation pattern of a different
type, as possibly hinted by "new" comet Kohoutek, whose preliminary data on preperihelion behavior rather suggested control by a
more volatile material like CO_2 (Delsemme 1975).

It is well known that, because of their random walk in orbital
energy due to planetary perturbations, "new" comets (that is,
comets that have never crossed the inner solar system) can be
only statistically defined as a group. The reason is that any
"older" comet may come backwards to the "new" comets' orbital
energy; however, the probability of the process is such that the
"new" comet sample is about 90% pure and only 10% polluted by
"older" comets.

Using the Marsden and Roemer's (1982) list of the original
binding energies of 220 "new" and long-period comets, I have

selected all comets studied by Bobrovnikoff, that were close to
the peak of binding energies of "new" comets. I found that,
from Bobrovnikoff's data, it was possible to reasonably reconstruct the light curves of those comets listed in Table III.
Table III also lists their absolute magnitude H_0, the characteristic distance r_0 found from Bobrovnikoff's light curves, a
column mentioning whether the light curve was established before
or after perihelion and the binding energy per unit mass of the

TABLE III

The Characteristic Distance r_0 of Eleven Comets

Comet	Name	H_0	r_0(A.U.)	Observat.	1/a(orig.)
periodic	Encke	10.0	1.7	B.P.	periodic
1900 II	Borelly	9.0	1.8	A.P.	+610
1899 I	Swift	6.5	2.1	B.P.	-109
1908 III	Morehouse	4.6	2.3	B.P.	+174
1903 IV	Borelly	6.0	2.4	B.P.	+ 33
1937 IV	Whipple	6.0	2.8	B.P.	+ 62
			2.8	A.P.	
1912 II	Gale	5.6	2.8 (min)	A.P.	+ 45
			2.8 (max)	A.P.	
1902 III	Perrine	6.5	2.8	B.P.	+ 27
1914 V	Delavan	2.2	2.8 (min)	B.P.	+ 29
			4.1 (outburst)	B.P.	
1973 XII	Kohoutek	5.0	4.4?(75 obs.)	B.P.	+ 20
			3.4 (2808 obs.)	B.P.	
			2.2 (2808 obs.)	A.P.	
1910 II	Halley	5.0	3.6	B.P.	periodic
			4.2?	A.P.	

Comparison with Sublimation Theory

Nature of Ices	Possible Range for r_0	
H_2O or clathrates	1.6 to 4.0 AU	
HCN	3.0 to 7.6 AU	(the range is set by
CO_2	5.2 to 13.2 AU	reasonable limits for
CH_2O	8.8 to 22 AU	the albedos; see text
CH_4	24 to 60 AU	for details)

Notes for Table III

H_0 is the absolute magnitude (magnitude reduced to $r = 1$ AU and $\Delta = 1$ AU).

r_0 is the characteristic distance separating the sublimation steady state from
the radiative steady state; on a light curve versus log r, it occurs 4 magnitudes
lower than the straight extrapolation of an inverse-square law (slope -2).

Under "observat.", B.P. means before perihelion and A.P. means after perihelion.

1/a (orig.) means the reciprocal of the semi major axis of the original orbit,
expressed in 10^{-6} AU^{-1}. The original orbit is the orbit before entering the
planetary system, referred to its barycenter ("new" comets are mainly <100).

For comet Delavan, the value $r_0 = 4.1$ refers to outbursts periodically observed
near 1.2, 1.6 and 3.3 AU (see Fig. 10).

For comet Kohoutek, the value $r_0 = 4.4$ refers to my early interpretation of the
light curve, when only 75 observations were available. This value should be
corrected to 3.4 before and 2.2 after perihelion, when all 2808 now available
observations are included.

For comet Halley, only the value 3.6 (B.P.) is secure; the value 4.2 (A.P.) is
based on inconsistent data with too many gaps.

original orbit, 1/a being expressed in 10^{-6} A.U.$^{-1}$. The new results are those for comets 1899 I, 1903 IV, 1912 II, 1914 V, 1902 III, 1937 IV and 1937 XII (2808 observations), as well as for comets 1900 II and 1908 III which are, strictly speaking, not quite "new", but whose binding energies imply that they are unlikely to have been perturbed by deep passages through the inner solar system.

The results for comets p/Encke and Kohoutek, 1973 XII (75 observations before perihelion) are from Delsemme (1975) and have been added for easy comparison. However, since Kleine and Kohoutek (1977) have published an analysis of 2796 visual observations of comet Kohoutek 1973 XII, in order to improve my 1975 assessment I have combined these observations with twelve crucial data obtained from 4.6 to 4.0 A.U., a few weeks after discovery (and that had not been used by Kleine and Kohoutek).

The new results from these 2808 observations, distributed in daily averages to smooth out the light curve, diminish somewhat r_o for comet Kohoutek before perihelion and bring it at the upper end, but still within the possible range for water ice (contrarily to my preliminary result, based on 75 observations only). For comparison, the possible range of r_o is given under Table III for different types of volatile snows. The range limits are set by the range of reasonable albedos, as in Fig. 2, namely $A_o \geq 0.1$ for $A_1 = 0.7$ and $A_o \leq 0.7$ for $A_1 = 0.1$ (0.7 is the Bond Albedo of ordinary snow or ice).

VI. DISCUSSION

Figures 3 through 10 show the light curves of the new and quasi-new comets found in Bobrovnikoff's data; their reduced magnitude M_Δ is shown versus the heliocentric distance expressed in logarithmic units. M_Δ is assumed to be in proportion to log Z (Z, production rate) therefore an inverse square law of the distance produces a straight line of slope - 2 on these diagrams, which are equivalent to those used on Figures 1 and 2.

Comet Encke's (Fig. 2) light curve before perihelion, fits in the best with a vaporization curve. This implies that all other phenomena that might perturb the steady state (see Table II) play a minor and diminishing part in its decaying activity. The present position of its spin axis can probably explain the small dissymmetry observed after perihelion (Whipple and Sekanina 1979); its r_o implies a very black nucleus.

Comet Borelly 1900 II (Fig. 3) has one of the shortest r_o, close to that of Encke, suggesting a very dark nucleus like Encke. Its original binding energy is too large to be a "new" comet.

Although the range of r covered by the observations is insufficient to give a good value of r_o, it would be difficult to imagine it larger than 2 AU.

Comet Swift 1889 I (Fig. 4) has a rather well defined r_o after perihelion in spite of two outbursts in the light curve. Its original orbit is nominally hyperbolic, but it is probably due to the neglect of its NGF.

Comet Morehouse 1908 III (Fig. 5) has a well defined light curve before perihelion; its $r_o = 2.3$ AU comes as a surprise because it implies that comet Morehouse's sublimation was completely controlled by water ice, in spite of the anomalously high abundance of its CO^+, identified in the 1908 spectra.

Comet Borelly's 1903 (Fig. 6) light curve does not extend far enough to give any accuracy to $r_o = 2.4$ A.U. Its outburst near 1.3 A.U. does not help the fitting, that could be extended to values to more than 3 A.U. as an upper limit.

Comet Whipple 1937 IV (Fig. 7) was observed only between 1.6 to 2.2 A.U., which is indeed the right position to observe a large slope, but the range was not large enough to give any accuracy. However, the extrapolation of the two light curves before and after perihelion, gives the same $r_o = 2.8$ A.U.

This $r_o = 2.8$ A.U. is also obtained twice for comet 1912 Gale (Fig. 8) in spite of four wide fluctuations producing four maxima and five minima in its light curve. This periodic outburst behavior suggests periodic mantle blow off (Table II).

Again $r_o = 2.8$ A.U. is obtained for comet Perrine 1902 III as well as for comet Delavan 1914 V, if a large outburst near 3 A.U. is neglected. If this outburst is given a large weight, $r_o = 4.1$ is obtained, at the price of a very poor fit between 1.5 and 2 A.U. (Fig. 9 and 10).

For completion and easy comparison on the same log-log diagrams, Fig. 11 and 12 give the light curves of comets Halley 1910 II and Kohoutek 1973 XII. Before perihelion, $r_o = 3.6$ A.U. is quite acceptable for Halley. The violent outburst at and just after perihelion, plus the gaps in the subsequent curve, make the fitting almost impossible, although the general drop of the light curve suggests an r_o near 4.2 after perihelion.

Comet Kohoutek's light curve before perihelion can reasonably be fitted to $r_o = 3.4$, if it is accepted that it reaches vaporization steady state between 1.0 and 0.5 A.U. In this range, the slope is indeed close to 2, however, for smaller r's, the slope grows again in the vicinity of 3, probably showing the result of

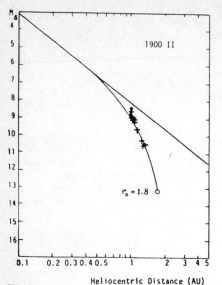

FIG. 3.-
Light curve of Comet 1900 II Borrelly. The dots are
observations before perihelion. The crosses are
observations after perihelion.

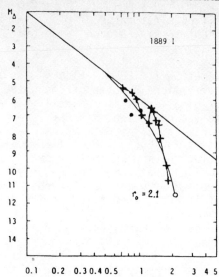

FIG. 4.-
Light curve of Comet 1889 I Swift. The dots are
observations before perihelion. The crosses are
observations after perihelion.

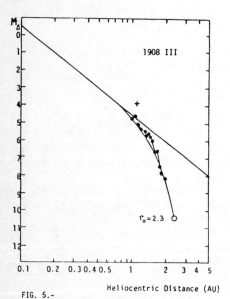

FIG. 5.-

Light curve of Comet 1908 III Morehouse.
The dots are observations before perihelion.
The cross is an observation after perihelion

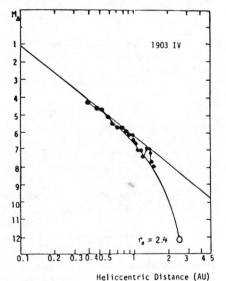

FIG. 6.-

Light curve of Comet 1903 IV Borrelly.
All the observations are before perihelion.

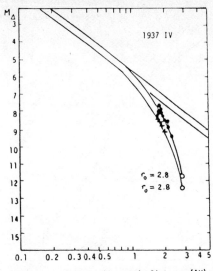

FIG. 7.-

Light curve of Comet 1937 IV Whipple. The dots
are observations before perihelion. The crosses
are observations after perihelion.

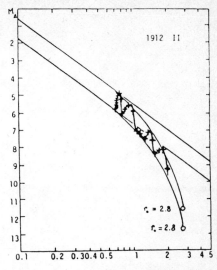

FIG. 8.-   Heliocentric Distance (AU)

Light curve of Comet 1912 II Gale. The dot is
an observation before perihelion. The crosses
are observations after perihelion.

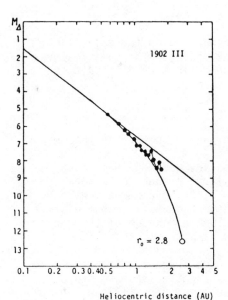

FIG. 9.-

Light curve of Comet 1902 III
Perrine; all the observations
are before perihelion.

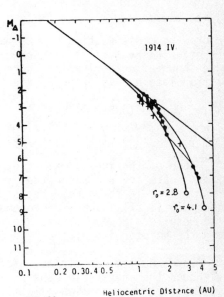

FIG. 10.-

Light curve of Comet 1914 V Delavan.
The dots are observation before perihelion.
The crosses are observations after perihelion

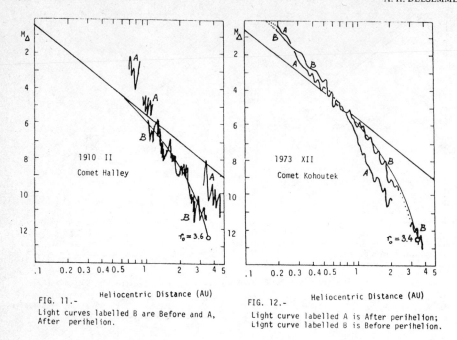

FIG. 11.-
Light curves labelled B are Before and A,
After perihelion.

FIG. 12.-
Light curve labelled A is After perihelion;
Light curve labelled B is Before perihelion.

a growing cross section of the nuclear region for solar light
(scattering by optically thick grain halo). The same features
are shown after perihelion; here an r_0 is more difficult to
extrapolate but is probably near 2.2 A.U.

VII. CONCLUSIONS

The conclusions of this discussion can be summarized this way:

a) None of the individual light curves deduced from Bobrovnikoff's data cover a range that is large enough to be totally
conclusive, in the sense that some of the phenomena mentioned
in Table II still interfere enough to hide or somewhat distort
the results. However, they are valuable because of the quality
and the homogeneity of the reduced data.

b) Comet Halley was already better observed for a longer time,
although the inevitable gaps in its light curve make the observations after perihelion almost useless. Fig. 11 is given only
for comparison purposes; complete data should be consulted in
Morris and Green (1982).

c) Comet Kohoutek's light curves cover one of the most complete
ranges before and after perihelion. My reduction of the magni-

tudes from 3.8 to 3.0 A.U. can always be disputed because the observations were made with large instruments; but the general range of r_o would not be changed enough to modify the general conclusions.

d) In spite of deficiencies in individual light curves, the message of the whole set of observations comes clearly out of the noise. It is already clearly apparent in the list of r_o's given in Table III, and it comes a big surprise: all "new" and "quasi-new" comets are not essentially different from short period comets in the sense that their sublimation--contrarily to the earlier belief--is also controlled by water ice only.

e) A possible interpretation of this surprising result is that none of the eight comets of Table III that have an original (1/a) smaller than 200 (in 10^{-6} AU^{-1} units) is a genuine "new" comet. All of these comets must only be very long-period comets that have already been several times through the outer solar system; they have lost already most of their ices more volatile than water ice, but the perturbations of Uranus and Neptune have not been large enough to change sizeably their binding energy to the Sun. Of course, if this is true, the puzzle is displaced. In a random sample of orbits with 1/a < 100, Oort's interpretation predicted 90% of new comets, and we do have a random sample here. Where are all new comets?

The answer to that question may be found in the fact that the Oort's cloud is five times as small as believed by Oort (40,000 AU instead of 200,000 AU), therefore the average velocity perturbation introduced by stellar passages is only of the order of 2 to 3 ms^{-1} per cometary revolution, much too small to bring a genuine "new" comet down to the vicinity of the Earth. New comets come down by steps into the inner solar system, and they need several revolutions to become observable from the Earth. Our results are the first empirical observations that confirm this.

We thank Orlando Naranjo who has drawn the light curves of the "new" comets on log-log diagrams, from the tabular data given by Bobrovnikoff. Grants from NSF-AST 82-07435 and from NASA-NSG-7381 are gratefully acknowledged.

REFERENCES

A'Hearn, M. F. and Millis, R. L., 1980, Abundance Correlations among Comets, Astron. J. 85, 1528; erratum 86, 802.
A'Hearn, M. F., Millis, R. L. and Birch, P. V., 1981, Comet Bradfield, the Gassiest Comet?, Astron. J. 86, 1559.
A;Hearn, M. F., Millis, R. L. and Thompson, D. I., 1983, The Disappearance of OH from Comet p/Encke; Icarus, 55, 250.
Bobrovnikoff, N. T., (1941) Investigations on the Brightness of

Comets I; Contr. Perkins Obs. No. 15.

Bobrovnikoff, N. T. (1942) Investigations on the Brightness of Comets II; Contr. Perkins Obs. No. 16.

Cochran, A. L., 1982, The Chemical Evolution of the Coma of Comet p/Stefan-Oterma; Univ. of Texas, Publ. Astron. No. 21.

Delsemme, A. H., 1972, Vaporization Theory and Non-Gravitational Forces in Comets, in "On the Origin of the Solar System", ed. H. Reeves, publ. CNRS Paris, pp. 305-310.

Delsemme, A. H., 1975, Physical Interpretation of the Brightness Variation of Comet Kohoutek, p. 195 in "Comet Kohoutek", ed. Gilmer, A. G., publ. NASA SP-355, Washington, D.C.

Delsemme, A. H., 1983, Ice in Comets; J. Physical Chem. 87, 4214.

Delsemme, A. H., Miller, D. C., 1971, Physico-Chemical Phenomena in Comets III: Planet. Space Sci., 19, 1229.

Delsemme, A. H., Rud, D. A., 1973, Albedos and Cross Sections for the Nuclei of Comets 1969 IX, 1970 II and 1971 I; Astron. Astrophys. 28, 1.

Delsemme, A. H., Swings, P., 1952, Hydrates de gaz dans les noyaux cométaires et les grains interstellaires, Annales Astrophys., 15, 1.

Festou, M. C., Feldman, P. D., 1981, The Forbidden Oxygen Lines in Comets; Astron. Astrophys. 103, 154.

Finston, M. L., Probstein, R. F., 1968, A Theory of Dust Comets I and II, 154, 327 and 353.

Hellmich, R., 1981, Influence of Radiation Transfer in Cometary Dust Halos on the Production Rate of Gas and Dust; Astron. Astrophys. 93, 341.

Keller, H. U., 1979, Feedback of the Dust Coma on the Evaporation Process, p. 57 in Proc. Comet Halley Micrometeoroid Hazard Workshop, ed. N. Longdon, ESA-SP-153, ESTEC Noordwijk, The Netherlands.

Keller, H. U., Lillie, C. F., 1974, The Scale Length of OH and the Production Rates of H and OH in Comet Bennett (1970 II) Astron. Astrophys. 34, 187.

Klinger, J., 1983, Classification of Cometary Orbits Based on Orbital Mean Temperature; Icarus 55, 169.

Malaise, D. J., 1970, Collisional Effects in Cometary Atmospheres; Astron. Astrophys. 5, 209.

Marsden, B. G., Sekanina, Z., Yeomans, D. K., 1973, Comets and Non-Gravitational Forces, Astron. J. 78, 211.

Morris, C. S., Green D. W. E., 1982, The Light Curve of Periodic Comet Halley 1910 II, Astron. J. 87, 918.

Spinrad, H., 1982, Red Auroral Oxygen Lines in Nine Comets, Publ. Astron. Soc. Pacific, 94, 1008.

Tatum, J. B., Gillespie, M. I., 1977, The Cyanogen Abundance of Comets, Astrophys. J. 218, 569.

Wallis, M. K., 1982, Are Comets Made of H$_2$O Ice?, p. 451 in Proceedings of Third European I.U.E. Conference, Madrid, Spain; ed. E. Rolfe, A. Heck, B. Battrick; ESA-SP-176, Publ. ESTEC, Noordwijk, The Netherlands.

Weaver, H. A., Feldman, P. D., Festou, M. C., A'Hearn, M. F.,
    1981, Water Production Models for Comet Bradfield, Astrophys.
    J. 251, 809.
Whipple, F. L., 1950, A Comet Model I; Astrophys. J. 111, 375.
Whipple, F. L., 1951, A Comet Model II; Astrophys. J. 113, 464.
Whipple, F. L., Sekanina, Z., 1979, Comet Encke: Precession of
    Spin Axis, Non-Gravitational Motion and Sublimation; Astron.
    J. 84, 1984.
Yeomans, D. K., 1977, Comet Halley and Nongravitational Forces;
    p. 61 in Comets, Asteroids, Meteorites, ed. A. H. Delsemme,
    publ. University of Toledo Bookstore.

DISCUSSION

P. Feldman - The behavior of comet Kohoutek could perhaps be explained by alteration of the surface layers by cosmic rays. Not
all "new" comets have necessarily the same previous history.

A. H. Delsemme - Your point is well taken. However, as you will
see in the final version of my paper, I have substituted a new
light curve based on 2808 observations of comet Kohoutek and the
puzzle introduced by the early light curve has disappeared. The
early assessment was made from inhomogeneous observations that
were not all of the same quality. The number of observations
itself is irrelevant, a few points may be sufficient if they
cover the right range, but it is essential that they be homogeneously reduced, if possible from a single observer's observations. This is the reason why I believe that most of the data
by Beyer or by Bobrovnikoff are more significant than many of the
light curve data found elsewhere.

P. Weissman - Dynamically we know that about 10% of apparent
Oort cloud comets have been within 3 AU of the sun before, and
about 50% have been within the orbit of Saturn. The average
comet has been within Neptune's orbit four times before we see
it. Thus, many if not all comets would be expected to lose much
of their volatile mantles prior to our observing them.

A. H. Delsemme - I wholeheartedly agree.   Since the Oort Cloud
is smaller than Oort believed in 1950, stellar perturbations are
accordingly smaller, therefore "new" comets perihelia do not
come from as far away as Oort believed. Your work has quantified
this fact and you just summarized your findings beautifully.
However, in the numerous comets we have observed so far, I would
expect that one or two would not be exactly average; for instance
when I started this work I was rather sure that comet Morehouse
would have a CO or CO_2 excess because of its anomalous CO^+
spectrum. It came as a surprise that it behaved exactly as the
other ones, that is as if its sublimation was controlled by

water ice only. If 50% of the "new" comets only have been within
the orbit of Saturn and if pristine comets contain CO_2 or HCN in
excess, then half of my sample of eight "new" comets should show
it: none does. I am less concerned with methane, carbon monoxide
or nitrogen, since, as my Fig. 1 combined with your remark implies, they would have been lost, at least from the outer layers
during their earlier perihelia passages within Neptune's orbit;
but where are the CO_2 and the HCN? I proposed with Swings,
thirty-two years ago, that all volatiles might be imprisonned in
clathrates of water. The explanation may still be valid.

Mayo Greenberg - What is the true depth of the cosmic ray pene-
trations effect? If only 0.1 m this would be gone early in a
first apparition. It was my impression that several meters is
a good possibility for cosmic ray modification. However, the
question is how much change is produced by cosmic ray processing
of already processed interstellar dust material. The answer
would appear to be that whatever volatiles were contained in the
original aggregate of interstellar dust would have been modified
towards nonvolatile organics, so that there would be a crust of
nonvolatiles and probably low visual near-ultraviolet albedo.

L. J. Lanzerotti - The penetration depth of cosmic rays grows
fast with their energy. A 1 MeV proton in H_2O ice is stopped in
less than 0.1 mm, whereas a 100 MeV proton would penetrate 10 cm.

V. Pironello - At still higher energies, one can roughly estimate that 1 GeV proton has a penetration in ice of a bit less
than 10 meters. Even if fluxes of so high-energy particles are
low, it is conceivable that a considerable irradiation of cosmic
rays has occurred in the first few meters of cometary nuclei,
during their long life outside the solar system.

A. H. Delsemme - This, of course, is the other possible answer
to my previous question: because of cosmic ray irradiation, all
materials more volatile than water have disappeared from the
outer layers of the comet. However, whatever the true depth of
the cosmic ray penetration effect, since we vaporize a few meters
of the outer layers per passage, we should see more volatile
material after 3, 10 or 30 passages. The fact that we don't
tells us something about the actual nature of the cometary stuff:
for instance, we could imagine that the accretion temperature of
comets was larger than 50 or 60 K, therefore, that neither CH_4
nor CO nor N_2 were condensed out of the early solar nebula. Even
if comets are made of interstellar grains, they may have been
later slightly processed by heat before accretion in the solar
nebula.

T. Owen - Could the problem with comet Kohoutek be solved if the
albedo were very high, about 1.0, instead of making the comet

rich in CO_2?

A. H. Delsemme - With the early light curve data shown in Nice, the answer was: no. No reasonably high albedo (like that of pure white snow, say 0.7) could have been an acceptable substitute for a more volatile stuff like CO_2. With the revised more homogeneous data used here, the need for CO_2 has disappeared (see my Table III and my answer to P. Feldman).

Gaetano Foti - The amorphous to crystal transition is, of course, a thermally activated process of the type $R_o exp(-E/kT)$. However, in amorphous Ge or Sb, it has been observed that the transition to crystalline phase may occur in an explosive way. The transition temperature which is usually near 500°C from amorphous to crystalline phase, may occur very fast at room temperature with a wavefront moving at meter-per-second velocity, when a vacuum-deposited amorphous phase is used. I submit that the proposed model of the cometary nucleus, where the amorphous to crystalline transition of water ice plays an important role, should take into account non-equilibrium processes, including the explosive crystallization that we have described in the literature (Leamy, Brown and Foti, Applied Physics Letters, 1981).

A. H. Delsemme - I am sure this remark has been heard and will be taken in consideration by Klinger. Since in cometary physics we have centuries or millenia to change amorphous ice into a crystalline phase, the real question rather is: What is the lowest possible temperature at which amorphous ice is going to change to crystalline ice? I do not know the proper answer to this question.

A. Mendis - One should be careful about identifying dust/gas ratios observed in the coma with dust/gas ratio in the nucleus. The dust in the coma is dragged off by the gas, therefore dust production rate is expected to be in proportion to gas production rate, independently of the dust/gas ratio in the nucleus.

A. H. Delsemme - I am quite aware of this problem, that I have discussed in my chapter of Wilkening's book "Comets" (1982). However, if we believe that comets are radially undifferentiated (see my discussion in same book) then the steady state implies that eventually, sooner or later, the two dust-to-gas ratios oscillate around an average value which is the same. The danger, of course, is that we are often faraway from any steady state!

CONDENSATION AND AGGLOMERATION OF COMETARY ICE: THE HDO/H$_2$O RATIO
AS TRACER

W.-H. Ip
Max-Planck-Institut für Aeronomie
D-3411 Katlenburg-Lindau
Federal Republic of Germany

ABSTRACT. The cosmochemical significance of the HDO/H$_2$O ratio in
cometary ice is discussed in the light of recent studies of H$_2$O
and HDO formation in interstellar clouds. In contrast to condensation processes in the cooling solar nebula, both gas phase
chemistry and surface chemistry on interstellar grains are expected to produce large isotope fractionation effects leading to
a HDO/H$_2$O ratio orders of magnitude higher than the value of the
D/H ratio in cosmic abundance. The deuterium abundance in cometary ice therefore could be very high if comets are formed by
direct agglomeration of interstellar grains. This idea is compared to the observational constraints set by the D/H ratios in
the upper atmospheres of the Jovian planets as well as the chemical composition of the interstellar icy grains derived from IR
observations and laboratory experiments.

The elemental and isotopic abundances in planetary atmospheres and comets are of basic importance in the study of the
formation of the solar system (1-5). The current development of
this interesting topic began with the direct measurements of the
^{3}He/H ratio in the solar wind (6,7). Geiss and Reeves (1) compared the inferred D/H ratio of the Sun with the corresponding
values found on Earth and in meteorites and proposed, for the
first time, that the observed deuterium enrichment by a factor of
about 10 in the terrestrial oceans and the meteorites must be due
to chemical fractionation at low temperatures in the solar
nebula. The subsequent discovery of an even higher deuterium
effect in the Chainpur meteorite with D/H ratios up to 9×10^{-4} by
Robert et al. (8) led to the specific idea that such high D/H

J. Klinger et al. (eds.), Ices in the Solar System, 389–395.

ratios could have their origin in the interstellar dark clouds -
as a result of ion-molecule chemistry (3). In particular, Geiss
and Reeves (3) drew attention to the observations of HDO in the
Orion nebula by Turner et al. (9) and the HDO/H_2O ratio of $\sim 10^{-3}$
so derived.

Being the most primitive bodies in the solar system, comets
represent the pristine sample of the early solar nebula; their
chemical abundance and isotope ratios are consequently of particular interest. Measurements of these quantities should provide
basic information on the relation between the interstellar medium
(dust and gas) and the cometary ices. Following the above mentioned studies on the D/H ratios in meteorites, our plan is thus
to first review the present ideas about the origin of comets and
the compositions of cometary ices and the interstellar icy
grains. Their implications on the D/H ratio of the cometary ice
are then compared to the pertinent observational data.

Recent results from IUE observations have indicated that
there is little difference in the chemical composition of the
volatile ices from comet to comet - with the exception of CO and
dust (14). The appearance of CO (or CO^+)-rich comets like comet
Morehouse 1908III nevertheless reminds us that caution must be
taken in accepting the idea of a homogeneous composition of the
condensed icy nuclei too soon. Note that Lacy et al. (15) have
matched the 4.61 μm absorption features due to icy grains observed in W33A and NGC 7538/IRS9 to CO and cynano group molecules
in solid phase. It is therefore possible that some comets could
be intrinsically CO-rich if they are formed by agglomeration of
the solid interstellar CO grains. Of course, the differences in
the abundances of dust and CO could be simply a matter of thermal
evolution of the cometary nuclei instead of basic variations in
the icy composition of comets of different origin. The point
could only be resolved after we have learnt more about the surfaces and interiors of comets from space missions to comets.

Even though there is good reason to believe that a large
number of icy planetesimals (or rather cometesimals) must have
been ejected into the Oort cloud region (10,11), our present lack
of knowledge of the initial conditions of the solar nebula does
not allow us to preclude the interesting possibility that km-
sized objects could have formed in the outer region of the pre-
solar nebula or/and in the collapsing interstellar cloud. Both
processes could contribute significantly to the cometary population (12,13). Thus the comets that come into our view could
actually have rather diverse origins.

In the context of ejection of comets from the accretion
zones of Uranus and Neptune it is interesting to note that the

D/H ratio ($\simeq 2 \times 10^{-5}$) measured for the upper atmosphere of Uranus
(5) might pose constraints on the idea of deuterium enrichment in
the solar nebula. It is possible that the interior of Uranus
could still be enriched in deuterium if there exists little in-
termixing between the atmosphere and the core (3,5). This is because the atmosphere of Uranus (and that of Neptune) might have
been acquired by gaseous accretion of H_2 gas in the solar nebula
after the formation of an icy core (J. Geiss, private communication, 1984). Note that the upper limit of the OD/OH ratio as determined from the observations of comet Seargent 1978XV is about
10^{-2} (16). Thus, if the D/H ratio is found to be large (i.e. D/H
$\approx 10^{-3}$-10^{-4}) in comet Halley, say, several interesting issues
would arise. First, this might mean that there is indeed no appreciable communication of the material between the core and the
atmosphere of Uranus. Second, the condensation of cometary ice
must have proceeded via non-equilibrium processes or other means.

As summarized before, Geiss and Reeves (3) have compared the
large D/H ratios found in meteorites to the corresponding values
observed in several interstellar molecules (i.e. HCN, NH_3, HCO^+,
N_2H^+, and H_2O). By analogy, an interstellar connection of the
cometary ice could thus be established. There are actually two
possible scenarios for the formation of the interstellar icy
grains:

(1) Assuming the water formation mainly occurs in the gas
phase, the icy grains condensed out of the gas phase should then
bear the same magnitude of deuterium enrichment. The amorphous
H_2O ice observed in interstellar clouds (17,18) therefore could
be enhanced in the HDO content; and if the interstellar icy
grains so formed have been later agglomerated into comets, a
large HDO/H_2O ratio may be expected from this point of view.

(2) More recently, Tielens and Hagen (19) and Tielens (20)
have considered surface chemistry between interstellar molecules
and dust grains in interstellar clouds with gas kinetic temperature and grain surface temperature both = 10°K and the hydrogen
number density $n(H_2) \simeq 10^3$-10^5 cm^{-3}. From such model calculations
the molecular composition of the interstellar grain mantles could
be estimated in a consistent manner. As discussed by Tielens
(20), at low grain surface temperature the evaporation time scale
from the grain surface for D atoms is much longer than that of
the H atoms, as well as the surface reaction time scale with
other radicals. This effect permits the efficient recombination
of the D atoms with the O_3 molecules (from surface recombination
of O and O_2) - after several intermediate steps - into HDO. Since
the D/H ratio ($\simeq 0.01$) in the gas phase for $n(H_2)$, as given in
the model of Tielens (20), basically determines the influx rates
of the D and H atoms to the grain surfaces, the resulting HDO/H_2O

ratio in the interstellar grain mantles follows closely the same enrichment factor.

Jones and Williams (21) have compared the formation time scales of ice mantles from direct H2O condensation and from the gas-surface reaction:

$$O_{gas} \Rightarrow O_{surface} \overset{2H}{=\!=\!=\!=\!=\!=\!\Rightarrow} H_2O_{surface} , \qquad (1)$$

and have reached the conclusion that surface reactions as given in equation (1) could be very significant in contributing to the ice formation in interstellar clouds. This study thus tends to reinforce the idea that a large D/H ratio is to be detected at comet Halley, if comets did form from the direct agglomeration of interstellar dust and icy grains.

One noteworthy result of the surface chemistry of ice forma-tion is simply that H_2CO could be nearly as abundant as H_2O; this should mean the production of deuterated H2CO also in the icy mantles (19,21). The recent positive detection of HCO and possi-ble detections of H2CO and DCO in comet IRAS-Araki-Alcock 1983d by Cosmovici and Ortolani (22) are therefore intriguing. On the other hand, the absorption features at 6 and 6.8 μm due to solid phase H_2CO have not been detected in interstellar clouds (23), thus care must be taken in interpreting the theoretical and ob-servational results.

Another example is that a number of studies have shown that the 3 μm feature observed for the Becklin-Neugebauer (BN) object (24) could be fitted satisfactorily by invoking large amorphous grains characterized by a mixture of H2O/NH3 ≈ 2 (25-27). Such mixing ratio has been predicted by the gas-surface interaction models described before (19,21). However, it has been pointed out most recently by Krätschmer (28) that the identification of the NH3 signature could be in error. Since a mixing ratio of $H_2O/NH_3 \approx 20$ would not be diagnosed by current IR observations (Krätschmer, private communication, 1984), it is still possible that interstellar icy grains contain 5% of NH_3.

It has been established by observations of cometary comae that the abundance of NH3 should be on the order of 10^{-3} as com-pared with that of H_2O (29,30). The discrepancy between the theo-retical models of interstellar icy grain formation and observa-tions of the compositions of the interstellar icy grains and cometary ice may be resolved by taking into consideration de-struction and/or reprocessing of the grain material. For in-stance, the NH3 ice could be preferentially sputtered away by

energetic charged particles (31). In fact, the recent detection
of S_2 molecules in the coma of comet IRAS-Araki-Alcock 1983d has
led A'Hearn and Feldman (32,33) to the opinion that these S_2
molecules could have been produced by irradiation effects of
small icy grains. The supposed irradiation could be in the form
of UV photolysis (34) or energetic particle bombardment (35) -
either in interstellar space or in the solar nebula during the
T Tauri phase of the young Sun.

These considerations indicate that intercomparison of the
chemical compositions of the interstellar icy grains and the
cometary ice is not straightforward; and one of the more reliable
signatures for cometary ice being of interstellar origin instead
of formation in the solar nebula would be the expected deuterium
enrichment effect. On the other hand, much more work is required
before we can quantify the D/H ratio to be derived in interstellar icy grains. The uncertainties in our understanding of the ice
condensation processes in interstellar space and in the solar
nebula are further accentuated by our lack of knowledge of the
physico-chemical structures of cometary surfaces and nuclei, not
to mention the inter-relation between comets and the outer planets (Uranus and Neptune). Thus we agree with Gautier and Owen (5)
in that: Observations of D/H and other isotopic ratios must be
accommodated by these models in ways that are not yet completely
clear. In any event, after the first in-situ measurements of the
D/H ratio in the coma of comet Halley in 1986, we might be able
to tell along which direction the future investigation should go.

ACKNOWLEDGEMENTS

We wish to thank Profs. M.F. A'Hearn, J. Geiss, and
Dr. W. Krätschmer for informative discussions and J. Geiss in
particular for his very useful comments.

REFERENCES

(1) Geiss, J. and Reeves, H., 1972, Astron. Astrophys. 18,
    pp. 126-132
(2) Black, D.C., 1973, Icarus, pp. 154-159
(3) Geiss, J. and Reeves, H., 1981, Astron. Astrophys. 93,
    pp. 189-199
(4) Reeves, H., 1974, in "On the Origin of the Solar System",
    ed. H. Reeves, CNRS Publication, pp. 51-53
(5) Gautier, D. and Owen, T., 1983, Nature 304, pp. 691-694
(6) Geiss, J., Eberhardt, P., Bühler, F., Meister, J., and
    Signer, P., 1970, J. Geophys. Res., pp. 5972-5979

(7) Geiss, J., Bühler, F., Cerutti, H., Eberhardt, P., and
    Filleux, Ch., 1972, Apollo 16 Preliminary Science Report,
    NASA SP-315, pp. 14.1-14.6
(8) Robert, F., Merlivat, L., and Javoy, M., 1979, Nature 282,
    pp. 785-789
(9) Turner, B.E., Zuckerman, B., Fourikis, N., Morris, M., and
    Palmer, P., 1975, Astrophys. J. 198, pp. L125-L128
(10) Safronov, V.S., 1972, "Evolution of the Protoplanetary Cloud
     and the Formation of the Earth and the Planets", Israel
     Program for Scientific Translation, Jerusalem
(11) Fernandez, J.A. and Ip, W.-H., 1981, Icarus 47, pp. 470-479
(12) Biermann, L. and Michel, K.W., 1978, Moon Planets 18,
     pp. 447-464
(13) Cameron, A.G.W., 1973, Icarus 18, pp. 407-450
(14) Feldman, P.D., 1982, in "Comets", ed. L.L. Wilkening, Univ.
     Arizona Press, pp. 461-479
(15) Lacy, J.H., Baas, F., Allamandola, L.J., Persson, S.E.,
     McGregor, P.J., Lomsdale, C.J., Greballe, T.R., and van de
     Bult, C.E.P., 1984, Astrophys. J. 276, pp. 533-543
(16) A'Hearn, M.F., Schleicher, D.G., Donn, B., and Jackson,
     W.M., 1981, in "The Universe at Ultraviolet Wavelengths: The
     First Two Years of IUE", pp. 73-81
(17) Hagen, W., Tielens, A.G.G.M., and Greenberg, J.M., 1981,
     Chem. Phys. 56, pp. 367-379
(18) Whittet, D.C.B., Bode, M.F., Longmore, A.J., Braines,
     D.W.T., and Evans, A., 1983, Nature 303, pp. 218-221
(19) Tielens, A.G.G.M. and Hagen, W., 1982, Astron. Astrophys.
     114, pp. 245-260
(20) Tielens, A.G.G.M., 1983, Astron. Astrophys. 119, pp. 117-184
(21) Jones, A.P. and Williams, D.A., 1984, Mon. Not. Roy.
     Astron. Soc. 209, pp. 955-960
(22) Cosmovici, C. and Ortolani, S., 1984, Nature 310,
     pp. 122-124
(23) Kitta, K., Krätschmer, W., 1983, Astron. Astrophys. 122,
     pp. 105-110
(24) Merrill, K.M., Russell, R.W., and Soifer, B.T., 1976,
     Astrophys. J. 207, pp. 763-769
(25) Knacke, R.F., McCorkle, S., Puetter, R.C., Erickson, E.F.,
     Krätschmer, W., 1982, Astrophys. J. 260, pp. 141-146
(26) Hagen, W., Tielens, A.G.G.M., and Greenberg, J.M., 1982,
     Astron. Astrophys. 117, pp. 132-140
(27) Léger, A., Gauthier, S., Defourneau, D., and Rouan, D.,
     1983, Astron. Astrophys. 117, pp. 164-169
(28) Krätschmer, W., 1984, Verhandlungen der Deutschen
     Physikalischen Gesellschaft, 7, pp. 1478-1491
(29) A'Hearn, M.F., Hanisch, R.J., and Thurber, C.H., 1980,
     Astron. J. 85, pp. 74-80
(30) Johnson, J.R., Fink, U., and Larson, S.M., 1984, Icarus, in
     press.

(31) Lanzerotti, L.J., Brown, W.L., Marcantonio, K.J., and Johnson, R.E., 1984, Nature 312, pp. 139-140

(32) A'Hearn, M.F., Feldman, P.D., and Schleicher, D.G., 1983, Astrophys. J. 274, pp. L99-L103

(33) A'Hearn, M.F. and Feldman, P.D., 1984, this volume

(34) Greenberg, J.M., 1982, in "Comets", ed. L.L. Wilkening, Univ. Arizona Press, pp. 131-163

(35) Strazulla, G., Purronello, V., and Foti, G., 1983, Astrophys. J. 271, pp. 255-258

FINAL NOTE:

Dr. Daniel Gautier has communicated to me the most recent determination of the D/H ratio on Uranus by C. de Bergh, B.L. Lutz and T. Owen (1985) from re-analysis of previous CH_3D measurements; the derived value of $(9^{+9}_{-4.5}) \times 10^{-5}$ would mean a possible deuterium enrichment by as much as a factor of 5-10 over the solar value. For further detail, see Gautier and Owen (in "Protostars and Planets", ed. T. Gehrels, Univ. Arizona Press, 1985).

AMORPHOUS AND POROUS ICES IN COMETARY NUCLEI

R. Smoluchowski

Department of Astronomy
University of Texas
Austin, Texas 78712   USA

ABSTRACT

The two main questions concerning the nature of cometary
nuclei are crystallinity of the ices and their porosity.
Various arguments lead to the conclusion that the water-ice
constituent of new comets is amorphous and its subsequent
crystallization would produce gaseous tails and flare-ups at
heliocentric distances about 70% larger than those expected for
crystalline ice nuclei. The effect should be weaker for older
comets, but it may be related to the behavior of the comet P/SW-1
if the layers of the crystallized ice peel off. An important
factor controlling the behavior of comets is the heat flux in and
out of the nuclei before and after perihelion. It turns out that
the surface temperature and the associated rate of evaporation
are lower than usually calculated and the effect is strongly
dependent on the porosity of the nucleus. The surface temperature
of the nucleus reaches a maximum after perihelion, in agreement
with the statistical studies of comae and tails. Delayed heating
of the CO_2-rich inclusions deep in the nucleus leads to outbursts
and splittings as far as 9 A.U. after perihelion also in
agreement with observation. At sufficiently high temperatures
heat transport in porous cometary nuclei is controlled by vapor
diffusion. In particular the presence of CO_2 vapor in the pores
significantly increases the heat flow in mixed nuclei at
temperatures above 140°-150°K. Thus for heliocentric distances
between 3 and 5 AU, the above mentioned phenomena should be
considerably enhanced in proportion to the admixture of CO_2-ices.
Closed pores move up the radial thermal gradient so that the
outside layer of a cometary nucleus should become denser. The
pore size distribution is altered by the alternating thermal

J. Klinger et al. (eds.), Ices in the Solar System, 397–406.
© *1985 by D. Reidel Publishing Company.*

gradient but these effects are, however, exceedingly slow, especially in water-ice.

1. INTRODUCTION

Essentially all our knowledge of cometary nuclei is indirect because it is either deduced from observations of cometary comae and tails and not of the nuclei themselves or it is based on theoretical arguments. Also, the answers to the fundamental question of the origin of these nuclei range from formation at the far outskirts of the primitive solar nebula, though not as far as the present Oort cloud, to formation within the orbits of the giant planets (followed by ejection) to capture from outside of the solar system. All these locations imply a very low ambient temperature so that the growth of amorphous rather than crystalline water-ice, which forms by condensation below 150 K, is favored. It is important that besides having very low thermal conductivity upon warming above 150 K, this ice crystallizes producing enough heat to raise its temperature by 37 degrees.

The presence of rather rough porosity in the nuclei can be inferred from the random nature of the process of accretion of solids and is favored by consideration of the apparent mechanical weakness of cometary nuclei (1). There is also an, admittedly tenuous, analogy to terrestrial glaciers and polar ice fields. The fact that water-ice is the least volatile and probably the most abundant of all ices in cometary nuclei suggests that the escape of those volatiles which have not been trapped as clathrates leaves behind a rather porous structure on a fine scale. A value of 0.3 to 0.5 for minimum local porosity seems reasonable; on a larger scale it could rise, however, to 0.8 and account for the apparent brittleness of some nuclei. The nature and structure of cometary ices becomes of crucial importance when the heat conduction and heat content of the nuclei is considered. Actually, it is only quite recently that the heat flux into a cometary nucleus and its role has been analyzed in some detail. The paper outlines the present state of our knowledge of this field and discusses some recent developments and new questions.

2. HEAT CONDUCTIVITY

The usual heat balance equation of a comet includes the captured solar heat, the heat radiated out into space by the surface at temperature T_s of the nucleus and the heat used in evaporating $Z(T_s)$ molecules, mostly H_2O, of the nucleus. The orientation of the rotation axis of the nucleus and the reflectivity and emissivity of the surface has to be taken into account. What is, however, usually omitted is the heat

$$K_s(dT/dr)_s \qquad \qquad (1)$$

diffusing into the nucleus, the justification being that it is
small compared to other terms. Although this approximation is
acceptable as far as the calculation of T_s is concerned, this
heat does play an important role (2) as discussed further below.
In Eq. (1), K_s is the thermal conductivity and $(dT/dr)_s$ is the
radial gradient of temperature both pertaining to the surface
layer of the nucleus. Since the radius of the nucleus is large
compared with the thickness of the layer here considered, one can
use a linear gradient $(dT/dx)_s$. The heat entering the nucleus is
important because it leads to solid state reactions and to mass
transport in the nucleus and also it gives the nucleus considerable
thermal inertia which affects the formation of comae, tails and
other phenomena.

If a nucleus contains amorphous water-ice which exists only
up to 150 K, then its thermal conductivity is a few orders of
magnitude lower (3) than if the ice were crystalline and it
increases with temperature which is the opposite trend of
crystalline ice. As a result, the thermal gradient below the
surface of the nucleus in Eq. (1) is much steeper than for a
crystalline nucleus. It is determined by the time-dependent
temperature profile

$$T(x,t) = x(2\pi\kappa_a)^{-1/2} \int_0^t T_s(t) \ (t-\tau)^{-3/2} \ \exp\{x^2/4\kappa_a(1-\tau)\}dt \qquad (2)$$

in which κ_a is the thermal diffusivity. For amorphous ice κ_a is
given by $v\lambda/3$ with v the velocity and λ the mean path of phonons.
When the initially amorphous ice crystallizes at about 150 K it
remains crystalline on subsequent cooling and its conductivity
is high.

Heat conduction of porous ice is complicated because at low
temperatures pores are an obstruction to heat flow while at higher
temperatures their vapor content increases the mean heat flow
through the ice (4). The transition between the two regimes
occurs near 150 K and near 200 K for CO_2 vapor in 1 mm and 0.1 mm
pores respectively, while for H_2O vapor the corresponding
temperatures are around 210 K and 250 K as shown in Fig. 1.

3. HEAT FLOW INTO THE NUCLEUS

Knowing the nature and structure of ice and using the above
equations, one can obtain T_s, the temperature profile $T(x)$ and
the heat flow into the nucleus as a function of t or of
heliocentric distance r_h. It turns out that the heat flow into

the nucleus is of the order of a few percent of the heat absorbed
by the surface so that T_S is lowered only by a few degrees.

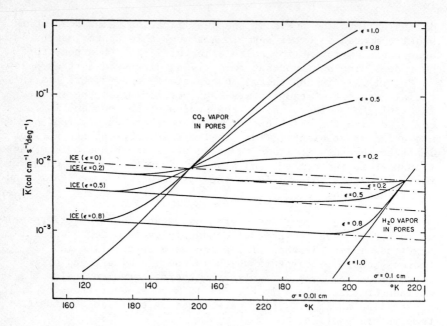

Fig. 1 Effective thermal conductivity of H_2O and H_2O-CO_2 ices
with porosity ε and pore size σ. (Reference 4)

The rate of evaporation $Z(T_S)$ depends, however, exponentially on T_S

$$dZ/dT_S = Z(T_S) \{ d(\log p)/dT - 1/2T_S \} \qquad (3)$$

so that the rate of evaporation of ices from the surface is
actually substantially lower than if it were calculated without
correction for the heat diffusing into the nucleus. It is
therefore strongly affected by porosity.

Another important aspect of the heat flow into the nucleus
during the pre-perihelion path of the comet is that as long as
T_S increases the radial temperature gradient below the surface
of the nucleus also increases. Just before the perihelion,
however, the solar heat incident on the comet gradually reaches
saturation, the thermal gradient reaches its maximum and then
begins to drop starting at the surface. This means that the rate
of increase of T_S is slowed down so that T_S reaches a maximum
some time after the perihelion. Because of its thermal inertia,
the nucleus is hotter than its surface and the heat flow is
reversed. As a result during the post-perihelion period T_S and

the rate of evaporation is higher than during the pre-perihelion
phase at the same r_h (Fig. 2).

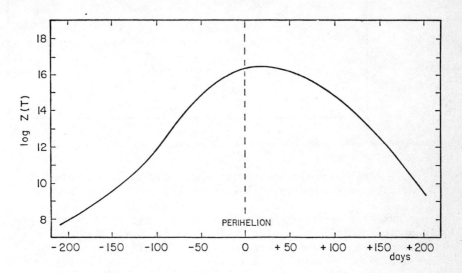

Fig. 2   Rate of vaporization $Z(T)$ in molecules cm^{-2} sec^{-1} of a
cometary nucleus near perihelion.   (Reference 2)

One concludes that, under normal circumstances, the coma and
the tail should be bigger after the perihelion than before
(2), the typical delay being up to 100 days for an almost
completely densified nucleus.   This normal behavior may be
masked or even reversed when there are anomalous structural
or compositional active regions on the surface as has been
suggested (5, 6, 7) for instance for Kohutek and Encke comets.

     As described below, a new porous comet has a low thermal
inertia but in the course of its subsequent passages its surface
layers and especially its interiors become denser.   As a result
these effects will progressively increase the asymmetry of
comae and tails with respect to the perihelia.   This conclusion
appears to be in agreement with statistical studies of cometary
brightness which show that for new comets this brightness
depends on heliocentric distance as r_h^{-n} with n - 2.5 but for
old, that is short-period, comets n - 4 before and n - 2.5 after
perihelion (7).   A similar delay is described further below
in connection with densification (Fig. 3).

4. PHASE CHANGES

     As mentioned in the introduction it is plausible that the

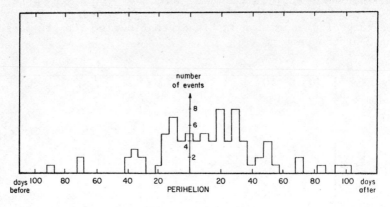

Fig. 3 Occurrence of outbursts and break-up of 87 comets in
proximity of perihelia. (Reference 11)

nucleus of a comet which is on its first approach to the Sun
contains amorphous H_2O-ice. If it is so, then, as soon as the
surface temperature of the subsolar point reaches 150 K the ice
crystallizes in an exothermic reaction to a cubic, also
metastable, form. Part of the heat released during this phase
change (8) is radiated out and part heats up the neighboring
layers of amorphous ice which in turn crystallize. The
steepness of the radial or tangential temperature gradient
below the surface determines the thickness of the ice layer
in which within a few tens of seconds the temperature rises
up to about 190 K and becomes crystalline. This rapid rise of
local temperature increases the rate of evaporation by several
orders of magnitude leading to a flare-up of the comet (2).
These flare-ups should occur at about 70 percent larger distances
from the Sun than those in crystalline H_2O-ice nuclei as has
been often observed.

     The rotation of nuclei may lead to a series of similar
flare-ups from other parts of the nucleus. This effect should
be pronounced for comets which have a slow rate of rotation
and are rather porous so that the heat does not spread rapidly.
For small perihelia, the freshly crystallized layers of the
nucleus may be entirely lost to rapid evaporation while at
larger perihelia they may remain in place slowing down the
flare-ups during subsequent orbits. A similar role may play
silicate impurities and other dust which will tend to form a
protective layer on the nucleus (7). The rather unusual
behavior of P/SW-1 which shows irregularly spaced flare-ups may
be the result of phase changes here described or of thermal
stresses discussed below.

5. DENSIFICATION OF COMETARY NUCLEI

There are two mechanisms of densification of porous icy bodies in a thermal gradient. The first one is the result of migration of roughly equiaxed closed pores up the thermal gradient that occurs because in each pore the H_2O or CO_2 vapor from the warmer and condenses on the colder end. The velocity of this motion increases with temperature, with thermal gradient and with the pore size so that clustering of pores is expected (9). As a result, the warmer side of the icy body becomes denser because upon reaching the surface the pores disappear as shown in Fig. (4).

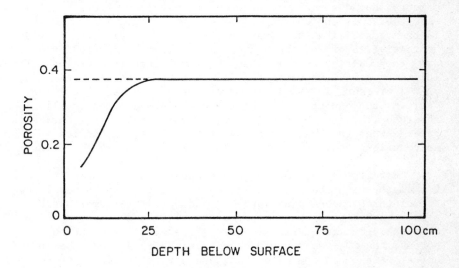

Fig. 4 Drop of porosity at the surface of one meter thick layer of CO_2-ice in a temperature gradient between 20 and 120 K after 10^6 sec. Initial porosity 0.386. (Ref. 15).

In cometary nuclei near the perihelion the temperatures and the thermal gradients are high but they last only for a rather short time. As a result, even for short period comets, this densification is negligible for H_2O-ice; it may be, however, significant for CO_2-ice if the ice is protected from rapid evaporation by a dense H_2O-ice layer or by a non-volatile dust crust. Because of the reversal of the thermal gradient during the orbital motion the pore size distribution will be non-uniform but the effect could be significant in short period comets.

The other densification process is the result of the

gradient of pressure along rather narrow but long pores which are parallel to the temperature gradient so that H_2O or CO_2 vapors diffuse towards the colder end where they condense. The rate of this process is controlled either by Knudsen or by viscous diffusion. Inasmuch as the vapor pressure of H_2O-ice is much lower than that of CO_2 the process is particularly important for cometary nuclei which contain a mixture of CO_2 and H_2O-ices. The progressive deposition of CO_2-ices in the pores in the cold interior of the nucleus increases its thermal conductivity and its thermal inertia. As a result, various effects such as the asymmetry of the heat balance on both sides of the perihelion mentioned above are enhanced. This densification of the nucleus is a continuous process so that its effects should be more pronounced in the older comets. Another result of this enrichment in CO_2 of the interior of a nucleus is the possibility that when in the course of time the outside protective layers of the nucleus are either evaporated or shed away the CO_2-rich area may become warmer than it ever was before. This increase of temperature will lead to an exponential rise of CO_2 pressure in the interior of the nucleus so that it may split. Such outbursts and break-ups have been occasionally observed (10) and they seem to occur mostly after the perihelion passage when the heat content of the nucleus reaches its maximum as described above (compare Figs. 2 and 3).

There is an interesting consequence of the instability of CO_2 clathrate hydrate below 120 K (11). At these temperatures, the clathrate decomposes into CO_2 and H_2O, the former having a much higher saturation vapor pressure than the latter which presumably becomes a very loose aggregate. If the nucleus is porous, the CO_2-vapor can escape from the area, condense in a colder part or escape altogether and not be available to re-form the clathrate hydrate when the temperature rises above 120 K. On the other hand in a dense, that is presumably in an old, cometary nucleus the CO_2 cannot escape but permeates easily the loose H_2O-ice structure and upon warming the clathrate hydrate can form again. It seems that presence of CO_2 clathrates in cometary nuclei should be rather rare. Other clathrates do not seem to show similar anomalies.

6. CONCLUSIONS

It is apparent from this brief survey of the phenomena and reactions which can take place in amorphous or porous ices in cometary nuclei that they are important for explaining various cometary phenomena and behavior. In general, however, the situation is quite complicated. In fact recent studies (12, 13, 14) indicate that the behavior of amorphous water-ice may be more complicated than it was initially believed. Even if a new cometary nucleus were initially chemically and structurally perfectly homogeneous, it would soon develop local surface anomalies and radial gradients of composition and density. These arguments lead to the important conclusion that the thermal conductivity of a cometary nucleus can be quite non-uniform and also that it changes with time. Thus the frequent assumption of a constant thermal conductivity of a uniformly crystalline or uniformly amorphous ice is erroneous. The big task ahead is to see to what extent the observations of the apparent differences among old and new comets can be accounted for by a suitable analysis of thermal and chemical processes taking place in the icy interiors of cometary nuclei.

Supported by NASA grant 7505 Sup. 4.

REFERENCES

1. Wetherill, G. W., and Re Velle, D. O., 1982, Relationships between comets, large meteors and meteorites. In "Comets", ed. L. L. Wilkening, U. of Arizona Press, Tucson, pp. 297-319.

2. Smoluchowski, R., 1981, Heat content and evolution of cometary nuclei. Icarus, 47, pp. 312-319.

3. Klinger, J., 1980, Influence of a phase transition of ice on the heat and mass balance of comets. Science, 209, pp. 271-272.

4. Smoluchowski, R., 1982, Heat transport in porous cometary nuclei. J. Geoph. Res., 87, Supp. A 422.

5. Sekanina, Z., 1982, The problem of split comets in review. In "Comets", ed. L. L. Wilkening, U. of Ariz. Press, Tucson, pp. 251-287.

6. Sekanina, Z., 1979, Fan-shaped coma, orientation of rotation axis and surface structure of cometary nucleus. Icarus, 37, pp. 420-442.

7. Whipple, F. L., 1978, Cometary brightness variation and nucleus structure. Moon and Planets, 18, pp. 343-359.

8. Ghormley, J. A., 1968, Enthalpy changes and heat capacity changes in the transformation from high surface area amorphous ice to stable hexagonal ice. J. Chem. Phys., 48, pp. 503-508.

9.   Smoluchowski, R., Marie, M. and McWilliam, A.,
     1984, Evolution of density in solar system ices.
     Earth, Moon and Planets, 30, pp. 281-288.

10. Campins, H. and Ferrin , I., 1978, Cometary out-
     bursts. Bull. Am. Astr. Soc., 10, 613 and private
     communication.

11. Miller, S. L., 1961, The occurrence of gas hydrates
     in the solar system. Proc. Nat. Acad. Sci., 47,
     pp. 1798.

12. Mayer, E. and Pletzer R., Polymorphism in vapor
     deposited amorphous solid water. In this book.

13. Whalley, E., 1985, The physics of Ice : some funda-
     mentals of planetary glaciology. In this book.

14. Dubochet, J. and McDowell, A. W., 1981, Vitrifica-
     tion of pure liquid water for electron microscopy.
     J. of Microsc., 128, RP3 and (1982) Electron mi-
     croscopy of frozen water and aqueous solutions,
     128, pp. 219.

15. Smoluchowski, R. and Mac William, A., Icarus, to be
     published.

COMPOSITION AND STRUCTURE OF THE COMET NUCLEUS AND ITS EVOLUTION ON A PERIODIC ORBIT

J.Klinger

Laboratoire de Glaciologie et de Géophysique de
l'Environnement
B.P. 96, 38402-St martin d'Hères Cedex - France

ABSTRACT

Present ideas on the chemical composition and structure of the comet nucleus are briefly discussed. In order to be able to estimate what changes the nucleus could undergo during its history, a formalism is developped that could be able to describe the thermal evolution of the interior.

INTRODUCTION

The difficulty of cometary physics and chemistry is that the whole cometary phenomenon is primarily controlled by the nucleus but the nucleus is, in the present state of observational technique, not directly accessible to investigation. So all our knowledge of the nucleus is based on observations of the coma, the dust and ion tails and on astrometric data. Even with this indirect information, we have now some ideas on size, rotation axis and rotation period, chemical composition and dust too.

What is still unknown is the internal structure and the thermal evolution of the structure and the chemical composition as a function of time. Starting with some present day concepts on the composition, structure and origine of comets, the author tries to point out what the evolution of a comet could be after capture on a periodic orbit.

J. Klinger et al. (eds.), Ices in the Solar System, 407–417.
© *1985 by D. Reidel Publishing Company.*

SOME CONSIDERATIONS ON THE STRUCTURE AND CHEMISTRY OF A PRISTINE
NUCLEUS

Virtually all modern concepts of the cometary nucleus are
based on Whipple's icy conglomerate model (1). For a great
number of comets, the light curves as well as the non
gravitational forces as a function of heliocentric distance can
be explained best by the evaporation of water ice (2,3). On the
other hand it is hopeless to try to fit the light curves of all
comets by a universal law, a power law for example. Most comets
show an asymmetry with respect to perihelion as well in the
light curves as in OH production. What defeats any attempts to
find a brightness law are irregular outburst that are shown by a
great number of comets. The origine of these irregularities in
the light curves is not completely clear until now and a great
number of mechanisms have been proposed. Attempts to explain
outbursts by extrinsic causes like collisions with boulders (4)
are difficult to hold in view of statistics. In fact why should
some comets collide with boulders several times during one
revolution while others in similar conditions do not ? This
evidently does not exclude that isolated outburst could be due
to collisions with boulders. But it is most likely that the
majority of outbursts are due to intrinsic mechanisms related to
the chemical composition and the physics of the nucleus. Several
of the "intrinsic" mechanisms proposed to explain the outbursts
need a threshold temperature to be reached at least locally in
or on the comet nucleus. This is in particular the case for the
onset of the evaporation of a particular type of ice. The
evaporation rate as a function of temperature rises in fact very
steeply with temperature. The same is the case for special
chemical or physical processes like recombinations of free
radicals or phase transitions of ices. Such kind of effects
have been proposed during the past thirty years.

In 1956, Haser (5) proposed that free radicals like C_2, C_3,
OH, NH, CN and CH could be initially present in the nucleus. The
recombination of such radicals could provide an energy source
for outbursts (6). These radicals, perhaps with the exception of
OH should recombine at temperatures considerably lower than 100
K. As recent observations of HCO and perhaps H_2CO strongly
support the interstellar origine of at least some comets (7),
laboratory simulations of the formations of interstellar grains
(8) could teach us a lot about the composition of comets. It
turns out that comets could contain rather complicated organic
molecules and that at least a part of the "missing carbon" (9)
could be hidden in an "organic refractory mantle" of
interstellar grains (8).

On the other hand amorphous ice has been identified in
interstellar clouds (10). If indeed H_2O is a major constituant

of the nucleus as suggested by the light curve as a function of heliocentric distance, we strongly suppose that water ice in comets is amorphous.Pataschnick et al. (11) suggested that the heat release during the phase transition from the amorphous to the cubic state of H_2O could provide an energy source for outbursts. Smoluchowski (12) and Klinger (13,14) developped this model further by considering that the transition will increase the thermal inertia of the nucleus. This effect could produce an asymmetry of the light curves with respect to perihelion. This evidently does not mean that other mechanisms related to the inclination of the roation axis (15) or the accumulation or blow-off of a dust mantle (16) should not be taken into consideration.

In 1952, Delsemme and Swings (17) suggested that comets could contain clathrate hydrates. The presence of clathrate hydrates in comets is indeed an attractive possibility to explain the simultaneous outgassing of substances with very different vapor pressures. It must be said that it is quite unlikely that clathrate hydrates form during the accumulation process of a pristine cometary body . For CO_2 clathrate hydrates the reason is very simple : in the phase diagramme of the system CO_2-H_2O we see that the equilibrium line of solid CO_2 + CO_2 gas crosses the line of the system ice + hydrate + CO_2 gas at a temperature of the order of 120 K (18). This means that the clathrate is not formed at temperatures lower than 120 K. At temperatures higher than 120 K confining pressures in the 0.1 millibar range are necessary. Such pressures can indeed be obtained by gravitational densification in a kilometer size body. But this needs a rather compact structure of the nucleus in order to avoid escape of CO_2 gas during the clathrate formation. Formation of CO_2 clathrate hydrates from CO_2 gas acting on ice have indeed been reported by several authors (19,20). The metastability of CO_2 clathrate hydrates at temperature lower than 120 K is not found for hydrates of other substances. But it seems quite unlikely that a clathrate structure can be formed at temperatures lower than 30 K. So if comets contain clathrate hydrates, those substances probably formed much later than the nucleus itself.

SECULAR EVOLUTION OF A PERIODIC COMET

Suppose that comets are undifferentiated pristine objects. What could be the evolution of its internal structure and internal temperature on a periodic orbit ? At the present state of knowledge we can only do some very simplistic modelling. The orbital mean temperature concept could be a rather useful tool for doing this (14,21). The basic principle is the following : let π be the period of the orbit and τ a time constant characterizing the penetration to the centre of the nucleus of a heat perturbation applied on the surface. If $\pi \ll \tau$ the temperature of the centre of the nucleus will tend versus a value T_m given by the relation :

$$\varepsilon\sigma T_m^{\,4} = \frac{1}{\pi} \int_0^\pi (P(t) - \varphi(t))dt \qquad (1)$$

with

   ε = emissivity
   σ = Stefan-Boltzman constant
   π = orbital period
   P = power input from solar radiation and other heat sources
   φ = power lost due to dissipative processes, sublimation for example

The time constant is given by

$$\tau = \frac{R^2}{a\pi} \qquad (2)$$

with

   $a = K/c$ = mean thermal diffusivity of the nucleus
       K = heat conduction coefficient
       c = heat capacity
       ρ = density

If $\pi \simeq \tau$ the temperature of the centre of nucleus will oscillate in an interval containing T_m. If $\pi \gg \tau$ the amplitude of this oscillation will be comparable to the temperature variations at the surface.

The idea is now the following : if the comet stayed on an orbit characterized by a value of T_m for a time span $n\pi \gg \tau$ where n is a positive number the centre will have reached a temperature only slighly lower than T_m and all volatile substances that evaporate in an appreciable manner at temperatures lower than T_m will be lost within a short time. But this implies that evaporation partially occurs in deeper layers and not exclusively on the surface as it is generally considered. The fluffy structure of the nucleus should be no obstacle to subsurface evaporation. Further if T_m of the orbit is higher than the threshold temperature of an exothermal effect (recombination of free radicals, phase transitions of ices etc..), this transformation will propagate through the whole nucleus in a nearly continuous manner.

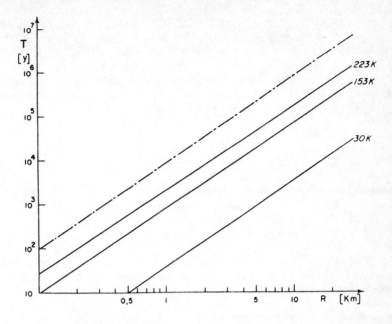

Fig.1 : Thermal time constants as a function of radius
              for compact ice spheres
full lines   :compact crystalline ice at
                             different temperatures
dashed-dotted line :compact amorphous ice.

    Figure 1 gives an idea of the orders of magnitude of   that
we have to expect in nuclei composed of compact ice. The time
constants found in this way should be considered as upper limits
even if in the case of subsurface evaporation of very volatile
ices the fluffiness could be partially compensated by heat
transport in the gas phase that fills up the pores (22). For
crystalline ice the heat diffusivity has been calculated using a
constant density $\rho = 0,9$ and phenomenological laws for the heat
conduction coefficient K and the specific heat c as (13,14) :

$$K = \frac{5.67}{T} \frac{W}{cmK} \qquad\qquad (3)$$

$$C = 7.49 \ 10^{-3} \ T + 9.10^{-2} \ \frac{WS}{gK} \qquad (4)$$

For amorphous ice the thermal diffusivity in a first order approximation is independent of temperature and can be calculate by the following relation (13,14) :

$$a = \frac{V\lambda}{4} \qquad (5)$$

with

$$v = 2.5 \ 15^{5} \ cm/s = \text{mean velocity of sound}$$

$$\lambda = 5.10^{-8} \ cm \qquad = \text{phonon mean free path}$$

For a typical nucleus built up of cubic ice with a radius of one kilometer we expect time constants between several hundreds and several thousands of years. For very cold cubic ice the time constant could be as small as some tenths of years. For amorphous ice we find values in the order of 10^{4} years.

"SEASONAL" AND "DIURNAL" TEMPERATURE VARIATIONS AND THEIR CONSEQUENCES ON THE STRUCTURE AND COMPOSITION OF THE SURFACE LAYER

A complete description of the chemical and structural modifications undergone by the subsurface layers of the nucleus during one rotation or revolution period needs a complete knowledge of the state of those layers at a moment to and the local temperature profile as a function of time in the nucleus. This means that even if we neglect tangential heat transfer, we have to solve, for any point of the surface of the nucleus, the following equation :

$$\frac{\partial T}{\partial t} = \frac{\partial}{\partial x} \left(a \ \frac{\partial T}{\partial x} \right) + \sum_{i=1}^{n} \Theta_i - \sum_{j=1}^{m} \phi_j \qquad (6)$$

with the boundary condition :

$$\frac{C_s}{r_h^2} \left(1 - \Delta \right) = \varepsilon \ \sigma \ T_s^4 + \dot{Z} L + K \left. \frac{dT}{dx} \right|_{-R} \qquad (7)$$

with :

θ_i = heat freed per unit of time due to internal heat sources

ϕ_i = heat lost due to dissipative processes in the nucleus, subsurface evaporation for example

C_S = solar constant

r_h = heliocentric distance

A = albedo

$\dot{Z}$ = evaporation rate on the surface

L = latent heat

K = heat conduction coefficient in the surface layer.

It must be stated that A and ε depend on the wavelength and angle of incidence of the incomming radiation as well as on the surface structure.$\dot{Z} \cdot L$ depends on the chemical composition and on temperature whereas K depends on chemical composition, temperature and on the porosity. Even feedback mechanisms may exist. K for example depends on the gas pressure in the pores (21) and thus on surface and subsurface evaporation. It is evident that in the present state of knowledge this problem cannot be solved in a satisfying manner. But in some simplified cases models of the surface temperature distribution have been obtained as a function of longitude, latitude and orbital position. For comet Halley, Weissman and Kieffer (23) took for the heat capacity, density and the heat conduction coefficient typical values for water ice at 273 K. Froeschlé and Rickman (24) used for C and K temperature dependance as : $C \propto T$ and $K \propto T^{-1}$, that are more realistic for crystalline ice. A further development of such kind of models can allow us to estimate the depletion of different volatile species in the subsurface layers. Further we can estimate at what depth physical or chemical phenomena needing a threshold temperature can be triggered. Periodic temperature variation in a semi-infinite heat conductor can be described using the so called "skin depth" defined as:

$$S = \left(\frac{a \; \pi}{\pi \tau} \right)^{1/2} \qquad (8)$$

This is the depth at which the amplitude of a sinusoïdal temperature variation applied at the surface is lowered by a factor $1/e$. This statement remains valid for our purpose even if the evolution of the surface temperature is not sinusoïdal. The reason is that the higher terms in the Fourier series of a periodic temperature variation applied on the surface are damped out very rapidly in the heat wave that penetrates into the

interior. We can consider that a depth level $x \simeq 5$ S the
amplitude of the seasonal temperature variation becomes
negligeable and for short period orbits the temperature will
establish, in the isothermal approximation at the orbital mean
temperature T_m within a few revolutions. For a rotating nucleus
the "internal temperature" for every latitude strip at $x \simeq 5$ S
evidently depends on the orientation of the rotation axis.
Values for the seasonal skin depths for short and intermediate
revolution periods are given in table I for compact amorphous
ice and for compact crystalline ice at 153 K.

Similar considerations can be applied to diurnal temperature
variations. Some values of the rotational skin depth are given
in table II.

Table I : Orbital skin depths as a function of
revolution period for compact amorphous ice (S_a)
and compact crystalline ice (S_c) at 150 K.

(years)	S_a (m)	S_c (m)
1	1.77	5.48
2	2.5	7.76
3	3.06	9.5
5	3.96	12.3
10	5.6	17.4
25	8.86	27.4
50	12.5	47.5
100	17.7	54.9

Table II : Rotational skin depths as a function of
rotation period Π_{rot} for compact amorphous (S_{ar}) and
compact crystalline ice (S_{cr}) at 150 K.

Π_{rot} (h)	S_{ar} (cm)	S_{cr} (cm)
1	1.9	5.9
2	2.7	8.3
10	6	18.5
20	8.5	26.2

Evidently the preceeding considerations can be strongly modified when evaporation is important on the surface and in the subsurface layers or when internal heat sources become active.

DISCUSSION

As stated by Delsemme (2) most of the "new" comets have already lost their most volatile ices on some kind of "transfer orbits" with perihelion distances beyond Jupiter's orbits before they are captured in orbits with small heliocentric distances from where they become visible. The present approach allows to estimate that for a pristine, kilometersize object this could happen at a time scale of 10^5 years. But if volatiles can be lost in this way this means that evaporation does not only occur on the surface of the nucleus.

For several comets lag angles of 90° or even higher between cometocentric noon and maximum activity have been reported (25,26). The best way to explain so high lag angles is to consider that evaporation takes place under a dust coverage or in a subsurface ice layer with different structure and/or com-position.

The fact that species with very different vapor pressures evaporate simultaneously is generally explained by the eva-poration of hydrates. An alternative explanation could be that due to the temperature profile in the subsurface layer different species evaporate simultaneously at different depths under the surface.

When the mean temperature of an orbit is higher than the threshold temperature of an exothermal process the whole nucleus will transform within a few revolutions. The formalism discussed in this paper has been applied to the phase transition of amorphous to cubic ice in order to establish a classification of periodic orbits (21). This classification reflects indeed qualitatively the behaviour of a great number of periodic comets.

Short period comets like Encke have orbital mean temperatures higher than the transition temperature of amorphous to cubic ice. Thus the thermal inertia of the nucleus is fairly important and the heat conduction to the interior of the nucleus is not negligeable in the heat balance equation.

The centre of intermediate period comets like Halley never reaches a sufficiently high temperature to allow a complete transformation of the nucleus. Only a small crust of crystalline ice will establish. But during the perihelion passage the thickening of this crust is in competition with surface evaporation . Further the crust and the dust coverage of the surface may peal off due to subsurface evaporation favored by the heat release at the transition layer.

A very interesting case is Schwassman-Wachmann 1. The orbital mean temperature of its orbit is only slightly higher than 100 K and thus too low to permit a transformation of all the water ice. But recent calculations (27) confirming the results of ref. (14) show that in some cases the surface temperature at the subsolar point could be high enough to trigger the phase transition and thus an outburst. As the orbital mean temperature is sufficiently low to allow the persistence of carbon dioxyde in deeper layers an extinction mechanism for outbursts is readily found when the heat wave reaches the CO_2 layers. As no phase correlation exist between the revolution, the rotation and the penetration of the heat wave into the nucleus, outburst occur in a random manner.

The preceeding statements show that the phase transition from amorphous to cubic ice seems to contribute to the thermal evolution of comets in a rather important manner. Evidently similar mechanisms could be important at greater heliocentric distances.

In order to get a more realistic picture of the internal structure of the nucleus a great number of observational data are necessary. The production rates of volatile species with very different vapor pressures as a function of heliocentric distance before and after perihelion for several comets could be a good tool to check our models.

Broad band radiometric data in the centimeter wavelength range obtainable from a spacecraft accompanying a comet on its orbit could give us most valuable information on subsurface thermal gradients.

ACKNOWLEDGEMENTS

This work has been supported by the French "Institut National d'Astronomie et de Géophysique" ATP n° 47-54. Numerous discussions with Dr H.Rickman from the Astronomical Observatory of Uppsala/Sweden are gratefully acknowledged.

REFERENCES

(1)   Whipple, F.L. 1950, Astrophys. J. 111, pp.375-394.
(2)   Delsemme, A.H. 1983, in :"Comets", Laurel L. Wilkening ed., The University of Arizona Press, Tucson Arizona, pp.85-130.

(3)   Delsemme, A.H. this book.

(4)   Hughes, D.W., Searle, T.M., Street, R.A. 1974, Nature, 252, pp.615-6126.

(5)   Haser, L. 1956, Comptes-rendus de l'Académie des Sciences, Paris 241, pp.742-743.

(6)   Shul'man, L.M. 1972, Proc. of IAU, Symp. 45eds., Chebotarev,G.A., Kazimirchak-Polonskaya E.I.,Marden,B.G., B.G., Reidel Publishing Co., pp.265-270.

(7)   Cosmovici, C.B., Ortolani, S. this book.

(8)   Greenberg, J.M. this book.

(9)   Delsemme, A.H., 1981, in :"Modern Observational techniques for Comets", Proc. of a Worshop held at Goddart space flight center, October 22-24, 1980, pp.5-13.

(10) Leger, A., Klein, S., de Cheveigne, S., Guinet, C., Defourneau, D., Belin, M. 1979,Astron. Astrophys. 79, pp.256-259.

(11) Patashnick, H., Rupprecht, G., Schuerman, D.W. 1974, Nature 250, pp.313-314.

(12) Smoluchowski, R. 1981, Astrophys. J. 244, pp.L 31-L 34.

(13) Klinger, J. 1980, Science 209, pp.271-272.

(14) Klinger, J. 1981, Icarus 47, pp.320-324.

(15) Cowan, J.J., A'Hearn, M.F., 1979, The Moon and the Planets 21, pp.155-171.

(16) Mendis, D.A., 1985, this book.

(17) Delsemme, A.H., Swings, P. 1952, Annales d'Astrophysique 15, pp.1-6.

(18) Miller, S.L. 1973, in :"Physics and Chemistry of Ice"eds. Whalley E., Jones.

(19) Adamson, A.W. and Jones, B.R., 1971, J. of Colloïd and Interface Sc. 37, pp. 831-835.

(20) Ocampo, J. and Klinger, J. 1982, J. of Colloïd and Interface Sc. 86, pp.377-383.

(21) Klinger, J. 1983, Icarus 55, pp.169-176.

(22) Smoluchowski, R. this book.

(23) Weissman, P.R. and Kieffer, H.H. 1981, Icarus 47, pp.302-311.

(24) Froeschlé, C. and Rickman, H. 1983, in"Asteroïds, Comets, Meteors"eds., Lagerkvist, C.I. and Rickman, H. University of Uppsala, pp.225-231.

(25) Sekanina, Z. 1979, Icarus 37, pp.420-442.

(26) Gerard, E. (private communication).

(27) Froeschlé, C., Klinger, J., Rickman, H. 1983, in : "Asteroids,Comets, Meteors", University of Uppsala, pp.215-224.

AMORPHOUS-CRYSTALLINE PHASE TRANSITION AND THE LIGHT CURVE OF COMET P/HALLEY

Hans Rickman

Astronomiska Observatoriet, Box 515, S-751 20
Uppsala, Sweden

Claude Froeschlé

Observatoire de Nice, B.P. 139, F-06003
Nice Cedex, France

Jürgen Klinger

Laboratoire de Glaciologie du CNRS, B.P. 68
F-38402 Saint-Martin-d'Hères Cedex, France

ABSTRACT

The comparison of the light curve of comet P/Halley with the results of theoretical model computations is briefly discussed. It is concluded that standard models, where sublimation proceeds from the surface of the nucleus and is fed only by absorbed sunlight, can not explain the observed gas production of this comet, especially at large distances from the Sun. We give preliminary discussion of the expected effects of the amorphous-cubic phase transition on the temperature structure of the nucleus of comet P/Halley, and thus of its influence on the gas production curve of the comet.

INTRODUCTION

The light curve of a comet, i.e., its observed total visual magnitude reduced to a standard distance (1 AU) from the Earth and plotted as a function of the distance from the Sun, yields a very easily available, yet highly diagnostic tool for revealing the

419

J. Klinger et al. (eds.), Ices in the Solar System, 419–428.
© *1985 by D. Reidel Publishing Company.*

composition and structure of the nucleus of the comet. Using
theoretical models the physical and chemical characteristics of
the nucleus can be linked to the production rates of the most
abundant molecules as well as dust particles. This modelling is
still in its infancy: there is evidence for a wide variety of
processes influencing these production rates, but so far each
model tends to concentrate on only one of the phenomena, thus
being complementary but not easily comparable to the other
existing models. The work to be described in the present paper
indeed fits into this tradition.

A major problem occurs for the vast majority of comets where
only visual light curves have been obtained and no direct observations of production rates are available. How can the visual
magnitudes of those comets be interpreted in terms of gas and
dust production rates ? There are methods to solve this problem,
but their reliability for discussing individual comets is open
to debate. Comet P/Halley is typical of this category. The only
data so far available, from which a light curve can be derived,
come from the 1910 apparition. Newburn's (18) semi-empirical
photometric theory of cometary gas and dust production and the
visual light curve of P/Halley in 1910 as compiled by Yeomans
(32) were used by Newburn (19) and by Newburn and Yeomans (20)
to compute the production rate of H_2O molecules as a function of
orbital position. This gas production curve has already been
referred to several times in presentations of theoretical model
calculations for P/Halley (e.g. Rickman and Froeschlé (23);
contribution by Weissman and Kieffer at the present colloquium).
We show in Fig. 1 the same comparison of "observations" and
theory as in the paper by Rickman and Froeschlé.

The curves plotted in Fig. 1 differ distinctly in two aspects :
a) the significant perihelion asymmetry of the "observed" curve
has no counterpart in the model curves;
b) the overall shape of the "observed" curve at r $\gtrsim$ 2 AU, especially after perihelion, is not well modelled, the theoretical
curve falling off too steeply with r. From the detection of a
6" coma in Sept. 1909 at r = 3.4 AU pre-perihelion (Bobrovnikoff
(2)) it appears likely that such a discrepancy exists before
perihelion as well.

In this connection it should also be noted that the recent observations of comet P/Halley and the preceding failures to detect
the object, notably by Felenbok et al. (6), provide an indication
that the comet was already active at 11 AU or more before
perihelion (see West (31)). Such an activity is very far from
being explicable by means of the theoretical model plotted in
Fig. 1.

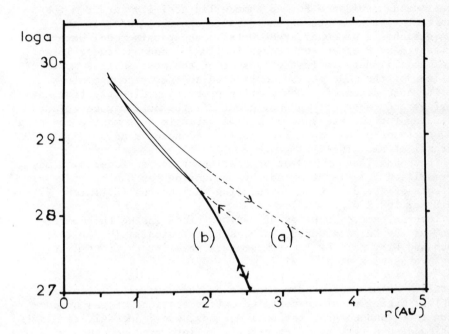

Figure 1. Logarithm of the production rate (a) of
H_2O molecules, per second, from the nucleus of comet
P/Halley plotted versus heliocentric distance (r).
Arrows indicate the sequence of time. Curve (a) shows
data given by Newburn ((19) : solid curve) and
Newburn and Yeomans ((20) : dashed extensions),
derived from observations in 1910. Curve (b) shows
the result of an H_2O-dominated thermal model computed
by Rickman and Froeschlé (23).

IMPROVEMENTS OF THE MODEL

While it can not be excluded that the discrepancy illustrated by
Fig. 1 may be due in part to the uncertainties of the photometric
theory underlying the "observed" curve, there does seem to be
clear enough evidence for an unmodelled activity far from the
Sun that improvements of the physico-chemical model of the
nucleus should be considered. Details about this model have been
presented by Rickman and Froeschlé (23,24) and by Froeschlé and
Rickman (7). Here we list only some of the most obvious modifications of the model to be considered in the present context.
a) The assumption of a sublimation proceeding directly from the
surface and constantly sweeping out all the dust may be relaxed,
allowing for the presence of a time-variable dust mantle covering
the surface (Mendis and Brin (16,17); Brin and Mendis (3);
Horanyi et al. (10)).
b) The conditions of radiative heating of the surface layer may
be modified to a large extent by accounting for the optical
effects of the dust coma surrounding the nucleus (Hellmich (9);
Weissman and Kieffer (30)).
c) Internal energy sources of various kinds inside the cometary
nucleus may be considered, such as free radicals (Haser (8);
Donn and Urey (5)) or low-temperature phase transitions (see
e.g. Patashnick et al. (21)).

A fully satisfactory understanding of the physics of cometary
nuclei, making possible detailed interpretations of gas production rates and nongravitational motions, would necessitate models
taking also a variety of other effects into account, such as
e.g. particle irradiation and micrometeorite bombardment (Brown
et al. (4); Johnson et al. (11)), latent energy transport from
day to night sides of the nucleus by "winds" in the cometary
atmosphere (Wallis and Macpherson (29); Wallis (28)) or localized
"active regions" of high sublimation flux (e.g. Sekanina (25,26).
Aiming eventually toward this goal, we are presently considering
the thermal effects of an internal heat source , namely, the
amorphous-crystalline phase transition. As yet, no quantitative
modellings of these effects have been published, but Smoluchowski
(27) and Klinger (12,13) have recently drawn attention to their
possible influence on important aspects of cometary evolution.
The present paper is a preliminary progress report aimed at describing only the general ideas upon which our work is based.

EFFECTS OF THE PHASE TRANSITION

The picture which seems most plausible for comet P/Halley is one
in which the pristine interior of the nucleus is made up of an
ice-dust mixture containing amorphous H_2O ice. This is surrounded

by a crust consisting of another ice-dust mixture where the
H_2O ice is crystalline.The reason for imagining this model
is that surface temperatures on the nucleus of comet P/Halley
reach values in excess of the transition temperature at 153K
(see e.g. Weissman and Kieffer (30); Froeschlé and Rickman (7)),
so at least in the surface layer the ice must be crystalline.
However, the internal temperature of P/Halley in a state of
thermal equilibrium as estimated by the orbital mean temperature
(Klinger (14,15)) is only appr. 80K, and using the different
estimate proposed by Bailey (1) an even lower value of 60K is
obtained. This is too far below the transition temperature to
allow the phase transition to spread explosively down to the
center of the nucleus. Such a phenomenon would rather occur for
comets of much shorter periods, such as the typical ones of the
Jupiter family (Klinger (15)).

If thus the downward propagation of the phase transition must be
triggered by heat conducted from the surface, there must be a
stable equilibrium condition governing the thickness of the
crystalline crust. Should the transition penetrate too deeply
into the nucleus, it will stop because the underlying ice is too
cold in order to be heated at once to the critical temperature.
In this situation the surface will progressively approach the
transition level due to sublimation. If it should approach too
closely, a large heat flux would reach the transition level from
above, and this would induce a rapid downward propagation of the
phase transition. This interplay of sublimation and transition
rates will certainly depend on the orbital motion of the comet,
so the thickness of the crust may be different at different
orbital positions. However, in the absence of a secular evolution
of the thermal state of the nucleus, the total displacements of
the surface and the transition level over one orbital revolution
must be the same.

It is interesting to note that the average thickness of the crystalline crust and the depth of the layer sublimated off the
nucleus during each orbital revolution can be estimated to be in
the same general range for comet P/Halley. A lower limit to the
crustal thickness is provided by the zero-order estimate of appr.
0.5m by Rickman and Froeschlé (23) as the maximum depth at which
the temperature in a standard thermal model reaches 153K. As an
approximate upper limit we may take the thermal skin depth for
the time scale of the orbital motion. For thermal inertias considered as reasonable for cometary material (10^2 to 10^3 SI units:
see Weissman and Kieffer (30), Rickman and Froeschlé (24)' we
find that this skin depth is several meters. Thus we estimate
the average thickness of the crust to be $\sim$ 1m. The mass loss
rate for comet P/Halley derived from gas production curves (see
Newburn and Yeomans (20)) yields an integrated loss of several

meters of surface material per orbital revolution, somewhat
larger than our estimate of the average crustal thickness. We
therefore expect that during the orbital motion of comet P/Halley,
significant variations of the thickness of the crust and hence
also significant variations of the influence of the internal heat
source on surface temperatures may occur.

Whether the gas production curve of the comet can thus be modified
to a noticeable extent, and whether the agreement between theoretical and observed curves can thus be improved with respect to
Fig. 1, remains to be investigated. Thermal models for comet
P/Halley including the amorphous-crystalline phase transition
should naturally be much more complicated than the models hitherto computed. As an example, the mere existence of a transition
level separating amorphous and crystalline ice with different
thermal properties implies that the usual simple spacing of depth
levels, increasing exponentially from the surface into the interior, can no longer be maintained. Furthermore, a variable
thickness of the crust means that the depth levels in the crust
will have to be redefined during the course of the computation.

MODELLING OF THE PHASE TRANSITION

The most important problem, however, is the choice of a proper
formalism for modelling the phase transition. Two different
approaches may be considered. In the first approach one imagines
a unique transition level representing a discontinuity in the
structure of the nucleus: the ice is regarded as completely amorphous below this level and completely crystalline above it.
According to this picture the phase transition occurs instanta-
neously and separates two different heat flows - one in the
crystalline crust and another in the amorphous interior. The two
usual boundary conditions at the surface and at very large depth
must be supplemented by appropriate intermediate conditions at the
transition level, connecting the two heat flows. Due to the
simple structure of the resulting model, standard analytical
solutions (e.g. Ribaud (22)) may be attempted at least for preliminary explorations of the possible motion of the transition
level.

The second approach leads to a more complicated structure but
appears physically more realistic, and we elaborate it somewhat
in the following. In this picture we consider a finite time scale
for the phase transition, and we thus introduce a reaction rate
(r) such that the number of molecules in the amorphous state (N)
at a given location varies with time (t) according to the
equation :

$$dN = - r N dt \tag{1}$$

This reaction rate is a function of temperature (T). The details of this function are not known, but it appears reasonable to use an exponential representation :

$$r(T) = r^* \exp\{-\frac{T-T^*}{\tau}\} \qquad (2)$$

T^* = 153K is the critical temperature of the amorphous-cubic phase transition. The other two parameters, r^* and τ, can only be roughly estimated at present. Further laboratory data are urgently needed.

The heat diffusion can now be formulated using two variables : the temperature (T) and the fraction of H_2O molecules in the amorphous state (f). Both these variables may be considered to depend on time (t) and depth (z) below some fixed reference level. f can only decrease with time, at a rate which is derived from Eq. (1). As f decreases, latent energy is released in the form of heat. We denote by H the amount of latent energy available per unit mass of amorphous ice. During the time interval Δt, f changes by : $\Delta f = df/dt.\Delta t$, and the amount of heat thus liberated per unit mass is : $\Delta E = -H\Delta f$. This must be included as a source term into the diffusion equation specifying $T(z,t)$. We thus obtain the equations :

$$\frac{\partial f}{\partial t} = -r(T).f \qquad (3)$$

$$C\rho \frac{\partial T}{\partial t} = \frac{\partial}{\partial z}(K\frac{\partial T}{\partial z}) + H\rho \; r(T).f \qquad (4)$$

A solution of these equations with proper boundary conditions at the surface and in the interior yields the structure $T(z,t)$ and $f(z,t)$ over one orbital period:$0 \leq t \leq P$. We note that Eq.(4)is indeed an ordinary diffusion equation with a source term (C is the heat capacity, ρ is the density and K is the thermal conductivity). In this picture we have only one heat flow for the whole nucleus, and we do not need to impose any intermediate boundary conditions. The resulting structure is characterized not by a single transition level but instead by a transition zone of finite extent, within which the ice is partly amorphous, partly crystalline ($0 < f < 1$). Both the depth and the thickness of this transition zone may be expected to vary with the orbital position of the comet.

Preliminary calculations indicate that at least over part of the orbit of comet P/Halley, for reasonable estimates of r^* and τ, the transition zone becomes very narrow. Hence at a given instant, $f(z)$ appears nearly as the step function considered in the discontinuous model, while $T(z)$ exhibits important variation in

a very narrow interval of z in contrast to the much slower variation outside this interval, thus simulating a discontinuous jump.

From Eqs. (1) and (2) it follows that strictly speaking this
model can not allow the ice to be completely amorphous or crystalline. There is always some change of phase going on everywhere
in the nucleus. However, in reality the value of τ is likely to
be quite small, so that the reaction rate varies rapidly with
temperature in the vicinity of the critical value T*. Therefore,
in practice we may define a limiting temperature T_0 such that
we may neglect the transition rate altogether for $T < T_0$. We may
thus distinguish between two possible states of the nucleus, which
we may call "active" and "passive". In the passive state the
temperature does not reach T_0 anywhere in the transition zone. No
latend energy is liberated, and the limits of the zone remain
fixed. In the active state, on the other hand, at least part of
the transition zone has $T > T_0$, so we have to consider a local
energy source together with a change of structure and in
general a motion of the zone.

There is certainly a part of the orbit of comet P/Halley, notably
near perihelion, where the nucleus is in the active state. On the
other hand, approaching aphelion it is highly probable that it
transites to the passive state, since the transition zone will
experience rapid cooling due to heat conduction twoward both the
surface and the interior. Because it is thus expected that the
internal heat source is concentrated to an orbital arc near perihelion, and since it takes some time for the released heat to
reach the surface of the nucleus, we expect that the phase transition causes an enhancement of the post-perihelion gas production
with respect to the pre-perihelion one. To the extent that there
was really a perihelion asymmetry of the gas production of comet
P/Halley in 1910 such as indicated by the observed magnitudes
(Fig. 1), the amorphous-cubic phase transition of H_2O ice appears
to offer one possibility of explaining this phenomeñon.

CONCLUSION

The light curve of comet P/Halley in 1910 together with some data
from recent observations indicates a twofold discrepancy with
respect to theoretical predictions of the gas production from
standard models of the cometary nucleus : a/ an unmodelled perihelion asymmetry, and b/ an unmodelled activity at heliocentric
distance $\gtrsim$ 3 AU. The internal heat source provided by the amorphous-crystalline phase transition is shown to provide a possible
explanation to these phenomena. In this preliminary report we
present arguments for the existence of a crystalline crust in the
nucleus of comet P/Halley, whose thickness may show significant
variation with orbital position but on the average is estimated

at $\sim$ 1 m. A formalism for a continous representation of the phase transition is suggested, leading to the existence of a finite transition zone below the crust. Differential equations for the time evolution of the structure of the nucleus in this model are presented. We continue our work in order to obtain numerical solutions to these equations.

This research was supported by grant nr. U-FR 3271-104 of the Swedish Natural Science Research Council in the framework of the exchange programme between Sweden and France.

REFERENCES

(1) Bailey, M.E. 1984, Mon. Not. Roy. Astron. Soc. (in press).
(2) Bobrovnikoff, N.T. 1931, Publ. Lick Obs. 17, pp.309-482.
(3) Brin, G.D. and Mendis, D.A. 1979, Astrophys. J. 229, pp. 402-408.
(4) Brown, W.L., Lanzerotti, L.J. and Johnson, R.E. 1982, Science 218, pp. 525-531.
(5) Donn, B. and Urey, H.C. 1956, Astrophys. J. 123, pp. 339-342.
(6) Felenbok, P., Picat, J.P., Chevillot, A., Guerin, J., Combes, M., Gerard, E.,Lecacheux, J. and Lelièvre G. 1982, Astron. Astrophys. 113, pp. L1-L2.
(7) Froeschlé, Cl. and Rickman, H. 1983, in "Asteroids, Comets, Meteors", eds. C.-I. lagerkvist and H. Rickman, Univ. of Uppsala, pp. 225-231.
(8) Haser, L. 1955, C.R. Acad. Sci. Paris 241, pp. 742-743.
(9) Hellmich, R. 1981, Astron. Astrophys. 93, pp. 341-346.
(10) Horanyi, M., Gombosi, T.I., Cravens, T.E., Kecskeméty, K., Nagy, A.F. and Szegö, K. 1983, in "Cometary Exploration", vol. I, ed. T.I. Gombosi, Hungarian Acad. Sci., pp. 59-73.
(11) Johnson, R.E., Lanzerotti., L.J., Brown, W.L., Augustyniak, W.M. and Mussil, C. 1983, Astron. Astrophys. 123, pp. 343-346.
(12) Klinger, J. 1980, Science 209, pp. 271-272.
(13) Klinger, J. 1981, Icarus 47, pp. 320-324.
(14) Klinger, J. 1983a, Icarus 55, pp. 169-176.
(15) Klinger, J. 1983b, in "Asteroids, Comets, Meteors", eds. C.-I. Lagerkvist and H. Rickman, Univ. of Uppsala, pp. 205-213.
(16) Mendis, D.A. and Brin, G.D. 1977, the Moon 17, pp. 359-372.
(17) Mendis, D.A. and Brin. G.D. 1978, the Moon and the Planets 18, pp. 77-89.
(18) Newburn, R.L. 1979, in "The Comet Halley Micrometeoroid Hazard" ed. N. Longdon, ESA SP-153, pp 35-50.
(19) Newburn, R.L. 1982, in "The Comet Halley Dust and Gas Environment", eds. B. Battrick and E. Swallow, ESA

           SP-174, pp. 3-18.
(20)   Newburn, R.L. and Yeomans, D.K. 1982, Annu. Rev. Earth
           Planet. Sci. 10, pp. 297-326.
(21)   Patashnick, H., Rupprecht, G. and Schurman , D.W. 1974,
           Nature 250, pp. 313-314.
(22)   Ribaud, G. 1960, "Conduction de la chaleur en régime
           variable", Gautier-Villard, Paris, pp. 62-64.
(23)   Rickman, H. and Froeschlé, Cl. 1983a, in "Cometary Explo-
           ration", vol. I, ed. T.I. Gombosi, Hungarian Acad.
           Sci., pp. 75-84.
(24)   Rickman, H. and Froeschlé, Cl. 1983b, in "Cometary Explo-
           ration", vol. III, ed. T.I. Gambosi, Hungarian Acad.
           Sci. pp. 109-126.
(25)   Sekanina, Z. 1981a, Annu. Rev. Earth Planet. Sci. 9, pp.
           113-145.
(26)   Sekanina, Z. 1981b, Astron. J. 86, pp. 1741-1773.
(27)   Smoluchowski, R. 1981, Astrophys. J. 244, pp. L31-L34.
(28)   Wallis, M.K. 1982, in proc. of "Third European IUE Confe-
           rence", eds. E. Rolfe, A. Heck and B. Battrick, ESA
           SP-176, pp. 451-453.
(29)   Wallis, M.K. and Macpherson, A.K. 1981, Astron. Astrophys.
           98, pp. 45-49.
(30)   Weissman, P.R. and Kieffer, H.H., 1981, Icarus 47, pp. 302-
           311.
(31)   West, R.M. 1983, The Messenger, No. 32, pp. 1-3.
(32)   Yeomans, D.K. 1981, "The Comet Halley Handbook : an
           Observer's Guide", 1st ed., JPL., Pasadena.

DISCUSSION

P. Feldman :
    One should be very careful in using visual light curves to
deduce the variation of gas production rate with heliocentric
distance, especially for a dusty comet like P/Halley.

L. Lanzerotti :
    The release of energy is discussed solely on the basis of
thermal "equilibrium", with no effects from the outer environment. However, particle irradiation can produce effects that
enhance sublimation at certain temperature. We have shown that
irradiating CO_2 at 10 K with energetic ions, then raising the
temperature to $\sim$ 50 K (without irradiation), causes significant
enhancement in the volatility at that temperature. Such enhancements must be considered when discussing energy balances in
the outer layer of a comet. See Brown, Lanzerotti and Johson,
1982, Science 218 pp.525; Johnson et al., 1983, Astron. Astrophys.
123 pp. 343.

MODEL FOR AN ICY HALO IN COMETS

J.F. CRIFO, C. EMERICH

L.P.S.P., BP 10, 91371 VERRIERES LE BUISSON CEDEX,
FRANCE

A B S T R A C T

In order to assess quantitatively the detectability of cometary
icy halos by remote sensing, a numerical model of the distribution of volatile and refractory grains in an active comet coma
has been constructed. The grains are assumed to be accelerated
outwards by cometary gas drag, and in the antisolar direction
by radiation pressure. In the case of volatile grains, the
change with time in their radius, temperature and radiation
pressure cross section is taken into account rigorously. The
optical emission from the grains, which is the sum of scattered
solar light and grain thermal emission, is then computed.
Spectra extending from visible to thermal infrared wavelengths
are presented and discussed, assuming flyby probe observing
conditions.

INTRODUCTION

The importance of detecting water ice in comets is unquestio-
nable in view of the central role attributed to ice in all
accepted comet models (see for instance the recent review by
Delsemme (1)). Direct spectroscopy of a comet nucleus being
impossible from Earth, ice can only be looked for under the
form of a halo of vaporizing grains dragged outwards by
cometary gasses. Such a halo is expected to alter the radial
emission profiles of Cometary molecular lines, since molecule
evaporation will occur throughout the halo, rather than at the

J. Klinger et al. (eds.), Ices in the Solar System, 429–442.

nucleus surface only. Similarly, the radial profile of the
scattered solar light continuum will be made steeper by scattering from ice grains, the size of which decreases with distance to the nucleus. Futhermore, this continuum will include
the characteristic absorption bands of water ice in the near-
infrared.

The identification of such halo features through Earth based
observations is apparently made difficult when icy grains
coexist with refractory dust, since the latter can build-up a
coma of much larger extent. Hanner (2), explained in this way
why no I.R. ice band has been detected in Comets at small
(≼ 2 A.U.) heliocentric distances, and a quantitative computation by Crifo (3), came to a similar conclusion. The only
uncontroversial detection of an icy halo to this date seems to
be the one in the almost dust-free Comet Burnham by Delsemme
and Miller (4), based on analysis of radial brightness
profiles.

The question then arises whether remote sensing from closer
distances (flyby missions) will allow to discriminate icy and
refractory grain emissions. We have in mind the forthcoming
infrared observations of Comet Halley that will be made between
2.5 and 12 μm with the French sounder "IKS" (5), mounted on the
Soviet "Vega" spacecrafts.

The essential difference between earth based and flyby experiments comes from the geometry of the observations : the former
collect light from practically the whole coma, while the latter
observe a cone through that coma ; in the case of "IKS", the
base of this cone in a plane containing the comet nucleus has a
diameter of only about 200 km (3), i.e. just the most probable
size of an icy halo. Due to the crucial role played by the
geometry, accurate numerical computations of the signals to be
expected by the instruments are needed to assess the detection
abilities. We present here preliminary results obtained from
such a work.

The first works dedicated to the physics of expanding volatile
grains (6, 4) were not developed into models capable of predicting in detail the radiative poperties of icy haloes. On the
other hand, sophisticated programs for the computation of the
optical properties of a coma of refractory grains have recently
appeared (7, 8, 3). Unfortunately, in most cases, they cannot
be used to predict the radiation of a coma of volatile grains,
for the two following reasons. First, in "refractory models",
the grain number density is derived assuming constant grain
mass ; in contrast, volatile grains have a mass which changes

considerably and in a complicated way with time. Second, computing the optical emission from a refractory coma is made easier from the fact that each grain keeps a constant scattering efficiency and a constant temperature during its motion ; quite to the contrary a volatile grain has changing parameters and even the simple assumption of single size production leads to the summation of optical contributions from a broad size distribution.

DYNAMICAL MODEL

A comet with water vapour production rate $\dot{M}_o$ (g/sec) and heliocentric distance D is assumed to produce isotropically refractory grains at a mass rate $\dot{m}_d$, and ice grains with initial radius a_o at a mass rate $\dot{m}_o$ (g/sec). It is further assumed that :

$$\dot{m}_d + \dot{m}_o \ll \dot{M}_o,$$

so that the grain acceleration by gas drag can be computed using Finson Probstein theory (9) even though the latter does not consider volatile grains. This grain acceleration stops at small altitudes and then the grains undergo radiation pressure acceleration.

We generalize the "fountain model" equations for the grain number density using a method first proposed by Schwehm and Kneissel (10), but without the restrictions used by these authors (constant parameters when grain size varies).

In a cometocentric rectangular coordinate system with the Ox axis pointing towards the Sun, the following symbols will be used :

F	Solar flux at heliocentric distance D
$v_o(a)$	Velocity of an ice grain with radius a at the end of the gas drag acceleration phase
$\gamma(t)$	Grain acceleration due to radiation pressure
$\phi\,(x,y,z)$	Azimuth of the grain velocity vector
$\alpha\,(x,y,z)$	Angle between the grain velocity vector and the yOz plane
$Q_{pr}\,(a)$	Efficiency factor for solar radiation pressure at radius a and heliocentric distance D

ρ_o Density of ice (0.9 gcm^{-3})

$a(t)$ Radius of an icy grain with initial radius a_o

$\dot{Z}$ (a) Vaporization rate of an ice grain at radius a and heliocentric distance D (in molecule/s)

μ_o Mass of a water molecule

We have :

$$\gamma\ (t) = \frac{3}{4}\ \frac{Q_{pr}\ (a(t))}{\rho_o \cdot a\ (t)}\ \cdot\ \frac{F}{c} \tag{1A}$$

where :

$$a\ (t) = a_o - \frac{\mu_o}{\rho_o}\ \int_o^t \dot{Z}\ (a\ (t))\ dt \tag{1B}$$

The grain motion obeys the equations :

$$x = v_o t\ \sin\alpha - \int_o^t dt \int_o^t \gamma\ (t)\ dt$$

$$y = v_o t\ \cos\alpha\ \cos\phi \tag{1C}$$

$$z = v_o t\ \cos\alpha\ \sin\phi$$

and the steady state coma grain number density is given by :

$$n\ (x,\ y,\ z) = \frac{3\ \dot{m}_o}{16\ \pi^2\ \rho_o a_o^3}\ \sum_i \frac{1}{v_o^2 t_i^2 (v_o - \sin\alpha_i\ \int_o^{ti} \gamma(t)dt)} \tag{2}$$

where $\alpha_i (x,y,z)$ and t_i (x,y,z) are solutions of the system of equations (1C), and the sum extends over all existing solutions. Eliminating α and ϕ from this system, t_i is obtained from the transcendental equation :

$$t_i = \frac{1}{v_o}\ \left(x^2 + z^2 + (y - \int_o^{ti}[\int_o^t \gamma(t)dt]dt)^2 \right)^{1/2} \tag{3}$$

Each solution t_i corresponds to one trajectory from the nucleus to the point (x,y,z), and the grains which have travelled along this trajectory have a radius :
$$a_i\ (x,y,z) = a\ (t_i)$$

OPTICAL MODEL

The emission of an individual ice grain is the sum of the solar light scattered by the grain and of the thermal emission from the grain at its equilibrium temperature $T(a)$.

The grains are assumed spherical and their scattering, absorption and radiation pressure efficiencies are computed by Mie theory. The optical constants of clean ice are the composite of the published experimental data used by Crifo (11). Impure ice is simulated by assuming a constant value k = 0.002 for the imaginary part of the complex index of refraction below λ = 1.5 μm, as in (2) and (11).

The flux received by the IKS focal plane package is computed by integration of the grain emissions over the volume of the coma intercepted by the instrument field of view of 1°. Details on this computation are given in (3).

THERMAL MODEL

The coma is assumed optically thin and the grain equilibrium temperature is computed in the classical way by balancing the solar energy flux absorbed by the grain, and the sum of radiative and vaporization losses. The thermodynamic properties of ice are those used in (11).

As examples of the evolution of icy grains, we show on Fig. 1 the variation of grain radius with distance to the nucleus for various ejection radii, and for clean and dirty ice. One sees that the shrinking of the grains is fast, except for the well known case of pure ice grains in the 10 μm size range.

COMPUTATIONAL PROCEDURE

To make the computation of the density fast enough to be usable in field-of-view integrations, the integrals occurring in equations (2) and (3) are evaluated, after changing the integration variables from t to a, by elementary formulae of the type :

$$\int_{a_1}^{a_2} f(a)\, da = F(a_2) - F(a_1)$$

The required primitive functions F (a) are tabulated once for all at the beginning of the program, as well as T (a), Q_{pr}(a), and the efficiency factors for absorption and scattering, Q_A (a, λ) and Q_S (a, λ) needed in the computation of the optical emission. In the course of the computation of the coma fluxes, all values of F, Q_{pr}, Q_A, Q_S, T are derived from the tables of these functions by linear interpolation.

Equation (3) is solved by a fast iteration procedure which finds out the number of roots and their values.

A single field-of-view integration is needed to derive the fluxes collected at a set of wavelengths $\{\lambda_i\}$ since the integrands differ only by a term involving Q_A, Q_S and T.

The computer program is modular and can be used to compute the near U.V. to far I.R. emissions from halos of various ices at various heliocentric distances.

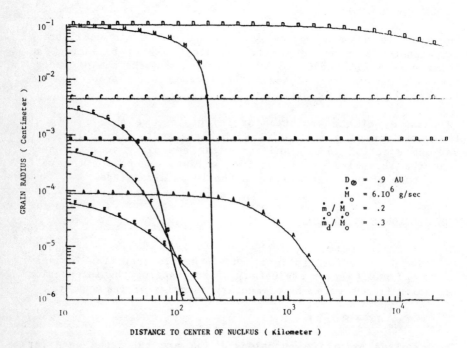

Figure 1 - Radius of the grains ejected by a model comet representative of Comet Halley at 0.9 A.U. post perihelion, as a function of their distance to comet nucleus. Curves A to D refer to clean ice grains of initial radii 0.9, 8, 45 and 1000 µm. E-H label the corresponding curves for dirty ice grains (k_v = .002). Notice that dirty grains have already appreciably melted at 10 km from nucleus. Intersection between different curves is possible because the expansion velocities differ.

RESULTS

The program was run using cometary parameters appropriate to Comet Halley at 0.9 A.U. post perihelion, derived by Newburn and Reinhard (12), except that we set :

$$\dot{m}_o/M_o = 0.2$$

Due to this assumption, the grain ejection velocities are slightly reduced with respect to the usually accepted values (which correspond to $\dot{m}_o = 0$).

The fluxes collected by the I.R. sounder IKS at its closest nominal approach to the comet from icy halos of grains ejected with initial radius a ~ 0.9, 9.0 or 34 μm, are presented on Figs. 2 (pure ice) and 3 (dirty ice). The instrument is assumed on the subsolar axis of the comet, i.e. the scattering angle is 180°, and at a distance 10^4 km from the nucleus. The computation wavelength step is $\Delta\lambda = 1$ μm, except between 2 and 4 μm where it is 0.1 μm.

For comparison, Fig. 4 presents the flux which is expected from the refractory grain component. This curve has been obtained using the simple radiative model for refratory grains described in (3) and does not include absorption or emission bands which, in real, may well be present.

All curves on Fig. 2-4 have the same general pattern consisting in a strong scattering maximum at visible wavelengths, and a secondary maximum in the thermal infrared corresponding to the grain thermal emission. Since refractory grains have equilibrium temperatures (~ 350 K) well above those of icy grains (~ 140 K for clean ice and ~ 185 K for dirty ice), their thermal emission is much greater.
As expected, all icy halo emissions display a conspicuous scattering dip near $\lambda = 2.9$ μm. This line is narrow (~ 0.4 μm FWHM), in contrast to the corresponding signatures observed on solid icy bodies in the solar system. Weaker scattering minima expected near $\lambda = 1.5$ μm and $\lambda = 2.0$ μm cannot be adequatly seen here due the coarse mesh in λ used. Other details in the spectrum, such as the ripple near $\lambda = 10$ μm in Fig. 2, curve a = 8.1 μm, may not be real and may be produced by the interpolation procedure and/or by the assumption of initial injection of single size spherical grains : this will be clarified in the future.

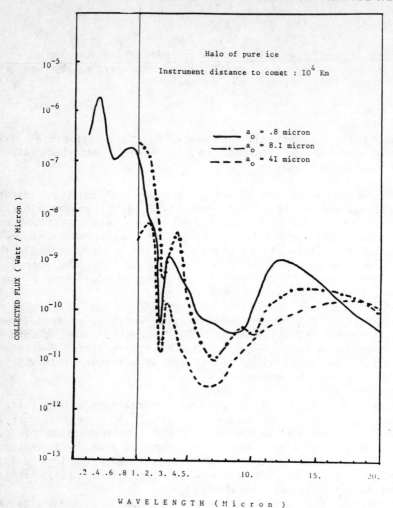

Figure 2 - Energy collected at focal plane of the "IKS" instrument (effective collecting area 32 cm ^{2} ; field of view 1° full angle) from icy halos of pure ice. The halos are formed by ejection of spherical grains with uniform initial size a_o, by an active comet meant to represent Comet Halley 0.9 A.U. post perihelion. The wavelength step in the computation was 0.1 μm between 2.0 and 4.0 μm, and 1 μm elsewhere. Data below λ = 1 μm are shown for one curve only with a mesh 0.2 μm. Notice that two different linear scales are used, one below and one above λ = 1 μm.

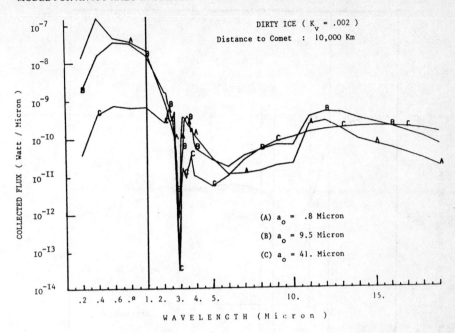

FIGURE 3 - Same as Fig. 2, except that dirty ice grains (K_v = 0.002) are considered.

An apparently surprising result is that there is not much difference —in order of magnitude— between the thermal fluxes of the six halos considered here, even though their geometrical size differs enormously as indicated by Fig. 1. Indeed, this is a consequence of the small field of view of the observation and small observing distance : the greatest part of the signal comes from grains very close to the comet surface, which have had no time to melt, and have a high number density because their expansion velocity is small. Furthermore, the higher temperature of impure grains somewhat compensates their faster melting. A consequence of this fact is that the detection of an icy halo by a flying-by instrument will probably not depend very much upon the halo parameters (ice purity, size distribution).

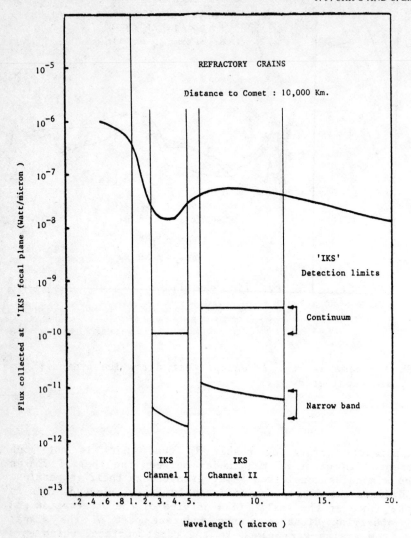

FIGURE 4 – Same as Fig. 2 and 3, but the halo here consists in "fancy" refractory grains with $a_o = 1$ μm, an albedo $A_v = 0.52$, and an infrared emissivity $\varepsilon = 0.3$. Except for emission bands, the curve adequately represents the most probable flux to be expected from refractory grains (See (3)). Also shown on the figure are "IKS" spectroscopic channel ranges and detection limits for continuum emission, and for lines with width $\Delta\lambda = \lambda/50$.

CONCLUSION

Comparison of Fig. (2-3) with Fig. 4 shows that there is no
hope to detect the thermal continuum from an icy halo with a
flying-by spectrometer. In this respect, the situation is
similar to earth based observations !

The 3 μm band, which is the most conspicuous signature of the
halo, lies one to two orders of magnitude below the refractory
grain continuum. This situation turns out to be very similar to
that of the fluorescence bands of the comet primary molecules
-including water vapour near 3 and 6 μm- as computed by
Crovisier and Encrenaz (13). But the "IKS" instrument was
designed with this situation in mind : it has the capability of
discriminating narrow spectral features against smooth continua
(5). The detection limits improve as the width $\Delta\lambda$ of the
spectral line decreases. On Fig. 4 these limits are indicated,
both for continuum and for $\lambda/\Delta\lambda = 50$ lines. One sees that the
3.3 μm line will be detectable in all cases presented here.

Finally, let us try to assess the validity of the above conclusions. The results of Fig. 2 and 3 are not very sensitive to
changes in the assumptions of isotropic gas and dust production
and in the parameters of the grain acceleration equations.
Similarly, the choice of optical ice constants and the assumption of grain sphericity are not believed to have critical
effects on the conclusions. The hypothesis of time stationarity
is certainly more critical : it may not be acceptable for a
very active comet like Halley, and it will have consequences on
the detection capabilities of IKS which are not yet evaluated.
A second hypothesis which may not be realistic and which has
strong effect on our conclusions, is that of optical thickness.
If the gas production rate of the comet is at its maximum
predicted value rather than at its nominal one, and/or if the
dust-to-gas ratio are higher, the vicinity of the nucleus will
become opaque at optical wavelengths. The equilibrium of the
expanding gas-grain mixture will strongly differ from what is
assumed here, and so will presumably the evolution of icy grain
size and temperature.

ACKNOWLEDGEMENTS

The computations were financed by CNES Contract 83/202 and by
CNRS Contract 83/RMB. Assistance received from Mr. F. BOUGNET
in operating the computer codes on CDC CYBER 760 and on NAS
9080 computers is greatly acknowledged.

REFERENCES

1.   DELSEMME, A.H., 1983, Ice in Comets, J. Phys. Chem.
        87, pp. 4214-4218.

2.   HANNER, M.S., 1981, On the detectability of icy grains
        in the comae of Comets, Icarus, 47, pp. 342-350.

3.   CRIFO, J.F., 1983, Analysis of the nucleus and
        circumnuclear area of Comet Halley with the "IKS"
        infrared sounder mounted on the "Vega" flyby probes,
        Adv. Space Res. 2, 12, pp. 203-206.

4.   DELSEMME, A.H. and MILLER, D.C., 1971, Physico
        Chemical Phenomena in Comets - III. The continuum of
        Comet Burnham (1960 II), Planet. Space Sci., Vol 19,
        pp. 1229-1257.

5.   ARDUINI, M., BIBRING, J.P., CAZES, S., COMBES, M.,
        CORON, N., CRIFO, J.F., ENCRENAZ, T., GISPERT, R.,
        HARDUIN, D., LAMARRE, J.M., MALAISE, D.,1983, The
        Comet Halley Flyby infrared sounder "IKS", Adv. Space
        Res. 2, 4, pp. 113-122.

6.   HUEBNER, W.F., WEIGERT, A., 1966, Eiskörner in der
        Koma von Kometen, Z. Astrophys. 64, pp. 185-201.

7.   DIVINE, N., 1981, Numerical models for Halley dust
        environments, Proc. Joint NASA-ESA Working Group
        Meeting, ESA SP-174, pp. 25-30.

8.   HELLMICH, R., 1981, The influence of radiation
        transfer in cometary dust halos on the production
        rates of gas and dust, Astron. and Astrophys. 93, pp.
        341-346.

9.   WALLIS, M.K., 1982, Dusty gas-dynamics in real Comets,
        in : Comets, Ed. L.L. Wilkening, University of Arizona
        Press., pp. 357-369.

10.   SCHWEHM, G.H., KNEISSEL, B., 1981, Optical and
        physical and dynamics properties of dust grains by
        Comet Halley, Proc. Joint NASA-ESA Working Group
        Meeting, ESA SP-174, pp. 77-84.

11.   CRIFO, J.F., 1983, Visible and infrared emissions from
       volatile and refractory cometary dust : a new
       interpretation of Comet Kohoutek observations, Proc.
       International Conference on Cometary Exploration,
       Hungarian Academy of Science, Ed. T.I. Gombosi,
       Vol. II, pp. 167-176.

12.   NEWBURN, R.L., REINHARD, R., 1981, The derivation of
       Halley parameters from observation, Proc. Joint NASA-
       ESA Working Group Meeting ESA SP-174, pp. 19-24.

13.   CROVISIER, J., ENCRENAZ, T. : 1983, Infrared
       fluorescence in Comets : the general synthetic
       spectrum, Astron. and Astrophys., 126, pp. 170-182.

DISCUSSION

H. CAMPINS :

Which albedo did you use in your computations ? Yours grains
must be very dark, isn't it ?

J.F. CRIFO :

We do not use albedo, but compute exactly absorption and
scattering using Mie theory. Doing so, we find for instance
that 0.1 μm grains absorb 0.2 % of the total solar flux if they
are clean, and 0.7 % when they are dirty. At the other extreme,
50 μm grains absorb 7 % of the solar flux if clean, and 84 % if
dirty.

M. GREENBERG :

What is the scattering angle ?

J.F. CRIFO :

It is full backscattering, the flyby probes will pass between
Sun and Comet.

A. LEGER :

In the preceeding presentations, it did not seem that the
3 μm line was the best choice for detecting water. Why do you
appear to show the contrary ?

J.F. CRIFO :

In general, people consider what they can measure. It appears
that many ground-based observations do not extend to 3 μm. On
the other hand, our instrument does not observe below
2,5 μm. But, when we made the choice of our observing
wavelength, a debate took place and several scientists, among
them Dr. KLINGER, recommended the choice of the 3 μm band.
Indeed when you look at the emission rate from individual
grains, you see that the 3 μm feature is by far the most
prominent one. The others are there, of course, but less
conspicuous.

ICE IN COMETARY GRAINS

Humberto Campins

Astronomy Program, University of Maryland

ABSTRACT

The presence of icy grains has important consequences on the physical and chemical processes which take place in the comae of comets. The search for the infrared signature of these icy grains is outlined. The absorption feature near 3 μm found in the coma grains of Comets Bowell [1] and Cernis [2] is interpreted as being due to water ice.

I. INTRODUCTION

The solid particles of ice and/or dust present in the coma and tail of comets originate in the nucleus where they are dragged away by expanding gasses. The dust particles have been observed in most comets using visual and infrared techniques. Icy grains or ice-coated dust grains, on the other hand, have proven to be much more difficult to detect.

The presence of frozen volatiles (H_2O, NH_3, CO_2 and others) as the major constituent of the nuclei of comets is the basis for Whipple's [3] icy conglomerate model. The ejection from the nucleus of pieces of the icy matrix was suggested by Huebner and Weigert [4] who postulated the existence of submicron-sized icy grains in the comae of comets, however, the size of the icy halo predicted by these authors was too small to be observable.

The existence of a larger icy halo which could be observed under certain circumstances was proposed by Delsemme and Wenger [5] based on their laboratory observations of the behavior of a

443

frozen hydrate of methane. As Delsemme and Miller [6] point out the existence of icy grains has important consequences on the physical and chemical processes which take place in the coma of comets, i.e., the release of parent molecules from icy grains in the coma, rather than directly from the nucleus, can lead to drastically different apparent lifetimes of the parent molecules and photometric profiles of the daughter radicals. Unfortunately, the observational evidence for the existence of these icy grains has been indirect and remains scarce [7].

Reflectance spectrophotometry has proven to be a very useful tool in the study of ices in the solar system [8]. A very good spectral window in which to search for cometary ices is the 1 to 4 µm region. There, many ices likely to be present in comets have characteristic absorption features; figure 1 [9] shows laboratory spectra of several such ices. In addition, the 1 to 4 µm spectra of bright comets, unlike their visual spectra, are virtually free of gaseous emission [10], this facilitates observations of the light scattered by the solid grains.

Of particular importance is the spectrum of frozen H_2O since this seems to be the dominant volatile species in most and maybe all comets; furthermore, water ice has been identified on several solar system objects using reflectance spectrophotometry of the 1 to 4 µm region [11,12,13], and references therein). The optical properties of water ice in the infrared have been discussed in detail by Irvine and Pollack [14] and most recently by Fink and Sill [9]. The 3 µm band (a fundamental absorption) is the deepest of the near infrared bands of water ice (see figure 1) and it is also the most diagnostic of the structure of the ice [9]. The bands shortward of 2.5 µm (overtones) are weaker and much harder to interpret [15]. Ironically, because the strength of the solar continuum decreases with increasing wavelength and the ambient thermal noise increases sharply beyond about 2.5 µm, observations of solar system objects shortward of 2.5 µm are considerably easier than observations near 3 µm and beyond.

II. PREVIOUS WORK

The first attempt to observe the absorption features due to ices in a comet was made by Oishi et al. [16] who obtained a spectrum of Comet West from 2.8 to 3.6 µm. At the time the comet was at 0.53 AU from the sun and the spectrum obtained showed only thermal emission from hot (T ~ 450 K) dust grains. In 1981 A'Hearn et al. [17] reported in their paper entitled "Where Is The Ice in Comets?" observations of the 1.4 to 2.6 µm spectrum of periodic comet Stephan-Oterma. Their observations showed no evidence for any absorption including the 1.5- and

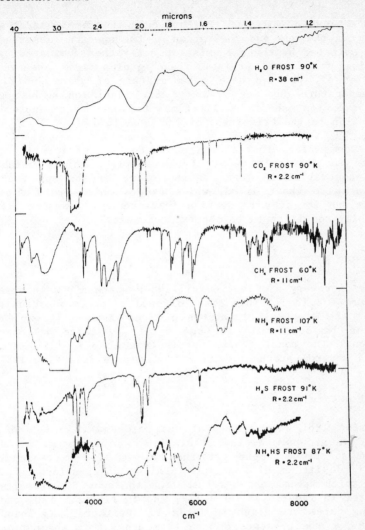

Figure 1: Reflection spectra of various frosts in the near infrared. The spectra are ratio spectra of the frost to a white, nonabsorbing standard. Intensity scale is linear with zero levels displaced as indicated. Each frost has its own distinctive spectral signature. Most of the absorptions in this region of the spectrum are combination and overtone bands. Solids with hydrogen bonding (H_2O and NH_3) have relatively broad absorption features. Fine structure near 1.7, 1.9, and 1.4 µm is due to incompletely cancelled H_2O vapor in the ambient laboratory atmosphere [9].

2.0-μm features of H_2O ice. In this case the comet was at about
1.6 AU from the sun and the region of the spectrum observed was
not contaminated by thermal emission but rather dominated by the
scattering from dust grains. These grains were observed by
Campins et al. [10] to have a temperature of about 265 K.   At
this temperature any icy grain or icy mantle on a dust grain
would have too short a lifetime to make any appreciable
contribution to the brightness of the coma [18].

It seems clear that the direct detection of ices in comets
has been so difficult mainly because bright comets which are
easily observable are generally so close to the sun that icy
grains are too short lived, while comets which are far enough
from the sun for ices to survive (heliocentric distance > 2 AU)
are usually too faint to be observed adequately.

III.   COMET BOWELL

The discovery of Comet Bowell 1980b sparked new interest in
the search for ices.   This comet offered a better opportunity to
look for the spectral signatures of ices because it was bright
enough to be observable in some detail at large heliocentric
distances.   The comet reached perihelion at 3.3 AU in March of
1982 and several groups searched for ices in it during 1981 and
1982.

A.   1981 Observations

During 1981 when the comet was between 4.5 and 5 AU from
the sun two groups, Campins et al. [19] and Jewitt et al. [20],
obtained spectrophotometry of the 1.5- to 2.5-μm region.   In
addition, Jewitt et al. obtained similar observations of Comet
Panther which was at about 2.1 AU from the sun.   The results are
summarized in figures 2 and 3 [19] and in figure 4 [20] and the
observing dates are given in Table 1.   Neither group found any
evidence for the presence of the 1.5- and 2.0-μm absorption due
to the H_2O ice.

Table 1:   1981 Observations of Comet Bowell (1980b)

Observers	Dates	R(A.U.)	Δ(A.U.)	Scattering Angle
Campins et al. [19]	4-10 to 5-15	4.6 to 4.4	3.6 to 3.7	177° to 169°
Jewitt et al. [20]	5-18	4.4	3.7	169°

Figure 2: The
1.25 to 3.5 µm
reflectance spectrum
of Comet Bowell in
1981 [19].

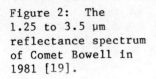
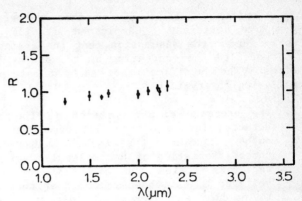

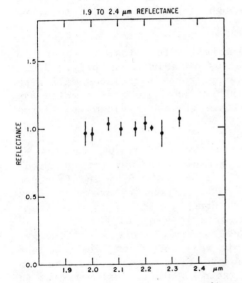

Figure 3: The 1.9 to
2.4 µm reflectance
spectrum of Comet Bowell
in 1981 [19].

Figure 4: The
1.5 to 2.5 µm
reflectance
spectrum of
Comets Bowell
and Panther
[20].

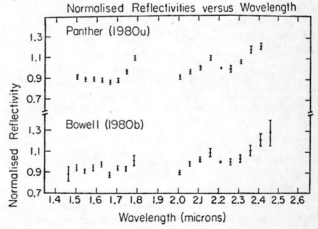

Jewitt et al. concluded that the grains in Comet Bowell were ice deficient. Campins et al. [19], on the other hand, worked under the assumption that the grains contained water ice and modelled the scattering by the grains to match the observations. The justificaton of such an assumption is based on the following observations:

1. The presence of OH emission in the coma of Comet Bowell detected in the ultraviolet by A'Hearn et al. [21]. These authors concluded that the OH in the coma was the result of photodissociation of H_2O molecules which had sublimated from the coma grains and not from the nucleus. The nucleus believed to have been too cold at that heliocentric distance to produce any appreciable OH; the coma particles, on the other hand, were warm enough ($\sim$ 150 K; [22]) to have produced the observed OH.

2. The disappearance of the 10-μm silicate emission feature in comets at heliocentric distances of about 2 AU [23]. Rieke [24] points out that this disappearance correlates well with the expected stability of clathrate hydrate grains which may act as glue holding the silicate grains in clumps too large to show the 10-μm feature.

After modelling the scattering by icy grains, Campins et al. [19] concluded that if the grains were dark enough (geometric albedo about 0.05) then the 1.5- and 2.0-μm ice features would be absent at the 5% level. How to make icy grains so dark was the next step. Based on work done by Clark [13] they found that adding only a small amount of dark dust to the ice would make the ice as dark as the dust itself. More specifically, they found that if one were to mix well only 1 to 10% by weight of dust particles like those observed in comets near the sun (1 to 5 μm in radius) in an ice grain of about 1 mm in radius (a size consistent with that found by Sekanina [25] from a dynamical analysis of the coma of Comet Bowell) then the albedo of the dirty ice grain would become as low as that of the dust itself.

The above analysis could explain the absence of the 1.5- and 2.0-μm features in the scattering by grains made mostly of ice, however, even in the darkest of grains modelled the 3 μm H_2O ice feature was very prominent.

B. The 1982 Observations

The following year Comet Bowell was about 1 AU closer to the sun and about a magnitude brighter. During this observing period priority was given to observations of the 3 μm area which

had not been possible the preceeding year. This time the groups
making the infrared observations were A´Hearn et al. [26] and
Campins et al. [1]. A strong absorption was found near 3 μm by
Campins et al. [1]. The detection was the result of four nights
of observation: April 23, June 11, 13 and 22 UT (all 1982),
using a filter centered at 3.25 μm and 0.45 μm wide. Standard K
(2.2 μm) photometry was interlaced with the longer wavelength
measurements to define the continuum at 2.2 μm. The combined
result is that the albedo at 3.25 μm was 0.52 ±0.06 times that
at 2.2 μm. This result is plotted in figure 5. Also in figure
5 are the results of the broadband photometry of the comet
obtained by A´Hearn et al. in April 1982; and although their
results are not by themselves enough to unambiguously determine
the presence of the 3 μm absorption, they are consistent with
the measurements by Campins et al. and they suggest that the
absorption is broad and the spectrum stays depressed towards
4 μm. The spectrum of Callisto [27] which has absorption
features thought to arise primarily from water frost, is also
shown in figure 5. This is to illustrate another case similar
to Comet Bowell´s in which the low reflectivity of the surface
has substantially decreased the relative strength of the 2.0 μm
water frost band while the 3 μm band remains quite prominent.

Of the ices which are likely to be found in comets, H_2O,
NH_3, H_2S, and CH_4 have absorption features near 3 μm (see figure
1); however only H_2O is a plausible identification at the
heliocentric distance at which Comet Bowell was then observed
(3.4 AU). From dynamical considerations, it is believed that

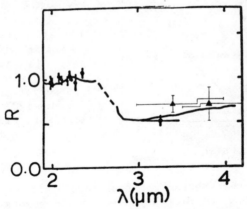

Figure 5: The 2 to 4 μm reflectance spectra of Comet
Bowell in 1982 (dots and triangles) and Callisto
(continuous curve). The dots are observations reported by
Campins et al. [1], the triangles are by A´Hearn et al.
[26]. The spectrum of Callisto is from Lebofsky [27].

the grains in this comet are $\geq$ 0.5 mm in diameter [25]. The grains had a temperature of approximately 170 K [28]. Under these conditions, it can be shown [29] that the lifetimes against evaporation are respectively on the order of one month, and one second for H_2O and NH_3 ice grains; the other candidate ices are even more volatile than NH_3. Therefore, the H_2O grains are stable enough to have produced the observed absorption, but all other candidate ices are too volatile by orders of magnitude.

Hydrated minerals also exhibit an absorption band near 3 μm which is usually narrower than those due to ice. As an example, a spectrum of the meteorite Orgueil, a CI carbonaceous chondrite with about 20% water (by weight) in the form of hydrated minerals [30] is plotted in figure 6 along with the observations of Comet Bowell. Notice that the point at 3.8 μm indicates a broader feature in the comet, in support of the interpretation that the feature is due to ice and not due to hydrated minerals. Furthermore, Hanner and Campins [28] conclude that the presence of icy grains, as opposed to hydrated minerals, in the coma of Comet Bowell in 1981 and 1982, is necessary to explain both the OH emission and the thermal (10- and 20-μm) emission from the coma grains.

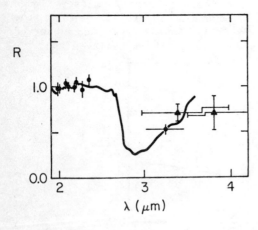

Figure 6: The reflectance spectrum of the meteorite Orgueil (continuous curve) a CI carbonacious chondrite with almost 20% water (by weight) in the form of hydrated minerals [30]. The observations of Comet Bowell in 1982 are plotted similarly to Figure 5.

It is clear that better spectral resolution is necessary to establish the shape of this 3 μm feature in comets. The shape and strength of this feature can be diagnostic of: (a) the structure of the ice (amorphous or crystalline [9], (b) the composition and size of the icy grains [31]; and (c) the presence of hydrated minerals in the grains. The 2.6 to 3.1 μm region (part of which is not observable from the ground) is the most diagnostic of the presence of hydrated minerals as opposed to water ice [32,33]

IV. COMET CERNIS

In 1983 there was the apparition of Comet Cernis (1983ℓ). This comet resembled Comet Bowell (1980b) in several ways. It was observed to have a well developed coma at a large helio-centric distance and it never came any closer to the sun than 3.2 AU. Because the comet was not discovered until perihelion its apparition was much shorter than that of Comet Bowell. This meant that advance planning was not possible for this object, however, it was observed in the infrared by Hanner [2] and in the UV by A´Hearn and Feldman [34].

Hanner found evidence for the 3 μm absorption feature based on three narrow filter observations made at 2.9, 3.0 and 3.2 μm. She also found that the thermal spectrum of this comet resembled closely that of Comet Bowell, indicating that the coma particles in both objects were similar. A´Hearn and Feldman found the gaseous emissions (from OH and CN) in this comet to behave like those in Comet Bowell where the OH seemed to come mainly from the icy grains.   Although Comet Cernis was not as widely observed as Comet Bowell there is enough evidence to confirm the findings in the earlier comet.

V. SUMMARY

An absorption feature near 3μm has been detected in the coma grains of Comets Bowell [1] and Cernis [2].   Additional indirect data make the identification of this feature with water ice by far the most plausible one. Nevertheless, observations of this feature with better spectral resolution are necessary to identify any other components and to determine the structure of the ice.

ACKNOWLEDGEMENTS

I am grateful to Dr. A. Dollfus and the organizing committee of this workshop for their travel support without which attendance to the meeting would not have been possible. I would like to thank Drs. M. A´Hearn and U. Fink for helpful discussions and for providing me with their original figures; and Dr. A. Delsemme and an anonymous referee for helpful comments. This work was supported by NASA grant NSG-7322.

REFERENCES

1. Campins, H., Rieke, G.H., and Lebofsky, M.J. 1983, Nature, 301, pp 405-406.
2. Hanner, M.S. 1984, Ap.J. Lett. 277, pp L75-78.
3. Whipple, F.L. 1950, Ap.J., 111, pp 375-394.
4. Huebner, W.F. and Weigert, A. 1966, Zs. Astrophys., 64, pp 185-201.

5. Delsemme, A.H. and Wenger, A. 1970, Planet. Space Sci., 18,
       pp 709-715.
6. Delsemme, A.H. and Miller, D.C. 1970, Planet. Space Sci., 18,
       pp 717-730.
7. Delsemme, A.H. 1985, these proceedings.
8. Clark, R.N. 1985, these proceedings.
9. Fink, U. and Sill, G.T. 1982, "Comets", ed. L.L. Wilkening,
       pp 164-202.
10. Campins, H., Rieke, G.H., and Lebofsky, M.J. 1982a, Icarus,
       51, pp 461-465.
11. Kuiper, G.P., Cruikshank, D.P., and Fink, U. 1970, B.A.A.S.,
       2, pp 235.
12. Lebofsky, L.A. 1977, Nature, 269, pp 785-787.
13. Clark, R.N. 1982, Icarus, 49, pp 244-257.
14. Irvine, W.M. and Pollack, J.B. 1968, Icarus, 8, pp 324-360.
15. Clark, R.N. 1981, J. Geophys. Res., 86, pp 3087-3096.
16. Oishi, M., Kaware, K., Koybayashi, Y., Maihara, T., Noguchi,
       K., Okuda, H., and Suto, S. 1978, Publ. Astron. Soc.
       Japan, 30, pp 149-159.
17. A'Hearn, M.F., Dwek, E., and Tokunaga, A.T. 1981, Ap. J. 248,
       pp 147-151.
18. Hanner, M.S. 1981, Icarus 47, pp 342-350.
19. Campins, H., Rieke, G.H., Lebofsky, M.J., and Lebofsky, L.A.
       1982b, A.J., 87, pp 1867-1873.
20. Jewitt, D.C., Soifer, B.T., Neugebauer, G., Danielson, G.E.
       and Matthews, D., 1982, A.J., 87, pp 1854-1866.
21. A'Hearn, M.F., Schleicher, D.G., Feldman, P.D., Millis, R.L.,
       and Thompson, D.T. 1984, A.J. 89, pp 579-591.
22. Hanner, M.S., Veeder, G.J. and Matson, D.L. 1981, B.A.A.S.
       13, pp 705.
23. Rieke, G.H. and Lee, T.A. 1974, Nature, 248, pp 737-740.
24. Rieke, G.H. 1977, Infrared and Submillimeter Astronomy, ed.
       G.G. Fazio, pp 109-120.
25. Sekanina, Z. 1982, A.J., 87, pp 161-169.
26. A'Hearn, M.F., Dwek, E., and Tokunaga, A.T. 1984, Ap.J. 282,
       pp 803-806.
27. Lebofsky, L. 1982, personal communication.
28. Hanner, M.S. and Campins, H. 1984, submitted to Icarus.
29. Lebofsky, L.A. 1975, Icarus, 25, pp 205.
30. Larson, H.P., Feierberg, M.A., Fink, U. and Smith, H.A. 1979,
       Icarus, 39, pp 257-271.
31. Greenberg, J.M. 1983, in "Asteroids, Comets, Meteors;
       Exploration and Theoretical Modelling", Uppsala, Sweden,
       pp. 259-268.
32. Lebofsky, L.A., Feierberg, M.A., Tokunaga, A.T., Larson,
       H.P., and Johnson, J.R. 1981, Icarus, 48, pp 453-459.
33. Pollack, J.B., Witteborn, F.C., Erickson, E.F., Strecker,
       D.W., Baldwin, B.J. and Bunch, T.E. 1978, Icarus 36,
       pp 271-303.
34. A'Hearn, M.F. and Feldman, P.D. 1985, these proceedings.

ULTRAVIOLET ALBEDO OF COMETARY GRAINS

P. D. Feldman
Johns Hopkins University

M. F. A'Hearn
University of Maryland

ABSTRACT

Several dusty comets, including two at heliocentric distances
greater than 3 AU (Bowell 1982 I and Cernis 1983ℓ), have been
observed with the International Ultraviolet Explorer.
Ultraviolet albedos have been determined for the wavelength range
2600-3050 A by comparison of the IUE data with both recent solar
flux measurements and with IUE observations of solar analogue
stars. All of the comets show a similar reddening within this
spectral range as well as with respect to the visible and
infrared. No spectral signatures of possible ices are evident in
the data. The ratio of gas to dust production rates, evaluated
for 15 comets observed by IUE, shows a well defined dependence on
heliocentric distance.

INTRODUCTION

Over the past few years nearly twenty faint and moderately bright
comets have been observed by the ultraviolet spectrographs of the
International Ultraviolet Explorer (IUE) satellite. These
observations have been mainly applied to studies of the
composition, chemistry and evolution of the gaseous coma (1).
Most of the comets observed by IUE also show a weak ultraviolet
continuum spatially concentrated near the center of the
spectrograph aperture (10" × 20") in contrast to the extended
emissions of the atoms and molecules of the coma. Two comets,
Bowell (1982 I) and Cernis (1983ℓ), were bright enough to be
observed close to their perihelia at ~ 3.3 AU, at which distance
the grain continuum predominates over the gas emissions.

J. Klinger et al. (eds.), Ices in the Solar System, 453–461.
© *1985 by D. Reidel Publishing Company.*

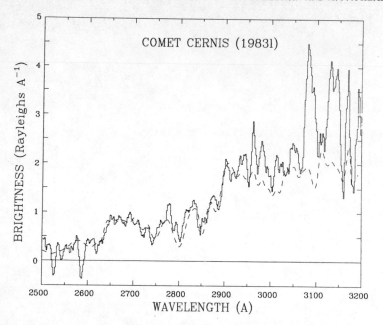

Figure 1. IUE long wavelength, low-dispersion spectrum
of Comet Cernis (1983ℓ) obtained on 7 September 1983.
The time of the exposure (LWR 16751) was 5.17 hours.
The dashed curve shows the solar spectrum of Mount and
Rottman (5) smoothed to 11 A resolution.

Hapke et al. (2) have suggested that the presence of icy grains
in a cometary coma could be detected through spectral
signatures in their ultraviolet reflectivity and presented
laboratory data for several candidate ice species. Information
about the chemical nature and the size distribution of the
cometary grains can also be inferred from near-simultaneous
determinations of the grain albedo from the ultraviolet to the
infrared (3). It is with these goals in mind that the IUE
continuum data are examined in this paper.

OBSERVATIONS

The spectra of a number of comets observed by IUE have been given
by Weaver et al. (4). For three of those comets, P/Stephan-
Oterma (1980 X), P/Borrelly (1981 IV) and Panther (1981 II), a
solar-type continuum is clearly distinguishable from the gaseous
emissions. In comets Bowell (1982 I) and Cernis (1983ℓ),
observed more recently, the spectrum is entirely continuum except
for very weak OH(0,0) emissions at 3085 A, as can be seen in

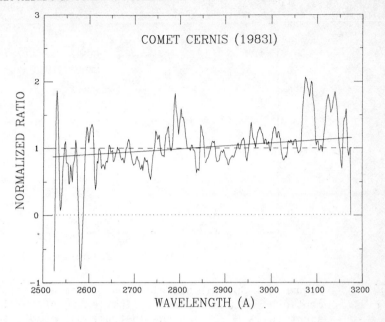

Figure 2. Ratio of the IUE spectrum of Figure 1 to the solar spectrum, normalized to the average value between 2600 and 3050 Å. The solid line is a least-squares best-fit to a linear variation of albedo with wavelength.

Figure 1. A solar spectrum, from Mount and Rottman (5), is also shown in the figure, smoothed to the resolution of the IUE spectrum. The relative albedo, as a function of wavelength, is obtained by taking the ratio of the cometary flux to the solar flux, as is shown for comet Cernis in Figure 2. For all five of these comets it is found that the albedo decreases by about 20% towards shorter wavelengths over the spectral region 2600-3050 Å. This behavior is similar to the ultraviolet reddening previously found for Comet West (1976 VI) (6). No spectral features, or signatures, are detectable in any of these spectral ratios but this may be the consequence of the large statistical and background uncertainties associated with the data. Most of these comets required very long exposures, greater than 5 hours for Comets Bowell and Cernis.

A'Hearn et al. (3) have combined the IUE spectra of Comet Bowell with ground-based visible and infrared observations to determine the grain albedo from 2600 Å to 2.25 μm. In order to normalize the various observations to the different apertures used, for each observation they derived the quantity Afρ, the product of

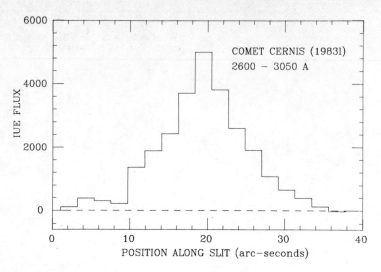

Figure 3. Spatial distribution of the ultraviolet
continuum of Comet Cernis. The length of the IUE large
aperture is ~ 22" and the instrumental resolution is
~ 5".

the grain albedo, filling factor and cometocentric radius of the
aperture. A'Hearn et al. show that this quantity is independent
of ρ if the spatial distribution of the dust continuum varies as
$ρ^{-1}$, which is the case for simple radial outflow at fixed
velocity, and, in fact, the ground-based data for Comet Bowell
support this conclusion. The spatial distribution of the
ultraviolet continuum within the 10" × 20" aperture of the IUE
spectrograph in Comets Bowell and Cernis, as seem in Figure 3, is
consistent with a $ρ^{-1}$ variation, at least for ρ small (near the
nucleus).

For comparison with Comet Bowell, values of Afρ were derived for
Comets Cernis and P/Stephan-Oterma using data from Hanner (7),
Hanner et al. (8) and Spinrad [private communication, 1983] in
addition to the IUE data. These are shown in Figure 4, where it
is seen that the albedos of the three comets follow the same
general behavior from the ultraviolet to the infrared. In fact,
the agreement is remarkably good, considering the difference in
observation dates for Comet Cernis, and implies a similar grain
size and composition for the three comets.

DISCUSSION

A'Hearn et al. (3) point out that Afρ is directly proportional to

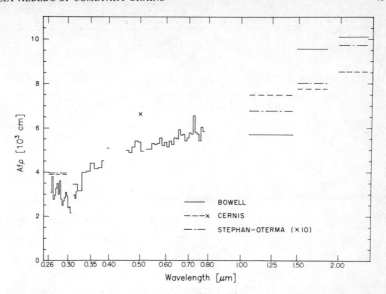

Figure 4. The product Afρ for Comet Bowell (1982 I) from A'Hearn et al. (3), together with IUE and ground-based data for Comets Cernis (1983ℓ) and Stephan-Oterma (1980 X).

the dust production rate, Q_{dust}. Thus, if the albedo, A, is the same in the ultraviolet for all of the comets observed by IUE, then the IUE data can be uniquely used to determine the ratio of gas to dust production rates. Weaver et al. (4) have demonstrated that the first several comets observed by IUE were all H_2O dominated. The derivation of the H_2O production rate from the emission brightness of the OH(0,0) band at 3085 Å was discussed by Weaver et al. (9). For the present purpose we are interested in Q_{gas}/Q_{dust}, which is taken to be proportional to $Q_{H_2O}/Af\rho$, and is evaluated in a consistent manner, using the continuum flux between 2900 and 3000 Å, for 15 comets observed by IUE.

The arbitrarily normalized ratio Q/Afρ is shown in Figure 5 as a function of the heliocentric distance, r, at the time of the observation. For those comets at r > 1.0 AU, r is generally not too different from the comet's perihelion distance. Surprisingly, the data show a very definitive trend with heliocentric distance suggestive of a power-law dependence. In fact, none of the points differs by more than a factor of two from a straight line corresponding to an $r^{-3.4}$ variation, strongly suggesting that the gas vaporization rate does not

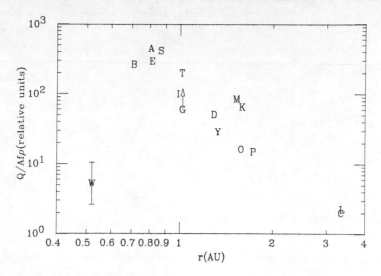

Figure 5. The normalized gas to dust production ratio,
Q/Afρ, for fifteen comets observed by <u>IUE</u> as a function
of heliocentric distance, r. An estimated value of the
same quantity for Comet West (1976 VI) is also shown.
Legend: A: Austin (1982 VI); Y: Borrelly (1981 IV); L:
Bowell (1982 I); B: Bradfield (1979 X); C: Cernis
(1983ℓ); D: D'Arrest (1982 VII); E: Encke (1980 XI); G:
Grigg-Skjellerup (1982 VI); I: IRAS-Araki-Alcock
(1983d); K: Kopff (1982k); M: Meier (1980 XII); P:
Panther (1981 II); S: Seargent (1978 XV); O: Stephan-
Oterma (1980 X); T: Tuttle (1980 XIII); W: West (1976
VI).

uniquely control the rate at which the grains are produced. For
comparison, Q/Afρ for Comet West (1976 VI) is also estimated
using the data of Smith et al. (10) and of Feldman and Brune
(11). The large uncertainty in this result is insignificant
compared to the two order of magnitude deviation from the
straight-line fit to the comets observed by <u>IUE</u> and demonstrates
that Comet West was uniquely different from the "typical" comet.
It is likely that the optically thick dust coma of this comet
effectively limited the gas production at small heliocentric
distances (12). In addition to having a much larger dust
production rate, this comet is also unique amongst recent comets
in having an extremely high (~ 0.2) ratio of CO to H_2O production
(13,14).

To support our conclusion, there are additional data in that
Comets Bradfield (1979 X) and Austin (1982 VI) were both observed

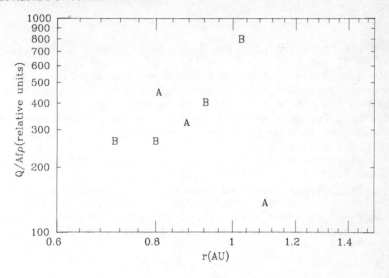

Figure 6. Variation of Afρ with heliocentric distance
for Comets Bradfield (1979 X) and Austin (1982 VI).

over a period of several weeks so that Q/Afρ can be determined
for each of these as a function of heliocentric distance. During
these observations, Comet Bradfield was receding from the sun
while Comet Austin was approaching. However, in both cases, as
Figure 6 demonstrates, the ratio of gas to dust production rates
appears to increase with time.

A more detailed analysis would take into account variations in
Afρ resulting from consideration of the variation of albedo with
scattering angle or the dependence of f on the dust size
distribution or outflow velocity. Most of the observations were
made at scattering angles from 80° to 150°, where little
variation is seen (15), except for Stephan-Oterma (174°), Bowell
(166°) and Cernis (167°). These comets would therefore show an
enhanced albedo of up to a factor of 2 (16) and hence a lower
value of Q/Afρ relative to those comets observed at smaller
scattering angle. The smaller expected dust outflow velocity at
larger heliocentric distances also suggests that Afρ probably
overestimates the dust production for those comets on the right-
hand side of Figure 5. Since this effect is dependent, probably
as a power-law, on heliocentric distance, the net result would be
to decrease the steepness of the slope of the $r^{-3.4}$ dependence
described above without increasing the scatter of the individual
data points.

CONCLUSION

Despite these caveats, the foregoing discussion demonstrates the
value of simultaneous observation of gas and continuum emission.
The data shown in Figures 5 and 6 are essentially ratios of the
gas and continuum fluxes corrected only for the observing
geometry and this correction is based on the observed spatial
distribution of these emissions. For the present study the
fluxes are assumed to depend in a linear way on the gas or dust
production rate so that detailed modeling is not necessary for a
determination of the trend with heliocentric distance. However,
these data may provide a valuable test of recent thermal models
of the cometary dust coma such as those proposed by Weissman and
Kieffer (17). Additionally, the data base for Figure 5 could be
increased with the inclusion of ground-based photometric or
spectrophotometric observations of other comets properly related
to the data presented here.

ACKNOWLEDGEMENTS

This work was supported by NASA grants NSG-5393 to the Johns
Hopkins University and NAG 5-279 to the University of Maryland.

REFERENCES

1. Feldman, P. D. 1983, Science 219, pp. 347-354.

2. Hapke, B., Wells, E., Wagner, J., and Partlow, W. 1981,
    Icarus 47, pp. 361-367.

3. A'Hearn, M. F., Schleicher, D. G., Feldman, P. D., Millis,
    R. L., and Thompson, D. T. 1984, Astron. J. 89, pp. 579-591.

4. Weaver, H. A., Feldman, P. D., Festou, M. C., A'Hearn, M.
    F., and Keller, H. U. 1981, Icarus 47, pp. 449-463.

5. Mount, G. H. and Rottman, G. J. 1981, J. Geophys. Res. 86,
    pp. 9193-9198.

6. Feldman, P. D. 1980, in "Solid Particles in the Solar System,"
    ed. I. Halliday and B. A. McIntosh, Reidel, Dordrecht, pp.
    263-266.

7. Hanner, M. S. 1984, Ap. J. (Letters) 277, pp. L75-L78.

8. Hanner, M. S., Tokunaga, A. T., Veeder, G. J., and A'Hearn,
    M. F. 1984, Astron. J. 89, pp. 162-169.

9. Weaver, H. A., Feldman, P. D., Festou, M. C., and A'Hearn, M. F. 1981, Ap. J. 251, pp. 809-819.

10. Smith, A. M., Stecher, T. P., and Casswell, L. 1980, Ap. J. 242, pp. 402-410.

11. Feldman, P. D. and Brune, W. H. 1976, Ap. J. (Letters) 209, pp. L145-L148.

12. A'Hearn, M. F., Thurber, C. H., and Millis, R. L. 1977, Astron. J. 82, pp. 518-524.

13. Feldman, P. D. 1978, Astron. Ap. 70, pp. 547-553.

14. A'Hearn, M. F., and Feldman, P. D. 1980, Ap. J. (Letters) 242, pp. L187-L190.

15. Ney, E. P. 1982, in "Comets", ed. L. L. Wilkening, Arizona, Tucson, pp. 323-340.

16. Millis, R. L., A'Hearn, M. F., and Thompson, D. T. 1982, Astron. J. 87, pp. 1310-1317.

17. Weissman, P. R. and Kieffer, H. H. 1981, Icarus 47, pp. 302-311.

COMMENT BY P. WEISSMAN

I want to clarify the statement that the dust chokes off the gas production. At high dust production rates the coma opacity becomes optically thick and the nucleus sees only a hot dust cloud around it, the energy of which varies as $1/r^2$. Thus, gas continues to be produced but the variation of the gas production with heliocentric distance is not as steep as for cases where the coma is optically thin (though not negligible).

S_2: A CLUE TO THE ORIGIN OF COMETARY ICE?

Michael F. A'Hearn

Astronomy Program, University of Maryland

and

Paul D. Feldman

Department of Physics and Astronomy
The Johns Hopkins University

ABSTRACT

   S_2 was recently discovered in Comet 1983d. It is argued that S_2 is a parent molecule, resident in the ices of the cometary nucleus. If so, this requires that cometary ice have accreted directly from interstellar grains with icy mantles which were never appreciably warmed before accretion.

INTRODUCTION

   In two recent papers (1,2) we have reported on the discovery of the sulfur dimer, S_2 in Comet IRAS-Araki Alcock 1983d and on its spatial and temporal variation in that comet. In this paper we will briefly summarize these findings, indicate some further details regarding the temporal variations and the cause of the observed outburst, and semi-quantitatively speculate about the source of the S_2. In summary, we will argue that the presence of S_2 implies that the ice in most comets arrived there in solid form directly from the interstellar medium, as previously proposed by Greenberg, and that the ice did not condense during the formation of the solar system.

J. Klinger et al. (eds.), Ices in the Solar System, 463–471.
© *1985 by D. Reidel Publishing Company.*

SUMMARY OF THE OBSERVATIONAL RESULTS

In (1) we showed that the S_2 was produced within at most 20 km of the nucleus and that the spatial distribution was consistent with S_2 being a parent molecule residing in the nucleus itself. The spatial distribution has the S_2 confined so closely to the nuclear region that it would not have been detectable at typical geocentric distances of comets. Thus it is not surprising that this species has not been seen in previously observed comets; even if it were present in all comets at the same relative abundance as observed transiently in comet 1983d, it would only have been barely detectable in the most favorable of our previous observations with IUE. In (1) we also reported on the abundance of S_2 and concluded that the production rate was of order 10^{-4} to 10^{-3} that of OH and comparable to that of CS.

In (2) we considered the temporal variations both of S_2 and of several other species in comet 1983d. We presented evidence indicating that the comet underwent an outburst on 1983 May 11 but that it also exhibited a periodic variation consistent with what one would expect from a heterogeneous surface and rotation period of order 1 or 2 days. The important points for our present discussion are that the S_2 was detectable in all our spectra as well as in spectra obtained at other times by Festou (private communication) and that it may be a parent molecule resident in the nucleus.

CAUSE OF THE OUTBURST ON MAY 11

Morris and Green (3) have pointed out that there was an outburst on May 11 as seen by visual observers, presumably the same outburst seen in our data with IUE at approximately UT 2000 on May 11. They noted a correlation with disturbed geomagnetic indices at that time. In order to examine the correlation between the outburst on May 11 and possible features in the solar wind, we have examined data from the IMP 8 satellite. The actual solar wind speed can be combined with the solar rotation rate to calculate that solar wind features should have arrived at the comet approximately 3 hours 50 minutes to 4 hours before arriving at Earth, i.e., at IMP 8. Figure 1 shows the speed of solar wind protons measured by IMP 8, shifted earlier in time by 4 hours, superimposed on the brightness variation that was presented in (2). It is clear that the sharp increase in brightness beginning near 20 hours UT is very well correlated with a sharp increase in the solar wind speed from approximately 650 km/s to more than 700 km/s. This was the last of three sharp increases which brought the solar wind speed up from a quiescent value of 450 km/s approximately 16 hours earlier. There was no change in the density of the solar wind at this time so the flux is directly proportional to the velocity. There was an increase in the thermal velocities of the protons but there was no major fluctuation in the magnetic field at this time. We note, incidently that there was a strong shock in the solar wind on the preceding day, at approximately May 10.6, and that there were major flares on the sun in the preceding period.

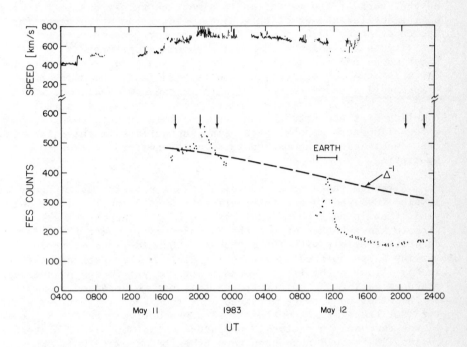

Figure 1: The light curve of comet 1983d as measured with the FES (Fine Error Sensor) of IUE as previously published (2). The mid-times of our spectra which show the presence of S_2 have been marked. Above the light curve is the newly available record of the speed of protons in the solar wind as recorded by IMP 8. These data have been shifted in time by 4 hours to allow for their arrival at the comet before their arrival at IMP 8. Although there is a data gap near 2000 hours on May 11, it is clear that there was a significant increase in the speed coincident with the onset of the outburst seen in the data from the FES of IUE. The two increases in speed of the solar wind earlier on May 11 are undoubtedly associated with the high starting level of the FES data shown here and with the visual outburst discussed by Morris and Green (3). The second sharp increase in the FES data, labelled Earth, is due to Earthshine when observing the comet very near the limb of Earth.

In (2) we pointed out that, independent of the outburst observed at 2000-2100 on May 11, the general level of molecular production was much higher on May 11.8 than on previous days when the production of OH was measured from the ground. Part of the variation is likely due to nuclear rotation (visible in Figure 1 as the slow curvature in the plot of FES counts on May 1) but another part of the variation is almost certainly due to the onset of the earlier phases of this high speed stream which began (at the comet) near 0330 UT on May 11. The correlation between activity in the solar wind and the cometary outburst is unambiguous and indicates that the time scale for response by the comet is of order minutes, not hours. In fact, the comet has completed its response and adjusted to the new conditions within an hour. W.-H Ip (private communication) has suggested that the coupling distance between ions and neutrals in this comet is only a few kilometers so that even mild fluctuations in the solar wind can squeeze electrons back down onto the nucleus to cause sputtering.

Although these data provide a unique opportunity to study the physics involved in the interaction between a comet and the solar wind, that is not the primary purpose of the current paper. We wish to know whether or not the S_2 observed in comet 1983d is associated solely with the outburst. The mid-times of spectra showing S_2 are marked in Figure 1. Clearly the spectra obtained on May 11, even the first, would be strongly affected by species produced during the outburst; the two spectra obtained on May 12, on the other hand, were obtained more than 200 S_2 lifetimes after the last stage of the outburst. The production rate for S_2 on May 12 was down only a factor 10 compared to that on May 11 whereas residual S_2 should have been down by a factor $\exp(-200)$. We therefore conclude that the S_2 which we observed was not primarily associated with the outburst except in so far as the difference in time scales between S_2 and other species made the emission features easier to detect during the outburst. Furthermore, Festou (private communication) obtained spectra of the comet with IUE on May 6 and May 8, well before the outburst, and reports that the S_2 is weakly visible then. Since the lack of previous observations of S_2 can be explained purely by geometrical effects, we infer that S_2 should be present in all comets.

THE SOURCE OF S_2

We ask now how the S_2 might have been produced in this comet and, by inference, in other comets. Much of this section will be speculative and we hope that it serves mainly to stimulate further work in this area. We must first ascertain whether or not the S_2 is truly a parent molecule rather than the daughter of a very short-lived parent. Formation in the gaseous phase by photo-

dissociation of a more complicated parent, for example, could
proceed if the parent has a sufficiently large cross section for
photodissociation. If the parent absorbed all photons shortward
of 300 nm, the upper limit of 20 seconds for the lifetime would
require a cross section (for all wavelengths shortward of 300 nm)
of a few 10^{-18} cm^2. Alternatively, the S_2 might be produced by
electron or proton impact on the nucleus. If the nucleus has a
radius of 4 km, consistent with other measurements, a quantum
yield of unity would imply an average current of a microamp per
square cm. Corresponding limits can be derived for the rate
constants of any 2-body gas phase reactions. Thermal dissociation
of a more complicated parent seems even less likely since the
temperature should be near 200 K at the surface and dropping
rapidly as the gas expands. We cannot totally rule out any of
these possibilities but the quantitative requirements must be met
by a plausible source before we can infer that the S_2 is not a
parent molecule. For the purpose of seeing the implications, we
will assume that the S_2 is a parent. Where did it come from?

The S_2 is unlikely to have condensed as a solid from the
gaseous phase. Sulfur atoms should combine primarily into a
variety of other species rather than into S_2, such as H_2S, CS_2,
and FeS. Many authors have considered the gas-phase chemistry of
the interstellar medium, perhaps the most recent being Mitchell
(4) who considered the chemistry behind shocks. He, like all
previous authors, found the abundance of S_2 to be only of order
10^{-3} that of other species containing S.

Duley et al. (5) have considered the effects of chemical
reactions on the surfaces of interstellar grains on which the
surface sites of OH^- have been converted to surface sites of SH^-
by the presence of neutral atomic S in the gasoues phase. In
order for their model to work, the grains must be very small and
more importantly the S must be depleted in the gaseous phase by at
least a factor 10 (6) and in fact Duley et al. assumed depletion
by a factor of 100. In this situation, S_2 can become the dominant
S-bearing species in the gaseous phase but most of the S, as
noted, is contained in other species on the grains.

Searches for S_2 at radio wavelengths in dark clouds by Liszt
(7) were unsuccessful and he set an upper limit for the amount of
S_2 at approximately 2% of the cosmic abundance of S. Since S has
only a moderate cosmic abundance, less than 10% that of C, the
amount of S_2 that can be produced in the gaseous phase relative to
the amount of C-bearing compounds is quite small.

Our observations showed that the production of S_2 was
comparable to that of CS, the only other S-containing species
observed in comets, and also comparable to the production of the
several species observed to contain C in most comets. Thus the

abundance of S_2 seems to be not only a major fraction of the total
S but even a major fraction of the C. This suggests that the C is
depleted from the volatile fraction of comets. The remainder of
the C may be in refractory grains such as graphite or the complex,
processed organic material suggested by Greenberg (8,9). The
abundance of S_2 further requires that the S_2 not come from the
interstellar medium in the form of S_2.

An alternative source of S_2 is irradiation of other S-containing species which are trapped in some sort of ice. As we have
previously pointed out (1), Fournier et al. (10) and Lee and
Pimentel (11) have made laboratory studies of the irradiation of
both OCS and CS_2 trapped in inert rare gas matrices at low temperatures. In their experiments, ultraviolet photolysis produced S in
the matrix and when the matrix was warmed to 10 or 20 K the S atoms
migrated and recombined to form substantial amounts of S_2. Earlier
experiments by Hopkins and Brown (12) suggested that S_2 disappeared
from matrices at temperatures above 30 K, possibly by chemical
reactions or by diffusion out of the matrix. Could such a process
have occurred in comets?

Although compounds of S have not been studied, Greenberg's
group (13,14,15) has carried out extensive experiments irradiating
mantles of simulated interstellar grains and shown that equivalent
chemical rearrangements do take place under ultraviolet
irradiation. Similar experiments by Moore (16,17) using high
energy protons to irradiate thin layers of icy material have shown
qualitatively similar effects. With both types of irradiation,
the chemical abundances of the various species change toward a new
equilibrium quite different from the original composition. For
example, if one starts with either pure CO or pure CO_2 mixed with
H_2O, the equilibrium mixture after extensive irradiation includes
comparable amounts of both CO and CO_2. Similar results might be
expected from S-bearing compounds so that a relatively large
percentage of S_2 would exist after extensive irradiation. The
important question is when the irradiation occurred.

Although it was our initial conclusion (1) that the
irradiation was produced by cosmic rays while the comet was in the
Oort cloud as discussed by Shul'man (18), Donn (19), and Whipple
(20), this suggestion is no longer tenable since the comet has
been found to be periodic. That process would have affected only
the outer few meters (if done by cosmic rays, microns if done by
ultraviolet light) and comet 1983d has surely lost much more than
its outer few meters of volatile material. Either the irradiation
occurred before the ice was accreted into comets (so that S_2 could
occur throughout the nucleus) or the S_2 is produced by irradiation
at the present time. The quantitative constraints on a process
for present-day irradiation have been discussed above and are
severe. A more natural place for the irradiation to take place is

in the interstellar medium. The thrust of Greenberg's work (8) has been to argue that interstellar grains have mantles which are highly processed by irradiation and that these grains have been incorporated into the solar system. If this is the case, then the presence of S_2 requires that the mantles of the grains never vaporized. More stringently it even requires that the grains remained extremely cold, no more than a few tens of K, so that the S_2 could neither diffuse out of the ice nor react with other species. This places rather tight constraints on the region of formation of comets.

Obviously much more work is required to test our suggestion. Specifically, the dose required to produce the observed amount of S_2 should be determined, the plausible S-bearing species need to be investigated, and the stability of S_2 in H_2O ices must be determined. Clearly more work is needed on alternative hypotheses. Can a satisfactory mechanism be found to produce the S_2 at the time of vaporization of the ices, by chemical reactions, by sputtering, or by photodissociation? More detailed modeling of the early solar nebula might include estimates of the amount of irradiation that can be accumulated then. Finally, it is important for S_2 to be sought elsewhere, both in other comets and in the interstellar medium.

ACKNOWLEDGEMENTS

This work has benefitted from informative discussions with many people including J. M. Greenberg, W.-H. Ip, W. M. Jackson, G. F. Mitchell, M. H. Moore, T. Hartquist, and A. T. Young. We particularly thank J. King and A. Lazarus for providing the data from IMP 8 on the magnetic field and flow of the solar wind. The work has been financially supported by NASA grants NAG 5-279 and NSG 5393.

REFERENCES

1. A'Hearn, M. F., Feldman, P. D., and Schleicher, D. G. 1983, Astrophys. J. 274, pp. L99-L103.

2. Feldman, P. D., A'Hearn, M. F., and Millis, R. L. 1984, Astrophys. J. 282, pp. 799-802.

3. Morris, C. S. and Green, D. W. E. 1983, Bull. Amer. Astr. Soc. 15, p. 803.

4. Mitchell, G. F. 1984, presented at Protostars and Planets Conference, Tucson, January 1984; also Astrophys. J. in press (15 Dec. 1984 issue).

5. Duley, W. W., Millar, T. J., and Williams, D. A. 1980, Mon. Not. Roy. Astr. Soc. 192, pp. 945-957.

6. Mitchell, G. F. 1978, Astron. J. 83, pp. 1612-1613.

7. Liszt, H. S. 1978, Astrophys. J. 219, pp. 454-457.

8. Greenberg, J. M. 1982, in "Comets", ed. L. L. Wilkening, Univ. Ariz. Press, Tucson), pp. 131-163.

9. Greenberg, J. M. 1983, in "Cometary Exploration", ed. T. I. Gombosi, Central Research Institute for Physics, Budapest, II pp. 23-54.

10. Fournier, J., Lalo, C., Deson, J., and Vermeil, C. 1977, J. Chem Phys. 66, pp. 2656-2659.

11. Lee, Y.-P. and Pimentel, G. C. 1979, J. Chem Phys. 70, pp. 692-698.

12. Hopkins A. G. and Brown, C. W. 1975, J. Chem. Phys. 62, p. 1598.

13. Greenberg, J. M., Yencha, A. J., Corbett, J.W., and Frisch, H. L. 1972, Mem. Soc. Roy. Sci. Liege, 6e Series, 3, pp. 425-436.

14. Hagen, W., Allamandola, L. J., and Greenberg, J. M. 1979, Ap. Sp. Sci. 65, pp. 215-240.

15. Greenberg, J. M., Allamandola, L. J., Hagen, W., van de Bult, C. E. P. M., and Baas, F. 1980, in "Interstellar Molecules", ed. B. H. Andrew, Reidel, Dordrecht, pp. 355-363.

16. Moore, M. H. 1981, Ph.D. Thesis, Univ. Maryland, College Park.

17. Moore, M. H., Donn, B., Khanna, R., and A'Hearn, M. F. 1983, Icarus 54, pp. 388-405.

18. Shul'man, L. 1982, in "The Motion, Evolution of Orbits and Origin of Comets", IAU Symp. 45, ed. G. A. Chebotarev et al., Springer-Verlag, NY, pp. 265-270.

19. Donn, B. 1976, in "The Study of Comets", NASA SP-393, ed. B. Donn et al., U. S. G. P. O., Washington, pp. 611-619.

20. Whipple, F. 1977, in "Comets, Asteroids, Meteorites", ed. A. H. Delsemme, Univ. Toledo Press, Toledo, pp. 25-35.

COMMENTS

L. Lanzerotti :

    One must look at the plasma conditions at the times of
comet outbursts (such as the May 11 outburst of IRAS-Araki-Alcock). The plasma conditions can vary considerably from one
interplanetary field condition to the other.

V. Pirronello :

    I should like to remark that not always solar burst have
measurable effects on comets also because particles they emit
could be quite highly beamed.

FORMALDEHYDE IN COMET IRAS-ARAKI-ALCOCK (1983 d). COSMOGONICAL IMPLICATIONS

C. B. Cosmovici[1] and S. Ortolani[2]

1) Space Research Group, DFVLR, 8031 Wesslin, FRG (cf. final note)
2) Asiago Astrophysical Observatory, 36012 Asiago (VI), Italy

ABSTRACT

Detailed investigations of spectra taken May 9.9 UT, 1983 with the grating spectrograph of the 182 cm Asiago Telescope led to the identifications of new cometary molecules. While the H_2CO, DCO, S_2 and NH_4-lines need further theoretical and laboratory investigations, the identification of HCO and H_2S^+ can be considered unambiguous. The new discoveries suggest an alternative place of origin for this comet (1).

I. INTRODUCTION

Cometary astronomers will maybe never have another opportunity in this century to observe a comet like IRAS-Araki-Alcock. Since 1770 (Comet P/Lexell 1770 I) no other observed comets came so close to the Earth ($4.5 \cdot 10^6$ km) thus giving for the first time the possibility to carry out observations with a spatial resolution up to 20 km.

In this way the nuclear region could be studied for the first time by means of photography, spectroscopy and radar technology and fundamental discoveries, especially in molecular spectroscopy, could be obtained from ground and from space. A'Hearn et al. (2) detected the S_2 molecules by means of observations with the IUE UV-satellite. This molecule was confined in a region of 100 km around the nucleus and had never been seen before in any astronomical object.

J. Klinger et al. (eds.), Ices in the Solar System, 473–485.
© 1985 by D. Reidel Publishing Company.

Altenhoff et al (3) report the detection of the parent molecules H_2O and NH_3 with the 100 m Effelsberg radiotelescope. Walker et al (4) made the first IR-observations of a comet from space with the IRAS satellite revealing morphology and physical details not detectable in the visible part of the spectrum. The Arecibo Radiotelescope was able to receive echo signals from the nucleus but no details or results have been reported up to now. Table 1 shows the status of our knowledge on cometary composition before and after Comet IRAS-Araki-Alcock (1983 d).

TABLE 1

Molecules and Atoms detected in Comets

COMA	IRAS-A-A (1983 d)
A) Molecules	
• Visible Spectrum (3000-10000 Å)	HCO*
$C_2, C_3, CH, CN, OH, NH, NH_2, CO*$	$(H_2CO*, DCO*, S_2*, NH_4*)$
• UV (1200-3000 Å)	
CO, CS, C_2	S_2
• Radio	
$(H_2O, HCN, CH_3CN)?, OH, CH$	H_2O, NH_3
• IR(1 - 5μ)	
OH, CN, C_2	
B) Atoms	
• Visible	
$H, O, Na, Ca, Cr, Co, Mn, Fe, Ni, Cu, V, Si, K$	
• UV	
H, C, O, S	
C) Grains	
Silicates : 10 μ	
TAIL	
• Visible	
$CO^+, CO_2^+, CH^+, N_2^+, H_2O^{+*}, OH^+, Ca^+$	H_2S^{+*}
• UV	
C^+, CO^+, CO_2^+	

* Discovered at the Asiago Observatory (see Refs. 6, 7, 18, 22, 23).

II. THE OBSERVATIONS

Both spectroscopic and photographic observations have been carried out on May 9.9 UT, 1983 with the 182 cm- and with the 90 cm- Schmidt telescopes at Asiago. Seven photographs and four spectra could be obtained. Observations during closest approach to the Earth (May 11) were not possible because of the weather conditions.

The analysis of the photographic plates was very puzzling, since the visible coma was directed toward the Sun with an inclination of about 30 °.

This was confirmed by the IUE- (5) and by the IRAS-satellite (4). From the physical point of view this behaviour is correct since the sunside of the nucleus sublimates much more material than the opposite part. This phenomenon could not be observed in previous comets because of the poor spatial resolution (Fig. 1).

The inclination of the nucleus - coma axis with respect to the Sun could be an indication for the nucleus rotation and the comparison with pictures taken on different days could permit the determination of the rotation period. During the time of observation (May 9.9 UT, 1983), the comet had a magnitude of 3, was at $10.6 \cdot 10^6$ km from the Earth and $152 \cdot 10^6$ km from the Sun.

The phase angle was $87.7°$ and the angle of the vector from the Sun to the comet was $171.3°$.

The spectra were taken on 103-a-D Kodak plates with exposure time varying from 5 to 30 min with the BC 600 + VARO grating spectrograph. Guiding was very difficult because of the high relative velocity of the comet.

The best spectrum taken with an exposure time of 30 min (Fig. 2) has about 450 lines, 50 of them still unidentified. The slit covered a region of 10 000 x 150 km centered on the brightest nuclear condensation. As in Fig. 1 also the spectrum clearly shows that the intensity distribution peaks toward the Sun. The spectral range was 3800 - 8500 Å and the resolution 6 Å.

III. SPECTROSCOPIC RESULTS

The Triplet and the Asundi System of CO discovered in Comet Bradfield 1980 t (6) (7), could be

SUN (North)

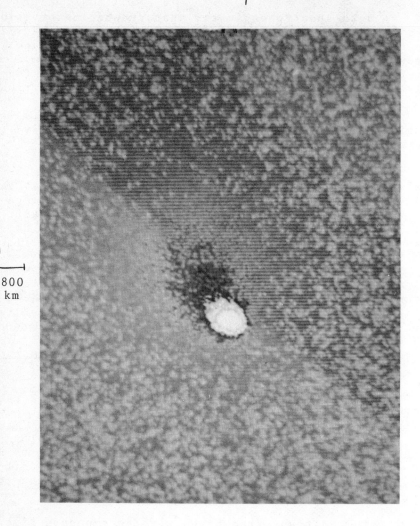

800
km

EST

Fig. 1   Plate taken with the 182 cm Asiago telescope
and processed with the PDS at ESO (Garching). Exposure
time 20 min. Different colours indicate different photon densities. The inclination of the nucleus – coma
axis with respect to the Sun is about 30°.

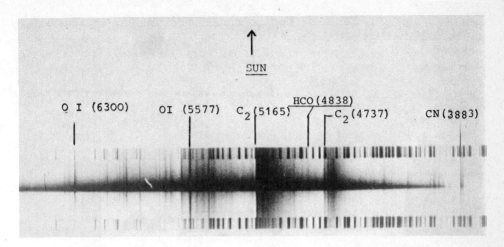

Fig. 2   Spectrum taken with 30 min exposure time. Slit 10000 x 150 km centered on the brightest part of the nuclear region in N - S direction.

unambiguously confirmed. Further the Ångstrom System (8) of CO ($B^1\Sigma$ - $A^1\pi$) could be detected for the first time.

Table 2 gives the new identified molecules. In this paper we will discuss only the HCO, H_2CO and DCO molecules which are the most important from the cosmogonical point of view.

The HCO lines tabulated in Table 2 are very clear and sharp (Fig. 3), whereas in the case of H_2CO and DCO the intensity is lower and the contamination probability in the respective wavelength range much higher at our resolution (6 Å). The intensity ratios between the different lines do not correspond to those observed in absorption by Johns et al. (9) and our strongest line appears at 4833 - 4838 Å (Fig. 3), but there could be some contribution of CO (Ångstrom system at 4835 A).

G. Herzberg (10) was somewhat concerned about our identification of HCO in emission at levels up tp v_2 = 15, because of the strong predissociation occurring

T A B L E   2
New Molecules in Comet IRAS-A-A (1983 d)

A) Identification positive

$(\nu_1,\nu_2,\nu_3)'$	$(\nu_1,\nu_2,\nu_3)''$	$\lambda(\overset{\circ}{A})$ observed.	$\lambda(\overset{\circ}{A})$ Laboratory	
			R	Q (Branch)
HCO (Ref. 9) 0 15 0	0 0 0	4838 - 9	4833.3 - 4838.39	
0 13 0	0 0 0	5199 - 5202	5195.6 - 5201.0	
0 11 0	0 0 0	Blended by C_2 5635.5 - 5628.6	5624.0 - 5629.2	
0 9 0	0 0 0	{ 6137 - 9 / 6143 - 6.	6138.0 - 6144.7	
0 7 0	0 0 0	{ 6763 - 4 / 6771 - 4	6766.3 - 6774.1	
H_2S^+ (Ref. 8) 0 3 0	0 1 0	4960 - 6	4962.4	
0 2 0	0 0 0	4819 - 21	4818.5	
0 4 0	0 1 0	4752 - 3	4755.7	
0 3 0	0 0 0	Blended by C_2 4684 - 4737	4692.4	
0 6 0	0 2 0	4618 - 21	4619.9	
0 4 0	0 0 0	4503 - 5	4507.2	
0 7 0	0 2 0	4423 - 5	4421.3	
0 5 0	0 0 0	4333 - 6	4336.4	
0 6 0	0 0 0	4176 ?	4172.5	

B) Strongly suspected

(identification to be confirmed by further laboratory and theoretical investigations)

DCO (Ref. 9) 0 19 0	0 0 0	4869 - 4872	4867.2 - 4870.8	
0 17 0	0 0 0	Blended by C_2	5149.9 - 5153.6	
0 17 0	0 1 0	5386 - 9	5385.0 - 5389.0	
0 15 0	0 0 0	Blended by C_2	5471.9 - 5475.9	
0 13 0	0 0 0	5843 - 4	5842.4 - 5846.7	
0 11 0	0 0 0	Blended by NH_2	6273.3 - 6277.9	
0 9 0	0 0 0	Blended by NH_2	6780.5 - 6785.5	
0 7 0	0 0 0	absent	7386.2 - 7391.8	
0 5 0	0 0 0	{ 8127 / 8130	8123.2 - 8129.8	
Diffuse bands			Maxima	
		5302	{ 5304.3	
0 16 0	0 0 0	5309 - 15	5312.5	
		5325 - 8	5317.4	
			5325.0	
0 16 0	0 0 0	5394 - 5	{ 5353 / 5395	
0 15 0	0 0 0	5488 - 96	{ 5491 / 5525	
0 14 0	0 0 0	5649	{ 5648.1 / 5656.9	
		5670 - 2	5665.0 / 5673.1	
0 14 0	0 0 0	5696	{ 5694	
		5753 - 6	5755	
0 13 0	0 0 0	5864 - 9 Blended by Na I	{ 5863.2 / 5898.8	
0 12 0	0 0 0	6053 - 9	{ 6057.2	
		6040 - 6	6043.7	
0 11 0	0 0 0	6322 - 4	6322.2	

T A B L E   2 (continued)

		$\lambda(\text{Å})$ observed	$\lambda(\text{Å})$ Lab.	I (Lab.)
H_2CO (Ref. 8)		4821	4821	3
		4678 - 4697	4673 - 95	5
		4556 - 65	4551 - 69	8
		4433	4434	8
		4348 - 9	4347	8
		Blended by CH + NH^+		
		4214 - 6	4220 - 40	1 0
		4240 - 5		
		4118 - 4121	4121	8
		4046 - 50	4044	6
		Blended by C_3 + CO_2^+		
		3950	3952	1 0
		Blended by C_3 + CO^+		
S_2 (Ref. 8)	Main System	4429 - 33	4433.4	6
		4650	4651.1	4
		4789 - 92	4790.6	6
		4934 - 39	4937.0	5
		5710 - 12	5710.1	4
		5837 - 43	5840.6	4
	Red System	6984	6983.8	
		7071	7069.6	
		7321	7319.5	

All other laboratory lines (Ref. 8) blended.

		$\lambda(\text{Å})$ observed	$\lambda(\text{Å})$ Lab.	I (Lab.)
NH_4 (Ref. 10)		6635	6633 (doublet with a spacing of 3 Å)	

Note : 1) The hypen - in the column "λ observed"   means the wavelength range in three different spectra.

2) The intensity of the HCO line at 4838 Å is of the order of 5.10^{-11} erg $cm^{-2}s^{-1}$. (Absolute flux of C_2 at 5135 Å kindly provided by M. A'Hearn).

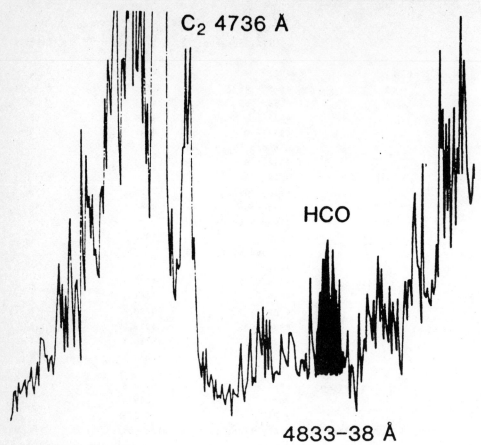

C₂ 4736 Å

HCO

4833-38 Å

Fig. 3   Part of the spectrum in Fig. 2 showing the strongest HCO line (Q and R branch). The absolute flux of this line was calculated by comparison with the absolute flux of the C_2 line at 4736 Å (see Table 2). The intensity scale on this figure is arbitrary.

at intermediate vibrational levels. But recent laboratory measurements (11) of HCO emission (v_2 = 9 and v_2 = 13) produced by laser photolysis of formaldehyde have demonstrated the detectability of those lines under particular excitation and environmental conditions. HCO is a well known interstellar molecule (12) (13) and was detected by radio astronomers at 3.46 mm. Its very low dissociation energy ($\sim$ 1 eV) makes its observation impossible in environments not protected from the UV radiation. Our Fig. 3 shows that HCO practically disappears at about 1000 km from the nucleus, where the dust and

gas intensity becomes insufficient for UV-screening.
Nevertheless, the strong line at 4838 Å was already observed in Comet Mrkos (1957 V) and given as unidentified
(14) (15). The other HCO lines (Table 2) were indeed
never been reported before. The HCO production, vibrational relaxation, chemical kinetics are well studied
(16) and the rate coefficients of the H_2CO - six possible decay products are given by Huebner and Carpenter
(17). The most probable production mechanism of HCO is
the photodissociation of formaldehyde

$$H_2CO + h\nu \longrightarrow HCO + H$$

(Rate coefficient : $85 \cdot 10^{-6} s^{-1}$, threshold : 3493 Å,
Photolysis Products Excess Energy : 0.5 eV).

Since HCO is a primary intermediate in hydrocarbon oxidation it would be very difficult to find it as
a parent in the icy nucleus of comets. The second peak
at 670 km in Sun direction (Fig. 4) shows that maybe at
this distance the dissociation of a parent molecule increases the HCO production. It is also not excluded that
formic acid detected in interstellar clouds at 18 cm
(12) could also be responsible for HCO production (17)

$$HCOOH \longrightarrow HCO + OH$$

(Threshold : 2500 Å , Rate coefficient : $490 \cdot 10^{-6} s^{-1}$,
Photolysis Products Excess Energy : 1.8 eV).

The excitation mechanism for the observed lines
in the cometary inner coma should be investigated in detail. It was demonstrated (7) (18) that fluorescence and
direct collisional excitation can not explain the observed emission from the triplet system of CO and that the
only plausible mechanisms were the dissociative recombination of HCO^+ and the Alfven mechanism (connected
with plasma instabilities) which could be responsible
for the production of hot electrons. If this second case
would apply to HCO has not yet been investigated.

Formaldehyde is also one of the most common interstellar molecules observed in the mm- and cm- range,
either in emission or in absorption (12). The dissociation energy is higher (D = 3.37 eV) than for HCO. Its
photochemistry is one of the best studied in laboratory
(19) because of its fundamental importance in chemistry,
astrophysics, polluted atmospheres and also in exobiolo-
gy. Biermann et al. (20) assume that the H_2CO abundance
in cometary nuclei of interstellar origin should vary

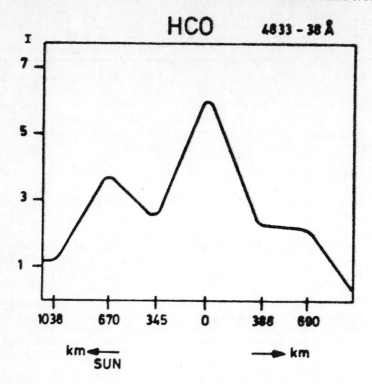

Fig. 4   PDS-profile of the HCO line at 4833 Å (published in Nature, Ref. 22), obtained directly from the
"TRACE" command of the IHAP system without any smoothing or filtering procedure (see IHAP, ESO Manual,
1983). Only the continuum subtraction has been made on
the original scan. The explanation of the "smooth" appearance is that the PDS scan aperture slit used was
100 x 5 μ with x and y steps of the same corresponding
size. In conclusion, the smoothing is the result of the
y size of the slit, much higher than the typical noise
frequency of the photographic emulsion grain.
On the y axis : arbitrary density units.
Km 0 represents the centre of the nuclear brightest region. The second peak at 670 km is probably due to dissociation of H_2CO (or HCOOH).

between 22 and 25 %.

The high percentage of HCO here observed and
also deduced in Comet Bradfield (7) (18) is not consistent with published results on the chemistry of dense
interstellar clouds and this shows that both new observations and new theoretical studies are greatly needed.

As suggested by Biermann (1) the place of the
origin of comets surrounding the solar system has not
necessarily been the outer fringes of the presolar nebula. Another nearby fragment of the same interstellar
cloud, in which the presolar nebula came into existence 4.5 billion years ago, appears to be an at least
equally probable place of origin. In this case only very diluted light from the newly born sun would influence the chemical evolution, such that the conditions
would be more like those in a normal dense interstellar
cloud.

Deuterated formyl (DCO) has been observed as ion
in interstellar clouds at 4.16 mm in emission (12). Because of the cosmic D/H ratio (2 · 10^{-5}) one would not
expect the presence of molecules containing deuterium
in a cometary nucleus. Thus our identification of the
reported DCO-lines was very surprising and we are still
very careful in the confirmation of such fundamental
discovery.

Otherwise recent calculations made by Ip (21)
have shown that both gas phase chemistry and surface
chemistry on interstellar grains can produce large isotope fractionation effects leading to $[HDO]$ / $[H_2O]$ ratios orders of magnitude higher than the $[D]$ / $[H]$ ratios.

On the basis of our spectroscopic results further detailed theoretical calculations are necessary in
order to establish the $[DCO]$ / $[HCO]$ ratio and to try
to determine the deuterium abundance in comets. This
would allow to establish a fundamental milestone in cometary research and cosmogony.

ACKNOWLEDGEMENTS

We are deeply grateful to Profs. G. Herzberg,
L. Biermann, F. Whipple, M. A'Hearn, P. Feldman,
W. Huebner, C. Arpigny and A. Delsemme for valuable dis-
cussions and suggestions.

REFERENCES

1. Biermann, L., 1981, Phil. Trans. R. Soc., London,
    A 303, pp. 351-352.

2. A'Hearn, M. F., Feldman, P. D. and Schleicher, D.
    G., 1983, Ap. J., 274, pp. L99 - L103.

3. Altenhoff, W. J., Batrla, W., Huchtmeier, W. K.,
    Schmidt, J., Stumpff, P., and Walmsley, M., 1983,
    Astron. Astrophys., 125, pp. L19 - L22.

4. Walker, R. G., Aumann, H. H., Davies, J., Green, S.,
    de Jong, T., Houck, J. R., and Soifer, B. T.,
    1984, Ap. J. 278, pp. L11-L14.

5. Festou, M., 1983, private communication.

6. Cosmovici, C. B., Barbieri, C., Bonoli, C., Borto-
    letto, F. and Hamzaoglu, E., 1982, Astron. Astro-
    phys. 114, pp. 373-387.

7. Cosmovici, C. B., Biermann, L. and Arpigny, C.,
    1982, Proc. ESO Workshop "The Need for Coordina-
    ted Ground-based Observations of Halley's Comet",
    pp. 131-147.

8. Pearse, R. W. B. and Gaydon, A. G., 1976, "The
    identification of molecular spectra", 4th Edition,
    Chapman and Hall, London.

9. Johns, J. W. C., Priddle, S. H., and Ramsay, D. A.,
    1963, Discussion Faraday Soc., 35, pp. 90.

10. Herzberg, G., 1983, private communication.

11. König, R. and Lademann, J., 1983, Chem. Phys. Lett.,
    94, pp. 152.

12. Cosmovici, C. B., Inguscio, M. Strafella, F. and
    Strumia, F., 1979, Astrophys. and Space Sci., 60,
    p. 475-491.

13. Hollis, J. M. and Churchwell, E., 1983, Astrophys.
    J., 271, pp. 170-174.

14. Greenstein, J. L., and Arpigny, C., 1962, Astro-
    phys. J., 135, pp. 892-905.

15. Woszczyk, A., 1962, "Le Spectre du Radical NH_2 ;
    son application aux spectres cométaires", Mém.
    Soc. Sci. Liège, Fasc. 6.

16. Reilly, J. P., Clark, J. H., Bradley Moore, C. and
    Pimentel, G. C., 1978, J. Chem. Phys., 69, pp.
    4381-4394.

17. Huebner, W. F. and Carpenter, C. W., 1979, "Solar
     Photo Rate Coefficients", LA-8085-MS, Los Alamos.

18. Arpigny, C., Biermann, L. and Cosmovici, C. B.,
     1983, Proc. Intern. Conference on Cometary Explo-
     ration, Budapest, Vol. I, pp. 185-189.

19. Houston, P. L. and Bradley Moore, C., 1976, J. Chem.
     Phys., 65, pp. 757-770.

20. Biermann, L., Giguerre, P. T. and Huebner, W. F.,
     1982, Astron. Astrophys., 108, pp. 221-226.

21. Ip, W. H., Condensation and agglomeration of Come-
     tary Ice, in This book.

22. Cosmovici, C. B. and Ortolani, S., 1984, Nature
     310, pp. 122-124.

23. Benvenuti, P. and Wurm, K., 1974, Astron. Astro-
     phys. 31, pp. 121-122.

DISCUSSION

Bar-Nun :

        Since there was much CO observed coming out,
could the observed HCO be produced, in the inner coma,
by 3-body recombination of H atoms with CO ?

Cosmovici :

        The known laboratory reactions are :

1) $H_2CO + h\nu \longrightarrow H + HCO$

2) $H_2CO + h\nu \longrightarrow H_2 + CO$ (see Chem. Phys. Lett. 94, 152,
                                                           1983)
For 1) the rate coefficient is $160 \cdot 10^{-6} s^{-1}$ and the
photolysis products excess energy (ΔE) is 0.5 eV.

For 2) we have $85 \cdot 10^{-6} s^{-1}$ and 0.5 eV. In the case of
$H+H+CO$ we would get $32 \cdot 10^{-6} s^{-1}$ and 3.0 eV.

NOTE added in proofs :

Present address of Dr. C. B. Cosmovici :

Istituto Fisica Spazio Interplanetario, C. N. R.,
00044 Frascati, Italy.

THE EFFECT OF DUST HALOS AND DUST MANTLES ON NUCLEAR OUTGASSING

D. A. Mendis

Dept. of Electrical Engineering & Comp. Sci. and
Center for Astrophysics and Space Science,
Univ. of Calif., San Diego, La Jolla, CA   92093   USA

ABSTRACT

As a cometary nucleus, conceptualized as an agglomerate of grains of ice and dust, approaches the sun, the ices sublimate and drag off some fraction of the nuclear dust. While solar radiation pressure eventually drives these dust grains in the antisolar direction to form the dust tail, much of the newly-emitted dust is concentrated around the nucleus to form an optically significant shell (or "halo"). The fraction of the dust that is not entrained by the outflowing gas would accumulate on the surface to form a gradually thickening "mantle," although under certain circumstances this could be "blown off" rather rapidly near perihelion.

Both the circum-nuclear dust halo and the dust mantle modify the solar radiation reaching the "evaporating surface" of the comet in different ways, and thereby regulate the nuclear outgassing.

While no study has yet been made of the simultaneous presence of dust halos and dust mantles on nuclear outgassing, several independent studies have been made of each effect separately.

Here I will present a brief summary of the findings of these studies. I will also discuss the chemical differentiation of the outer layers of a comet during perihelion passage, and show how it may explain a number of observations pertaining to the relative production rates of various cometary molecules.

J. Klinger et al. (eds.), Ices in the Solar System, 487–504.
© *1985 by D. Reidel Publishing Company.*

#1. INTRODUCTION

As a cometary nucleus, visualized as a "dirty snowball",
approaches the sun, the sublimating gasses drag off some
fraction of the nuclear dust. While solar radiation pressure
eventually drives these dust grains in the antisolar direction
to form the dust tail, much of the newly-emitted dust is
concentrated around the nucleus to form an optically significant
shell (or "halo"). The fraction of the dust that is not
entrained by the outflowing gas would accumulate on the surface
to form a gradually thickening "mantle." This is illustrated
schematically in Figure 1.

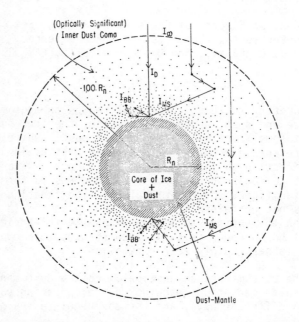

Fig. 1   Schematic representation of the dust halo and the dust
mantle around a dirty-ice cometary nucleus. The direct
(attenuated) solar radiation field I_D, as well as the diffuse
radiation fields due to multiple scattering, I_{MS}, and thermal
re-radiation, I_{BB}, by the hot circum-nuclear dust, are also
shown.

Both the circum-nuclear dust halo and the dust mantle
modify the solar radiation reaching the "evaporating surface" of
the comet in different ways, and thereby regulate the nuclear
outgassing.

While no study has yet been made of the simultaneous presence of dust halos and dust mantles on nuclear outgassing, several independent studies have been made of each effect separately.

Here I will present a brief summary of the findings of these studies.

#2. DUST MANTLES

The existence of an insulating layer of dust on the surface of a rotating icy cometary nucleus was first proposed by Whipple (1) in order to produce the outgassing time lag necessary to explain the non-gravitational perturbation to the revolution period of short-period comets. Subsequently he also suggested that an insulating crust, purged near perihelion, but reformed on the outgoing branch of the orbit, perhaps in a spotty fashion may explain certain cometary brightness variations with heliocentric distance (2).

Brin and Mendis (3) developed a quantitative model of the dust release and dust mantle development in a comet to explain the observed assymetry between the inbound and outbound brightness curves of certain comets. In that model this brightness hysteresis is basically explained in terms of the varying thickness of the dust mantle, which consists of the nuclear material (envisaged as a uniform admixture of ice and dust) evacuated of its icy component. While the sublimation of the ice tended to increase the thickness of this mantle, the dust erosion due to the drag of the sublimating gasses tended to decrease its thickness. These authors assumed that dust erosion took place throughout the mantle as all particles smaller than some critical size (given by the balance between the gas drag force and nuclear gravity) escaped.

The basic equations, defining the core-mantle model, assuming quasi-steady conditions, represent the energy balance at the mantle surface (i.e., the net absorption by the surface is equal to the thermal re-radiation into space plus the thermal conduction into the mantle), the energy flow through the mantle, and the energy balance at the core-mantle boundary (viz., that the energy reaching it by conduction through the mantle is utilized entirely in sublimating the volatile ice in a thin layer). This is supplemented by an assumption about the dust mass spectrum and the dust-to-gas ratio.

The variation of the mantle thickness Δ and the temperatures of the outer (mantle) surface, T_s, and the core

surface, T_e, are shown in Figure 2 for Halley's comet (assumed
to have a radius R_n = 2.5 km, and a dust-to-gas ratio of 1). An
interesting aspect of this, seen in this figure, is the purging
of the mantle at about 1.5 AU on the inward leg, and its
formation at around 2 AU on the outward leg, confirming the
essential validity of the suggestion by Whipple (2). Between
1.5 AU and perihelion (inbound) and perihelion and 2 AU
(outbound), the nucleus is "bald."

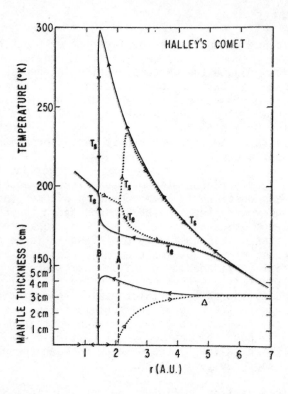

Fig. 2   Variation of the mantle thickness Δ (lower curves) and
T_s and T_e (upper curves) with heliocentric distance. Solid
lines represent inward path while dotted lines represent
outward path (From (3)).

     The variation of the brightness of the C_2 (0-0),
5165Å emission band, which is among the most dominant in the
visual region (corrected for geocentric distance variation), is
shown in Figure 3. Other emissions (e.g., CN at
3883Å, OH at 3090Å) are also expected to follow the same

brightness curve in the case of a clathrate hydrate nucleus that
is assumed here. It is seen that the comet is brighter after
perihelion than before, at the same perihelion distance, up to
about 5 AU. This is due to the fact that the mantle is thinner
after perihelion than before. The maximum brightness assymetry
is between about 1.5 AU and 2 AU. This is due to the fact that
the nucleus is bald in this range, after perihelion, while it
has a thick mantle before perihelion. A brightness hysteresis
of this type was indeed exhibited by Comet Halley during its
1910 apparition (4,5).

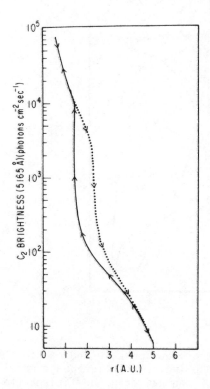

Fig. 3   Variation of the brightness of the C_2 (0-0)
5165Å emission band (corrected for geocentric distance
variation) with heliocentric distance. Solid line represents
inward path while dotted line represents outward path (From
(3)).

Horanyi et al (6) introduced some essential improvements to
the core-mantle model of Brin and Mendis (3). In their so-
called "friable sponge" model, they make the more plausible

assumption that the dust is lost entirely from the surface
rather than from all throughout the mantle, as assumed by Brin
and Mendis. They also assume that the size distribution of the
dust grains in the mantle remain constant with time. This
requires that particles of all sizes can be removed from the
surface, which in turn implies that the larger grains be broken
up into smaller ones (i.e., the assumption of "friability").

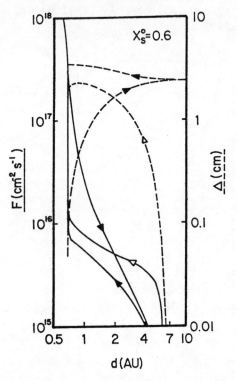

Fig. 4 Variation of the production rate and mantle thickness
with heliocentric distance in the Friable Sponge model for a
relatively large value of the gas drag coefficient (to which
X_s^0 has an inverse co-relation). (From (6)).

Finally, they assume that the mass loss rate of dust is
proportional to the momentum flux of the gas:

$$\frac{dM_d}{dt} = \beta \cdot 4\pi R_n^2 \, F \, m \, u \tag{1}$$

where β is the proportionality constant. It is really the drag
coefficient, which depends most strongly on the grain size
(7). Consequently, if the dust size distribution remains

constant, as assumed, β would indeed be approximately
independent of the heliocentric distance.
     The variation of the mantle thickness, Δ, and the
production rate of the gas, F, (in this case H_2O) for Comet
Halley is shown in Figure 4 for a particular value of X_s^o
(= 0.6) where X_s^o is related to β by

$$X_s^o = \frac{1}{1 + u_o \beta} \tag{2}$$

where u_o is the gas flow speed at 1 AU, when $\Delta = 0$. Once again,
a rapid purging of the mantle is observed close to perihelion;
in this case around 0.7 AU. Also, since the brightness is
proportional to the production rate of H_2O in the quasi-steady
state, the comet is brighter after perihelion.

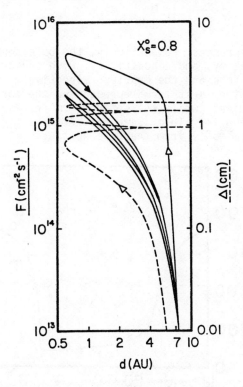

Fig. 5   Same as Fig. 4, but for a relatively smaller value of
the gas drag coefficient. (From (6)).

     The variation of Δ and F for a larger value of X_s^o (and
therefore a smaller value of the drag coefficient β), is shown
in Figure 5. Here one sees a continuing increase of Δ and a

corresponding continuous decrease of F at any given heliocentric
distance. In this case the comet will get dimmer and dimmer
with each revolution, and will ultimately become "asteroidal."

Horanyi et al (6) also solve the time-dependent heat
conduction equation through the mantle and core of a rotating
nucleus with appropriate boundary conditions, and allowing for
the diurnal heating and cooling (see also 8). In this case the
appropriate time-dependent conduction equations for the mantle
and the core, respectively, are:

$$-\rho_m C_m \frac{\partial T}{\partial t} + \frac{\partial}{\partial y}\left(K_m \frac{\partial T}{\partial y}\right) = -\frac{3}{2}k\dot{z}(T_1)\frac{\partial T}{\partial y} \qquad (3)$$

and

$$-\rho_c C_c \frac{\partial T}{\partial t} + K_c \frac{\partial^2 T}{\partial y^2} = 0 \qquad (4)$$

Here ρ, C and K represent, respectively, the bulk density,
specific heat capacity, and conductivity, with subscripts c and
m designating the core and the mantle, respectively; K_m has the
form $A + BT^3$ with the first term representing the normal contact
conduction coefficient, while the second term represents a
radiative conduction term (e.g., see 3). Also, $\dot{z}(T_1)$ represents
the flux of gas ($cm^{-2}s^{-1}$) from the core surface.

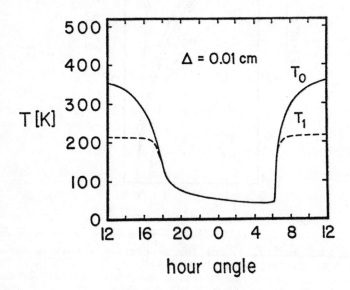

Fig. 6   Diurnal variations of the equatorial/surface
temperature and core-mantle interface temperature for a rotating
core-mantle nucleus at 1 AU, with a mantle of thickness 0.01 cm
(from (6)).

The variations of the mantle surface temperature (T_0) and
the core surface temperature (T_i) at the equator, with hour
angle, for two different values of Δ, are shown in Figure 6 and
7. It is seen that when the mantle thickness is small
(Δ = 0.01 cm), the diurnal variation of the core surface
temperature tends to follow that of the mantle surface.

However, when the mantle thickness is substantial
(Δ = 1 cm), the temperature of the core temperature is low and
shows no diurnal variation. This is clearly due to the large
heat capacity of the thick mantle. Notice also that, in this
case, while T_0 > T_i on the day side, the opposite is true on the
night side. So while heat flows into the core during the day,
heat flows out during the night. It is clear that a thick
mantle ensures a more or less isothermal core, and thus a more
spherically symmetric outflow of gas.

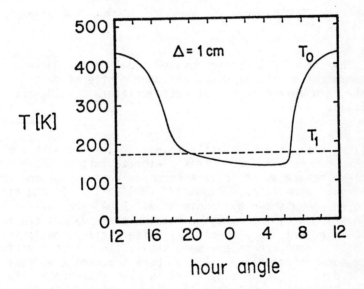

Fig. 7   Same as Fig. 6, but with a mantle thickness 1 cm.
(From (6)).

3. CHEMICAL DIFFERENTIATION OF THE NUCLEUS: DUST AND ICE
MANTLES

The prevailing view that the chemical composition of the
volatile component of the cometary nucleus is a clathrate
hydrate, where all the other volatile species are entrapped in
the H_2O-ice lattice (9) explains a number of cometary phenomena

(10). However, since the clathrate hydrate can hold a maximum
of one guest molecule for every six H_2O molecules, the
production rate of any guest molecule $\dot{Q}$ (Guest) $< 0.17 \dot{Q}$
(H_2O).

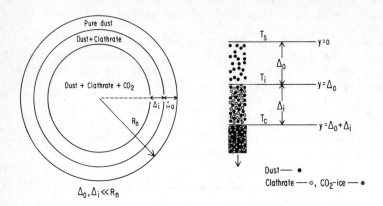

Fig. 8 The chemically differentiated model of the nucleus
having an outer mantle of dust, an inner mantle of dust +
clathrate, and an inner core of dust + clathrate + CO_2-ice
(From (11)).

        Mixing more of the volatile (e.g., CO_2 or CO) with the
water than can be trapped in this clathrate lattice would make
the nucleus behave as if it were composed entirely of the more
volatile substance (i.e., $\dot{Q}$ (Guest)/$\dot{Q}$ (H_2) $\gg$ 1). On the other
hand, it has been shown by Houpis et al[2] (1984) that values
of $\dot{Q}$ (Guest)/$\dot{Q}(H_2O)$ intermediate between these values can be
understood in terms of a chemically differentiated nucleus (see
Figure 8). Indeed, there are some observations that seem to
indicate such intermediate values. Here too, while we have an
evacuated dust mantle on top, there is also a second mantle of
dust plus clathrate ice, where the excess volatile ice (CO_2 in
this case) that could not be accommodated in the clathrate
lattice has also escaped. In this situation, since the volatile
CO_2 is partially thermally insulated by the clathrate mantle, it
will sublimate at a lower temperature ($T_c < T_i$) than this H_2O-
ice in the second mantle. Consequently while $\dot{Q}(CO_2)/\dot{Q}(H_2O)$ can
be $\geqslant$ 0.17 it need not be $\gg$ 1.
        The variations of Δ_o and Δ_i with heliocentric distance for
such a comet (assumed to have equal mass fractions of dust,
clathrate and CO_2) are shown in Figure 9, while the heliocentric
variation of $\dot{Q}$ $(H_2O)/\dot{Q}$ (CO_2) is shown in Figure 10.
Interestingly it is seen that this ratio becomes larger than 1,
inbound only when d $\leqslant$ 1.5 AU . It then reaches a maximum just

before perihelion and decreases again, becoming less than 1
outbound around 0.75 AU. This is essentially due to the fact
that T_i drops faster than the surface temperature T_c of the
highly insulated core.

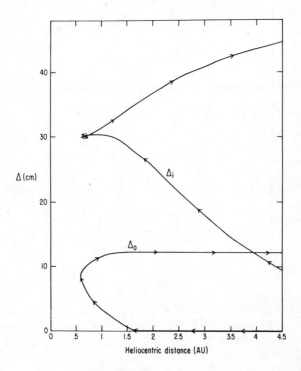

Fig. 9 Heliocentric variation of the thickness of the outer
dust-mantle (Δ_o) and the inner dust + clathrate mantle Δ_i, in
the chemically differentiated nucleus (From (11)).

These results are merely meant to display the plausibility
of such a chemically differentiated model in explaining values
of $\dot{Q}$ (volatile) / $\dot{Q}$ (H_2O) $\geqslant 0.17$ but not $\geqslant 1$.   The present
quantitative results are not to be taken too seriously. In
fact, in this present model, we have, for simplicity, assumed a
constant (0.8) blow-off rate of the dust mantle. This is why it
does not exhibit catastrophic blow-off near perihelion. Houpis
et al (11) are presently refining this model by assuming that
the dust erosion rate is proportional to the momentum flow of
gas, as was assumed by Horanyi et al (6). Such a model would
give qualitatively different results.

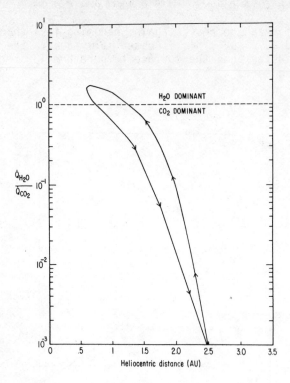

Fig. 10 Heliocentric variation of the production rate ratio
$\dot{Q}_{H_2O}/\dot{Q}_{CO_2}$ in the chemically differentiated dusty nucleus.
(From (11))

#4. DUST HALOS

Several authors (12,13,14) have considered the effects of
dust halos on the development, dynamics and thermodynamics of
cometary atmospheres.

Recently Marconi and Mendis (14) developed a self-
consistent multi-fluid model of the dynamical and thermal
structure of a dusty gas atmosphere of a comet. The diffuse
radiation fields due to multiple scattering of the solar
radiation, as well as thermal re-radiation by the circum-nuclear
dust halo, were also included by introducing a three-stream
approximation based on the shadowing geometry of the nucleus. A
similar investigation was earlier carried out by Hellmich (12)
who used, however, the simplifying assumption that the expansion
of the gas dust mixture was adiabatic.

The radial profiles of the normalized mean intensities of the direct and the diffuse radiation fields, calculated by Marconi and Mendis (14) for Comet Halley at 0.89 AU, assuming a dust-to-gas ratio of 0.5, and a single dust size (r_g = 0.5μm), along the sun comet axis is shown in Figure 11. It[9] is seen that while the mean intensity of the direct solar radiation decreases monotonically towards the nucleus, the intensity of both diffusive radiation fields (the first due to multiple scattering and the second due to thermal re-radiation from the dust) increases from small values at r = 100 km, to maxima close to the nucleus (r ≈ 4-6 km) before decreasing towards the nuclear surface (r = 3 km). It is also seen that the mean contribution to the intensity close to the nucleus comes from the thermal re-radiation by the dust. Also, the contribution from the diffuse radiation field due to multiple scattering is larger than that due to direct solar radiation when r < 10 km. It is further seen that the total intensity of radiation inwards of r ≈ 200 km is larger than that near the top of the atmosphere.

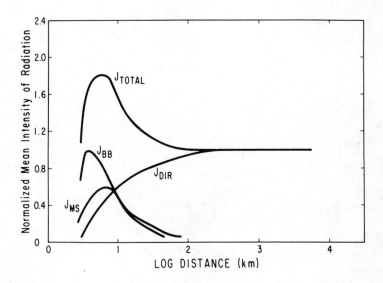

Fig. 11 The radial profiles of the mean intensity of radiation along the comet-sun axis. The subscripts DIR, MS and BB refer to the direct solar radiation, the multiply scattered solar radiation, and the black body thermal re-radiation by the dust for Comet Halley at 0.89 AU (From (14)).

   The reason for the maxima in the diffuse radiation fields
very close to the nucleus is also clear. This is due to the
fact that the shadow cone of the nucleus intercepts an
increasingly larger fraction of the grains on its night side as
we approach the nucleus. Qualitatively similar results have
been obtained by Weissman and Kieffer (13; see also 12),
although no quantitative comparison can be made due to different
assumptions regarding the properties of the grains and the
nucleus.

   Since the thermal modelling of a rotating cometary nucleus
with and without dust halos is dealt with at length elsewhere
(15), I will not dwell on that subject here.   I will merely
show the isotherms on the surface of the comet obtained by
Weissman and Kieffer (13) without (Figure 12) and with (Figure
13) the appropriate dust halo at 1 AU. As I mentioned earlier,
it is clear that the nucleus is much closer to being isothermal
in the latter case. In fact, the minimum temperature in this
case is only 191.6° K.

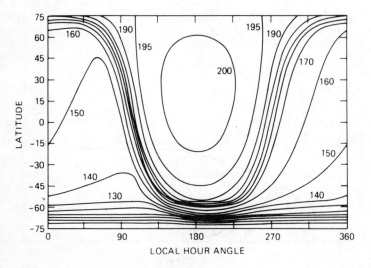

Fig. 12 The isotherms on the surface of Comet Halley at 1 AU,
with no dust halo. Temperature contours are every 10°K below
170°K and every 5°K above (From (13)).

5. CONCLUSIONS

   As we have pointed out, existing studies show that while
dust halos tend to increase the nuclear outgassing, dust mantles
do the opposite. Both have implications for the heliocentric

brightness curves. While dust halos tend to increase the
steepness of the brightness curve near perihelion
(i.e., $B \propto d^{-n}$, $n > 2$), dust mantles also produce a hysteresis
in the brightness curve.

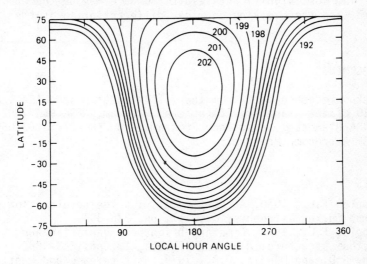

Fig. 13 The same as Fig. 12, but with the dust halo.
Temperature contours are every 1°K. The minimum temperature is
191.6°K (From (13)).

What is lacking is a study of the simultaneous presence of
dust halos and mantles on nuclear outgassing, since these are
expected to co-exist. Also, the claim that both dust mantles
and dust halos produce more uniform nuclear outgassing has to be
treated with caution because this is based on the assumption
that both dust mantles and dust halos are spherically symmetric
in the first place.

The small kilometer-sized cometary nucleus is unlikely to
be even close to being spherical. Nor are its surface
properties likely to be uniform. Mantle formation due to
nuclear outgassing from such a patchy surface could produce both
more uniform outgassing and less uniform outgassing, depending
on various conditions, particularly the variation of the
volatile-to-dust ratio. Such a non-uniform surface could give
rise to sunward-directed fan-shaped dust jets, as are often
observed, rather than spherically symmetric dust halos. These
jets, by focussing more (diffuse) radiation on to the regions
from which they originate, could produce even less uniform
outgassing, thereby reinforcing the jet effect.

Therefore, the present models, while correctly emphasizing the physical processes that are assoicated with both dust mantles and dust halos, cannot be taken as reliable quantitative guides to what may take place around a "real" cometary nucleus. The most significant results to date may be the ones associated with the evolving chemical differentiation of the nucleus.

ACKNOWLEDGEMENTS

I acknowledge support from the following grants: NASA NGR-05-009-110 of the Planetary Atmospheres Program, NASA NSG 7102 of the NASA Planetary Geoscience Program, and NSF AST 83 03105 of the NSF Astronomy Program.

REFERENCES

1. Whipple, F.L., 1950, A comet model. I. The acceleration of Comet Encke, Astrophys. J., 111, pp. 375-394.
2. Whipple, F.L., 1978, Cometary brightness variation and nucleus structure, Moon and Planets, 18, pp. 343-359.
3. Brin, G.D. and Mendis, D.A., 1979, Dust release and mantle development in comets, Astrophys. J., 229, pp. 402-408.
4. Jacchia, L.G., 1974, The brightness of comets, Sky and Tel., 47, pp. 216-220.
5. Bortle, J. E., Brooks, W. R., and Morris, C. S., 1984, Brighter prospects for Halley's comet, Sky and Telescope 67, pp. 9-12.
6. Horanyi, M., Gombosi, T.I., Cravens T.E., Korosmezey A., Kecskemety K, Nagy, A.F., and Szego K, 1984, The friable sponge model of a cometary nucleus, Astrophys. J., 278, pp. 449-455.
7. Probstein, R.F., 1968, The dusty gas dynamics of comet heads, in "Problems of Hydrodynamics and Continuum Mechanics" (Ed. M.A. Laurentiev), SIAM (Philadelphia), p. 568-583.
8. Dobrovol'skj, O.V., and Markovich, M.Z., 1972, On nongravitational effects in two classes of models for cometary nuclei, in "The Motion, Evolution of Orbits and Origin of Comets" (Eds. G.A. Chebotarev, E.I. Kazimirchak-Polonskaya and B.G. Marsden), D. Reidel Pub. Co., pp. 287-293.
9. Delsemme, A.H. and Swings, P., 1952, Hydrates de gaz dans les noyaux cometaires et les grains interstellaires, Ann. Astrophys., 15, pp. 1-6.
10. Delsemme, A.H., 1973, Gas and dust in comets, Space Science De., 15, pp. 89-101.

11. Houpis, H.L.F., Ip, W-H and Mendis, D.A., 1984, The
     chemical differentiation of the cometary nucleus: The
     process and its consequences, (in preparation).
12. Hellmich, R., 1981, The influence of radiation transfer in
     cometary dust halos on the production ratio of gas and
     dust, Astron. Astrophys., 93, pp. 341-346.
13. Weissman, P.R. and Kieffer, H.H., 1981, Thermal modelling
     of the cometary nucleus, Icarus, 47, pp. 302-311.
14. Marconi, M.L. and Mendis, D.A., 1984, The effects of the
     diffuse radiation fields due to multiple scattering and
     thermal re-radiation by dust on the dynamics and
     thermodynamics of a dusty cometary atmosphere Astrophys.
     J. (submitted).
15. Weissman, P. R. and Kieffer, H. H., 1985, Thermal modelling
     of cometary nuclei (this volume).

DISCUSSION

Klinger:

Yeomans states that the non-gravitational forces have
remained constant during several thousands of years. Could you
explain such a behavior with your model without making an
assumption on the intrinsic dustiness? (The phase transition
model is able to explain the constant activity without
considerations other than orbital parameters).

Mendis:

Yeomans' statement is compatible with those comets that
both build and purge mantles during each orbit, as is the case
for those comets that are not too dusty intrinsically. It would
not be the case for those that are intrinisically very dusty
because they would develop continuously deepening dust mantles,
which would eventually choke off the sublimation. Therefore,
yes, I agree with you; one has to make an assumption about the
intrinsic dustiness of the nucleus in order to explain the long-
term constancy of the non-gravitational force, as is the case
for Comet P/Halley. On the other hand, there are other comets
(e.g., P/Encke) whose non-gravitational forces appear to have
decreased substantially over periods less than a thousand years.

Fanale:

Your qualitative description of mantle cycling seems
intuitively reasonable. However, I have trouble accepting the
absolute values. In particular, your model predicted that while
reaching from 2 to 5 AU, Halley's comet sublimes many

centimeters of ice through many centimeters of very fine soil
cover. Calculations recognizing the inhibition of the
sublimation rate by a static (no longer fluidized) soil cover
through which the water must pass by Knudsen diffusion (since
the mean free path is greater than the pores) suggest rates
orders of magnitude lower. The mantle should be considered as
both a thermal barrier and as a barrier to the sublimation of
underlying ice.

Mendis:

Estimates made during our earlier calculations seemed to
indicate that the main effect of the dust mantle was the partial
thermal insulation of the icy core rather than providing a
physical barrier to the gas flux. Of course, this effect
clearly depends on the assumed porosity of the mantle, and more
quantitative work needs to be done. It is indeed possible that
the mantle thicknesses calculated by Brin and myself were
overestimates, although similar thicknesses have been obtained
more recently by the "friable-sponge" model of Horanyi et al.
In any case, I think the main purpose that all these models
serves at this time is to draw attention to some plausible
physical processes rather than provide reliable quantitative
results. Besides the point you make, there are several others
that need to be taken into account; in particular, the non-
uniformity of the nuclear surface and the simultaneous presence
of dust halos and dust mantles.

WHAT WE DO NOT KNOW ABOUT COMETARY ICES: A REVIEW OF THE INCOMPLETE EVIDENCE

A. H. Delsemme

The University of Toledo, Toledo, OH   43606   USA

Only two features seem rather well established about the cometary nucleus: it is an icy conglomerate that contains a large amount of volatile ices, and its major constituent seems to be water ice. The evidence on all the minor constituents is incomplete and fragmentary and there are still major difficulties in establishing a complete list of the parent molecules from the numerous radicals, ions and atoms observed in the coma. In particular, the case of the forbidden lines of oxygen is discussed here in detail; new arguments are shown to contradict the need for CO_2 as a major source of $O(^1D)$. The physical nature of the nucleus is likely to be very complex, geometrically as well as kinetically; some progress may have been recently achieved on the crystal structure of the ices by circumstantial consideration, but their crystallography as well as their mineralogy will remain in doubt until the first cometary rendez-vous by a spacecraft.

I. THE EXISTENCE OF ICES IN COMETS

The existence of volatile ices in comets is reasonably well established although it has been reconstructed from partial and circumstantial evidence only.

The best arguments for the presence of an icy conglomerate in the cometary nucleus are probably still those developed by Whipple (1950, 1951) and brought in sharp focus somewhat later (Whipple 1963).

The major argument is based on the interpretation of the non-gravitational forces (NGF) as resulting from the jet effect of

J. Klinger et al. (eds.), Ices in the Solar System, 505–517.
© *1985 by D. Reidel Publishing Company.*

the asymmetrical sublimation of the nucleus. This interpretation
has been given a more quantitative approach (Delsemme 1972,
Marsden et al. 1973) by using the steepness variation of the
dependence law of the NGF on distance.

Whipple's second argument was based on the large gaseous production rates needed to explain the NGF, combined with the large
number of observed passages to perihelion before a serious
dimming of the cometary brightness; this implied that the volatile
fraction was at least of the same order of magnitude, and even
possibly larger, than the dusty (refractory) fraction. A reasonable measure of the gas-to-dust ratio in the outer layers of the
nucleus has been obtained since in two comets, by Finson and
Probstein's (1968a) technique based on the fluid dynamics of
the dust dragged away. The technique was used by Finson and
Probstein (1968b) and by Sekanina and Miller (1973) for comets
Arend-Roland and Bennett respectively. Reduced to uniform modern
values of albedos, of molecular masses, etc., the two dust-to-gas
mass ratios are 0.8 ± 0.2 and 0.6 ± 0.4 respectively (Delsemme
1982).

Six days before passage to perihelion, comet Arend-Roland had an
outburst during which the dust-to-gas mass ratio went up for 3
days to values larger than 4. Delsemme (1982) interprets this
occurence as the blowoff of a mantle of outgassed dust that had
accumulated during the previous weeks, according to the model
developed by Mendis and Brin (1977). He concludes that the
asymptotic value of 0.8 for the dust-to-gas mass ratio in the
vaporizing gases is related to pristine material and probably
reflects fairly well the mean composition of the outer layers.
In the same paper, Delsemme (1982) also reviewed arguments that
are all in favor of a radial undifferentiation of the nucleus.
For this reason, 0.8 seems to be the best assessment available
today for the refractory-to-ice ratio of the icy conglomerate
present in the nucleus: a value still very close to Whipple's
predictions.

Whipple's third argument was the non-negligible nature of the
cohesive forces existing in the cometary nucleus, for some comets
to survive as an entity after perihelion, in spite of the tidal
forces developed during sun grazing passages. Although still
very strong, this argument has not been developed in a more
quantitative way because tidal forces are critically dependent
on the nuclear density, and we have never been able to measure
with any accuracy, neither the density nor the mass of any
cometary nucleus. In particular, we still have no idea on the
degree of compaction of the ices (or snows?) present in the icy
conglomerate. With a gravity field in the range of 10^{-3} to 10^{-4}
g, any density is possible, including very fluffy aggregates of
snow down to the nuclear core, since in the core the weight of

the outer layers could still be less than a few hundredths of one atmosphere.

With these rather uncertain data on the elusive nucleus, it is good to know that the reality of its existence and even its approximate size have been confirmed by a direct detection of its radar echo (Kamoun et al. 1982). The size of comet Encke's nucleus found by this (yet marginal) detection is entirely compatible with that deduced by Delsemme and Rud (1973), using the theory of vaporization and the observed vaporization rates.

Another source of information on ices is the possible existence of a halo of icy grains surrounding the cometary nucleus. Of course, icy grains in the nuclear region are not a direct proof of the existence of ice in the nucleus: a theory on the origin of the icy halo must be first established.

The first conceptual mention of an icy halo seems to be in Huebner and Weigert's (1966); however, the optical depth of micron-sized grains implied that the proposed halo was too small to be observed. The idea of a halo of large ice grains was derived by Delsemme and Wenger (1970) from their observation that, in laboratory experiments studying the behavior of the solid hydrate of methane, large ice grains (sizes 0.1 to 1.0 mm) were stripped away from the sublimating snows. A model of this type of halo was developed by Delsemme and Miller (1970, 1971) and used to explain the continuum of comet Burnham 1960 II; this comet was almost dust-free. However, its spectrum showed a narrow continuum, disappearing with puzzling large gradients in the two opposite directions. This vanishing continuum could not be explained by permanent dust, but could be easily explained by icy grains vanishing from sight because of their sublimation. The icy halo model fitted the large gradients quite well with a mean halo radius of 25,000 km, implying a distribution tail of very large grains (1 to 3 cm). More recently, Hobbs et al. (1975, 1977) interpreted the microwave radiation of comets Kohoutek and West as produced by an ice halo. The tails of several comets at large heliocentric distances also suggest that these comets eject icy grains (Sekanina 1975). However, A'Hearn et al. (1981) were surprised not to detect the infrared absorption bands of ice in the spectrum of comet Stefan Oterma. Hanner (1981) showed that it is unlikely that the 1.5 or 2.0 μm ice band could ever be detected. However, the 3 μm ice band could possibly be detected at large heliocentric distances. She also showed that the ice halo is probably limited to smaller sizes than previously believed: the minimum rate of sublimation is reached for particles near 20 μm, therefore they survive longer. As predicted by Hanner, the 3 μm absorption band, presumably due to water, was finally found in comet Bowell at 3.4 AU (Campins et al. 1983). However, it is unclear whether this water band belongs to a water of hydration (for instance in a silicate) or comes from water ice; the same band was later observed by Hanner in comet Cernis (unpublished results).

As a conclusion, the existence of water ice in a halo of icy grains is reasonably probable; however, infrared identifications are marginal so far and cannot be used as a proof that water either is a major constituent, or controls the sublimation of the icy conglomerate.

II. THE PREVALENCE OF WATER ICE

In another chapter, (Delsemme 1984) I have reviewed the dependence law of the NGF on distance; it shows a steepness variation compatible with water ice only, for all those short-period comets with enough observational data. I have also shown that the brightness curves of "new" comets seem to be compatible with water ice only. All available clues suggest therefore that water ice is the major volatile constituent, and that it controls the sublimation of the icy conglomerate of most comets, pristine or not, including some whose behavior had been suspected as meaning the presence of a stuff more volatile than water; an example is comet Morehouse, which in spite of its anomalously high production of CO^+, is found to be controlled by water ice; another example is comet Kohoutek, whose infamous light curve seems to be an extreme case still controlled by water, and not by CO_2 as I had suggested earlier (Delsemme 1975).

All other data invoked to justify the existence of water in comets usually stop short at the existence of water vapor as a parent molecule; they do not try to prove that this water vapor comes indeed from the sublimation of water ice in the nucleus. The only proper evidence for the prevalence of water ice in comets is that which has been reviewed hereabove.

However, it is useful to mention that the production rates of H and OH in the coma are completely consistent with the vaporization rate of a few square kilometers of water ice near 200°K in the nucleus. The production rates of H, OH and other species have been observed in the coma of the bright comets of the 1970's, Keller and Lillie (1974), Feldman et al. (1974), Feldman and Brune (1976), Opal and Carruthers (1977) and Smith et al. (1978); these authors have confirmed in general that OH and H are indeed the most abundant molecular fragments observed in cometary atmospheres.

Further evidence, including the resonance line of oxygen, was also obtained from the set of comets observed later in the vacuum ultraviolet by the International Ultraviolet Observer (Feldman 1982).

III. THE CASE OF THE FORBIDDEN LINES OF OXYGEN

The forbidden lines of oxygen may have a great significance in
cometary chemistry, at least if their formation mechanism is
understood in detail, because their mere intensity implies that
they do not come from a fluorescence mechanism. Twenty years
ago, Bierman and Trefftz (1964) showed that their origin in
comets was likely to be the photodissociation of a (yet undeter-
mined) molecule into fragments, producing, in particular, an
oxygen atom in an already excited singlet state: ^1D producing
the two red lines at 6300 and 6364 A or $^1S_0$ cascading to ^1D by
producing first the green line at 5577 A. The early literature
on observations of the green and the red lines in comets is
confusing, because of the blends with the same airglow lines in
the terrestrial atmosphere. Twenty-five years ago the "higher
resolution" spectra of comet Mrkos still had a dispersion of
27 A/mm; they did not allow Swings and Greenstein (1958) to
separate airglow lines from cometary lines by Doppler shift;
their 1958 argument was based on the higher intensity of the line
blend near the nucleus; However, Swings (1967) reported later
that he doubted very much of the existence of the green line in
comets because an unsuspected vignetting had been found to reinforce the central part of the spectra taken at Palomar with a
dispersion of 27 A/mm. Since then, the oxygen green line has
never been reported with great confidence in comets, because if
it is present, it is always faint and blended with some structure
of the (1-2) C_2 band. For this reason we will not discuss the
green line any further.

Only recent very high dispersion spectra have been able to
separate the Doppler shifted cometary red lines from the airglow,
for instance, by Fabry Perot spectrometry (Huppler et al. 1975)
in comet Kohoutek. However, Delsemme and Combi (1979) have
published spatial brightness profiles of the brightest red forbidden line in Comet Bennett, showing that near the nucleus they
reach intensities more than 20 times as bright as the airglow
level; this level can therefore be discounted easily from the
profiles, without requesting any spectral resolution. A more
difficult problem is the blend of the 6300 A line with many faint
lines of NH_2. Delsemme and Combi (1983) have measured the spatial
brightness profile of NH_2 at a different wavelength; they deduced
that the NH_2 contribution falls down to values much smaller than
10% of that of the oxygen line in the immediate vicinity of the
nucleus. The intensity of the two red lines of oxygen are therefore real, and they indicate that in comet Bennett, oxygen in
the singlet D state corresponds roughly to 25 to 50% of the
amount of water. However, I had shown (Delsemme 1980) that at
most 12% of H_2O can be dissociated by solar light into $H_2 + O(^1D)$,
(through a predissociated band irradiated by the solar ultraviolet between 1000 and 1250 A, assuming a large contribution of

a strong Lyman alpha). Since Lyman alpha is variable, Festou
(1981) finds 7% only for the quiet sun, in agreement with my
results. I concluded that another source of $O(^1D)$ has to exist
in order to explain comet Bennet's observations. I proposed CO_2
(Delsemme 1980) because $67\% \pm 21\%$ of CO_2 dissociates into
$CO + O(^1D)$; the 21% uncertainty comes from the gaps in the known
absorption cross-section of CO_2, mainly between 900 and 1100 A.
At any rate, not a large amount of CO_2 would be needed to
explain the extra brightness of the two red lines. Since then,
Spinrad (1982) has published observations of the red forbidden
lines for several comets, finding that for six comets $O(^1D)$ is
roughly between 4 and 6% of H_2O, therefore probably explainable
from water only, whereas for comets Meier and Kohoutek $O(^1D)$ is
between 18 and 25% of H_2O putting these comets in the same
class as our observations of Bennett.

However, a new possible channel for the production of $O(^1D)$ has
been very recently identified (Singh, van Dishoek and Dalgarno
1983); it comes from the further dissociation of OH, which is
one of the molecular fragments from water. OH can indeed be
photodissociated by the absorption of solar Lyman alpha into
higher energy doublet states ($^2\Delta$ and $^2\Pi$) yielding repulsive
potential curves to $H + O(^1D)$. The intensity of solar Lyman
alpha is of course variable, unfortunately it is not monitored
daily, but it does not have to reach extremely high values to
explain all or almost all of the anomalous intensities of the
forbidden red lines in comets Meier, Kohoutek and Bennett. We
have combined here the two channels yielding $O(^1D)$; the rates
in the parentheses are due to the branching ratios, the ranges
in the rates are not error bars; they come mainly because of
the known variation of Lyman alpha.

First step: 100% $\begin{cases} H_2O \rightarrow OH + H & (88\text{-}93\%) \\ H_2O \rightarrow H_2 + O(^1D) & (\ 7\text{-}12\%)\ (1) \end{cases}$

Second step: about 90% OH $\begin{cases} OH \rightarrow H + O(^3P) & (70\text{-}85\%) \\ OH \rightarrow H + O(^1D) & (\ 5\text{-}20\%)\ (2) \end{cases}$

Total $O(^1D)$: (from (1) and (2)): 12-32% of H_2O

The total production rate of $O(^1D)$ may become even more variable
when the change in radial velocity of the comet is taken into
account for the solar excitation.

When Singh, van Dishoek and Dalgarno's(1983) data on OH dissociation are combined with that of H_2O and with the observations of the red forbidden lines in comets, the puzzle of the
parent molecule of $O(^1D)$, submitted by Delsemme and Combi (1979)
and emphasized by Spinrad (1982) seems almost to vanish. However, the lifetime of OH against photodissociation is of the
order of 10^5 sec, therefore the OH contribution to $O(^1D)$ is produced in an extended source of the order of 10^5 km (Delsemme

1980) whereas the strong intensity peak of $O(^1D)$ was observed
in comet Kohoutek in a zone smaller than 10^3 km (Delsemme and
Combi 1983). The disagreement with OH as the only parent of
$O(^1D)$ remains therefore very strong for comets Kohoutek, and
probably as strong for comets Bennett and Meier. However, because of the newly identified channel to produce $O(^1D)$ from OH,
it is possible that an amount of CO_2 as small as 15% of the
amount of water could explain the extra intensity of the oxygen
red line. The smaller intensity of the green forbidden line
has been measured quantitatively for the first time by Cochran
(1984) in comet Iras Araki Alcock (1983d). He concludes from
the surprizingly small ratio $O(^1S)/O(^1D) \leq 0.03$ that neither CO
nor CO_2 are present and that H_2O is the primary parent in this
particular comet. Because of the cascading effect through ^1D
from ^1S in CO and CO_2, the two lines would be of almost equal
intensities.

The 2-order of magnitude discrepancy mentioned earlier between
photodissociation lifetimes and the observed concentration of
the source of $O(^1D)$ near the nucleus, is as large for CO_2 as for
OH, and even much larger for CO.

This suggests that photodissociation is not the major mechanism
and that collisional reactions with fast electrons play an
important role, at least in large comets like Bennett, Meier
and Kohoutek.

The collision of the cometary plasma with the solar wind still
is rather poorly understood, and it is still uncertain whether
fast electrons could penetrate down into the collision zone.
However, all observed cometary ions are violently ejected
towards the tail; the behavior of a plasma implies that the
electrons would be strongly coupled to such a motion and, therefore, the forbidden line of oxygen would show a strong dissymetry.
It does not. The problem of the production of $O(^1D)$ in an extremely small zone near the nucleus has therefore not be solved.

However, present clues suggest that CO_2 does not have to be more
abundant than 10-15% of water, even in comets Bennett, Meier
and Kohoutek: an important consequence for the chemistry of the
nuclear snows, since CO_2 could still be imprisonned in clathrate
hydrates, as other minor constituents.

IV. THE MINOR CONSTITUENTS IN THE VOLATILE ICES

As far as the chemical nature of the volatile ices is concerned,
we are left with hard evidence for one single major constituent
only, namely water; the minor constituents must be deduced from
a rather large number of molecular fragments which are observed

in the coma, (Table I) with abundances in the range of less than

TABLE I

Observed in Cometary Spectra (until January 1984)

Organic	Inorganic	Metals	Ions
C	H	Na	$C+$
C_2	NH	K	$CO+$
C_3	NH_2	Ca	CO_2+
CH	NH_3	V	$CN+$
CN	O	Cr	$CH+$
CO	OH	Mn	H_2O+
HCO	H_2O	Fe	$OH+$
H_2CO	S	Co	$Ca+$
CS	S_2	Ni	N_2+
HCN	*silicates*	Cu	
CH_3CN			

1% of water. For instance, A'Hearn (1982) has reviewed the molecular abuandances in terms of the abundance of OH (that can probably be assumed to be that of water). He gives:

C_2 0.4%, CH 0.13%, CH $\geqslant$ 0.05%, C_3 0.015% and $NH_2 \geqslant$ 0.1%

I believe that it is still too early to make a significant chemical model for any of the individual comets studied so far. However, since no spectacular difference has emerged among comets, I have built (Delsemme 1982) a heuristic model giving the elemental ratios for the average of the recent bright comets. Normalizing to the oxygen abundance defined at unity, I find:

H = 1.5,   C = 0.2,   N = 0.1,   O = 1.0   and   S = 0.003

This, which includes dust as well as gas, shows that comets are really very depleted in hydrogen; not only in respect to cosmic abundances (depletion factor of 1000) but also in respect to oxygen: there is barely enough hydrogen to make enough water; therefore hydrocarbons (like methane) or any hydrogenated compound (ammonia) must also be rather depleted in order to explain the bulk of water. It is true that the ratio H/O is not known with a good accuracy; it is also true that it was found closer to 2 or 2.5 than to 1.5 in some comets and that present models are imperfect; but even so, the redox ratio implies that carbon is rather in CO, CO_2 or H_2CO than in methane or other hydrocarbons; and that unsaturated hydrocarbons are more likely than paraffins. The model predicts the oxygen abundance close to cosmic abundances, whereas the carbon is depleted by a factor of 4, at least for the fraction present in compounds volatile enough to be

dragged away by the sublimation of water. In order to reach
cosmic abundances, carbon should be in rather non-volatile compounds and should therefore represent 30% of the dust in mass.
This is the gist of the interpretation proposed by Mayo
Greenberg's model.

Newburn and Spinrad (1984) have recently reported the dependence
on heliocentric distance of different ratios in the production
rates, namely CN/[OI], C_2/[OI], C_3/[OI], C_2/CN. They emphasize
that (assumedly because of the large scattering of the data) the
results indicate trends only. They offer guidance to studies of
detailed mechanisms, but in no way do they replace such studies.
The lines through their data points indicate only one subjective
interpretation of the general trends, and no physical interpretation is offered. For this reason, we assume that they concur
with our general conclusion that it is too early to make significant chemical models.

V. THE RECONSTRUCTION OF THE PARENT MOLECULES

In spite of so many and so beautiful cometary spectra, how is it
possible that we know so little on the other molecules present
in the cometary nucleus? The reason is that we observe only
those molecular fragments that can be excited by a complex fluorescence mechanism due to the solar radiation. Most of the parent
molecules cannot be observed, at least at visual wavelengths, because they cannot fluoresce in the solar light. The fluorescence
mechamisms are all slightly different and they must be understood
each time in detail, often including the solar flux effects at
many different wavelengths, in the infrared for the populating
of the vibration and rotation levels, in the ultraviolet for some
electronic transitions that often become important just because
of a wavelength coincidence with one of the bright emission lines
of the solar corona. The production rates derived from light
intensities also depend on the uncertain lifetimes of the observed molecular fragments against photodissociation and/or
photoionization; whereas their scale lengths in the observed
coma depend on their uncertain (non-radial) velocities due to
the energy balance sheet of their production mechanisms. Finally,
these fragments come from parent molecules that may have been
reshuffled by ion-molecular reactions.

The collisional models of the coma chemistry: Huebner et al.
(1982), Mitchell et al. (1981), Cochran (1982), must all be taken
with a grain of salt for different reasons. First, half of the
rate constants of the hundreds of charge-exchange and photolytic
reactions that are needed to make a realistic model are not known
with good enough an accuracy; since one path may dominate the
chemistry at one place, another one at another place, one single

rate constant that has not been properly assessed may ruin a whole model. Second, we cannot compute backwards the flow of entropy, therefore we cannot start from the coma and compute backwards where it comes from. Models start from an assumed nuclear composition and follow its evolution; therefore they can demonstrate the consistency, but never the necessity of the assumed composition. Finally, the smallest and the brightest comets show a surprising homogeneity in their spectra. No clear evidence has emerged of any radical, any ion, any molecule that appears in large comets only, in spite of the fact that the smallest comets have practically no collision zone. This fact speaks strongly in favor of the assumption that ion-molecular reactions do not play any significant role in comets. This fact may indeed come from the origin of the cometary molecules: if they were formed in interstellar space, the ion-molecular re-actions whose kinetics is fast, may have occurred already in interstellar space some five billion years ago.

VI. NUCLEAR COMPLEXITY, MINERALOGY AND CRYSTALLOGRAPHY

The physical nature of the nucleus is difficult to study because it remains in the best cases an elusive point of light not clearly separated from the nuclear region of the cometary head. Its spectrum shows a bland continuum without features; whereas the dust in the coma sometimes show the emission signature of silicates in the infrared.

Circumstantial evidence suggests that the nucleus is a complex object, probably irregular in shape, spinning around its axis, with regions of different nature and different albedos because of observed differences in activity (jet structure and halos in inner coma). The axis is itself precessing and secularly changes the sunshine exposition of different zones at different points of the orbit. Zones can become more or less depleted in volatiles, yielding outgassed grain mantles that vary in position. The nucleus may be surrounded by an unsteady halo of dusty and icy grains, that changes its effective cross section for solar light and, through multiple scattering, its total absorption of solar light.

The mineralogy and crystallography of the ices is also totally derived from circumstantial considerations. Some progress seems to have been achieved recently (Smoluchovski 1977, 1984, Klinger 1983) in respect to the existence and the behavior of amorphous versus crystalline water ice in comets. The models are essen-tially based on the assumption, not only that water is the major constituent in comets, but also that this water is indeed in the form of ice. We have just reviewed the facts that support this contention; however, it is clear that we must wait for the first comet rendez-vous with a spacecraft to be able to know much more about the cometary nucleus.

NSF grant AST 82-07435 and NASA grant NSG-7381 are gratefully acknowledged.

REFERENCES

A'Hearn, M. F. (1982), Spectrophotometry of Comets at Optical Wavelengths, p. 433 in "Comets", ed. L. L. Wilkening, Univ. of Arizona Press, Tucson.

Bertaux, J. L., Blamont, J. E., Festou, M. (1973), Interpretation of Lyman Alpha Observations of Comets Bennett and Encke, Astron. Astrophys. 25, 415.

Biermann, L. and Trefftz, E. (1964), Ueber die Mechanismen der Ionisation und der Anregung in Kometenatmospheren., Z. Astrophys. 59, 1.

Campins, H., Rieke, G. H., Lebovsky, M. J. (1983) Water Ice IR Band Detected in Comet Bowell, Nature, 301, 405.

Cochran, A. L. (1982), The Chemical Evolution of the Coma of Comet P/Stefan-Oterma, Univ. of Texas, Publ. in Astronomy No. 21.

Cochran, W. D. (1984) Detection of [OI]^{1}S-^{1}D in Comet IRAS-Araki-Alcock; Icarus 58, 440.

Delsemme, A.H., 1972, Vaporization theory and non-gravitational forces in comets, in "On the Origin of the Solar System", ed. H. Reeves, Publ. C.N.R.S. Paris, pp. 305-310.

Delsemme, A. H. (1975), Physical Interpretation of the Brightness Law of Comet Kohoutek, p. 195 in "Comet Kohoutek" ed. G. A. Gary, NASA-SP 355, Washington.

Delsemme, A. H. (1980), Photodissociation of CO_2 into $CO+O(^1D)$; p. 515 in "Spectres des Molécules Simples", 21st Colloq. Intern. Astrophys. Liège (Univ. Liège).

Delsemme, A. H. (1982), Chemical Composition of Cometary Nuclei, p. 85 in "Comets", ed. L. L. Wilkening, U. of Ariz. Press.

Delsemme, A. H. (1985), The Surface Temperature of the Cometary Nucleus; (this book).

Delsemme, A. H. and Combi, M. R. (1979), $O(^1D)$ and H_2O+ in Comet Bennett, Astrophys. J. 228, 330.

Delsemme, A. H. and Combi, M. R. (1983), Neutral Cometary Atmospheres IV; Astrophys. J. 271, 388.

Delsemme, A. H., and Miller, D. C. (1970, 1971), Physico Chemical Phenomena in Comets II, III, IV; Planet. Space Sci., 18, 717; 19, 1229 and 1259.

Delsemme, A. H. and Rud, D. A. (1973), Albedos and Cross-Sections for the Nuclei of Comets 1969 XI, 1970 II and 1971 I., Astron. Astrophys. 28, 1.

Delsemme, A. H. and Wenger, A. (1970), Physico-Chemical Phenomena in Comets I., Planet. Space Sci. 18, 709.

Feldman, P. D. (1982), Ultraviolet Spectroscopy of Comae, p. 461

in "Comets" ed. L. Wilkening, Univ. of Ariz. Press, Tucson.

Feldman, P. D. and Brune, W. H. (1976), Carbon Production in
Comet West, Astrophys. J. 209, L 145.

Feldman, P. D., Tanacs, P. Z., Fastie, W. G., Donn, B. (1974)
Rocket U.V. Spectrophotometry of Comet Kohoutek, Science
185, 705.

Finson, M. L. and Probstein, R. F. (1968a), A Theory of Dust
Comets I. Models and Equations., Astrophys. J., 154, 327.

Finson, M. L. and Probstein, R. F. (1968b), A Theory of Dust
Comets II. Results for Comet Arend-Roland, Astrophys. J.
154, 353.

Hanner, M. S., (1981) On the Detectability of Icy Grains in
Comets; Icarus, 47, 342.

Hobbs, R. W., Maran, S. P., Brandt, J. C., Webster, W. J.,
Krishna Swami, K. S., (1975) Microwave Continuum Radiation
from Comet Kohoutek: Emission from an Icy Grain Halo?,
Astrophys. J. 201. 749.

Hobbs, R. W., Brandt, J. D., Maran, S. P. (1977) Microwave
Continuum from Comet West; Astrophys. J. 218, 573.

Huebner, W. F., Giguere, P. T., Slattery, W. L. (1982) Photo-
Chemical Processes in Inner Coma, p. 496 in "Comets", ed.
L. L. Wilkening, Univ. of Arizona Press, Tucson.

Huebner, W. F., and Weigert, A. (1966) Eiskoerner in der Koma
von Kometen; Zs. Astrophys. 64, 185.

Huppler, D., Reynolds, R. J., Roesler, F. L., Scherb, F.,
Trauger, J. (1975), Observations of Comet Kohoutek with a
Fabry-Perot Spectrometer, Astrophys. J., 202, 276.

Kamoun, P. G., Pettengill, G. H., Shapiro, I. I. (1982) Radar
Dectability of Comets, p. 288 in "Comets", ed. L. L. Wilkening,
Univ. of Arizona Press, Tucson.

Klinger, J. (1983), Classification of Cometary Orbits Based on
the Concept of Orbital Temperature, Icarus, 55, 169.

Marsden, B. G., Sekanina, Z., and Yeomans, D. K. (1973), Comets
and Non-Gravitational Forces V., Astron. J. 78. 211.

Mendis, D. A. and Brin, G. D. (1977), Monochromatic Brightness
Variations of Comets II. Core-Mantle Model, Moon, 17, 359.

Mitchell, G. F., Prasad, S. S., Huntress, W. T. (1981),
Chemical Model Calculations of C_2, C_3, CH, CN, OH and NH_2
Abundances in Cometary Comae; Astrophys. J. **244**, 1087.

Newburn, R. L., Spinrad, H. (1984), Spectroscopy of Seventeen
Comets; Astron. J. 89, 280.

Opal, C. B., Carruthers, G. R. (1977), Lyman Alpha Observations
of Comet West, Icarus, 31, 503.

Sekanina, Z. (1975), A Study of the Icy Tails of Distant Comets;
Icarus 25, 218.

Sekanina, Z. and Miller, F. D. (1973), Comet Bennett 1970 II;
Science 179, 565.

Singh, P. D., van Dieshoeck, E. F. and Dalgarno, A. (1983), The
Dissociation of OH and OD in Comets by Solar Radiation, Icarus,
56, 184.

Smith, A. M., Stecher, T. P., Casswell, L. (1980) Production of
  Carbon, Sulfur and CS in Comet West; Astrophys. J. 242, 402.
Smoluchovski, R. (1977), Ice Grains in Space; p. 47 in "Comets,
  Asteroids, Meteorites", ed. A. H. Delsemme, publ. Univ. of
  Toledo Bookstore.
Smoluchovski, B. (1985), Amorphous and Porous Ices in Cometary
  Nuclei; (in this book).
Spinrad, H. (1982), Observations of the Red Auroral Oxygen Line
  in Nine Comets; Publ. Astron. Soc. Pacific, 94, 1008.
Swings, P. (1967), Forbidden Lines of Oxygen in Comets, Oral
  Communication at Commission 15 Meeting, Internat. Astron.
  Union General Assembly, Prague.
Swings, P. and Greenstein, J. L. (1958) Présence des Raies
  Interdites de l' Oxygène dans les Comètes; Comptes Rendus
  Acad. Sci., Paris, 246, 511.
Whipple, F. L. (1950), A Comet Model I, Astrophys. J. 111, 375.
Whipple, F. L. (1951), A Comet Model II, Astrophys. J. 113, 464.
Whipple, F. L. (1963), On the Structure of the Cometary Nucleus,
  p. 639 in "The Moon, Meteorites and Comets", ed. B. Middlehurst
  and G. P. Kuiper, Univ. of Chicago Press, Chicago.

Part IV

Ices on Mars

THE MARTIAN POLAR CAPS : A REVIEW

Philippe L. MASSON

Laboratoire de Géologie Dynamique Interne - UA 730
Université Paris-Sud, bât. 509
F - 91 405 ORSAY Cedex

ABSTRACT

Each martian pole is covered by ice cap. Each cap consists of a seasonal canopy of CO_2 which extends 50-65° latitudes during winter, and of a summer remnant cap that extends a few degrees across. The northern polar cap consists of water-ice and contains large amounts of dust. The southern polar cap upper layer is mainly CO_2 and dust-free. Both martian polar caps are spiraled in appearance which may be the result of the Coriolis forces and of preferential frost removal. Layered deposits are observed at the poles. These deposits are up to 5 km thick, and they extend out to about 80° latitudes. They probably consist of ice cemented dust. The layered deposits are partly eroded, being transected by numerous curvilinear valleys and escarpments. Erosional processes that contribute to the formation of these valleys and escarpments, are probably related to the planet's precession, obliquity, and eccentricity.

J. Klinger et al. (eds.), Ices in the Solar System, 521–534.

INTRODUCTION

        Two planets in the Solar System are known to have polar
ice caps : the Earth and Mars. The martian polar ice caps were
identified by telescopic observations, and they were among the
first features of the planet surface to be recognized from Earth.
Since the discovery of the south polar cap in the later part of
the 17th century, the composition of the martian polar caps has
been a matter of controversy. By analogy with Earth, the caps
were long assumed to be composed of water-ice. William Herschel
thought that they were thin deposits of snow and ice. He also
attributed the annual growth and shrinkage of the polar caps to
the changing seasons. In the mid-1960s CO_2 was recognized as the
main component of the martian atmosphere (95,3 %) (Table I).

                        - TABLE I -

GAS	PROPORTION	
	MARS	EARTH
Carbon dioxide	95.32 %	0.033 %
Nitrogen	2.7 %	78.084 %
Argon	1.6 %	0.934 %
Oxygen	0.13 %	20.946 %
Carbon monoxide	0.07 %	
Water vapor	0.03 %	(x)
Neon	2.5 ppm	18.2 ppm
Krypton	0.3 "	1.2 "
Xenon	0.08 "	0.9 "
Ozone	0.03 "(x)	

(x) variable

TABLE I : Compositions of the martian and terrestrial atmospheres
          (from Owen et al.), (1).

        In 1965 the Mariner 4 measurements of the cap
temperatures found them to be $-80°$ C, too cold to be frozen
water. Such temperature indicated dry ice, e.g. frozen carbon
dioxide (2), (3). The presence of CO_2 ice was subsequently
confirmed by Mariner 7 (4). Leighton and Murray (2) also

predicted that the seasonal variations in the cap dimensions could produce variations in atmospheric pressure because a large fraction of the atmospheric CO_2 freezes into the winter cap. The Viking temperature and humidity measurements of the martian poles showed an important difference between them. The southern cap is assumed to be mainly frozen carbon dioxide (possibly a mixture of CO_2 and H_2O ices) whereas the summertime remnant of the north polar cap is water ice. Each winter it is covered by layers of carbon dioxide that freezes out of the atmosphere. Thin layers of CO_2 ice constitute the seasonal parts of the caps that retreat rapidly during the spring. About 30 % of the atmospheric CO_2 condense out to form a seasonal cap during the winter in one hemisphere. Then it vaporizes during the spring, and it freezes out again at the opposite pole during the winter in that hemisphere. Based upon the measurements of atmospheric pressure variations from northern winter to summer, it was established that the amounts of CO_2 frozen in the southern polar cap is sufficient to cover the maximum extent of the cap with an average thickness of more than 22 cm of CO_2 ice (5). Most of the water on Mars appears to be tied up in the permanent polar caps, ground ice, and as bound water in the regolith. Farmer et al (6) estimated that the water ice thickness at the north pole ranges between 1 m and 1 km. The caps may contain up to 10 000 times as much water as the present martian atmosphere.

During two centuries the polar cap's advance and retreat were monitored. Due to eccentricity of the planet's orbit, winter is longer in the southern hemisphere. Consequently the southern cap extends much farther than the northern cap. At their maximum size, the polar caps reach down to a latitude of about 60°. The maximum southern cap reachs a width of about 65°, while the northern cap maximum extent does not usually exceed 50°. The northern cap center and the rotational pole coincide whereas the southern cap center is offset about 5°. During the spring, the caps retreat rapidly. According to telescopic observations (7), (8), the southern cap width has 5° radius just before it desappears under the polar hood (fig. 1a). The northern cap extends up to 8° in width at the end of the summer in the northern hemisphere (fig. 1b). The smaller size of the southern cap is because Mars is closer to the sun during the southern hemisphere summer.

The Mariner 9 images (fig. 2) revealed that the remnant summer cap of the southern pole was surrounded by a thick sequence of layered deposits resting unconformably on old cratered terrains. These deposits appeared to be mixtures of volatiles and wind blown dust. Their layering suggests cyclic variations in the deposition rate that could be related to climate changes and seasonal fluctuations of solar energy. The

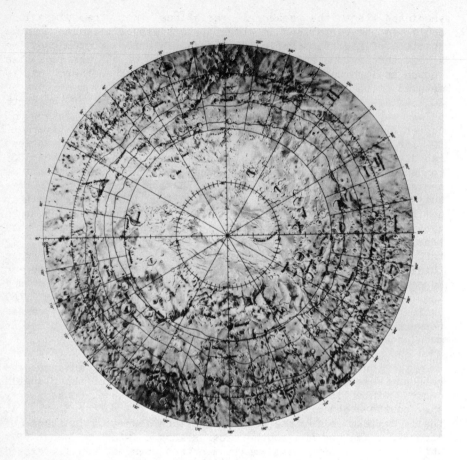

Fig. 1a : Topographic map of the south pole showing
the polar cap at its minimum extent (e.g. residual
summer cap). The layered terrains are better seen
around the south than around the north pole (because
of the permanent frost cover), (from USGS), (9).

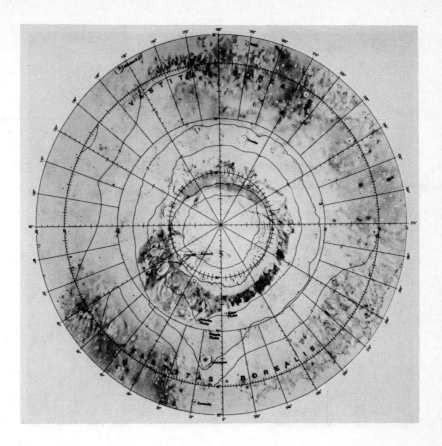

Fig. 1b : Topographic map of the north pole showing
the polar cap at its minimum extent (e.g. residual
summer cap), (from USGS), (9).

Fig. 2a : Mariner photomosaic of the northern polar
ice cap of Mars (NASA-JPL).

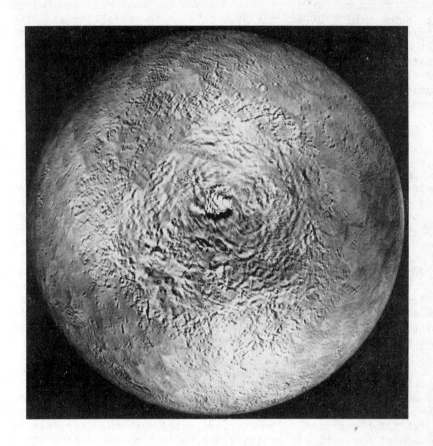

Fig. 2b : Mariner photomosaic of the southern polar
ice cap of Mars (NASA-JPL).

climate changes influence both surface temperatures and winds
that subsequently affect the stability of the volatiles and the
mobility of the wind transported fine materials. The seasonal
fluctuations of solar energy are related to the variations in the
planet's rotation and orbit. The Viking images revealed that
layered terrains also exist at the north pole and that they are
relatively young (less than 10^3 years) as indicated by their
sparcely cratered surfaces.

STRUCTURE AND DYNAMICS OF THE SEASONAL CAPS

        Mars is close to aphelion during southern winter, and
the southern summer occurs close to perihelion. Consequently
winters in the southern hemisphere are colder and longer than
those in the north, and the south polar cap is therefore more
important. The southern summer is relatively hot, causing the
southern cap to retreat further (350 km across) than the northern
summer cap (1000 km across). The southern polar cap's retreat is
non uniform in longitude but similar from year to year. At its
maximum extend the cap is roughly circular around the pole.
Numerous frost outliers are observed along the cap edge. These
outliers are mostly associated with craters. The outliers appear
generally to be slope-controlled, as shown by preferential frost
retentions just within the north walls of the craters (10). The
frost appears also to be removed by wind.

        When the southern cap retreats to its minimum extend,
its position becomes more asymmetrical with respect to the pole.
Because of mixed dust, the seasonal cap has a slightly reddish
color, whereas the remnant cap is less red and slightly brighter.
Terrestrial observations, however, reported darkness increases of
the remnant cap surface, probably due to dust concentration after
frost removal of the seasonal cap (7). As the seasonal cap
retreats, the characteristic spiral pattern of the permanent cap
appears. This pattern is due to frost removal preferentially on
the north-facing slopes in valleys and on escarpments of the
layered terrain (10). The spiral pattern turns in a clockwise
direction and is centered on the 80° W longitude that does not
coincide with the geographic pole or with the cap center.

        The spring retreat of the northern cap leaves also
outliers that appear to be more local than the southern outliers,
because of the less rugged terrains across which the cap
retreats. Most of the outliers are associated with craters. The
northern cap's retreat appears to be more symmetric than the
southern cap's retreat. Because the northern remnant's extension
approximates the layered terrain's extension, the northern
layered deposits are overlaid by a discontinuous frost cover. Due
to the same process (e.g., preferential defrost of the valley

Fig. 3 : Oblique view of the residual cap, showing
the spiral pattern (NASA, Viking Orbiter Image 710
A 74).

walls and escarpments of the layered terrain), the northern
residual cap shows the same spiral pattern as the southern
residual cap (fig. 3). Based upon the seasonal variations of
atmospheric pressure, the average thickness of the seasonal CO_2
cap was estimated a minimum value of 23 cm, possibly 50 cm (11).

COMPOSITIONS AND EVOLUTIONS OF THE RESIDUAL CAPS

The northern remnant is probably water-ice. The
brightness temperatures registered by the Viking spacecrafts
during the northern summer, were near -70° C (e.g., close to the
frost point of water) whereas darker areas were warmer (-40° C).
It seems therefore unlikely that solid CO_2 could exist, because
the sublimation temperature of CO_2 at 6.1 mb is -125° C (12). In
addition these temperatures are consistent with the observation
of large amounts of vapor water in the atmosphere over the poles
at the summertime (6). During the southern summer, the
temperatures are substantially colder (-113° C) than the northern
summer temperatures, despite the closer proximity of Mars to the
sun, and the level of water vapor remains the same as elsewhere
on the planet at this season (13). Because the -113° C
temperature could be caused by CO_2 ice at its frost point (-125°
C), it was concluded that the upper surface of the southern
remnant is predominantly CO_2 (14).

These main differences between the northern and southern
residual caps are attributed to the effects of global dust storms
(14). Dust storms that initiate in the southern hemisphere during
summer, raise large amounts of dust in the atmosphere and partly
shields the residual southern cap from the sun's radiation. The
CO_2 that condenses out at the north pole at the same time,
nucleates around the dust grains thus retrieving dust from the
atmosphere (15). In winter, when the southern cap forms, the
atmosphere is clear, and little dust is incorporated into the
seasonal cap. Subsequently, the southern cap is mainly CO_2 ice
dust-free, whereas the northern cap is dirty water ice. Because
the southern cap is partly shielded by dust from the sun, whereas
the atmosphere is clear over the northern remnant, the
temperatures are warmer on the northern cap than on the southern
during summer. These substantial differences in temperatures
contribute to complete CO_2 volatilization at the north pole and
CO_2 retention at the south pole during summer (10).

It is assumed that the caps alternate with the
precessional cycle (10), and that the actual conditions (northern
residual cap mainly consisting of dirty water-ice, and southern
residual CO_2 ice cap) will be reversed in 25,000 years. For
example the dust storms will occur during the northern summers,
and dust will deposit on the southern cap. Such cycle could be

also responsible for the asymmetric distribution of dune fields
around each pole. Kieffer and Palluconi (14) suggested that the
actual distribution, e.g. large fields of dunes surrounding the
north pole and limited dunes localized within craters at the
south pole, could result from the precipitation of the
atmospheric dust by CO_2 condensing at the north pole in winter.
The CO_2 dissipation in spring would release most of the dust.
According to this precessional model, dunes of the north pole are
no more than 25,000 years old, and similar dune fields should
surround the south pole within 25,000 years from now.

MORPHOLOGY AND ORIGIN OF THE LAYERED TERRAIN

        Both poles exhibit layered deposits that lie on the old
cratered terrain in the south and in plains in the north within
the 80° latitude circles (fig. 1a and b). The layered terrain has
a smooth and crater free surface which upperpart is incised by
numerous valleys and low escarpments.   These   valleys   and
escarpments show swirl patterns oriented counter clockwise in the
north and clockwise in the south.   These   depressions   are
approximately equally spaced. Their walls range in height from
100 m to 1000 m, and their slopes range up to 6° (16). Their
minimum thicknesses were estimated at 1-2 km in the south and 4-6
km in the north (16).

        The origin of the swirl pattern of the layered terrain
remains uncertain. Wind appears to have played a significant role
in   sculpting the polar landscape (17), (18). Differential
sublimation and condensation may also have played a role in
sharping the layered deposits (18), (19), (20).

        The layers appear as fine horizontal bandings that extend
for hundreds of kilometers (21), (22), (23) (fig. 4). The average
thickness of individual layer ranges from 10 m to 50 m (16).
Adjacent layers seem to maintain a continuity in relative
thickness over long distances, and their disruption by impacts
are rare that indicates a relative young age of these layered
terrain.

        The layered terrain are probably the result of volatile
and dust accumulations which could modulated by climatic
variations produced by the orbital and rotational motions of the
planet (21), (22), (24), (25). Dust accumulations at the north
and south poles follow the 25,000 years cycles mentionned in the
previous section, e.g. the precession of the orbital and
rotational axes. However the eccentricity of the martian's orbit
and the planet's obliquity affect the sun's insolation and the
energy balance at the poles that modify both the dust deposition
rates and the processes of the CO_2 adsorption-desorption.

Fig. 4 : Detailed view of the layered terrain in the north, showing fine layering on partly defrosted slopes (frame is 65 km across and centered at 80° W, 347° W), (NASA, Viking Orbiter Image 56 B 84).

Consequently, cycles of 95,000 years and 2×10^6 years (variations in eccentricity) and cycles of 51,000 years (precession of the orbital and rotational axes) would control the periodic variations of the dust deposition rate at the poles. Pollack and co-workers (15) estimated that the present rate of dust deposition at the poles is 4×10^{-2} cm/yr.

CONCLUSION

        As a consequence of precession of the orbital and rotational axes and variations in eccentricity and obliquity, the martian climate changes periodically. The general atmospheric circulation, volatile stability, and energy rate must be affected by these changes that may also cause complex erosional and depositional processes. At the present time, the north polar cap is composed by permanent dirty water-ice covered by a thin CO_2 seasonal cap retreating almost uniformily in summer, whereas the southern cap could be mainly CO_2 dust free ice covered by a thin CO_2 seasonal cap retreating non uniformly in summer. These present conditions will probably be reversed in 25,000 years because of the precessional cycle of Mars. Both polar caps are surrounded by layered terrain that corresponds to different cycles of volatile and dust accumulations induced by climate changes. Accumulations may be currently occurring at the north pole whereas the accumulation of dust at the southern cap may be less effective. Because the precession conditions will be reversed in 25,000 years, the south pole will accumulate the dust raised by the storms.

REFERENCES

1. Owen, T., 1977, J. Geophys. Res. 82, pp. 4635-
    4639.

2. Leighton, R. B. and Murray, B. C., 1966, Science
    153, pp. 136-144.

3. Leovy, C. B., 1966, Science 154, pp. 1178-1230.

4. Neugebauer, G. et al., 1971, Astron. J. 76, pp.
    719-728.

5. Snyder, C. W., 1979, NASA Conf. Publication 2072,
    pp. 76-79.

6. Farmer, C. B. et al., 1976, Science 194, pp. 1339-
    1341.

7. Dollfus, A., 1965, Ann. Astrophysique 28, pp. 722-
    747.

8. Dollfus, A., 1973, Icarus 18, pp. 142-155.

9. USGS, 1976, Topographic Map of Mars, 1 : 25.000,
    Map I-961.

10. Carr, M. H., 1981, Yale University Press.

11. Hess, J. E. et al., 1979, J. Geophys. Res. 84, pp.
    2923-2927.

12. Kieffer, H. H. et al., 1976, Science 194, pp. 1341-
    1344.

13. Farmer, C. B. and Doms, P. E., 1979, J. Geophys.
    Res. 84, pp. 2881-2888.

14. Kieffer, H. H. and Palluconi, F. D., 1979, NASA
    CP-2072, pp. 45-46.

15. Pollack, J. B. et al., 1979, J. Geophys. Res. 84,
    pp. 2929-2945.

16. Dzurisin, D. and Blasius, K. R., 1975, J. Geophys.
    Res. 82, pp. 4225-4248.

17. Cutts, J. A., 1973, J. Geophys. Res. 78, pp. 4211-
    4221.

18. Sharp, R. P., 1973, J. Geophys. Res. 78, pp. 4222-
    4230.

19. Howard, A. D., 1978, Icarus 34, pp. 581-599.

20. Cutts, J. A. et al., 1979, J. Geophys. Res. 84, pp.
    2975-2994.

21. Cutts, J. A., 1973, J. Geophys. Res. 78, pp. 4231-
    4249.

22. Murray, B. C. et al., 1972, Icarus 17, pp. 328-345.

23. Soderblom, L. A. et al., 1973, J. Geophys. Res. 78,
    pp. 4197-4210.

24. Cutts, J. A. et al., 1976, Science 194, pp. 1329-
    1337.

25. Sharp, R. P., 1974, J. Glaciology 13, pp. 173-183.

MARS: LONG TERM CHANGES IN THE STATE AND DISTRIBUTION OF H_2O

Fraser P. Fanale, James R. Salvail, Aaron P. Zent, and
Susan E. Postawko
Planetary Geosciences Division, Hawaii Institute of
Geophysics, Univ. of Hawaii, Honolulu, Hawaii 96822

Abstract

A model for H_2O distribution and migration on Mars has been
formulated which takes into account: 1) thermal variations at
all depths in the regolith due to variations in obliquity,
eccentricity and the solar constant; 2) variations in
atmospheric partial pressure of water (PH_2O) caused by
corresponding changes in polar surface insolation; and 3) the
finite kinetics of H_2O migration in both the regolith and
atmosphere. Results suggest that regolith H_2O transport rates
are more strongly influenced by polar-controlled atmospheric
PH_2O variations than variations in pore gas PH_2O brought about
by thermal variations at the buried ice interface. The
configuration of the ice interface as a function of assumed soil
parameters and time is derived. Withdrawal of ice proceeds to
various depths at latitudes $<50°$ and is accompanied by filling
of regolith pores at latitudes $>50°$ and transfer of H_2O to the
polar cap. The transfer has a somewhat oscillatory character,
but only $<1g/cm^2$ is shifted into and out of the regolith during
each obliquity cycle. The net irreversible and inexorable
transfer of H_2O to higher latitudes involves between 1×10^6
km^3 and 1×10^7 km^3 of H_2O over the history of Mars for most
reasonable sets of assumptions. This mass is comparable to the
amount of material removed from deflated terrain at mid and low
latitudes and to the mass of the polar cap. We conclude that
this process combined with periodic thermal cycles played a
major role in development of the fretted terrain, deflationary
features in general, patterned ground, the north polar cap and
the layered terrain.

J. Klinger et al. (eds.), Ices in the Solar System, 535–564.
© *1985 by D. Reidel Publishing Company.*

A. INTRODUCTION

An abundance of cold climate surface features on Mars suggests that a substantial amount of subsurface water ice may have existed in the porous regolith throughout Martian geologic history. These features include debris flows, patterned ground, thermokarst pits, table mountains and rampart craters (e.g., Carr and Schaber, 1977 [1]; Rossbacher and Judson, 1981 [2]). The geologic evidence for the widespread occurrence of subsurface ice is in agreement with theoretical estimates of the global inventory of water on Mars. The work of McElroy et al., (1977) [3] and Pollack and Black (1979) [4] suggests that a layer of ice in excess of 100 m, averaged over Mars' surface, may have been degassed while studies of the fate of H_2O and the geochemical sinks for it suggest that most of that H_2O may have been stored as ground ice (Fanale, 1976 [5]; Rossbacher and Judson, 1981 [2]).

Mars has experienced great insolation variations (especially near the poles) as the results of changes in its obliquity and orbital eccentricity (Ward, 1974 [6]; Ward et al., 1979 [7]). Changes in the solar constant (D.O. Gough, 1977 [8]) may have played a role as well. The consequences for CO_2 distribution and climate change have been investigated and appear to be profound, including periodic atmospheric pressure changes of a factor of 100 and exchange of very large quantities of CO_2 ($\sim 10^{20}$ g) between the regolith and caps (Fanale and Cannon, 1979 [9]; Toon et al., 1980 [10]; Fanale et al., 1982 [11]). This paper presents preliminary results of an attempt to examine the possible effects of these changes on the state and distribution of regolith, polar cap H_2O, and atmospheric H_2O.

In considering those long term effects which require not only in situ changes in H_2O state (e.g., melting and freezing) but also migration of H_2O molecules through the regolith, we must also take into account the sometimes severe kinetic limitations on migration of the H_2O molecule through a cold, adsorbing porous medium. Clifford and Hillel (1983) [12] constructed an analytical model to determine the lifetime of a 200m layer of subsurface ice overlaid by 100m of ice free regolith for the latitudes between $\pm 30°$ on Mars. They assumed that the regolith is isothermal in depth and in time, and their gas diffusion model accounts for the possibility of both molecular and Knudsen diffusion. They obtained results for twelve soil types, representing different pore sizes and porosities, and for various assumed values of the iso-thermal regolith temperature. Their analytical model does not quantitatively include (1) solar insolation changes due to various causes, (2) geothermal heating and heat flow, (3) temperature variations with depth and time caused by (1) and

(2), and (4) variations in atmospheric concentration of H_2O with location and time. We find that the last may be a particularly important consideration. Toon et al., (1980) [10] constructed a model with many similarities to Clifford and Hillel (1983) [12], but it was primarily seasonal in scope using various chosen values of obliquity and eccentricity.

In this paper we develop an analytical model that treats the same basic problem in that we consider the factors that would cause changes in any initial distribution and state of H_2O through Mars' history. Starting from an initial uniform ice configuration at the beginning of Mars' geologic history, we calculate the changes in ice layer thickness throughout geologic time for the entire range of latitudes. This will yield the present global depth distribution of subsurface ice and/or the lifetime of the ice layer for various latitudes as appropriate. The model quantitatively takes into account (1) obliquity and eccentricity variations, (2) long term changes in the solar luminosity, (3) albedo changes at higher latitudes due to seasonal condensation of CO_2, (4) geothermal heating and heat flow, (5) temperature variations in the regolith as a function of depth, time and latitude due to (3) and (4), (6) molecular and Knudsen diffusion, (7) annual average water concentrations and fluxes at the pole, and (8) variations in the atmospheric concentration of water vapor with latitude and time.

In computing regolith temperatures we have neglected the small amount of convected heat, and the latent heat from the subliming ice. For our atmospheric transport calculations we have used a simple one dimensional diffusion model, which does not consider atmospheric circulation. The reasons for this simplification are discussed. We have used a simplified soil geometry model that uses an average porosity and pore size, rather than the complex pore size and porosity distributions used by Clifford and Hillel (1983) [12]. Given the limitations in our knowledge of deep regolith properties, we are convinced that the attainment of accuracy in this problem depends considerably more on a realistic thermodynamic model, rather than on great detail in soil geometry, primarily due to the exponential dependence of H_2O concentration on temperature. To account for the effect of soil geometry, we leave pore size as a free parameter and use values that seem to be reasonable bounds for this important property.

B. MODEL DESCRIPTION

We will now describe the method and assumptions which characterize our model. Detailed mathematical description and derivations will be presented elsewhere, together with a more

extensive array of results (Fanale et al., 1985) [13]). The
problem ultimately involves computation over three different
time scales. The largest time scale is the entire duration of
Mars' history: 4.5 billion years. This is due to the changing
solar luminosity, which has been postulated to be about 40%
greater now than at the beginning of Mars' history (D.O. Gough,
1977 [8]). The second time scale is due to the obliquity and
eccentricity cycles. Ward (1979) [14] has obtained data
concerning obliquity and eccentricity variations from
theoretical considerations. The mean value of Mars' obliquity
is ~24.4° with a maximum amplitude of ~13.6° and maximum and
minimum obliquities of 38.0° and 10.8°. The period for the
short term oscillations is ~1.2 x 10^5 years. These are
superimposed on longer term oscillations of 1.2 x 10^6 years.
Also included in the model are pre Tharsis obliquity conditions
as determined by Ward et al. (1979) [7]. Their results indicate
that prior to the Tharsis uplift, the mean value of the
obliquity was ~27.5° with maximum and minimum obliquities of
~46° and 9°. Wise et al. (1979) [15] have concluded that the
Tharsis uplift occurred between 3.5 and 4.0 billion years ago.
For the purpose of this work, 3.5 billion years ago will be used
as the transition point for the obliquity conditions. The mean
value of the eccentricity has been determined to be .07 with
maximum and minimum values of .141 and .004. There are also two
characteristic periods. Relatively rapid oscillations take
place with a period of 9.5 x 10^4 years. These are superimposed
over a longer term variation of 2 x 10^6 years. We have used
Fourier analysis to obtain series formulations that very closely
describes these complicated obliquity and eccentricity
variations.

The third time scale is the annual cycle. Although we are
more concerned about average annual conditions, we have found it
necessary to consider annual variations for two reasons. First,
the water transport through the regolith will be seen to depend
strongly on the annual average atmospheric H_2O concentration at
the poles. This annual average H_2O concentration does not
correspond to the annual average temperature because of the
exponential dependence of vapor pressure on the temperature.
Secondly, the albedo at latitudes 50° and above varies
considerably throughout the year due to seasonal condensation of
CO_2. The albedo of CO_2 ice is ~.65, and the average planetary
albedo is .25. To obtain results that agree fairly closely with
the Viking Thermal Model of Kieffer et al. (1977) [16] it was
necessary to use an annual average albedo, which is a function
of latitude. Such seasonal averaging should be satisfactory for
our purposes, since our critical boundary condition is the mean
annual temperature.

The one dimensional, transient heat diffusion equation was

used to obtain near surface temperature profiles and surface temperatures throughout a Martian year. The surface boundary condition was a heat flux condition that included the daily averaged solar insolation, infrared thermal emission and latent heat due to condensation of CO_2. The daily averaged insolation is a complicated function of planetocentric latitude, solar latitude, argument of perihelion, obliquity, true anomaly and eccentricity and was provided by Ward (1974) [6]. Annual surface temperatures were initially obtained for each 10^0 latitude using an average planetary albedo of .25 for latitudes below 80^0 and an albedo of .40 for latitudes 80^0 and 90^0, since a permanent water ice cap is thought to exist near the north pole. At each point in the annual cycle where the surface temperature was calculated and at each latitude, the corresponding CO_2 vapor pressure was calculated. If the CO_2 vapor pressure was found to be less than the atmospheric pressure, CO_2 was assumed to condense; and the albedo at that time and latitude was set equal to .65. Then an annual averaged albedo was calculated for each latitude using the albedos given above and the fraction of the year that CO_2 ice was present at that latitude. The above calculations were repeated for each of the 100 time steps we used in the long term obliquity cycle to obtain an array of annual averaged albedos for each latitude and obliquity. At the north pole H_2O vapor pressures were calculated for each point in the annual cycle using the vapor pressure equation, and the atmospheric H_2O concentration was calculated using the ideal gas equation. This array of annual H_2O vapor densities was used to calculate an annual averaged H_2O vapor density for each obliquity. This was used later to calculate the surface boundary condition for water vapor at each latitude.

The one dimensional, transient heat diffusion equation was again used to calculate annual averaged regolith temperatures, including surface temperatures. The surface boundary condition was similar to that used for computing annual temperatures except that an annual averaged insolation was used. This is a complicated function of latitude, solar longitude, obliquity and eccentricity and was also provided by Ward (1974) [6]. The annual averaged albedos, discussed above, were used in the surface boundary condition. The lower boundary condition was assumed to be the internal planetary heat flow deduced by Fanale (1976) [5], ~30 ergs/cm^2-sec. A heat generation term was also included in the heat diffusion equation. Using a time step of 1.2×10^4 years, regolith temperature profiles were obtained as a function of time/obliquity for each 10^0 latitude.

Regolith temperatures were calculated for two different depth domains for different purposes. The regolith was assumed to extend to a constant depth of 1km for the entire planet. In fact, the permeable regolith is probably much thicker near the poles than in low latitude. Temperature profiles were calculated for this domain for the purpose of calculating atmospheric pressures. Temperatures were also calculated from the surface to the instantaneous depth of the ice layer for use in the H_2O diffusion calculations to be described later. Since this depth is constantly changing due to the sublimation of the ice, the depth step was made variable with time.

A knowledge of atmospheric pressures throughout a long term obliquity cycle is required to determine the time and latitude of the seasonal CO_2 cap and to calculate H_2O fluxes through the CO_2 background gas, both in the atmosphere and in the porous regolith. We use the results of our previous work (Fanale et al. 1982 [11]) to calculate CO_2 atmospheric pressures at each point in the long term obliquity cycle assuming a regolith of ground basalt. For details see the above reference. For the total CO_2 inventory we use an amount which yields the currently observed atmospheric pressure at the present time. We also assume that the soil pores are not completely filled with ice and that the ice does not effectively prevent CO_2 transport between the regolith and the atmosphere on the time scales considered.

A simple one dimensional diffusion model was used to calculate the H_2O atmospheric concentration as a function of latitude. This will be used to provide approximate annual average values of H_2O atmospheric concentration to be used as the surface boundary condition for the problem of H_2O diffusion in the regolith. For this purpose it is assumed that the atmospheric CO_2 pressure is the same at all surface locations at any given time. It is also assumed that the passage of H_2O through the CO_2 is always horizontal and orthogonal to latitude lines and that there are no winds. We are aware that several models of Mars' atmospheric circulation have shown that H_2O transport in the atmosphere is mostly convective rather than diffusive. The one dimensional diffusion model has been used as a mathematical device to obtain an annual average H_2O density variation with latitude and time. We have used a diffusional model rather than a more realistic circulation model for the following reasons: 1) We have not as yet developed a circulation model; 2) Only rough annual average values of H_2O density are desired, since the variation at the north pole over an obliquity cycle is at least two orders of magnitude more than the variation with latitude at a given time; 3) A global circulation model would not be economically feasible to use in this situation, since it would have to be used repeatedly in an already long-running program; and 4) Our current diffusion constant gives a minimum value for the actual e folding time between 10° latitudinal zones of 18 days. This is only a factor of three longer than e folding times calculated by Jakosky (1983) [17] considering wind transport. Further, atmospheric transport is not the slowest step in the H_2O cycle. Hence, this convenient procedure provides a reasonably good fit to the annual averaged Viking data, and at the same time it provides a convenient way to relate past polar sublimation rates to past atmospheric H_2O concentrations at lower latitudes.

The H_2O flux through the atmosphere was expressed quantitatively in terms of horizontal H_2O atmospheric concentration and temperature gradients and a diffusion coefficient that is evaluated in terms of molecular speed and mean free path, which is itself a function of atmospheric temperature and pressure. We assumed that the near surface atmospheric temperature was equal to the local annual averaged surface temperature as a rough approximation. The H_2O flux expression was combined with the conservation of mass principle to obtain a linear partial differential equation with variable coefficients. This equation was solved iteratively, using the annual average H_2O atmospheric density at the pole, as discussed earlier and an equatorial boundary condition to obtain the H_2O atmospheric density profile as a function of latitude. In the above analysis we included a transient term for the sake of generality. However, because of the large time scale inherent in the problem, it can be viewed as a steady state problem. The atmospheric H_2O density, as a function of latitude, is used as the surface boundary condition for the regolith H_2O diffusion problem to be described next. It was necessary to obtain only approximate values for the local atmospheric H_2O density because the annual average atmospheric H_2O density at the pole varies by three orders of magnitude over an obliquity cycle. Observations of Mars have shown that atmospheric H_2O concentrations are generally in the same order of magnitude over the northern latitudes and that annual average values for the northern latitudes vary by less than a factor of two. Therefore, the difference between the atmospheric H_2O density at different latitudes at a given time is very much less than the difference between the annual average H_2O density at the pole at different times in the obliquity cycle.

The flow rate of H_2O through the porous regolith is calculated using a one dimensional model which includes the effect of vapor density gradients, temperature gradients and gravitational effects. In addition, the model allows for two modes of flow. If the mean free path of H_2O in the CO_2 is less than a soil pore diameter, molecular diffusion will occur. If the mean free path is greater than a pore diameter, flow will be governed by Knudsen diffusion. Unfortunately, current Mars atmospheric pressure and pore size are such as to place the transport regimes near the boundary between the two, so both must be treated.

For the case of Knudsen diffusion, the diffusion coefficient is determined by the porosity, pore size and tortuosity of the soil matrix. The mass flux of H_2O vapor through the regolith is formulated in terms of H_2O density and temperature and their depth gradients, the acceleration of gravity and the diffusion coefficient. For molecular flow the mass flux is formulated similarly except that the diffusion coefficient contains the diameter of the CO_2 molecule instead of the soil pore diameter. The conservation of mass principle is then used to obtain transient, linear partial differential equations with variable coefficients for both modes of flow. The H_2O density boundary condition at the ice interface is obtained from the temperature calculated earlier and the vapor pressure and ideal gas equations. The surface boundary condition was discussed in the previous section. The governing equation for each flow regime, together with the boundary conditions, is solved interactively as a function of depth and time to obtain the H_2O density profile as a function of depth for each value of the obliquity. The mass fluxes can then be calculated in a direct manner. The thickness of ice consumed during a time step is easily calculated from the H_2O mass flux, the density of solid ice, the soil porosity and the time step. The total change in thickness during a long term obliquity cycle is found by summing the results for each time step. The calculations could not be performed continuously over Mars' history, since this would require an unacceptable amount of computer time. Therefore, Mars' history was divided into 100 epochs, during each of which the solar luminosity was assumed to remain constant. Then the consumption of ice was extrapolated over each epoch based on the results obtained for the obliquity cycle at the beginning of each epoch. The results were summed for each epoch to give the current depth distribution or lifetime of ground ice for each latitude.

C.   RESULTS

Before discussing the extent of long term variations in Mars' regolith temperature and atmospheric H_2O pressures and considering migration of H_2O through geologic time, we shall review the implications of current Viking-observed conditions for the state and distribution of Martian H_2O. Figure 1 (adapted from Fanale, 1976 [5]) shows an approximate sketch of the zones where ice can currently exist in equilibrium with the atmosphere and caps, where ice could exist if protected from equilibration with the atmosphere and cap cold traps, where brines could exist (with the same caveat) and where even pure water could exist. The assumptions used to construct the diagram include an estimate of Mars' heat flow and regolith

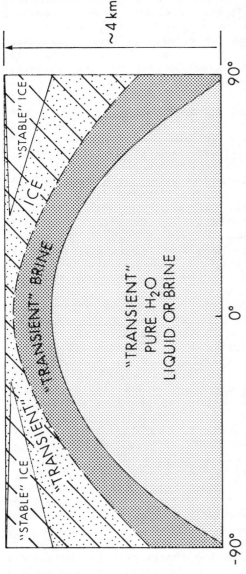

FIGURE 1 A schematic representation of a possible distribution of important thermal boundaries for H_2O on Mars. This is the configuration that the boundaries would have if current conditions of insolation were representative of all of Mars' history. The wedges represent zones where ice could always exist in equilibrium with the atmosphere. That is, the temperature is $<200°K$. The zone below could harbor ice since it is $<273°K$, and the zone below that could harbor sulfate rich brines since it is above $235°K$. However, no ice or brine below the two wedges could be in equilibrium with the atmosphere. Such ice or brine would constantly be a source of H_2O molecules until they were depleted. Nonetheless, ice and brines could, once emplaced, persist for geological periods of time here. The degree to which ice or brines populate the pores in any part of the regolith depends on the simple availability of H_2O molecules as well as the temperature and time for H_2O migration. A "chondritic" heat flow, Q, for Mars, and a regolith conductivity of hard frozen limonite is used to construct this diagram. Values are $Q = 30$ erg cm^{-2} sec^{-1} and $K = 8 \times 10^4$ erg cm^{-1} sec^{-1} K^{-1}.

conductivity, as discussed below. Despite these assumptions, we
believe that the figure at least qualitatively represents the
implications that the Viking results, which include atmospheric
H_2O pressures ranging from ~5-100μm and mean annual temperatures
as reported by Kieffer et al. (1977) [16], would have for Mars'
H_2O distribution if all these conditions were constant
throughout Mars' history. It should be borne in mind however,
that Figure 1 ignores many of the climate variations discussed
in this paper. Figure 2 (Fanale and Clark, 1983) [18]
represents the most obvious direct effect of the obliquity
variation on mean annual temperatures. The latter are displayed
as a function of latitude, depth and phase of the obliquity
cycle. Assumptions are similar to those in Figure 1. The most
obvious points are: (1) the greatest mean annual temperature
variations are near the poles; they are small near the equator
and almost zero at mid latitude; (2) although diminished by
penetration into the thermal capacitor represented by the deep
Mars' regolith, we expect that a substantial portion of the
obliquity driven thermal wave will penetrate to depths of up to
1 km. This portrayal does not include the effects of variable
albedo, which are included in the subsequent results. We will
now examine the model-predicted potential effects of the thermal
variations in Figure 2 as well as postulated variations in the
solar constant on both the atmospheric H_2O pressure and the H_2O
pressures and H_2O stability zones in the regolith.

The calculations described in the previous section were
carried out for each ten degrees of latitude from the equator to
the north pole and for two values of the pore or capillary
radius of the porous regolith. We assumed uniform soil
properties, including the capillary radius, porosity and
tortuosity, rather than the complicated and more realistic
distributions used by Clifford and Hillel (1983) [12]. Since so
little is known about the properties of the Martian regolith,
and since there is such a great variation in the properties of
Earth soils, we felt that using intuitively reasonable uniform
soil properties would yield results that were just as accurate
on the average as those obtained using pore size distributions.
We did not use different values of porosity and tortuosity,
since the possible range of variation of these properties is
relatively small. This is not so for the pore or capillary
radius; therefore, we attempted to bound this important
parameter using values of 1 and 10 microns. Results were not
obtained for the southern latitudes because we believe that the
depth distribution of subsurface ice is roughly symmetric about
the equator. Approximate values of the annual average surface
temperature were obtained for both northern and southern
latitudes using the results from the Viking Thermal Model
(Kieffer et al., 1977) [16]. It was found that the annual
average surface temperatures were nearly symmetrical about the

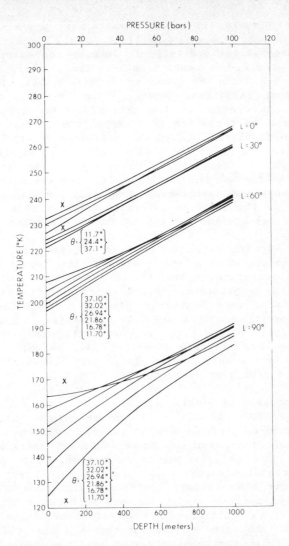

FIGURE 2   The Mars regolith thermal regime
below ~2m (Fanale and Clark, 1983) [28]. Temperatures are given as a function of latitude,
depth and obliquity. Assumptions are similar to
those in Figure 1. Albedo variations at high
latitudes are not taken into account in this
figure, but are in the calculations represented
by Figures 3-5.

equator. This would also be true for annual average temperatures and H_2O vapor densities at the subsurface ice interface. Observations presented in Jakosky and Farmer (1982) [19] show that the H_2O concentration over the southern latitudes is on the average about an order of magnitude less than that over the northern latitudes. However, for latitudes between $\pm$ 40^0 the H_2O vapor density at the subsurface ice interface is usually from one to three orders of magnitude greater than the atmospheric H_2O density. The annual average H_2O vapor density at the pole also varies by about three orders of magnitude over an obliquity cycle. Therefore, a variation of atmospheric H_2O density by a factor of two or three, on the average, between northern and southern latitudes, by itself, should cause the subsurface ice in the southern latitudes to be only of the order of ~1% less and deeper than for the northern latitudes.

Calculations were performed in the sequence indicated in the previous section. For each latitude the results yielded annual surface temperatures and annual average albedos, annual average regolith temperatures including those for the surface and ice interface, annual average atmospheric H_2O densities, regolith H_2O density profiles and mass flow rates, and the depth of subsurface ice. Obliquities, eccentricities and atmospheric (CO_2) pressures were obtained globally. All calculations were performed for 100 evenly spaced long term obliquity cycles spanning Mars' 4.5 billion year geologic history.

We now present a small sample cross section of our preliminary results. A much more comprehensive presentation will be made in a complete paper, which will soon be submitted. Figures 3a and b give polar cap temperature vs. solar longitude at the high, middle and low points in the obliquity cycle for different eccentricity conditions. The isothermal part of the curves represents the northern hemisphere winter season during which the higher northern latitudes are in total darkness, and the temperature is maintained at a constant value by the condensing CO_2. The value of the isothermal temperature is the temperature at which the vapor pressure of CO_2 is equal to the atmospheric pressure, which varies throughout the obliquity cycle. The results show that the polar temperatures are highest for the largest obliquities. The opposite is true for latitudes below ~40^0. The results shown also depend on the eccentricity cycle. As stated previously, the obliquity cycle has a period of ~1.2×10^6 years while the eccentricity cycle has a period of 2×10^6 years in the long term. At different points in Mars' history the obliquity and eccentricity cycles have different phase relationships. The obvious way to model these relationships exactly is to calculate continually over Mars' history. Alternatively, it may be possible to calculate over a composite cycle consisting of a number of obliquity and

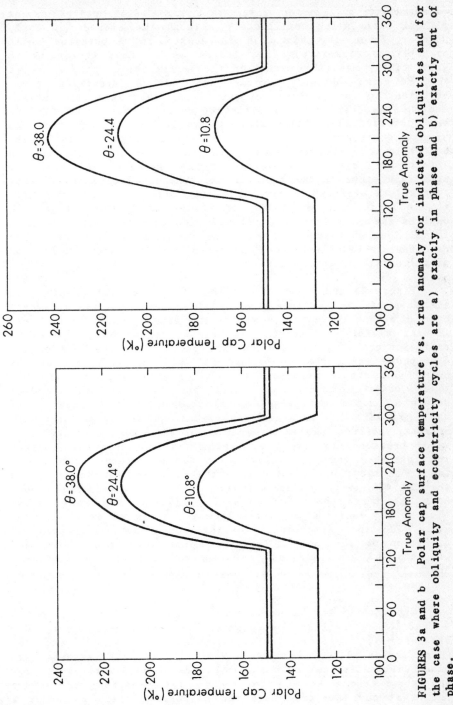

FIGURES 3a and b Polar cap surface temperature vs. true anomaly for indicated obliquities and for the case where obliquity and eccentricity cycles are a) exactly in phase and b) exactly out of phase.

eccentricity cycles for which the phase relationships repeat without calculating continuously through the history of Mars. However, even in this case the costs of computation would probably be excessive. The alternative we have chosen is to artificially make the period of the eccentricity cycle equal to that of the obliquity cycle and use a specified phase relationship for a given computer run. The phase relationship can then be changed for a different run. This is done for the two most different phase relationships to see the maximum effect of the eccentricity cycle. In Figure 3a, the obliquity and eccentricity cycles are exactly in phase. In this case the two effects work against each other in the northern hemisphere and with each other in the southern hemisphere. The result is that temperature differences for different obliquities are minimized at the north pole and maximized at the south pole. Figure 3b depicts the case when the obliquity and eccentricity cycles are totally out of phase. In this case, temperature differences for different obliquities are maximized at the north pole and minimized at the south pole.

Figure 4a and b depict the log of H_2O partial pressure at the base of the atmosphere and at the subsurface ice interface over an obliquity cycle for relatively recent conditions. These figures vividly illustrate the driving force that propels the diffusion process. The H_2O partial pressure at the base of the atmosphere is controlled by and follows the variations at the pole. The annual average H_2O pressure at the pole varies by three orders of magnitude over an obliquity cycle, which accounts for the very large excursions shown in the figure. The much smaller H_2O pressure excursions at the subsurface ice boundary reflect the much smaller local temperature and H_2O vapor pressure variations at depth. The higher the subsurface ice interface curve is relative to the atmospheric curve, the faster is the H_2O diffusion from the regolith. Figure 4a shows that at 40° latitude H_2O diffuses out of the regolith except near the three points of highest obliquity. Figure 4b shows that at 50° latitude the subsurface ice curve is near the mean value of the atmospheric oscillation, indicating periodic diffusion of H_2O out of and into the regolith. Below 40° latitude H_2O is constantly transported out of the regolith. Above 50° to 60° latitude H_2O ice can always be found near the surface.

Our results for annual average atmospheric H_2O pressure as a function of latitude were compared to the annual average H_2O pressures for selected latitudes obtained from the graphical data presented in Jakosky (1983) [17]. Our results are practically identical to the annual average values obtained from the observations at 40° latitude. The greatest discrepancy occurs at the equator, where the calculated value is about a

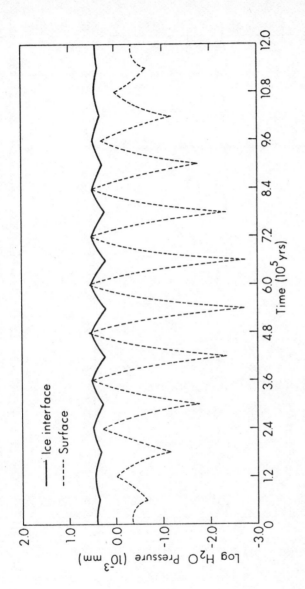

FIGURE 4a   Log of H_2O pressure at the base of the atmosphere and at the sub-surface ice interface vs. time for one 'current' obliquity cycle at 40° latitude.

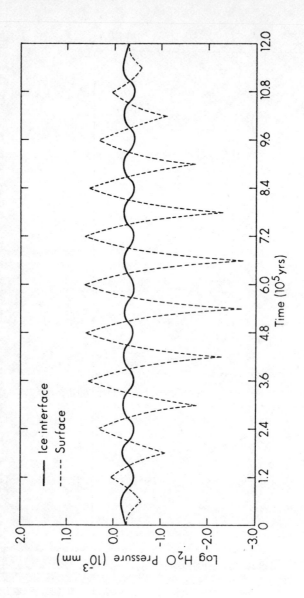

FIGURE 4b   Log of H_2O pressure at the base of the atmosphere and at the sub-surface ice interface vs. time for one 'current' obliquity cycle at 50° latitude.

factor of 3 less than the value obtained from observations. Going toward the north pole, our results show H$_2$O pressures that are increasingly greater than the values obtained from observations up to a factor of 2 at the pole. However, our calculations show that the change in the H$_2$O density gradient between the subsurface ice and the atmosphere caused by the differences between predicted annual average H$_2$O pressures and annual average H$_2$O pressures obtained from observations is at most only about 1.4% for latitudes below 40°. A similar difference was found for latitudes 70° and greater. At 60° latitude the difference in the H$_2$O density gradient is about 50%, but in either case the ice front is advancing toward the surface at nearly all points in the obliquity cycle. Therefore, it is seen that the diffusion of H$_2$O through the regolith is not sensitive to the atmospheric H$_2$O density at most latitudes. However, near 50° latitude the annual average H$_2$O density at the subsurface ice boundary is comparable to the annual average H$_2$O density at the base of the atmosphere. A relatively small change in the atmospheric H$_2$O density can cause a reversal in the density gradient and, thus, in the direction of H$_2$O vapor transport. Therefore, our results at 50° latitude may depart significantly from actual conditions although predicted and observed values differ only by about 25%. The ice front may actually be slowly retreating. However, it is certain that near 50° latitude there is a latitude at which there has been no net transport of water throughout Mars' geologic history.

Figure 5a gives the depth of the subsurface ice layer as a function of latitude for five points in Mars' history using a pore/capillary radius of 1µm. The numbers on the right are in units of billions of years with the lowest curve depicting present conditions. The slopes of the curves are near zero at the equator, indicating symmetry between northern and southern latitudes. A crossover point for all curves occurs between 45° and 50° latitude. For latitudes lower than the crossover point there is a net transfer of water vapor from the regolith into the atmosphere, and the ice front is receding from the surface. For latitudes higher than the crossover point there is a net transfer of water vapor in the opposite direction, and the ice front is advancing toward the surface. The figure shows that the ice has already reached the surface at essentially all latitudes above 50°. Figure 5b is a plot of subsurface ice depth as a function of latitude for a pore/capillary radius of 10µm. The general characteristics are the same as in Figure 5a except that diffusion occurs at a much faster rate. For this case, both Knudsen diffusion and molecular diffusion occur, with molecular diffusion taking place at higher temperatures. Subsurface ice has been removed from all latitudes lower than 50°, and this removal was completed relatively early in Mars' history, at least for latitudes $\leq$40°. For latitudes 60° and

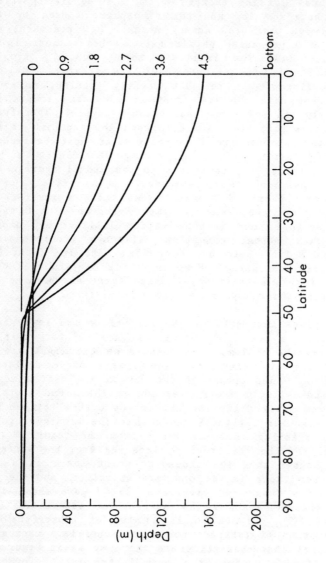

FIGURE 5a Depth of subsurface ice vs. latitude for indicated times from the beginning of Mars' history in billions of years, using a pore/capillary radius of 1μm.

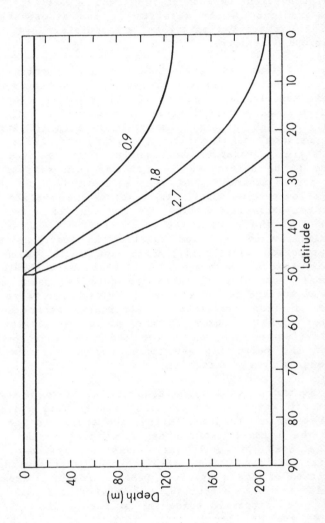

FIGURE 5b Depth of subsurface ice vs. latitude for indicated times from the beginning of Mars' history in billions of years, using a pore/capillary radius of 10μm.

greater near surface ice has existed since very early in Mars'
history. Near 50° latitude the results for both pore sizes show
that near surface ice exists. However, this result may not
depict actual conditions because of the great sensitivity of the
H_2O density gradient in the regolith on the atmospheric H_2O
density and the inherent limitations in accurately calculating
the atmospheric H_2O density.

When the ice front has advanced to within approximately a
few tens of centimeters from the surface, the mass flux of H_2O
will oscillate rapidly into and out of the regolith along with
the seasonal thermal wave. However, the mass flux of H_2O is
much too small and the seasonal time scale is much too short to
cause the position of the ice front to oscillate on a
macroscopic scale. Therefore, the position of the ice front
will continue to be determined by the long term variations in
temperature, and an advancing ice front will continue to advance
slowly in geologic time until it reaches the surface. The
program does not explicitly follow the movement of water vapor
after it has left the regolith at lower latitudes. However, a
water molecule, having left the regolith at lower latitudes,
will be transported alternately north and south, generally,
along with the seasonal water wave. Each year there should be a
very small net deposition of water ice onto the polar cap so
that the annual average concentration of atmospheric water vapor
remains more or less constant. A particular water molecule
would migrate slightly closer to the polar cap during each
annual oscillation to compensate for the very small net ice
deposition on the polar cap and to make room for new water
molecules emerging from the regolith.

Figures 5a and b can also be used to calculate the volume
of the permanent water ice cap. The figures were integrated
graphically between the constant initial ice depth and the curve
representing the present profile using a panel width of 2.5°
latitude. The area of each panel was multiplied by the
circumferential distance along the latitude representing the
midpoint latitude of the respective panel. Positive areas are
those where ice has been evacuated, and negative areas are those
where ice has been deposited below the surface. The volume of
regolith where ice has been removed is much greater than the
volume of regolith where ice has been deposited. Therefore, the
final result does not depend significantly on the assumed
initial ice depth of 10m. Performing the above calculations and
also taking into account the assumed effective porosity of 0.5,
the volume of the permanent water ice cap was found to be ~2.75
x 10^6 km^3 for Figure 5a and 5.66 x 10^6 km^3 for Figure 5b. This is
the amount of H_2O ice which would be expected to migrate to high
latitudes if ice were initially uniformly distributed. These
estimates are not greatly dependent on the initial thickness of

overlying ice free regolith or the initial ice distribution in general, but they are dependent on the pore size assumed. Even in the case of pore size, the difference in the predicted near polar water mass is only a factor of two using 1.0μm and 10.0μm pore sizes.

D. DISCUSSION

Figure 1 (modified from Fanale, 1976) [15] shows the distribution of the zones which may potentially harbor ice in equilibrium with the present atmosphere (at least providing that insolation has always been equal to that observed by Viking). It also shows zones where ice could exist out of equilibrium with the atmosphere. Such ice could be regarded as being "metastable" relative to the cold trap represented by the current atmosphere and polar caps. Below the 235°K isotherm brines of likely Martian composition (sulfate rich) could exist. This means they could exist not far below the surface near the equator. Brines are always "metastable", according to Figure 1, because they must be at temperatures greater than 235°K whereas, to be "stable" relative to the atmosphere, ice must be <200°K (the atmospheric or polar cap cold trapping temperature). In no case is it specified that the species of H_2O indicated actually occupies the pores; only that, if H_2O were available, it could exist in that state. Obviously the later a "deposit" of a metastable species is emplaced, the more likely it is to have survived to the present. Also the deeper it is, the greater the soil cover and the slower the loss to the cold trap. The problem of depletion of such "metastable" species has been treated in greatest detail by Clifford and Hillel (1983) [12].

Although useful, Figure 1 should be regarded as merely a "starting point" in modeling the state and distribution of H_2O on Mars. There are several reasons other than the planetary availability of H_2O why the actual distribution of H_2O species on Mars is likely to be quite different from that shown in Figure 1. They are:

1)   The general recession of H_2O from the equatorial zone, as indicated by the work of Clifford and Hillel (1983) [12] and this work, is sure to have taken place to great depth. This conclusion is reinforced by the observation that the numerous changes in atmospheric and regolith conditions considered in detail in this work do not mitigate or slow that recession and that the extent of the recession appears to be great for all average pore sizes >1.0μm. This will be discussed in more detail below.

2)   The thermal changes caused by the obliquity variations
     will affect the direction of H_2O transport at depth in
     an oscillatory manner, expecially at high latitudes.

3)   The periodic variation in atmospheric H_2O everywhere
     produced by obliquity/eccentricity driven variations
     in polar sublimation rates will not directly penetrate
     to the deep regolith, but will effect somewhat the
     position   of   the   boundary   in   that   the   average
     atmospheric conditions would not necessarily be the
     same as those observed by Viking or those used to
     construct Figure 1.

Figure 2 shows variations in regolith temperature as a
function of three parameters: depth, latitude and phase of the
obliquity cycle. These variations (Ward, 1974 [6], Ward et al.,
1979 [7]) have been shown to be capable of producing periodic
variations of a factor of 100 in atmospheric CO_2 pressure and
effecting transfer of tens of atmospheric masses of CO_2 between
the regolith and polar caps (Fanale and Cannon, 1979 [9]; Toon
et al., 1980 [10]; Fanale et al., 1982 [11]). This is so
because the characteristic depth to which a thermal "wave"
penetrates is proportional to the square root of the period; for
reasonable assumptions as to thermal conductivity and for a 10^5
year period, the thermal "wave" could be felt to a depth of
hundreds of meters. It is tempting to immediately consider the
effect of such large thermal variations on the state of trapped
regolith H_2O, especially near the poles. However, we defer this
since this effect cannot be considered in isolation from
variations in atmospheric PH_2O produced by surface thermal
insolation variations on the polar caps.

Figures 3a and b show seasonal variation in surface polar
cap temperatures as a function of obliquity and eccentricity.
These temperature variations are sufficiently large to cause
corresponding variations in summer PH_2O near the caps of over
two orders of magnitude during the obliquity/eccentricity
compound cycle. It should be pointed out that the variations
shown at "zero depth" (the ordinate) in Figure 2 do not
correspond to those shown in Figure 3. Figure 2 (taken from a
previous publication) shows the thermal variation for a bare
patch of dirt subjected to the predicted insolation variations.
It does not take into account the effect of either albedo
changes or latent heat effects associated with seasonal or
astronomically driven changes in polar cap extent, as does the
present model.

The changes in polar cap seasonal thermal curves with
obliquity/eccentricity shown in Figure 3 are reflected in Mars'
wide changes in seasonal abundances of atmospheric basal PH_2O at

all latitudes. However, quantification of the relationship between the former and the latter requires an understanding of how the large near polar summer concentrations of H_2O are propagated seasonally across Mars. A similar problem was addressed by Jakosky (1983) [17] in attempting to model atmospheric H_2O column abundances on the basis of a polar source and either an opposite polar or regolith sink. Characteristic times deduced for propagation of the polar water were deduced by these investigators by taking into account global wind patterns and estimates of diffusion rates between cells. Values for e-folding times between reservoirs of ~7 days per 10° latitude band were deduced. We used, in our computations, an effective diffusion constant which was expressed in terms of the atmospheric parameters so as to represent the actual diffusion rate of H_2O through a static atmosphere. Although slower than the most likely rate of dispersion by winds, the predicted diffusion through a static atmosphere is only a factor of three slower. Since the latter is sure to occur and since it is easier to predict diffusion for other atmospheric pressures than to predict global weather patterns for those conditions, we choose to use a static diffusion model even though wind transport is somewhat dominant. Static diffusion would probably dominate at lower obliquities (hence pressures) in any event. Finally, the choice is not so critical since of the three processes involved in changing the abundance of atmospheric H_2O at lower latitudes--sublimation at the poles, atmospheric transport, and equilibration with regolith or polar sinks--the first is probably the slowest.

Figures 4a and b show predicted consequences of these obliquity/eccentricity variations for atmospheric PH_2O at mid latitudes (40° and 50°). We chose this latitude because under virtually all circumstances the boundary between stable and metastable ice varies between 30° and 60°. The variations shown in Figure 4 are between one and three orders of magnitude during the cycle. The curve with smaller oscillation represents the oscillations in PH_2O at the subsurface ice interface (here we chose the current depth of the ice interface). This is produced by the changes in the pore gas saturation PH_2O at the ice interface produced by the attenuated surface thermal wave at the interface.

From Figures 4a and b we can draw an important conclusion: the most important cause of variations in PH_2O gradients in the regolith, hence in the direction and magnitude of H_2O transport, is the atmospheric H_2O pressure variation brought about by surface temperature changes at the poles, not the effect of thermal changes at depth. This is so because a) thermal changes at the poles are greater than thermal changes elsewhere and are reflected in H_2O abundances at other latitudes, and b) thermal changes near the surface are obviously greater than the attenuated corresponding changes at depth. The fact that the curves in 4b cross periodically indeed means that net transport occurs from the atmosphere to the regolith and vice versa with a reversal time of ~10^5 years. However even though a portion of the regolith is alternately refilled with and dessicated of ice, the amount of ice involved is small. Our model suggests that the amount typically transferred to the interface in 10^4-10^5 years is <1g/cm^2. Hence profound morphological effects may not result. In some cases, however, exchange of H_2O at shallow depth could possibly aid some mechanical weathering processes, such as the recession of steep walls.

Figures 5a and b show the global recession of ice at low latitudes <40^0 - 50^0 and the deposition of ice at latitudes >50^0. The dessication of the equatorial zone to some significant depth was predicted by the work of Clifford and Hillel (1983) [12]. Several interesting new aspects emerge: First, note that the recession is predicted here despite the fact that our model takes into account all the intermediate term thermal variations in PH_2O at both the base of the atmosphere and at the ice interface as well as the thermal variations at the poles and at depth at all latitudes. Those variations, although they give a slight oscillatory character to the recession, do not affect it greatly. The atmospheric PH_2O variations constitute the main effect. These variations do cause a change in the surface H_2O density boundary condition, while the boundary condition at the ice interface remains relatively constant, being affected only slightly by the attenuated thermal wave (see Figure 4a and b). At latitudes <40^0 for most of Mars' history, the net flow of H_2O is clearly outward. However, as shown in Figure 4a, there are times (at highest obliquity) when the atmospheric H_2O pressure is so high that it is close to the H_2O pressure at the interface and the flow is greatly slowed. At somewhat higher latitudes (e.g., Figure 4b) the atmospheric H_2O pressure is alternately higher and lower than the H_2O pressure at the interface, and some periodic reinsertion and withdrawal does result. The amount transported per obliquity cycle is probably <1g cm^{-2} however, as indicated above.

Our results indicate that the average annual Mars atmospheric PH_2O at any place on Mars varies by up to three orders of magnitude during the obliquity and eccentricity cycles. This suggests caution in using Viking data to establish boundary conditions for global H_2O evolutionary models. The extremes are, however, 1.0 order of magnitude above and 2.0 orders of magnitude below the current value (~3 x 10^{-4} mm). Thus the effective average PH_2O which serves as the boundary condition for regolith ice stability is only a factor of 5-10 higher than that suggested by Viking results. The agreement is somewhat fortuitous, and is due to the fact that Mars' current obliquity is near its average value. Therefore, the current ice distributions suggested by Fanale (1976) [5] may not require serious revision on those grounds alone. Similarly, despite all the complex behavior of this system, the global removal of H_2O ice at low latitudes proceeds inexorably as in the model of Clifford and Hillel (1983) [12].

North of 50°, water is constantly added. In Figures 5a and b this is depicted as a filling in of originally ice free soil in the top ten meters. The existence of this soil in Figure 5 is purely an arbitrarily stipulated initial condition in our model. All such soil at latitudes >50° can, however, harbor stable ice. Note that H_2O mass is not conserved in Figure 5. The missing H_2O mass migrates to the polar cap. The amount of mass transferred to the cap during geologic time is surprisingly independent of initial conditions or assumptions. It is also rather independent of the amount of the initial overlying soil cover and even pore size, since the process is fairly efficient so long as the average pore size is >1μm. The concept of pore size may not be particularly meaningful at low latitudes since there appears to be a sharp concentration of true megaregolith at very high latitudes, while at lower latitudes any large reservoirs of megaregolith may--at least at present--be quite thoroughly infiltrated with consolidated intrusive and extruded igneous material (Guiness et al., 1984) [20].

For an average pore size of 1μm the amount transported poleward is ~3 x 10^6 km^3 and for 10μm it is 6 x 10^3 km^3. Several studies have called attention to morphological evidence of net removal of material from mid- and low-latitudes. These investigators have indicated that H_2O removal may have played an important role in shaping Mars' terrain, but caution that it is often difficult to distinguish between H_2O removal, removal of unconsolidated silicates, and magma withdrawal [e.g., Sharp (1973) [21] and McCauley (1973) [22]. The amount of material removed has been estimated to be between 10^6 and 10^7 km^3, while the mass of the north polar cap is ~3 x 10^6 km^3 (Masursky et al., 1980) [23]. We therefore speculate that the general correspondence among these values suggests that the global long term H_2O transfer process depicted in Figure 5 has played a major role in development of deflationary features at mid-latitudes and low latitudes and in the development of depositional features at high latitudes including the north polar cap and the layered terrain. In the case of the layered terrain, the water transported to the polar regions in excess of that currently in the cap is mixed with silicate in unknown proportion.

All the global generalization indicated by our model are subject to regional perturbations which may be localized in space and time. The most prominent of these is intrusion and extrusion of igneous material which can greatly alter the thermal structure. However, it has been shown that these effects are extremely limited in time and that the ambient global thermal regime is quickly reestablished (Coradini and Flamini, 1979) [24]. A more telling effect is the emplacement of juvenile H_2O or the remobilization and reemplacement of H_2O, the effects of which may persist locally for hundreds of millions of years. Nonetheless, the global picture developed in the model should be somewhat reflected in landforms, and morphological conflicts with the model may be relatable to regional igneous activity during Mars' history. Future studies should carefully identify the age of the suspected ice-related features, the minimal physical-chemical conditions required for their formation and the effect that the terrain type has on their apparent latitudinal distribution. Such comparisons between limits on physical conditions as a function of time and space suggested by morphological status vs. those suggested by models are extremely useful, but beyond the scope of this study. We note as an example however, two recent papers by Lucchitta in which polygonally fractured ground (Lucchitta, 1983) [25] and fretted terrain development (Lucchitta, 1984) [26] were related to the presence of ice. Our model is compatible with the suggestions of Lucchitta in the case of the polygonal ground, since the requirements are only for the presence of ice and significant thermal changes at depth. The first requirement (the features referred to occur at 47° N) is clearly satisfied (Figure 5) and so is the second (Figure 2). We note, however, that ice plays no role at all in Pechman's (1980) [27] exclusively tectonic scheme for polygonal ground formation. This suggests caution in using morphological studies to "test" models of H_2O distribution. In the case of the fretted terrain, Lucchitta suggests that an apparently higher concentration of ice at moderate depths at midlatitudes is suggested. This is attributed by Lucchitta to higher past obliquities. In this case, our model (see especially Figure 5) leads us to suggest that any such age effect is more easily attributed to the simple irreversible global water transfer which will gradually increase the depth to the ice interface at midlatitudes throughout Mars' history.

E.   CONCLUSIONS

1)   Variations in Mars' obliquity and eccentricity have produced great variations in two boundary conditions which drive H_2O transport in the regolith and H_2O transfer between the regolith and the polar caps. These two conditions are: a) the annual average H_2O

pressure at the base of the atmosphere and b) the H_2O pressure at the ice interface, controlled by astronomically driven deep regolith thermal variations. The first, which involves average atmospheric water abundance variations of a factor of $>10^2$ at all latitudes, is more important, especially at the critical mid latitude.

2) Despite the oscillatory variations in the water vapor gradient suggested above, inexorable and irreversible withdrawal of H_2O from the equatorial zone occurs throughout Mars' history as suggested by Clifford and Hillel (1983) [12] and is accompanied by transfer to any available regolith pores at latitudes $>45 - 50°L$ and to the polar cap and surrounding layered terrain. The amount of transferred H_2O is probably between 10^6 and $10^7 km^3$.

3) The dynamic processes described in the preceding conclusions played a major role in the development of the fretted terrain (probably) patterned and polygonal ground, the north polar cap and the layered terrain.

REFERENCES

1. Carr, M.H., and G.G. Schaber (1977). Martian permafrost features. J. Geophys. Res. **82**, pp. 4039-4054.

2. Rossbacher, L.A., and S. Judson (1981). Ground ice on Mars: Inventory, distribution and resulting landforms. Icarus **45**, pp. 39-59.

3. McElroy, M.B., T.Y. Kong, and V.L. Yung (1977). Photochemistry and evolution of Mars' atmosphere: A Viking perspective. J. Geophys. Res. **82**, pp. 4379-4388.

4. Pollack, J.B., and D.C. Black (1979). Implications of the gas compositional measurements of Pioneer Venus for the origin of planetary atmospheres. Science 205, pp. 56-59.

5. Fanale, F.P. (1976). Martian volatiles: Their degassing history and geochemical fate. Icarus **28**, pp. 179-202.

6. Ward, W.R. (1974). Climate variations on Mars, 1, Astronomical theory of insolation. J. Geophys. Res. **79**, pp. 3375-3386.

7. Ward, W.R., J.A. Burns, and O.B. Toon (1979). 2, Past obliquity oscillations of Mars: Role of the Tharsis uplift. J. Geophys. Res. **84**, pp. 243-259.

8. Gough, D.O. (1977). Theoretical predictions of variations in the solar output. in "The Solar Output and Its Variation". (O.R. White, ed.), Colorado Assoc. Univ. Press, Boulder, CO.

9. Fanale, F.P., and W.A. Cannon (1979). Mars: CO_2 adsorption and capillary condensation on clays: Significance for volatile storage and atmospheric history. J. Geophys. Res. **84**, pp. 8404-8414.

10. Toon, O.B., J.B. Pollack, W. Ward, J.A. Burns and R. Bilski (1980). The astronomical theory of climate change on Mars. Icarus **44**, pp. 552-607.

11. Fanale, F.P., J.R. Salvail, W.B. Banerdt, and R.S. Saunders (1982). Mars: The regolith-atmosphere-cap system and climate change. Icarus **50**, pp. 381-407.

12. Clifford, S.M., and D. Hillel (1983). The stability of ground ice in the equatorial region of Mars. J. Geophys. Res. **88**, pp. 2456-2474.

13. Fanale, F.P., J.R. .Salvail, A.P. Zent, and S.E. Postawko (1985). Distribution and Migration of Water on Mars. Submitted to Icarus.

14. Ward, W.R. (1979). Present obliquity oscillations of Mars: Fourth order accuracy in orbital E and I. J. Geophys. Res. **84**, pp. 237-241.

15. Wise, D.U, M.P. Golembleck, and G.E. McGill (1979). Tharsis province of Mars: Geologic sequence, geometry, and a deformation mechanism. Icarus **38**, 456-472.

16. Kieffer, H.H., T.Z. Martin, A. R. Peterfreund, and B. M. Jakosky (1977). Thermal and albedo mapping of Mars during the Viking primary mission. J. Geophys. Res. **82**, pp. 4249-4291.

17. Jakosky, B.M. (1983). The role of seasonal reservoirs in the water cycle II: Coupled models of the regolith, the polar caps, and atmospheric transport. Icarus **55**, pp. 19-39.

18. Fanale, F.P., and R.N. Clark (1983). Solar System Ices and Mars Permafrost, in "Permafrost: Fourth International Conference Proc." ISBN 0-309-03435-3 National Academy Press, Washington, D.C.

19. Jakosky, B.M., and C.B. Farmer (1982). The seasonal and global behavior of water vapor in the Mars atmosphere: Complete global results of the Viking atmospheric water detector experiment. J. Geophys. Res. **87**, pp. 2999-3019.

20. Guiness, E.A., C.E. Leff, and R.E. Arvidson (1984). On the Latitudinal Distribution of Debris in the Northern Hemisphere of Mars, in "Water on Mars", Proc. of the Workshop on Water on Mars, Pub. Lunar and Planetary Sciences Institute, Houston, Texas.

21. Sharp, R.P. (1973). Mars: Fretted and chaotic terrains. J. Geophys. Res. **78**, pp. 4073-4083.

22. McCauley, J.F. (1973). Mariner 9 evidence for wind erosion in the equatorial and mid-latitude regions on Mars. J. Geophys. Res. **78**, pp. 4123-4133.

23. Masursky, H., A.L. Dial, and M.H. Strobel (1980). Polar ice inventory - A progress report. In "Reports of Planetary Geology Program, 1979-1980" (NASA Technical Memorandum 81776), NASA, Washington, D.C. (Abstract)., pp. 399-401.

24. Coradini, M., and E. Flamini (1979). A thermodynamic study
     of the Martian permafrost. J. Geophys. Res. **84,** pp.
     8115-8130.

25. Lucchitta, E.K. (1983). Permafrost on Mars: Polygonally
     fractured ground. in"Permafrost: 4th Intl. Conf. Proc."
     Natl. Acad. Press, Washington, D.C.

26. Lucchitta, B.K. (1984). Ice and debris in the fretted ter-
     rain, Mars. J. Geophys. Res. **89,** (suppt.), pp. B409-B418.

27. Pechman, J.C., (1980). The Origin of Polygonal Trosyho on
     the Northern Plains of Mars: Icarus, **42,** pp. 185-2310.

28. Fanale, F.P. and R.N. Clark (1983). Solar System Ices and
     Mars permafrost.in"Proc. Fourth International Conference on
     Permafrost."

29. Fanale, F.P., and W.A. Cannon (1974). Exchange of adsorbed
     H_2O and CO_2 between the regolith and atmosphere of Mars
     caused by changes in surface isolation. J. Geophys. Res.
     79, pp. 3397-3402.

30. Farmer, C. B., D.W. Davies, and D.D. LaPorte (1976). Mars:
     Northern summer ice cap-water vapor observations from Vik-
     ing 2. Science **194,** pp. 1339-1341.

31. Farmer, C.B., D.W. Davies, A.L. Holland, D.D. LaPorte, and
     P.E. Doms (1977). Mars: Water vapor observations from the
     Viking Orbiters. J. Geophys. Res. **82,** pp. 4225-4248.

32. Farmer, C.B., and P.E. Doms (1979). Global seasonal varia-
     tion of water vapor on Mars and the implications for per-
     mafrost. J. Geophys. Res. **84,** pp. 2881-2888.

33. Hess, S.L. (1976). The vertical distribution of water va-
     por in the atmosphere of Mars. Icarus **28,** pp. 269-278.

34. Smoluchowski, R. (1968). Mars: Retention of ice. Science
     159, pp. 1348-150.

35. Walker, J.C.G. (1977). Evolution of the Atmosphere, Mac-
     millan Co., New York.

SUBSURFACE ICE AND PERMAFROST ON MARS

Duwayne M. Anderson

Texas A&M University

ABSTRACT

Terrestrial permafrost varies widely in its physical and mechanical properties and behavior. Ice content, for example, may range from 0 to 100 % by volume.
The types of subsurface ice are numerous and the crystal structure of terrestrial subsurface ice is variable.
Most subsurface ice is hexagonal, Ice-I ; clathrate
structures are known, however. The ice content of permafrost is only a fraction, albeit the predominant one, of
the water present. A significant portion of the water
present exists in an unfrozen state and is distributed
throughout the pore space and in interfacial areas. The
proportion of ice to unfrozen water varies, in a characteristic manner, with temperature and solute concentration. These basic facts are important in determining the
strength and deformation properties of permafrost and
also its hydrological and electrical properties. Reliable relationships among these properties are derivable
from basic thermodynamic theory and from empirical relationships recently established on the basis of laboratory and field data.

Permafrost exists at all latitudes on Mars and
subsurface ice probably is abundant. The temperatures
and pressures characteristic of each location or region
determine, to a large extent, the depth and distribution
of permafrost. Together with ground water salinity, they
control the ice content, strength and deformation characteristics, in addition to other physical and electrical properties of local permafrost. Calculations based
on the Viking Mission Data indicate that permafrost

J. Klinger et al. (eds.), Ices in the Solar System, 565–581.

thicknesses range from about 3.5 km at the equator to approximately 8 km in the polar regions. The depths to the bottom of Martian permafrost are more than three times the depth characteristic of permafrost in terrestrial polar locations.

Martian permafrost, in general, is much colder than terrestrial permafrost. Consequently, the proportion of unfrozen water to ice generally is much lower. This, however, probably is somewhat offset by a significantly higher salinity of the Martian permafrost. The combination of low temperatures and great thicknesses of Martian permafrost, coupled with the low atmospheric pressure and very small snowfall, enhance the stability of the Martian surface. The "active layer" on Mars is extremely thin compared to that of terrestrial permafrost, making Martian permafrost more resistant to deformation and abrasion than is the case on Earth. The occurrence, quantities and behavior of subsurface ice, currently a matter of speculation and conjecture, is important in many respects. Its determination has been an objective of high priority in the exploration of Mars.

The presence of water-ice on Mars was inferred many years ago, mainly from spectroscopic data. These data indicated that Martian temperatures are cold enough to provide the necessary environmental conditions for stability and the absorption spectra of the polar caps were sufficiently similar to the spectra observed for ordinary ice on earth to justify the conclusion that ice is common on Mars (1-2-3-4). Data obtained from the Mariner Missions and with instruments on Viking Landers 1 and 2 verified this conclusion. The meteorological instruments confirmed the existence of environmental conditions necessary for the permanent presence of water-ice. The gas chromatograph/mass spectrometer instrument confirmed the presence of water in the soil materials scooped from the Martian surface (5-6) and the video cameras recorded the episodic appearance of frost and snow (7-8-9). At the same time video cameras on the Mariner Missions, particularly Mariner 9, and the Viking Orbiters mapped extensive deposits of snow and ice in the north and south polar regions (10-11).

Many questions remain unanswered, however. Although there is evidence that solid carbon dioxide is deposited periodically with water-ice on the polar caps, the mode of deposition and the manner in which these substances are intermixed is not known. Admixtures of various proportions are possible, especially at depth where pressure and temperature conditions are known to be sufficient for the formation of clathrate and interstitial forms (12). Although natural gas hydrates are known to occur widely at depth on Earth (13-14-15-16), they are not likely to be present in similar abundance on Mars because of a general lack of biogenic compounds on Mars (5). Abiogenic origins are possible, however, so the possibility that they may occur at depth cannot be completely ruled out.

Fine grained, mineral matter frequently is known to be brought into the Martian atmosphere by sustained, strong winds. It is then deposited, partially removed and redeposited again and again over virtually the entire surface of the planet. Imagery of the Martian polar caps showing banded layers in the Martian polar ice together with a varying albedo indicate the presence of wind transported mineral matter in widely varying proportions in the polar snow and ice (11). At other latitudes, although accumulations of surface ice may be absent, temperatures everywhere generally are low enough to sustain subsurface ground ice.

Maximum and minimum surface temperatures for virtually all locations on the Martian surface were obtained from radiometer-derived temperatures recorded by the two Viking Orbiters (9). Information on the thermal conductivity of the Martian surface materials and the underlying regolith together with information on the thermal flux from the interior of the planet permit estimates of the geothermal gradient for Mars. An estimate of 4.4° K per 100 meters recently was published (17). To arrive at this figure, a thermal conductivity characteristic of an iron-oxide rich soil material containing about 60% ice was assumed together with a surface heat flow of 0.035 W m^{-2} (derived by Toksoz & Hsui (18) on theoretical grounds). This estimate, together with the average surface temperatures observed by the Viking Orbiters, implies depths to the bottom of the Martian permafrost ranging from about 1 km at the equator to about 3 km at the poles.

These estimates are conservative, however. It is unlikely that an iron-oxide rich soil containing 60% ice is very representative of the Martian Regolith, spatially and to depths of several kilometers. If it is assumed that the regolith is composed mainly of basalt at lower ice contents with a thermal conductivity of approximately 2.5 W m^{-1} $°K^{-1}$ instead of the 0.8 W m^{-1} $°K^{-1}$ employed by Rossbacker and Judson, an estimate of about 1.5 °K per 100 meters is obtained. This implies distances to the bottom of Martian permafrost ranging from a little over 3 km at the equator to between 7 and 8 km at the poles.

Estimates such as these are very useful, but only to illustrate general features and to begin introductory discussions of Martian permafrost. Actual circumstances and occurrences surely must be very complex and the degree of

variability very large. For example, the thermal flux on Earth
commonly varies from location to location by as much as a
factor of four. No doubt variations of similar magnitude occur
on Mars. Even wider variations may exist between areas of
current or recent volcanic activity and less active locations.
Values of thermal conductivity must vary widely from place to
place also, just as is the case on Earth, but the data presently
available are insufficient to support any except general
conclusions and hypothetical and speculative discussions
(19-20-21-22).

On Earth the depth to the bottom of permafrost in the
North Polar Region, where investigations have been
concentrated, can be as much as 1.5 km in certain locations in
Siberia and North America. There it is quite thick and almost
continuous. At more southerly locations, where permafrost
often is found in discontinuous, isolated masses, it may be
few meters in thickness. Mars is a much colder planet than
Earth. It is reasonable, therefore, to expect a much thicker
and more extensive cryosphere there than exists on Earth.
Observations generally bear this out. Another consequence of
the generally colder temperatures on Mars is to cause the
"Active Layer," (the thickness of the surface layer that rises
annually above 0°C) to be much thinner than is characteristic of
permafrost on Earth. It ranges from about 2 cm at the poles to
about 10 cm at the equator. This is the layer that might be
said to be seasonally "thawed". Actually it is kept dry or
almost dry by evaporation and sublimation. In this state it
provides the finely divided mineral matter that commonly becomes
entrained in the atmosphere during periodic atmospheric
disturbances and that causes the extensive dust storms
frequently observed on the planet.

The properties and behavior of permafrost become important
when one considers the possibility of establishing research
outposts or more elaborate settlements on Mars. Its properties
and behavior wherever it may be found are determined by complex
interactions between ice in its various forms and mineral
matter in its various forms and states of subdivision. The
purpose of this paper is to review our understanding of the
most important factors that govern these interactions and to
consider some of the consequences as they have a bearing on the
strength and rheological behavior of frozen ground.

The surface of ice in contact with air is known to support
a liquid-like transition layer (23-24-25-26-27). When ice
crystals are placed in contact with the surfaces of silicate
minerals this liquid-like transition zone is enlarged. The
thickness of the liquid-like, interfacial water separating the
ice crystals from mineral surfaces depends sensitively on

temperature. From the freezing point down to about -5°C it ranges from 5 x 10^{-7}cm or more to about 9 x 10^{-8}cm. Below -5°C and down to -150°C the thickness of this interfacial layer diminishes from about 9 x 10^{-8}cm to about 3 x 10^{-8}cm. In extremely fine grained materials such as clays which have very large specific surface areas, the quantity of unfrozen water in the interfacial layer separating the ice from the silicate surfaces ranges from 5 to 10% by weight at extremely low temperatures, to approximately 25% at temperatures between 5°C and -10°C. At temperatures of -1°C to a few tenths of a degree below freezing it may amount to 100% or more by weight.

If soluble substances are present, the quantities of unfrozen interfacial water are larger at any given temperature and pressure. The effect of strong brines is to increase the thickness of the unfrozen, interfacial zone. Increasing pressure at any given temperature is known to have a similar effect, although the effect of pressure on unfrozen water contents has ye to be investigated systematically (28). The relationship between the quantity of unfrozen interfacial water and temperature is illustrated in Figure 1 where experimental data for some representative materials is displayed. When direct measurement is not possible the unfrozen water content, w_u, may be estimated from the relationship

$$\ln w_u = a + b \ln S + c S^d \ln \theta \quad (1)$$

if specific surface area of the mineral matter is known. In Equation (1) S is the specific surface area in square meters per gram, is in degrees below 0°C and a, b, c and d are 0.2618, 0.5519, -1.449, -0.264, respectively (19).

At temperatures below -10°C the mobility of the unfrozen interfacial water becomes much reduced. Low temperature scanning calorimeter data for nontronite are shown in Figure 2. Cooling curves for nontronite at different water contents with and without the presence of a calcium chloride and sodium sulphate brine are shown in three dimensional plots (29). These curves are interpreted as follows: As the temperature is lowered from 280° K through the normal freezing point of water, to about 266° K nothing is observed even though the sample has become cooled below its normal freezing point. Spontaneous nucleation occurs at about 265° K in the case of the clay with the highest water content (about 40% water by weight). Nucleation is accompanied by the release of latent heat. As it appears, it is registered in arbitrary intensity units by the calorimeter detector. In this case three separate peaks are observed, suggesting that three distinct nucleation events occurred within the sample. This leads, as usual, to the

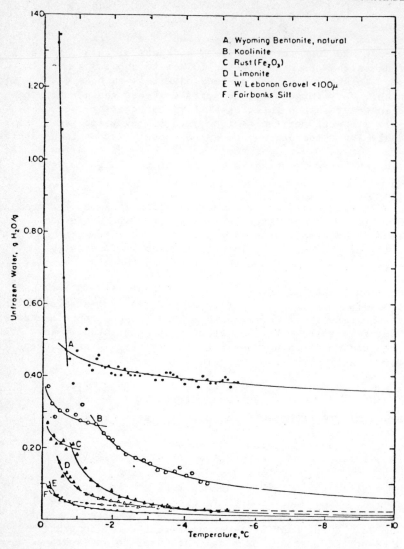

Figure 1.   Water-ice phase-composition curves for
            2 representative clays, A, B; 2 repre-
            sentative soils, E, F; limonite and rust
            (27).

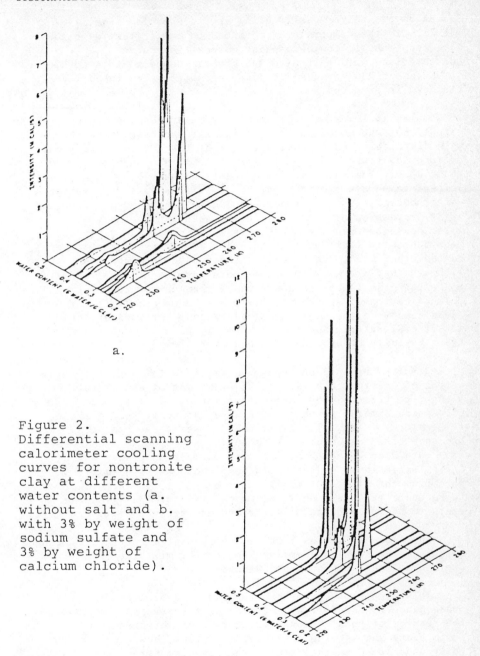

a.

Figure 2.
Differential scanning
calorimeter cooling
curves for nontronite
clay at different
water contents (a.
without salt and b.
with 3% by weight of
sodium sulfate and
3% by weight of
calcium chloride).

b.

formation of polycrystalline ice within the clay mineral
matrix. As the temperature continues to fall, freezing becomes
more complete, following a phase composition curve
characteristic of nontronite and almost identical to A in Figure
1. At about 236° K an additional, relatively low intensity
exotherm is produced.

In samples at lower water contents, similar results are observed
except that the magnitude of the initial exotherm diminishes
with diminishing water content. All of the curves exhibit one
or more low temperature exotherms at temperatures between 230° K
and 240° K. The results of many analyses have shown that the
initial temperature of nucleation diminishes with diminishing
water content, in general correspondence to the normal freezing
point depression of the water in the specimen at the various
water contents plus 4 to 6° K corresponding to the super cooling
required to initiate nucleation. The low temperature exotherms,
on the other hand, are virtually independent of water content,
although they are seen to become more complex at very low water
contents. The low temperature phase changes may result from the
solidification of a eutectic mixture of some complexity or be
associated with the freezing of additional portions of the
interfacial water. Additional investigation is required in
order to fully understand this behavior.

Results obtained on the same sample containing 3% by
weight of calcium chloride and 3% by weight of sodium sulfate
brought to various water contents are shown in Figure 2b. The
curves are similar to those of Figure 2a except that the
temperature of initial nucleation is lower because of the
presence of these solutes. The curves are smoother, however,
and contain no evidence of multiple nucleation events during
initial freezing. A distinct second exotherm is seen to
accompany the first but trailing it by 12° to 15° K. The low
temperature exotherms observed in the case of the salt-free
clays to be independent of water content are not visible.

The data of Figures 1 and 2 are presented to illustrate and
emphasize the complex freezing behavior of mixtures of fine
grained silicate minerals and water.

This complex behavior has important rheological
consequences. However, the effects of interfacial phenomena,
the results of phase changes brought on by variations in
temperature and varying levels of heat flux, however, are not
fully understood at this time. Cursory discussions and
generalizations are available but additional study is required
before the mechanical behavior and the mechanical properties of
frozen and thawing earth materials can be fully described
(30). Partly because of the presence of unfrozen interfacial

water and partly because of the normal response of ice to the imposition of stress, which ranges from pressure melting to brittle fracture, continuous deformation due to creep under sustained stress is the dominant geotechnical feature of permafrost and frozen earth materials. The rate of creep under a given load is dependent principally on temperature, unfrozen water and ice content and the mineralogy and physical state of the mineral matter present.

The reaction of polycrystalline ice to the stresses of loading has been extensively investigated. A great quantity, mostly uncorrelated, of stress-strain behavior has been observed and reported. From this, it is apparent that during creep the two most important destructive processes that occur are displacement along crystal interfaces (which leads to structural damage), the redistribution of impurities (which further weakens the material), and the eventual formation of "slip-bands" and cracks. According to Michel (31), the formation of "slip-bands" is the main mechanism of shear. At low and intermediate stresses a small elastic deformation, followed by a transient creep that eventually reaches steady state often is observed. This steady state creep may accelerate to a rate that is about a factor of ten higher than the inital rate of steady state creep. Cracks usually appear, eventually, even at relatively low stresses and their number increases rapidly at higher stresses, leading to rapid failure.

The phenomenon of regelation, on the other hand, is a healing mechanism that often is very effective. The interrelationship of these destructive and healing processes has been effectively illustrated by Pusch (32). He consolidated published data in a plot of log strain rate versus log time. The result is shown in Figure 3. In his diagram it is seen that three of the curves, those corresponding to the highest stresses and rates of strain, all eventually proceed to failure (defined as accelerating rates of strain). Those for lower stresses and rates of strain usually reach a condition of state creep (constant rate of strain). The shapes and positions of these curves support the view that polycrystalline ice fails by cracking at high stresses. At low stresses and low rates of strain the healing processes of regelation are thought to allow the formation of slip-bands. These then act to facilitate deformation while retaining continuity and strength. At rates of strain and for times below the dotted line in Figure 3, the Andrade relation, (log strain rate proportional to -2/3 log time) often applies.

The strength and deformation characteristics of permafrost are similar. As mentioned above, they depend upon the nature of the mineral matter, the applied stress levels, the

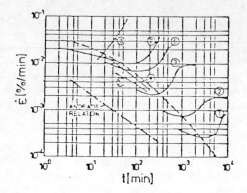

1) σ= 0.7 MPa, T = -4.8°C, Granular ice (Ref.33)
2) σ= 0.9 MPa, T = -4.8°C, Granular ice (Ref.33)
3) σ= 1.2 MPa, T = -4.8°C, Granular ice (Ref.33)
4) σ= 1.5 MPa, T = - 10°C, Columnar ice (Ref.34)
5) σ= 1.0 MPa, T = - 10°C, Columnar ice (Ref.34)
6) σ= 1.3 MPa, T = -4.5°C, Fine-granular ice (Ref.35)
7) σ= 1.0 MPa, T = -4.5°C, Fine-granular ice (Ref.35)
8) σ= 1.6 MPa, T = - 10°C, Columnar ice (Ref.34)

Figure 3.   Uniaxial creep of ice (Ref. 32)

temperature below freezing and the amounts of ice and unfrozen water present. A very wide range of behavior is observed as these factors are varied. At sufficiently high ice contents, the healing process of regelation leads to behavior like that illustrated in Figure 3, except that much higher stresses may be applied because of the additional strength contributed by the rigid granular material. Typical values of unconfined compressive strength as a function of temperature for several frozen soil materials are shown in Figure 4.

Three curves are shown for ice, illustrating the wide variation in its behavior. This variation in strength is a reflection of variation in ice crystal size, air content, homogeneity, and purity. It is significant that the fine grained soil materials show increasing strength down to very low temperatures. This no doubt is related to the diminishing quantities and changes in the properties of the unfrozen interfacial water in these materials. These effects are observed in frozen sand down to about -50°C. Below that temperature the quantity of unfrozen interfacial water in this relatively coarse material is completely negligible and the strength of the frozen sand becomes comparable to that of ice.

It is evident from the data presented in Figure 4 that very heavy loads can be supported by permafrost as long as it is protected from rising temperatures. This may be accomplished by thermally isolating all load-bearing structures from the permafrost. Loads may be increased by lowering the temperature of permafrost. This may be accomplished in a variety of ways well known now in permafrost engineering. Martian permafrost generally is colder than -50°C. Consequently, the very high loading densities of Figure 4 may be safely employed in preliminary design considerations. The possible presence of saline conditions must be anticipated, however. The presence of soluble substances in the Martian permafrost will act to increase the unfrozen water content of Martian permafrost and thus lead to a reduction in strength. The influence of soluble substances on the amount of water remaining unfrozen at any given temperature may be estimated by the method described by Banin and Anderson (39). Methods of predicting the strength and deformation characteristics of saline permafrost from unfrozen water contents are not yet available, although investigations designed to accomplish this are in progress.

It is quite possible that experimental research stations can be established on Mars by the end of this century or in the early part of the 21st century. Unlike the moon, which is almost totally devoid of water, the Martian surface can be expected to contain ice and unfrozen water or brine in varying

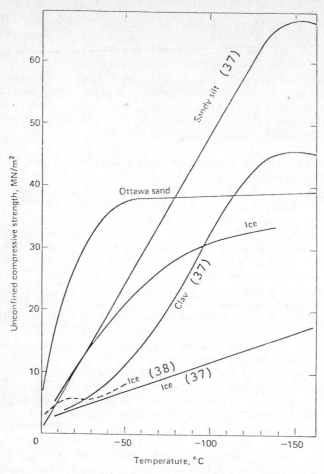

Figure 4. The unconfined compressive strength
           of several frozen soils and ice as
           a function of temperature (Refs 30-
           36-37-38)

proportions. The ability of this surface to sustain long term
loads will depend upon the creep characteristics of the Martian
permafrost as they are determined by the characteristics of the
mineral fabric, the temperature of the frozen ground, the ice
and unfrozen water contents and the presence or absence of
soluble materials. Sufficient information on the properties
and behavior of terrestrial permafrost is available now for
preliminary design purposes. The additional information needed
to assure a reliable final design can be obtained from
experiments and measurements that could be conducted by
unmanned soft landers or ballistic bodies instrumented and
configured for this purpose. From the preceding discussion, it
is apparent that critical measurements are:

1. Gross resistance of the Martian surface to deformation
    and penetration.

2. Subsurface temperature profiles.

3. Ice and unfrozen water contents as a function of
    depth.

4. Quantities and identity of water soluble substances
    present.

5. Grain size distribution of the unconsolidated
    materials.

Other measurements, such as local environmental conditions,
topography etc., are possible and will be useful. All of these
measurements can be made from unmanned spacecraft. Soft
landers such as Vikings 1 and 2 are capable of very extensive
surface and subsurface investigations. Ballistic bodies
designed for deep penetration of the Martian surface also are
capable of accomplishing these determinations at a much lower
cost per impact. Consequently, more extensive spatial coverage
is feasible. A manned landing would permit even more extensive
measurements and investigations.

The relatively favorable Martian environment and the
presence of available water at the surface of Mars justifies an
objective of establishing an experimental station there. The
additional information required to design the necessary
components of this station, insofar as the bearing capacity and
other properties of the Martian surface is concerned, are
clearly defined. The technological feasibility of the overall
endeavor has been demonstrated and is no longer a serious
question. One or more additional space missions to Mars,
however, is required for site selection and site
characterization. A Mars penetrator spacecraft is well suited
for this task.

REFERENCES

(1)   Sinton, V. M. and Strong, J. (1960). Radiometric
          Observations of Mars. Astrophys. J., 131, pp.
          459-469.

(2)   Leighton, R. B., and Murray, B. C. (1966). Behavior of
          Carbon Dioxide and other Volatiles on Mars. Science,
          84, pp. 136-144.

(3)   Morrison, D., Sagan, C. and Pollack, J. B. (1969).
          Martian Temperatures and Thermal Properties. Icarus,
          11, pp. 36-45.

(4)   Neugebauer, G., Munch, C., Chase Jr., S. C., Hatzenbeler,
          H., Miner, E. and Schofield, D. (1969). Mariner
          1969: Preliminary Results of the Infrared Radiometer
          Experiment. Science, 166, pp. 98-99.

(5)   Biemann, K., Oro, J., Toulmin, III, P., Orgel, L. E.,
          Nier, A. O., Anderson, D. M., Simmonds, P. G., Flory,
          D., Diaz, A. V., Rushneck, D. R., Biller, J. E., and
          Lafleur, A. L. (1977). The Search for Organic
          Substances and Inorganic Volatile Compounds in the
          Surface of Mars. Journal of Geophysical Research,
          82, pp. 4641-4658.

(6)   Anderson, D. M. (1978). Water in the Martian Regolith.
          Comparative Planetology, Academic Press, pp.
          219-224.

(7)   Farmer, C. B., Davis, D. W., Holland, A. L., LaPort, D.
          D., and Doms, P. E. (1977). Mars: Water Vapor
          Observations from the Viking Orbiters. Journal of
          Geophysical Research, 82, pp. 4225-4248.

(8)   Farmer, C. B. and Doms, P. E. (1979). Global Seasonal
          Variations of Water Vapor on Mars and the
          Implications for Permafrost. Journal of Geophysical
          Research, 84, pp. 2881-2888.

(9)   Kieffer, H. H., Martin, T. Z., Peterfreund, A. R.,
          Jakosky, B. M., Miner, E. D., Palluconi, F. D.
          (1977). Thermal and Albedo Mapping of Mars During the
          Viking Primary Mission, Journal of Geophysical
          Research, 84, pp. 4249-4291.

(10) Murray, B. C. and Malin, M. C. (1973). Polar Volatiles on
        Mars - Theory Versus Observation. Science, 182, pp.
        437-443.

(11) Cutts, J. A., Blasius, K. R., Briggs, G. A., Carr, M. H.,
        Greeley, R., and Masursky, H. (1976). North Polar
        Region of Mars: Imaging Results From Viking 2.
        Science, 194, pp. 1329-1337.

(12) Miller, S. L., and Smythe, W. D. (1970). Carbon Dioxide
        Clathrate in the Martian Ice Cap. Science, 170, pp.
        531-533.

(13) Judge, A. (1982). Natural Gas Hydrates in Canada.
        "Proceedings Fourth Canadian Permafrost Conference",
        National Research Council, Ottawa, Canada, pp.
        320-328.

(14) Weaver, J. S. and Stewart, J. M. (1982). In Situ Hydrates
        Under the Beaufort Sea Shelf. "Proceedings Fourth
        Canadian Permafrost Conference", National Research
        Council, Ottawa, Canada, pp. 312-319.

(15) Makogon, Y. F. (1982). Perspectives of the Development of
        Gas-Hydrate Deposits. "Proceedings Fourth Canadian
        Permafrost Conference", National Research Council,
        Ottawa, Canada, pp. 299-304.

(16) Kvenvolden, K. A. (1982). Occurrence and Origin of Marine
        Gas Hydrates. "Proceedings Fourth Canadian Permafrost
        Conference", National Research Council, Ottawa, Canada,
        PP. 305-311.

(17) Rossbacker, L. A. and Judson, J. (1981). Ground Ice on
        Mars: Inventory, Distribution, and Resulting
        Landforms. Icarus, 45, pp. 39-59.

(18) Toksoz, M. N. and Hsui, A. T. (1978). Thermal History and
        Evolution of Mars. Icarus, 34, pp. 537-547.

(19) Anderson, D. M., Gatto, L. W. and Ugolini, F. (1973). An
        Examination of Mariner 6 and 7 Imagery for Evidence
        of Permafrost Terrain on Mars. "International
        Conference on Permafrost, 2'd Yakutsk, Siberia, N.
        American Contribution". National Academy of Science
        Pub., pp. 449-508.

(20) Gatto, L. W. and Anderson, D. M. (1975). Alaskan
        Thermokarst Terrain and Possible Martian Analog.
        Science 188, no. 4185, pp. 255-257.

(21) Coradini, M. and Flamini, E. (1979). A Thermodynamical
        Study of the Martian Permafrost. Journal of
        Geophysical Research, 84, pp. 8115–8130.

(22) Fanale, F. P., Salvail, J. R., Banerdt, W. B. and
        Saunders, R. J. (1982). Mars: The
        Regolith-Atmosphere-Cap System and Climate Change.
        Icarus, 50, pp. 381–407.

(23) Hosler, C. L., Jenson, D. C. and Goldschlak, L. (1957).
        On the Aggregation of Ice Crystals to Form Snow. J.
        Meteorol., 14, pp. 415–420.

(24) Jellinek, H. H. G. (1967). Liquid-Like (Transition) Layer
        on Ice. J. Colloid Interface Sci., 25, pp. 192–205.

(25) Jellinek, H. H. G. and Ibrahim, S. H. (1967). Sintering
        of Powdered Ice. J. Colloid Interface Sci., 25, pp.
        245–254.

(26) Hobbs, P. V. and Mason, B. J. (1964). The Sintering and
        Adhesion of Ice. Phil. Mag., 9, pp. 181–197.

(27) Anderson, D. M. and Morgenstern, N. R. (1973). Physics,
        Chemistry and Mechanics of Frozen Ground.
        "International Conference on Permafrost, 2'd Yakutsk,
        Siberia, N. American Contribution". National Academy
        of Sciences Pub., pp. 257–288.

(28) Anderson, D. M. (1967). The Interface Between Ice and
        Silicate Surfaces. Journal of Colloid and Interface
        Science, 25, pp. 174–191.

(29) Anderson, D. M. and Tice, A. R. (1980). Low Temperature
        Phase Changes in Montmorillonite and Nontronite at
        High Water Contents and High Salt Contents. Cold
        Regions Science and Technology, 3, pp. 139–144.

(30) Andersland, O. B. and Anderson, D. M. (1978).
        Geotechnical Engineering for Cold Regions.
        McGraw-Hill.

(31) Michel, B. (1977). A Mechanical Model of Creep of
        Polycrystalline Ice. Canadian Geotechnical Journal,
        15, pp. 155–170.

(32) Pusch, R. (1980). Creep of Frozen Soil, A Preliminary
      Physical Interpretation in "Proceedings Second
      International Symposium on Ground Freezing."
      Norwegian Institute of Technology, Trondheim, Norway,
      pp. 190-201.

(33) Steinemann, S. (1958). Experimentelle Untersuchungen zur
      Plastizität von Eis. Beiträge zur Geologie der
      Schweiz. Hydrologie No. 10. Kommissionsverlag
      Kummerly & Frey Ag., Geographischer Verlag, Bern.

(34) Gold, G. W. (1960). The Cracking Activity in Ice During
      Creep. Can. J. Phys., 38, pp. 1137-1148.

(35) Ting, J. M. and Martin, R. T. (1979). Application of the
      Andrade Equation to Creep Data for Ice and Frozen
      Soil. Cold Regions Sci. and Technology, 1, pp.
      29-36.

(36) Sayles, F. H. (1966). Low Temperature Soil Mechanics.
      U.S. Army Cold Reg. Res. Eng. Lab. Tech. Note,
      Hanover, N. H.

(37) Wolfe, L. H. and Thieme, J. O. (1967). Physical and
      Thermal Properties of Frozen Soil and Ice. Soc. Pet.
      Eng. J., 4, pp. 67-72.

(38) Butkovich, T. R. (1954). Ultimate Strength of Ice. U.S.
      Army Res. Rep. 11.

(39) Banin, A. and Anderson, D. M. (1974). Effects of Salt
      Concentration Changes During Freezing on the Unfrozen
      Water Content of Porous Materials. J. Water
      Resources Research, 10, pp. 124-128.

DISCUSSION

E. Whalley :

The low-temperature, transitions that you attributed to a
liquid-to-glass transition in a thin layer of water occurred with
an absorption of heat, whereas a glass transition occurs with a
reduction of heat capacity but no heat absorption. It looks more
like the heat effects that occur at an eutectic transition. Could
that be the explanation ?

D. Anderson :
      Yes ; it is definitely another possibility.

GEOMORPHOLOGIC EVIDENCE FOR GROUND ICE ON MARS

Baerbel K. Lucchitta

U. S. Geological Survey, 2255 North Gemini Drive,
Flagstaff, Arizona 86001

ABSTRACT

For ground ice to exist on Mars, two conditions
have to be met. One is the presence of permafrost ; the
second is the availability of water. Because the mean
temperature of Mars'surface is - 80° C., permafrost 1-3
km thick occurs over the entire planet. Remote-sensing
measurements suggest that water presently exists in the
atmosphere and in the polar caps ; frost has been observed at the Viking landing sites. Landforms that support
the contention that ground ice exists on Mars include lobate ejecta from craters, small valley networks, and numerous features that may be attributed to thermokarst.
Chaotically collapsed terrain, chain craters, irregular
depressions, and valleys tributary to canyons all appear
to be related to loss of ground ice. Patterned ground has
dimensions similar to that on Earth as well as dimensions
of giant size. Masswasting features are common on scarps.
Even though individual features may be explained as results of other processes, the combined evidence indicates that a large reservoir of ground ice has existed on
Mars.

INTRODUCTION

Since the seventeenth century, when Cassini and
Huygens sketched bright patches at the Martian poles, we
have known that Mars has polar caps. In the next century,
Herschel noticed their waxing and waning, and first suggested that Mars may have polar caps comparable to those

J. Klinger et al. (eds.), Ices in the Solar System, 583–604.

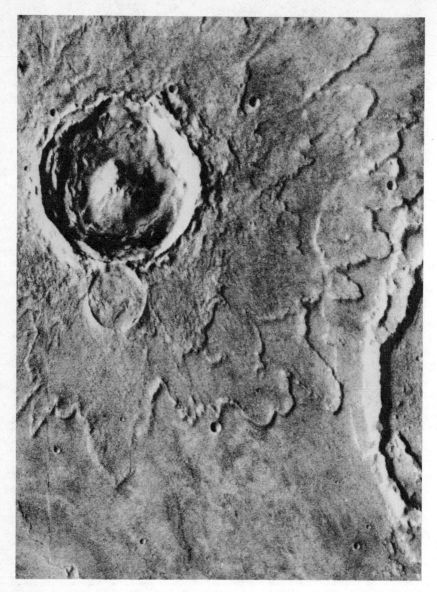

Fig. 1   Crater with lobate ejecta. Unlike crater
ejecta on the Moon and Mercury, which are radially arranged and ballistically emplaced, crater ejecta on
Mars tend to have lobate margins and appear to be emplaced by ground-hugging flow. Picture about 50 km wide ; north toward upper left, illumination from left
(lat 22° N., long 34° W. ; Viking Orbiter image 3A07).

on Earth. Now we know that the Martian north-polar cap
is composed of dirty water ice, and that the south cap
most likely contains water ice with frost of CO_2 (1). Is
the ice contained in the polar caps the only ice on Mars,
or does Mars contain ice elsewhere, which is not as easily accessible to remote sensing measurements ? Lewis
(2) predicted that Mars'initial water content should be
six times per unit mass that of the Earth on the basis
of equilibrium models of the Solar System, but measurements obtained from the Martian surface since 1974 appear to be at odds with his prediction ; except for indicating winter frost at the Viking II lander site (3),
these measurements give no indication of surface ice in
the Martian equatorial areas and middle latitudes.

However, other remote-sensing observations do
suggest that Mars contains much ice. These observations
are based on images. Imaging does not provide readily
quantifiable data, yet it is an important tool, and provides constraints on theoretical models. Deductions based on images rely on comparisons with observations founded in our terrestrial experience, and no single observation necessarily proves that similar processes operated on other planets where environmental conditions were
and are different. Yet, if large numbers of observations
and comparisons all point towards the same condition or
process, the probability becomes high that the inferred
condition indeed existed, or that the inferred process
indeed took place on other planets.

Images returned from Mars show numerous landforms that resemble those produced by ground ice on Earth.
For ground ice to exist on Mars, two conditions have to
be met. One is the presence of permafrost ; the second
is the availability of water.

THE MARTIAN ENVIRONMENT

Mars'mean annual temperature is - 80° C overall,
- 60° C at the equator (4). Rossbacher and Judson (5)
calculated that permafrost would be 1 km thick near the
equator and 2-3 km thick near the poles, assuming the
heat-flow values of Toksöz and Hsui (6) and a surface
layer composed of hard-frozen limonitic soil with a high
(60 %) ice content. Near the equator, temperatures may
reach 15° C, but night-time temperatures remain far below 0° C (7). Freeze-thaw processes, therefore, are theoretically possible, however, the formation of an active
layer is unlikely because the equatorial near surface
soil is likely to be desiccated (4).

Fig. 2   Small valley networks. Integrated system suggests that the valleys were carved by a fluid, which was
most likely derived by sapping of ground ice. Picture
180 km wide ; north toward upper left, illumination from
upper right (lat. 42° S., long 93° W. ; Viking Orbiter
image 63A09).

Mars'atmosphere is composed dominantly of carbon dioxide at a pressure of approximately 600 Pa (.006
bar). Water vapor in the present atmosphere is scarce ;
Farmer and Doms (8) measured generally less than 10 precipitable microns except near the North Pole in summer,
where the amount rose to as much as 50 precipitable microns. Because of the thin atmosphere and low temperatures, water vapor is in equilibrium with the ground poleward of about lat. 40° ; equatorward, water vapor migrates from the ground into the atmosphere (4) unless fine
soil builds a diffusion barrier that permits ice to remain in the ground, possibly for billions of years (9).

The presence of water is also documented by the
Viking landers. Samples were lighter after mass-spectro-
meter measurements than when introduced into the measuring chamber ; apparently, carbon dioxide and water (an
estimated 2 %) were driven off (10). This estimate does
not include capillary water, physically adsorbed water,
and ice, and so the total water content of the samples
may have been higher. Clark (11) estimated on theoretical grounds that the content of ice taken up in the soil
may have been as high as 80-120 % the weight of the soil.
Seasonal frost observed on the Viking 2 lander site (3)
probably also consisted of water ice or clathrates, because the temperature of the site was too high for carbon dioxide ice to remain on the ground without evaporating (12).

Water, though scarce now, may have been much more abundant on Mars in the past. As mentioned above, Lewis (2) proposed that initially Mars had much more water
per unit mass than the Earth. Viking atmospheric measurements have changed this assessment : Anders and Owen
(13) projected that Mars never contained more than about
10 meters of water, uniformly distributed over its surface (on the basis of argon-isotope ratios and chondritic and terrestrial abundances) ; however, McElroy et al.
(14) suggested that Mars may have had as much as 160-200
meters of water (on the basis of nitrogen-isotope ratios).
By comparison, the equivalent value for the Earth is
about 3 km of water (15). Thus, even though water on Mars
is not as abundant as on Earth, sufficient water is available to form ice-rich ground over the entire surface.
The extent and depth of this ice is controlled by local
rock conditions and their porosity, by the atmospheric
pressure and water vapor at the surface (permitting dessication near the surface in equatorial areas), by geothermal gradients (establishing the lower boundary of

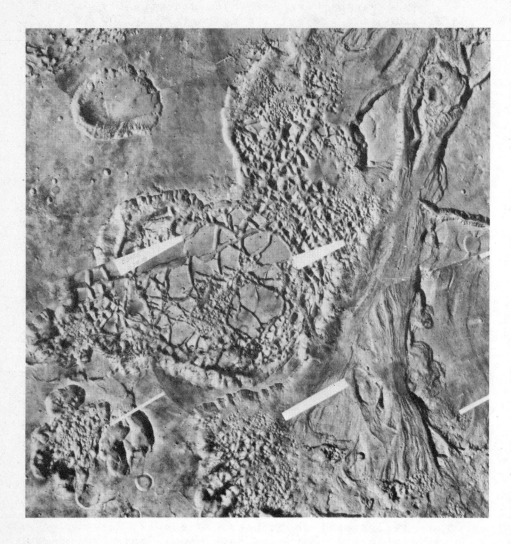

Fig. 3   Chaotically collapsed ground. Note large
channel on right emerging from chaos near bottom. Picture 225 km wide ; north toward upper right, illumination from right (lat. 3° N., long 28° W. ; part of Viking mosaic 211-5556).

permafrost), and by the composition of the water (brines would depress the freezing temperature).

CRATERS

One of the most cogent arguments for the existence of volatiles in the Martian subsurface comes from
craters and their ejecta. Craters on bodies lacking an
atmosphere, such as the Moon and Mercury, have ejecta
blankets that are ballistically emplaced, resulting in
a radial arrangement of ejecta materials. By contrast,
many ejecta blanckets of craters on Mars show flow lobes (16) (Fig. 1). In places, the lobes are diverted
around obstacles, indicating that the flow was a ground-
hugging surge. Most workers (17) feel that this type of
flow is due to the release of volatiles from the subsurface upon impact. Recent discovery of similar lobes on
the icy surface of Ganymede (18), which lacks an atmosphere, suggests that the lobes indeed result from impact
into water- or ice-rich ground.

A study analysing the minimum diameters of flow-
lobe craters over the Martian surface (19) led to an estimation of the minimum depth at which craters intersected the ice-rich layer, and thus an estimation of the
thickness of the desiccated zone. It was found that the
minimum crater size of flow-lobe craters systematically
decreased from the equator toward the poles, giving an
approximate depth to the top surface of an ice-rich
layer of 300-350 meters in the equatorial area and 100-
150 meters near 40° latitude. The study was prepared,
however, on a rather limited data set, and could be refined considerably with presently available information.
Nevertheless, the study agrees with theoretical estimates of a deeper desiccated zone in the vicinity of the
equator and a diminished thickness toward the poles.

VALLEYS

A major observation suggesting that ground ice
exists on Mars is the presence of small valley networks.
The networks are ubiquitous in ancient highland terrain
(20), and are composed of either short, solitary channels or integrated channel systems (Fig. 2). The integration makes it plausible that the systems were produced by a fluid, most likely water. However, the origin
of these valleys remains enigmatic. Even though super-
ficially similar to terrestrial drainage networks, when
analyzed in detail the valleys have less dendritic di-

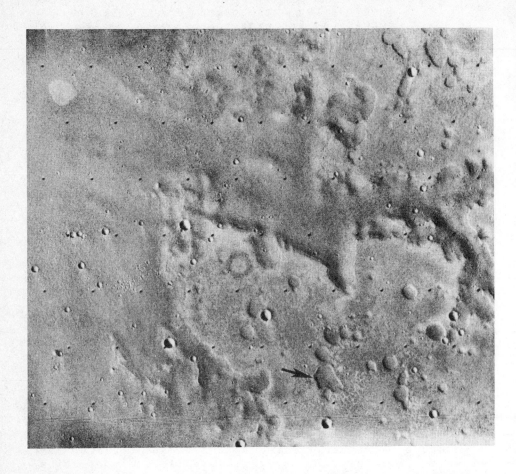

Fig. 4   Mesas (right center and upper right) bounded
by scalloped scarps. Note irregular depressions on top
surfaces of mesas. Also note light-colored polygonal
pattern surrounding depression (arrow) near bottom center. Scallops and depressions resemble alases formed by
thermokarst and are similar in morphology and size to
Alaskan thaw lakes. Picture 47 km wide ; north toward
upper left, illumination from lower left (lat. 23° N.,
long 35° W. ; Viking Orbiter image 8A74).

versification, the tributaries tend to be parallel to
one another, and the junction angles are more acute
than those in drainage networks on Earth (21). Moreover, the valleys closely resemble valleys formed by sapping on Earth, showing characteristic alcove-shaped valley heads. Therefore, Pieri (21) suggested that their
origin is best explained by seepage of water from ground
ice, resulting in collapse of the headwall of the valleys and headward migration.

THERMOKARST

Thermokarst occurs where ground collapses due
to loss of subsurface volatiles. Numerous collapse features are observed on Mars, and most are best explained
by loss of ice.

Enormous expanses of chaotic terrain (Fig. 3)
were first observed on Mariner images. Chaotic terrain
consists of irregular depressions and appears to have
formed by collapse. Sharp (22) attributed the chaotically collapsed ground to the loss of ground ice, but cau-
tioned that it would be difficult for lenses of segregated ice to form on such a scale. However, if the ice
accumulated by some other means, he felt that thermokarst was the most reasonable explanation for the formation of the chaotic terrain. Because chaotic terrains
tend to occur in low areas and at places at the mouths
of channels, the massive ice responsible for later collapse may have come from water frozen in lakes or trapped in deltas of flood plains (23). Futhermore, the
most extensive chaotic terrains occur in the source region of the large Martian outflow channels, an observation that corroborates that water was abundant in chaotic terrain areas.

Large chain craters are strung out like beads
along faults and grabens that parallel the large equatorial tectonic troughs on Mars: the Valles Marineris.
The lack of obvious deposits on the rims of these craters makes it unlikely that they are of volcanic origin ; the prevalent hypothesis is that they formed by
collapse (24). Their location along faults may be significant : increased heat flow and volatile content
along these tectonic avenues may have lead to preferential melting of ground ice that drained and caused collapse of the ground near the surface.

Collapsed ground on Earth that is attributed to

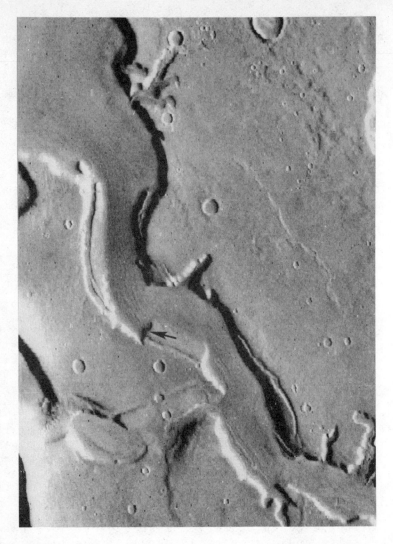

Fig. 5   Bent and tilted margins of valley near northern highland boundary. Material appears to have been
removed from the subsurface and caused the more resistant surface layers to collapse. Vague flow lines apparently emerge from under collapsed layer (arrow). Picture 45 km wide ; north toward top, illumination from
right (lat. 37° N., long 344° W. ; Viking Orbiter image 230S35).

loss of ground ice forms alases or thaw lakes. Alases
are irregular-shaped, flatfloored, and steep-sided depressions that may have interior lakes. In the Alaska
coastal plain thaw lakes occupy depressions whose shape
and size are similar to depressions on Mars located on
a plateau near the northern highland scarp (25, 26, 24)
(Fig. 4). The similarity strongly suggests that the
ground in this area of Mars was underlain by ice.

Tributaries and alcoves in the banks of large
terrestrial rivers flowing across ground-ice terrain in
Siberia have been studied by Jahn (27). He concluded
that, because of the melting of ground ice, short tributaries in the banks of the Lena River formed rounded
alcoves with narrow V-shaped outlets. The length profile of the tributaries is convex upward. Many of the tributaries and alcoves in the walls of the troughs in Valles Marineris have similar shapes and convex profiles,
suggesting that there, also, ground ice may be present.

The northern highland boundary on Mars forms a
prominent scarp that is locally dissected into steep-
sided, flat-floored valleys, scarps, and outlying mesas.
The valley floors are covered with longitudinally grooved material that apparently moved downvalley ; the mesas are surrounded by debris aprons that apparently flowed away from the scarps. The flow is thought to have
been facilitated by interstitial ice (28). The provenan-
ce of this ice has been a matter of controversy ; a derivation from atmospheric precipitates that seeped into
the ground was suggested by Squyres (28). However, the
precipitated amounts are insufficient (29). Other evidence points toward derivation of the lubricating ice
directly from the ground : collapsed ground can be seen
locally where upper layers sag or bend downwards upon
withdrawal of material from underneath (Fig. 5). The removed material appears to have flowed out, suggesting
that the debris on the valley floors and surrounding the
mesas came from collapsed and mobilized ice-rich ground.

The northern plains of Mars provide further evidence that the ground may contain ice. A lobate flow,
most likely lava, is sunk below the general ground level
in the vicinity (Fig. 6). The edge of the flow coincides
with the edge of the sunken area ; the coincidence suggests that the heat from the flow melted the underlying
ground ice and the flow sank accordingly. This type of
flow is marked by small, light-colored spots that are
locally aligned into subparallel curvilinear stringers

Fig. 6   Lobate flow in northern plains. The lobate
shape suggests that the flow is lava ; the flow is sunk
beneath the general ground level in the vicinity. The
observation suggests loss of material, most likely ice,
from the subsurface. Picture 150 km wide ; north toward
upper left, illumination from left (lat. 45° N., long
352° W. ; Viking Orbiter image 673B40).

like those in fingerprint patterns. The spots may well
be related to the outgassing of volatiles forming fuma-
roles or pseudocraters ; the curved patterns may be due
to the venting of gasses along curvilinear flow patterns.

Small mounds with central depressions are abundant in the northern plains (30). They could be pseudocraters, resulting from steam eruptions where lava flowed over ice-rich ground (31). On the other hand, the
small cratered mounds have the correct size for ice-co-
red mounds called pingoes. Pingoes tend to have collapsed tops that have formed craters. However, pingoe craters are irregular in shape, whereas the craters on
mounds on Mars are very regular. Alternatively, the
mounds may be cinder cones or small eroded impact craters having inverted relief, and may have no bearing
whatsoever on the existence of ground ice.

To obtain massive ground collapse, ice must be
segregated into lenses. On Earth, segregrated ice forms
where coarse grained aquifers supply water, and where
fine-grained sediments favor the concentration of ice
(32). Do such conditions prevail on Mars ? Mars'surface
is probably underlain by megabreccia from the accretionary period early in Martian history, or contains megabreccia under superficial lava caps. The breccia could
provide the necessary aquifers and fine-grained sediments. Wind-blown dust is also prevalent and may form
fine-grained blankets. Moreover, the most extensive
chaotic terrains are situated in a regional low where
ancient flood-plain deposits may have accumulated (23),
forming alluvium of gravel and silt, an ideal environment for the formation of ice lenses.

PATTERNED GROUND

On Earth, polygonally fractured ground is most
prevalent in areas underlain by ground ice. It forms in
polar areas where frozen, water-rich ground cools rapidly during the cold season and contraction cracks form
in the ice (33, 34). Such patterned ground on Earth
forms polygons up to about 100 meters in size. On Mars,
high-resolution images locally show polygons that measure 50-300 meters across, similar in size to those on
Earth (Fig. 7). In one place, polygons occur on the rim
of a collapse depression, suggesting association with
ground ice (35). On Mars, ice is in equilibrium with the
atmosphere (8) at the latitudes where small polygons are
observed. Presence of ground ice is also indicated by the
disintegration of some polygonal ground into fields of hummocks ;

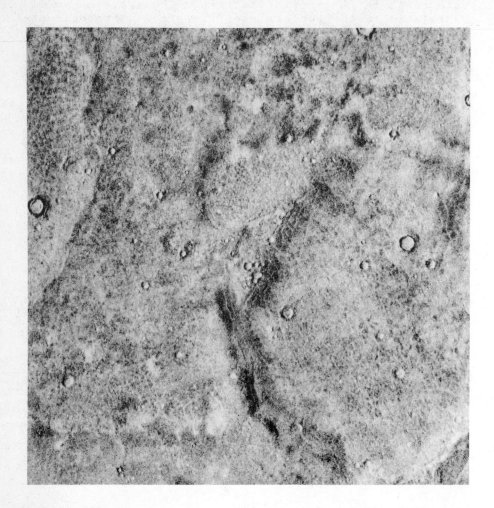

Fig. 7   Ground fractured by small polygons near northern highlands margin. Polygons are similar in size to
those on Earth. Polygons occur on flat terrain and on
gently sloping wall of ancient crater rim. Picture 13
km wide ; north toward upper right, illumination from
left (lat. 47° N., long 346° W. ; Viking Orbiter image
458B67).

Fig. 8   Trough traversing Viking 2 Lander site. It resembles troughs over polygonally arranged cracks in Antarctica. Massive boulder in right center is about 1 m across (Viking Lander 2 event, number 21A024).

the disintegration suggests loss of material. In addition to thermal contraction in ground ice, polygons on Mars may have formed by desiccation. Climatic cycles may have permitted ice to remain in the ground only for the colder part of the cycle (36). Nevertheless, desiccation implies that the ground was wet before the desiccation occurred, indicating that ground ice was present for at least part of the cycle.

Polygons may be present at the Viking 2 Lander site, where small troughs traverse the ground (37) (Fig. 8). Some trough segments are at angles to one another. Again, the hypothesis of ground ice is supported. The size of these troughs is similar to that of troughs of polygons in Antarctica (38).

Areas of patterned ground with polygons 2-20 km in diameter surround the Martian north-polar cap. The troughs forming the polygons have upturned margins (Fig. 9) similar to those of actively developing ice-wedge polygons on Earth (39). The large size of these polygons makes it questionable that they are contraction cracks (40), but their association with the polar cap suggests ice-related mechanisms. Neither lava-contraction cracks (41) nor tectonic fractures (42) are satisfactory alternatives to explain the origin of these polygons. Contraction or desiccation due to climatic cycles of 10^5 to 10^6 years rather than annual seasonal cycles have been suggested for their origin (43). Long-term climatic cycles could result in polygons two to three times the size observed on Earth (44) but a precise mechanism for growth to the size observed on Mars has not yet been advanced.

MASS WASTING

Mass wasting occurs ubiquitously on Earth, but is especially efficient in polar areas because freeze-thaw cycles and thermal contraction in ice loosen rock. Except for wind erosion, mass wasting is the most conspicuous erosional process on Mars. Talus slopes occur on all high steep slopes and resemble talus slopes found in terrestrial polar regions.

Landslides are also common in the Martian Valles Marineris, where scarps are steep and as high as 10 km. Landslides are not in themselves indicative of ground ice ; most large terrestrial landslides are dry. However, gigantic landslides inside the Valles Marineris troughs differ from terrestrial slides in some important

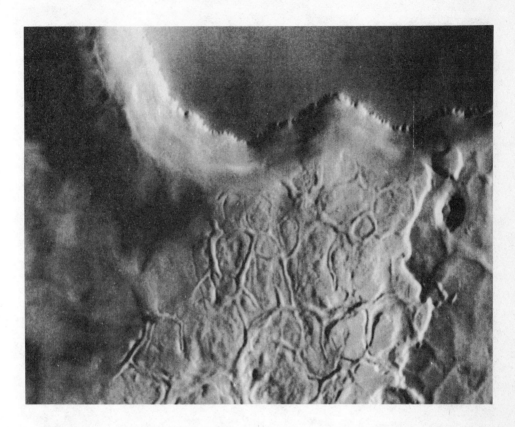

Fig. 9   Ground fractured by large polygons at edge of
north-polar ice cap (top). Troughs have raised margins
similar to active ice-wedge polygons on earth. Picture
about 60 km wide ; north toward upper right, illumination from left (lat. 81° N., long 63° W. ; Viking Orbiter image 560B42).

Fig. 10   Outwash deposit from landslides in Candor
Chasma. Note doughnut-shaped ridges at bottom and bottom
left center that resemble kettle holes in terrestrial
moraines. Kettle holes are formed by the melting of ice
blocks. The observation suggests that the landslide outwash contained blocks of ice. Picture 13 km wide, north
toward top, illumination from upper left (lat. 6° S.,
long 72° W. ; Viking Orbiter image 815A50).

respects. Their debris blankets bear longitudinal rather than transverse ridges which are indicative of reduced frictional resistance affecting the sliding mass (45, 46). On Earth, longitudinal ridges are observed only in landslides that flow over ice, such as the Sherman landslide in Alaska (47). On Mars, reduced frictional resistance may come from lubricating water or ice. The idea that the large landslides contained water is suggested by outwash that flowed from the landslide masses and apparently debouched into a level, flat area similar to a flood plain or pond. This "pond" occurs in the lowest part of the central Valles Marineris (23).

Presence of ice in the landslide debris is further suggested by a hummocky outwash deposit that contains doughnut-shaped ridges, similar to kettle holes in terrestrial moraines that result from the melting of ice blocks (Fig. 10). This proposed ice in the landslide deposit could come only from the collapsed trough walls at the head of the slide, suggesting that ground ice was present in the subsurface there. It is important to note that the Valles Marineris landslides are located in the equatorial area and occurred relatively recently on a Martian time scale. This observation implies that Mars was not desiccated completely in the equatorial areas ; at least some wall segments intersected by the landslide scars must have contained water or ice.

SUMMARY

We can explain some of the individual features described above—craters with flow lobes, small valley networks, thermokarst features, patterned ground, and mass wasting—by factors other than ground ice, though for some only with difficulty. However, the combined evidence of these features indicates that a large reservoir of ground ice has existed on Mars. The ice occurred in equatorial as well as polar areas and may have persisted into relatively recent times.

REFERENCES

1. Kieffer, H. H., Martin, T. Z., Peterfreund, A. R., Jakosky, B. M., Miner, E. D. and Palluconi, F. D., 1977, J. Geophys. Res. 82, pp. 4249-4271.

2. Lewis, J. L., 1974, Science 186, pp. 440-443.

3. Wall, S. D., 1981, Icarus 47, pp. 173-183.

4. Fanale, F. R., 1976, Icarus 28, pp. 179-202.

5. Rossbacher, L. A. and Judson, S., 1981, Icarus 45,
    pp. 39-59.

6. Toksöz, N. M. and Hsui, A. T., 1978, Icarus 35,
    pp. 537-547.

7. Michaux, C. M. and Newburn, R. L., 1972, Jet Prop.
    Lab. Docu. 606-1.

8. Farmer, C. B. and Doms, P. E., 1979, J. Geophys.
    Res. 84, pp. 2881-2888.

9. Smoluchowski, R., 1968, Science 159, pp. 1348-1350.

10. Anderson, D. M. and Tice, A. B., 1978, Proc. Se-
    cond Coll. on Planet. Water and Polar Processes,
    U. S. Army Cold Regions Research and Engineering
    Laboratory, Hanover, New Hampshire, pp. 55-61.

11. Clark, B. C., 1978, Icarus 34, pp. 645-665.

12. Jones, K. L., Bragg, S. L., Wall, S. D., Carlston,
    C. E. and Pidek, D. G., 1979, Science 204, pp.
    799-806.

13. Anders, E. and Owen, T., 1977, Science 198, pp.
    453-465.

14. McElroy, M. B., Kong, T. Y. and Yung, Y. L., 1977,
    J. Geophys. Res. 82, pp. 4379-4388.

15 Turekian, K. K. and Clark, S. P., 1975, J. Atmos-
    pheric Sci. 32, pp. 1257-1261.

16. Mouginis-Mark, P. J., 1979, J. Geophys. Res. 84,
    pp. 8011-8022.

17. Carr, M. H., Crumpler, L. S., Cutts, J. A., Gree-
    ley, R., Guest, J. E. and Masursky, H., 1977, J.
    Geophys. Res. 82, pp. 4055-4065.

18. Horner, V. M. and Greeley, R., 1982, Icarus 51,
    pp. 549-562.

19. Kuzmin, R. O., 1980, Lunar and Planet. Sci. XI,
    The Lunar and Planet. Sci. Institute, Houston, TX,
    pp. 585-586.

20. Carr, M. H. and Clow, G. D., 1981, Icarus 48, pp. 91-117.

21. Pieri, David, 1980, Science 210, pp. 895-897.

22. Sharp, R. P., 1973, J. Geophys. Res. 78, pp. 4073-4083.

23. Lucchitta, B. K. and Ferguson, H. M., 1983, Proc. Thirteenth Lunar and Planet. Sci. Conf., Part 2 : J. Geophys. Res. 88, Suppl., pp. A553-A568.

24. Lucchitta, B. K., 1981, Icarus 45, pp. 264-303.

25. Gatto, L. W. and Anderson, D. M., 1975, Science 188, pp. 255-257.

26. Carr, M. H. and Schaber, G. G., 1977, J. Geophys. Res. 28, pp. 4039-4054.

27. Jahn, Alfred, 1975, Polish Scientific Publishers, Warsaw, 223 p.

28. Squyres, S. W., 1978, Icarus 34, pp. 600-613.

29. Lucchitta, B. K., 1984, Proc. Fourteenth Lunar and Planet. Sci. Conf. Part 2 : J. Geophys. Res. 89, Suppl. pp. B409-B418.

30. Allen, C. C., 1979, J. Geophys. Res. 84, pp. 8048-8059.

31. Frey, H., Lowry, B. L. and Scott, A. C., 1979, J. Geophys. Res. 84, pp. 8075-8086.

32. Washburn, A. L., 1973, Periglacial processes and environments : St. Martin's, New York, 320 p.

33. Lachenbruch, A. H., 1962, Geol. Soc. Am. Spec. Paper 70, 69 p.

34. Black, R. F., 1976, Quat. Res. 6, pp. 3-26.

35. Brook, G. A., 1982, Repts. of Planet. Geol. Prog. 1982, NASA TM 85127, pp. 265-267.

36. Lucchitta, B. K., 1983, Proc. 4th Int. Conf. Permafrost, pp. 744-749.

37. Liebes, S., Jr. 1982, NASA Contractor Rept. 3568,

289 p.

38.   Morris, E. C., Mutch, T. A. and Holt, H. E., 1972,
      U. S. Geol. Surv. Interagency Rept. Astrogeology
      52, 156 p.

39.   Péwé, T. L., 1975, U. S. Geol. Surv. Prof. Paper
      835, 143 p.

40.   Black, R. F., 1978, Proc. Second Coll. on Planet.
      Water and Polar Processes, U. S. Army Cold Re-
      gions Research and Engineering Laboratory, Hano-
      ver, New Hampshire, pp. 127-130.

41.   Morris, E. C. and Underwood, J. R., 1978, Repts.
      Planet. Geol. Prog. 1977-1978, NASA TM 79729,
      pp. 97-99.

42.   Pechmann, J. C., 1980, Icarus 42, pp. 185-210.

43.   Coradini, M. and Flamini, E., 1979, J. Geophys.
      Res. 84, pp. 8115-8130.

44.   Helfenstein, P. and Mouginis-Mark, P. J., 1980,
      Lunar and Planet. Sci. 11, pp. 429-431.

45.   Lucchitta, B. K., 1978, Geol. Soc. Am. Bull. 89,
      pp. 1601-1609.

46.   Lucchitta, B. K., 1979, J. Geophys. Res. 84, pp.
      8097-8113.

47.   Shreve, R. L., 1966, Science 154, pp. 1639-1643.

COMMENT :
COMPARED DISTRIBUTION OF H_2O ON MARS AND THE EARTH

André CAILLEUX

9, avenue de la Trémouille, F - 94100   FRANCE

On the Earth, hydrologists consider 7 main H_2O
subsets. Let us ordinate them, from the greatest volume
to the least (Table 1) and apply to them post-ordinant

T A B L E   1

VOLUMES OF H_2O (WATER, ICE, VAPOR) [1]

	On the Earth km^3 [2]	Rsq	On Mars at present
1 Seas	1 370 000 000		0
		51,0	
2 Glaciers	27 000 000		polar
		3,4	
3 Underground	8 000 000		very abundant
		91,0	
4 Fresh water lakes	88 000		0
		4,6	
5 Soils	19 000		small amount
		1,6	
6 Atmosphere	12 000		small amount
7 Rivers	1 200	10	0

(1) For ice and vapor, volume of water having the same
    mass.
(2) According to hydrologists.

605

J. Klinger et al. (eds.), Ices in the Solar System, 605–606.
© 1985 by D. Reidel Publishing Company.

operations (1). The span of variation, expressed by the extremal ratio R_{ex} (greatest value divided by the smallest), is enormous : more than 1 million. The sequential ratio R_{sq} (quotient of each value by the following one) is mostly small (1 to 10), with two noteworthy exceptions, two very sudden jumps : one from seas to glaciers (51), the second, even stronger (91), from underground water to lakes. Those two jumps define a clearcut family, grouping glaciers (Antarctica, Greenland, etc...) and underground H_2O (liquid, solid and vapor) combined in the interstices of the rocks.

On Mars, there are presently neither seas, nor lakes, nor rivers, and only a small amount of H_2O in the atmosphere and in the superficial soils. H_2O mainly exists in the form of polar glaciers, stratified (niveo-eolian and glacio-eolian) terrain and underground ice (above) and water (lower). It is noteworthy that the great abundances occur in the two subsets (glaciers and underground) which, as on the Earth, form a well defined family.

Therefore, Table 1 shows clearly that the seven main natural H_2O subsets are not ordinated on Mars as on the Earth. Thus, our planet contrasts, not only with Mars but also with all other bodies in the Solar System, by its essential surface characteristic, i. e. overwhelming abundance of oceanic water, which most impressive consequence is its original geodynamic action.

REFERENCE

1.   Cailleux, André, 1983, Les opérations post-ordinannantes, 157 pp., 12 figs., 33 tables, Guérin., 4501 Drolet, Montréal and Eska, 30 rue de Domrémy 75013 Paris.

HYDROLITHOSPHERE AND PROBLEMS OF SUBSURFACE ICE IN THE EQUATORIAL ZONE OF MARS

BATTISTINI René

Université d'Orléans, France

ABSTRACT

It is assumed that on Mars liquid water (lower hydrolithosphere) is confined under the icy upper lithosphere (upper hydrolithosphere), which is indeed a rigid and practically impervious carapace. The mobility of the lithospheric water and probably also of a part of the stocking rocks under the permafrost probably originates by sapping some important surface features, for instance subcircular depressions of increasing sizes, with steep slopes, aligned along the main fractures. This process is probably primordial in the origin of the Valles Marineris system. If the oscillations of the cryomarge (basis of the permafrost) are able to make thin enough the icy upperlithosphere until a break out point, the disruption of the aquifer to the surface is possible, generating high artesian pressure in the lowest regions of the planet (supposed process of the outflows according with the Carr's model). It seems likely that more or less important ice caps existed in some places in the equatorial zone of Mars, with the same origin (subsurface water). Three possibilities are suggested to explain the variations of the cryomarge and the thinning of the permafrost : modifications of the geothermal gradient ; climatic oscillations ; progressive heating of the water during the filling of the aquifers.

INTRODUCTION

The largest part of the water stored in Mars is

607

J. Klinger et al. (eds.), Ices in the Solar System, 607–617.
© 1985 by D. Reidel Publishing Company.

subsurface water included in pervious rocks, either as
ice (permafrost) in the upper hydrolithosphere, or as
liquid water below the lower level of frost (cryomarge)
in the lower hydrolithosphere (1). The icy upperpart of
the hydrolithosphere is probably the most important phenomenon to be emphasized for the understanding of many
aspects of the surface geomorphology, especially slopes
evolution, origin of various types of depressions, and
mechanism of the outflows. Liquid water is confined under the icy upper lithosphere, which is indeed a rigid
and practically impervious carapace. The study of geomorphological features of the surface of Mars provides
only possibilities to obtain data concerning volume, site and action (for instance formation of underground cavities) of liquid water under the permafrost.

1 - DISTRIBUTION OF THE HYDROLITHOSPHERE (Fig. 1)

        Craters with fluidized ejecta are a good tool
for the study of the hydrolithospheric ice and deep water. It is indeed admitted that the ejecta fluidity is
related to the water or to the subsurface ice melted by
impact (2). Inventory and mapping of the fluidized ejecta craters provide a method to point out the existence
of significant amount to subsurface lithospheric water
(or ice) in pervious rocks. The highest porosity of the
rocks (> 0.1) is probably located in volcanic ashes and
lapillis, and mostly in eolianites and accumulations
built up by large flows (outflow channels). After Carr
(2) grained rocks or lavas microbroken up by tectonics
or by impacts have a smaller porosity (< 0.1). Refering
to this method, it is possible to identify very large
areas with a continuous hydrolithosphere (Coprates,
Lunae Planum, Chryse Planitia, Syrtis Major Planitia :
these areas also display wrinkle ridge fields), and regions without hydrolithosphere, or with discontinuous
hydrolithosphere within depressions which are often
great craters, corresponding to the oldest cratered terrains (1). The upper part of the hydrolithosphere is often capped with a crust 500-1000 m thick : on the walls
of Valles Marineris, for instance, it is possible to observe in the upperpart of the outcrops a crust apparently hard, some hundreds of meters thick, overhanging by
place. In the large volcanic Tharsis and Elysium areas
the top of the hydrolithosphere deepens ; there, the meteoritic impacts had to go through the "bulk" cover of
recent lava flows - more than 4 kilometers thick in some
places - to reach the lower beds where water is stored.
The use of selected large craters with fluidized ejecta

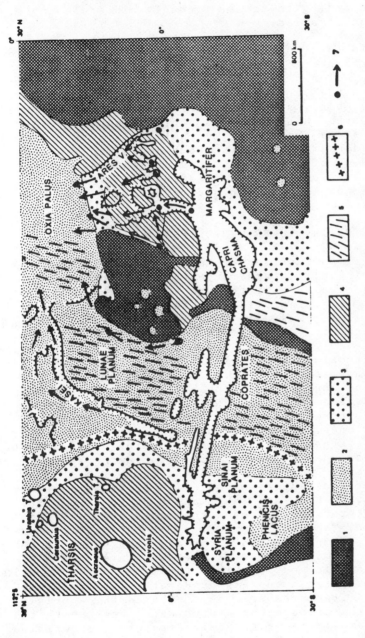

Fig. 1   Distribution and depth of the hydrolithosphere at the lowest latitudes of Mars, between 0° and 112°, 5 longitude. 1) discontinuous hydrolithosphere or lack of hydrolithosphere (generally old heavily cratered terrains). 2) to 4) areas with continuous hydrolithosphere : 2) depth of the top of the hydrolithosphere = 1-km (± 500 m.) ; 3) depth of the top of the hydrolithosphere = between 1 and 2 km ; 4) depth of the top of the hydrolithosphere = more than 2 km. 5) fields of wrinkle ridges. 6) lava flows cover limit of the Tharsis bulge volcanic region. 7) point of outflows and flow direction. The scale is approximate.

also permits to evaluate in each region the lower level
of the hydrolithosphere, therefore its thickness, and
allows to estimate roughly its water content, which is
different according to areas, probably since the porosity of the rocks varies.

It is interesting to compare the distribution of
the hydrolithospheric areas to the distribution of various geomorphologic features. For instance, the comparison between the cartography of hydrolithospheric terrains, on one hand, and the distribution of wrinkle ridges, on the other hand, shows a strict geographic coincidence. It seems that their morphological similitude
with Moon ridges is due to the same dynamical genesis,
the compression of a thin crust in disconformity with
underlying terrains. But the relationship is, on Mars,
not with volcanism, but with underground ice and water
(3).

2 - MOBILITY OF THE LOWER HYDROLITHOSPHERE

Some important geomorphologic features indicate
a mobility of the lithospheric water and probably also
the mobility of a part of the stocking rocks under the
permafrost : large landslides at the foot of the great
slopes (4, 5), for instance in Valles Marineris, creating subsidence faults, as well as collapsed depressions
of great size, with low gradient, and subcircular depressions of increasing sizes, with steep slopes aligned
with the main fractures (Fig. 2). For instance, sapping
by underground aquifers is probably the initial reason
for Ganges Catena formation (NE Ophir Chasma), created
by the coalescence of smaller subcircular depressions
with steep slopes, aligned along EW fracture (Fig. 2).
This process is probably primordial in the origin of the
larger depressions in the vicinity (including Valles Marineris - D and E in Fig. 3). In the following stages
the slopes of these depressions evolved out by backvas-
ting process, melting and sublimation of ice. In other
heavily fractured areas around the Tharsis bulge, for
instance Alba, Tantalus (C in Fig. 3), Mareotis (A and
B in Fig. 3), Tempe, Memnonia Fossae, subsidence features are also common. Narrow micrograbens are initially
of tectonic origin, but probably their morphology results
mostly from sapping by underground water. It is assumed
that the main fractures are, as on the Earth, a privile-
ged way for the circulation of the water underneath the
ground. Especially along the main fractures, material of
the basis of the hydrolithosphere has been probably exposed since millions of years to diagenetic alteration

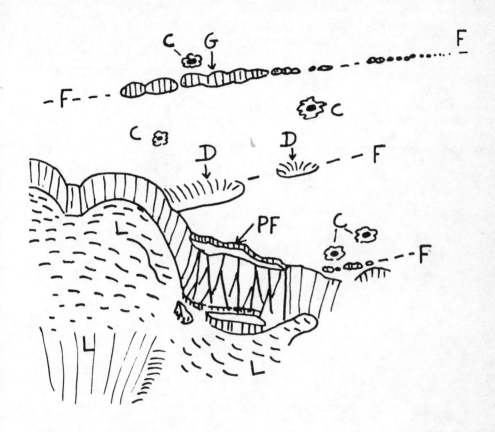

Fig. 2   Forms in relationship to the mobility of the
lower hydrolithosphere in the vicinity of Ophir Chasma
(Valles Marineris).
L   : landslides at the foot of the great slopes.
P F : panamean faulting.
F   : main E-W fractures.
D   : collapsed depressions with low gradient.
G   : Ganges Catena.
C   : main craters with fluidized ejecta.

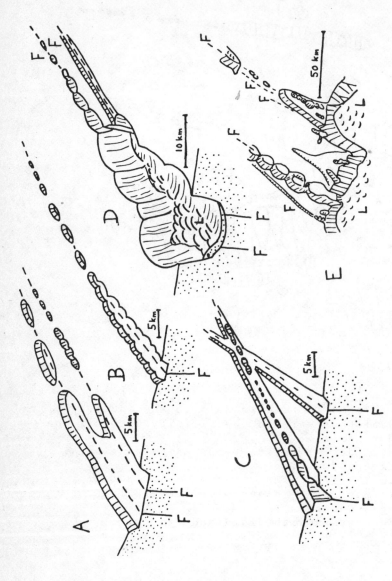

Fig. 3 Geomorphologic features (schematic figuration) probably in relationship to
the mobility of the lower hydrolithosphere along the main fractures, in various areas.
A, B : Mareotis Fossae (Arcadia SE) ; C   : Tantalus Fossae (Arcadia SW) ; D, E :
Valles Marineris ; F : fracture ; L : landslides.
The scale is approximate.

and corrosion by water. We postulate the possibility of
cavities (underground karst)along the main fractured
areas, where pressurized liquid water is stored.

The lack of hydrolithosphere in the oldest heavily cratered terrains (except in discontinuous areas)
does not exclude the possibility of fossile aquifers of
underground water or ice, located in underground cavities along the fractures ; an indication of this possibility is the maltiplicity, in some areas, of collapsed
depressions, with a great variety of forms, along major
and even minor fractures.

3 - THE SUPPOSED PROCESS OF THE OUTFLOWS

The outflow channels are probably exclusively
Martian (6). We agree with Carr's model (7) concerning
the outflows. Carr supposed confined aquifers in the lower part of the megaregolith, under the permanent ice
layer of the permafrost with the possibility of high
pressures in relation with the topography. In this artesian system the permafrost plays the most important role
by its ubiquity and its imperviousness. The disruption
of the aquifer to the surface under high pressure occurs
in very localized zones, generally characterized by chaotic topography. After Carr's model, the location of outflow channels around the Chryse Basin is in relation
with the topography : this area in one of the lowest regions on the planet and its margins were ideal locations
for generating high artesian pressures. It is necessary
to imagine underground water stored in a network of open
cavities of very great areal extend, and suddenly released, to understand the important quantities of water related to the outflows formation.

An important question arises concerning the origin of the variations of the cryomarge (basis of the permafrost), able to make the icy upperlithosphere thin
enough to create a break out point generally located in
the lowest regions. Three possibilities are suggested :

1) modifications of the geothermal gradient ;
2) climatic oscillations ;
3) progressive heating of the water during the
filling of the aquifers.
The morphology of the outflow channels with in places
deep canyons supposes that the catastrophic outflow phenomena is very repetitive. Consequently the second or the
last hypothesis seems the most probable. Many craters
are superimposed on the channels and it has been conclu-

ded (6) that they must be old (generally > 1 billion years). The outflow phenomena is then neither today nor recent phenomenon. The explanation of this fact is not easy : the cause is probably the loss by the planet of an early dense atmosphere, and the impossibility now to reconstitute the aquifers, only by juvenile water. That is perhaps combined by a decrease of the temperatures and a thickening of the permafrost.

4 - GLACIAL CAPS IN THE EQUATORIAL ZONE

Morphological features, looking like morainic ones, are to be found in the equatorial zone of Mars, in the following sites : W. of Arsia Mons ($3° - 11°$S, $123°-130°$W) ; N. N. W. of Pavonis Mons ($0° - 6°$N, $114° - 118°$W) ; W. of Ascraeus Mons ($11° - 13°$N, $107° - 109°$W) ; N. W. part of Kasai Vallis ($21° - 28°$ N, $74° - 79°$W) ; N. W. part of Aeolis ($3°$S - $6°$N, $204° - 222°$W). It seems likely that more or less important ice caps existed in these places. This cannot have been glaciers to be linked with atmospheric precipitations : actually their origin is to be traced not on high reliefs, but in the thoughs of depressions. In all likelyhood these ice caps are due to subsurface water, as in the case with outflows. The ice caps near Arsia Mons, Pavonis Mons, and Ascraeus Mons, are probably related to volcanism (high geothermal gradient in the vicinity of these high volcanoes, having created break out points through the permafrost).

REFERENCES

1. Battistini, R., 1984, L'utilisation des cratères météoritiques à ejecta fluidisés comme moyen d'étude spatiale et chronologique de l'eau profonde (Hydrolithosphère) de Mars. Revue de Géomorphologie Dynamique, 33, pp. 25-41.

2. Carr, M. H., Crumpler, L. C. S., Cutts, J. A., Greeley, R., Guest, J. E., Masursky, H., 1977, Martian impact craters and emplacement of ejecta by surface flow, J. Geophys. Res., 82, pp. 4055-4065.

3. Battistini, R., 1984, Morphology and origin of ridges in low-latitude areas of Mars, Earth, Moon, and Planets, 31, pp. 49-61.

4. Lucchitta, B. K., 1978, A large landslide on Mars, Geol. Soc. Am. Bull., 89, pp. 1601-1609.

PLATE 1

Three stages in the formation of Valles Marineris
   (12°–18°S, 57°–62°W)

(1) subcircular little depressions of increasing sizes aligned
    along the main fractures, probably created by sapping by under-
    ground aquifers through the permafrost;
(2) coalescence of these depressions;
(3) wide depressions, with steep slopes evolving out probably by
    melting and sublimation of ice. Some meteoritic craters have
    fluidized ejecta (F); on the bottom of the largest depressions,
    landslides (L) at the foot of the great slopes.

The scene is 250 km across (610 A 07) N.A.S.A. photograph

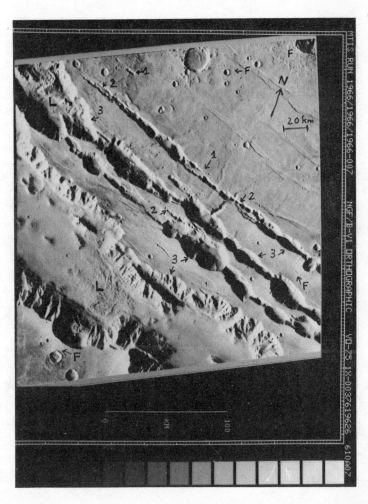

PLATE 2

Morphological features in the NW part of Kasai Vallis
    (19°–23° N, 74°–78° W)

Longitudinal lineaments in the central part (A), and sopposed la-
    teral morainic accumulation or eolian accumulation on the NW
    limit (B) of the presumed glacier.
Arrows show the postulate ice flow direction.
Meteoritic craters, some with fluidized ejecta (C), and a wrinkle
    ridge (R) are older than the glacier (fine striations cutting
    across the ejecta and the ridge).

The scene is 210 km across (519 A 05) N.A.S.A. photograph

5. Lucchitta, B. K., 1978, Morphology of Chasma walls,
    J. Res. U. S. Geol. Survey 6, pp. 651-662.

6. Baker, V. R., 1982, The Channels of Mars. University
    of Texas Press, pp. 198.

7. Carr, M. H., 1979, Formation of Martian flood fea-
    tures by release of water from confined aquifers,
    J. Geophys. Res., 84, pp. 2995-3007.

Part V

Rings, Icy Satellites and Pluto

ICY SATELLITES, RINGS and PLUTO

J.Klinger

Laboratoire de Glaciologie et de
Géophysique de l'Environnement
B.P. 96, 38402-St Martin d'Hères Cedex - France

Abstract

The present state of knowledge on the occurence of ices on
the satellites of Jupiter, Saturn, Uranus and Neptune, on ring
particles and on Pluto is briefly reviewed. An overview is given
of the role of ices in the geological evolution of the above
mentioned bodies.

INTRODUCTION

It has been stated in the introduction of this book that the
orbit of Jupiter is beyond the limit where evaporation of water
ice is sufficiently small in order to allow the survival of this
substance for time spans comparable to the age of the solar
system. We thus can establish a natural limit between the inner
solar system where the geological evolution of planets and
satellites is dominated by rock material and the outer solar
system where the history of a great number of bodies should be
dominated by moderatly volatile substances as defined by Whalley
(1). This two regions are separated by the asteroïd belt.

Since the pioneer work of Kuiper (2) the above mentioned
rather intuitive affirmations have been progressively
confirmed by infrared reflexion spectroscopy

Table I summarizes the ices that have been identified on
outer solar system bodies or that are likely to occur there in
view of the local physical conditions. For water ice, the phases
that are compatible with pressures and temperatures on the surface

621

J. Klinger et al. (eds.), Ices in the Solar System, 621–629.
© *1985 by D. Reidel Publishing Company.*

and in the interior of satellites or on ring particles are also indicated .

Table I : Ices in Satellites, Rings and Pluto.
Water ice phases compatible with the physical conditions in the body are indicated too.

Object	Radius R(km)	Average density (Mg/m^3)	Albedo	Varieties of ices identified or inferred	Ref.
J1 Io	1819	3.53	0.6	SO_2	(3,4)
J2 Europa	1563	3.03	0.6	$H_2O(I)$	(5,6)
J3 Ganymede	2638	1.93	0.4	$H_2O(I,II,V, VI,VII,$ amorphous in polar caps ?)	(5,6)
J4 Callisto	2424	1.79	0.2	$H_2O(I,II,V, VI,VII)$	(5,6)
Rings of Saturne				$H_2O(I,$ amorphous?)	(7,8, 9)
S1 Mimas	196	1.190	0.6	$H_2O(I)$	(10,6)
S2 Enceladus	250	1.2	0.9	$H_2O(I,$ formerly amorphous?) clathrate hydrates ? ammonia hydrates ?	(10,6)
S3 Tethys	530	1.21	0.8	$H_2O(I)$	(11,12, 6)
S4 Dione	560	1.43	0.62	$H_2O(I)$	(11,12, 6)
S5 Rhea	765	1.33	0.65	$H_2O(I,II)$	(11,12, 13,6)
S6 Titan	2575	1.88	0.2	$H_2O(I,II,V, VI,VII)$ clathrate hydrates?	(6)
S7 Hyperion	205x130x110		0.3	$H_2O(I)$	(14)

S8	Iapetus	730	1.16	$0.5/0.05$	$H_2O(I)$	(11,12, 6)

Small inner satellites S10-S17	$H_2O(I)$?	(15)
Rings of Uranus	modified CH_4?	(16)
U5 Miranda	$H_2O(I)$	(17)
U1 Ariel	$H_2O(I)$	(17,18)
U2 Umbriel	$H_2O(I)$	(17,18)
U3 Titania	$H_2O(I)$	(17)
U4 Oberon	$H_2O(I)$	(17)
N1 Triton	CH_4	(19,20)
Pluto 1200-1800	CH_4	(20,21 22,23)

It must be said that even in the cases where table I indicates only ice I for H_2O containing satellites, high pressure phases may be present in the ejecta of impact craters. These high pressure ices are formed during the impact event. At sufficiently low temperatures they are metastable after relaxation of the impact pressure. It is possible to identify high pressure ices by spectroscopical methods (24).

THE GALILEAN SATELLITES OF JUPITER

The Galilean satellites of Jupiter show a remarkable evolution of the density as a function of the distance from Jupiter, the closest (Io) being the densest and the farthest (Callisto) the less dense one. Probably this decrease in density corresponds to an increase in ice content, thus indicating that the satellites close to Jupiter formed at higher temperature or accreted with higher impact velocities (25).

The only intermediately volatile solid that has been identified on Io's surface is SO_2 (4). The deposit of SO_2 is presumably due to the volcanoes that are active on this planet. H_2O ice has been identified on the surface of the other Galilean Satellites. Among these ice bearing satellites the closest to Jupiter (Europa) has the smoothest surface (only three impact

craters > 10 km in radius have been identified (26)) and Callisto
the most cratered one. This sequence probably indicates that
Europa experienced a more drastic geological evolution than
Ganymede and Callisto even if it is likely that the smooth surface
of Europa is partially due to sputtering by ions of Jupiter's
magnetospheric plasma (27).

According to model calculations by Consolmagno and Lewis
(28), Europa is a well differentiated satellite with a silicate
core and a mantle of liquid water covered by a thin crust of
Ice I

The question whether under present day conditions Europa
contains liquid water under its icy crust has not yet received a
final answer (29,30,31). The results of this kind of models
strongly depend on the strength of the internal heat source
(radioactive and tidal heating) and on the heat transport
mechanism. The tidal dissipation is rather difficult to estimate
as it depends on the dissipation function used. As far as the heat
transport mechanism is concerned, most models disregard the heat
transport by solid convection. Whether or not heat transport by
solid convection is significant depends on the rheological
properties of the solids in the satellite.

The surface of Ganymede shows at least two types of terrains
with different crater density : cratered terrain and grooved
terrain (32). Polar caps have been identified as well by earth
based observations (33,34) as from Voyager images (26). Since it
is most likely that these polar caps are vapor deposits and since
vapor deposits on icy satellites shoud be amorphous (35,36),one
can conclude that the polar caps of Ganymede are composed of
amorphous ice.

At present, it is very difficult to give a final answer
concerning the degree of differentiation of Ganymede which is
likely to contain ice VI, which has a very low viscosity (37).
Thus solid convection could have prevented the satellite from a
complete differentiation. Whether the furrows visible on
Ganymede's surface are due to tectonism or to local events (38,39)
is still controversial.

Callisto is the most cratered of the Galilean satellites and
thus is thought to have undergone very little geological
evolution. Despite the fact that its surface indicates heavy
meteoritic bombardment in the early stage of the solar system,
there is some evidence for tectonism (40).

Voyager images of Callisto did not reveal any important
differences in impact cratering between the leading and the

trailing hemisphere. On the other hand, earth based optical polarimetry showed a striking dissymetry (41,42) : the trailing hemisphere has the polarimetric characteristics of clean rock whereas the leading hemisphere shows a behavior comparable to lunar fines. The explanation given by Mandeville et al. (42) is the following : the impacting objects which produce the asymmetry are ancient planitesimals that remained after the formation of the Jovian satellite system. These bodies moove on ellipses having pericenters close to Jupiter. The movement is in the same sense as Callisto's. As the orbital velocity of particles is smaller than that of Callisto these bodies can fall on its leading hemisphere. The impact craters produced in this way are too small to be within the limit of resolution of Voyager images.

THE SATURNIAN SYSTEM

The most spectacular features of the Saturnian system are the rings which have been discovered by Galileo in 1610 and correctly interpreted by Huygens in 1656. As we know now, these rings are composed of a great number of individual particles on which water ice has been identified (7,8). Considerations on heating curves after eclipses as well as the fact that the material of the ring particles seems to be replenished by vapor deposition at rather short time scales strongly suggest that the water ice on ring particles is amorphous (9,16,43).

The satellites of Saturn do not show any systematic evolution of density as a function of distance from the mother planet as Jupiter's satellites do. Water ice has been detected on a number of Saturnian satellites and is likely to be present on all of them. With the exception of Titan and Rhea, the Saturnian satellites have too low internal pressures for the formation of high pressure phases of H_2O in their interiors. Titan, the biggest of Saturn's satellites, is the only one to have an atmosphere and an "ocean" of liquid methane. The simultaneous presence of ice and liquid hydrocarbons probably gave rise to clathrate hydrates (44,45).

Hyperion as well as the smallest of Saturn's satellites have irregular shapes. The shape of these objects may provide interesting clues to the rheological properties of the ices that composes them and/or to their collisional history (46).

Despite their rather small mass, the "intermediate size",regularly shaped satellites show a great variety of geological evolution (47,48). Whereas Mimas is heavily but not uniformely cratered, Enceladus, its immediate neighbor has undergone striking geological evolutions. Voyager 2 images show at least four different types of terrain (49,50). If we consider these satellites as spheres of pure crystalline water ice mixed with rock material it is easy to show that radioactive heating is not sufficient to produce melting in such a body. Even tidal heating under present day conditions is not sufficient for melting partially a water ice body of the size and density of Enceladus unless we use unrealistic values for the dissipation function (51). An other attempt to explain the geological activity of Enceladus by tidal heating is due to Yoder (52). This author proposes that tidal heating occurs not continuously but episodically. Several authors suggested that H_2O ice could be mixed with NH_3 (53,54), which would lower the melting point making tidal heating sufficient to produce the observed resurfacing . Others authors attribute the geological activity of the satellites of Saturn to the decomposition of clathrate hydrates (55,44). A further possibility to explain the resurfacing of Enceladus is perhaps the phase transition between amorphous and cubic ice (56). If we consider that a body accretes from pieces of amorphous ice and if we use the approximation that the impact velocity of the accreting body we see that the amorphous ice of a body of the size of Enceladus is likely to survive the accretion phase without being transformed into cubic ice. If now the interior of the satellite is heated by radioactivity or tidal dissipation the phase transition may start and the excess heat freed during this phase transition may contribute to the resurfacing of the satellite.

At the present state of knowledge it is rather difficult to check the different models. More observational data on the Saturnian system and more laboratory studies on thermal and rheological properties of ices are necessary .

THE URANIAN SYSTEM, TRITON AND PLUTO

The rings of Uranus have been discovered only a few years ago (57) during a stellar occultation. It has been suggested that the Uranian rings where initially composed of CH_4-ice that has been altered due to photochemical reactions and bombardment by magnetospherioc particles (58).

As on the satellites of Jupiter and Saturn, water ice has been identified on the satellites of Uranus (17,18). So far no data from space missions to Uranus are available so that all our knowledge comes from earth based observations, spectroscopy and spectrophotometry which is very little as compared to the huge amount of data available for Jupiter's and Saturn's satellites and their ring systems. Nevertheless it has been possible to constrain the radii of those satellites as well as their densities. It was possible to show that the differences in reflectivity between the Uranian satellites and other ice bearing bodies are due to contamination by dark material (59).

Methane ice has been discovered on the surfaces of Triton and Pluto (20). These objects are too distant and too small for direct measurement of their diameter so that the available estimates of size and albedo are very uncertain (60). Thus it is not possible to speculate on their geological evolution and on the composition.

CONCLUSION

The study of icy objects like satellites and rings that condensed in the outer part of the solar system can give us important information on the chemistry and physical conditions in the solar nebula during planet formation.

For a long time all our knowledge on ice bearing objects in the outer solar system was due to earth based observations. Even so we where able to get some ideas on the chemical composition and texture of the surfaces of these bodies.

Although the recent Voyager missions gave us a huge amount of data relevant to the geological evolution of satellites of Jupiter and Saturn and about the planetary rings. In particular the imaging results gave a good basis for extensive geomorphological work. What is still a problem is the degree of internal evolution of the Jovian and Saturnean satellites. One of the most important information that we need in this context are some data on the mass distribution in the satellites' interior, leading to a better knowledge on the degree of differentiation. The most promising way to get these data is to measure the moment of inertia of the satellites during future space missions.

Spectroscopic indentification of high pressure phases of ice in ejecta of impact craters may help to characterize the impacts. Recent vapor deposits of H_2O could be detected when amorphous ice will be spectroscopically identified. This information might become available from further space missions and from improved earth based and earth orbiting observatories. More laboratory data of thermal, optical and rheological properties H_2O, NH_3, CO_2, CH_4 and their hydrates will be needed for further work.

ACKNOWLEDGEMENTS

This study has been supported by the French "Institut National d'Astronomie et de Géophysique", ATP "Planètologie" n°47-54. Numerous suggestions by Dr Benest, Prof. Dollfus and Prof. Smoluchowski are gratefully acknowledged.

REFERENCES

(1)   Whalley,E., 1985, this book.
(2)   Kuiper, G.P., 1957, the Astron. J., 62, p.245.
(3)   Fanale, F.P., Brown, R.H., Cruikshank, D.P. and Clark, R.N.,
      1979, Nature, 280, pp. 761-763.
(4)   Cruikshank, D.P., Howell, R.R., Geballe, T.R. and Fanale,
      F.P., 1985, this book.
(5)   Pilcher, C.B., Ridgway,S.T.and Mc Cord, T.B., 1972, Science
      178, pp.1087-1089.
(6)   Poirier, J.P., 1982, Nature 299, pp.683-688.
(7)   Pilcher, C.B., Chapman, C.R., Lebofsky, L.A. and Kieffer,
      H.H. 1970, Science 167, pp.1372-1373.
(8)   Clark, R.N. and Mc Cord,T.B. 1980, Icarus 43, pp.161-168.
(9)   Smoluchowski, R. 1978, Science 201, pp.809-811.
(10) Stone, E.C. and Miner, E.D., 1981, Science 212, pp.159-162.
(11) Fink, U., Larson, H.P., Gautier III, T.N. and Treffers, R.R.,
      1976, Astrophys. J 207, pp.L63-L67.
(12) Morrisson, D., Cruikshank, D.P., Pilcher, C.B. and Riecke,
      G.H., 1976, Astrophys. J.207, pp.L213-L216.
(13) Clark, R.N. and Owensby, P.D., 1981, Icarus, 46, pp.354-360.
(14) Cruikshank, D.P. and Brown, R.H. 1982, Icarus, 50, pp.82-87.
(15) Cruikshank, D.P., 1985, this book.
(16) Burns, J.A., 1985, this book.
(17) Cruikshank, D.P., 1982, in :"Uranus and the outer planets,"
      Proceedings   of   the   IAU/RAS,   Colloquium   n°60,
      Cambridge University Press, pp. 193-210.
(18) Cruikshank, D.P. and Brown, R.H., 1981, Icarus 45,
      pp.607-611.
(19) Cruikshank, D.P. and Apt, J. 1984, Icarus 58, pp.306-311.
(20) Cruikshank, D.P., Brown, R.H. and Clark, R.N., 1985, this
      book.
(21) Cruikshank, D.P., Pilcher, C.B. and Morrison, D. 1976,
      Science 194, 835-837.
(22) Cruikshank, D.P. and Silvaggio, P.M., 1980, Icarus 41,
      pp.96-1202.
(23) Soifer, B.T., Neugebauer, G. and Matthews, K., Astron. J.,85,
      pp. 166-167.
(24) Gaffney, E.S. and Matson, D.L. 1980, Icarus 44, pp.511-519.
(25) Ahrens, T.J. and O'Keefe, J. 1985, this book.
(26) Smith, B.A. and the Voyager imaging team, 1979, Science 206,
      pp.927-950.
(27) Johnson, R.E., Boring, J.W., Reimann, C.T., Barton, L.A.,
      Sieveka, E.M., Garrett, J.W., Farmer, K.R., Brown, W.L. and
      Lanzerotti, L.J. 1983,Geophysical Research Letters, 10,
      p.892-895.
(28) Consolmagno, G.J. and Lewis J.S. 1975, In :"Jupiter",T.A.
      Gehrels, Ed., University of Arizona Press, Tucson,
      pp.1035-1051.

(29) Cassen, P., Reynolds, R.I. and Peal, S.J. 1979, Geophysical Research Letters 6, pp.731-734.

(30) Cassen, P., Peale, S.J. and Reynolds, R.T., 1980, Geophysical Research Letters 7, pp.987-988.

(31) Squyres, S.W., Reynolds, R.T., Cassen, P.M. and Peale, S.J. 1983, Nature 301, pp.225-226.

(32) Bianchi, R. and Casacchia, R. 1985, this book.

(33) Lyot, B. 1953, L'Astronomie 67, pp.1-19.

(34) Dollfus, A. and Murray, J.B. 1974, in :"Exploration of the planetary system" eds.A. Wosczyk and C.Iwaniszewska, IAU Symp. n°65, pp.513-525.

(35) Smoluchowski, R. 1983, Science 222, pp.161-163.

(36) Smoluchowski, R. and Mc Williams, A. 1984, Icarus 58, p.282-287.

(37) Poirier, J.P., Sotin, C. and Peyronneau, J. 1981, Nature 292, pp.225-227.

(38) Lucchitta, B.K. 1980, Icarus 44, pp. 481-501.

(39) Forni, O.P., Thomas, P.G. and Masson, P.L., 1985, this book.

(40) Thomas, P.G. and Masson, P.L. 1985, this book.

(41) Dollfus, A. 1975, Icarus 25, pp.416-431.

(42) Mandeville, J.C., Geake, J.E. and Dollfus, A.1980, Icarus 41, pp.343-355.

(43) Smoluchowski, R. 1985, this book.

(44) Owen, T. 1985, this book.

(45) Lunine, J.I. and Stevenson, D.J. 1985, this book.

(46) Farinella, P., Milani, A., Nobili, A.M., Paolicchi, P. and Zappala, V. 1985, this book.

(47) Plescia, J.B. and Boyce, J.M. 1982, Nature 295, pp.285-293.

(48) Boyce, J.M. and Plescia, J.B. 1985, this book.

(49) Smith, B.A. and the Voyager Imaging Team, 1982, Science 215, pp.504-536.

(50) Passey, Q.R. 1983, Icarus 53, pp.105-120.

(51) Poirier, J.P., Boloh, L. and Chambon, P. 1983, Icarus, 55, pp.218-230.

(52) Yoder, C.F. et al., 1981, EOS, 42, p.939.

(53) Stevenson, D.J. 1982, Nature 298, pp.142-144.

(54) Squyres, S.W., Reynolds, R.T., Cassen, P.M. and Peale, S.J. 1983, Icarus 53, pp.319-331.

(55) Miller, S.L. 1961, Proc. Nat. Acad. Sci. U.S.A. 97, pp.1798-1808.

(56) Klinger, J. 1982, Nature 299, p.41.

(57) Elliot, J.L. and Nicholson, P.D. 1984, in :"Planetary Rings" R.Greenberg and A.Brahic Ed., Univ. of Arizona Press Tucson, pp.25-72.

(58) Lanzerotti, L.J., Brown, W.L. and Johnson, R.E. 1985, this book.

(59) Brown, R.H. 1985, this book.

(60) Morrison, D., Cruikshank, D.P. and Brown, R.H. 1982, Nature 300, pp. 425-427.

SHOCK VAPORIZATION AND THE ACCRETION OF THE ICY SATELLITES OF JUPITER AND SATURN

Thomas J. Ahrens and John D. O'Keefe

Seismological Laboratory, California Institute of Technology, Pasadena, CA 91125

Abstract. Shock wave data, thermodynamic and phase diagram data for ice, porous ice, and water are taken together with a Rice-Walsh-Bakanova equation of state to define the shock pressures and impact velocities required to induce incipient melting (IM) (6 GPa), complete melting (CM) (10 GPa), and passage through the vapor-liquid critical point (CP) upon isentropic release (22.5 GPa). Upon expanding along the isentrope which passes through CP $\sim$0.61 kPa (6.1 mbar) is achieved. Below this pressure, ice sublimates and $\sim$0.4 mass fraction H_2O gas is in equilibrium with ice I. The minimum impact velocity required to induce IM, CM, and isentropic release through CP is 2.1, 3.0, and 4.5 km/sec for silicate impactors. For icy projectiles, Hugoniot states achieved in icy targets or projectiles depend only weakly on initial temperature of ice. The IM, CM, and CP isentropes are achieved upon impacting with an icy projectile an icy surface at velocities of 3.4, 4.4, and 7.2 km/sec, respectively. We observe that at a partial H_2O pressure below 0.61 kPa and temperatures below 273K, ice partially vaporizes and requires $\sim$3000 kJ/kg of heat of vaporization for complete sublimation. We examine the hypothesis that the smaller satellites of Saturn having mean densities in the 1.1 to 1.4 Mg/m^3 range represent primordial accreted planetesimal condensates (60% (wt.) H_2O, and 40% (wt.) silicate) formed in the proto-Jovian and Saturnian accretionary planetary discs. These densities are in the range expected for water-ice/silicate mixtures constrained to the solar values of O/Si and O/Mg atomic ratios. If the large satellites accreted from the same group of planetesimals which formed the small Saturnian satellites, impact vaporization of water upon accretion into a porous regolith at low H_2O partial pressure can account for the increase in mean planetesimal density from 1.6 Mg/m^3 (43% H_2O + 57% silicate) to a mean planetary density of 1.9 Mg/m^3 for Ganymedean-sized water-silicate objects. If impact-volatilization of initially porous planetesimals is assumed, we demonstrate that starting with planetesimals composed of 54% H_2O and 40% silicate (1.35 Mg/m^3) partial

J. Klinger et al. (eds.), Ices in the Solar System, 631–654.

devolatilization upon accretion will yield a Ganymede-sized planet, having a
radius of 2600 km and a density of 1.85 kg/m³, similar to that of Ganymede,
Callisto, and Titan.

1. INTRODUCTION

We apply knowledge of the properties of water and ice over a wide range of
pressures and temperatures and describe constraints on the shock vaporization
processes for water and ice in the solar system. In particular, we examine the
role of impact vaporization acting during the formation of the Jovian and
Saturnian satellites in an attempt to explain the observed density in terms of
composition of these rock and ice objects. The imaging instruments on the
Voyagers I and II spacecraft (25,35-37) have provided definitive knowledge of
the volumes, and hence, densities, as well as, the surface characteristics of at
least 12 satellites of Jupiter and Saturn having planetary radii $\gtrsim$ 100 km
(Table 1).

Table 1. Characteristics of Major Jovian and Saturnian Satellites[a]

Satellite	Mean Orbital Radius (10^5 km)	Radius (km)	Mass (10^{20} kg)	Mean Density (Mg/m^3)	Satellite Class.[b]
J5 Amalthea	1.8	(96)	(0.13)	(3.5)	s
J1-Io	4.2	1815±10	892	3.55	s
J2-Europa	6.7	1569±10	487	3.04	s
J3-Ganymede	11	2631±10	1490	1.93	d
J4-Callisto	19	2400±10	1075	1.83	d
S1-Mimas	1.9	195±5	0.37	1.44±0.18	p
S2-Enceladus	2.4	250±10	0.72	1.16±0.55	p
S3-Tethys	2.9	525±10	6.06	1.21±0.16	p
S4-Dione	3.8	560±10	10.30	1.43±0.06	p
S5-Rhea	5.3	765±10	24.38	1.33±0.09	p
S6-Titan	12.2	2560±26	1335.25	1.88±0.01	d
S8-Iapetus	35.6	720±20	18.76	1.16±.09	p

[a] Data compiled from Morrison (1982), Stone and Miner (1982), Lunine and Stevenson (1982).
[b] s = silicate; d = (partially) devolatilized; p = primordial

Pollack and Reynolds (29), and more recently, Lunine and Stevenson (18)
and Pollack and Fanale (28) have discussed the origin of the larger Jovian
satellites within a framework of condensation from a "protostellar" disc of
gaseous and particulate material rotating around a newly formed proto-
Jupiter (of approximately solar composition). This disc is much like the larger
scale protosolar disc from which the sun and the planets themselves form.

Like the solar system, itself, this type of model of the Jovian satellite system implies a higher temperature close to Jupiter and thus ultimately less relative condensation of volatiles (e.g. H_2O, NH_3) in the satellites close to the planet as compared to the outer satellites. The Jovian system clearly suggests such a formational history, as the inner satellites (Amalthea, Io, and Europa) have higher mean densities, 3.0 to 3.6 Mg/m^3, and in an analogy to the terrestrial planets) contain a higher fraction of silicates than do the outer larger (lower density) satellites (Ganymede and Callisto). The latter have densities of 1.8 to 1.9 Mg/m^3 (Fig. 1). In such formation models of the Jovian satellites, it is usually assumed that these accreted from small solid objects which formed via gravitational or other instabilities in the proto-Jovian disc. The inner satellites accumulated lower inventories of volatiles as a result of either or both of the following two effects:

(1) Because of the higher temperatures in the vicinity of the early Jupiter, relatively less volatiles can condense. For example, in the Lunine and Stevenson model the temperatures and pressures at the orbits of Amalthea and Callisto are 1350K and 13 MPa and 150K and 0.01 MPa, and

(2) Because the Keplerian velocity around Jupiter of the proto-satellites and hence the speed with which they overtake and accrete smaller solid objects embedded in the gaseous nebula have a lower encounter velocity with increasing radius from Jupiter, less impact vaporization can occur during accretion.

The "miniature solar system" model for the formation of the Jovian satellite system is expected to always yield relatively more volatile material condensing with increasing radius from Jupiter. This will give rise to high density inner satellites versus low density outer satellites. In the case of the Saturnian system, the satellite mean density versus orbital radius relation is irregular, as can be seen in Fig. 1.

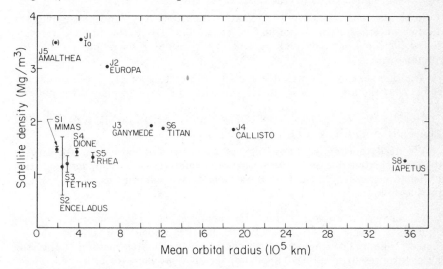

Figure 1. Mean orbital radius versus satellite density for the Jovian and Saturnian systems.

Examination of Table 1 suggests that one could classify the Jovian and Saturnian satellites into three groups according to mean density: silicate (s), primordial (p), and partially devolatilized (d) (e.g. 9). By silicate density we mean the density of terrestrial planetary mantles (3-3.5 Mg/m^3). An estimate of a primordial density can be obtained by assuming the solar O/Si atomic ratio of 15.5 (33). We assume that this ratio determines the abundance of H_2O which is in turn controlled by the abundance of oxygen relative to silicon and hence the abundance of Si. For example, if we consider SiO_2 as being controlled by the abundance of Si then we infer that an atomic ratio is required of 5.16 moles of H_2O to 1 mole of SiO_2. This ratio thus implies a density of 1.23 Mg/m^3 for a water-ice and quartz mixture. Assuming that the most abundant anhydrous silicate in the solar system is probably enstatite $MgSiO_3$ and utilizing the solar atomic ratio of O/Mg of 17.38 and assuming the H_2O/MgO ratio controls the density of primordial low temperature volatile-silicate condensate mixture, a density of 1.08 Mg/m^3 is inferred by the same reasoning. The additional small amounts of oxidized iron will, of course, increase this density. We assume that densities in the above range, correspond to a combination silicate plus some oxidized iron with primordial volatile condensate. This is seen to be representative of densities of the group of the smaller satellites in the Saturnian systems indicated by (p) in Table 1.

Previously Lunine and Stevenson (18) predict final satellite density on the basis of differing environment for accretion as a function of satellite radius from Jupiter. We assume <u>all</u> the satellites in the Jovian and Saturnian system (except J5, J1, and J2) accreted from the same composition of planetesimal material and that the least affected samples of these planetesimals are the smaller satellites of Saturn and possibly also the ring materials of both planets. The partially devolatilized satellites whose density we attempt to predict (Ganymede, Callisto, and Titan) are then assumed to be partially devolatilized, indicated as (d) in Table 1.

In the present paper we first examine the shock wave and phase diagram data for water, solid and porous water-ice and define the impact velocity required for shock-induced melting and vaporization of water and ice at various temperatures.

We then examine a possible model of accretion of icy satellites which predicts that the amount of ice devolatilization is related to planetary size.

2. SHOCK WAVE EQUATION OF STATE AND THERMODYNAMIC DATA FOR WATER, WATER-ICE, AND POROUS ICE.

We need to combine knowledge of the Hugoniot curves for ice targets and the calculated release isentrope curves to determine the impact velocities required to generate sufficient entropy to melt and vaporize ice. The train of logic is as follows:

(a) Given the measured Hugoniot shock velocity, U, versus particle velocity, u, data, shock pressure versus shock particle velocity curves may be constructed. These allow prediction via the impedance match method (22) of the peak shock pressures induced upon impact of a specific projectiles (e.g. rock or ice) against solid or porous ice or water targets. The amount of material subjected to this peak pressure is in the order of 10^0 to 10^1 times the projectile mass (27).

(b) The single shock data must then be combined with data obtained by double shocking water and/or porous ice to the same pressure but at lesser and greater temperatures (and hence lesser and greater specific volumes) as well as thermodynamic data to calculate the isentropes in pressure-volume temperature space. The tactic we employed is that of Ahrens and O'Keefe (2). We utilize pressure-volume-entropy-enthalpy data at a single moderate pressure (1 GPa) and increasing (high) temperatures to define a series of states with increasing entropy to specify the foot of isentropes.

(c) These isentropes are then projected to higher pressures using a complete equation of state formalism like that proposed by Bakanova et al. (5). Once the intersections with the water and ice Hugoniots are found, we can infer the entropy along the Hugoniot from the intersection of the various isentropes with the Hugoniot. Then we can determine whether the entropy density is sufficiently high in a given shock state such that upon pressure release shock induced melting or vaporization will occur. Of special interest are the isentropes which define the shock pressures required to just produce incipient melting (IM), complete melting (CM), and bring water to the liquid-vapor (steam) critical point (CP).

(d) When the calculated release isentropes are taken together with the low pressure thermodynamic data for the water-ice-steam system, as for example, summarized in the form of a Mollier diagram (7), the mass fraction of vapor produced by a given entropy density as a function of partial pressure of water vapor in the planetary environment may be inferred.

(1) Shock wave data.

Because ice covers the surface of more than 20 of the Jovian and Saturnian satellites (except for J5, J1, and S6) and is a major constituent of the interior of these objects (and it of course covers 10% of the earth) it is important to determine the projectile impact velocities and hence shock pressures required to induce melting and/or vaporization.

The available Hugoniot data for water-ice and porous-ice in terms of shock velocity, U, particle velocity, u, relations of the form

$$U = C_o + \lambda_1 u + \lambda_2 u^2 \tag{1}$$

is given in Table 2. Shock pressure and shock density (or specific volume, V) can be obtained from the constants in Table 2 via the Rankine Hugoniot shock equations for pressure, P, density, ρ, and internal energy, E, or enthalpy, H, respectively.

$$P-P_o = \rho_o u U \tag{2}$$

$$\rho = \rho_o U/(U-u) \text{ or } V = V_0 (U-u)/U \tag{3}$$

$$E-E_o = (P_o+P)(V_{00}-V)/2 \text{ or } H-H_o = (P_o+P)(V_{00}+V)/2 \tag{4}$$

In equations 2-4, P_o, ρ_o, V_0, E_o, and H_o are the pressure, crystal density, specific crystal volume, internal energy, and enthalpy in the initial state. In the present paper, energies and enthalpies (and entropy) are measured relative to conditions at 1 bar and 0 K. V_{00} is the pre-shock initial specific volume

Table 2. Shock-Particle Velocity Data for Water, Ice, and Porous Ice

Initial Material	Initial Density (Mg/m^3)	C_o (km/sec)	λ_1	λ_2 (sec/km)	Range of u (km/sec)	Ref.
Water (294-298K)	0.9979 ± 0.0004	2.393	1.333	-	$1.5 < u < 7.1$	(a)
Water (293)	$0.99823^{(b)}$	1.50	2.0	-0.107	$0 < u < 4.0$	(c)
Water (293)	$0.99823^{(b)}$	3.2	1.144	-	$u > 4.0$	(c)
Ice (258)	0.915	1.57	1.465	-		(c)
Ice (263)	$0.9230^{(b)}$	1.317	1.526	-		(d)
Porous Ice (258)	0.60	0.74	1.425	-		(c)
Porous Ice (258)	0.35	0.0	1.425	-		(c)

[a] Mitchell and Nellis (1982)
[b] Nominal value Weast (1982)
[c] Bakanova et al. (1976)
[d] present fit to 18 data points (8 data, Bakanova et al., 1976; 8 data, Anderson, 1968; 1 point, Larson et al., 1972; 1 point, Gaffney and Ahrens, 1980).

which in the case of porous ice will be greater than V_0. For no porosity $V_0 = V_{00}$. In the case of liquid water, the shock wave data are described via a set of bilinear equations by Mitchell and Nellis (24) who assume a high pressure regime extending from 4.4 to 83 GPa and Bakanova et al. (5) who defined the high pressure regime from 8 to 55 GPa. Double (reflected) shock experiments on water have also been carried out by Walsh and Rice (40), Bakanova et al. (5), and Mitchell and Nellis (24), the latter to peak pressures of 230 GPa. In addition, a large number of shock wave data for solid ice initially at temperatures of 258 to 263 K are available over a pressure range from 0.6 to 50.3 GPa (Fig. 2). These data are also closely fit by a linear shock velocity-particle velocity relation. Below 3 GPa, shock waves in solid ice demonstrate interesting structure involving intermediate shock states due to phase transitions. These are described by Gaffney (13) in this volume.

When the final shock state data summarized in Table 2 are cast into the pressure-particle velocity plane (Figure 3) via equation 2, it can be seen that liquid water (on account of its greater density) initially has a slightly higher shock impedance (larger value of shock pressure at a fixed particle velocity) than ice and that this small impedance contrast reverses (on account of more shock heating in ice) above $\sim$95 GPa at a particle velocity of 7.7 km/sec. Porous ices with lower initial densities have correspondingly lower shock impedance.

The representation of Hugoniot curves in the pressure-particle velocity plane via the impedance match method (22) allows straightforward determination of projectile-target peak shock pressures from the impact velocities. To demonstrate how this is carried out we indicate in Fig. 2 the escape velocity of water-bearing objects in the solar system along the abscissa (from 0.16 km/sec for Mimas to 11.19 km/sec for the earth). In general, the infall velocity is the minimum velocity for which a projectile can impact the planetary surface in the case where no atmosphere exists or the projectile is large enough not to be affected by its passage through the atmosphere. This minimum velocity then

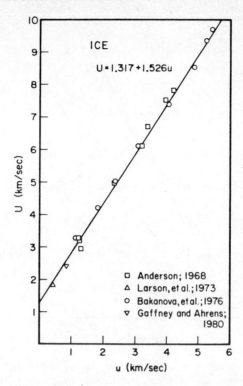

Figure 2. Shock velocity versus particle velocity for ice at 258 and 263 K.

can be increased substantially in the case of the satellites of the major planets by the following three factors which are listed in order of importance.

(1) The effect of gravity focusing by the mass of the planet.

(2) The effect of heliocentric velocity of the planet, and

(3) The effect of the orbital velocity about the planet.

The effect of gravitational focusing by the high infall velocity (escape velocity of Jupiter and Saturn, 60 and 36 km/sec, respectively) of objects on their planetary satellites is demonstrated with the mean impact velocity and flux concentrations given in Table 3. Table 3 assumes that for the Jovian system objects have a mean encounter velocity of 8 km/sec. Shoemaker and Wolf (34) suggest that in addition to co-orbiting objects at Jupiter's heliocentric distance, many of the objects which have impacted the Jovian satellites were also cometary.

The pressure-particle velocity Hugoniot of the impactors is plotted backward centered at the impact velocity. Thus shown in Fig. 3, a projectile impacting at 11.19 km/sec having the pressure-particle velocity characteristics of serpentinite, will induce a shock pressure of 41, 63, 90, and 90.5 GPa in 0.35 and 0.6 Mg/m^3 porous ice, solid ice, and water, respectively. In contrast to the major and terrestrial planets the minimum infall velocities of the Jovian and Saturnian satellites are much lower as indicated in Fig. 3. For

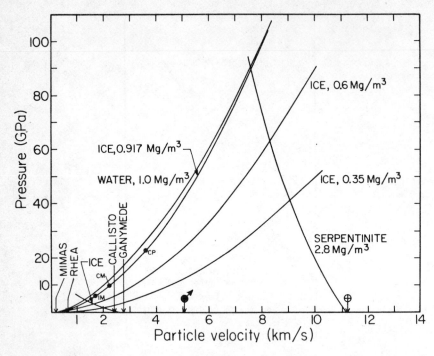

Figure 3. Shock pressure versus particle velocity for water, ice, and porous ice. Curve
for serpentinite, representative of planetary silicates after Marsh (1980). Escape velocity for satellites and planets indicated along the particle velocity axis. Shock pressures required for incipient melting (IM), complete melting (CM), and the critical point
isentrope (CP) for ice are indicated.

Table 3. Relative cratering rates on the Galilean satellites for mean
encounter velocity of 8 km/sec (Smith et al., 1979b)

Satellite	Flux concentration factor by Jupiter gravity	Mean Impact Velocity (km/sec)	Relative Crater Production Rate
Callisto	3.3	14	1
Ganymede	5.0	18	2.0
Europa	7.3	21	2.8
Io	11.1	26	5.3

solid ice infalling at the escape velocity of Ganymede (2.74 km/sec) a peak
pressure of only 4.2 GPa is induced in the projectile and target.

(2) Equation of state for water and ice.

It was recognized as early as 1957 when Rice and Walsh (32) carried out
the first extensive analysis of single and reflected shock data for water that

the formulation of a complete thermal equation of state according to the Mie-Gruneisen equation --- so effective for metals (e.g. 31) was infeasible for application to water and other molecular fluids. Instead Rice and Walsh proposed on the basis of a series of reflected shock data in which different enthalpy states were achieved at a given pressure that the thermodynamic quantity $(\partial H/\partial V)_P$ depends strongly on P (and is effectively independent of V or T at constant P). They then defined a thermodynamic quantity ξ (P) which we will call the Rice-Walsh parameter, as

$$\xi(P) \equiv C_p/(\partial V/\partial T)_P = \left[\frac{\partial H}{\partial T}\right]_P \Big/ \left[\frac{\partial V}{\partial T}\right]_P \tag{5}$$

Eq. (5) is analogous to the Mie-Gruneisen equation of state, where a state at a given E can be calculated for a given pressure, P, by comparing it to the internal energy (E_r), and pressure (P_r) at the same volume V. Here E_r and P_r refer to the reference thermodynamic path such as the principal Hugoniot. The Mie-Gruneisen equation-of-state is written as:

$$E = E_r + V/\gamma (P-P_r) \tag{6}$$

where

$$\gamma \equiv V (\partial P/\partial E)_v \tag{7}$$

In analogy with the Mie-Gruneisen parameter, the Rice-Walsh parameter is used to relate enthalpy H and volume, V, at the same pressure, P, to that along the reference path. The reference path chosen here is the principal Hugoniot of water denoted by the subscript 1. Thus the Rice-Walsh equation is

$$H(P, V) = H_1 (P) + \xi(P) [V-V_1(P)] \tag{8}$$

This treatment satisfactorily describes the available Hugoniot and reflected shock data to 25 GPa and is also concordant with static compression data for water at high temperature as well at high pressure and relatively low temperature on the properties of steam. Upon obtaining additional reflected shock data to 120 GPa and porous ice Hugoniot data, Bakanova et al. (5) utilized an equivalent and more convenient relation than equation 8 specifying internal energy at constant pressure. The Bakanova et al. parameter is defined as

$$\eta (P) \equiv P (\partial V/\partial E)_P \tag{9}$$

and is related to the Rice-Walsh parameter by applying the relation

$$(\partial H/\partial V)_P = (\partial E/\partial V)_P + P \tag{10}$$

to obtain

$$\eta (P) = P/(\xi-P) \tag{11}$$

From equation 9 it follows that

$$E(P, V) - E_1(P) = \frac{P}{\eta(P)} (V-V_1(P)) \tag{12}$$

where E_1 and V_1 are again the reference internal energy and volume at pressure, P.

To obtain an equation for an isentrope we substitute for E_1 (P)

$$E_1 = E_o + (V_o - V) P/2 \tag{13}$$

using the Rankine Hugoniot energy equation (relative to the internal energy, E_0 at STP) into equation 12 and solve for V. This yields

$$V = V_1 (1+\eta/2) - \eta \left[\int_{V_{io}}^{V} PdV - E_{io} + E_o + PV_0/2 \right] / P \tag{14}$$

where we have also substituted for E

$$E = - \int_{V_{io}}^{V} PdV + E_{io} \tag{15}$$

Here V_{io} and E_{io}, and T_{io} are the specific volume, internal energy and temperature state which defines the foot of the isentrope at pressure P_{io}.

An equation for the temperature along the isentrope can be obtained from the differential relation

$$0 = TdS = T (\partial S/\partial T)_P \, dT + T (\partial S/\partial P)_T \, dP \tag{16}$$

substituting in the first and second term, respectively, and using the thermodynamic relations

$$C_P/T = (\partial S/\partial T)_P \tag{17a}$$

$$(\partial S/\partial P)_T = -(\partial V/\partial T)_P \tag{17b}$$

yields the differential equation

$$dT/T = dP(\partial V/\partial T)_P/C_P \tag{18}$$

which upon substituting for ξ (P) from equation 5 and integrating yields

$$T_a = T_{io} \exp \int_{P_{io}}^{P} \frac{dP}{\xi(P)} = T_{io} \exp \int_{P_{io}}^{P} \frac{\eta dP}{(1+\eta)} \tag{19}$$

In order to calculate the P-V-T-E isentropic path from equations 14 and 19 the function $\xi(P)$ or $\eta(P)$ must be determined. Rice and Walsh examined four reflected shock data to 23.5 GPa and fit these to

$$\log_{10} \xi = 1.17943 + 0.030338 \, P(GPa) \tag{20}$$

whereas Bakanova et al. fit both porous ice data and water double shock data and fit these to the relation

$$\eta \, (P) \, GPa = 0.07 \, [1 - \exp(-0.08436P)] + 0.044095P \exp (-0.048202 \, P) \tag{21}$$

As can be seen in Fig. 4 the $\eta(P)$ function from both studies are similar. The peak in η in the 15 to 20 GPa range is attributed to the increase pressure-induced ionization which occurs in water and is reflected in the rapid increase in electrical conductivity (24) of shocked liquid water. To further demonstrate the adequacy of the Bakanova equation of state in describing the thermodynamic properties of water to high pressures, we apply it to calculating shock

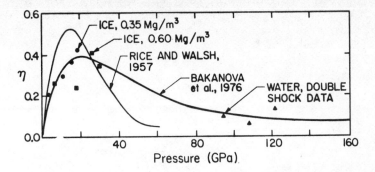

Figure 4. Bakanova parameter, versus, pressure.

temperatures for water. Comparison of calculated and measured shock temperatures is carried out in Appendix A.

In order to calculate the specific release isentrope P-V paths for H_2O other than the paths given by Bakanova et al. (5), Fig. 5, we can utilize both the phase diagram of water (Fig. 6) and the Mollier diagram (Fig. 7). The phase diagram of water sketched in Fig. 6 demonstrates that upon shocking either liquid water or ice above ~20 GPa it achieves states above the vapor-liquid critical point (647K, 22.1 MPa, 3.1 m^3/Mg). It should be noted that a well-defined critical shock pressure marking the onset of incipient vaporization upon release cannot be given. Shock-induced vaporization depends strictly on the partial pressure of H_2O in the environment into which shocked water or ice expands. If the partial pressure of H_2O is sufficiently low (below the ice-I vapor or liquid phase line) shocked water ice will always be vaporized upon release to that environment.

The existence of the ice I-liquid-vapor triple point at 273K and 0.61 kPa implies that shocked ice need not melt first but can bypass the liquid field and sublimate provided the pressure is below the ice I-sublimation line which is given explicitly in Dorsey (11) as

$$\log_{10} P \text{ (torr)} = -\frac{2445.5646}{T(K)} + 8.2312 \log_{10} T - 1677.006 \, (10^{-5}) \, T +$$
$$120514 \, (10^{-10}) \, T^2 - 6.757169 \tag{22}$$

The Mollier diagram for water, ice, and steam shown in Fig. 7 defines the critical isentrope (vertical lines in Fig. 7) which can be defined for incipient melting (IM), complete melting (CM), and the isentrope which passes through the vapor-liquid critical point (CP). These critical isentropes in the pressure-volume plane are plotted in Fig. 8 for ice at 70 and 263K, the range of temperatures appropriate for the surfaces of the Jovian and Saturnian icy satellites. Whereas the isentropes corresponding to IM and CM produce little vapor until relatively low pressures are achieved, the release isentrope passing through the CP gives rise to ~0.5 mass fraction vapor upon release through the critical point and upon release to pressures below 0.61 kPa some 0.47 mass fraction remains vaporized (Fig. 7).

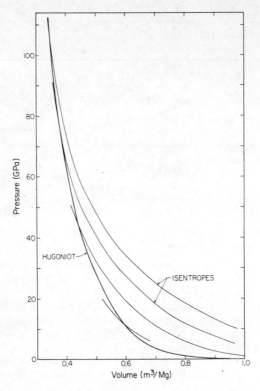

Figure 5. Principal Hugoniot and release isentropes for water in the pressure specific volume plane after Bakanova et al. (1976).

The critical isentropes of Fig. 8 and Tables 4 and 5 are also indicated where they cross the Hugoniot in the pressure-particle velocity plane for solid ice in Fig. 3. Examination of where the IM, CM, and CP points fall on the ice Hugoniot (Fig. 3) demonstrates that the minimum infall velocities for ice onto an icy satellite so as to produce melting or vaporization can be specified. Silicate projectiles impacting at greater than 2.1 km/sec or icy projectiles impacting 3.4 km/sec are required to induce IM of ice. CM of ice requires silicate projectiles impacting at least 3 km/sec or ice projectiles impacting at 4.4 km/sec. To obtain 0.4 to 0.5 mass fraction vapor upon passing through the CP requires ~4.5 km/sec for silicate impactors or 7.2 km/sec for pure ice impactors. The latter velocities are in approximate agreement with those of Smoluchowski (38).

Thus we conclude that although copious melting and vaporization of water ices on icy satellites will occur at meteoroid and cometary impact velocities on the Jovian and Saturnian satellites (Table 3), vaporization processes during accretion will not be significant unless the partial pressure of water is below 0.61 kPa during accretion of these objects. We explore the consequences of accretion of the icy satellites in this environment in the next section.

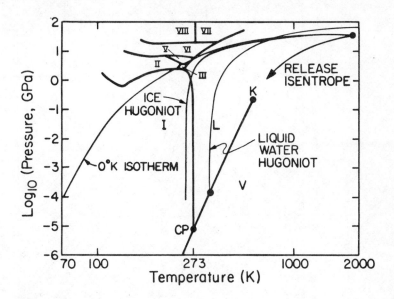

Figure 6. Log_{10} pressure (GPa) versus temperature phase diagram for H_2O. Hugoniot for ice and liquid water and release isentrope passing through liquid-vapor critical point (K) and water-vapor-ice I, triple point (CP) are indicated.

3. ACCRETION OF ICY SATELLITES

We have demonstrated the importance of the existence of low partial pressure of H_2O (below 0.61 kPa or 6.1 millibar) for vaporization to be important in impact processes during accretion. For accretion under low water partial pressure conditions and low temperatures, impact vaporization is, however, readily accomplished because the liquid stability field of water is entirely bypassed. Examination of the Mollier diagram of water (Fig. 7) in this low temperature region indicates that an enthalpy gain of ∼3000 kJ/kg is required to completely vaporize Ice I at pressures below the vapor-liquid ice I triple point. Rather than assume that the Jovian satellites accreted in a gaseous nebula in which the partial pressure of water was at the saturation point at temperatures varying from (150 to 230K) as assumed by Lunine and Stevenson (18), in the present paper we will examine the consequences of creating icy satellites via infall of planetesimals in a dry, undersaturated environment.

We assume as an extreme case once material is vaporized upon impact, it is lost from the planet. The more complicated problem of precipitation versus thermal loss of H_2O and the general thermal state of the accreting planetary atmosphere is not addressed.

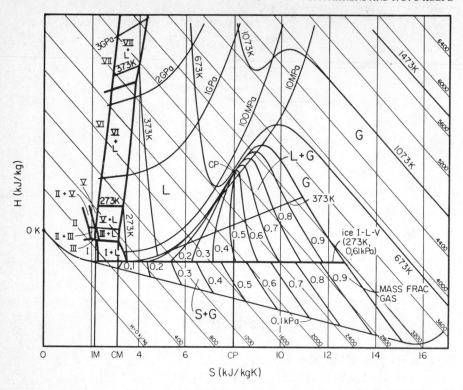

Figure 7. Simplified Mollier enthalpy entropy diagram of water, after Bosnjakovic, Renz, and Burow (1970). Note coordinates, enthalpy, H, and entropy, S, are nonorthogonal and enthalpy values along negative sloping lines are given along left edge of figure. Isentropes are vertical lines and the critical ones are labeled. Stability fields of ice I-VII are indicated as are also solid (S), liquid (L) and gas (G).

We shall first examine the case of accretion of a pure icy planet at low pressures. If we assume that a porous icy regolith forms in this situation, we can model the fraction of ice impact vaporized, f, as being equal to:

$$f = V_i^2/(2 \cdot \Delta H_{subl}) \tag{23}$$

where V_i is planetesimal impact velocity.

We conclude that it is most likely that at low temperatures and at undersaturated conditions a porous ice regolith some $\sim 10^0$ to 10^2 km thick would be retained on accreting icy satellites because: (1) Porous regoliths are known to adhere to objects as small as $\sim 10^2$ km diameter asteroids in the case of Phobos and Deimos. (2) At low temperatures, ice retains $\sim 50\%$ porosity until overburden lithostatic stresses of ~ 5 MPa are exerted (Anderson and Benson, 1963).

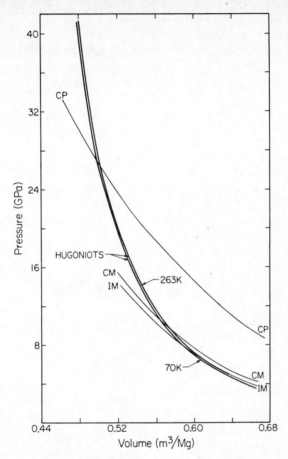

Figure 8. Hugoniot for ice centered at 70 and 263 K and release isentropes for incipient melting (IM), complete melting (CM), and critical point (CP) isentropes in pressure volume plane. Hugoniot of ice centered at 70 K inferred from thermal expansion data for ice collected in Fletcher (1970).

At any point in the accretion of a planetary satellite of uniform density, ρ_0, the minimum infall velocity, $V_{i,min}$ is equal to the escape velocity, V_{esc} which is

$$V_{esc} = \sqrt{8\pi \, G \, \rho_0/3} \; R \tag{24}$$

where G is the gravitational constant. We now assume that

$$V_i = V_{i,min} = V_{esc} \tag{25}$$

and calculate for a pure water planet how much incident material accreted is retained versus planetary mass (Fig. 9). We observe that at planetary masses of $\sim 10^{24}$g the infall velocity is so slow (0.5 km/sec) that little impact vaporization of H_2O occurs. However, as the pure H_2O satellite accretes in this

Table 4. Critical Isentropes in the Water System

| Critical Isentrope | Determining State | Entropy* (kJ/kgK) | Phase | State at 1 GPa | | | |
				Temperature (K)	Enthalpy* (kJ/kg)	Volume (Mg/m³)	Internal Energy* (kJ/kg)
Incipient Melting (IM)	Triple Point Ice III-Ice V-Liquid	2.21	Ice VI	400	1140	0.735	405
Complete Melting (CM)	Triple Point Ice I-Ice III-Liquid	3.16	Ice VI+ Liquid	400	1400	0.736	614
Liquid-Vapor Critical Point (CP)	Liquid-Vapor	7.95	Liquid	1017	4080	1.085	2995
1 GPa and 1273K	none	8.76	Vapor	1273	5022	1.2532	3769

* referenced to 1 bar and 0K

Table 5. Critical Release States, Ice Hugoniots

Isentrope[a]	S(kJ/kgK)	70K			263K		
		T(K)	P(GPa)	V(m³/Mg)	T(K)	P(GPa)	V(m³/Mg)
IM	2.21	378	7.6	0.594	363	6.2	0.616
CM	3.16	410	10.8	0.562	298	9.6	0.576
CP	7.95	1756	25.5	0.509	1779	26.7	0.500

(a) referenced to 1 bar and 0K

environment an increasing fraction of the mass delivered is impact vaporized and becomes lost to the object. As can be seen in Fig. 9 when the planet mass is $\sim 10^{26}$g, the minimum infall velocity is $\sim$2 km/sec and only $\sim$0.44 mass fraction of the incident planetesimal material has been retained. For a pure H_2O planet as massive as Titan or Callisto only $\sim$15% of the infalling mass is not impact vaporized.

If now the same calculation is carried out for 0.4 mass fraction of silicate, 0.6 mass fraction water, planetesimals typical of material condensing from the primordial Jovian and Saturnian nebulae, and, we assume a silicate density of 3.0 Mg/m^3 for the silicate fraction, the water vaporization is less extreme as also indicated in Fig. 9. Planetesimals with this composition when accreting to form a planet retain $\sim$ 65% of the incident mass for an object when the object becomes massive as Titan or Ganymede.

We sought to examine the question of whether accretion of a water-silicate mixture, initial density 1.27 Mg/m^3 which approximates the mean density of the smaller satellites (Mimas, Enceladus, Tethys, Dione, Iapetus and Rhea) results in sufficient impact vaporization to obtain a mean density of >1.8 Mg/m^3 corresponding to larger satellites Callisto, Titan, and Ganymede. As indicated in Fig. 10 accretion of such a mixture results in a mean planetary density of only $\sim$1.5 Mg/m^3. Moreover, also shown in Fig. 10, the situation is not especially improved if one assumes a 10^3 km silicate core for the larger satellites. In this case the increase in the average final density is only slight. The interesting minimum density at 1.55 Mg/m^3 is caused by the process of initially reducing the mean density of the growing planet with a silicate core by accreting water-bearing material. However, as the planet grows, an increasing fraction of the water-bearing material contributed is shock-vaporized and lost, and finally at a radius of $\sim$2300 km the mean density begins to increase again. We conclude from these calculations that the simple process of impact vaporization in a regolith and a loss of all the shock volatilized water cannot make planets with densities as high as Callisto, Titan, and Ganymede from a protosolar mixture of 60% (mass) water and 40% silicate which both correspond to a primordial water silicate mixture as well as approximates the average density of the smaller satellites of Saturn. As indicated in Fig. 10 if one relaxes the constraint on the planetary composition slightly, and assume 43% H_2O and 57% silicate, it is possible to produce objects with exactly the density of Callisto, Titan, and Ganymede (which correspond to ~40% water and 60% silicate) by the impact volatilization process outlined above. Finally, if instead of assuming a porous regolith, we make a slightly more drastic assumption and assume porous planetesimals (and restrict impact vaporization to the mass of planetesimal material <u>only</u>) it is possible to slightly enhance the impact volatilization process to achieve objects close to the radius of Callisto, Titan, and Ganymede with material very close to the mean density of the smaller satellites of Saturn. In this case our calculations demonstrate that starting with 54% H_2O and 46% silicate composition (starting density 1.35 Mg/m^3) planets having the Ganymede radius of 2600 km and a density of 1.85 to 1.9 can be achieved.

The present results suggest that impact volatilization can give rise to the greater density of the larger Jovian and Saturnian satellites as a result of vaporization of porous, water ice (and possibly also NH_3) ice during accretion

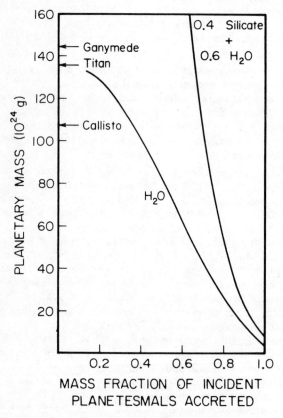

Figure 9. Mass fraction of incident planetesimal material accreted, versus, planetary mass for pure H_2O and 0.4 silicate + 0.6 H_2O planet. A porous, ideal locking (Linde and Schmidt, 1966) solid (similar to snow) is assumed on the surface of the growing planet. All the gravitational energy of accretion is deposited in this layer and gives rise to partial vaporization of ice at temperatures below 273.16 K and 0.61 kPa (6.108 mbar).

and all but the essentially silicate satellites can have accreted from the same primordial solar composition material. The final density of Jovian and Saturnian satellites then may just be the consequence of the amount of material in orbit around Jupiter and Saturn at a particular radius, and hence, ultimate satellite size.

4. SUMMARY AND CONCLUSIONS

Shock wave data for polycrystalline ice, porous ice, and liquid water are available over a wide range of pressure. The peak shock state achieved upon accretionary impacts up to 2.6 km/sec, as well as impacts of 20-30 km/sec for cometary encounters can be described using the impedance match method. A key calculation is to determine the intersection of the isentrope with the Hugoniot curve of water and ice so as to determine the entropy function along the Hugoniot. Knowledge of the entropy gain along the Hugoniot of ice determines whether rarefaction from a peak shock pressure will result in incipient melting, complete melting, or partial vaporization. The latter will occur upon passage of the isentropes through thermodynamic states near the vapor-liquid critical point (at 640 K, 221 MPa and 3.1 m^3/Mg). For ice, the shock pressure required to induce incipient melting is 6 GPa; to achieve complete melting, is 10 GPa; and to pass through the vapor-liquid critical point, is 22.5 GPa.

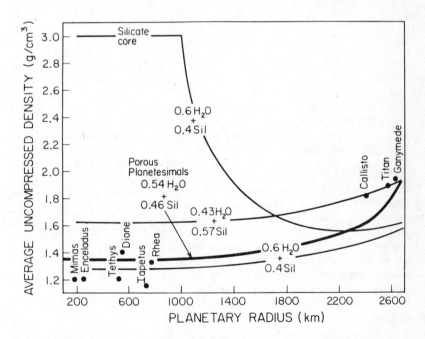

Figure 10. Average uncompressed density of an icy planet versus planetary radius. Light curves are for accretion of solid planetesimals and a regolith-covered planet. Heavy curve is for accretion of porous planetesimals and a solid planetary surface.

These pressure values depend only slightly on the initial temperature of the water-ice. Upon release along the isentrope passing through the critical point, the mean fraction of water vaporized is ~0.5 which decreases to 0.47 upon cooling and expansion to below the vapor-liquid-ice I triple point.

The vapor-liquid-ice I triple point at 0.61 kPa (6.1 mbar) and 273 K occurs at a point where the condensed phase density is 1 Mg/m^3 and the vapor phase density is 5 x 10^{-6} Mg/m^3. It plays an important role in impact vaporization in an environment with low partial pressure of H_2O. The minimum impact velocity onto an icy satellite surface required to induce the incipient melting is 2.1 and 3.4 km/sec for silicate and icy projectiles, respectively. Complete melting of ice requires silicate or icy impactors traveling at ~3 and 4.4 km/sec, respectively. To obtain 0.4 to 0.5 mass fraction vapor upon passing through the liquid-vapor critical point requires 4.5 to 7.2 km/sec impact velocities for silicate and ice projectiles, respectively.

The Rice-Walsh-Bakanova equation of state in which internal energy or enthalpy is specified along a pressure-volume reference curve can be used to calculate other thermodynamic states such as along isentropes, at the same pressure, for different volumes and temperatures from the thermodynamic parameters $(\partial H/\partial V)_P$ or $P(\partial V/\partial E)_P$. These parameters, which are assumed to be only a function of P, are fit to the singly or doubly shocked Hugoniot data for water, ice, and porous ice.

The satellites of Jupiter and Saturn are classified into three groups according to density. The inner, silicate (s), satellites of Jupiter (Amalthea, Io, and Europa) the partially devolatilized (d) satellites (Ganymede, Callisto, and Titan), and the smaller primordial (p) satellites (Mimas, Enceladus, Tethys, Dione, Rhea, and Iapetus). The latter two groups have mean densities of 1.8 to 1.9 Mg/m^3 and 1.0 to 1.4 Mg/m^3. We infer that the densities of the p group also correspond to the densities of the particles in the rings of Saturn and Jupiter. We suggest that the p satellites are composed largely of primordial solar ratio material controlled by the oxygen/silicon ratio (15.5) and the oxygen/magnesium ratio (17.4) of the primitive Jovian and Saturnian nebulae. The primordial solar ratios, if these correspond to objects having silicate and water ice compositions, would have densities of 1.1 to 1.2 Mg/m^3. We have examined the hypothesis that the large d satellites achieved their higher densities (1.8 to 1.9 Mg/m^3) upon partial impact devolatilization during accretion starting with planetesimals of material similar in density, 1.27 Mg/m^3, (60% ice, 40% silicate) to that of the smaller satellites. Accretion of planetesimals of such a composition, assuming impact volatilization at partial pressures of H_2O below 0.61 kPa in a porous regolith, yields Ganymedean-sized planets with mean planetary densities of only 1.5 Mg/m^3. However, starting with planetesimals having densities of 1.62 Mg/m^3 (43% H_2O and 57% silicate) give rise to mean satellite densities of 1.9 Mg/m^3 for Ganymedean-sized objects.

If impact volatilization of initially porous planetesimals is assumed, our calculations demonstrate that starting with planetesimals of composition 54% H_2O and 46% silicate (mean density 1.35 Mg/m^3) and partial devolatilization occurring during accretion leads to objects having a Ganymedean radius of 2600 km and a mean density of 1.85 Mg/m^3.

The present results suggest a mechanism for the formation processes of the higher density partially devolatilized, satellites of Saturn and Jupiter. The accretion-devolatilization process depends on the relative amount of material present in different zones orbiting in the nebula around each proto-planet and not on the assumed mean radial temperature at a given distance from the planet.

Acknowledgments. Research supported under NASA grants NSG 7129 and NAGW-205. We have benefited from the previous work of Mark Morley who first looked at the accretion problem from a different perspective, as well as the technical discussions with D. Stevenson and J. Lunine. We are grateful to E. Gaffney and J. Klinger for helpful comments on the manuscript, as well as, A. Dollfus and R. Smoluchowski for inviting us to present this paper to a critical audience in Nice, France. Contribution #4107, of the Division of Geological and Planetary Sciences, California Institute of Technology, Pasadena, California.

APPENDIX A. SHOCK TEMPERATURES AND THE WATER EQUATION OF STATE

A stringent test of the complete thermal equation of state for fluids such as water are to compare predicted sound speeds and Hugoniot temperatures to measurements. Previously Bakanova et al. (1976) very favorably compared their predicted sound velocity along the water Hugoniot with measurements to ∼60 GPa. Shock temperatures were also calculated and compared with previous relatively low pressure data of Kormer (1968). Other work by Lyzenga and Ahrens (1980), Lyzenga et al. (1983), Ahrens et al. (1982), Nellis et al. (1984), and Boslough et al. (1984) for a variety of solids and fluids demonstrate that upon comparing calculated and measured shock temperature T_H versus shock pressure, P, the absolute value of T_H at a given pressure was very sensitive to the total energy budget of the material (e.g. occurrence of phase changes) whereas a slope of the P versus T_H relation is sensitive to specific heat.

In Fig. 1A the available H_2O shock temperature data of Kormer (1968) and the more recent higher temperature and pressure data of Lyzenga et al. (1982) are shown relative to previous shock temperature calculations. As is evident from the figure, the calculation of Ree based on a repulsive parameter equation of state which does not take into account, the already mentioned, extensive pressure-induced ionization (Mitchell and Nellis, 1982) yields temperatures which are too high by ∼600K, especially at higher pressures (above 50 GPa). Rice and Walsh's calculations are too high largely because the data range they studied, up to ∼25 GPa, did not encompass as much of the higher pressure range where subsequent measurements demonstrate that the Hugoniot curve is more compressive (especially above 30 GPa) (Mitchell and Nellis, 1982) than would be anticipated from simple extrapolation of the lower pressure data. Both the curve labeled Cowperthwaite and Shaw (1970) and a present calculation agree well with Kormer's 1968 lower pressure results. Cowperthwaite and Shaw (1970) use a thermal equation of state which was derived from a careful fit to the lower pressure thermodynamic data, and took into account the strong temperature dependence of specific heat at constant volume. For the present calculation curve we employed the form of the

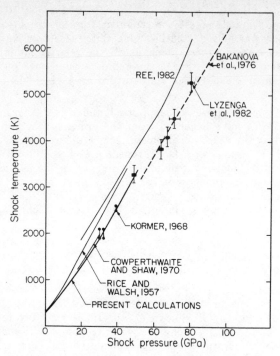

Figure 1A. Shock temperature versus shock pressure. Theoretical calculations and experimental measurements.

Table 1A. Shock Temperature, Liquid Water

Isentropic Centered State, 1 GPa			Hugoniot State			
E(kJ/kg)	V(m³/Mg)	T(K)	E(kJ/kg)	V(m³/Mg)	T(K)	P(GPa)
3166[a]	1.11[a]	1073[a]	8766	0.484	1964	31.5
3769[b]	1.25[b]	1273[b]	10,987	0.460	2447	37.9
(c)	0.868[d]	428[d]	(c)	0.595	584	10.8
(c)	1.014[d]	1286[d]	(c)	0.443	2555	44
(c)	1.19[d]	2143[d]	(c)	0.382	4703	75
(c)	1.381[d]	3000[d]	(c)	0.344	6859	107.4

(a) Bosnjakovic et al., 1970
(b) from Burnham et al., 1969
(c) not calculated
(d) Bakanova et al., 1976

Bakanova equation of state and used equation 21 for specifying $\eta = P/(\partial V/\partial E)_P$ as a function of P. Instead of using a theoretical value of E_{io}, T_{io}, and V_{io} to specify the foot of a series of isentropes at $P_{io} = 1$ GPa in the low temperature and high pressure regime as do Bakanova et al. (1976), we used the experimentally based thermodynamic and equation of state data specified in Table 1A. Hence it is not surprising that our results agree so closely with those of Cowperthwaite and Shaw (1970). Finally we note that over the range from 65 to 80 GPa the Bakanova et al. (1976) equation of state is in excellent agreement with the recently measured shock temperatures. We thus conclude that this form of this equation of state provides an accurate description of the properties of shocked water and presumably ice up to 50 GPa when constrained by the thermodynamic data extending to $\sim$1 GPa and $\sim$1273K. At higher pressures $>$ 50 GPa the theoretical description of the high temperature - low pressure properties of water proffered by Bakanova et al. (1976) fits both the shock temperature data, the sound speed, and the reflected shock, and porous ice data.

REFERENCES

1 Ahrens, T.J., Lyzenga, G.A., and Mitchell, A.C. 1982, in *High Pressure Research in Geophysics*, ed. by S. Akimoto and M.H. Manghnani, Center for Academic Publications, Japan, pp. 579-594.

2 Ahrens, T.J. and O'Keefe, J.D. 1972, The Moon 4, pp. 214-249.

3 Anderson, D.L. and Benson, C.S. 1963, in *Ice and Snow*, ed. by W.D. Kingery, MIT Press, pp. 391-411.

4 Anderson, G.D. 1968, U.S. Army CRREL Res. Rept. 257.

5 Bakanova, A.A., Zubarev, V.N., Sutulov, Yu.N., and Trunin, R.F. 1976, Sov. Phys. JETP. 41, pp. 544.

6 Boslough, M.B., Ahrens, T.J., and Mitchell, A.C. 1984, J. Geophys. Res., 89, B9, pp. 7845-7851.

7 Bosnjakovic, F., Renz, U., Burow, P. 1970, *Mollier Enthalpy, Entropy Diagram of Water*, GZH, Zagreb, Hemisphere Publishing Corp., Washington, D.C.

8 Burnham, C.W., Holloway, J.R., and Davis, N.F. 1969, The Geological Society of America special paper #132, 99 pp.

9 Cole, G.H.A. 1984, Quart. J. Roy. astr. Soc. 25, pp. 19-27.

10 Cowperthwaite, M. and Shaw, R. 1970, J. Chem. Phys. 53, pp. 555-560.

11 Dorsey, N.E. 1940, *Properties of Ordinary Water-Substance*, Reinhold Publ. Corp., N.Y., 673 pp.

12 Fletcher, N.H. 1970, *The Chemical Physics of Ice*, Cambridge Univ. Press, 271 pp.

13 Gaffney, E.S. 1985, Proc. NATO Workshop on Ices in the Solar System, Nice, France, Jan 16-19, 1984.

14 Gaffney, E.S. and Ahrens, T.J. 1980, Geophys. Res. Lett. 7, pp. 407-409.

15 Kormer, S.B. 1968, Sov. Phys. Usp. 11, pp. 229-254.

16 Larson, D.B., Bearson, G.D., and Taylor, J.R. 1973, *No. Amer. Contrib. 2nd Int. Conf. Permafrost, Yakutsk*, pp. 318-325.

17 Linde, R.K. and Schmidt, D.N., 1966, J. Appl. Phys., 37, pp. 3259-3271.

18 Lunine, J.A. and Stevenson, D.J. 1982, Icarus 52, pp. 14-39.

19 Lyzenga, G.A, Ahrens, T.J., and Mitchell, A.C., 1983, J. Geophys. Res., 88, pp. 2431-2444.

20 Lyzenga, G.A., Ahrens, T.J., Nellis, W.J., and Mitchell, A.C. 1982, J. Chem. Phys. 76, pp. 6282-6286.

21 Lyzenga, G.A. and Ahrens, T.J. 1980, Geophys. Res. Lett. 7, pp. 141-144.

22 McQueen, R.G., Marsh, S.P., Taylor, J.W., Fritz, J.N., and Carter, W.J. 1970, in *High Velocity Impact Phenomena*, ed. by Kinslow, R., Academic Press, New York, pp. 294-419.

23 Marsh, S.P. 1980, LASL Shock Hugoniot Data, University of California Press, Berkeley, pp. 327.

24 Mitchell, A.C. and Nellis, W.J. 1982, J. Chem. Phys. 76, pp. 6273-6281.

25 Morrison, D. 1982, in *Satellites of Jupiter*, ed. by D. Morrison, U. Ariz. Press, Tucson, pp. 3-43.

26 Nellis, W.J., Ree, F.H., Trainor, R.J., Mitchell, A.C., and Boslough, M.B. 1984, J. Chem. Phys. 80, pp. 2789-2799.

27 O'Keefe, J.D. and Ahrens, T.J. 1977, Proc. Lunar Sci. Conf. 8th, pp. 3357-3374.

28 Pollack, J.B. and Fanale, F. 1982, in *Satellites of Jupiter*, ed. by D. Morrison, Univ. Ariz., Tucson, pp. 872-910.

29 Pollack, J.B. and Reynolds, R.T. 1974, Icarus 21, pp. 248-253. 30 Ree, F.H. 1982, J. Chem. Phys. 76, pp. 6287-6302.

31 Rice, M.H., McQueen, R.G., and Walsh, J.M. 1958, Solid State Phys. 6, pp. 1-63.

32 Rice, M.H. and Walsh, J.M. 1957, J. Chem. Phys. 26, pp. 824-830.

33 Ross, J.E. and Aller, L.H. 1976, Science 191, pp. 1223-1229.

34 Shoemaker, E.M. and Wolf, R.F., 1982, *Satellites of Jupiter*, ed. by D. Morrison, pp. 277-339, U. Ariz. Press.

35 Smith, B.A., Soderblom, L.A., Johnson, T.V., Ingersoll, A.P., Collins, S.A., Shoemaker, E.M., Hunt, G.E., Masursky, H., Carr, M.H., Davies, M.E., Cook, A.F. II, Boyce, J., Danielson, G.E., Owen, T., Sagan, C., Beebe, R.F., Veverka, J., Strom, R.G., McCauley, J.F., Morrison, D., Briggs, G.A., and Suomi, V.E. 1979a, Science 204, pp. 945-972.

36 Smith, B.F., Soderblom, L.A., Beebe, R., Boyce, J., Briggs, G., Carr, M., Collins, S.A., Cook, A.F., Danielson, G.E., Davies, M.E., Hunt, G.E., Ingersoll, A., Johnson, T.V., McCauly, J., Masursky, H., Owen, T., Sagan, C., Shoemaker, E.M., Strom, S., Suomi, V.E., and Veverka, J. 1979b, Science 206, pp. 927-950.

37 Smith, B.F., Soderblom, L.A, Beebe, R., Boyce, J., Briggs, G., Bunker, A., Collins, S.A., Hansen, C.F., Johnson, T.V., Mitchell, J.L., Terrille, R.J., Carr, M., Cook, A.F., Cuzzi, J., Pollack, J.B., Danielson, G.E., Ingersoll, A., Davies, M.E., Hunt, G., Masursky, H., Shoemaker, E.M., Morrison, D., Owen, T., Sagan, C., Veverka, J., Strom, J., Suomi, V.E. 1981, Science 212, pp. 163-191.

38 Smoluchowski, R. 1983, Science 222, pp. 161-163.

39 Stone, E.C. and Miner, E.D. 1982, Science 215, pp. 499-504.

40 Walsh, J.M. and Rice, M.H. 1957, J. Chem. Phys. 26, pp. 815-823.

41 Weast, R.C. 1982, *CRC Handbook of Chemistry and Physics*, 63rd Ed., Chemical Publ. Co., Cleveland.

THE COMPOSITION AND STRUCTURE OF PLANETARY RINGS

Joseph A. Burns

Cornell University

The properties of planetary ring systems are summarized herein ; emphasis is given to the available evidence on their compositions and to their dynamical attributes. Somewhat contaminated water ice makes up the vast expanse of Saturn's rings. Modified methane ice may comprise Uranus'rings while silicates are the likely material of the Jovian ring. Saturn's rings form an elaborate system whose characteristics are still being documented and whose nature is being unravelled following the Voyager flybys. Uranus'nine narrow bands display an intriguing dynamical structure thought to be caused by unseen shepherd satellites. Jupiter's ring system is a mere wisp, probably derived as ejecta off hidden parent bodies.

INTRODUCTION

Enormous advances have been made during the past decade in our information about, and understanding of, planetary ring systems. An excellent reference which summarizes the latest knowledge and background physics on the subject is the volume Planetary Rings (1). The growth of this field has been accomplished through observations from the ground as well as from spacecraft, and by impressive theoretical developments in our understanding of the operation of ring systems.

In regard to the specific topic of this volume, somewhat contaminated water ice is omnipresent throughout Saturn's rings (and in fact this ring system provides the largest exposure of ice in the entire solar

J. Klinger et al. (eds.), Ices in the Solar System, 655–679.

system), methane ice may comprise the Uranian rings,
whereas the Jovian ring is believed (from relatively
crude measurements) to not contain any ice. The opportunity to look "inside" solar system matter and the relative ease of observing Saturn's rings makes them ideal
for studying solar system ice. In addition ring particles have a distribution of sizes but most are small
and therefore relatively unmodified by internal processes.

The three ring systems will be described separately in the remaining sections of this paper ; for each
case the ring's individual morphology will be first presented, followed by a description of the particle compositions and sizes, and finally by a brief consideration
of the possible origin and/or operation of the systems.
Before doing this we first make some general remarks
about planetary rings.

Planetary ring systems provide a contemporary
dynamical laboratory in which to witness processes that
are thought to have operated when the solar system originated and similar mechanisms that may be at work in
spiral galaxies today. Gravity governs many of these
processes and therefore only mass - whether due to ice
or rock - matters in their operation. However, collisions are common in most rings and the precise nature
of these collisions may determine the specific structural endpoint ; clearly the particular composition of the
colliding ring members will be an important determinant
of the outcome of the event. In addition, interactions
of the rings with other components of the circumplanetary space (e. g., magnetospheric particles (2, 3) and
interplanetary micrometeoroids) can produce atmospheres about rings and can influence the lifetime of some
ring particles (3, 4) ; here again the composition of
the ring may alter the end result. Conversely, properties of the ring material can depend upon the specific
ring structure ; for example, the nature of the ice sur-
faces-whether the molecular arrangement is amorphous or
crystalline (5, 6) as well as whether very volatile components survive - will be influenced by the relative importance and ferocity of internal ring collisions versus
external projectiles, and this relies on the ring's optical depth and the ring particles random velocities.

A planetary ring is composed of innumerable objects orbiting individually about their planet in a thin,
flattened disk. These objects occasionally jostle one

another due to small random motions about their nearly
circular orbits ; typical collision times are of the order of half the orbital period divided by the normal optical depth τ (in which $e^{-τ}$ is the fractional diminish-
ment of light passing perpendicular to the ring plane).
Planetary rings are near their primary planet because
tidal forces (effectively, the difference in gravity
across an object) which compete against the agglomeration of material are largest there. In the most detailed treatment of this physics, Weidenschilling et al.
(7) have suggested that Saturn ring particles are ephe-
meral over a period of a few weeks ; they grow rapidly
through collisions but, once large enough, they fragment owing either to tides or unusually energetic impacts.

        Small satellites - which some regard as merely
the largest of the ring particles - seem intimately associated with each of the three ring systems (8). These
objects presumably share the composition and physical
properties of ring particles (9, 10, and 11). For the
Jovian case, two satellites about 30-40 km in diameter
orbit near the outer extremus of the main ring and determine most of the little observable structure ; the
larger Amalthea and Themis are embedded in a newly-found
tenuous ring. Many distinctive features of the Saturn
system are engendered at the locations of resonances
with satellites that orbit just beyond the ring edge.
Unseen objects the size of small stellites have been
implicated by the waviness of the edges of Encke's division (12), which is reminiscent of the braiding and kin-
king seen in Saturn's F ring (13). Almost all the unique
structure of the Uranian rings is thought to result from
the action of small, unseen "ring-moons" (14, 15, 16).

THE RINGS OF SATURN

        The most elaborate and most studied ring system
is that of Saturn. In connection with the topic of this
conference, it is also the one ring system having unequivocal evidence for ice, specifically water ice. Although investigated from the ground for more than three
and a half centuries after its first sighting by Galileo
in 1610, the information base about this ring system ex-
ploded with the Voyager flybys (17, 18). Given the ban-
quet of information now available, we will only be able
to provide a taste of the Saturn system ; for a fuller
meal, the reader is referred to several recent review
articles (8, 19, and 20).

Structure

The various rings of Saturn (see Figs. 1 and 2) are designated alphabetically in the order of their discovery : these "rings" are in fact circular bands of circumplanetary space with distinctive characteristics, each containing innumerable ringlets. Most information on Saturn's rings concerns the very substantial A and B rings which are separated by the Cassini division, a region relatively (but not completely) devoid of material. The A ring (2.02 - 2.27 R_s) has modest optical depth (0.5 $\lesssim \tau \lesssim$ 0.7) and, particularly near its outer perimter, a series of regularly-spaced brightness oscillations. The B ring (1.52 - 1.95 R_s) is more opaque (1.2 $\lesssim \tau \lesssim$ 1.8) and, with a few exceptions, has an irregular and partially time-variable structure that occurs over all length scales down to the resolution limit of any particular image (19). Both the C ring (1.23 - 1.52 R_s) and the Cassini division contain broad plateaus of optically thin regions punctuated by well-ordered, optically thick ringlets, several of which are elliptical and/or isolated by clear gaps. The D ring (1.11 - 1.23 R_s) is much fainter than the C region and consists of many distinct narrow ringlets.

Fig. 1   A voyager image of Saturn's rings. At the right is the narrow F ring and a shepherd satellite. Interior to this lies the A ring ; the faint curves are mainly resonances but the darkest gap is the Encke division. The Cassini division is the band of four bright ringlets (each 500 km across) bounded by empty gaps, the innermost of which is the Huygens gap which sits at the Mimas 2 : 1 resonance. The smooth and fuzzy B ring breaks down into innumerable ringlets at higher resolution ; spokes are seen in its midsection. The C ring, bounded by two banded regions, forms the inner quarter of this ring image.

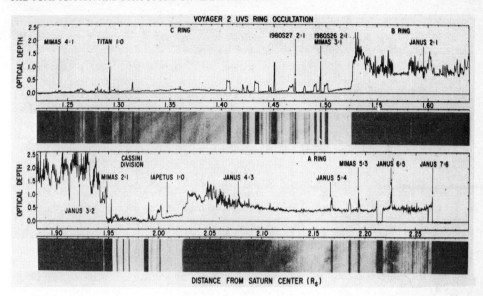

Figure 2   The Voyager 1 ultraviolet spectrometer optical depths are compared to Voyager 1 images of the unil-
luminated side of the rings, where the rings are viewed
in diffusely transmitted light (19). The brightest portions of the images correspond to intermediate optical
depths while the dark portions represent either very large or low optical depths. Various important satellite
resonances are indicated by arrows. R_s = 60,330 km in
the bottom distance scale.

        Several airy rings lie outside the main ring sys-
tem. The E ring ($\lesssim$ 3 - $\lesssim$ 8 R_s) (20, 21) is a diffuse
sheet of low optical depth with some vertical extent ;
it may be associated with the enigmatically-young satellite Enceladus. The F ring (17, 18, 16), which was found
by Pioneer 11, is a narrow ribbon (2.324 R_s) whose properties are both spatially - and temporally - variable
(8, 20). Depending on the observation, its width apparently varies from about 100 (17) to about 500 km (18)
while the number of its constituent strands changes from
three to at least five. Various parts of the ring are
clumped, braided, and kinked over periodic length scales
$\sim 10^4$ km, probably because of the action of two shepherding satellites on eccentric orbits (13, 16). Very faint
material has been located spanning the gap between the A
ring and the tiny moon Atlas'orbit (22), and then even

more tenuously extending out to the F ring (23). The G
ring (18, 24, 20) is a gossamer ring of bounded width
($\sim 10^3$ km) and appreciable thickness ($\sim 10^2 - 10^3$ km)
centered at 2.82 R_s ; little is known of its properties.

Several other features of the Saturn ring system
are notable. Spokes are elongated, roughly radial albedo ($\pm$ 10 %) patches that develop in the middle of the
dense B ring. From their photometric behavior, these
structures contain unusually large numbers of micron
particles that are suddenly liberated. The boundaries of
these patches generally move at Keplerian speeds but,
when some are forming, move at the synchronous rate.
This property and the fact that the spokes are narro-
west (or even nonexistent) at synchronous orbit, and the
small particle sizes implicate electromagnetic forces.
At one time, Saturn's electrostatic discharges (bursts
of broadband microwave radiation with a periodicity of
$\sim$ 10h 10m) were thought to come from this same region
but now are believed to be generated in atmospheric
thunderstorms.

Saturn's rings also exhibit an east/west brightness asymetry, particularly in the inner A ring, in
which brightness enhancements ($\lesssim$ 15 %) are present in
those quadrants immediately preceding orbital conjunction with the observer (Earth or Voyager). Usually this
effect is ascribed to the gravitational wakes of large
ring particles that organize streamlines ; another idea
is that some ring particles are aligned.

A total of about fifty features of several types are located at resonances, those portions of the
ring where orbital periods are nearly commensurate with
the orbital periods of major satellite perturbers (either those nearby or massive (15, 12, 19). The outer edge of the A ring is located at Janus'7 : 6 resonance and
the outer edge of the B ring at Mimas'2 : 1 resonance ;
the cause of the inner edges is not known. Elsewhere,
mainly in regions of modest optical depth located in
the outer A ring, spiral density waves are engendered
at some forty or so resonances. At a few, vertical bending waves are also noted in the ring and may help account for the ring's thickness. The ring's surface density determines the wavelengths of these features ; typical values are $10 - 10^2$ g cm^2, implying a total ring
mass of $\sim 5 \times 10^{-8}$ M_s. The damping rates of these waves
allow estimates of ring kinematic viscosities which can
be converted to random particle velocities of tenths of
cm sec^{-1}, corresponding to ring thickness of $\sim 10 - 50$m.

Other ways of estimating the ring thickness give comparable values, but considerable interest remains in this
quantity in order to distinguish between two classical
models for interpreting data, a monolayer and a many-
particle-thick ring (8, 15). A final important issue regarding density waves concerns the timescales over which
angular momentum should be exchanged between the rings
and the causative satellites ; this process pushes satellites away from the rings while simultaneously causing the ring to collapse inward toward Saturn. These
lifetimes are embarrassingly short ; for example, Atlas
should have moved out from the A ring's edge to its current position in only $\sim 3 \times 10^6$ yr. Either the rings
are recent, the mechanism is incorrectly described, or
something (such as a resonance tie to another satellite) restrains the satellites.

The B ring displays structure on all spatial
scales down to the resolution limit. No clear gaps are
located in this region, and therefore hidden embedded
moonlets are not suspected. Instead a diffusional instability has been suggested ; this relies on a particular form for the coefficient of restitution as a function of impact velocity, for which preliminary experiments on water ice with velocities relevant to Saturn's
rings have been carried out (25). Some of the brightness variations in this region seem to be due to changes in albedo, rather than τ, and may result from an
extraordinarily large amount of dust (8).

Sizes and Compositions of the Particles

The approximate particle sizes known from ground-
based measurements (mainly radio and radar) were confirmed and refined by Voyager. This mission allowed photometric observations from many phase geometries, high-
resolution stellar occultation studies in the visible
and ultraviolet, and, most importantly, coherent radio
occultations at 3.6 and 13 cm (19, 26). The emerging
picture is that the cumulative distribution of ring
particles between 1 cm and 10 m satisfies a power law
with an index roughly -3 and that an upper size cutoff
of about 5 m radius exists. In such a distribution most
of the mass resides in the largest particles while the
smallest provide substantial area. The cutoffs at large and small sizes, as well as the precise exponent in
the power law may vary from place to place in the ring.
The upper cutoff in particle size is probably just a
steepening of the index in which case hundreds of km-
size objects could be present in the ring. Particle si-

ze estimates of 10 - 100 μm based upon near-infrared
reflectance bands and far infrared spectral variations
are now believed to characterize the granulation on the
surfaces of individual ring particles.

Micron grains are abundant in a few restricted
areas of the ring, and may serve as a tracer of an unusual electromagnetic environment or of energetic collisions. The spokes in the middle of the B ring, as well
as several zones in the outer A ring are the primary sites for small particles in the main rings. On the other
hand, the ethereal E and G rings are largely micron-
sized-dust, and the F ring contains many micron and submicron particles in addition to a complement of larger
objects (20).

The rings exhibit a strong opposition effect
(i. e., they brighten appreciably at small phase angles)
that is generally interpreted as due to a many-particle-
thick ring. The individual particles in the main rings
are primarily back-scattering at visual wavelengths,
which implies rough surfaces much larger than a visual
wavelength. Pang et al. (21) have suggested that the E
ring particles must be spherical because of that ring's
strong backscatter peak, but such is not necessarily
the case (8). The polarization properties of the main
rings (a weak negative branch and limited polarization
at large phase angles) are consistent with those of a
bright object with a grainy surface like a snowbank (19).

In broad outline, the A and B rings are bright
and red (see Fig. 3). At least some of the reddening is
caused by multiple scattering, so that the particles
themselves could possibly have colors like that of Europa, except in the UV where the ring spectrum seems flat-
ter. The C and Cassini region particles are distinctly
darker (albedo $\sim$ 0.2) and apparently less red, although
some of this is a multiple scattering effect. Surprisingly the E ring particles are blue, and in fact form the
only blue, icy outer solar system target ; this is, however, likely to be a size effect rather than indicative of composition. Radar reflectivities are also available but provide few compositional constraints.

The infrared reflectivity is most diagnostic of
composition. Water ice absorption features are seen at
1.6, 2.0, and 3.0 μm (see Fig. 4) ; the detailed shape
of the first two of these are well-matched by laboratory spectra of medium-grained water frost at 100° K (53,

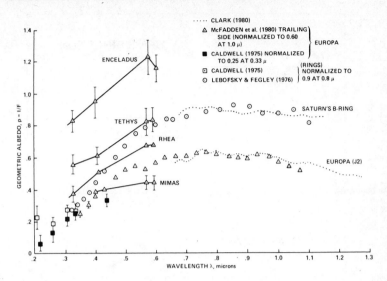

Fig. 3   Spectral reflectivity of Saturn's A and B rings from 0.2 μm compared with that of several icy satellites. The rings are seen to be even redder than most other nearby objects. Some contamination is implied. From Esposito et al. (19). Some of the comparison data come from Caldwell (57), Clark (58), and McFadden et al. (59).

54). It may be critical, however, to recall that this identification is non-unique since near-infrared spectra cannot distinguish pure water ice from various clathrate-hydrates of ammonia or methane. Further evidence for the presence of water ice is the predominance of hydrogen and oxygen throughout the Saturnian magnetosphere ; presumably these atoms are knocked off the icy rings and moons embedded in the magnetosphere. The reddening of the rings implies an impurity whose spectral signature may include the slight absorptions seen at 0.6 and 0.85 μm (55). These, as well as the overall blue absorption and the uniformly high reflectivity between 0.9 and 1.5 μm are consistent with various silicates ; sulfur impurities have also been suggested but they do not satisfy the high albedo found near a micron. The low ring reflectivity at 3 μm (56) implies that the impurity is wellmixed.

Far-infrared and millimeter observations provi-

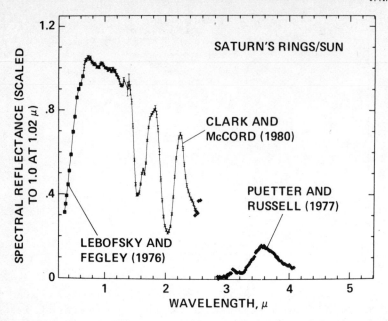

Fig. 4   A composite reflectance spectrum of the rings
of Saturn scaled to unity at 1.02 μm. From Clark and
McCord (54).

de some compositional information. Eclipse measurements
show that particles have very low thermal inertias, and
suggest porous surfaces or perhaps an amorphous phase
of water ice. Low microwave temperatures are not consistent with the classically-used absorption coefficient
for hexagonal-phase (Th) ice at 85° K. Groundbased radar and radio observations restrict the mass fraction
of silicates to $\lesssim$ 10 %, or perhaps twice as much, depending upon one's choice for the ice absorption coefficient.

        Apparent spatial variations in the composition
of ring material are of considerable interest in connection with possible ring origins and the subsequent transport of material through the system. The material in
the A and B rings seems to differ from that in the C
ring and Cassini division by being redder and much brighter, and containing many more small particles ; upper
size cutoffs and power-law indices may also vary. At
least to some extent, the different properties of these

regions reflect the environments : residents of the Cassini division and C ring are more exposed to external
influences than are denizens of the more crowded main
ring. However, variations in particle properties (e. g.
albedo) appear to be present over small scales in the B
ring (8) ; their cause is uncertain.

THE URANIAN RINGS

Structure

        The nine narrow rings encircling Uranus were
discovered in 1977 during a stellar occultation by the
planet (see Fig. 5 ; 27). About fifteen subsequent stel-

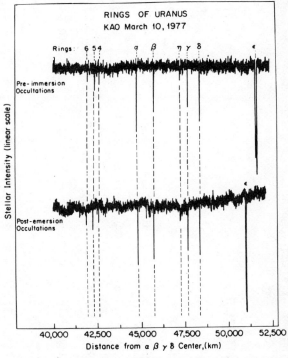

Fig. 5   The star's brightness as traced during the
stellar occultation in which the Uranian rings were discovered (27). The pre-emersion profiles are plotted
using a common distance scale from the center of Uranus
in the ring plane. The nine confirmed rings are labeled.
The abruptness and the opacity of the rings are apparent. The uniqueness of the ε ring is obvious : it is
most opaque, broadest, and most eccentric.

TABLE 1.   PROPERTIES OF THE URANIAN RINGS

Ring	Semimajor Axis(km)	Optical Depth τ	Radial Width(km)	Eccentricity $e \times 10^3$	Inclination $i \times 10^3$(rad)
6	41,877.3±16.6	≥0.5	0.4-2	1.01±0.10	1.15±0.21
5	42,275.2±16.6	≥1.0	0.8→2	1.85±0.08	0.87±0.17
4	42,609.6±16.8	≥0.8	0.7→2	1.15±0.04	0.38±0.09
α	44,758.3±16.4	0.7→∿1.4	5→10	0.78±0.02	0.30±0.05
β	45,701.0±16.5	0.35→∿1.5	5→11	0.43±0.02	0.10±0.03
η	47,214.9±16.5	≥0.6	0.5-2 broad(55)	(0.03±0.04)	(0.05±0.06)
γ	47,666.3±16.4	≥1.5	∿3	(0.04±0.02)	0.10±0.03
δ	48,338.7±16.5	≥1.5	2-3	0.06±0.02	0.21±0.05
ε	51,188.1±17.0	?→>1.2	20→96	7.94±0.02	(0.05±0.05)

Based on Elliot and Nicholson (28)

lar occultations have allowed a precise description of
the system's structure and kinematics to the point that
rms errors of occultations are now about 0.1 sec. Kine-
matical models (28) currently solve for five orbital
elements (a, e, i, ω, Ω ; see Table 1) for each of the
rings, as well as for Uranus'rotation pole and the planet's gravitational moments J_2 and J_4.

   The Uranian rings lie from 1.6 R_U to 1.9 R_U and
generally have widths between 2 and 12 km ; the dense
outermost ε ring is unique in that its width attains
nearly 100 km ; the only other broad ring component
($\sim$ 55 km) is of low τ ($\sim$ 0.1) and surrounds a narrow
core to the η ring. Seven of the nine rings are elliptical with typical eccentricities $\sim 10^{-3}$; again the
ring stands out since its eccentricity is about an order of magnitude larger than the others. The widths of
the α, β, and ε rings vary linearly with orbital radius,
implying that both the inner and outer ring edges are
elliptical ; in all these cases, the outermost edge is
more elliptical than the inner with $\delta e/\delta a/a \sim 0.5$. These elliptical shapes appear to rotate rigidly in the
planet's oblate gravity field, and this can be explained if ring self-gravity is important (29). Most rings
are also slightly inclined (typically $\sim 10^{-3} - 10^{-4}$ ra-
dians). Ring inclinations seem crudely correlated with
eccentricities : both decrease roughly linearly with
distance from the planet. Once more the ε ring is unusual : it has no measurable inclination.

Composition and Particle Size

   The Uranian rings with a geometric albedo of only 2-3 % between 0.89 and 3.9 μm are the darkest material known in the solar system. The spectrum, shown in
Fig. 6, is essentially flat between 2.0 and 4.0 μm (30).
This, coupled with the overall low albedo, rules out water or ammonia frost. The lack of an absorption feature
centered at 2.9 or 3.0 μm also places strong limits on
the amount of bound or absorbed water. The apparent absence of water ice in the Uranian ring is somewhat surprising in light of its known predominance on the surfaces of the Uranian satellites (57). Most would argue
that carbonaceous material form the ring surfaces. Cheng
and Lanzerotti (31, 32, 33) suggest that the Uranian
ring particles may originally have been methane ice
which has been polymerized and darkened through its bombardment by ultraviolet radiation and magnetospheric
particles. The Uranian case would be different from Sa-

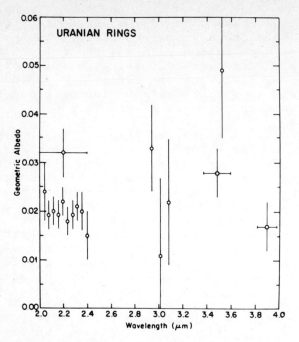

Fig. 6   The observed geometric albedo for the Uranian rings between 2.0 and 3.9 µm assuming an average
ring width of 85 km. Circles correspond to circular variable filter data (resolution 0.03 to 0.05 µm), squares represent narrowband filter measurements at 3.48
and 3.89 µm ; the diamond is a broadband 2.2 µm measurement (28).

turn's situation for two reasons : colder solar nebula
temperature at Uranus would allow methane to be incorporated primordially into water ice but then magnetospheric particles must be more able to penetrate and act
throughout the less substantial Uranian ring system.
This radiation would preferentially eliminate the water
ice and darken the leftover methane. Others (e. g. M.
A'Hearn and P. Feldman, private communications, 1984)
see similarities with cometary grains which also are
dark and have presumably formed in comparably cold environments.

       Several limits can be placed on typical sizes
of Uranian ring particles. First, the rings cannot have a significant complement of micron-sized grains for,
if they did, optical depth profiles in the near-infrared
would vary with wavelength, contrary to observations.

Accordingly, if all particles were a single size, that radius would be larger than 4 µm. In addition, their uniformly low albedo limits the percentage of Rayleigh scatterers : so any reasonable power law can contain very few particles $\lesssim$ 0.1 µm (34). On the other hand, large grains (r $\gtrsim$ 100 µm) would normally produce a noticeable opposition effect ; since the ring apparently does not, most light may come from a relatively limited size range although other processes might account for the absence of an opposition surge. Dynamical models (14, 39) imply that the average ring surface density is $\sim$ 25 g cm^{-2} in the ε ring but only $\sim$ 1 g cm^{-2} elsewhere. These models then suggest typical particle sizes of 1-10 cm. If both parts of the above argumentation are correct, a bimodal size distribution seems required.

Discussion

        The confinement of narrow rings with abrupt edges (16) can be explained by the presence of as-yet unseen moonlets (radii of a few km) between the rings. According to the shepherding mechanism that Goldreich and Tremaine (14) introduced to explain this feature, satellites repel nearby disks of matter ; in competition with collisions, they also generate ring inclinations and eccentricities (15, 35). The torque that they apply $\sim$ m^2 Δa^{-4}, where m is the satellite mass and Δa is the distance from the ring. Fits to the sets of Uranian ring occultation data are now accurate to $\sim$ 1 km, and the effects of such moonlets have been sought, so far unsuccessfully. The widths of the narrow, nearly circular, Uranian rings do however vary significantly and are thought to result from Lindblad resonances (36). Clearly a major objective of the Voyager 2 flyby of Uranus expected in January 1986 will be to search for these shepherd satellites. Predictions as to other discoveries only depend on the limits of one's imagination.

THE JOVIAN RING SYSTEM

        Jupiter's ring, discovered by Voyager 1 (37) and explored by Voyager 2 (38) in 1979, is the least substantial and the least understood of the three planetary ring systems. Preliminary attempts to comprehend this system have been made (39, 40, 41). A recent reanalysis of the Voyager data that has helped to elucidate more clearly the nature of this ring system (20, 42, 43) will be summarized here.

Structure

        The three components of the Jovian ring - the
"bright" main ring, the vertically extended halo, and
the exterior "gossamer" ring - are sketched in Fig. 7.
The main ring starts at an illdefined inner edge ($\sim$
122,000 km or 1.71 R_J) and extends to a fairly abrupt
outer boundary at 129,130 $\pm$ 100 km, about 150 km beyond
the small satellite Adrastea's orbit. It is thin (t $\lesssim$
30 km) and its outer edge is circular (to better than 1
part in 10^3). Beyond the breakdown into three components,
little structure can be discerned in the ring, owing
partly to image smear. Showalter et al. (43) identify
three brightness enhancements centered at 127,600 km,
128,250 km, and 128,850 km ; the outermost and innermost appear associated with the moons Adrastea and Metis, respectively, each of which resides roughly midway

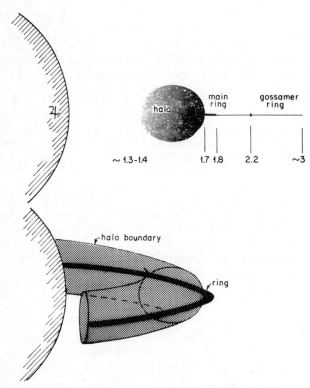

Fig. 7   A sketch of that illustrates the three components of the Jovian ring system (20, 42, 43). As seen in
the bottom panel, the halo arms overlap one another as
well as the main ring.

in a region where the brightness increases with decreasing radius.

Inward of the main ring is the toroidal halo, which stretches from the main ring's inner edge to about halfway to Jupiter's cloudtops ($\sim$ 1.3 - 1.4 R_J), where it is no longer identifiable. The halo is symmetric about the ring plane, and extends about 10^4 km above and below it. This diaphanous structure becomes fainter with distance off the equatorial plane and as the planet is approached.

Showalter et al. (44, 23) have discovered a very faint component, which reaches outward from the main ring to $\sim$ 220,000 km in the vicinity of Thebe's orbit. Its intensity decreases with distance from the main ring with an average value $\sim$ 5 % of the main ring. The only feature noted in this gossamer ring is a $\sim$ 20 % enhancement at synchronous orbit, thought to be caused by reduced plasma drag and no vertical forces there (44, 45).

Sizes and Composition of the Particles

The light forward-scattered by the main ring mostly comes from particles that range in size from several tenths to several tens of microns, and that satisfy a power-law size distribution [$n(r)dr = C(r/\mu m)^{-p}$ $(dr/\mu m)$ with p = 2.5 $\pm$ 0.5] ; the distribution is expected for collisional ejecta which then evolve inward at a rate $\sim r^{-1}$ such as by plasma drag. In this distribution, particles less than r = 100 μm provide an optical depth τ of 1-5 x 10^{-6}, depending on p. In addition to these forward-scatterers, larger particles are also required in order to absorb charged species, as noted by Pioneer experimenters (46), and to provide the measured back-scattered signal. If these macroscopic particles have a 5 % geometrical albedo, τ = 3 x 10^{-6}.

According to Voyager measurements in four wavelength bands (0.4 - 0.6 μm), the larger Jovian ring grains are red and have a photometric function between phase angles α = 2.5° and 18° that is reminiscent of Amalthea (43). Likely all surfaces in the inner Jovian magnetosphere are contaminated with the sulfur Io spews into the magnetosphere and also are altered through impacts with micrometeoroids and especially magnetospheric material. Groundbased measurements performed in the optical (0.55 to 1 μm) (17)   and the infrared (1.7 to 2.4 μm) (47) show an absence of absorption features.

Ices of water, methane, and ammonia may thus be eliminated as possible materials, and accordingly a silicate
or carbonaceous composition has been inferred. This is
not unreasonable in light of the composition of nearby
Amalthea and Io (48).

The halo particle size distribution and its integrated τ are not markedly different from those of the
micron-sized component of the main ring. The ring and
the halo each show an unexplained $\sim$ 10 % brightness excess of the far arm over the near arm. The ring particles making up the exterior gossamer ring are probably
somewhat smaller ($\sim$ 1.5 μm in radius) than particles in
the main ring since this ring brightness by a factor of
$\sim$ 2 between scattering angles of 7° and 6°.

Nature of the Jovian Ring

If the ring is a permanent feature of the solar
system, a continual resupply of the ring is needed because the observed ring particles are so small that
they have limited lifetimes. Burns et al. (20) estimate
inward orbital evolution times of 2 x $10^{2\pm1}$ yr due to
plasma drag ; if not lost in this manner, particles are
sputtered away by magnetospheric particles in $10^{3\pm1}$ yr.
The Jovian ring is believed to be debris lost through
collisions of interplanetary micrometeoroids (49) or
perhaps Io dust (50) into unseen parent bodies. The location of the main ring must in part be determined by
the presence of the satellites Adrastea and Metis which
are likely ring sources given the brightness enhancements they apparently generate. Other parent bodies are
required in addition and their overall distribution may
determine the placement of the main ring.

Resonances may account in some yet unspecified
way for much of the Jovian ring structure. Burns et al.
(45, 56) point out that the 5 : 3 eccentricity and inclination resonances with Amalthea both overlap the
ring's outer edge and Adrastea's orbit to within the
available resolution. They also call attention to Lorentz resonances, in which particles experience oscillatory radial and vertical electromagnetic perturbations
with periods that nearly match the particle's orbital
period. These resonances, which would cause large radial and vertical excursions, are located near the boundaries of the halo and less clearly at the outer part
of the gossamer ring. The inner resonance with quadru-
polelike terms is at 1.41 R_J, the inner "octupole" re-

sonance is at 1.71 R_J and the outer dipole resonance at 3.56 R_J, while higher order components accumulate at synchronous orbit. The thickened nature of the halo probably results from electromagnetic forces, which have an out-of-plane component due to Jupiter's tilted magnetic field, and from the absence of a vertical damping mechanism (40, 51, 20, 52).

CONCLUDING REMARKS

Water ice is the principal compound in Saturn's rings while a modified methane ice composition could account for the unique blackness of Uranus'rings. However, the makeup of these rings is not known in the detail necessary to fully understand the operation of the rings. Additional observations and modelling will be necessary to constrain the physical properties of the rings. Such work should allow us to better appreciate the nature of ice in the solar system.

ACKNOWLEDGMENTS

Preparation of this manuscript was supported by NASA Grant NAGW-310, while some travel expenses were provided by NATO and Cornell University. I thank M. Talman for manuscript preparation.

REFERENCES

1. Greenberg, R. and Brahic, A., Eds., 1984, "Planetary Rings", U. of Arizona Press, Tucson.

2. Burns, J. A., 1985, Planetary rings, in "Dust-Magnetosphere Interactions", Ed. G. Morfill, COSPAR, Pergamon Press.

3. Morfill, G., 1985, "Dust-Magnetosphere Interactions", COSPAR Proceedings, Pergamon Press.

4. Durisen, R. H., 1984, Transport effects due to particle erosion mechanisms, in "Planetary Rings", Eds. R. Greenberg and A. Brahic, U. Arizona Press, pp. 416-446.

5. Smoluchowski, R., 1983, Solar system ices : Amorphous or crystalline, Science 222, pp. 161-163.

6. Smoluchowski, R., 1985, Ices in planetary rings. This volume.

7. Weidenschilling, S. J., Chapman, C. R., Davis, D.
    R. and Greenberg, R., 1984, Ring particles : Col-
    lisional interactions and physical nature, in
    "Planetary Rings", Eds. R. Greenberg and A. Bra-
    hic, U. Arizona Press, pp. 367-415.

8. Cuzzi, J. N., Lissauer, J. J., Esposito, L. W.,
    Holberg, J. B., Marouf, E. A., Tyler, G. L. and
    Boischot, A., 1984, Saturn's rings : Properties
    and processes, in "Planetary Rings, Eds. R. Green-
    berg and A. Brahic, U. Arizona Press, pp. 73-199.

9. Cruikshank, D. P., 1985, The small, icy satellites
    of Saturn. This volume.

10. Thomas, P., Veverka, J. and Dermott, S., 1985,
    Small satellites, in "Satellites", Eds. J. A.
    Burns and D. Morrison, U. Arizona Press, in press.

11. Farinella, P., Milani, A., Nobili, A. M., Paolic-
    chi, P. and Zappala, V., 1985, Shapes and stren-
    gths of small icy satellites. This volume.

12. Cuzzi, J. N. and Scargle, J., 1985, Wavy edges :
    Evidence for moonlets in Encke's division, Astro-
    phys. J., submitted.

13. Showalter, M. R. and Burns, J. A., 1982, A numeri-
    cal study of Saturn's F ring, Icarus 52, pp. 526-
    544.

14. Goldreich, P. and Tremaine, S., 1979, Towards a
    theory for the Uranian rings, Nature 277, pp. 97-
    99.

15. Goldreich, P. and Tremaine, S., 1982, The dynamics
    of planetary rings, Ann. Rev. Astron. Astrophys.
    20, pp. 249-283.

16. Dermott, S. F., 1984, Dynamics of narrow rings, in
    "Planetary Rings", Eds. R. Greenberg and A. Bra-
    hic, U. Arizona Press, pp. 589-637.

17. Smith, B. A. and Reitsema, H. J., 1980, CCD obser-
    vations of Jupiter's ring and Amalthea, in"Procee-
    dings of I. A. U. Colloquium 57".

18. Smith, B. A. and the Voyager imaging team,1982, A
    new look at the Saturn system : The Voyager 2 ima-

ges, Science 215, pp. 504-537.

19. Esposito, L. W., Cuzzi, J. N., Holberg, J. B., Marouf, E. A., Tyler, G. L. and Porco, C. C., 1984, Saturn' rings : Structure, dynamics and particle properties, in "Saturn", Eds. T. Gehrels and M. S. Matthews, U. Arizona Press, pp. 464-545.

20. Burns, J. A., Showalter, M. R. and Morfill, G., 1984, The ethereal rings of Jupiter and Saturn, in "Planetary Rings", Eds. R. Greenberg and A. Brahic, U. Arizona Press, pp. 200-221.

21. Pang, K. D., Voge, C. C., Rhoads, J. W. and Ajello, J. M., 1984, The E ring of Saturn and satellite Enceladus, J. Geophys. Res. 89, pp. 9459-9470.

22. Graps, A. L., Lane, A. L., Horn, L. J. and Simmons, K. E., 1984, Evidence for material between Saturn's A and F rings from the Voyager 2 photopolarimeter experiment, Icarus 60, pp. 409-415.

23. Burns, J. A., Schaffer, L. E., Greenberg, R. J. and Showalter, M. R., 1985, Lorentz resonances and the structure of the Jovian ring, Nature, submitted.

24. Van Allen, J. A., 1983, Absorption of energetic protons by Saturn's G ring, J. Geophys. Res. 88, pp. 6911-6918.

25. Bridges, F. G., Hatzes, A. and Lin, D. N. C., 1984, Structure, stability and evolution of Saturn's rings, Nature 309, pp. 333-335.

26. Marouf, E. A., Tyler, G. L., Zebker, H. A. and Eshleman, V. R., 1983, Particle size distributions in Saturn's rings from Voyager 1 radio occultation, Icarus 54, pp. 189-211.

27. Elliot, J. L., Dunham, E. and Mink, D., 1977, The rings of Uranus, Nature 267, pp. 328-330.

28. Elliot, J. L. and Nicholson, P. D., 1984, The rings of Uranus, in "Planetary Rings, Eds. R. Greenberg and A. Brahic, U. Of Arizona Press, pp. 25-72.

29. Goldreich, P. and Tremaine, S., 1979, Precession of the ε ring of Uranus, Astron. J. 84, pp. 1638-1641.

30.   Nicholson, P. D. and Jones, T. J., 1980, Two-micron
      spectrophotometry of Uranus and its rings, Icarus
      42, pp. 54-67.

31.   Cheng, A. F. and Lanzerotti, L. J., 1978, Ice sput-
      tering by radiation belt protons and the rings of
      Saturn and Uranus, J. Geophys. Res. 83, pp. 2597-
      2602.

32.   Cheng, A. F., Haff, P. K., Johnson, R. E. and Lan-
      zerotti, L. J., 1985, Interactions of planetary
      magnetospheres with icy satellite surfaces, in
      "Satellites", Eds. J. A. Burns and D. Morrison,
      U. Arizona Press, in press.

33.   Lanzerotti, L. J., Brown, W. L. and Johnson, R. E.,
      1984, Laboratory studies of the ion irradiations
      of water, sulfur dioxide and methane ices. This
      volume.

34.   Pang, K. D. and Nicholson, P. D., 1984, Composi-
      tion, particle size and age of the Uranian rings,
      BAAS 16, p. 678.

35.   Borderies, N., Goldreich, P. and Tremaine, S.,
      1984, Excitation of inclinations in ring-satelli-
      te systems, Astron. J. 284, pp. 429-434.

36.   French, R. G. and Elliot, J. L., 1984, The widths,
      radii and optical depths of the rings of Uranus,
      BAAS 16, pp. 677-678.

37.   Smith, B. A. and the Voyager imaging team, 1979, The
      Jupiter system through the eyes of Voyager 1,
      Science 204, pp. 951-971.

38.   Smith, B. A. and the Voyager imaging team, 1979,
      The Galilean satellites and Jupiter : Voyager 2
      imaging science results", Science 206, pp. 927-
      950.

39.   Owen, T., Danielson, G. E., Cook, A. F., Hansen,
      C., Hall, V. L. and Duxbury, T. C., 1979, Jupi-
      ter's ring, Nature 281, pp. 442-446.

40.   Jewitt, D. C., 1982, The rings of Jupiter, in "Sa-
      tellites of Jupiter, Eds. D. Morrison, U. Arizona
      Press, pp. 44-64.

41.   Jewitt, D. C. and Danielson, G. E., 1981, The Jo-

vian ring, J. Geophys. Res. 86, pp. 8691-8697.

42. Showalter, M. R., 1984, Jupiter's ring system re-
     solved : Physical properties inferred from the Vo-
     yager images. Ph. D. dissertation. Cornell U.,
     213 + xiii pp.

43. Showalter, M. R., Burns, J. A., Cuzzi, J. N. and
     Pollack, J. B., 1985, Reexamining Jupiter's ring.
     In preparation.

44. Showalter, M. R., Burns, J. A., Cuzzi, J. N. and
     Pollack, J. B., 1985, The discovery of a tenuous
     Jovian ring, Science, submitted.

45. Burns, J. A., Schaffer, L. E., Greenberg, R. J.
     and Showalter, M. R., 1985, Lorentz resonances
     and the structure of the Jovian ring, Nature, sub-
     mitted.

46. Fillius, W., 1976, The trapped radiation belts of
     Jupiter , in "Jupiter", Ed. T. Gehrels, U. Arizo-
     na Press, pp. 896-927.

47. Neugebauer, G., Becklin, E. E., Jewitt, D., Terri-
     le, R. and Danielson, G. E., 1981, Spectra of the
     Jovian ring and Amalthea, Astron. J. 86, pp. 607-
     610.

48. Clark, R. N., Fanale, F. P. and Gaffey, M. J.,
     1985, Surface composition of natural satellites,
     in "Satellites", Eds. J. A. Burns and D. Morri-
     son, U. Arizona Press, in press.

49. Burns, J. A., Showalter, M. R., Cuzzi, J. N. and
     Pollack, J. B., 1980, Physical processes in Jupi-
     ter's ring : Clues to its origin by Jove !, Ica-
     rus 44, pp. 339-360.

50. Grün, E., Morfill, G., Schwehm, G. and Johnson,
     T. V., 1980, A model of the origin of the Jovian
     ring, Icarus 44, pp. 326-338.

51. Consolmagno, G. J., 1983, Lorentz forces on the
     dust in Jupiter's ring, J. Geophys. Res. 88, pp.
     5607-5612.

52. Schaffer, L. and Burns, J. A., 1984, Dust motion
     in Jupiter's tilted magnetic field, in "Dust-Ma-

gnetosphere Interactions", COSPAR, Pergamon Press,
in press.

53. Pilcher, C. B., Chapman, C. R., Lebofsky, L. A.
    and Kieffer, H. H., 1970, Saturn's rings : Identi-
    fication of water frost, Science 167, pp. 1372-
    1373.

54. Clark, R. N. and McCord, T. B., 1980, The rings of
    Saturn: New near-infrared reflectance measurements
    and a 0.326-4.08 µm summary, Icarus 43, pp. 161-
    168.

55. Lebofsky, L. A. and Fegley, M., 1976, Laboratory
    reflection spectra for the determination of chemi-
    cal composition of icy bodies, Icarus 28, pp. 379-
    387.

56. Puetter, R. C. and Russell, R. W., 1977, The 2-4
    micron spectrum of Saturn's rings, Icarus 32, pp.
    37-40.

57. Brown, R. H., 1984, Physical properties of the Ura-
    nian satellites, In "Uranus and Neptune", Ed. J. T.
    Bergstrahl, NASA Conf. Pub. 2330, pp. 437-461.

58. Caldwell, J., 1975, Ultraviolet observations of
    small bodies in the solar system by OAO-2, Ica-
    rus 25, pp. 384-396.

59. Clark, R. N., 1980, Ganymede, Europa, Callisto,
    and Saturn's rings : Compositional analysis from
    reflectance spectroscopy, Icarus 44, pp. 388-409.

60. McFadden, L. A., Bell, J. and McCord, T. B., 1981,
    Visible spectral reflectance measurements 0.3 -
    1.1 µm of the Galilean satellites at many orbital
    phase angles 1977-1978, Icarus 44, pp. 410-430.

DISCUSSION

R. E. Johnson :

The difference between Uranus and Saturn may be
simply a temperature difference. That is, if you can
trap methane (e. g. as a clathrate) at Uranus but not
at Saturn, then U. V. and charge particle conversion of
methane into a solid form of carbon will only occur at
Uranus. This process can occur in the presence of water

ice and since it is very exothermic you would lose the water.

L. J. Lanzerotti :

   Saturn magnetosphere cut off at or inside Enceladus ; hence plasma does not intersect with the A- and B-rings, only with the E - ring. Such a situation may or may not exist at Uranus : magnetosphere could engulf the rings, rather than be cut-of as at Saturn.

ICES IN PLANETARY RINGS

R. Smoluchowski

Department of Astronomy
University of Texas
Austin, TX 78712 USA

ABSTRACT

Understanding the structure and behavior of Saturnian rings
in terms of properties of ices is basic for evolutionary
planetology. The available information indicates the presence
of quite pure, probably amorphous water ice in the form of medium-
grained frost in a fairy castle structure with a low thermal
inertia. Tidal forces and interparticle collisions lead to
continuous break-up and re-formation of loose aggregates of
smaller particles. Micrometeoroid bombardment and proton
irradiation are important for explaining the mechanical and optical
properties of the surfaces of ring particles. The transfer of
angular momentum and mass among the rings should homogenize the
chemical and structural characteristics of the ring ices. Uranian
rings may be made of carbon-covered methane ice particles.

#1. INTRODUCTION

Although three of the outer planets have rings, only the
well-known rings of Saturn and its E-ring are made primarily of
water-ice. Those of Jupiter are probably non-icy ejecta from
satellites while those of Uranus seem to be covered by carbon (1)
produced by irradiation of methane. Ever since the spectacular
Pioneer and Voyager investigations, we know a great deal about
the Saturnian rings. Nevertheless, we would like to know much
more because there are good reasons to believe that understanding
the formation, evolution and properties of these rings will help
to unravel some of the mysteries concerning the formation of
planetary systems. The question thus arises, "To what extent can

681

J. Klinger et al. (eds.), Ices in the Solar System, 681–689.
© 1985 by D. Reidel Publishing Company.

we use what we know about the Saturnian rings to understand all
ring systems rather than only those made of ices?" Water, in its
various forms, is the most widespread compound in the solar
system, and presumably in the universe, but in many ways it is ·
unique. For these reasons, an interpretation of the facts about
the Saturnian rings in terms of properties of water ice, or ices,
is of special importance.

#2. PURITY

Studies of albedo and spectral reflectivity of Saturn's
rings leave little doubt that they contain at least 90 percent
H_2O ice. A comparison of the somewhat asymmetric shape of the
bands in the 1 to 2.5 μm region with those studied in the
laboratory indicates that the surfaces of the ring particles
resemble medium fine-grain frost rather than frost deposited on
ice (2). The geometric albedo (3) of the rings for λ > 0.7 μm is
around 0.9 and falls to very low values near 0.2 μm in a manner
similar to the albedo of Europa which is believed to be
occasionally re-covered with fresh ice. Although Europa's albedo
is 25 to 30 percent lower, one should remember that
albedo of micron-size particles is strongly dependent on their
radii (4). The unusual brightness of the rings agrees with the
suggestion discussed further below that the surfaces of the ring
particles are continuously refreshed. The temperature of the
surface of the rings (5, 6, 7) is 78+5K but the brightness
temperature shows a rather complicated behavior with a minimum of
less than 60K between 100 and 400 μm, a peak near 80K at 500 μm,
and a gradual drop at longer wavelengths. It has been speculated
that this behavior may indicate (8) a slightly lower temperature
in the interior of the larger particles of the rings.

Differences between certain optical properties of the rings
and those of pure H_2O fine-grain frost suggests that the most
likely contaminants are very small amounts of carbonaceous
chondrites and of silicates. Both of these traces would be a
natural contribution from impacting micrometeorites. It is worth
noting, however, that the optical properties can be best
reproduced if these impurities do not lie on the surface of the
icy particles but are embedded in water frost or slightly covered
by it (5). The precision of the measurements is not sufficient
to indicate whether any clathrates are present in the rings,
though actually one would not expect to find them there. One
reason for this conclusion is that the usually most abundant
CO_2-clathrate is not stable (9) below 120 K and decomposes into
CO_2 and H_2O. As far as the other clathrates, such as those
containing NH_3 or CH_4 are concerned, it is very unlikely that
they would survive the continuous sputter, re-deposition and
bombardment of the icy surfaces of the ring particles as

discussed further below.

#3. STRUCTURE OF ICES

In view of the low temperature of the rings, it is natural
to suspect that the H_2O-ice is not crystalline but amorphous (10)
because the latter forms when water vapor condenses below 150 K.
Unfortunately, at present, there is no direct observational
evidence in favor of, or against, the presence of amorphous ice
in the rings. In laboratory the absorption band at about 3 μm,
which corresponds to the O-H stretching mode, is broader in
amorphous ice than in crystalline ice, it has a sharp central
peak and two rather pronounced shoulders. The studies of
reflectivity of the rings in this wavelength region show only a
rather weak featureless band so that no definite conclusion can
be drawn. Also, for crystalline ice the position of the peak and
the width of the band appear to depend on the length of the
optical path and on particle size (10) which further complicate
a diagnostic comparison between observation and laboratory data
(Fig. 1). There may be also an effect of temperature.

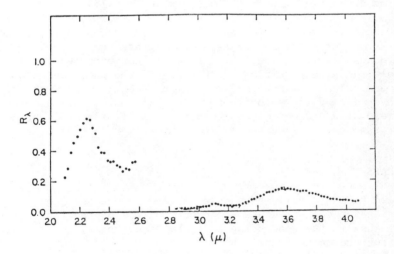

Fig. 1   The IR spectrum of Saturn's ring showing the weak 3 μ
peak (Ref. 27).

It is worth pointing out that an insight into the nature of
crystallinity of ices on Saturnian rings could be obtained if a
spacecraft were to make observations of the sunlight penetrating
the rings. In the space between the ring particles there should
be a considerable concentration of small icy fragments broken

off during interparticle collisions or by micrometeoritic
bombardment. If these particles are crystalline then a halo
around the Sun at about 22 degrees and another, weaker one at
about 47 degrees should be observable just as they are known to
occur in high altitude crystalline clouds in our atmosphere. Of
course an absence of such halos would not constitute a strong
argument in favor of amorphous ice.

Amorphous ice on the rings could form by condensation of
water molecules present in the atmosphere which is known to
surround the rings (12). Although observations indicate only
that this atmosphere contains hydrogen, one can surmise that
there are also hydroxyl groups and water molecules sputtered from
the rings by protons, electrons and by micrometeorites (13). At
the low temperature of the rings, the condensation process is
random and there is negligible surface mobility of the molecules
so that the result should be a very porous structure often
described as "fairy castles". Such a surface would show a low
thermal inertia in accord with observations (14) made at the
entrance and emergence of the rings from the planet's shadow.
Furthermore, such a structure would explain the opposition
effect of the rings which is too strong to be interpreted solely
in terms of multiple interparticle scattering (15).

Finally, it should be mentioned that, although pores in ices
in a thermal gradient are expected to migrate to surfaces (16)
and produce densification, this effect is not expected to play
any role for ring particles. First of all the particles are
small and probably isothermal apart from a shallow surface effect
due to shadowing and secondly the motion of pores in water ice is
extremely slow and would require periods many orders of magnitude
longer than the unperturbed lifetime of a ring particle.

#4. MASS TRANSFER

Two recent developments have an important bearing on
understanding the structure and nature of ice on the ring particles.
One of them is the conclusion that the larger particles are not
solid but are rather weakly bonded agglomerates of much smaller
particles (17, 18). By accretion these agglomerates can grow to
particles of the order of meters which may become sufficiently
large to be sensitive to tidal forces so that they break up. The
old idea that the large particles are solid, though undoubtedly
porous, chunks of ice implied that the influence of various
external effects such as proton, electron and micrometeoroid
bombardment and deposits of fresh ice were limited to the
surfaces of these large particles. The new model of loosely
bound and continuously changing agglomerates indicates that the
effect of the various surface processes are fairly uniformly

distributed throughout the large particles (Fig. 2).

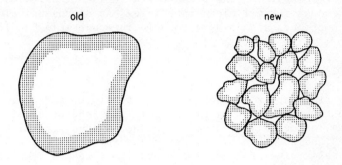

Fig. 2   Schematic cross section of a large Saturnian ring
particle showing the distribution of damage close to the surface
according to the old and new model.

The larger particles have therefore not only low thermal inertia
surfaces as discussed above but also their interiors should have
much lower thermal conductivity than was previously assumed.
Depending on wavelength of the incident radiation, the large
agglomerates will show higher reflectivity and less forward
scattering.

    The second important new insight into the nature of the icy
particles is based on the idea that as a result of collisions
there is considerable transfer of angular momentum and of
particles of all sizes between the various rings (19, 20).   This
transfer leads to radial differences in ring thickness and to a
small continuous loss of mass.   The process presumably also
equalizes any compositional differences between the rings which
may have existed or developed in the course of time.   It is clear
that an important aspect of this accretion, mass transfer and
tidal break-up is the strength of the bonding between the small
particles.   The mechanical interaction between two ice particles
which collide with relative velocity of the order of a fraction
of cm s^{-1} is quite weak.   If these particles had a solid elastic
ice surface the chances of adhesion at the few points of contact
would be very small.   On the other hand, particles covered with
a fragile, and at least partly amorphous, fairy castle layer may
adhere much easier because the weak impact will not be elastic
but will fracture the convoluted surface layer and produce many
points, if not a whole area, of contact.   Another factor which
may help adhesion is the possibility that the weak impact would
lead to formation of local bonds and even perhaps to local
crystallization with the attendant evolution of heat (21).   In
favor of this mechanism is the low thermal conductivity of the

irregular surface and, in particular, the low specific heat of
amorphous ice both of which would facilitate a highly localized
release of energy. At the present time a quantitative estimate
of the process is well nigh impossible. As discussed below
radiation effects also play a role in interparticle bonding.

#5.   RADIATION EFFECTS

    The ring particles are subject not only to micrometeoroids
but also to bombardment by plasma particles and by atomic
particles, especially secondary protons, produced by high energy
primaries and observed by Voyager spacecraft (22, 23, 24). The
flux of solar wind protons is comparatively weak. From the point
of view of formation of atomic defects in ring particles, the
most effective are protons around 100 kev which have a range of
the order of microns. These protons produce the well-known
atomic sputtering (12), the less known evaporation sputtering (25)
and finally, thermal fatigue fracture (26). The last one is based
on the fact that in a rather good thermal insulator such as ice,
the impinging particle heats up the solid along its path forming
the so-called thermal spike. This heat produces rapid local
expansion which cannot be accomodated without some local plastic
deformation. Upon cooling these plastic effects do not reverse,
and are cumulative, so that additional thermal spikes produced by
other nearby incident particles lead to progressively increasing
highly localized lattice deformation and strain (Fig. 3).

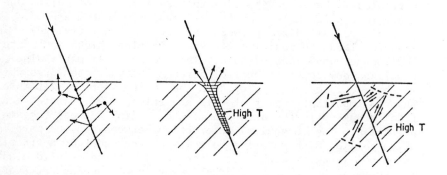

Fig. 3   Displacement, thermal and plastic deformation spikes
along a path of an energetic particle penetrating an insulator.

Such a highly strained surface layer (which is of the order of a
micron thick) is very brittle and can splinter during the frequent
though weak interparticle collisions. This mechanically weakened
surface facilitates adhesion in a manner similar to that of
"fairy castles" mentioned above. Those pieces of ice which

become loose contribute to the layer of ice dust which
cover the ring particles and influence the formation of
radial spokes (26).

#6. CONCLUSION

        The major new conclusion based on a variety of
observations and theoretical studies of rings and of
ring particles is that, contrary to previous models,
ring particles (and the ice which covers them) are continuously undergoing changes. These changes are brought
about by bombardments an by interparticle collisions.
It is very likely, that the ice on the ring particles
is amorphous rather than crystalline but this conclusion still awaits observational check. The surface of
ring particles should be very porous, irregular and mechanically weak which would enhance interparticle adhesion, lead to low thermal inertia and to characteristic optical effects.

        Supported by NASA Grant NGS7505 supp. 4.

REFERENCES

1.   Pang, K. D. and Nicholson, P. D., 1984, Composi-
     tion and Size of Uranian Ring Particles, Lunar
     and Planetary Science, 15, in print.

2.   Clark, R. N., 1980, Ganymede, Europa, Callisto and
     Saturn's Rings : Compositional Analysis from Re-
     flectance Spectroscopy, Icarus, 44, pp. 388-409.

3.   Esposito, L. W., 1984, Photometry and Polarimetry
     of Saturn's Rings, in "Natural Satellites",
     I. A. U. Coll. 77, Icarus, in print.

4.   Greenberg, J. M., 1984, private communication.

5.   Esposito, L., O'Callaghan, M. and West, R. A., 1983,
     The Structure of Saturn's Rings : Implications
     from the Voyager Stellar Occultation, Icarus, 56,
     pp. 439-452.

6.   Fink, U. and Sill, G. T., 1983, The Infrared Spec-
     tral Properties of Frozen Volatiles, in "Comets",
     ed. L. L. Wilkening, U. of Ariz. Press, Tucson,
     pp. 164-202.

7.   Daniel, R. R., Ghosh, S. K., Iyengar, K. V. V.,

Rengajaran, N., Tandon, S. N. and Verma, R. P.,
1982, Far-Infrared Brightness Temperature of
Saturn's Disk and Rings, Icarus, 49, pp. 205-212.

8.   Cuzzi, J., 1983, Planetary Ring Systems, Rev. Geoph.
     and Space Physics, 21, pp. 173.

9.   Miller, S. L., 1961, The Occurrence of Gas Hydrates
     in the Solar System, Proc. Nat. Acad. Sci., 47,
     pp. 1798.

10. Smoluchowski, R., 1978, Amorphous Ice on Saturnian
     Rings and Icy Satellites, Science, 201, pp. 809-
     811.

11. Lucey, P. and Clark, R. N., 1984, Spectral Proper-
     ties of Water Ices and Contaminants, in this book.

12. Cuzzi, J. N., Esposito, L. W., Hollberg, J. B.,
     Marouf, E. A. and Boischot, A., 1984, Saturn's
     Rings : Properties and Processes, in "Planetary
     Rings", ed. R. Greenberg and A. Brahic, Univ. of
     Ariz. Press, Tucson.

13. Brown, W. L., Lanzerotti, L. J., Poate, J. M. and
     Augustyniak, W. M., 1978, Sputtering of Ice by
     MeV Light Ions, Phys. Rev. Lett., 40, pp. 1027-
     1030.

14. Froidevaux, L., Matthews, K. and Neugebauer, G.,
     1981, Thermal Response of Saturn's Ring Particles
     During and After Eclipse, Icarus, 46, pp. 18-26.

15. Pang, K. D. and Rhoads, J. W., 1984, Interpreta-
     tion of Integrated Disk Photometry of the Uranian
     Satellites, in "Natural Satellites", I. A. U.
     Colloq. 77.

16. Smoluchowski, R., Marie, M. and McWilliam, A., 1984,
     Evolution of Density in Solar System Ices, Earth
     Moon and Planets, 30, pp. 281-288.

17. Weidenschilling, S. S., 1983, "Dust in a Turbulent
     Solar Nebula", Bull. Am. Astr. Soc., 15, pp. 811
     (abstract), and in "Planetary Rings", ed. R. Green-
     berg and A. Brahic, U. of Ariz. Press.

18. Greenberg, R., Davis, D. R., Weidenschilling, S.S.
     and Chapman, C. R., 1983, The Dynamic Ephemeral

Bodies in Saturn's Rings , Bull. Am. Astr. Soc.,
15, pp. 812-813 (abstract) and in "Planetary Rings",
ed. R. Greenberg and A. Brahic, U. of Ariz. Press.

19. Durisen, R. H., 1984, Transport Effects Due to
     Particle Erosion Mechanisms , in "Planetary Rings",
     ed. R. Greenberg and A. Brahic, U. of Ariz. Press.

20. Lissauer, J. J., 1984, Ballistic Transport in Sa-
     turn's Rings : An Analytic Theory , Icarus, 57,
     pp. 63-71.

21. Ghormley, J. A., 1968, Enthalpy Changes and Heat
     Capacity Changes in the Transformation from High-
     Surface-Area Amorphous Ice to Stable Hexagonal
     Ice, J. Chem. Phys., 48, pp. 503-508.

22. Chenette, D. L., Cooper, J. F., Eraker, J. H., Pyle,
     K. R. and Simpson, J. A., 1980, High Energy Trap-
     ped Radiation Penetrating the Rings of Saturn ,
     J. Geoph. Rev., 85, pp. 5785-5792.

23. Van Allen, J. A., Randall, B. A. and Thomsen, M.
     F., 1980, Sources and Sinks of Energetic Elec-
     trons and Protons in Saturn's Magnetosphere ,
     J. Geoph. Res., 85, pp. 5679-5694.

24. Krimigis, S. M. and Armstrong, T. P., 1982, Two-
     Component Proton Spectra in the Inner Saturnian
     Magnetosphere , Geoph. Res. Lett., 9, pp. 1143.

25. Ollerhead, R. W., Bottiger, J., Davies, J. P.,
     L'Ecuyer, J., Haugen H. K. and Matsunami, N.,
     1980, Evidence for a Thermal Spike Mechanism in
     the Erosion of Frozen Xenon , Rad. Eff., 49, pp.
     203.

26. Smoluchowski, R., 1983, Formation of Fine Dust on
     Saturn's Rings as Suggested by the Presence of
     Spokes , Icarus, 54, pp. 263-266.

27. Puetter, R. C. and Russell, R. W., 1977, Icarus,
     32, pp. 37-40.

THE SMALL, ICY SATELLITES OF SATURN

Dale P. Cruikshank

Institute for Astronomy
University of Hawaii
Honolulu, Hawaii 96822 USA

ABSTRACT

At least nine small satellites exist in the inner regions of
the Saturn system. From their high geometric albedos and by
inference from the larger satellites, at least the surfaces of
the small satellites are dominated by water ice. The bulk of our
limited information on these bodies comes from analysis of images
from the Voyager 1 and 2 spacecraft.

INTRODUCTION

The presence of water ice in the Saturn system has been
known or strongly suspected since Kuiper's original infrared
spectrophotometric measurements of the rings which were reported
only in a brief abstract (Kuiper (1)). Verification of water
ice in the rings came when infrared spectra of higher resolution
were compared with low-temperature ice in the laboratory (Pilcher
et al. (2) ; Kuiper et al. (3) (4)). The detailed story of the
1970 work will be an interesting episode for future historians to
disentangle, but the outcome is simple enough; the infrared spectrum of the rings shows the very strong water ice spectral signatures in the region 0.9-2.6 μm. The water ice bands are stronger
in the ring spectrum than in any other solar system body so far
studied.

Water ice was also found on the larger satellites of Saturn
(except Titan) by near-infrared photometry (Johnson et al. (5) ;
Morrison et al. (6)) and by near-infrared spectroscopy (Fink et
al. (7)). The most recent and detailed ground-based investiga-

691

J. Klinger et al. (eds.), Ices in the Solar System, 691–697.

tion of the surface compositions of the larger satellites of
Saturn has been made by Clark et al. (8). A review of the
compositions and microstructures of all the satellites of Saturn,
including both ground-based and Voyager spacecraft data, has been
published by Cruikshank et al. (9). A complimentary study of
the geological properties of the same bodies has been presented
by Morrison et al. (10).

In the Cruikshank et al. (9) review, the basic information on the optical properties of the small, icy satellites, the
subject of the present paper, was assembled and analyzed chiefly
by Drs. P. Thomas, J. Veverka, and B. Buratti, working with
Voyager spacecraft data.

THE SMALL SATELLITES

Ground-based observations of Saturn at the times of ring-
plane passage (1966 and 1980), together with observations from
Voyagers 1 and 2 (1980/1981) revealed several small satellites in
the inner part of the Saturnian system. A selection of Voyager
images of these bodies, reproduced at the correct relative
scales, is given in Figure 1. It can be seen that most of the
objects are irregular in shape, and some are heavily cratered.
Available spacecraft data indicate that the small satellites are
in synchronous spin states with the largest axes pointed toward
the planet.

Photometric studies of the spacecraft data show that the
normal reflectances (and presumably the geometric albedos) are
high, ranging from about 0.4 to at least 0.8. The reflectances
and colors cover the range observed for the larger satellites of
Saturn known to be ice-covered; it is thus reasonable to conclude
that the smaller satellites are also ice-covered even though
there is no specific infrared (or other) spectroscopic evidence
to confirm this conclusion. Thomas et al. (11), (12)) discuss the
possibility that there is contamination of the icy surfaces by
small amounts of dark, opaque material. Contaminated ices are
known on many other satellites throughout the solar system,
though the nature of the contaminants is presently unknown.

The essential results of the study of ground-based and
Voyager spacecraft data on the small satellites are summarized in
Table I from Cruikshank et al. (9). Precise geometric albedos
for these bodies are difficult to determine because of the low
quality of ground-based photometry and the lack of spacecraft
data below phase angle = 24°. The values given in Table I are
based upon an extrapolation to opposition of the phase behavior
observed at large phase angles (details in Thomas (11)). For
two satellites in this group the phase coefficients can be deter-

Figure 1. Seven of the inner satellites of Saturn are shown here
in their correct relative sizes. The left-hand column, top to
bottom, shows the larger coorbital (Janus), Calypso (leading
Tethys lagrangian), 1980 S26 (outer F-ring shepherd), 1980 S27
(inner F-ring shepherd). The right-hand column shows 1980 S6
(Dione lagrangian), Epimetheus (smaller coorbital), and Telesto
(trailing Tethys lagrangian). NASA photograph courtesy Jet
Propulsion Lab.

Table I. Properties of Small, Inner Satellites of Saturn[a]

Satellite			Orbital Radius ($\times 10^3$ km)	Cross Sectional Radius at Elongation A x C (km)	Geometric Albedo	$+\Delta m$ (Relative to Mimas)	Mean Opposition Magnitude V_o
No.	Name	Type					
S17	Atlas	A-Ring Shepherd	138	20 x 10	0.4	--	--
S16	1980 S27	Inner F-Ring Shepherd	139	70 x 37	0.49-0.68[b]	3.1-2.7[b]	15.6-15.2[b]
S15	1980 S26	Outer F-Ring Shepherd	142	55 x 33	0.52	3.6	16.1
S10	Janus	Larger Coorbital	151	110 x 80	0.63	1.6	14.1
S11	Epimetheus (1980 S3)	Smaller Coorbital	151	70 x 50	0.42-0.60	3.0-2.6	15.5-15.1
S1	Mimas	Regular	186	196 x 196	0.60 ± 0.1	0	12.5
S14	Calypso	Tethys Leading Lagrangian	295	12 x 11	0.53-0.75	6.3-5.9	18.8-18.4
S13	Telesto	Tethys Trailing Lagrangian	295	15 x 8	0.81-1.15	6.0-5.6	18.5-18.1
S12	1980 S6	Dione Leading Lagrangian	377	17 x 15	0.46-0.66	5.7-5.3	18.2-17.8

[a]Reproduced from Cruikshank et al. (9).
[b]Where a range is given, the two estimates correspond to two different assumptions about the phase coefficient below $\alpha = 30°$: (a) intrinsic phase coefficient of 0.0092 mag/deg (derived for Outer F Shepherd), and (b) intrinsic phase coefficient of 0.022 mag/deg (derived for Larger Coorbital). In this table, we have assumed that the geometric albedo equals the normal reflectance (i.e., no limb darkening at $\alpha = 0°$, or k = 0.5). If k = 0.6 were more appropriate, the geometric albedos in the table should be decreased by about 10% and the values of V_o increase by 0.1 mag.

mined well. The larger coorbital (Janus) and the outer F-Ring shepherd satellites have intrinsic phase coefficients of 0.02 and 0.092 mag/deg, respectively. For comparison, the same parameter for Mimas, Enceladus, Tethys, Dione, and Rhea ranges from 0.007 to 0.018 mag/deg (Buratti et al. (13); Buratti (14)). Among these bodies there appears to be no clear dependence of the phase coefficient on surface reflectance.

Table I also lists the mean opposition magnitudes derived from Voyager data. These values result from photometry of spacecraft images using Mimas as the primary reference source. The work of Buratti and Veverka (15) indicates that the mean opposition magnitude of Mimas is +12.5, and not +12.9 as has been previously quoted. The values of V_O in the table are thus referenced to the new magnitude of Mimas. The Voyager data are in good agreement for the magnitude ($V_O = 18.5 \pm 0.5$) of the ground-based value of the Tethys leading lagrangian satellite determined by Larson et al. (16).

The existing spacecraft data on the small satellites of Saturn have been exploited nearly to their fullest extent. It appears, however, that new techniques used with large ground-based telescopes may shed further light on the physical properties of these bodies, even though they are very difficult to observe because of their proximity to Saturn and the bright rings (e.g., Veillet (17)).

ACKNOWLEDGMENTS

Our understanding of the small satellites of Saturn is due chiefly to the work of J. Veverka, P. Thomas, and B. Buratti, whose contributions to this paper are gratefully acknowledged.

REFERENCES

1. Kuiper, G. P., 1957, Infrared observations of planets and satellites, Astron. J. 62, p. 245 (abstract).

2. Pilcher, C. B., Chapman, C. O., Lebofsky, L. A. and Kieffer, H. H., 1970, Saturn's rings : Identification of water frost, Science 167, pp. 1372-1373.

3. Kuiper, G. P., Cruikshank, D. P. and Fink, U., 1970, The composition of Saturn's rings, Bull. Amer. Astron. Soc. 2, pp. 235-236 (abstract).

4. Kuiper, G. P., Cruikshank, D. P. and Fink, U., 1970, Composition of Saturn's rings, Letter to Editor, Sky and Telescope

39, p. 80.

5. Johnson, T. V., Veeder, G. J. and Matson, D. L., 1975, Evidence for frost on Rhea's surface, Icarus 24, pp. 428-432.

6. Morrison, D., Cruikshank, D. P., Pilcher, C. B. and Rieke, G. H., 1976, Surface compositions of the satellites of Saturn from infrared photometry, Astrophys. J. 207, pp. L213-L217.

7. Fink, U., Larson, H. P., Gautier, T. N. and Treffers, R. R., 1976, Identification of water ice on Iapetus, Rhea, Dione, and Tethys, Astrophys. J. 207, pp. L63-L67.

8. Clark, R. N., Brown, R. H., Owensby, P. D. and Steele, A., 1984, Saturn's satellites : Near-infrared spectrophotometry (0.65-2.5 μm) of the leading and trailing sides and compositional implications, Icarus 58, pp. 265-281.

9. Cruikshank, D. P., Veverka, J. and Lebofsky, L. A., 1984, Satellites of Saturn : The optical properties. In "Saturn", ed. T. Gehrels, Univ. of Arizona Press, pp. 640-667.

10. Morrison, D., Johnson, T. V., Shoemaker, E., Smith, B. A., Soderblom, L., Thomas, P. and Veverka, J., 1984, Satellites of Saturn : Geological Perspective. In "Saturn", ed. T. Gehrels, Univ. of Arizona Press, pp. 609-639.

11. Thomas, P., Veverka, J., Morrison, D., Davies, M. and Johnson, T. V., 1983, Phoebe : Voyager 2 observations, J. Geophys. Res. 88, pp. 8736-8742.

12. Thomas, P., Veverka, J., Morrison, D., Davies, M. and Johnson, T. V., 1983, Saturn's small satellites : Voyager imaging results, J. Geophys. Res. 88, pp. 8743-8754.

13. Buratti, B., Veverka, J. and Thomas, P., 1982, Voyager photometry of Saturn's satellites. In "Reports of Planetary Geology Program" (NASA TM-84127), pp. 41-43.

14. Buratti, B., 1983, Ph. D. dissertation. Cornell University, Ithaca, NY.

15. Buratti, B. and Veverka, J., 1984, Voyager photometry of Rhea, Dione, Tethys, Enceladus, and Mimas, Icarus 58, pp. 254-264.

16. Larson, S. M., Fountain, J. W., Smith, B. A. and Reitsema, H. J., 1981, Observations of the Saturn E ring and a new satellite, Icarus 47, pp. 288-290.

17, Veillet, C,, 1984, Satellites of Saturn, IAU Circular n° 3940.

COMMENT by J, Plescia

       S, Synott of J, P. L. has made a survey of many of the
Voyager images at the Saturn area and discovered several small
(few km) diameter bodies. However, only a small fraction of space
was viewed and many more satellites could exist.

THE SHAPES AND STRENGTHS OF SMALL ICY SATELLITES

P. Farinella[1], A. Milani[1], A. M. Nobili[1],
P. Paolicchi[2] and V. Zappala[3]
1) Dipartimento di Matematica, Università di Pisa, Italy
2) Osservatorio Astronomico di Brera, Merate, Italy
3) Osservatorio Astronomico di Torino, Pino Torinese,
   Italy

ABSTRACT

   In the set of small icy satellites of Saturn
recently imaged by the Voyager probes, we can observe
the transition from irregularly shaped, strength-dominated objects to larger, gravity-dominated bodies with
shapes roughly fitting the theoretical equilibrium figures. The transition occurs at a radius of 100 ± 50 km,
corresponding to a typical material strength on the order of 10^7 dynes/cm^2. We discuss briefly the cases of
Mimas, Enceladus, Hyperion, Phoebe and the small coorbital and F-ring shepherding moons, comparing them with
other bodies of similar size like Amalthea and the intermediate-sized asteroids. We show that an analysis of
the shape data can often provide interesting clues on
the physical properties, origin and collisional history
of these objects.

I. INTRODUCTION

   Due to some basic properties of two fundamental
physical interactions – the electromagnetic force giving rise to the solid-state strength and the gravitational force –, an important transition occurs among the
small bodies of the solar system. Below a critical size,
the shape of the bodies can be defined as 'irregular',
depending mainly on the origin and evolutionary history
of the various objects. On the other hand, beyond the
critical size self-gravitational forces overcome the
solid-state rigidity of the material, and the shapes

J. Klinger et al. (eds.), Ices in the Solar System, 699–710.
© *1985 by D. Reidel Publishing Company.*

approximate roughly (apart from minor surface topography) the figures of gravitational equilibrium consistent with the rotational state. These figures are different depending on whether the body is isolated or is a satellite distorted by the gravitational tides of a primary ; moreover, the internal density distribution affects the surface figure significantly. For homogeneous objects, the classical theory (reviewed by Chandrasekhar (1)) specifies directly the shape parameters once the density and the spin rate are known (the satellites are always assumed to be 'corotating', i. e., with synchronized spin and orbital periods). In all the cases of interest for us, the equilibrium figures are ellipsoids. Thus, given the semiaxes a ⩾ b ⩾ c, the shapes are determined unequivocally by the two ratios c/a and b/a.

On the other hand, when the solid-state strength is large enough to prevent relaxation towards equilibrium, the shape of the bodies retains an important record of their previous history. For instance, many small bodies of the solar system have been subjected to a heavy external bombardment by impacting objects, and their present properties frequently arose as a result of a process of collisional fragmentation. In this case, the resulting shapes are complex and irregular, depending on the details of the fragmentation history itself. The analysis of some laboratory impact experiments (2-3-4) has confirmed the large variability in shape of the fragments, but has shown that, statistically, the shape distribution can be characterized by the following values of the axial ratios : c/a = 0.50 ± 0.14 ; b/a = 0.73 ± 0.13 (here the axial lengths are defined as the dimensions of the objects along three perpendicular directions). This result is remarkable in that it does not depend on the type of collision nor on the shape and composition of the target ; hence it can be used to assess whether a given body is a likely collisional fragment or not.

What is the critical size for the transition described above ? From a theoretical point of view, we can evaluate it by requiring that the material strength S is high enough to sustain the local gravitational stress caused by topography of some fixed height h (5). In this case S is approximately given by

$$S = \frac{4\pi}{3} G\rho^2 h R, \qquad\qquad (1)$$

where G is the gravitational constant and ρ and R are the density and radius of the body, assumed to be near-ly spherical and homogeneous. Alternatively, we can re-quire that S is equal to the hydro-static internal pres-sure on the spherical surface of radius αR (α being a coefficient smaller than 1), obtaining

$$S = \frac{2\pi}{3} G\rho^2 R^2 (1 - \alpha^2). \qquad (2)$$

Let us define a 'regular' body as one whose surface to-pography is lower than, for instance, 10 % of the ra-dius, implying h < 0.1 R and α > 0.9. Then the critical radius is

$$R_{cr} = (15 \; S/2\pi G\rho^2)^{1/2}, \qquad (3)$$

which obviously depends on the density and strength of the material constituting the object. In particular S is highly variable and difficult to estimate reliably for the materials forming the small bodies of the solar system (see the discussion in Farinella et al. (6)). If we assume that these materials, accumulated presumably under conditions of low gravitational compression, are mostly weakly consolidated aggregates, then $S \simeq 10^7$ dynes/cm^2 could be a reasonable estimate. With $\rho = 1,5$ g/cm^3, this yields $R_{cr} \simeq 130$ km. This result is consis-tent with the presently available observational data : several recent papers have discussed the shape data for asteroids (7-8), Phobos and Deimos (9) and Amalthea (10), and these studies seem to confirm that, perhaps within a factor two, the critical transition radius is of the order of 100 km.

Conversely, if we can measure the maximum height of the topography on a given body, this gives a lower bound to the strength of the material forming the surfa-ce layer of the object. This lower bound is close to the real strength when the surface shows no sign of endoge-nic evolution (e. g., tectonic and vulcanic activity, melting phenomena, etc.) capable of affecting substan-tially the topographic relief, and when we can exclude any significant creep deformation which would lead to relaxation towards equilibrium. The relaxation timesca-le of a nonequilibrium equatorial bulge of a viscous bo-dy is

$$\tau = 57\eta/8\pi G\rho^2 R^2 \qquad (4)$$

(remember that the viscosity η is defined as the shear

stress - of the order of the surface gravity divided by the area of the body - needed to keep a flow of a given velocity V per length D ; and in our case τ = D/V). τ is longer for topographic features of wavelength λ smaller than R : for instance, if $\lambda \sim R/4$, τ is increased by a factor ~ 2.5 (5). The viscosity has a strong temperature dependence, whose form is given by the empirical formula

$$\frac{\eta(T_1)}{\eta(T_2)} = \exp\left[- a \left(\frac{T_m}{T_2} - \frac{T_m}{T_1}\right)\right], \tag{5}$$

where T_m is the melting temperature (whose dependence on pressure can be neglected for bodies with $R \lesssim 1000$ km) and a is an experimentally determined constant close to 18 for many materials, including ice (11). The viscosity of ice for $T = T_m$ is of the order of 10^{10} p (12), but this estimate must be considered uncertain by at least plus or minus one order of magnitude, implying a corresponding uncertainty of τ. Table 1 gives the relaxation times (in years) for topographic features of wavelength $\sim R$ and $\sim R/4$, for R = 200 and 1000 km and for T = 95° K (close to the temperature of the Saturnian moons measured by the Voyagers' infrared observations) and T = 110° K (in order to check whether a modest thermal evolution could have caused extensive viscous relaxation). We have assumed always $\eta(T_m) = 10^{10}$ p and $\rho = 1.5$ g/cm^3.

T A B L E   1

τ	T = 95° K	T = 110°K
R = 200 km $\lambda \sim R$	5.3×10^9	4.6×10^6
R = 200 km $\lambda \sim R/4$	1.3×10^{10}	1.1×10^7
R = 1000 km $\lambda \sim R$	2.1×10^8	1.8×10^5
R = 1000 km $\lambda \sim R/4$	5.3×10^8	4.6×10^5

The values of τ shown in Table 1 indicate that the smaller, geologically inactive moons can have retained substantially irregular shapes, contrasting with larger and/or thermally evolved objects for which we expect relaxation of large-scale topographic features (like megaimpact craters). These results, together with the estimate of R_{cr} given by Eq. (3), support the idea that the small satellites of Saturn recently imaged by the Voyager probes, which span the range of sizes $10 \gtrless R \gtrless 10^3$ km, are an almost ideal sample of objects to see the transition from strength-dominated to gravity-dominated bodies. Thus, for the larger moons we should be able to obtain from the equilibrium shapes information on their physical properties and internal structure (as suggested originally by Morrison and Burns, (13)), while the smaller objects should display irregular shapes as a consequence of their presumably intense collisional history. In the following, we shall discuss briefly these items for some individual satellites, based of course on the imaging data provised by the Voyager mission (14-15).

II. MIMAS AND ENCELADUS

These satellites, with radii of 195 (Mimas) and 255 km (Enceladus), are the first couple of solar system bodies in this size range for which detailed images are available (notice, for instance, that the few largest asteroids after Ceres have comparable sizes). Their shape is definitely 'regular', with topographic features (craters, peaks, ridges, etc.) not higher than 5 to 10 km for Mimas and 1 to 2 km for Enceladus. Mimas presents a remarkable giant crater (130 km across) whose bowl-shaped bottom with a large central peak is clearly unrelaxed to the shape of the surrounding surface, indicating that no viscous flow has occurred on this satellite (a similar giant crater on Tethys has been largely flattened, and this is consistent with the relaxation timescales shown in Table 1 provided Mimas' surface has remained 'cold' since the formation of the crater). Mimas' topography implies a lower bound for the strength of about 0.5 to 1 x 10^7 dynes/cm^2 ; this lower bound is probably close to the actual strength, because it seems plausible that the main factor in limiting the impact-generated topography has been the failure of the crust once gravitational loads get too large. The same conclusion is not valid for Enceladus, whose thermal history has clearly affected the surface features and reduced the topographic relief. At any rate, even if thermal mechanisms have reduced the topography in the

past, this should cause a "better" fit to the equilibrium
figures ; thus, there can be no doubt that the global
shape of both the satellites has been molded by gravitational forces, due to the bodies themselves and (to a
lesser degree) to Saturn.

    If the satellites had nearly homogeneous inte-
riors, the equilibrium figures would be Roche ellipsoids (Chandrasekhar, (1) Ch. 8) ; using the densities
derived by Tyler et al. (16) from the Voyager radio
tracking data, i. e., 1.44 $\pm$ 0.18 g/cm³ for Mimas and
1.13 $\pm$ 0.55 g/cm³ for Enceladus, the equilibrium theory yields an (a-c) difference of 15 $\pm$ 2 km and $11^{-3.5}_{+10.5}$
km for Mimas and Enceladus, respectively (the large uncertainty of the latter value is due to the poorly determined density of Enceladus). The corresponding values of (b-c) are too small for a meaningful comparison with the Voyager images, because of their limited
resolution and of the 'noise' due to topography. However, the Roche theory is inadequate if the bodies are
differentiated, with a denser silicate core and an icy
mantle ; in this case, the tidal deformations with respect to sphericity are smaller than in the homogeneous
case. The differences between semi-axes are reduced by
a factor H which can be computed by applying Dermott's
(17) theory. In a previous paper (18) we have analyzed
different models for the interior of the satellites,
and concluded that H can be as small as 0.7 if a dense
rocky core of radius more than 1/2 of the total radius
is present.

    The Voyager images allow us to compare the pre-
dicted (a-c) values with the actual shapes of the satellites. A systematic fit of a geodetic network of 'control points' on the surface (19) has given (a-c) =
6 $\pm$ 3 km for Mimas and 8 $\pm$ 5 km for Enceladus. The former value leads to the surprising conclusion that Mimas'
shape is clearly (at the 3-sigma level) less elongated
than predicted by the Roche theory. This cannot be due
to 'irregular' topography sustained by the strength of
the surface layer, which in any case could only explain
an irregular shape but not a surface closer to sphericity. Neither can this problem be resolved by the tidal
evolution of Mimas' orbit, because its semimajor axis is
presently increasing, and thus if the 'freezing' of the
surface shape occurred a long time ago, the tidal distortion would be larger than that corresponding to the
present orbital distance (and not smaller as is observed). The unescapable conclusion appears to be that
Mimas has a differentiated core-mantle structure, cau-

sing a H value close to 0.7 and hence an (a-c) value of
about 10 km. This is higher than Davies and Katayama
'nominal' value, but consistent with their stated uncertainty. We remark that a differentiated structure for
Mimas has far-reaching implications on the formation
process and/or thermal history of this satellite, whose
discussion is outside the purposes of this paper (20).

　　　　For Enceladus the conclusions are much less
sharp. Enceladus should be expected to fit the equilibrium shape even better than Mimas, because of its lower topography and because of the extensive resurfacing
which indicates important thermal evolution. Unfortunately for Enceladus neither the density nor (a-c) have
been determined with great accuracy ; therefore, we can
only state that the available data is consistent with a
homogeneous internal structure, although the presence of
a core (possibly smaller and/or lighter than that of
Mimas) cannot be ruled out.

III. HYPERION

　　　　This satellite probably had a peculiar collisional history connected with its orbital resonance locking with Titan, as discussed in detail by Farinella
et al. (21). Voyager observations give a $\simeq$ 205 km,
b $\simeq$ 130 km and c $\simeq$ 110 km, with a 'mean' radius $(abc)^{1/3}$
of about 143 km. The strongly irregular shape suggests
clearly a catastrophic impact event (or sequence of
events) : Hyperion is probably a piece of the core of a
larger primordial body, collisionally disrupted with no
subsequent reaccumulation of the ejected fragments
(which possibly contributed to the cratering record of
the inner satellites). This hypothesis is supported by
the fact that Hyperion's axial ratios (c/a $\simeq$ 0.54,
b/a $\simeq$ 0.63) are typical for fragments produced in catastrophic collisions, as shown by the laboratory experiments (see Sect. 1). The irregular shape implies that
the least strength consistent with Eq. (1) is fairly
high (S $\underset{\sim}{>}$ 4 x 10^7 dynes/cm^2 for h $\simeq$ 50 km) and that no
significant creep has occurred since Hyperion's forma-
tion ; on the basis of Eqs. (4) and (5), this means that
the satellite's temperature has never exceeded $\simeq$ 100° K.
The high strength and the relatively low geometric albedo ($\simeq$ 0.3, compared with $\simeq$ 0.6 for other icy satellites) could suggest that Hyperion contains a substantial fraction of rocky material ; however, the overall
shape of the body should be controlled by the strength
of the weakest component (ice).

IV. PHOEBE

This small satellite (R ≃ 110 km) has been imaged at low resolution by Voyager 2, and displays a number of interesting properties : (a) a retrograde, eccentric and inclined orbit at very large planetocentric distance ; (b) a very dark surface (mean albedo of ∿ 0.06) with strong albedo markings, corresponding to a lightcurve amplitude of ∿ 0.2 mag ; (c) a spin period of about 9 hr, not synchronized with the orbital period ; (d) a nearly spherical, fairly regular shape. All these properties make plausible the hypothesis that Phoebe is a captured, asteroid-like object (we recall that light-curve data suggest that most asteroids with R > 100 km have nearly equilibrium figures : see ref. 8). Indeed, several C-type main-belt asteroids are similar to Phoebe both in size and in rotational properties. We remark that such a distant and non-corotating satellite should present an equilibrium figure determined only by the balance between self-gravitational and centrifugal forces (with no significant tidal deformation). Provided the interior is nearly homogenous, this figure must be a Maclaurin axisymmetric and oblate ellipsoid (see Chandrasekhar, (1) Ch. 5). Unfortunately ellipsoidal fits to the actual shape of Phoebe are not yet available ; since the rotational period is known, even a rough determination of c/a might allow one to discriminate between icy (low density) and rocky (high density) compositions. For ρ = 1.0 and 2.5 g/cm^3, the equilibrium values of c/a are 0.82 and 0.93, respectively. Hence the shape data could provide an intersting clue about the origin of this puzzling object.

V. F-RING SHEPHERDING AND COORBITAL SATELLITES

These four small satellites represent a sample of icy bodies having mean radii in the range from 50 to 100 km, that is just in the range where the transition from completely irregular to nearly equilibrium shapes has to begin for S ∿ 10^7 dynes/cm^2. We note also that regular shapes should be expected if any of these satellites has experienced collisional fragmentation followed by reaccretion at low relative velocities, as assumed by Smith et al. (15) for the major inner satellites of Saturn's system. This could explain the contrast with respect to Hyperion's irregular shape, because for Hyperion no reaccumulation was possible (21) ; another irregular object of similar size is probably the Jovian moon Amalthea (10), whose rocky constitution presumably allows a higher strength. In Table 2, we have tried to compare the observed axial ratios (with uncertainties corresponding to the resolution of the Voyager images) with the axial ratios of suitable Roche ellipsoids, chosen by varying the density as a free parameter ; the densities giving a reasonable fit between observed and 'theoretical' shapes are also indicated.

T A B L E   2

Satellite	$(c/a)_{obs}$ $(b/a)_{obs}$	$(c/a)_{Roche}$ $(b/a)_{Roche}$	ρ_{Roche} (g/cm^3)
1980 S27	0.53 ± 0.15 0.71 ± 0.15	0.61 0.64	0.82
1980 S26	0.60 ± 0.18 0.76 ± 0.20	0.67 0.71	0.85
Janus	0.73 ± 0.08 0.86 ± 0.08	0.81 0.84	1.09
Epimetheus	0.71 ± 0.15 0.81 ± 0.19	0.81 0.84	1.09

It is not easy on the basis of the available data to decide whether the fit is significant and the satellites are, at least roughly, shaped as Roche ellipsoids, or they are simply irregular bodies, possibly formed as fragments from catastrophic collisions with no subsequent reaccretion. The coorbitals, however, which are the largest bodies in the sample with mean radii of about 95 and 60 km, show several features favouring the hypothesis of quasi-equilibrium shapes : (a) a rather smooth general appearance, contrasting with Hyperion and with the smaller satellites (Atlas, Telesto, Calypso, etc.), which are just tens of km across ; (b) Janus and Epimetheus, at the same orbital distance, have similar shapes : this is an obvious consequence of the Roche model, while otherwise it would be due only to chance for two irregular fragments ; moreover, the observed shapes are significantly less elongated than typical fragments ; (c) the density derived by the Roche model (~ 1.1 g/cm^3) is very plausible for an icy body, and is just in the range of the observed values for most larger satellites

(16) : this would be again a coincidence for bodies with
irregular shapes. If the coorbitals have really nearly
equilibrium figures, this would imply that the strength
of the icy material forming the inner satellites of Saturn is not higher than 10^6 to 10^7 dynes/cm^2, in good
agreement with the estimate based on the surface relief
of Mimas. For the two F-ring shepherding satellites, on
the other hand, we favour the idea of irregular shapes,
because their sizes are smaller (with mean radii of
about 40 and 50 km) and their shapes are much more frag-
ment-like.

In conclusion, the shape data on the small satellites of Saturn suggest that the transition from irregular, strength-dominated objects to gravity-dominated
bodies with nearly equilibrium shapes occurs in the range of R between $\sim$ 50 and $\sim$ 150 km, the same range in
which the transition appears to occur among asteroids
(7) ; the corresponding material strength is on the order of 10^7 dynes/cm^2, with perhaps a factor 10 of variability.

ACKNOWLEDGMENTS

We thank J. A. Burns and J. Veverka for helpful
discussions and suggestions. This work was partially
supported by the National Research Council of Italy
(C. N. R.).

REFERENCES

1.   Chandrasekhar, S., (1969). *Ellipsoidal Figures of
     Equilibrium*. Yale University Press, New Haven and
     London.

2.   Fujiwara, A., Kamimoto, G. and Tsukamoto, A. (1978).
     Expected shape distribution of asteroids obtained
     from laboratory impact experiments. Nature 272,
     pp. 602-603.

3.   Lange, M. A. and Ahrens, T. J. (1982). Impact frag-
     mentation of ice-silicate bodies. Lunar Planet.
     Sci. XIII, pp. 417-418.

4.   Capaccioni, F., Cerroni, P., Coradini, M., Farinel-
     la, P., Flamini, E., Martelli, G., Paolicchi, P.
     and Zappalà, V. (1984). On the shape of collisio-
     nal fragments and asteroids. Nature 308, pp. 832-
     834.

5. Johnson, T. V. and McGetchin, T. R. (1973). Topo-
    graphy on satellite surfaces and the shape of as-
    teroids. Icarus 18, pp. 612-620.

6. Farinella, P. Ferrini, F., Milani, A., Nobili, A.
    M., Paolicchi, P. and Zappalà, V. (1982). The sha-
    pe of small solar system bodies : Gravitational
    equilibrium vs. solid-state interactions. In *The
    Comparative Study of the Planets* (A. Coradini and
    M. Fulchignoni, Eds.), Reidel, Dordrecht, pp. 71-
    77.

7. Farinella, P., Paolicchi, P. and Zappalà, V. (1982).
    The asteroids as outcomes of catastrophic colli-
    sions. Icarus 52, pp. 409-433.

8. Catullo, V., Zappalà, V., Farinella, P. and Paolic-
    chi, P. (1984). Analysis of the shape distribu-
    tion of asteroids. Astron. Astrophys., in press.

9. Soter, S, and Harris, A. W. (1977). The equilibrium
    figures of Phobos and other small bodies. Icarus
    30, pp. 192-199.

10. Ferrini, F., Milani, A. and Nobili, A. M. (1981).
    On the shape of Amalthea. Advances Space Res. 1,
    pp. 191-197.

11. Weertman, J. (1970). The creep strength of the
    Earth's mantle. Rev. Geophys. Space Phys. 8, pp.
    145-168.

12. Pounder, E. R. (1965). *The Physics of Ice*. Perga-
    mon Press, Elmsford, New York.

13. Morrison, D. and Burns, J. A. (1976). The Jovian
    satellites. In *Jupiter* (T. Gehrels, Ed.). Univer-
    sity Of Arizona Press, Tucson, pp. 991-1034.

14. Smith, B. A., and 26 other authors (1981). Encoun-
    ter with Saturn : Voyager 1 imaging science re-
    sults. Science 212, pp. 163-191.

15. Smith, B. A., and 28 other authors (1982). A new
    look at the Saturn system : The Voyager 2 images.
    Science 215, pp. 504-537.

16. Tyler, G. L., Eshleman, V. R., Anderson, J. D., Le-
    vy, G. S., Lindal, G. F., Wood, G. E. and Croft,

T. A. (1982). Radio science with Voyager 2 at Saturn : Atmosphere and ionosphere and the masses of
Mimas, Tethys and Iapetus. Science 215, pp. 553-
558.

17.   Dermott, S. F. (1979). Shapes and gravitational moments of satellites and asteroids. Icarus 37, pp.
575-586.

18.   Farinella, P., Milani, A., Nobili, A. M., Paolicchi,
P. and Zappalà, V. (1984). The shape and structure
of Mimas. Earth, Moon, and Planets 30, pp. 39-42.

19.   Davies, M. E. and Katayama, F. J. (1982). The control networks of Mimas and Enceladus. Icarus 53,
pp. 332-338.

20.   Consolmagno, G. L. and Lewis, J. S. (1977). Preliminary thermal history models of icy satellites.
In *Planetary Satellites* (J. A. Burns, Ed.). University of Arizona Press, Tucson, pp. 492-500.

21.   Farinella, P., Milani, A., Nobili, A. M., Paolicchi,
P. and Zappalà, V. (1983). Hyperion : Collisional
disruption of a resonant satellite. Icarus 54,
pp. 353-360.

ICY SATELLITES OF URANUS

Robert Hamilton Brown

Planetary Geosciences Division
Hawaii Institute of Geophysics
University of Hawaii
2525 Correa Road
Honolulu, Hawaii 96822

Recent work on the satellites of Uranus has revealed many of their basic physical properties. Radiometric measurements have shown that the Ariel, Umbriel, Titania and Oberon have diameters which range from 1630 to 1110 km and albedos which range from 0.30 to 0.18. Spectrophotometric observations of Miranda suggest that it may have the highest albedo of the known Uranian satellites and a diameter of about 500 km. Near-infrared measurements show that Ariel, Titania and Oberon have the largest known opposition surges. All five known satellites of Uranus have surfaces which are composed of water ice contaminated with small amounts of dark material. The dark material on the surfaces of Ariel, Umbriel, Titania and Oberon is spectrally bland and has spectral similarities to carbon black, charcoal, carbonaceous chondritic material and other dark, spectrally neutral materials. Recent density determinations suggest that there may be large density differences among Ariel, Umbriel, Titania and Oberon, with density increasing with distance from Uranus.

INTRODUCTION

Rapid advancement in the technology of electro-optical detector systems, especially detectors optimized for the near-infrared, have resulted in useful, groundbased studies of the physical

711

J. Klinger et al. (eds.), Ices in the Solar System, 711–729.
© 1985 by D. Reidel Publishing Company.

properties of the satellites of Uranus. The Uranian satellite
system is an interesting system for many reasons, not the least of
which being the unusual orientation of its angular momentum and
the regularity of the orbits of the five known satellites. The
satellites of Uranus also comprise the most distant of the regular
satellite systems and as such may have formed under much different
conditions than their warmer counterparts in the Jovian and Saturnian systems. All five known satellites have orbits that fit the
criteria for a regular system. In order of distance from Uranus,
the five satellites are U5 Miranda, U1 Ariel, U2 Umbriel, U3
Titania and U4 Oberon.

One dynamical aspect of the Uranian system is particularly
interesting, that being the inclination of the rotation axis of
Uranus with respect to the plane of its orbit (and also to the
plane of the ecliptic). Uranus' axial inclination of 98° with
respect to the ecliptic pole is the one of the most extreme examples of axial tilt among the planets. It has been suggested that
the origin and evolution of the Uranian satellites may have been
strongly affected by the events which are responsible for the
observed axial tilt (1). The 1984 aspect of the Uranian system as
seen from the Earth is essentially polar and this has simplified
the derivation of some orbital parameters from astrometric observations, as well as simplifying other observations such as photometry. The polar aspect does, however, frustrate groundbased
observations which would search for such properties as albedo
asymmetries with respect to the leading and trailing sides of the
Uranian satellites.

Because the initial groundbased reconnaissance of the Uranian
satellites is reasonably mature, a review of their physical properties is timely. Specific topics discussed for the bodies in
this paper will be size, surface compositions, photometric properties and densities, as well as problems for further study from
both the ground and spacecraft. A compilation of the known properties of the satellites of Uranus prior to 1982 has been published
by Cruikshank (2), so this paper will concentrate primarily on
work done since the Cruikshank review.

ORBITAL PROPERTIES

Studies of the dynamics of satellite systems have three general goals: the first is a complete characterization of the orbital properties of the satellites in order to accurately predict
their positions, the second is a determination of the mass and the
gravitational moments of the central body from the observed orbital parameters and the third is to determine the satellite masses
from observation of those orbital parameters which are modified by
the gravitational interactions of the individual satellites.
Several studies have been made of the dynamics of the Uranian
satellite system and thorough discussions of the theory and

observations prior to 1979 have been published by Greenberg (33–
35). An important study of the dynamical properties of the
Uranian satellites has recently been completed by Veillet (24) in
which all the astrometric observations of the Uranian satellites
since their discovery have been analyzed. Veillet's work
represents the most complete characterization of the orbital
parameters of the five Uranian satellites to date, and from precise measurements of their mutual perturbations he has determined
the masses of Miranda, Ariel, Umbriel, Titania and Oberon. From
Table I, which is adapted from Veillet's work, it can be seen that
the orbits of the Uranian satellites are regular. All the satellites' orbits have low eccentricities and all but that of Miranda
have essentially zero inclination. In one sense the Uranian system seems to be the most regular satellite system in the solar
system because all its known members have approximately circular,
low-inclination orbits. This is in contrast to the Jovian and
Saturnian satellite systems which have a few members with highly
irregular orbits.

Table I: Orbital Properties of the Uranian Satellites			
	P (days)	A (km)	e	i (deg)
Miranda	1.41347925 (14)	129390 (135)	0.0027 (6)	4.22 (16)
Ariel	2.52037935 (10)	191020 (90)	0.0034 (3)	0.31 (11)
Umbriel	4.1441772 (2)	266300 (90)	0.0050 (3)	0.36 (8)
Titania	8.7058717 (3)	435920 (75)	0.0022 (1)	0.142 (31)
Oberon	13.4622389 (5)	583530 (60)	0.0008 (1)	0.101 (24)

Table I: Orbital properties of the Uranian satellites. P is the
semi-major axis in km, A is the siderial period in days, e is the
eccentricity and i is the inclination of the satellite orbits in
degrees. Uncertainties (one standard deviation of the mean) in the
last 1 or 2 digits are given in parentheses for the siderial
periods. All data in this table are adapted from Veillet (24).

SURFACE COMPOSITIONS

Most of what is known about the surface compositions of
satellites in the outer solar system is derived from observations
of their reflectance spectra. Absorption features characteristic
of the surface mineralogy of a planetary body can be observed in
the entire region of the solar spectrum where there is detectable
reflected light (~0.1 to 5 μm). Observations of the reflectance of
icy bodies are particularly diagnostic of surface composition in

the near infrared where several cosmochemically important molecules (e.g. H_2O, NH_3 and CH_4) have vibrational transitions which result in absorptions seen in spectra of their diffuse reflectance. This technique has been applied to the Uranian satellites by many researchers and has resulted in a reasonable characterization of their surface compositions.

The first study of the near-infrared spectral reflectance of Titania and Oberon was published by Cruikshank (3). In his spectra appear absorptions at 1.5 and 2.0 μm characteristic of the presence of water ice or frost on the surfaces of Titania and Oberon. In a follow-up study of Ariel and Umbriel, Cruikshank and Brown (4) also found water-ice absorptions in the near-infrared reflectance spectra of these satellites. Due to its faintness and proximity to Uranus, reflectance spectra of Miranda in the near-infrared have been extremely difficult to obtain using available telescopes and detector systems. Nevertheless, Brown and Clark (5) succeeded in obtaining a spectrum of Miranda in the 1.6-2.4 μm spectral region which clearly shows a deep absorption at 2.0 μm characteristic of water ice. The spectrum of Miranda from Brown and Clark overlaid with a spectrum of water frost from Clark (6) is displayed in Figure 1. It is now clear that all five Uranian satellites have water-ice surfaces.

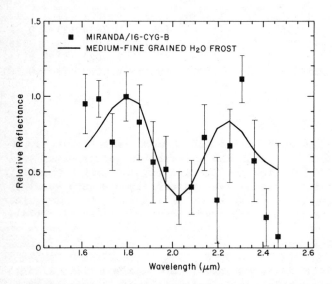

Fig. 1. Plotted is the reflectance spectrum of Miranda from Brown and Clark(5). The Miranda spectrum is normalized to 1.0 at 1.79 μm and is overlaid with a spectrum of fine-grained water frost from Clark (6). The laboratory frost spectrum has been convolved to the resolution of the Miranda spectrum and similarly normalized.

Soifer et al.(7) obtained reflectance spectra of Umbriel, Titania and Oberon which confirmed the discovery of water ice and raised the possibility that these satellites had lower albedo than are characteristic of pure water-ice surfaces heavily gardened by meteoritic infall. That some of the Uranian satellites have relatively low albedos was firmly established by the radiometric measurements of Brown et al.(8) who found the visual geometric albedos of Ariel, Umbriel, Titania and Oberon to be roughly intermediate to the visual geometric albedo range for most solar system bodies (~0.03 to 0.7). This does not seem to be the case for Miranda, however, because the depth of the 2.0 μm absorption in its reflectance spectrum implies that the water ice on its surface is relatively pure (5). This suggests that Miranda's albedo is near the upper end of the range (Table II) and that its surface may be similar to those of some of the icy satellites of Saturn. With their p_v range of 0.2 to 0.3, the nearest albedo analogs for the large Uranian satellites among other icy satellites are Callisto and Hyperion whose p_v are 0.19 and 0.28 respectively (9,10). As we shall see below, Hyperion may be similar to Ariel in surface composition as well as albedo.

Because the relatively low albedos of the Uranian satellites suggest the presence of a dark contaminant either on or in their surfaces, some recent studies of the near-infrared reflectance of the Uranian satellites have concentrated on its identification. Brown and Cruikshank (11) and Brown (12,19) have obtained reflectance spectra of Ariel, Umbriel, Titania and Oberon in the 0.8- to 2.6-μm spectral region which indicate that the non-water component of the surfaces of these satellites has a relatively bland reflectance spectrum. They have further noted that the non-water component has spectral similarities to substances such as charcoal, carbon black, carbonaceous chondritic material and other neutrally colored, low-reflectance materials. Composite spectra of the Uranian satellites are displayed in Figure 2. All spectra in Fig. 2 show the strong H_2O absorption at 2.0 μm and some show the 1.5 μm absorption as well. A laboratory spectrum of a sample of fine-grained water frost obtained by Clark (6) is displayed in Figure 3 to illustrate this point.

Brown (12) has investigated areal mixtures of water frost with isolated patches of dark, opaque, spectrally neutral material as potential analogs for the surfaces of Ariel, Umbriel, Titania, Oberon and Hyperion. Some of his results are displayed in Figure 4. The simulated spectra were constructed by the linear superposition of two laboratory spectra: that of fine-grained water frost, and that of an intimate mixture of 30 wt % charcoal and 70 wt % water ice. Reasonable matches constructed using this method demonstrate the consistency of the Uranian satellite spectra with areal mixtures of water frost and isolated patches of dark spectrally neutral materials. The spectral matches shown in Fig. 4 are,

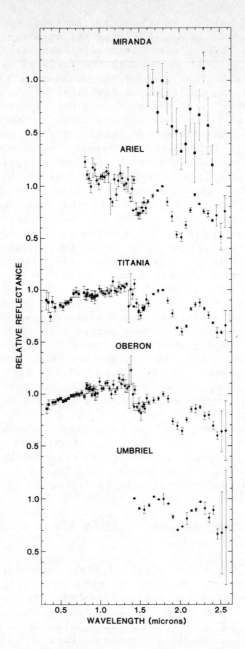

Figure 2: Plotted are composite spectra, from several sources, of the five satellites of Uranus (5,11,12,32). All spectra are normalized to 1.0 near 1.8 μm.

of course, not unique, but serve to demonstrate a large class of
spectral analogs capable of approximately matching the depths of
the absorption bands and the overall reflectance level of the
Uranian satellite spectra. It is not possible, however, to determine conclusively using available data whether the dominant state
of dispersal of the dark component of the Uranian satellite surfaces is voluminal or areal.

While looking for spectral analogs for the surfaces of the
Uranian satellites, Brown and Cruikshank (11) noticed that the
0.8- to 2.6-μm reflectance spectrum of Saturn's satellite Hyperion
(10) is very similar to that of Ariel. This is demonstrated in
Figure 5 where the 1.5- to 2.6-μm spectra of Ariel and Hyperion
have been normalized to 1.0 at 1.79 μm and overlaid. The spectral
similarity extends to shorter wavelengths as well (Fig. 4) and is
supported by the similarity of the two bodies visual geometric
albedos (0.30 for Ariel and 0.28 for Hyperion). The exact reason
for this spectral similarity is not clear, but it may result from
a similarity in the distribution and spectral characteristics of
the dark components contaminating the water ice on their surfaces
(12). A problem with this interpretation is the fact that the
spectrum of Hyperion is redder than those of Titania and Oberon in
the spectral region 0.3 to 0.8 μm. This would suggest that the
dark component of Hyperion's surface is different from that of
Ariel, or that Ariel's dark surface component is different from
that of the other Uranian satellites.

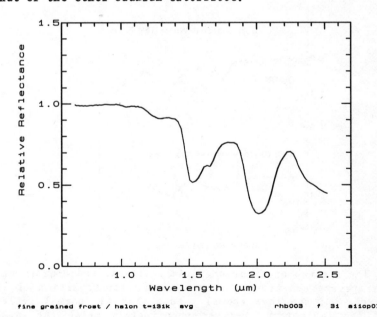

fine grained frost / halon t=131k avg   rhb003   f   31   a11op013n

Figure 3: A spectrum of fine-grained water frost from Clark (6).

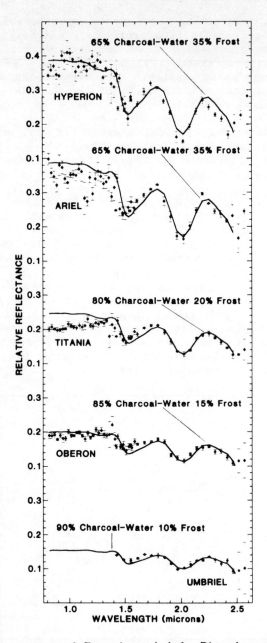

Figure 4: The spectra of Hyperion, Ariel, Titania and Oberon over-laid with spectra constructed from a linear superposition of lab spectra (ref. 12; see text above). Note that the normal reflec-tances of the laboratory spectra approximately match the satel-lites geometric albedos.

Many theories of the formation of bodies in the outer solar system predict the incorporation of volatiles such as ammonia, methane and carbon monoxide into the surfaces of the Uranian satellites (13-18,31). Conclusive evidence for the presence of such volatiles in the Uranian satellite surfaces has not yet been uncovered, but there are some interesting features of the reflectance spectra of Ariel and Hyperion which might result from the presence of NH_3, CH_4 or CO. The spectral feature in question is subtle, but amounts to a depression of the continuum at 2.25 μm in the Ariel and Hyperion spectra relative to that of pure water frost (see Figs. 3, 4 and 5). The effect results in the spectra of Ariel and Hyperion peaking at 2.19 μm instead of at 2.25 μm as is normal for pure water ice or a mixture of water ice and small amounts of spectrally neutral material. Spectra of Ariel obtained during the 1983 apparition of Uranus (Brown and Clark, unpublished) confirm the spectral feature seen in the data of Brown and Cruikshank (11). Ammonia, methane and carbon monoxide all have strong absorptions in the 2.2- to 2.3-μm region, but the low-resolution and precision of the existing data prevents a positive identification of which of these compounds, if any, is present on the surfaces of Ariel or Hyperion.

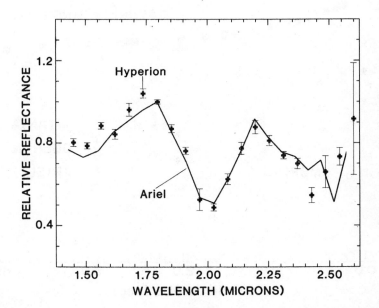

Figure 5: Plotted are the spectra of Ariel and Hyperion, both normalized to 1.0 at 1.79 μm. The data are from Brown and Cruikshank (11). To facilitate comparison, no error bars are shown for the Ariel spectrum though they are comparable in size to those of the Hyperion spectrum.

OPPOSITION SURGES

A non-linear increase in logarithmic brightness approaching zero-degrees solar phase angle (opposition surge) has been observed to exist for many solar system bodies, but recent observations indicate that the Uranian satellites opposition surges are unusual. The near-infrared opposition brightness surges of the Uranian satellites are the largest known over the 3° of solar phase angle which can be observed from Earth (11,19). In Figure 6 are displayed the near-infrared brightness versus solar phase angle of Ariel, Umbriel, Titania and Oberon as well as broadband visual data on Saturn's rings, which, until recently, had the largest known opposition surges. As can be seen from Figure 6, the opposition surges of at least three of the Uranian satellites is 0.5 mag or more.

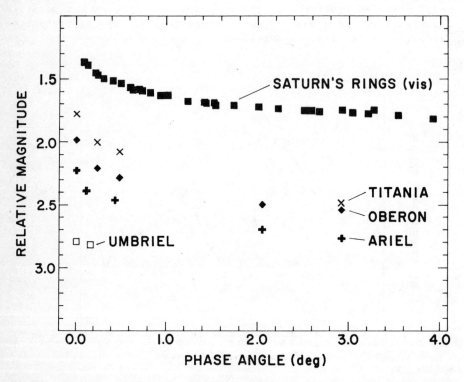

Figure 6: Near-infrared opposition brightness surges of Ariel, Umbriel, Titania and Oberon. The data are from Brown and Cruikshank (11). Also plotted are the visual opposition surge data for Saturn's rings from Franklin and Cook (30). The data for the Uranian satellites contain an arbitrary offset to facilitate comparison to the data for Saturn's rings.

A follow-up study of the broadband visual (V) opposition surges of the Uranian satellites (20) tentatively shows comparable results, but more observations are required to fully characterize the visual phase curves. It is not clear what surface properties of the Uranian satellites might be responsible for the large opposition surges, but some work points to surfaces composed of particles with highly back-scattering phase functions (21). Hapke (22) has been able to model the very large and narrow opposition surges of the Uranian satellites by requiring that the density of the surface layer of scattering particles increase with depth from near zero to close packed over a distance of about 30 times the mean particle size. Veverka and Gradie (23) argue that grain size distribution, grain shape and packing texture are some of the several mechanisms that can be invoked to explain why the surfaces of icy satellites, in general, are not lambert scatterers.

RADII AND DENSITIES

Mean density deduced from measurements of mass and radius is an important physical property to determine in the study of the origin of a satellite system. Until recently, however, this quantity could only be estimated for the satellites of Uranus. The distance from Earth and faintness of the Uranian satellites makes direct determination of their radii nearly impossible using groundbased techniques. For the same reasons, observations from which accurate masses of the Uranian satellites can be derived are very difficult. Nevertheless, recent studies by Brown et al.(8), Brown and Clark (5), and Veillet (24) have made considerable progress in the determination of the densities of the satellites of Uranus. Brown et al.(8) have determined the radii of Ariel, Umbriel, Titania and Oberon by measuring the 20-μm thermal fluxes from the satellites and combining them with measurements of the satellites' broadband visual fluxes. These measurements were then incorporated into a version of the standard radiometric model (25-27), from which were deduced the radius and albedo for each satellite (Table II).

Because the photometric/radiometric technique is a model dependent determination, it is useful to digress slightly to discuss the limits of validity of the Brown et al.(8) measurements. The radiometric technique is model dependent in general and, for the specific case of the Uranian satellite diameter measurements, values were assumed for some of their photometric properties. The phase integrals of the Uranian satellites have never been determined because only $\pm$ 3° of solar phase angle can be covered by observations from the ground. Therefore, Brown et al. assumed phase integrals for each of the four large Uranian satellites by analogy with other icy satellites of similar albedo whose phase integrals are known. Fortunately, the effect of a rather large

Table II: Radii, Masses, Densities and Albedos			
	radius (km)	mass (10^{23} g)	density (g cm^{-3})	p_v
Miranda	250 ± 110	1.7 ± 1.7	~ 3	~ 0.5
Ariel	665 ± 65	15.6 ± 3.5	1.3 ± 0.5	0.30 ± 0.06
Umbriel	555 ± 50	10.0 ± 4.2	1.4 ± 0.6	0.19 ± 0.04
Titania	800 ± 60	59.0 ± 7.0	2.7 ± 0.6	0.23 ± 0.04
Oberon	815 ± 70	60.0 ± 7.0	2.6 ± 0.6	0.18 ± 0.04

Table II: Masses and densities are drawn from Veillet (24), radius and visual geometric albedo (p_v) are drawn from Brown et al.(8), except the radius and albedo of Miranda which are drawn from Brown and Clark (5).

uncertainty in the phase integral on the derived radius is quite small for the case of the Uranian satellites (e.g. the assumption of 0.8 ± 0.1 for the phase integral of Ariel results in a change in the derived radius of ± 3%). Brown et al. also found it necessary to assume a value of V(1,0) for each of the Uranian satellites from measurements of their brightness made at a solar phase angle of 0.01° (V(1,0) is a linear extrapolation of visual magnitude to 0° solar phase angle from phase angles greater than 6° with normalization to a distance of 1 AU from Earth and Sun). This was necessary because the standard linear extrapolation to zero—phase brightness can not be made from groundbased measurements. The uncertainty in V(1,0) of ± 0.2 mag specifically used by Brown et al., translates into about a ± 5% uncertainty in derived radius. This is not the case for the derived albedos because they depend almost entirely on the amount of light reflected by the body being measured. Therefore, rather large systematic errors in V(1,0) can translate into equally large errors in derived albedo. An additional uncertainty in the radiometric radii of the Uranian satellites derives from the fact that the model is based on monochromatic fluxes derived from broadband measurements. The derivation of monochromatic fluxes requires a knowledge of the atmospheric transmission function above the observing site as well as the convolved filter transmission and instrument response function. For the Brown et al. measurements, atmospheric transmission was determined from measurements of the column density of H_2O derived from observations of the 1.4-μm telluric absorption obtained contemporaneously with the 20-μm thermal measurements. Uncertainties of ± 0.1 mag in the correction of broadband fluxes to monochromatic fluxes were folded into the results of Brown et al. to maintain conservative error estimates.

Perhaps the greatest potential problem with any radius

derived using the radiometric technique is one basic assumption--
that the surface of the body is in instantaneous thermal equilibrium with absorbed insolation. If that assumption is seriously
violated (i.e. the entire surface of the body is isothermal), it
is possible to underestimate the radius of the body by as much as
a factor of root two. This is not likely to be a problem with the
Brown et al. measurements because the Uranian satellites are
essentially pole-on with respect to the Sun (unless, of course,
the bulk composition of the satellites is a material with very
high thermal conductivity or they all have substantial
atmospheres--neither of which are likely). Though it is possible
that there may be some residual systematic errors in the Brown et
al. radii, they are not likely to be larger than the quoted random
errors unless the surfaces of the Uranian satellites have highly
unusual thermal properties.

As can be seen from Table II, the satellites of Uranus are
comparable in size to the largest of Saturn's icy satellites
Dione, Iapetus and Rhea which have diameters of 1120, 1460 and
1530 km respectively (28). They are therefore among the largest
satellites in the solar system, but are considerably smaller than
the giant satellites Ganymede, Callisto, and Titan whose diameters
are approximately 5000 km.

As was mentioned previously, Veillet (24) has derived masses
for all the Uranian satellites from observations of their mutual
orbital perturbations. He has combined six years of his own observations of the Uranian satellites with all other astrometric
observations made since the discoveries of the satellites,
redetermining all their orbital parameters and mutual perturbations. Combining his mass measurements with the radii measured by
Brown et al.(8), Veillet has derived densities for Ariel, Umbriel,
Titania and Oberon. The masses of all the Uranian satellites as
well as their densities and albedos are listed in Table III (The
density of Miranda listed in Table II was derived from the diameter estimate of Brown and Clark (5) and the mass estimate from
Veillet). As can be seen from Table II, the Uranian satellites
seem to form two distinct density groups. Given that all the
Uranian satellites are known to have water-ice surfaces, and that
the densities of Ariel and Umbriel are comparable to those of
several of Saturn's satellites whose bulk compositions are thought
to be about 40% silicates and 60% water ice by weight, one can
conclude that the bulk compositions of Ariel and Umbriel may be
similar. In contrast, Titania and Oberon have densities which
indicate that much larger fractions of their bulk compositions are
of high density materials such as silicates. Since they are large
enough to have undergone melting and at least partial differentiation (29), densities of 2.6 to 2.7 g cm^{-3} for Titania and Oberon
suggest that they may have a large core which is mostly rock with
a skin of water ice composing their crusts and mantles. If the
apparent density variations are real, then we have a system whose

density gradient is opposite to that which one would expect if
primordial heat from the accretion and contraction of Uranus
determined what materials were available for incorporation into
the satellites. This might be seen to favor origins of the
Uranian satellites which are closely connected to the catastrophic
events hypothesized to be responsible for the present axial orientation of the Uranian system (1). Nevertheless, large uncertainties in the densities of the Uranian satellites council restraint
with regard to speculations about the satellites' origins based on
their apparent density differences. Further thoughts regarding
the origins of the Uranian satellites can be found in Stevenson et
al.(18).

SUMMARY

The Uranian satellites comprise a system of five regular
satellites, all having water ice surfaces of varying degrees of
purity. Ariel, Umbriel, Titania and Oberon have opposition surges
which are among the largest in the solar system and have low
albedos relative to those typical of relatively pure, heavily gardened water-ice surfaces. Present with the water ice on the surfaces of Ariel, Umbriel, Titania and Oberon is a dark spectrally
neutral component which has spectral characteristics similar to
those of carbon black, charcoal, carbonaceous chondritic material,
and other neutrally colored, low-reflectance materials. Compounds
more volatile than water ice (e.g. methane, ammonia and carbon
monoxide) have not yet been conclusively shown to exist on any
surfaces in the Uranian system, though there are as yet unidentified spectral features in the spectrum of Ariel which might be the
result of the presence of the hydrate of one of these compounds.
The Uranian satellites are comparable in size to the largest of
Saturn's icy satellites while density measurements suggest that
the bulk compositions of Ariel and Umbriel are quite different
than those of Titania and Oberon. Ariel and Umbriel have densities
which are similar to the icy Saturnian satellites and may have
similar bulk compositions. Titania and Oberon have densities which
suggest that they have a much higher proportion of silicates or
other high-density materials than do Ariel and Umbriel. If the
apparent increase in satellite density with increase in distance
from Uranus is real, it may perhaps result from catastrophic
events hypothesized to be responsible for Uranus' axial orientation.

PROSPECTS FOR FURTHER STUDY

Certainly some of the most important data on the Uranian
satellites will be obtained when Voyager 2 arrives in January,
1986. As the trajectory is now planned, Voyager's closest

satellite encounter in the Uranian system will be with Miranda, resulting in the highest resolution imaging of the five satellites, and, despite the non-ideal trajectory (the spacecraft will pass through the Uranian system on a trajectory roughly perpendicular to Uranus' equatorial plane), good imaging should also be obtained for Ariel, Umbriel, Titania and Oberon. High-quality images will likely yield clues to the relative importance of endogenic and exogenic processes in the modification of the surfaces of the Uranian satellites and may also help determine the source of the satellites' low albedos. The prospects are encouraging for highly accurate measurement by Voyager of all five satellites' radii, with more accurate density measurements being an important result. The Voyager 2 spacecraft, however, is not well equipped for detailed remote sensing of the surface compositions of satellites and, until another spacecraft arrives at Uranus with the proper instrumentation, groundbased research will be the primary source of further information of the composition of the Uranian satellite surfaces. Groundbased studies of the surface compositions of the Uranian satellites have been difficult in the past, but the new technology of infrared array detectors promises to allow significant further progress. As was mentioned in the section on surface compositions, a particularly interesting problem with regard to the surface compositions of the Uranian satellites, as well as other icy satellites, concerns the presence or absence of hydrated and anhydrous ammonia (NH_3), methane (CH_4), nitrogen and carbon monoxide (CO). A search for these compounds in the surfaces of the Uranian satellites is important, because the result has profound consequences for our understanding of the origins of icy bodies within the context of currently accepted theories.

REFERENCES

1.   Singer, S. F., 1975. When and where were the sa-
      tellites of Uranus formed ? Icarus, 25, pp. 484-
      488.

2.   Cruikshank, D. P., 1982. The satellites of Uranus.
      In "Uranus and the Outer Planets", ed. G. E. Hunt,
      Cambridge University Press, pp. 193-216.

3.   Cruikshank, D. P., 1980. Near-infrared studies of
      the satellites of Saturn and Uranus. Icarus, 41,
      pp. 246-258.

4.   Cruikshank, D. P. and Brown, R. H., 1981. The Ura-
      nian satellites : Water ice on Ariel and Umbriel.
      Icarus, 45, pp. 607-611.

5.   Brown, R. H. and Clark, R. N., 1984. Surface of
      Miranda : Identification of water ice. Icarus, 58,
      pp. 288-292.

6.   Clark, R. N., 1980. Water frost and ice : The near-
      infrared spectral reflectance 0.65-2.5 microns.
      J. Geophys. Res., 86, pp. 3087-3096.

7.   Soifer, B. T., Neugebauer, G. and Matthews, K.,
      1981, Near-infrared spectrophotometry of the sa-
      tellites and rings of Uranus. Icarus, 45, pp. 612-
      617.

8.   Brown, R. H., Cruikshank, D. P. and Morrison, D.,
      1982. Diameters and albedos of satellites of Ura-
      nus. Nature, 300, pp. 423-425.

9.   Squyres, S. W. and Veverka, J., 1981. Voyager pho-
      tometry of surface features on Ganymede and Cal-
      listo. Icarus, 46, pp. 137-155.

10.   Cruikshank, D. P. and Brown, R. H., 1982. Surface
      composition and radius of Hyperion. Icarus, 50,
      pp. 82-87.

11.   Brown, R. H. and Cruikshank, D. P., 1983. The Ura-
      nian satellites : Surface compositions and opposi-
      tion brightness surges. Icarus, 55, pp. 83-92.

12.   Brown, R. H., 1983. The Uranian satellites and Hy-
      perion : New spectrophotometry and compositional

implications. Icarus, 56, pp. 414-425.

13. Cameron, A. G. W., 1973. Formation of the outer planets. Space Sci. Rev., 14, pp. 383-391.

14. Cameron, A. G. W., 1973. Elemental and isotopic abundances of the volatile elements in the outer planets. Space Sci. Rev., 14, pp. 392-400.

15. Lewis, J. S., 1971 . Satellites of the outer planets : Their physical and chemical nature. Icarus, 15, p. 174-185.

16. Lewis, J. S., 1972. Low temperature condensation from the solar nebula. Icarus, 16, pp. 241-252.

17. Lewis, J. S., 1973. Chemistry of the outer solar system. Space Sci. Rev., 14, pp. 401-411.

18. Stevenson, D. J., Harris, A. W. and Lunine, J. I., 1984. Origins of satellites. In "Natural Satellites", Chapter 2. University of Arizona Press, in preparation.

19. Brown, R. H., 1982. The satellites of Uranus : Spectrophotometric and radiometric studies of their surface properties and diameters. Ph. D. dissertation, University of Hawaii, Honolulu, Hawaii.

20. Goguen, J., Hammel, H. and Brown, R. H., 1984. V photometry of Titania, Oberon and Triton. Icarus, submitted.

21. Pang, K. D. and Rhoads, J. W., 1983. Interpretation of disk integrated photometry of the Uranian satellites, in "Natural Satellites", in preparation.

22. Hapke, B., 1983. The opposition effect. Bull. Amer. Astron. Soc., 15, pp. 856-857 (Abstract).

23. Veverka, J. and Gradie, J., 1983. Why don't icy satellites scatter like model snow-covered planets ? Bull. Amer. Astron. Soc., 15, p. 853 (Abstract).

24. Veillet, Ch., 1983. De l'observation et du mouvement des satellites d'Uranus. Ph. D. dissertation, University of Paris, Paris, France.

25. Morrison, D., 1973. Determination of radii of sa-
     tellites and asteroids from radiometry and phot-
     metry. Icarus, 19, pp. 1-14.

26. Morrison, D. and Lebofsky, L. A., 1979. Radiometry
     of asteroids. In "Asteroids", ed. T. Gehrels, Uni-
     versity of Arizona Press, pp. 184-205.

27. Brown, R. H., Morrison, D., Telesco, C. M. and
     Brunk, W. E., 1982. Calibration of the radiome-
     tric asteroid scale using occultation diameters.
     Icarus, 52, pp. 188-195.

28. Smith, B. A. et al., 1982. A new look at the Sa-
     turn system : The Voyager 2 images. Science ,
     215, pp. 504-536.

29. Lewis, J. S., 1971 . Satellites of the outer pla-
     nets : Thermal models. Science, 172, p. 1127-1128.

30. Franklin, F. A. and Cook, A. F., 1965. Optical pro-
     perties of Saturn's rings II. Two-color phase cur-
     ves of the two bright rings. Astron. J., 70, pp.
     704-720.

31. Lunine, J. I. and Stevenson, D. J., 1983. The role
     of clathrates in the formation and evolution of
     icy satellites, in "Natural Satellites". In pre-
     paration.

32. Bell, J. F., Clark, R. N., McCord, T. B. and
     Cruikshank, D. P., 1979. Reflection spectra of
     Pluto and three distant satellites. Bull. Amer.
     Astron. Soc., 11, p. 572 (Abstract).

33. Greenberg, R., 1975. The dynamics of Uranus'satel-
     lites. Icarus, 24, pp. 325-332.

34. Greenberg, R., 1976. The Laplace relation and the
     masses of Uranus'satellites. Icarus, 29, pp. 427-
     433.

35. Greenberg, R., 1979. The motions of Uranus'satel-
     lites : Theory and application. In "Dynamics of
     the Solar System", ed. L. Duncombe, Reidel, pp.
     177-180.

COMMENT

L. J. Lanzerotti :

    Six years ago, Andy Cheng and I proposed the
dark nature of Uranus'rings was produced by magnetosphere particle bombardment of primordial CH_4 on the
rings. Our laboratory results show clearly that

$$\text{ions} + CH_4 \longrightarrow \text{"black gunk"}$$

with a resultant abundance H/C $\sim$ 2.

THE ATMOSPHERES OF ICY BODIES

Tobias Owen

Department of Earth and Space Sciences
State University of New York at Stony Brook
Stony Brook, New York 11794 U.S.A.

Abstract

Ices are considered as the major volatile-carrying solid in
the outer solar system. As such, they play a major role in forming atmospheres of icy satellites and the comas of comets and
they have probably contributed to the enrichment of heavy elements
in outer planet atmospheres. A possible role for clathrate hydrates in causing the activity of distant comets and sub-surface
processes on icy satellites seems worthy of further study.

1. INTRODUCTION: THE SIGNIFICANCE OF CLATHRATE HYDRATES

The Voyager observations of Titan have produced a large
amount of new information about the composition of the atmosphere
of this icy satellite. The major constituent of Titan's atmosphere is molecular nitrogen (80-99%) with methane the next most
abundant identified gas (1-6%). About 0.02% is molecular hydrogen, while at least 11 other trace constituents are also present
(Table 1). Up to 15% of the atmosphere may be primordial argon
(^{36}Ar and ^{38}Ar); uncertainties in the data preclude an exact
determination (1).
    One item of particular interest is the apparent absence of
neon; the UV spectroscopy sets an upper limit of $\sim$ 1%. This
means that the atmosphere of Titan was not simply captured from
the proto-Saturnian (or primordial solar) nebula (2). If it had
been captured, the abundance of neon would be similar to the
atomic abundance of nitrogen, which would make neon the most
abundant gas. Thus Titan's atmosphere must be secondary in
origin, produced by devolatilizing of the solid materials making

731

J. Klinger et al. (eds.), Ices in the Solar System, 731–740.

up the bulk of the planet. As ice can carry more gas than rocky
material can, it is the ices we need to study. The same thing
should be true of the atmospheres found on Triton and Pluto,
since the bulk composition of these bodies again seems likely to
include a high proportion of ice. Thus on each of these bodies
we would expect to find atmospheres similar to Titan's, with
variations depending on which gases were captured by the ices,
what kinds of chemistry took place after their release, and the
present surface temperatures.

We also have plenty of evidence for gases frozen or trapped
in cometary ices. Delsemme and Swings (3) suggested that CO_2,
CH_4, etc. clathrate hydrate formation was required in order that
the observed near constancy in the ratios of gas abundances would
be preserved with changing solar distance, despite the differences in the vapor pressures of the gases themselves. Miller (4)
pointed out that the original proposal required modification in
that a mixed hydrate rather than a mixture of pure hydrates would
be produced. Given that comet tail spectra show evidence of both
CO^+ and N_2^+, it seems reasonable to suggest that clathrate hydrates of these two gases (in un-ionized form) may also be present
in the comet nuclei.

Finally, it should be pointed out that the present compositions of the atmospheres of the giant planets also lead to a
requirement for the incorporation of gases in primordial ices.
Even giant Jupiter reveals an excess of carbon to hydrogen compared with the solar value of this ratio (5). On the other
giant planets, the discrepancy is even greater. Since hydrogen
cannot escape from these planets, the logical way to achieve
this enrichment of carbon is to increase the abundance of methane.
This tends to support Mizuno's (6) proposal for a two-stage process of giant planet formation in which a core forms first and
then induces the collapse of an envelope of solar composition
gas from the surrounding solar nebula. During the formation of
the core, a secondary atmosphere rich in nitrogen (probably as
ammonia) and methane would be produced. The mixing of that
atmosphere with the collapsing envelope of gas from the solar
nebula would then produce the atmosphere we find today (7).

There are thus three distinct end points for primordial
gas-carrying ices: the icy satellites, the comets, and the
cores of the giant planets. Only in the case of the comets do
we have rather good grounds for assuming that clathrate hydrates
are involved; in the other examples, one could invoke the
formation of ices of (e.g.) methane and ammonia directly. But
thermodynamics favors clathrate formation, provided there is no
kinetic barrier (4). Furthermore, the comets should represent
the smallest units of primordial icy material, from which the
larger bodies formed. If the comets contain clathrates, we can
reasonably expect that the icy satellites and the giant planet
cores were formed from similar material.

2. VARIATIONS IN INCORPORATED AND RELEASED GASES

All ices that formed in the outer solar system did not necessarily have the same composition. Which gases got incorporated and in what proportions will depend on the local temperature and pressure at which the ices form. This explains the absence of atmospheres for Ganymede and Callisto, for example, two bodies that are very similar in size and density to Titan. Evidently the ices that are incorporated in these two satellites formed at temperatures too high for clathrates or for the condensation of ammonium hydrate. This difference in the primordial environment compared with Saturn could have been caused by the greater heat liberated by Jupiter as it formed (8). It is not a result of less degassing on the Jovian satellites, since they are warmer than Titan. Models including liquid water mantles have been developed for all three bodies, again suggesting a structural similarity and similar opportunities for degassing (9).

Local variations in the gas content of ices can also be expected. As Prinn and Fegley (10) have shown, the values of CO/CH_4 and N_2/NH_3 in a protoplanetary cloud can be quite different from those found in the primordial nebula itself. High temperatures near the forming planets will favor smaller abundances of CO and N_2. The question then centers on the efficiency of radial mixing within the protoplanetary nebula (10). Experience with the marked inhomogeneities revealed by inclusions and other aspects of meteorite mineralogy suggest that such mixing was distinctly incomplete in the solar nebula (11). There is no reason to assume that it was markedly better around the planets.

3. IMPLICATIONS FOR TRITON AND PLUTO

If the N_2 now in Titan's atmosphere was indeed brought in as clathrate hydrate, then we would expect N_2 to be present on Triton and Pluto as well. The protoplanetary cloud around Neptune at the position of Triton's formation should have been reasonably similar to the proto-Saturnian nebula. Otherwise such a large satellite would not have formed. The discovery of an absorption feature at 2.16 μm in Triton's spectrum by Cruikshank (12) appeared to support this prediction, since this wavelength corresponds to the forbidden quadrupole absorption of N_2. Subsequent study of laboratory spectra by Cruikshank and collaborators (13) has raised the possibility of liquid N_2 oceans on the surface of this satellite. Methane is also present (13). The model predicts that CO and Ar should be in the atmosphere as well, but these gases are much more difficult to detect remotely.

In the case of Pluto, only methane has been detected, and it is not clear that gaseous methane is present (13). Pluto is evidently cold enough for methane to freeze on the surface. While this is true on Triton as well, the near-constancy of

methane absorption in the spectrum of Triton as the satellite
follows its orbit around Neptune suggests that a substantial
amount of methane is in fact in the atmosphere (13).

There may be an opportunity here to put a constraint on the
location in which Pluto formed. There is still speculation that
Pluto originated as a satellite in orbit around Neptune and was
later expelled by an encounter with a third body, instead of
forming as a separate planet. If Pluto in fact began its existence in the solar nebula, the lower pressure and temperature
(compared with a protoplanetary cloud) could prevent the formation of N_2, CO, and Ar clathrates while permitting CH_4 clathrate
(Figure 1 - after Miller [4]). If future observations made with
greater precision confirm the apparent absence of N_2 on Pluto
and its presence on Triton, this difference in the environment
in which the two bodies formed might provide the explanation.
We need laboratory information on the kinetics of clathrate
formation at low temperature in order to decide whether or not
this is a useful discriminant.

4. IMPLICATIONS FOR COMETS AND SMALL SATELLITES

One of the big surprises of the Voyager observations of
Saturn's small, icy satellites was the evidence for internal
geological activity on some of them (14). Enceladus was particularly striking in this respect, with a large fraction of the
surface showing no evidence of impact craters and a uniform
reflectivity close to 100% (14). The apparent association of
Enceladus and Saturn's E ring strongly suggests that the former
is the source of the latter, although no convincing mechanism
for producing ring particles from the satellite has yet been
proposed (15,16). The prominent cracks with bright rims on the
surface of Dione are another example of sub-surface activity.
On Dione, the process(es) responsible for the cracks presumably
occurred long ago, but the special puzzle associated with
Enceladus is the fact that the E-ring must be continually resup-
plied with particles. If Enceladus is indeed populating this
ring - as the circumstantial evidence suggests - it must be
doing so now (16,17).

A possible analogue to this behavior is offered by the
activity manifested on still smaller icy bodies - the comets.
We have not yet seen any details on comet nuclei, but the fact
that these tiny objects have some way of releasing energy from
within is nevertheless well known. Perhaps the most famous
example is P. Schwassman-Wachmann I (S-WI), but many others have
been observed (18).

The orbit of S-WI has the lowest eccentricity of any known
comet (0.105) and lies entirely between the orbits of Jupiter
and Saturn, at a heliocentric distance ranging from 5.5 to
7.3 AU. In its quiescent phase, the comet is of magnitude 18 or

19, and appears nearly stellar, with just a hint of a coma. Outbursts occur rapidly and apparently randomly, with an increase in brightness of five magnitudes or more taking place in less than a day. At its brightest, the comet exhibits a rather sharply bounded, expanding disk of nebulosity. As the nebulosity continues to expand, at a velocity of one to a few tenths of a kilometer per second, the brightness of the comet decreases. If no further outbursts occur, S-WI regains its quiescent appearance within three weeks to a month. This is also an unusually large comet. The likely low albedo coupled with the magnitude at minimum light suggests a diameter on the order of 70 Km (19).

Spectroscopic observations of the comet during both active and quiet phases reveal a solar spectrum in which CO^+ has only recently been identified (20). Evidently, both the faint coma and the expanding nebulosity consist predominantly of a mixture of dust and ice grains. If the comet were in orbit around a planet instead of around the sun, this material could contribute to the formation of a planetary ring.

5. A MODEL FOR ACTIVITY IN ICY OBJECTS

Given that comets can exhibit internal activity at distances on the order of 6 AU, is it possible that the process(es) responsible could be the same in the case of the icy satellites? Several authors have suggested that the transformation of ice from amorphous to cubic might be the common cause (21). It seems useful to review the energies involved in an alternative model involving clathrate hydrates.

Suppose that CO, N_2, CH_4, Ar, etc. form a mixed hydrate in the ices that form comets or accrete to form satellites. These gases could be released below the surfaces of such objects as the result of some dissipation of energy such as solar heating or tidal stresses. An outburst is caused when this gas collects in a region from which it is unable to escape, ultimately achieving sufficient pressure to overcome the strength of the surrounding ices. In the case of a small comet, the pressure may only be on the order of an atmosphere. This might also suffice near the surface of a satellite. The reason only Enceladus still exhibits this behavior (as far as we know) must be that this satellite formed with an unusually high content of low temperature volatiles that are still available to be "degassed" by an appropriate source of energy. Perhaps its surface or a subsurface layer was "sealed" early in its history. In this context, we should bear in mind that the smoothness of large regions of the surface of Enceladus may have resulted from local melting caused by the early dissipation of tidal energy when Enceladus was in resonance with one of the co-orbital satellites or their parent body (22). At present, not enough tidal energy is available to melt the satellite, and we should consider other

processes that might lead to the expulsion of icy grains. This
is where a possible analogy with comet outbursts becomes
attractive.

We can appreciate the scale of the cometary phenomenon by
considering S-WI. As we shall see, the characteristics of the
outbursts on this comet make them an appropriate model for gas-
driven ejection of material from Enceladus. Whitney (23) has
estimated that a typical outburst involves 10^{12} gm of material.
The average kinetic energy calculated by Whitney for the expanding halos of S-WI is 10^{21} ergs. These outbursts occur at a rate
that is not well-defined, but may average as high as 2/year. In
Whitney's model, the energy for these bursts is provided by
absorbed sunlight, which would be 3×10^{19} ergs s^{-1}. There are
two other potential sources of energy that will be dissipated in
the _interior_ of Enceladus, radioactivity and tidal heating.
Estimates of the tidal heating experienced by Enceladus as a
result of its 2:1 resonance with Dione range from 10^{14} to 3×10^{15}
ergs s^{-1}; we adopt 5×10^{14} ergs s^{-1} as a reasonably conservative
value (24). This is comparable to the heating rate from radioactivity if Enceladus contains only a few per cent of lunar-like
rocky material (24).

We shall assume that the outbursts on Enceladus are of a
similar magnitude to those on S-WI. The energy liberated within
the satellite is divided between the dissociation of clathrates
and dissipation through deforming and fracturing of the water ice.
The subsequent explosive release of gas propels grains of ice
into space. Assuming 1 per cent efficiency for this process, the
necessary energy is delivered every 2×10^8 seconds and the amount
of gas produced is $\sim 3 \times 10^{12}$ gm. Thus enough energy is available
in this system to produce one comet-like burst every 8 years from
Enceladus. The escape velocity of Enceladus is 0.2 km/sec, at
the low end of the range of expansion velocities observed in the
nebular disks produced by outbursts of S-WI. Thus most of the
material produced during an outburst will leave the satellite.
Material that doesn't escape simply falls back and coats the
surface maintaining its remarkably high surface brightness
($p_v = 1.0 \pm 0.1$).

It is more difficult to evaluate the effect of sunlight
since we need an additional model for the transfer of absorbed
energy at the surface to subsurface layers. It would seem that
gases near the surface would have been thoroughly vented long
ago. On the other hand, if the explosive release of ice grains
occurs over long enough intervals, darkening of the surface by
UV irradiation and solar wind sputtering in between these events
will lead to an increase in subsurface temperature that will have
a quasi-periodic effect on gas release. The arguments presented
here simply show that the available energy can produce the necessary mass through a plausible mechanism in an appropriately short
period of time. The expulsion of $\sim 10^{15}$ gm every $\sim 10^6$ years

results in a net loss to the satellite of less than 10^{-4} of its present mass over the lifetime of the solar system.

6. OBSERVATIONAL TESTS FOR THE PRESENCE OF CLATHRATE HYDRATES

The importance of clathrate hydrates in storing and releasing gases in the outer solar system can be tested in several ways. For example, if the N_2 on Titan was originally incorporated as a clathrate hydrate, then the present atmosphere should contain several per cent of primordial argon. The reason behind this assertion is fairly obvious. At temperatures and pressures assumed for the proto-Saturnian nebula at Titan's distance from the planet (T = 60°K, P = 0.1 bar), neither of these gases will condense, but both will form clathrates with very similar dissociation pressures (Figure 1). It is then simply a question of consulting a table of solar abundances to determine the relative proportions in the resulting atmosphere. Conversely, if the N_2 on Titan was produced by the incorporation of frozen ammonium hydrate followed by the release of NH_3 to the atmosphere and the photolytic destruction of NH_3 to yield N_2; then the amount of argon that is present could be very much smaller.

In the case of gases from ices in outer planet cores, atmospheric hydrogen, helium, and neon should exhibit solar relative abundances, while the other elements will be enhanced. In particular, primordial argon should show the same enrichment as carbon and nitrogen, if all three gases were brought in as a mixed clathrate.

It seems even more likely that argon should be present in comets. In this case one may be able to find lines of Ar^+ in the spectra of a bright comet's tail. Alternatively, Ar may be detected by a mass spectrometer on one of the spacecraft investigating Halley's Comet in 1986. But one would not expect neon, since this gas does not condense at T > 27°K, and the small atomic size and low polarizability of neon also mitigate against clathrate formation.

If the intervals between outbursts on Enceladus are very short, one test of the "comet-burst" hypothesis would be to search for the gas released with the particles. If it is only methane, the possibilities of detection are poor, although CH_4^+ may contribute to the ions of mass 16 ± 4 reported by Wolfe et al. (25) in the plasma beyond the orbit of Enceladus. If it is carbon monoxide and nitrogen, emission lines of CO^+ and N_2^+ might conceivably be detectable.

Acknowledgements: I am grateful to W. A. Baum, S. Dermott, and C. Yoder for helpful remarks. This research was supported in part by NASA contracts 953614 and NGR 33015141.

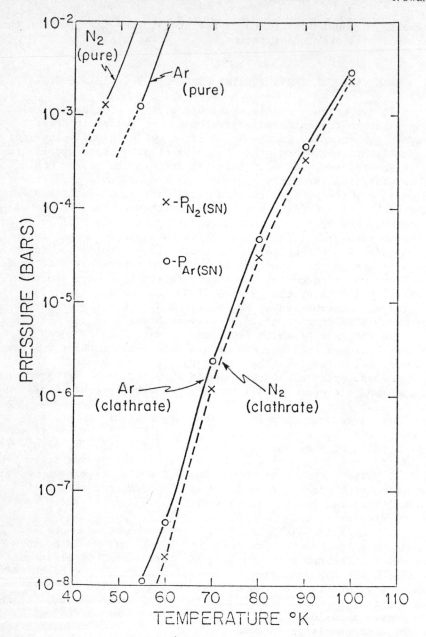

Figure 1.   A plot of the dissociation vapor pressures of argon
and nitrogen clathrate hydrates, compared with the
vapor pressures of the gases. The two points mark
the abundances expected in the proto-Saturnian Nebula.

THE ATMOSPHERE OF TITAN

Trace Constituents (parts per million)

Hydrogen (H_2)	2000 ppm

Hydrocarbons

Acetylene (C_2H_2)	2
Ethylene (C_2H_4)	0.4
Ethane (C_2H_6)	20
Diacetylene (C_4H_2)	0.1 – 0.01
Methylacetylene (C_3H_4)	0.03
Propane (C_3H_8)	20

Nitrogen Compounds

Cyanogen (C_2N_2)	0.1 – 0.01
Hydrogen Cyanide (HCN)	0.2
Cyanoacetylene (HC_3N)	0.1 – 0.01

Oxygen Compounds

Carbon Monoxide (CO)	50 – 150
Carbon Dioxide (CO_2)	0.0015

REFERENCES

1. Lindal, G.F., Wood, G.E., Hotz, H.B., Sweetnam, D.N., Eshleman, V.R. and Tyler, G.L. 1983, Icarus 53, p. 348.
2. Owen, T. 1982, J. Planet. Space Sci. 30, p. 833.
3. Delsemme, A.H. and Swings, P. 1952, Ann. Astrophys. 15, p. 1.
4. Miller, S.L. 1961, Proc. Nat. Acad. Sci. 47, p. 1798.
5. Gautier, D., Bezard, B., Marten, A., Baluteau, J.P., Scott, N., Chedin, A., Kunde, V. and Hanel, R. 1982, Astrophys. J. 257, p. 901.
6. Mizuno, H. 1980, Progr. Theor. Phys. 64, p. 544.
7. Gautier, D. and Owen, T. 1983, Nature 304, p. 691.
8. Pollack, J.B. and Consolmagno, G. 1984, in "Saturn" ed. T. Gehrels and M.S. Matthews (Tucson: Univ. of Arizona Press) p. 811.

9. Consolmagno, G.J. and Lewis, J.S. 1978, Icarus 34, pp. 280–293.

10. Prinn, R.G. and Fegley, B., Jr. 1981, Astrophys. J. 249, p. 308.

11. Papanastassiou, D.A., Lee, T. and Wasserburg, G.J. 1977, in "Comets, Asteroids, Meteorites" ed. A.H. Delsemme (Toledo: Univ. of Toledo Press) p. 343.

12. Cruikshank, D.P., 1983, private communication.

13. See Cruikshank, D.P., paper published in this volume.

14. Smith, B.A. and Voyager Imaging Team 1982, Science 215, p. 504.

15. Baum, W.A., Kreidl, T., Westphal, J.A., Danielson, G.E., Seidelmann, P.K. and Pascu, D. 1981, Icarus 47, p. 84.

16. Pang, K.D., Voge, C.C., Rhoads, J.W. and Ajello, J.M. 1984, J. Geophys. Res. 89, pp. 9459–9470.

17. Burns, J.A., Showalter, M.R. and Morfill, G.E. 1984, in "Planetary Rings" ed. R. Greenberg and A. Brahic (Tucson: Univ. of Ariz. Press) p. 200.

18. Hughes, D.W. 1974, Quart. J. Roy. Astron. Soc. 16, p. 4.

19. Roemer, E. 1966, in "Nature et Origine des Comètes" ed. P. Swings (Belgium: Univ. of Liége) p. 23.
    Hartmann, W.K., Cruikshank, D.P. and Degewij, J. 1982, Icarus 52, p. 377.

20. Larson, S.P. 1980, Astrophys. J. 238, p. L47.
    Cochran, A.L., Barker, E.S. and Cochran, W.D. 1980, Astron. J. 85, p. 474.

21. Patashnick, H., Rupprecht, G. and Schuerman, D.W. 1974, Nature 250, p. 313.
    Klinger, J. 1982, Nature 299, p. 41.

22. Lissauer, J.J., Peale, S.J. and Cuzzi, J.N. 1984, Icarus 58, pp. 159–168.

23. Whitney, C.A. 1955, Astrophys. J. 122, p. 190.

24. Yoder, C.F. 1979, Nature 279, p. 767.
    Pearle, S.J., Cassen, P. and Reynolds, R.T. 1980, Icarus 43, p. 65.

    Dermott, S. 1982, private communication.

25. Wolfe, J.H., Mihalov, J.D., Collard, H.R., McKibben, D.D., Frank, L.A. and Intrilligator, D.S. 1980, Science 207, p. 403.

EVOLUTION OF TITAN'S COUPLED OCEAN-ATMOSPHERE SYSTEM AND INTERACTION OF OCEAN WITH BEDROCK

Jonathan I. Lunine* and David J. Stevenson

Division of Geological and Planetary Sciences,
California Institute of Technology, Pasadena,
California 91125 USA
*Now at Lunar and Planetary Laboratory,
University of Arizona, Tucson, Arizona 85721 USA

ABSTRACT

A recent model for the surface state of Titan proposes a liquid ethane-methane-molecular nitrogen layer of order one kilometer thick which because of stratospheric methane photolysis has become increasingly ethane-rich with time (1). We explore the interaction of such an ocean with the underlying "bedrock" of Titan (assumed to be water-ice or ammonia hydrate) and with the primarily nitrogen atmosphere. It is concluded that although modest exchange of oceanic hydrocarbons with enclathrated methane in the bedrock can in principle occur, it is unlikely for reasonable regolith depths. The surprisingly high solubility of water ice in liquid methane (2) implies that topographic features on Titan of order 100 meter in height can be eroded away on a time scale $\lesssim 10^9$ years. The large solubility difference of N_2 in methane versus ethane implies that the ocean composition is a strong determinant of atmospheric pressure; a simple radiative model of the Titan atmosphere is employed to demonstrate that significant surface pressure and temperature changes can occur as the oceanic composition evolves with time.

1. INTRODUCTION

Recently a model for a liquid hydrocarbon ocean on Saturn's satellite Titan was proposed by Lunine et al. (1) to reconcile Voyager data on the lower atmosphere with current understanding

741

J. Klinger et al. (eds.), Ices in the Solar System, 741–757.
© 1985 by D. Reidel Publishing Company.

of methane photochemical cycles in the upper atmosphere. The spacecraft and ground-based data sets bearing on Titan are reviewed thoroughly in Hunten et al. (3).

The primary goal of this paper is to explore the coupled evolution of the ocean and atmosphere, and the interaction of this ocean with the underlying material. In the remainder of this section we review the basic model. For details the reader is referred to (1). Section 2 examines two possible interactions of the ocean and water ice + ammonia hydrate bedrock beneath: exchange of hydrocarbons between ocean and clathrate hydrate and erosion of bedrock topography. Section 3 presents a simple model for the coupled evolution of ocean and atmosphere, as the former becomes more ethane·rich with time.

The model for the surface state of Titan is shown in Figure 1a, superimposed on a plot of temperature versus altitude in the atmosphere from Lindal et al. (4). The atmosphere is essentially pure N_2 at a surface pressure of 1.5 bars and temperature 95 K (the possible presence of 12 mole percent argon in the atmosphere does not affect any of the arguments presented in the paper). The ocean methane composition of 25 mole percent is chosen to be consistent with the Voyager radio-occultation upper-limit of 3% methane in the atmosphere. More detailed modeling by Flasar (5) permits $\lesssim$ 9% methane in the lowermost atmosphere which would correspond to a 70% methane ocean. The calculation of the amount of dissolved nitrogen is described in section 3. The low vapor pressure of ethane ($\sim 10^{-5}$ bars at 95 K) allows the atmosphere above to follow an essentially dry adiabat as required by Voyager data (4). A nearly pure methane liquid is not consistent with the data.

The source of the oceanic ethane and the underlying solid acetylene sediment is stratospheric photolysis of methane. The depth of the ocean and sediment are calculated from the inferred rate of ethane and acetylene production (6) integrated over the age of the solar system. A 25% methane ocean has a depth of $\sim$ 800 meters; a 70% methane ocean corresponds to 1.8 kilometers depth. The sediment layer is $\sim$100-200 meters thick. (The relative production rates of ethane and propane are somewhat uncertain; the two are sufficiently similar thermodynamically that we do not distinguish between them in this paper.) The ocean is thus the source and sink of methane photolysis and becomes more ethane-rich with time, as illustrated schematically in Figure 1b. In the absence of an ocean, the maximum methane retainable by the atmosphere would be destroyed by photolysis in $\sim 10^7$ years; our model provides sufficient methane for $\gtrsim 10^9$ years. The ultimate source of the methane forms the basis for another study and is not discussed here.

We note here two details which are of relevance to calculations in later sections. Referring to Figure 1b, two methane molecules are photolytically destroyed for each ethane molecule produced. This decrease in number of molecules in the

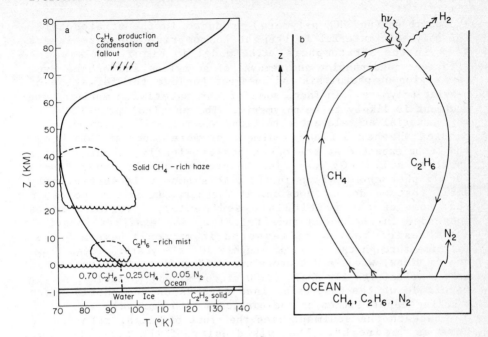

Figure 1. (a) Model of Titan's near-surface and atmospheric structure from Lunine et al. (1), with altitude versus temperature data from Lindal et al. (4). Below the surface the ocean temperature is assumed to follow an adiabat. Between 10 and 20 kilometers coexisting solid and liquid haze may be present. (b) Schematic model of evolution of ocean due to methane photolysis. Methane evaporated from the ocean is photolyzed in the stratosphere, producing hydrogen and ethane, acetylene and other products observed by Voyager IRIS (3). Hydrogen escapes from the atmosphere; the supersaturated ethane condenses and falls to the ocean. As the ocean enriches in ethane, N_2 comes out of solution. Note one molecule of ethane (or acetylene) is produced for every two methane destroyed. Only the methane-ethane-N_2 portion of the cycle is shown for clarity.

ocean with time, coupled with the rather different solubilities of N_2 in methane and ethane, forms a strong driver for atmospheric evolution as described in section 3. With regard to the acetylene layer, there is some disagreement in the literature as to the relative amounts of acetylene versus higher complex hydrocarbons (and HCN polymers) produced in the stratosphere. The polymeric material is thought to comprise the haze layers seen in the stratosphere. The model in Figure 1a assumes efficient condensation and removal of acetylene and ethane from the region of photolysis and does not include the formation of heavier polymers. In fact, some of the material comprising the sediment is likely to be polymeric. The physical properties of this material at relevant conditions are not well enough known to predict whether a flaky sediment or more coherent "blacktop" should be expected at the ocean base. Finally, about 1 meter worth of solid CO_2 may also be present in the layer from a CO-CO_2 photochemical chain (7) ; this amount is sufficiently small that we do not consider it further. It is also possible that some of the particulate sedimentary material remains suspended in the ocean indefinitely. For example, spherical particles of radius r (measured in microns) have a Stokes falltime through the ocean of roughly 10^2 r^{-2} years. Ocean flows due to tides, winds and temperature differences are larger than Stokes velocities for small ($\lesssim 100$ micron-sized) particles, so the magnitude of the suspension load depends on the poorly known sticking capabilities of the particles with each other.

Beneath the sediment lies the crust of Titan, referred to here as "bedrock". The bulk density of the satellite and cosmochemical considerations predict water ice probably at least partially transformed into clathrate, and perhaps ammonia hydrate, as the primary bedrock material (3). A small amount of silicate material may be present but does not affect any of our considerations because it is chemically inert at $\sim$100 K.

2. INTERACTION OF OCEAN AND BEDROCK

In this section we consider two types of interaction between the ocean and water ice bedrock: 1) exchange of hydrocarbons between ocean and clathrated bedrock, and 2) erosion of water-ice topography by solution in the ocean.

The possibility of enclathrated methane present on the Titan surface has been previously considered (8) but never in connection with an ocean. Clearly, a significant exchange of photochemically produced ethane in the ocean with methane in underlying clathrate can only occur if the ocean can establish thermodynamic equilibrium with a substantial thickness of the underlying ice. This thickness is estimated to >100 meters.

The methane (or ethane) clathrate is a modified water ice lattice enclosing guest molecules and with an approximate formula $X \cdot 6H_2O$ where X is methane and/or ethane. A full description of the thermodynamics of clathrate hydrates and application to solar system bodies is presented by Lunine and Stevenson (9), the results of which are used here.

At ambient conditions on Titan's surface, ethane, methane and

N_2 should fully incorporate in clathrate hydrate, with the result that no ocean could be present and the N_2 atmospheric pressure would be $\sim 10^{-3}$ bars. This situation is prevented by the inability of these species to come into contact with large quantities of water ice. Such contact requires that the ice not only have substantial permeability but that the fluid channels (fissures, cracks, etc.) penetrate the ice to the submicron level, since diffusion of methane or ethane into water ice is extremely slow even over a distance of one micron (9). The existence of porosity (and methane-ethane "aquifers") is certainly possible since the hydrostatic head and temperature may not be sufficient to deform the ice and squeeze out enclosed methane, even at 10 km depth (temperature ~ 120 K?, pressure ~ 200 bars; pressure difference between methane and ice columns of ~ 100 bars). Fluid penetration to a submicron scale is unlikely, however, since the ice is likely to have undergone annealing at some stage in its history. (It might have formed by freezing from a primordial NH_3-H_2O ocean, for example, or been subjected to impact. In either case, porosity would not extend to the submicron level.) Impact causes formation of a regolith as deep as a kilometer (10) which consists of an extreme diversity of particle sizes (11); only the uppermost 100 meters (or less) is likely to be stirred sufficiently frequently to allow efficient conversion to clathrate. We thus conservatively assume the equivalent of a 100 meter, finely fragmented ice layer. The relative fraction y of C_2H_6 to CH_4 incorporated in the clathrate is given by the rough formula

$$y \sim \frac{P^c(CH_4)}{P^c(C_2H_6)} \frac{f(C_2H_6)}{f(CH_4)} \tag{1}$$

where f(q) is the fugacity of species q in the ocean and $P^c(q)$ the dissociation pressure of clathrate containing pure species q. $P^c(q)$ is calculated from molecular properties of q in (9). f(q) is very nearly $X(q) \cdot P(q)$, X(q) the mole fraction of q in the ocean, P(q) the vapor pressure of pure q at system temperature. Inserting values from (9) equation (1) becomes

$$y \sim \frac{1}{3} \frac{X(C_2H_6)}{X(CH_4)} \quad .$$

Thus, as the ocean C_2H_6 composition is increased with time due to CH_4 photolysis, the underlying clathrate becomes progressively enriched in C_2H_6.

Consider now a starting state consisting of a pure methane ocean, depth ~ 1.4 km (calculated from a present ocean of 25% CH_4 and 0.8 kilometer depth), overlying a 100 meter regolith of methane clathrate hydrate. We assume for now the acetylene sediment layer does not prevent ocean-regolith contact and is

kinetically inhibited from enclathrating because it is in the
solid phase. Also, N_2 can be neglected because it is a poor
incorporator in clathrate compared to CH_4 or C_2H_6. As the ocean
composition becomes more ethane rich, some of the ethane
exchanges with the methane in the clathrate producing a negative
feedback on ocean composition. Figure 2 plots the fraction of
oceanic methane derived from clathrate as a function of the
ethane mole fraction in the ocean. By the time the present ocean
composition is reached, only 20% of the methane in the ocean is
derived from clathrate. Only ∿6% of the ethane produced has been
incorporated in the clathrate; the remainder resides in the
ocean. Arguments presented above suggest that this is an upper
limit to the amount of methane and ethane exchanged between ocean
and regolith. Moreover, the basal sediment layer of acetylene
and polymeric material could severely inhibit ocean-regolith
contact. We conclude that transfer of oceanic and regolith
hydrocarbons is probably not an important control on ocean
evolution, and may be neglected in what follows.

We next examine the ability of the ocean to erode bedrock
topographic features. Such a process could be particularly
efficient for newly formed features created by impact or
ammonia-water volcanic processes and which are not yet covered
with a layer of sediment. The surprisingly high solubility of
water ice in non-polar cryogenic liquids (6×10^{-5} mole fraction
in methane at 112 K (2)) permits substantial removal of ice from
topographic highs, provided saturation of the entire ocean is

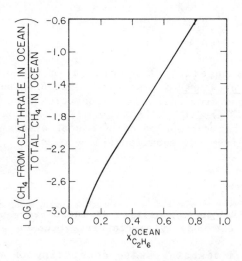

Figure 2. Logarithm of the fraction of oceanic methane derived
from clathrate as a function of ethane mole fraction in the
ocean. Starting state is pure methane ocean underlain by 100
meters of methane clathrate hydrate.

avoided. Tidal currents generated by Titan's non-zero free eccentricity (12) may provide a mechanism for dissolving and removing water.

Consider a topographic feature ("mound") of height h and length L, encountering an ocean current of velocity v (Figure 3a). Assume, for this example, that the fluid is highly undersaturated in H_2O. At the ocean-mound interface a diffusional boundary layer of thickness δ develops. If the boundary layer remains laminar then $\delta \simeq (Dy/v)^{1/2}$, where y is the distance measured downstream from the upstream tip of the mound. Since a saturated layer of thickness $\sim(DL/v)^{1/2}$ detaches from the mound and flows downstream at velocity v, it follows that the erosion time scale is:

$$\frac{h}{\dot{h}} = \frac{hL}{v\delta X} = \frac{hL}{v[D(L/v)]^{1/2}X} = \frac{h}{X}\left(\frac{L}{Dv}\right)^{1/2} \qquad (2)$$

where X is the saturation mole fraction of water ice in the ethane-methane liquid and D is a diffusion coefficient. For a dissolved constituent moving through the liquid we estimate D 10^{-4} cm^2/s; X at 95 K is conservatively estimated from the methane-water ice data to be $\sim 10^{-8}$. The flow velocity v ~ 1 cm/s is taken from the tidal model of Sagan and Dermott (12); the scale length of a topographic feature is estimated by analogy with large impact structures on Galilean satellites to be 10 km.

Figure 3b plots $h/\dot{h}$ versus h. Note that a 100 meter-high feature could be eroded over the age of the solar system. Unless the topographic feature is small, turbulence is likely to be important. In this case, one might argue that the boundary layer "breaks" and is assimilated into surrounding fluid once it achieves a downstream extent ~ 1 where vl/ν $\sim 10^3$, a critical Reynolds number. This leads to an enhancement of the erosion rate by a factor $\sim (L/l)^{1/2}$. For the predicted l $\sim$ 1-10 cm and L $\sim$ 10 km, this is an enhancement factor of several hundred. More realistically, the solution and dissolution of water and accumulated hydrocarbon sediment in the CH_4-C_2H_6 liquid depends on the details of the ocean circulation and temperature structure. One could imagine cavernous chemical erosion, similar to Karst topography on Earth (Karst derives its name from the northeastern shore of the Adriatic Sea where Yugoslavia and Italy meet, and is characterized by an irregular limestone terrain with many small depressions, which evolved from the dissolving power of the groundwater). The equivalent of stalactites or stalagmites might also form.

The vertical scale of impact and tectonic features on large icy satellites is 1 km (13); this scale may apply to volcanic edifices if their height is determined by isostasy as suggested for terrestrial volcanoes by Ben Avraham and Nur (14). It is

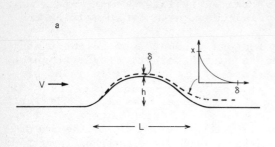

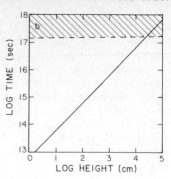

Figure 3. (a) Water ice topographic feature of height h and extent L submerged in ocean with current velocity v. δ is boundary layer thickness over which water mole fraction decreases from saturation to dilute value. Schematic plot shows decrease in mole fraction of dissolved water in boundary layer as a function of distance from feature. Vertical scale of feature and boundary layer thickness (the latter increasing in downstream direction) are schematic. (b) Time in seconds to erode a feature of height h versus h in centimeters. Parameter values given in text. Stippled area is t > age of solar system. As argued in text, the plotted erosion time may be a substantial overestimate.

thus possible that significant modification of submerged crater- and tectonic-forms has occurred on Titan. Radar profiling of the submerged ice structure, taking advantage of the low microwave absorption characteristics of hydrocarbons and significant difference in ice and hydrocarbon dielectric constants could reveal such degradation and would be a fascinating experiment.

3. COUPLED EVOLUTION OF OCEAN AND ATMOSPHERE

Lunine et al. (1) compute the solubility of N_2 in their nominal (25% methane) ocean under 1.5 bars N_2 pressure to be ∿5 mole percent, implying a dissolved N_2 mass ∿1/4 of the atmospheric mass. A pure methane ocean of similar mass under the same conditions would contain >20% N_2 or ∿ twice the atmospheric mass of N_2. Clearly the relatively high solubility of N_2 in methane, and low solubility in ethane implies that the ocean composition must exert a strong control on the N_2 atmospheric pressure, and as we show below, temperature. Titan's ocean-atmosphere interaction is thus intermediate between that of Earth's, for which the dissolved oceanic N_2 is <1% of the atmospheric mass (solubility data from (15) p. 256) and Triton, which may possess large amounts of N_2 (16) as either liquid or solid, and which would then be overlain by a N_2 saturated vapor atmosphere. In this section we construct a model for the evolution of Titan's surface temperature and pressure as the ocean becomes more ethane-rich with time, and discuss the implications for the original composition and timing of surficial emplacement of the ocean.

The model assumes the ocean and atmosphere to be a closed system as far as the total carbon and nitrogen budgets are concerned. Conversion of methane to ethane and acetylene proceeds at the rates given in Yung et al. (6); two molecules of methane are converted to one molecule of ethane or acetylene, with loss of hydrogen from the atmosphere. The ocean mass decreases with time, primarily because methane photolysis leads to the formation of insoluble acetylene. Hydrogen is also formed and escapes, but is less important for the mass balance. The solubility of N_2 in CH_4 and C_2H_6 separately is assumed to obey Henry's law (a valid assumption so long as the mole fraction of $N_2 \equiv X(N_2) \lesssim 0.1 - 0.2$):

$$P(N_2) = K(q) \ X(N_2) \tag{3}$$

where $K(q)$ is the Henry's law constant for N_2 dissolved in component q and $P(N_2)$ is the partial pressure of N_2 ($\approx$ total pressure). The Henry's law constant at a given temperature for the mixed CH_4-C_2H_6 liquid is ((17), p. 373):

$$\log K \sim X(CH_4) \ \log K(CH_4) + [1-X(CH_4)] \ \log K(C_2H_6) \tag{4}$$

since the cryogenic ethane-methane system is roughly ideal (18). Then the relationship between atmospheric pressure P and total N_2 mass $M(N_2)$ (ocean + atmosphere) is:

$$P = \frac{M(N_2)}{\dfrac{4\pi R_T^2}{g} + \dfrac{M_o}{(K - P)} \dfrac{\mu(N_2)}{\mu_o}} \tag{5}$$

where R_T = radius of Titan's surface, g = gravitational acceleration at R_T, M_o = mass of the ethane + methane component of the ocean, and μ_o = mean molecular weight of ethane-methane ocean mixture. $M(N_2)$ is computed from the inferred present-day ocean composition (25% methane) given in Lunine et al. (1) and the observed 1.5 bar atmospheric pressure; we also do a calculation for a 60%-methane ocean. The ocean mass is a decreasing linear function of ethane mole fraction tied to the present state for which $\sim 20\%$ of the converted methane is in the form of acetylene, and the presumed primordial state consisting of 100% methane.

The compositional and temperature dependences of the Henry's

law constant K require some comment. If N_2 dissolves in the
ocean in an ideal sense, $K = P_s$, the saturation vapor pressure
over pure condensed N_2 at the given temperature. At 92 K, P_s =
4.3 bars (19) while $K(CH_4) \approx 7.0$ bars and $K(C_2H_6) \approx 45$ bars (20).
The N_2-CH_4 and N_2-C_2H_6 systems both exhibit positive deviations
from ideality; however, the different sizes of the N_2 and C_2H_6
molecules in particular strongly inhibit solution of the N_2 in
C_2H_6. It is the difference in solubility of N_2 in CH_4 and C_2H_6
and the changing number of moles of hydrocarbon in the ocean
which drive the evolution of the atmosphere with changing ocean
composition in the model presented below.

If the solutions were ideal, the temperature dependence of K
would follow that of the N_2 vapor pressure curve, $\sim 10^{-|A|/T+B}$, A
and B constants. Data for N_2-CH_4 from 105 to 92 K (20, 21)
suggest this dependence, with a somewhat smaller $|A|$ than for
pure N_2 vapor, but a linear fit would do just as well. Moreover,
at very low temperatures near the solidus, the temperature
dependence may become very weak, since the data show increasing
non-ideality with decreasing temperature. Hence we adopt two
extreme temperature dependences for K, one going as $10^{-|A|/T+B}$
and the second constant with T at the 95 K value. As shown
below, the two extremes predict somewhat different relationships
between T and ocean composition; the qualitative implications for
our nominal ocean model, however, are similar. Thus, case "A"
involves the following equation for K:

$$\log_{10} K = -\frac{140}{T} + 3.16 - X(CH_4) \cdot \left(\frac{52}{T} + 0.22 \right) \tag{6}$$

where due to the lack of N_2-C_2H_6 solubility data at more than one
temperature point below 100 K, the temperature dependence of the
N_2-C_3H_8 system from Cheung and Wang (20) was used, C_3H_8 being
very similar thermodynamically to C_2H_6. For case "B", a
temperature-independent K was used:

$$\text{Log}_{10} K = 1.66 - 0.719 \ X(CH_4) \tag{7}$$

Both case A and case B yield $X(N_2) = 0.05$ for present atmospheric
conditions and the nominal ocean model.

The ocean evolution model is coupled to a conceptually simple
model of the Titan atmosphere. This atmosphere may be broadly
divided into three regions according to the mode of energy
absorption and emission (22). Haze particles in the stratosphere
absorb essentially all the violet end of the visible solar
spectrum, heating altitudes above 50 kilometers (see Figure 1a).
The red spectral end is primarily scattered to the ground; hence
an optically thick troposphere in radiative equilibrium is
present. A thin convective region also occurs below 3.5
kilometers. As Samuelson (22) notes, the atmosphere is dominated

by radiative processes; we consider the lower atmosphere to be grey and in radiative equilibrium with an effective temperature T_e fixed by ambient surface conditions. Hence the temperature at the surface, $T(R_T)$, is

$$T^4(R_T) = \frac{1}{2} T_e^4 \left(1 + \frac{3}{2} \tau (R_T) \right) \tag{8}$$

where τ, the optical depth, is assumed due primarily to pressure-induced N_2 opacity:

$$\tau = \frac{\bar{A}}{2} \frac{1}{k_B M_p \mu(N_2) g} \frac{1}{n^2} \frac{P^2}{T} \tag{9}$$

$\bar{A}$ = absorption coefficient in cm^{-1} amagat $^{-2}$, n = Loschmidt's number, M_p = mass of proton, and k_B = Boltzmann constant. The relative importance of collision-induced gas versus cloud opacity remains controversial (22). Using a value for A similar to that in (23), we find

$$\tau = 141 \frac{P^2}{T} \tag{10}$$

or an effective temperature of 72 K for the present surface conditions (95 K, 1.5 bars). T_e is low compared to the global value of 86 K given in Hunten et al. (3) since most of the solar energy is absorbed and radiated by the upper stratospheric haze. The haze does re-radiate some energy to the lower atmosphere (22) but to first order we can decouple the lower atmosphere and determine its evolution as the ocean composition changes using the simple model of equations (5), (6) or (7), (8), and (10). This approach may not be valid for the earliest ocean compositions (high methane), when the atmosphere is nearly optically thin according to our results, and backwarming by the haze could be important. Note that radiative equilibrium dictates the ground temperature to be larger than the atmospheric value just above it given by equation (8). In practice, convection redistributes the heat and narrows the discontinuity to $\lesssim$ several degrees (4). Here we make the simplifying assumption that the ocean temperature is given by equation (8).

Figure 4 presents results for the nominal ocean model, plotted as surface temperature versus oceanic methane mole fraction. The associated surface pressure and time before present are plotted as auxiliary scales. Both solubility cases A and B (equations (6) and (7)) are plotted. Superimposed on the figure is the ethane-methane solid-liquid phase diagram from Moran (24). The N_2 mole fraction in the ocean is $\lesssim 15\%$ throughout

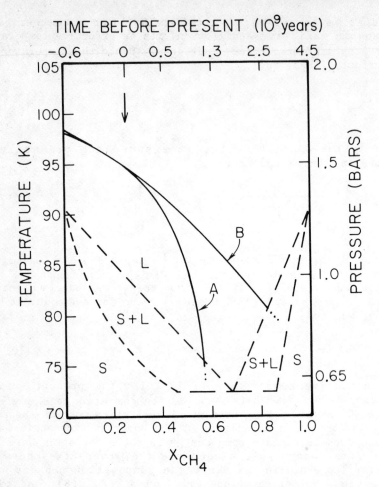

Figure 4. Surface temperature versus mole fraction of methane in ocean, nominal model. Two cases are shown: A, $K \propto 10^{-1/T}$ and B, K constant with temperature. Two auxiliary scales are plotted: atmospheric pressure in bars corresponding to temperature and time before present corresponding to methane mole fraction, assuming constant photolysis rate. Time is nonlinear in $X(CH_4)$ due to decrease in number of moles of oceanic hydrocarbon with ethane and acetylene production. Arrow points to present state. Dotted lines are ethane-methane solid-liquid phase diagram from Moran (24), with L = all liquid, S + L = coexisting liquid + solid and S = all solid. Total N_2 mass for this model = 1.1×10^{22} g, total present day ocean mass = 3.7×10^{22} g.

the evolution and hence lowers the phase boundaries at most $\sim$5 K.

Both cases exhibit monotonic decreases in temperature as the ocean is changed from the present state back to more methane-rich compositions. The strong positive temperature feedback for case A results in a more rapid evolution. The calculation becomes invalid as the curves cross the liquidus. This occurs for case A at $\sim$0.55 mole fraction methane; using the photolysis rate of Yung et al. (6) this corresponds to a time before present of $\Delta t \sim$ 1.1×10^9 years. Case B crosses the liquidus at X $\sim$ 0.8 corresponding to $\Delta t \sim 2.5 \times 10^9$ years (note Δt assumes constant photolysis rate, initial ocean is pure methane, and no bulk injection of methane or ethane from sources other than photolysis).

Figure 5 plots N_2 pressure and oceanic mole fraction versus temperature for case A along with the N_2 saturation vapor pressure. At T < 75 K, the atmospheric pressure begins to approach the saturation vapor pressure value. Since polar surface temperatures even at present may be several degrees below the equatorial value (3), our model could require some pure N_2 condensation in the polar regions at high methane ocean concentrations, until the global pressure is reduced below the saturation pressure at the poles.

We examined the sensitivity of our results to the variation of several parameters. Modeling of N_2 escape processes suggests that $\sim$20% of the present N_2 atmosphere could have been lost over the age of the solar system (3). Using a value of $M(N_2)$ 20% higher than for the Figure 4 calculation pushes the temperature up for a given oceanic methane composition, but the shape of the

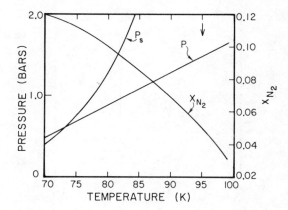

Figure 5. Surface pressure in bars and ocean N_2 mole fraction versus temperature in Kelvin for nominal model. Arrow points to present state. Pure N_2 saturation vapor pressure from Jacobsen and Stewart (19) is also plotted.

T-X evolution curve and the Δt needed to reach optically thin
conditions is roughly unchanged. Comparable modification of the
starting ocean mass has the opposite effect, but similar
magnitude. Lowering the N_2 absorption coefficient in the opacity
calculation makes the surface temperature less sensitive to
changing methane mole fraction as expected, but even a 50% change
has a rather minor effect on the curves in Figures 4 and 5.

Figure 6 shows model runs for a present day ocean containing
60% methane. Case B yields results roughly similar to the
nominal case, with the pure solid field crossed at $X(CH_4)$ ∿ 0.85.
Case A, however, shows a dramatically more rapid evolution. The
large mass of CH_4 and N_2 in the ocean relative to the N_2 in the
atmosphere, coupled with the strong temperature dependence of
solubility, yields an unstable situation: a slight increase in
oceanic methane strongly increases the amount of N_2 which can be
drawn out of the atmosphere, decreasing T and further increasing
the solubility so that the temperature drops. The same effect is
seen in Figure 4 case A at $X(CH_4) = 0.6$. We consider, however,

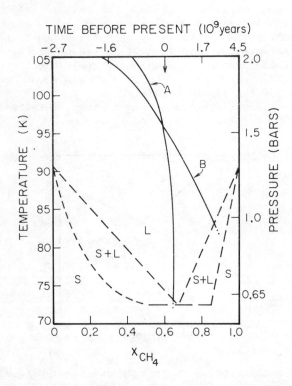

Figure 6. Same as Figure 4, high methane model. Total N_2 mass
1.6×10^{22} g, total present day ocean mass 5.8×10^{22} g.

the 60% methane ocean to be somewhat of an extreme case, as far as <u>present-day</u> surface conditions are concerned.

With the exception of the high methane case A result, the models explored evolve from low temperature states to the present-day over long time scales, $\sim 10^9$ years. Continuing the evolution to a pure ethane ocean composition leads to a modest temperature and pressure rise from the present day state (Figure 4). The nominal model would completely exhaust its methane on a time scale of $\sim 6 \times 10^8$ years, implying Titan's atmosphere-ocean system is and has been in a long term, slowly evolving condition.

The earliest evolution of the system is of great interest; however, we are limited at present to rather qualitative speculation. Two complications prevent us from using the above model to trace the evolution back past the point at which the liquidus is crossed: (1) the behavior of the ternary N_2-CH_4-C_2H_6 solid-liquid system is sufficiently uncertain that we cannot reliably predict, for example, the partitioning of N_2 between solid, liquid and gaseous phases, and (2) at T $\lesssim$ 75 K, the atmosphere in our model becomes optically thin, and significant radiative contribution from the stratospheric haze may dominate. A further complication is that some N_2 could have been frozen out at the poles at T $\lesssim$ 75 K, controlling the global atmospheric pressure, as indicated in Figure 5. Clearly more involved atmospheric and thermodynamic modeling is required.

We speculate briefly on some possible starting points for the ocean-atmosphere system. N_2 and CH_4 volatiles, extruded together from the interior could have coated the surface as a high albedo material, maintaining a low temperature. Production of dark (albedo ~ 0.2) haze material which subsequently covered the surface could have raised the temperature above 85 K, allowing N_2 to vaporize and regions containing methane and photochemically produced ethane ice to liquify and initiate an ocean in places. A complex trade-off between the albedo and emissivity properties of haze-covered and ocean-covered regions would have ensued, with a global ocean eventually forming and the atmosphere-ocean system running onto an evolution curve similar to those calculated above.

Other calculations by Lunine and Stevenson (25) suggest a hot formation scenario for an ammonia-hydrate rich Titan which could have allowed several bars of N_2 to be produced photochemically. As the surface temperature dropped below 180 K, and a solid surface formed, most of the N_2 would have remained in the atmosphere. The subsequent extrusion of methane from the interior could have had two effects: if all of the methane inferred to be the source of the present ocean were suddenly outgassed, and temperatures were low enough to allow it to condense, most of the atmosphere would dissolve into the ocean and freezing would ensue, with subsequent evolution as suggested above. If the extrusion were slow or episodic, so that photolytic conversion of CH_4 to C_2H_6 moderated the N_2 solubility,

the atmosphere could have slowly merged onto an evolution curve
similar to Figure 4's as the mass of methane extruded reached the
inferred present amount. As yet we cannot distinguish between
the various options, including the possibility that some ethane
or propane was also extruded from the interior. Further modeling
of the ocean evolution process, application of tidal constraints
(12), and predictions of atmospheric abundance of noble gases (9)
which could ultimately be measured and are diagnostic of
satellite formation and early history may constrain mechanisms
for the origin of the ocean.

As for Triton, the apparent detection of both condensed CH_4
and N_2 on the surface (16) along with the high solubility of CH_4
in N_2 liquid or solid (26) suggests that photochemical conversion
of CH_4 to heavier hydrocarbons could also strongly couple
atmospheric and surface evolution. Uncertainties regarding
relative and absolute abundances of CH_4 and N_2 on the surface and
in the atmosphere, the phase of condensed N_2 (liquid or solid)
and degree of areal mixing of those constituents renders modeling
a poorly constrained venture at present. The possibility that
Triton could represent a permanent analog for the frozen early
state of Titan's surface is an exciting one which should be borne
in mind as the Voyager close flyby of Triton (1989) draws near.

ACKNOWLEDGEMENT

This work was supported by NASA Grant NAGW-185. Contribution
number 4057 from the Division of Geological and Planetary
Sciences, California Institute of Technology, Pasadena,
California 91125 U.S.A.

REFERENCES

1)   Lunine, J.I., Stevenson, D.J., and Yung, Y.L. 1983, Science
     222, pp. 1229-1230.
2)   Rebiai, R., Rest, A.J., and Scurlock, R.G. 1983, Nature 305,
     pp. 412-413.
3)   Hunten, D.M., Tomasko, M.G., Flasar, F.M., Samuelson, R.E.,
     Strobel, D.F., and Stevenson, D.J. 1984, in "Saturn" (T.
     Gehrels, Ed.), U. of Ariz. Press, pp. 671-759.
4)   Lindal, G.F., Wood, G.E., Hotz, H.B., Sweetnam, D.N.,
     Eshleman, V.R., and Tyler, G.L. 1983, Icarus 53, pp. 348-363
5)   Flasar, F.M. 1983, Science 221, pp. 55-57.
6)   Yung, Y.L., Allen, M.A., and Pinto, J.P. 1984, Astrophys. J.
     Suppl. Ser., in press.
7)   Samuelson, R.E., Maguire, W.C., Hanel, R.A., Kunde, V.G.,
     Jennings, D.E., Yung, Y.L., and Aikin, A.C. 1983, J. Geophys.
     Res. 88, pp. 8709-8715.

8) Lewis, J.S. 1971, Icarus 15, pp. 174-185.
9) Lunine, J.I. and Stevenson, D.J. 1985, Astrophys. J. Suppl. Ser., in press.
10) Hartmann, W.K. 1973, Icarus 18, pp. 634-636.
11) Kawakami, S.-I., Mizutani, H., Takagi, Y., Kato, M., and Kumazawa, M. 1983, J. Geophys. Res. 88, pp. 5806-5814.
12) Sagan, C. and Dermott, S.F. 1982, Nature 300, pp. 731-733.
13) Squyres, S.W. 1981, Icarus 46, pp. 156-168.
14) Ben-Avraham, Z. and Nur, A. 1980, J. Geophys. Res. 85, pp. 4325-4335.
15) "International Critical Tables" 1928, Volume III, McGraw-Hill, New York.
16) Cruikshank, D.P., Brown, R.H., and Clark, R.N. 1984, Icarus 58, pp. 293-305.
17) Prausnitz, J.M. 1969, "Molecular Thermodynamics of Fluid-Phase Equilibria", Prentice-Hall, Englewood Cliffs, New Jersey.
18) Miller, R.C. and Stavely, L.A.K. 1976, Adv. Cryogenic Eng. 21, (K.D. Timmerhaus and D.H. Weitzel, Eds.), Plenum Press, New York, pp. 493-500.
19) Jacobsen, R.T. and Stewart, R.B. 1973, J. Phys. Chem. Ref. Data 2, pp. 757-922.
20) Cheung, H. and Wang, D.I.-J. 1964, I and EC Fundam. 3, pp. 355-361.
21) Chang, S.-D. and Lu, B.C.-Y. 1967, Chem. Eng. Progr. Symp. Ser. 63(81), pp. 18-27.
22) Samuelson, R.E. 1983, Icarus 53, pp. 364-387.
23) Hunten, D.M. 1978, in "The Saturn System", (D.M. Hunten and D. Morrison, Eds.), NASA CP 2068, pp. 127-140.
24) Moran, D.W. 1959, Dissertation, Univ. of London.
25) Lunine, J.I. and Stevenson, D.J. 1982, Bull. Amer. Astron. Soc. 14, p. 713.
26) Omar, M.H., Dokoupil, Z., and Schroten, H.G.M. 1962, Physica 28, pp. 309-329.

IMPORTANCE OF THE TECTONIC MOTIONS ON GANYMEDE

O.P. FORNI, P.G THOMAS, P.L. MASSON

Laboratoire de Géologie Dynamique Interne. Bat 509.
UA 730. Université Paris XI. 91405 Orsay Cédex. France.

ABSTRACT

Two icy satellites, Ganymede (Satellite of Jupiter) and
Enceladus (Satellite of Saturn), exhibit features that seem to be
unique in the solar system : the grooved terrains.
Local observations conducted at various scales, are
presented and discussed in order to constrain the importance of
the tectonic motions on Ganymede and of the grooved terrains
origin. Some of these observations were conducted with the help
of an image enhancement system. The observations lead us to
assume that the grooved terrains are very superficial and that
the tectonic motions in ice, if present, are very limited.

INTRODUCTION

The Voyager 1-2 spacecrafts launched in 1977, encountered
the Jovian system in 1979. These two encounters allowed close
observations of the Galilean satellite surfaces. For example,
Ganymede was observed at a 62,000 km distance by Voyager 2
cameras that obtained images with a 1 km resolution per line
pair. These images and the other scientific data collected by the
Voyager spacecrafts allowed to discover the surface characters of
each Galilean satellite and especially those of Ganymede's.
Ganymede is the largest Galilean satellite (5,270 km in
diameter). It has a moderately bright surface containing an
average of 90 % frozen water by mass (1). Voyager close-up views
reveal some provinces of old, cratered, dark terrains that are
broken into polygons and cut through by light stripes of grooved
terrains (2), (3).

759

J. Klinger et al. (eds.), Ices in the Solar System, 759–766.
© *1985 by D. Reidel Publishing Company.*

The crater density on the grooved terrains is variable, but averages about 1/10 that on the cratered terrains. The grooved terrains are therefore younger than the cratered terrains, but may have developped over an extended period of time. The grooved terrains are a mosaic of discrete systems of grooves. The grooves of one system end abruptly against the grooves of adjacent systems. When the grooved terrains are in contact with the cratered terrains, the grooves are often parallel to the contact. Each system consists of a few to several tens of grooves. Individual grooves are ca. 5-15 km wide and a few hundreds meters deep. Systems of grooves are ca. 10-100 km wide and range up to ca. 1,000 km in length. They are generally arcuate in plan and have sharp bands. It has been proposed that the grooves may have been formed by normal faulting that was produced by extensional tectonics followed by resurfacing processes (4), (5), (6), (7), (8). Similar structures are recognizable on Enceladus, other icy satellite, but are not present on Callisto though it has also an icy composition.

However, the origin and the significance of groove formation remain unclear and controversial. In order to contribute to a better knowledge of these structures, a detailed study of this terrain unit i.e., grooved terrains, and of its relationships with the other units was undertaken   . This paper aims to present and to discuss some of our local observations. This study was performed at various scales because these features seem to relate at least to two main scales of processes :
          - Minor displacements concerning only local features i.e., large scale processes.
          - Major displacements   concerning usually the major evolution of the surface i.e., small scale processes.

1. LARGE SCALE OBSERVATIONS

The feature   centered at Lat. -16°, Long. 172° (Fig. 1a) was interpreped as a crater truncated by a furrow and spread apart (9). It was therefore presented as one of the best evidence of extensional tectonics on Ganymede. However, our study of this feature allows us to propose another interpretation based on the following observations.

First, it is noticed that the northern part of the crater's rim crosscuts the limits of the furrow, meanwhile the southern part does not transsect this limit (Fig. 1b). In addition, the two parts of the crater do not have the same curvature radius : the smallest i.e., the northern part, would have a 20 km radius and the largest i.e., the southern part, would have a 27 km radius (Fig. 1c).

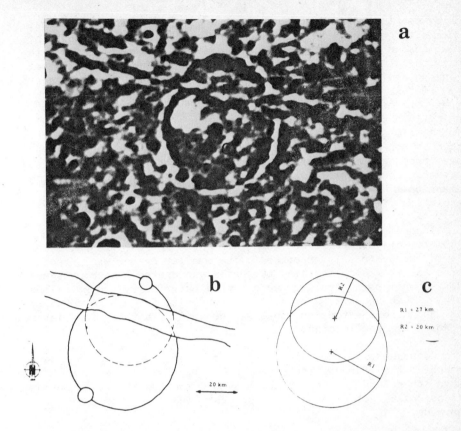

Fig. 1. (a) Detail of image 395J2-001 (Filtered) showing two
parts of craters (C1, C2) truncated by a furrow (F).
    (b) Sketch map illustrating the crosscutting of the
furrow by the crater's rim (In full line are drawn the
observations, in dashed line the virtual rims).
    (c) Interpretation in term of two distinct craters.

According to these observations, it seems that these two
crater parts do not belong to the same crater. On the contrary,
they should belong to two different craters. The smallest part
should belong to the older crater because its rim do not appear
inside the limits of the larger crater, as it should be if it has
been formed later on. Either one, the radius size difference
cannot be explained by vertical movements along the furrow. If
important vertical movement occured along the furrow, their
topographic expression should be important enough to be observed
on the Voyager high resolution images which is not the case.

Consequently, it seem unlikely that the furrow's formation

could be interpreted as the result of an important extensional
process. Moreover, in its northern limit, the crosscutting
relations of the furrow with the crater's rim seem to indicate
that the furrow is only a superficial feature. An additional
argument against an important extensional movement is provided by
the virtual part of the largest crater's rim. This virtual rim
appear to be tangent to the furrow's northern limit (Fig. 1b).
Taking into account the hypothesis of spreading mechanism, this
virtual rim should not be tangent to the furrow's northern limit,
which is not the case. Nevertheless, a limited extension that may
have produced a graben like feature, may have played a role.
However, according to our observations, the effect of such
extensional process should be smaller than the Voyager image
resolution (i.e. below 1.4 km per pixel).

According to the fact that an extensional process has
probably played a minor role along the furrow, the following
event succession is proposed : the smaller crater was created
first and it was obliterated by the larger crater. Later on these
two superimposed craters were partly burrried by the so called
furrow that appears to be only a superficial feature eventually
due to minor extension, that was covered by a thin ice layer
produced by overflooding.

Consequently the spreading mechanism, which should have
produced the furrow, seems very unlikely. The furrow cannot be
taken therefore as an argument to demonstrate that important
extensional tectonics did exist, at least in this region of
Ganymede.

2. SMALL SCALE OBSERVATIONS

The relationships that could exist, on one hand, between the
furrows observed in the terrain of Galileo Regio and the grooved
terrains, on the other hand, between two sets of grooved
terrains, were analyzed at a smaller scale than the previously
reported observations.

The relations between the furrows of Galileo Regio and the
grooved terrains that are observed in its immediate vicinity, are
presented first. For the purpose of this study, image centered at
Lat. -3°, Long 130° (Fig. 2a) was processed with an image color
enhancement system (Fig. 2b) that was used for directional
filtering (see appendix).

$$\text{Filtering based on a convolution matrix} \begin{matrix} 0 & 1 & 2 \\ -1 & 0 & 1 \\ -2 & -1 & 0 \end{matrix} \text{ enhanced}$$

the features oriented NW-SE. This image processing showed that
most of the furrows in this area seem to extend through the

grooved terrains (Fig. 2b, 2c). It was noticed additionaly that

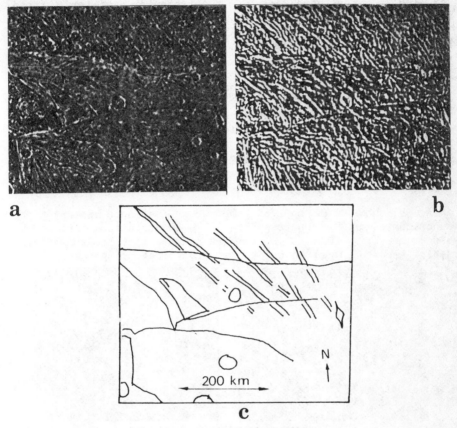

Fig. 2. (a) Detail of image 550J2-001 (Filtered) showing the
cratered terrains of Galileo Regio with its furrow system (GR)
and the grooved terrains in its vicinity (GT).
    (b) Processed image.
    (c) Sketch map illustrating the observations of the
processed image and showing the furrows extending through the
grooved terrains.

none of the other bi-dimensional analyses showed any other
preferential orientation which seems to indicate that the
structures observed through the grooves are not random features
nor artefacts. Moreover, because of their orientation and of
their apparent continuity with the furrow system of Galileo
Regio, it is deduced that these stuctures are furrows overlaid by
the grooved terrains. This implies that the grooves did not
affect significantly the underlying furrows. Consequently the
grooved terrains in this area appear to be superficial features,

else the underlying stuctures should not be apparent. Moreover, if the grooves were related to tectonic processes, these processes should have limited geometric effects i.e., relative displacements, because the underlying formations do not seem to be affected by these processes at the image resolution i.e., 800 meters per pixel.

The relations that could exist between two sets of grooved terrains located at Lat. $-30°$, Long. $129°$ (Fig. 3a) were analyzed. On the basis of crosscutting relationships, it is believed that the two sets were formed separetely in time.
The same type of image processing as previously described, was performed for the purpose of this analysis. In that particular case a convolution matrix $\begin{smallmatrix} 2 & 0 & 0 \\ 2 & 0 & -2 \\ 0 & 0 & -2 \end{smallmatrix}$ was used. The processed image (Fig. 3b) shows that one of the two grooved sets appears to be superimposed on its counterpart which seems to extend through the upper set (Fig. 3c). It is therefore deduced that the superficial set of grooves cannot be due to important tectonic process.

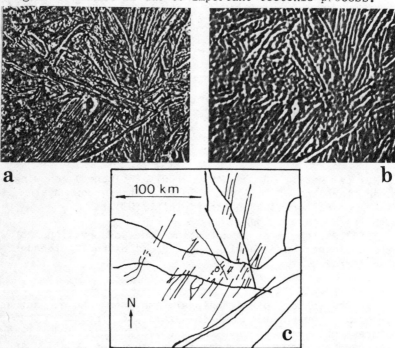

Fig. 3. (a) Detail of image 564J2-001 (Filtered) showing two sets of grooves, one (S1) seeming older than the other (S2).
        (b) Processed image.
        (c) Sketch map illustrating the processed image and showing the grooves of set S1 continuing through set S2.

CONCLUSIONS

Based on the observations conducted at various scales, it is noticed that the best evidence for extensional tectonics and spreading mechanism i.e., crater truncated by a furrow, should be taken into account carefully. The formation of this feature could be also explained by superimposition of two different craters and the truncating furrow could perhaps be due to a little graben with minor displacement accompagnied by overflooding of a thin ice cover. In addition, the grooved terrains that were studied in the vicinity of Galileo Regio do not seem to be mainly due to important extensional tectonics but to minor tectonic processes followed by superficial overflooding. It appears that tectonic processes, at least in the studied areas, if playing a role in the formation of local or regional features, are very limited. If these observations and interpretations were to be confirmed by extensive studies conducted on a more global scale, they would constrain the hypotheses of important extension on Ganymede's surface. These conclusions agree with those of (10) who show that post-grooves basins reveal the presence of a primitive "grid" pattern that is not disturbed by the grooved terrains.

It appears, so, that extentional tectonic processes, though their wide distribution and occurracy in the Solar System, are very limited on Ganymede perhaps due to its icy composition.

APPENDIX

The directional analysis is a bi-dimensional filtering applied on a 512x512 pixels digitalized image. Filtering is obtained by a 3x3 convolution matrix : each pixel of the original image is calculated again by accounting the pixel value and the values of its eight neighbours. The mathematical expression of the filtering process comes as follows :

$$V = 2^{-P} \sum_{i=1}^{9} V_i \cdot A_i + T$$

where V is the new value of each pixel, P is a constant varying from 1 to 8, V_i are the values of each previous pixel (V_5) and of its eight neighbours, A_i is the 3x3 convolution matrix which coefficients can vary from -39 to 39, T is a constant varying from 0 to 256.

ACKNOWLEDGMENTS

We are grateful to S. Equilbey from the C.D.S.I. for his help. This work was supported by Institut National d'Astronomie et de Géophysique du CNRS (ATP de planétologie) under research contract No 47.24.

REFERENCES

1. Clark, R. N., 1981, Icarus 44, pp. 388-409.

2. Smith, B. A. and the Voyager imaging team, 1979 , Science 204, pp. 951-972.

3. Smith, B. A. and the Voyager imaging team, 1979 , Science 206, pp. 927-950.

4. Allison, M. L., Head, J. W. and Parmentier, E. M., 1980, in"The satellites of Jupiter". Abstracts submitted to IAU colloqium No 57. May 13-16, 1980, Kailua-Kona, Hawaii, Sessions 6-7.

5. Head, J. W., Allison, M. L., Parmentier, E. M. and Squyres, S. W., 1981, Lunar and Planet. Science XII, pp. 418-420.

6. Parmentier, E. M., Squyres, S. W., Head, J. W. and Allison, M. L., 1982, Nature 295, pp. 290-293.

7. Squyres, S. W., Parmentier, E. M. and Head, J. W., 1981, Lunar and Planet. Science XII, pp. 1031-1033.

8. Shoemaker, E. M., Lucchitta, B. K., Wilhelms, D. E., Squyres, S. W. and Plescia, J. D., 1982, in"The Satellites of Jupiter", ed. D. Morrison, University of Arizona Press. Tucson. pp. 435-520.

9. Lucchitta, B. K., 1980, Icarus 44, pp. 481-501.

10. Thomas, P. G., Forni, O. P. and Masson, P. L., 1984, Earth, Moon and Planets.

SOME REMARKS ON THE GEOLOGY OF GANYMEDE

R. Bianchi and R. Casacchia

I.A.S. – Reparto di Planetologia, Roma, Italy

ABSTRACT

The icy jovian satellite Ganymede shows complex morphological and tectonic features whose interpretation is to be performed in the framework of the physical and chemical conditions in which the satellite evolved. Water-ice is a major constituent of Ganymede bulk composition. Two major geologic units have been recognized: cratered terrain and grooved terrain. These two units differ in albedo, crater density and in the tectonic framework. In this paper we report the preliminary results of a detailed geologic study of some of the largest areas of cratered and grooved terrains. Aspects of the geologic history of the icy surface of Ganymede are also discussed.

1. INTRODUCTION

Voyager images of the jovian galilean satellites have revealed that these bodies experienced an intensive geologic activity. The same processes that modelled the surfaces of the terrestrial planets, impact cratering, volcanism and tectonics, have occurred and, in some cases, are still active on Io, Europa, Ganymede and Callisto. In particular Ganymede's geologic history seems

J. Klinger et al. (eds.), Ices in the Solar System, 767–779.
© *1985 by D. Reidel Publishing Company.*

to have been characterized by at least three major e-
vents: an intensive extensional tectonic regime in the
early stage of its history, a period of heavy meteoritic
bombardment and resurfacing of large portions of its sur
face. All these events occurred in an environment chemically and physically different than that in which the
terrestrial planets evolved.

Water ice is a major constituent of Ganymede bulk
composition. Reflectance spectra data allowed the detection of H_2O on the surface of the satellite (1) and its
low mean density (1.92 g/cm^3) suggests that large portions of the interior could be constituted by volatiles
such as water-ice. Furthermore the morphologies observed
on Ganymede are unique among the bodies of the Solar Sys
tem imaged so far and suggest a complex geologic history
of the satellite.

The key to reconstruct the geologic history of Gany
mede's surface is provided by the detailed analysis of
its geologic features. Preliminary studies on Ganymede
images made by Smith et al. (2) allowed the recognition of two major units: heavily cratered areas and grooved terrain. The cratered terrain consists of low albedo surfaces (0.35) dissected by strips or belts of bright
grooved terrain (albedo 0.43), (3). Cratered and grooved
terrains differ also in the density of impact craters
and in the tectonic framework. The tectonic pattern of
large portions of cratered terrain is represented by fur
row systems extending for several hundreds up to a few
thousands of kilometers. The grooved terrain is character
ized by a mosaic of grooves domains varying in orientation and density. Crater densities along with the transecting relationships between the cratered and the grooved terrains indicate that the cratered terrain is the
oldest surface on Ganymede.

The cratering record of Ganymede shows a variety
of morphologies from fresh craters to palimpsests, impacts features with a flattened topography and an albedo comparable to that of the grooved terrain. Depth/di-
ameter ratios suggest that the overall morphology of
Ganymede's craters is generally shallower than that of
craters formed on the silicate surfaces of the inner
planets (4). This is probably due to the different re-

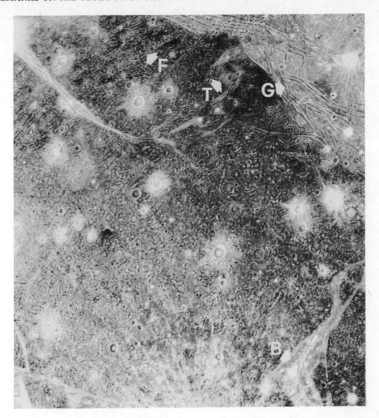

Figure 1. North-central part of Marius Regio confining
          with the southern Uruk Sulcus (grooved ter-
          rain). Furrows (F) on Marius Regio occur in
          the northern part decreasing in sharpness and
          density southward. In the picture are also in-
          dicated: sets of grooves (G), bright smooth
          terrain (B), transitional terrain (T).
          (Voyager 2 image centered at 8°N, 183°W; PICNO
          382J2-001).

sponse of an icy surface to the impact process resulting
in a higher rate of relaxation (5) of the original crater morphology or to different populations of impacting
bodies (6).
    This study presents the preliminary results obtain-
ed in mapping two of the largest areas of heavily cra-

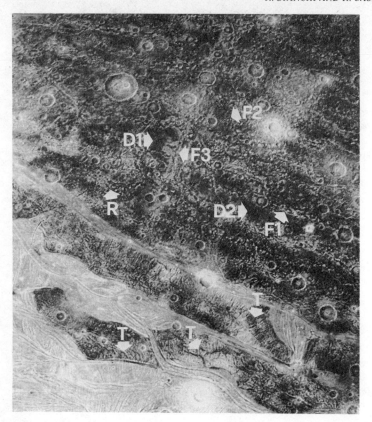

Figure 2. Southern Galileo Regio bounded by grooved ter-
          rain. Transitional terrain (T) occurs at the
          boundaries between the two units. N-S struc-
          tures (F2) intersect the main NW-SE furrow
          system (F1) with no appreciable offset. Smooth
          (D1) and rough (R) terrains show a sharp roun-
          ded boundary. Structure belonging to the NE-SW
          furrow system (F3); dark smooth terrain (D2)
          associated to F1. (Voyager 2 image centered
          at 14°N, 158°W; PICNO 461J2-001).

tered terrain, Galileo and Marius Regios, and Uruk, Tia-
mat, Mashu and Anshar Sulci (grooved terrain) on Ganymede. The mapped features have been subdivided in types of
terrains including both features related to impact craters and morphologic units identified on the basis of

albedo, degree of roughness and the geometric pattern of the terrain.

2. CRATERED TERRAIN

The cratered terrain has been interpreted as the oldest surface of Ganymede. It probably consists of ice with a mixture of rocky material which could be responsible for the lower albedo of the unit. Although cratered terrain appears similar on a global scale, a close examination of this unit on Galileo and Marius Regios leads to the recognition of different morphologic features on the two areas. Marius Regio shows almost uniformly a rough, irregular surface with a high density of small impact craters (center of Fig. 1), less than 10 km in diameter, while on the Galileo Regio two main morphologic units have been identified: rough terrain and dark smooth terrain (R and D in Fig. 2), (7). The first appears quite similar to that of Marius Regio. On both the areas the elements of roughness are constituted by arcuate, randomly oriented ridges, which could be the remnants of old impacts and of ancient tectonic features.

The dark smooth terrain on Galileo Regio is located mostly in the southern portion of the area and is spatially related to the most prominent tectonic structures of the region. Their albedo does not differ from that of the whole region and they are represented by smooth surfaces with low density of small impact craters. The boundaries between the rough and the dark smooth terrains on Galileo Regio are rarely sharp: these two units generally fade one into the other creating several inter mediate terms. In a few cases the dark smooth terrain shows circular boundaries with rough terrain (D1 in Fig. 2) resembling the remnants of disrupted crater rim and there fore it could be the floor of ancient impact crater flood ed by endogenic material. Furthermore dark smooth terrain appears related to the main tectonic structures of the region (D2 in Fig. 2).and looks younger than rough terrain. This suggests that the dark smooth terrain is probably fluid material extruded along structural weak-

nesses of the crust after rough terrain formation. The
absence of dark smooth terrain on Marius Regio could be
interpreted as the result of different tectonic history
of this area compared to the Galileo Regio.

In both regions the dominant tectonic framework
is a broadly arcuate furrow system NE-SW oriented on
Marius Regio (F in Fig. 1) and NW-SE oriented on Galileo
Regio (F1 in Fig. 2). Furthermore, on Galileo Regio the
furrow system extends all over the surface area covering
a range in latitude of 60°; on Marius Regio it is concen
trated mostly in the north-western portion of the area
and decreases progressively in sharpness and density of
the structures up to disappear southward. In addition to
this on Galileo Regio two more systems of furrows, respec
tively oriented N-S (F2 in Fig. 2) and NE-SW (F3 in Fig.
2) have been observed. Transecting relationships between
the three systems of furrows indicate that the NE-SW
system is the oldest, while the N-S appears to be the
youngest (7).

It is interesting to observe that the furrow systems on Galileo Regio intersect without or with little
displacement and that they all predate the formation of
any recognizable impact crater larger than 10 km in diameter. The global morphology of the furrow systems apparently resemble that of the grabens formed on the silicated surfaces of the terrestrial planets. For this rea
son they have been interpreted as the result of extensive
tectonic regimes occurred early in Ganymede's history.
Individual furrows can show different morphologies. In
the southern Galileo Regio they show raised rims border-
ing flat-floored valleys and have regular width and spac
ing. In the north-central portion of the same area furrows are more irregular with a width ranging from about
6 to 20 km and their morphology is represented also by
lineated hummocks or subparallel, discontinuous ridges.

Furrows on Marius Regio globally show a lower state
of preservation than on Galileo Regio and are more dense
ly spaced. The differences between the two areas could
be attributed to differences in the stresses that formed
these structures or to a different rheological behaviour
of the two regions.

3. GROOVED TERRAIN

Grooved terrain covers about the 60% of the imaged Ganymede's surface. The wide distribution of this unit testify that, after the solidification of the outermost layers, Ganymede was largely resurfaced. The grooved terrain is constituted by a mosaic of areas (or domains) characterized by different grooves orientation and density (G in Fig. 1) and areas totally smooth (bright smooth terrain; B in Fig. 1).

Grooved terrain has been observed for the first time on a planetary surface: its peculiar morphology is the result of surface deformations determined by tectonic stresses that occurred in an icy crust. The boundaries between the single domains can be sharp (graben like structures), or faint, just showing changing in groove orientation. On the grooved terrain we observe that, similarly to the cratered terrain, the main tectonic structures appear extensive in origin. At present, we cannot speculate about the mechanism which led to the formation of the different domains of grooves. Nevertheless a recent study by Bianchi et al. (8), concerning the azimuthal frequency and the cumulative lenght distributions of the tectonic features of the Uruk Sulcus region, suggests a close relationship between the orientations of the main graben-like structures and of the grooves. The former are largely concentrated in a range of 30° about a main NW-SE orientation. The grooves are distributed according to two prevalent orientations: NW-SE and NNE-SSW. NW-SE domains cover about the 35% of Uruk Sulcus, they show a higher grooving density and their structures are shorter than those of the NNE-SSW domains, which appear less densely grooved and occupy the majority of the region. The preliminary results of this statistical approach in studying the tectonics of the grooved terrain seems to indicate that the domains of grooves were not due to local stresses but are somehow related to the formation of the main structures of the grooved areas.

4.   TRANSITIONAL TERRAIN

This unit generally occurs at the boundaries between the cratered and the grooved terrains (T in Figs.
1 and 2). This terrain has been defined transitional because it exhibits features typical of the two major u-
nits. Its albedo is generally similar to that of the cra
tered terrain, but it could be of the same order of magnitude of the grooved terrain. Although few impact craters are superimposed on the transitional terrain its
morphological aspect can be sometimes similar to the old
est unit. Transitional terrain also shows graben-like
features and systems of densely spaced subparallel grooves, similar to those of the grooved terrain. The structures belonging to the domains of grooves extend even
to the transitional terrain.

This unit has to be considered as a structural unit.
Its morphology indicates that it is constituted by
edges of ancient cratered terrain, probably subjected
to the same stresses responsible for the formation of
the grooved terrain.

Finally the peculiar morphology of the transitional
terrain indicates that portions of the ancient cratered
unit has probably been modified by tectonic stresses due
to internal activity, thus representing a resurfacing
process of an icy surface. Therefore extrusion of water
ice material with a low rock content (bright terrain)
is not the only resurfacing process on Ganymede.

5.   CRATERING RECORD

The morphology of the impact craters on Ganymede
has been widely described by Passey and Shoemaker (4).
In a recent paper, Casacchia and Strom (7) pointed out
that several morphological transitions can be observed
on fresh craters on the Galileo Regio in relation to
their diameter. A bowl-shaped morphology is characteristic of objects smaller than 10 km in diameter. Central
peaks and flat floors are shown by craters in the range
10-20 km, while at larger diameter peaks are replaced
by pits generally associated with domed floors. Ejecta

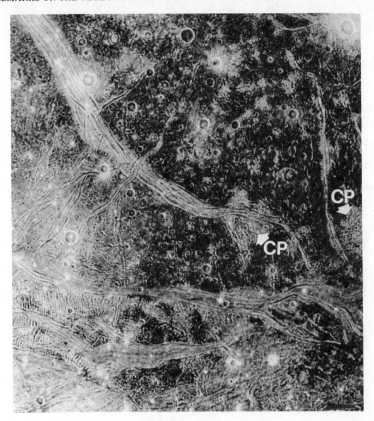

Figure 3. Southern Marius Regio. Cratered palimpsests
         (CP) are shown. These features are considered
         the oldest impacts structures on Ganymede.
         (Voyager 2 image centered at 18°S, 158°W;
         PICNO 482J2-001).

deposits are typically confined around the crater rim
in circular patterns extending for about 1-2 crater radii. The freshest craters show bright ejecta deposits;
on the cratered terrain the ejecta can show an albedo
similar to that of the palimpsests. On the grooved terrain the fresh craters often exhibit dark floors and
dark ejecta deposits. Rayed craters are also common on
Ganymede's surface and appear to be distributed mainly
on the grooved terrain rather than on the cratered terrain.

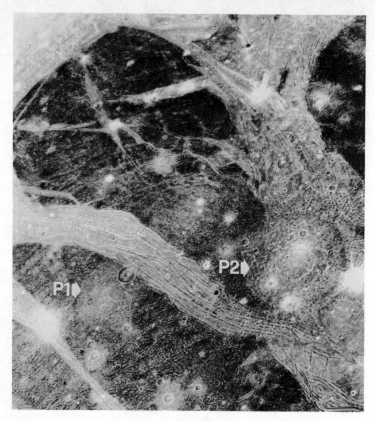

Figure 4. Northern Marius Regio. Pre-grooved palimpsests
         (P1) and post-grooved palimpsests (P2) are
         shown. (Voyager 2 image centered at 31°N, 188°
         W; PICNO 370J2-001).

    Globally impact craters formed onto an icy surface
show a more flattened morphology than their counterparts
formed onto a silicatic target and the crater population
observed on Ganymede appears depleted in small craters
compared to the cratering record of the terrestrial plan
ets (5). These observations are consistent with the rheological properties of icy crustal material the character
istics of which can change even for little variations of
temperature and pressure (9). The viscous relaxation of
the original impact topography on a icy crust is controlled by the thermal evolution of the satellite. This can

account for the observed crater flattening. Moreover,
a morphometrical analysis of a particular class of impact craters on Ganymede (domed central pit craters)
showed a possible correlation between crater morphometry
and the evolution of the crust which has changed during
its history in thickness, density and in the ratio rock
to ice.(10).

The striking aspect of Ganymede's cratering record
is the occurrence, both on the grooved and the cratered
areas, of crater palimpsests, bright circular spots,
showing a multiring structure around a central smooth
area. Palimpsests also show a wide range of morphologies
and many attempts to classify them have been carried out
by several authors. Passey and Shoemaker (4) discriminate between palimpsests and penepalimpsests of type I
and II; Casacchia and Strom (7) on the Galileo Regio
have mapped palimpsests and central pit palimpsests. We
classified three categories of palimpsests on the basis
of their global state of preservation and the overlapping relationships with the cratered and the grooved
terrains: cratered palimpsests (CP in Fig. 3), pre-grooved terrain palimpsests (P1 in Fig. 4) and post-grooved
terrain palimpsests (P2 in Fig. 4).

On the cratered regions palimpsests postdate the
main tectonic structures and overlap both the rough and
the dark smooth terrains. Their occurrence on the grooved terrain with morphologies similar to those of the
palimpsests on the cratered terrain suggests that they
do not represent a time-horizon marker in Ganymede's
crustal evolution. Theoretical and experimental studies
have been carried out in order to explain the origin of
the palimpsests. Two possibilities seem to be consistent
with our observations: they can either be structures
formed on thermally active ice surface or by high impact
velocity objects, different from those responsible for
the formation of the fresh craters (11; 12).

6.   SUMMARY

Geologic mapping of Galileo and Marius Regios shows
that the dominant morphologic unit is represented by
rough, irregular surfaces disrupted by intense extensive

tectonics. This unit, defined as rough terrain, has been
interpreted as the oldest unit and probably represents
the pristine crust of Ganymede. On Galileo Regio dark
smooth areas, spatially associated with the main tectonic structures indicate that this area has been locally
resurfaced by fluid material extruded along crustal struc
tural weaknesses. The tectonic activity occurred on the
cratered areas after the emplacement of the rough terrain resulting in the formation of arcuate furrow systems. The tectonic framework of the Galileo Regio is rep
resented by systems of furrows with three main directions
and different ages; on Marius Regio furrows posses one
main trend and appear discontinuous or totally absent.
This suggests differences in the stresses that determined the formation of the tectonic framework of the two
areas or a different rheological behaviour of Galileo
and Marius Regios when tectonic disruption occurred.

As suggested by the overlapping relationships between tectonic features and impact structures heavy bombardment of Ganymede surface occurred after the formation of the furrow systems. Crustal fracturing was still
active towards the end of the heavy cratering stage.
Crater densities on grooved and transitional terrains
are lower that on the cratered units but still characterized by a considerable number of impacts.

The occurrence of bright terrain indicates a complete breakdown of the dark unit and the replacement of
the oldest terrain by presumably water-ice material with
a low rock content. Locally ice was emplaced with a relatively smooth surface even though most of the bright material was modified in grooves by tectonic stresses that
probably produced plastic deformations. Whatever the ori
gin of the grooved terrain its structural pattern reveals a complex geologic history. Grooves on Uruk Sulcus
are distributed according to main trends, respectively
parallel and perpendicular to the trend of the main struc
tures of the area (NW-SE). This correlation seems to favour the hypothesis that grooves formation is somehow
connected with the emplacement of the main structures
and therefore they are probably due to regional tectonic
regimes rather than to local randomly oriented stresses.

Finally transitional terrain apparently indicates

that in same areas the breakdown process of the dark terrain was incomplete as revealed by the morphology and the albedo of the unit.

ACKNOWLEDGMENT

We thank J.E. Guest and B.K. Lucchitta for useful discussions.

REFERENCES

(1) Pilcher, C.B., Ridgway, S.T. and McCord, T.B. 1972, Science 178, pp. 1087-1089.
(2) Smith, B.A. and the Voyager Imaging Science Team 1979, Science 206, pp. 927-950.
(3) Squyres, S.W. 1981, NASA TM-84412, p. 356.
(4) Passey, Q.R. and Shoemaker, E.M. 1982, Satellites of Jupiter (D. Morrison, Ed., Tucson Arizona), pp. 379-434.
(5) Parmentier, E.M. and Head, J.W. 1981, Icarus 47, pp. 100-111.
(6) Woronow, A., Strom, R.G. and Gurnis, M. 1982, Satellites of Jupiter (D. Morrison, Ed., Tucson Arizona), pp. 237-276.
(7) Casacchia, R. and Strom, R.G. 1984, Jour. Geoph. Res. 89, Suppl., pp. B419-B428.
(8) Bianchi, R., Casacchia, R. and Pozio, S. 1984, XV Lunar and Plan. Sci. Conf., pp. 54-55.
(9) Cassen, P.N., Peale, S.J. and Reynolds, R.T. 1982, Satellites of Jupiter (D. Morrison, Ed., Tucson Arizona), pp. 96-128.
(10) Bianchi, R. and Pozio, S., 1984, Annales Geophysicae, in press.
(11) Greeley, R., Fink, J.H., Gault, D.E. and Guest, J.E. 1982, Satellites of Jupiter (D. Morrison, Ed., Tucson Arizona), pp. 340-378.
(12) Croft, S.K. 1983, Proc. XIV Lunar and Plan. Sci. Conf., pp. 72-89.

TECTONICS OF VALHALLA BASIN ON CALLISTO

P.G. THOMAS and P.L. MASSON,

Laboratoire de Geologie dynamique Interne, UA 730,
Bat. 509, Université Paris Sud, 91405 Orsay cedex,
FRANCE.

ABSTRACT

Callisto and Ganymede, the two largest known icy bodies
in the solar system exhibit concentric structures (impact basins)
which seem be unique in the solar system. Their "special" appearance is probably due to the physical and geometrical properties
of the icy lithosphere. But the geometric analysis of Valhalla
basin (detailed shapes or global distibution) seems to indicate
that the multi-ridges, scarps and furrows system around Valhalla
is the result of "Valhalla motions", that occured on a prefrac-
tured icy lithosphere. The direction of this old fracturation
may suggest that a planetary grid pattern existed on Callisto at
the time of Valhalla formation. Such a grid discovered on Callisto (icy body) exists also on almost all silicated bodies
(exept the Earth) despite the difference of lithospheric chemistry.

INTRODUCTION

The surface morphology of the outer Galilean Satellite,
Callisto, observed in 1979 by the Voyager missions, is dominated
by two large concentric patterns of bright ridges, scarps and
furrows named Asgard and Valhalla. Asgard and Valhalla basins are
respectively about 1500 and 4000 km in diameter. These structures
are the only tectonic features described on Callisto. The
multiring basins are recognized as major geological features
since the end of 19 th century. Such basins were observed on most
of the solid planetary bodies. These intensively studied
structures were interpreted as large impact basins. In generally,
the formation of the outer scarps have been interpreted as the

781

J. Klinger et al. (eds.), Ices in the Solar System, 781–790.
© *1985 by D. Reidel Publishing Company.*

result of radially motions directed toward the center of the cavity ("mega-slumping"). But the geometry and the morphology of the two ganymedean multi-ring basins differ in many aspects from their silicated counterparts.

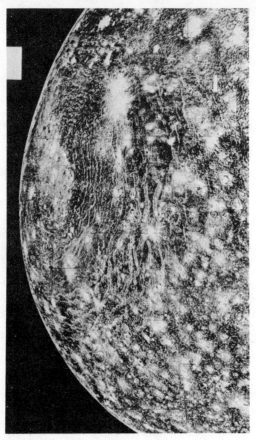

Fig 1) Mosaic of Voyager I images (n° 21285) of Callisto showing the 4000 km wide system of ridges, scarps, and furrows, and the bright center that defines the Valhalla structure.

The Valhalla structure (fig.1)

The center of Valhalla structure is a bright, relatively smooth and apparently ring free area, which is about 350 km in radius. Series of bright, roughly concentric ridges, scarps and furrows surround the central area of the structure and form rings extending outward from the edge of the bright interior. From 400 to 600 km of the center, the ridges are more or less subdued, closely spaced and more or less sinuous. The zone between 600 and 800 km is sparcely occupied by subdued ridges.

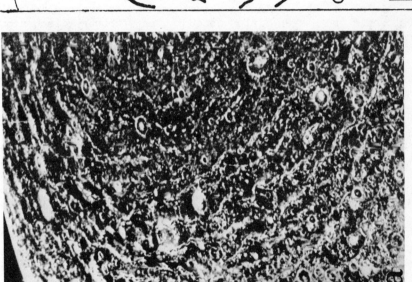

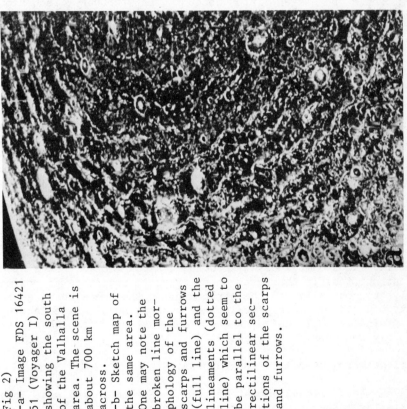

Fig 2)
-a- Image FDS 16421
51 (Voyager I)
showing the south
of the Valhalla
area. The scene is
about 700 km
across.
-b- Sketch map of
the same area.
One may note the
broken line mor-
phology of the
scarps and furrows
(full line) and the
lineaments (dotted
line) which seem to
be parallel to the
rectilinear sec-
tions of the scarps
and furrows.

Beyond 800-900 km, and as far as 2000 km, the scarps and furrows
are very well observed, and it is noticed that their spacing is
wider than near the center of the structure.

Several explanations were proposed to explain these
numerous ridges and scarps around Valhalla. The "Tsunami-like
oscillations" model (1, 2) was proposed for Callisto by Smith et
al (1979), (3). It was also proposed by Mc Kinnon and Melosh
(1980), (4), that impact craters penetrating a thin and weak
lithosphere overlaying a relatively fluid asthenosphere would
induce an asthenospheric flow pattern. This flow pattern produces
concentric fault scarps. These two hypothesis involve a thin and
weak icy lithosphere at the time of the impact, this greatest
thinness of the lithosphere being probably a consequence of its
icy nature.

A third explanation was proposed by Wood (1981), (5),
with regard to terrestrial morphological analogs of Valhalla
basin found in icelandic glaciers: a diapiric instability of a
thick and dense lithosphere into the water mantle inducing a
subsidence cauldron. The purpose of this paper is not to discuss
the origin of the Valhalla stucture, but to propose new data
which could be taken into account for further discussions
concerning the origin of this structure, or the thermal and
rheological models of the icy mantle of Callisto.

Detailed geometry of the scarps and furrows

We only undertook the study of the outer, scarps and
furrows beyond 700 km from the center. These rings are approxi-
matively concentric to the basin. But their circular extend is
discontinuous and they are irregular in plan. They are not
sections of circular arcs but exhibit a broken line morphology
(fig. 2). The rectilinear sections of these broken lines are
about 50 km in length and they often show a two by two
paralellism. Somewhere, the rectilinear sections continue as
faint but distinguishable lineaments, that extend beyond one or
both of their ends. Somewhere else, a lineament extends between
two rectilinear sections of two different ridges.

Such a pattern would be interpreted in term of inheri-
ted structures: the radial stress due to the formation of
Valhalla induced motions which occured along local preexisting
discontinuities in the areas where they are oriented approxima-
tively tangentially to the structure. In the areas where appro-
ximatively tangential preexisting discontinuities are lacking,
the "Valhalla motions" induced new structures, which are
perfectly concentric.

Global geometry of scarps and furrows.

As the local geometric patterns seems to indicate local
preexisting patterns, we undertook a global geometrical and

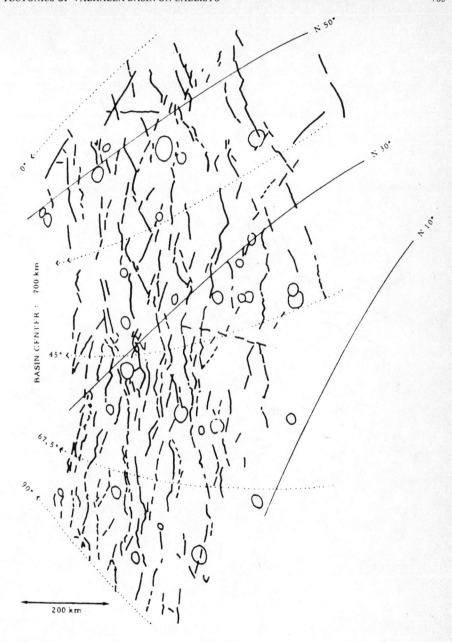

Fig 3) Non rectified structural map of the northern outer part
of Valhalla basin. One can note the broken line morphology of the
scarps and furrows (full line). The dotted lines are the limits
of the angular sectors used for fig 5.

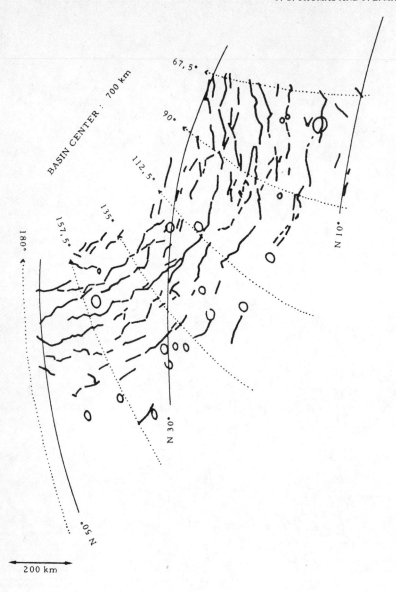

Fig 4) Non rectified structural map of the southern outer part of Valhalla basin. One can note the broken line morphology of the scarps and furrows (full line) The dotted lines are the limits of the angular sectors used for fig 5.

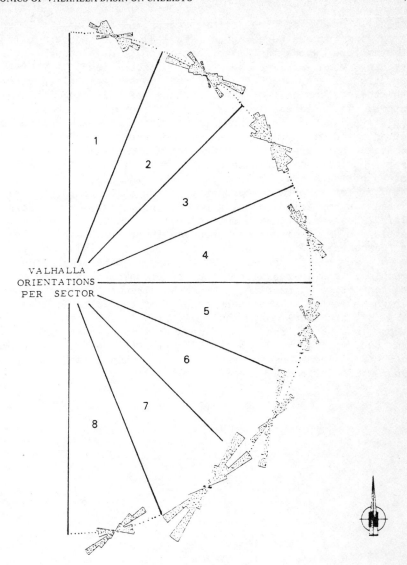

Fig 5) Scarps and furrows azimuthal rose diagrams per 22.5°
sectors for the eastern outer part of Valhalla basin. One can
note the non perfect concentricity of the structures, the
presence of the NW-SE direction in all the northern sectors and
the presence of the NE-SW direction in all the southern sectors.

azimuthal study of the outer scarps and furrows. All these
scarps and furrows were mapped (fig 3-4) in the eastern half of
the structure, where the image resolution is high enough. Optical
streching were applied to the oblique wiews. Arithmetical
corrections were used to avoid the images scale discrepancies.
Then the eastern half of the structure was divised into 22.5°
sectors. The ridges, scarps and furrows orientations and
cumulative length according the location of each sector were
plotted, and a systematic search for trends were conducted for
each sectors. Where the ridges are broken-line in plan, the
directions of the rectilinear sections were taken into account,
instead of the average direction of the ridges.

        The important concentric orientation of the scarps and
furrows examplified on figure 5 indicates their "Valhalla
origin". However, this concentric orientation is not perfect. For
example the NW-SE direction observed in sector 3 corresponds to
the concentric direction. But it is also observed in sectors 2
and 4, where it makes a 22.5° angle with the concentric
direction, and in the sector 1 where it makes a 45° angle with
the concentric direction. The SW-NE direction is also observed
everywhere in the southeastern quarter of the Valhalla area.

        The global azimuthal distribution of these ridges and
scarps in the entire mapped area shows a bimodal distribution
with two maxima: N.50° and N.150° (fig. 6). One can note that
these two directions are similar to the main directions of the
"lunar grid" defined by Strom (1964), (6).

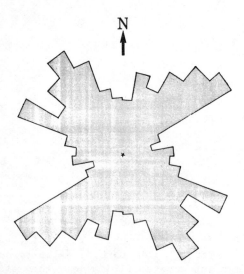

Fig 6) Global azimuthal rose diagram of the Valhalla scarps and
furrows length. We can note the bimodal distribution of these
structures and the dominant NW-SE and NE-SW directions.

It seems unlikely that such an unequal distribution could be only the result of an impact mechanism, which would have created an approximative circular pattern without any preferential trend dominance for one half of the structure.

The "sunlight direction effect" which emphasize ridges perpendicular to the sunlight direction would only produce a unimodal distribution. Moreover, this effect is here very low, because at the time the pictures were obtained, Valhalla was near the subsolar point.

Even if the bimodal distribution could be the consequence of the non circularity of the ridges and scarps, the variation of the curvature radius of the ridges being the result of lithospheric properties (variation in thickness, ...), the broken line morphology of the ridges, the presence of lineaments apparently "used again" and the coincidence of the Moon and Callisto's directions could not be explained.

It is proposed therefore that this global geometry is the result of the reactivation of an old global pattern -a grid of Callisto- by the Valhalla formation. This grid has two dominant directions: $N.50^{\circ}$ and $N.150^{\circ}$. The old pattern was intensively reactivated in the sectors where its orientation fitted with Valhalla stresses. This hypothesis would explain both the detailed morphology and geometry, and their global azimuthal distribution. The coincidence of the "Callisto and Moon grid's" directions would indicate a possibly common origin, i.e. tidal flexing or despinning.

CONCLUSION AND SUMMARY

These observations and their interpretation have two important implications on Callisto's history:
-Callisto is not a "tectonic-free" planetary body, as usually though, but, as most of silicated bodies observed so far in the solar system, and despite its icy nature, it exhibits a planetary grid pattern, that probably results from tidal flexing or despinning. We can note that such a global pattern was recently discovered on Ganymede, the largest icy satellite of Jupiter (7) and on Dione, an $NH3-H2O$ icy satellite of Saturne (8).
-At the time of Valhalla event, the icy lithosphere of Callisto was thick and strong enough to preserve the existance of a fracture pattern during the time interval between the despinning or flexing and the Valhalla event. This result should be taken into account in thermal and rheological models of the icy mantle of Callisto.

AKNOWLEDGMENT: This work was supported by Institut National d'Astronomie et de Geophysique du CNRS (ATP de Planetologie) under research contract n° 4724.

REFERENCES

1. Van Dorn, W. G., 1968, Tsunamis on the Moon, Natu-
    re 220, pp. 1102-1107.

2. Baldwin, R. B., 1974, On the origin of the mare ba-
    sins, Proc. 5th Lunar Sci. Conf. 1, pp. 1-10.

3. Smith, B. A., et al, 1979, The Jupiter system
    through the eyes of Voyager 1, Science 204, pp.
    951-972.

4. Mc Kinnon, W. B. and Melosh, H. J., 1980, Evolu-
    tion of planetary lithospheres : evidence from
    multiringed structures on Ganymede and Callisto,
    Icarus 44, pp. 454-471.

5. Wood, C. A., 1981, Possible terrestrial analogs of
    Valhalla and other ripple-ring basins, in "Multi-
    ring Basins", Proc. of Lunar and Plan. Sci. 12 A,
    pp. 173-180.

6. Strom, R. G., 1964, Analysis of lunar lineaments :
    tectonic maps of the Moon, Com. Lunar and Plane-
    tary Lab. Univ. of Arizona 39, pp. 205-216.

7. Thomas, P. G., Forni, O. P. and Masson, P. L.,
    1984, Geology of large impact craters on Ganyme-
    de : implications on thermal and tectonic histo-
    ries, submitted to Earth, Moon and Planets.

8. Moore, J. M., 1984, The Tectonic and Volcanic his-
    tory on Dione, Icarus 59, pp. 205-220.

CHRONOLOGY OF SURFACE UNITS ON THE ICY SATELLITES OF SATURN

Joseph M. Boyce[1] and Jeffrey B. Plescia[2]

1. NASA Headquarters, Code EL-4, Washington,
    D. C. 20546, U. S. A.
2. U.S. Geological Survey, 2255 North Gemini Drive,
    Flagstaff, Arizona 86001, U. S. A.

ABSTRACT

Because the impact flux history in the Saturnian system has been similar in its general form to that for the inner solar system and Jupiter it is possible to develop an absolute calibration model for the relative ages derived from the density of impact craters. This has been done by identifying the density at which the crater populations change characteristics from that during post-accretion heavy bombardment to a later population of low but continuous flux. The transition between these two populations occurred at a known time throughout the solar system, 3.9 by, and provides an absolute calibration point on which to construct a calibration curve. Results based on calibrating relative age data to this absolute flux model suggest that most surfaces are ancient > 3.9 by (or within several 100 million years). Only Enceladus and Dione have had prolonged activity that has resulted in resurfacing. The activity on Enceladus was probably driven by tidal forces whereas the mechanism for driving resurfacing on Dione has yet to be identified.

INTRODUCTION

The establishment of the sequence of formation of geologic units on a planet weather composed of silicate, ice, or a mixture of the two is paramount to understanding the geologic history of that body. While relative age dating is important, our understanding is most significantly increased when the sequence is put on an

791

J. Klinger et al. (eds.), Ices in the Solar System, 791–804.
© *1985 by D. Reidel Publishing Company.*

absolute time scale.

        The densities of impact craters superimposed on
planetary surfaces have been widely used as estimate of
the relative age. A knowledge of the history of the cratering rate on the planet can be used to "calibrate" the
crater density (relative) ages to the absolute time scale. This has been done for the Moon, where radiometric
sample ages were compared to observed crater densities
at the Apollo and Luna Landing sites (1, 2, 3, 4, 5).
These studies showed that early in lunar history, the
cratering rate was very high but decreasing exponentially until about 3.9 b. y. ago. Thereafter the flux became relatively low and nearly constant.

        A knowledge of the current population of small
bodies such as planet crossing asteroids and comets in
the solar system has been used to extrapolate the lunar
flux history to Mars (6, 7, 8, 9, 10, 11) and to the Galilean satellites (12). These studies suggest that the
flux histories in other parts of the solar system behaved in a manner similar to the moon. While the relative
contribution from various potential debris sources,
which are responsible for the craters, may have varied
and overall rates may have been different, the time scale and general characteristics of the sweep up has had a
similar character in both the inner (Mercury, Moon, Mars)
and outer (Jupiter, Saturn) parts of the solar system.

        The similar overall nature of impact cratering
histories result from the very rapid rate at which material is perturbed into planet crossing orbits or is ejected from the solar system. Because the probability of encounter of this material by the planets is so great, the
early sweep up of this material is expected to be rapid
with a time scale independent of position in either the
inner or outer solar system. After this early rapid
sweep up, only long period comets and the "Uranus-Neptune planetesimal cloud" material (13) appear to have orbital lifetimes longer than the age of the solar system.
It is because of these long lifetimes that Shoemaker and
Wolfe argue that comets (both extinct and active) have
been the dominate crater forming objects, particularly
in the outer solar system, over the last 3.5 b. y.

        In this paper, the impact flux history observed
and modeled for the inner solar system and Jupiter is extrapolated farther into the outer solar system, to the
Saturnian system. A model impact flux history, similar

to that of the inner solar system, was combined with
information regarding changes in the population of impact craters, to provide a basis for the modeled impact
flux history of the Saturnian system. The proposed model has been used to estimate absolute ages for surfaces on the Saturnian satellites from the observed crater densities (relative ages). Hence, resurfacing histories of the Saturnian satellites can been developed
based on this model.

IMPACT CRATERING HISTORY OF THE SATURNIAN SATELLITES

        Impact craters were counted for most observed
surface on the Saturnian satellites that were imaged at
high resolution by Voyagers 1 and 2 (14, 15, 16, 17).
The density of impact craters was found to vary by two
orders of magnitude from the least to the most cratered
surfaces (Table 1). The cumulative size-frequency distribution curves, constructed from the counts, showed a
systematic difference in slope as a function of crater
density (time). Heavily cratered surfaces typically have distribution curves with slopes of approximately -2.
Such distributions have been termed Population I (14,
15). Lightly cratered surfaces, typically having distribution curves with a -3 slope, are termed Population II.
A few surfaces have crater statistics which are intermediate between the heavily and lightly cratered surfaces
having distributions with slopes between -3 and -2.
Smith et al. authors (14, 15) have theorized that a
third population, Population III, exists. This third population includes only the youngest impact craters, most
of which are too small to be seen in Voyager images. Population III was thought to be responsible for the obli-
teration of the wispy terrain on the leading edges of
Dione and Rhea as theorized to result from a low flux of
impacts in the outer solar system after about 3.5. b. y.

        Smith et al. (14, 15) have discussed the ori-
gin of the different populations of craters and crater
forming objects recorded on the Saturnian satellites.
Population I is thought to have been caused by the impact of debris present near the end of the heavy bombardment, early in solar system history. Shoemaker and
Wolfe (13) agreed with Smith et al. (15) on the timing
and suggest that Population I craters were also formed
in part by material perturbed inward from the vicinity
of Uranus and Neptune.

        The origin of and, hence, evolutionary history
of Population II craters is more controversial. Smith

T A B L E   1

SATELLITE	O.C.D.	F.	N.C.D.	AGE B. Y.
Mimas	900 ± 197	0.05	45 ± 10	3.9 ± 0.05
	380 ± 113		19 ± 5	3.8 ± 0.05
	160 ± 59		8 ± 3	3.0 ± 0.2
Enceladus	168 ± 60	0.08	13 ± 5	3.7 ± 0.1
	75 ± 88		6 ± 7	2.9 ± 0.8
	32 ± 56		3 ± 4	1.3 ± 1.6
	0 + 0		0	0.8
Tethys	700 ± 66	0.18	129 ± 12	4.0 ± 0.05
	170 ± 36		31 ± 5	3.9 ± 0.05
	125 ± 34		24 ± 6	3.8 ± 0.05
	120 ± 50		22 ± 9	3.8 ± 0.3
Dione	275 ± 51	0.29	80 ± 15	3.9 ± 0.05
	72 ± 21		21 ± 6	3.8 ± 0.3
	16 ± 16		5 ± 5	2.4 ± 1.2
	7 ± 14		2 ± 4	0.8 ± 1.0
Rhea	380 ± 120	0.53	200 ± 63	4.0 ± 0.05
	274 ± 49		145 ± 26	4.0 ± 0.05
	200 ± 50		105 ± 26	3.95± 0.05
	160 ± 55		84 ± 29	3.9 ± 0.05
	15 ± 13		10 ± 8	3.5 ± 0.35
Hyperion	150 ± 58	0.44	66 ± 25	3.9 ± 0.05
Iapetus	740 ± 30	1.00	740 ± 30	4.0 ± 0.05

O. C. D. : Observed·crater density $\geq$ 20 km in diameter
            per 10^6 km^2.
F.   : Normalization factor, scales crater densities
            relative to Iapetus.
N. C. D. : Normalized crater densities from column O.C.
            D. using F.
AGE   : Age in billions of years estimated from mo-
            del.

et al. (14) suggested that Population II craters were "secondaries", resulting from ejecta derived from large Population I impacts on the satellites or the debris resulting from the breakup of smaller co-orbiting satellites. Therefore, they suggest that Population II craters were generated during or at the end of the formation of Population I craters and entirely within the Saturn system. Such a "secondary" origin has been used to explain the absence of the predicted leading edge - trailing edge crater density gradient (18). However, Plescia and Boyce (19) suggest that both Population II and III are actually the same, with Population III simply being the small size fraction of Population II. Plescia and Boyce further indicate that Population II represents a lower, but uniform flux of planet crossing objects which continued to impact after sweep-up of the last of the accretionary debris that formed Population I craters.

To support the contention that Population II craters are not generated predominantly as secondary debris internal to the Saturnian system, Plescia and Boyce (19) noted the severe timing problems in the geologic histories of the satellites that the suggestion of Smith et al (14, 15) imply. In order for Population II craters to be produced from locally derived debris, all Population II surfaces must have formed between the time of the production of secondary debris by the last Population I crater and the time that debris is reaccumulated to form the Population II craters. The time required for reaccumulation of debris is very short, on the order of 10^1 to 10^4 years (20, 21). Therefore, all Population II surfaces must have been formed within less thant 10^4 years, between the formation of the last large Population I craters and the sweep-up of debris they produced. Such ad hoc circumstances for the formation of all Population II surfaces is improbable and requires identical resurfacing histories for nearly all of the Saturnian satellites.

Further support for an external origin of Population II craters is provided by the probability of disruption of small satellites, as determined from the density of craters observed on the surfaces of the satellites. The statistical probability of destruction of co-orbiting bodies to produce Population II craters is very low (19). At the end of the heavy bombardment the probability, as determined from crater densities on Population I surfaces, was so low that for a 30 km diameter satellite, there was less than a 0.1 chance of des-

truction. Disruption and fragmentation of larger satellites, such as Hyperion or Phoebe, is even less probable because of the requirements for enormous craters to
cause fragmentation (the crater size-frequency distribution is inversely proportional to the square or the
cube of the crater diameter).

However, this is not to say that such co-orbiting satellites were never fragmented or destroyed, but
rather, that it occurred early, well before the end of
Population I time. The most heavily cratered regions of
the major satellites have crater densities that provide
a sufficient probability to suggest that fragmentation
of co-orbital satellites could have occurred. However,
such events would have occurred in Population I time,
too early to effect Population II cratering. As a result, the debris from these impacts could be incorporated into Population I craters.

For the above reasons, the most likely source
of material to produce Population II craters is not locally derived debris from Population I impacts, but long
period objects, such as that recorded on the moon after
about 3.8 b. y. ago. We agree with Smith et all. (14,
15) and Shoemaker and Wolfe (13) that Population I is
equivalent to the crater population of the lunar highlands and that both terminated at about 3.9 b. y. ago.

Therefore, we suggest that the impact flux history at Saturn was similar throughout most of the solar
system ; an early period of heavy bombardment that terminated about 3.9 b. y. ago and then a low, nearly constant flux of impacts to the present. This conclusion
forms the basis of our model of the impact flux history
in the Saturnian system.

BASIS FOR ABSOLUTE CALIBRATION OF RELATIVE AGE DATA

As mentioned, the crater size-frequency distributions determined by Smith et al. (14, 15) and Plescia
and Boyce (16, 17, 19), can be assigned to either Population I or II with a few exceptions which exhibit a
transitional (mixed) population. The presence of several areas with crater statistics intermediate between
Populations I and II on several satellites suggests that
the transition was not geologically instantaneous. Had
the transition been rapid, surfaces would display only
Population I or II craters. This situation is similar
to that for the light upland plains on the Moon and further suggests a similarity in history. The crater den-

sity which separates Population I from Population II distributions most likely represents the density corresponding to the time when the heavy bombardment was terminating (a point like that at 3.9 b. y. on the knee in the lunar flux curve). At Iapetus, the transition density transition is 25 craters/10^6 km^2 with diameters greater than or equal to 20 km. Alternatively, G. Neukum (personal communication) has suggested that there was continuous and gradual change (selective sweep up of large or high velocity bodies) in the characteristics of a single debris population and that these transitional areas merely record a segment of this continuous change.

Because the gravitation focusing effect of projectiles by Saturn (15) varies the effective cratering rate on each satellite, direct comparison and correlation of the crater density data on different satellites is not immediately possible. Shoemaker and Wolfe (18) have calculated the relative effects of gravitational focusing by Saturn on the crater populations of each satellite. They indicate that this effect increases the cratering rate with decreasing orbital radii in the Saturn system. The result is a cratering rate at Mimas which is about 20 times that at Iapetus (Table 1). Using the data of Shoemaker and Wolfe (18), the crater density data can be normalized to a common rate. We have done this and chosen to normalize the cratering rates relative to Iapetus (Table 1).

For comparisons of crater densities between satellites and for use in the model, the densities of craters > 20 km in diameter were chosen. This size was selected as it represents a compromise between image resolution and cratering statistics. The resolution of Voyager images commonly prevented the consistent observation and counting of craters less than 20 km in diameter. The use of a diameter much greater than 20 km was also unsuitable because of the limited number of such craters on some sparsely cratered surfaces. Crater density data, (for craters > 20 km/10^6 km^2) that was previously collected by Plescia and Boyce (16, 17, 19) is shown in Table 1. Also included in this table are the data after normalization to the cratering rate at Iapetus. The normalized density of impact craters at the Population I - Population II transition, at the knee in the flux curve, is about 25 craters/10^6 km^2 with diameter > 20 km.

We have used this value of 25 craters > 20 km/10^6 km^2 as a normalized calibration point at 3.9 b.

y. ago, and the previous arguments concerning the similarity of the history of impacts in the solar system,
to develop a model for the impact flux history for the
saturnian system. This model flux can be used to estimate the absolute ages of surface on the Saturnian satellites using crater density data. We have done this
in the next section and discuss the implications to geologic history of the satellites.

   The flux predicted here, and by Plescia and
Boyce (19), is about an order of magnitude higher than
the flux calculated by Shoemaker and Wolfe (18), who based their estimates on the observed small body population of the outer solar system. However, the difference in expected flux between these two models may be the
result of invalid assumptions on the part of Plescia
and Boyce (19), or of a sampling peculiarity brought
about by either short period cycles in comet flux caused by astronomical effects (22, 23) or by the high
statistical uncertainties due to the small sample analyzed by Shoemaker and Wolfe (18).

GEOLOGIC IMPLICATIONS

   Most of the observed surfaces on the Saturnian
satellites are heavily cratered and exhibit Population
I type craters and high crater densities, suggesting
that these surfaces are ancient, > 3.9 b. y. old. Surfaces so characterized include large parts of Rhea,
Diona, Hyperion and Mimas and all of the bright terrain
on Iapetus. Because the impact flux was exponentially
decaying before 3.9 b. y. ago, the large range in density (over two orders of magnitude) of craters on this
ancient surface probably only represents a short period of time ($\sim 10^8$ years).

   However, a few surfaces exhibit Population II
type distributions with low crater densities. These
surfaces formed after about 3.9 - 3.8 b. y. ago. Such
surfaces include all surfaces imaged at high resolution
on Enceladus, as well as parts of Mimas, Rhea and Dione. Using the flux history discussed in the previous
section and by Plescia and Boyce (19), an absolute
chronology of these surfaces can be estimated. Figure
1 shows the normalized crater density data from Table
1 plotted on the modeled cummulative impact flux curve
(discussed previously) and suggests some surprising results. Rather than static unevolved bodies, some of the
Saturnian satellites have had prolonged, dynamic geologic histories. Specifically, Figure 1 indicates that:

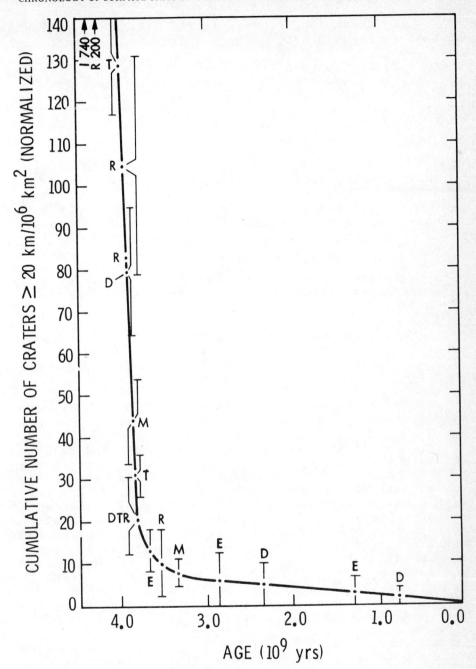

Fig. 1   Modeled flux history at Saturn, normalized to Iapetus. Also shown are crater densities for various surfaces on the satellites from Table I.

1) A mantling deposit in the equatorial region of Rhea is about 3.5 ± 0.3 b. y. old.

2) The south polar region of Mimas lacks craters larger than 30 km in diameter and is about 3.4 ± 0.25 b. y. old.

3) A smooth plains unit on the leading edge of Dione is about 2.4 ± 1.2 b. y. old.

4) A lobate deposit on the leading edge of Dione (24) is about 0.9 ± 1.0 b. y. old.

5) All density distributions observed on Enceladus are Population II type. The oldest unit is located in the north polar region and is about 3.7 ± 0.1 b. y. old.

6) Two sparsely cratered smooth plains units occur at lower latitudes on Enceladus and are about 2.9 ± 0.8 b. y. and 1.3 ± 1.6 b. y. old.

7) A smooth plains unit on Enceladus with no observable craters (larger than the limit of resolution of the ∿ 2 km) is less than 0.8 b. y. old and could be more recent.

8) All other surfaces observed, the heavily cratered areas, are ancient 3.9 b. y. or older.

The restricted size of the surfaces and the relatively low crater density of the younger surfaces result in large statistical errors associated with the data (an estimate of one standard deviation). Therefore, within the statistical limits of error many of these surfaces could be very young or ancient. More notable is that there are several surfaces even taking into account the associated uncertainty, which are relatively young : less than 3.0 b. y. old. In particular the lobate deposit on Dione, the sparsely cratered smooth plains and the uncratered smooth plains on Enceladus are predicted here to be no older than approximately 1.9 b. y., 2.9 b. y. and 0.8 b. y. old respectively.

Long lived internal activity in small low temperature icy bodies is unexpected. Most lack obvious energy sources that could provide the power to drive internal melting for extended periods (several billion years). Thermal history models (25, 26) for these satellites ba-

sed on radioactive decay as a heat source suggest that
due to the presence of ice clathrates and a surface regolith some internal melting might be expected for 10^8
to 10^9 years after satellite formation. Enceladus is
the exception. Yoder (27) has suggested that the long
lived activity on Enceladus is probably sustained by
tidal heating rather than from radioactive decay.

However, neither of these energy sources explain
the extended period of activity on Dione. Current models for internal melting due to radioactive heating
(28) suggest that Dione should not be active for the several billion years suggested here, even though Dione
contains the highest silicate fraction. In addition, the
current orbital dynamical models do not identify a candidate resonance that would provide tidal heat to drive
the prolong activity of Dione (29).

Therefore, Dione stands as a particular mystery
with young resurfacing processes. Dione is also peculiar in that it appears to modulate Saturn's kilometric
radio emission (30, 31) much like Io modulates Jupiter
decimetric radio emissions. There may be a cause and
effect relationship between these two observations, such
as the interior structure of Dione is arranged in a manner that would cause such a modulation or Dione is currently active.

SUMMARY

Most of the surface units of the small icy regular satellite of Saturn examined are ancient, 3.8 b.
y. and older. Only Enceladus and Dione have surfaces
that are significantly younger. Geologic characteristics of these young units suggest that the resurfacing
mechanism was driven by some internal processes. For
Enceladus the energy source to facilitate internal heating is probably tidal stresses. However, for Dione, no
energy source has been identified that to drive the prolonged activity indicate here.

ACKNOWLEDGEMENTS

The authors are grateful to the reviewers for
their constructive comments. Work on this research was
carried out at the Jet Propulsion Labs and the U. S. Geological Survey and supported by NASA through grants from
the Planetary Geology and Geophysics Program, Outer Planets Data Analysis Program and Voyager Project.

REFERENCES

1. Hartmann, W. K., 1970, Preliminary note on lunar
    cratering rates and absolute time scales, Icarus,
    12, pp. 131-133.

2. Hartmann, W. K., 1972, Paleocratering of the Moon :
    Review of post-Apollo data, Astrophys. Space Sci.,
    17, pp. 48-64.

3. Soderblom, L. A. and Lebofsky, L. A., 1972, Tech-
    nique for rapid determination of relative ages of
    lunar areas from orbital photography, J. Geophys.
    Res., 77, pp. 279-296.

4. Neukum, G., Konig, B., Fechtig, H. and Storzer, D.,
    1975, Cratering in the Earth-Moon system : conse-
    quences for age determinations by crater counting,
    Proc. Lunar Sci. Conf. VI, pp. 2597-2620.

5. Boyce, J. M., Schaber, G. G. and Dial, A. L., Jr.,
    1977, Age of Luna 24 mare basalts based on crater
    studies, Nature, 265, pp. 38-39.

6. Hartmann, W. K., 1971, Martian cratering 4, Mari-
    ner 9 initial analysis of cratering chronology,
    J. Geophys. Res., 78, pp. 4096-4116.

7. Hartmann, W. K., 1977, Relative crater production
    rates on planets, Icarus, 31, pp. 260-276.

8. Soderblom, L. A., Condit, C. D., West, R. A., Her-
    mon, B. M. and Kreideler, T. J., 1974, Martian
    planetwide crater distributions : Implications
    for geologic history and surface processes, Ica-
    rus, 22, pp. 239-263.

9. Soderblom, L. A., 1977, Historical variations in
    the density and distribution of impacting debris
    in the inner solar system : Evidence from plane-
    tary imaging, in "Impact and Explosion Cratering",
    eds. D. J. Roddy, R. O. Pepin, and R. B. Merrill,
    Pergamon, New York, pp. 629-633.

10. Neukum, G. and Wise, D. U., 1976, Mars : A stan-
    dard crater curve and possible new crater time
    scale, Science, 194, pp. 1381-1387.

11. Neukum, G. and Hiller, K., 1981, Martian ages, J.
    Geophys. Res., 86, pp. 3097-3121.

12. Shoemaker, E. M. and Wolfe, R. F., 1982, Cratering time scales for the Galilean satellites, in "Satellites of Jupiter", ed. D. Morrison, Univ. Arizona Press, pp. 277-339.

13. Shoemaker, E. M. and Wolfe, R. F., 1984, Evolution of the Uranus-Neptune planetesimal swarm, Lunar Planet. Sci. Conf. XV Abs., pp. 780-781.

14. Smith, B. A., et al., 1981, Encounter with Saturn : Voyager 1 imaging science results, Science, 212, pp. 163-191.

15. Smith, B. A., et al., 1982, A new look at the Saturn system : The Voyager 2 images, Science, 215, pp. 504-537.

16. Plescia, J. B. and Boyce, J. M., 1982, Crater densities and geologic histories of Rhea, Dione, Mimas, and Tethys, Nature, 295, pp. 285-290.

17. Plescia, J. B. and Boyce, J. M., 1983, Crater numbers and geological histories of Iapetus, Enceladus, Tethys and Hyperion, Nature, 301, pp. 666-670.

18. Shoemaker, E. M. and Wolfe, R. F., 1981, Evolution of the Saturnian satellites : The role of impact ; Lunar Planet. Sci. Conf. Abs. XII, Supplement A, pp. 1-3.

19. Plescia, J. B. and Boyce, J. M., 1985, Impact cratering history of the Saturnian satellites, J. Geophys. Res., in press.

20. Harris, A., 1984, personal communication.

21. Neukum, G. and Pozio, S., 1984, The cratering record of Ganymede, XV Lunar Planet. Sci. Conf. Abs., pp. 601-602.

22. Smoluchowski, R. and Scalo, M., 1984, Galactic gravitational shock waves and the extinction of the species, Bull. Amer. Astron. Assoc., 16, p. 493.

23. Smoluchowski, R. and Torbett, M., 1984, Orbital stability of the unseen solar companion linked to periodic extinction events, Nature, submitted.

24. Plescia, J. B., 1983, The geology of Dione, Icarus,

      56, pp. 255-277.

25.   Stevenson, D. J., 1982, Volcanism and igneous pro-
      cesses in small icy satellites, Nature, 298, pp.
      142-144.

26.   Ellsworth, K. and Schubert, G., 1983, Saturn's icy
      satellites : Thermal and structural models, Ica-
      rus, 54, pp. 490-510.

27.   Yoder, C. F., 1979, How tidal heating in Io drives
      the Galilean orbital resonance locks, Nature, 279,
      pp. 767-770.

28.   Stevenson, D. J., 1984, personal communication.

29.   Cuzzi, J. and Lissauer, J., 1984, personal commu-
      nication.

30.   Desch, M. D. and Kaiser, M. L., 1981, Saturn kilo-
      metric radiation : Satellite modulation : Nature,
      292, p. 739.

31.   Kurth, W. S., Gurnett, D. A. and Scarf, F. L.,
      1981, The control of Saturn's kilometric radio
      emission by Dione : Nature, 292, p. 742.

DISCUSSION

P. Weissman :

        The flux of objects at Saturn curren is at least
twice that at the Earth assuming either an Oort cloud
or trans-Neptunian planetesimal source (or some combination of the two). How then can you say that the current cratering note is an order of magnitude less than
the terrestrial value ? Are you talking about absolute
crater size or are you including energy scaling ? Even
so, I can't see that kind of cratering rate difference
assuming similar sources.

J. Plescia :

        The statement that the current cratering rate at
Saturn is ∿ 10x less than at the Earth is based on a
comparison of the terrestrial rates of Shoemaker and
those for Iapetus from our work. The cratering rate obviously increases with decreasing satellite orbital radius. The cratering rate on Mimas is 20x that of Iapetus
and therefore twice the terrestrial one.

SULFUR DIOXIDE ICE ON IO

D. P. Cruikshank and R. R. Howell

Institute for Astronomy
University of Hawaii
Honolulu, Hawaii 96822 USA

T. R. Geballe

Kapteyn Astronomical Institute
Rijksuniversiteit te Groningen
and United Kingdom Infrared Telescope
900 Leilani Street
Hilo, Hawaii 96720 USA

F. P. Fanale

Planetary Geosciences Division
Hawaii Institute of Geophysics
University of Hawaii
Honolulu, Hawaii 96822 USA

ABSTRACT

The 4-μm band system in the spectrum of Io, as well as the ultraviolet absorptions, are caused by SO_2 ice or frost distributed over a major fraction of the satellite. The spectral contribution of any adsorbed gas component of the surface cannot be discriminated from the ice absorption band with the data now available. The 4-μm band is strongest on the leading hemisphere of Io and weakest on the trailing. No temporal changes are seen in the six-year interval in which the infrared data have been studied. Individual spectral features attributed to SO_2 with various combinations of the ^{32}S, ^{34}S, ^{16}O, and ^{18}O isotopes are seen in the 4-μm region.

805

J. Klinger et al. (eds.), Ices in the Solar System, 805–815.
© *1985 by D. Reidel Publishing Company.*

Infrared spectra of Io during one of its eclipses by Jupiter
in 1983 revealed particles or flakes of SO_2 snow in the volcanic
eruption plume above the Loki volcano. Calculations of the eruption rate of SO_2 from Loki from our observations are consistent
with independent determinations of the deposition rate made from
other data. Spectroscopic studies in combination with photometric observations in the thermal infrared are providing a satisfactory way in which to monitor the activity of the volcanoes on
Io from ground-based observations.

INTRODUCTION

The discovery of sulfur dioxide on Io in the gaseous and
solid forms and the identification of its sources in the active
volcanism on that body are very recent developments in planetary
science. These discoveries, together with the development of our
understanding of the anomalous heat flow from the satellite and
its probable origin in the tidal competition between Jupiter and
the other three Galilean satellites, represent an advance of
enormous importance in our understanding of planetary processes.
Many aspects of the subject have been reviewed in the book by
Morrison (1), and more recently at the Ithaca conference on
planetary satellites (IAU Colloquium 77, July 1983; a conference
volume edited by J. A. Burns and D. Morrison is scheduled for
publication in (2), Univ. of Arizona Press). At that conference, Nash et al. (3) reviewed the evidence for SO_2 as a gas,
a solid, and as an adsorbed species permeating the soil of Io.

In this paper, we discuss certain aspects of the presence of
SO_2 ice on Io and defer to Nash et al.'s (3) major review for
a study of the other phases of this molecule.

SPECTROSCOPIC EVIDENCE FOR SO_2

The 4-μm region of Io's spectrum is the only part observable
from the Earth-based observatories in which distinct spectral
bands are found. Spectrophotometry was first obtained, approximately simultaneously, by Cruikshank et al. (4) with a ground-
based telescope and by Pollack et al. (5) with the Kuiper
Airborne Observatory. Both sets of observations showed a deep
absorption band centered between 4.0 and 4.1 μm, and a weaker
feature at 3.75 μm. These features were identified independently
as condensed sulfur dioxide by Fanale et al. (6), Smythe et
al. (7), and Hapke (8). In an analysis of spectra from
Voyager 1, Pearl et al. (9) found the ν_3 fundamental of SO_2
gas at 7.4 μm near one of Io's volcanic plumes. Further support
for solid SO_2 on Io came from the identification of a 0.32-μm
absorption edge in the satellite's spectrum as sulfur dioxide by

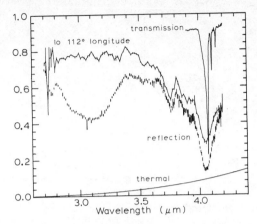

Figure 1. The geometric spectral albedo of Io, the reflectance of SO_2 frost, the transmission of a film of SO_2 frost (Barbe et al. (10)), and the thermal component of the Io spectrum. The transmission spectrum was obtained with much higher spectral resolution than the Io and reflectance data. The 3-μm band in the reflection spectrum is due to traces of H_2O in the sample; the surface of Io appears to be completely anhydrous. Reproduced from Howell et al. (11).

Nelson et al. (12). Using IUE and ground-based spectra, Nelson et al. have shown that the ultraviolet absorption in Io's spectrum is variable with the satellite's orbital (rotational) phase angle.

In Figure 1 we show spectra of Io obtained with a large ground-based telescope, and spectra of SO_2 frost or ice observed in transmission and diffuse reflection. The close coincidence among these data sets constitutes the primary basis for the identification of sulfur dioxide frost on Io. We note that central wavelength of the SO_2 gas absorption band ($v_1 + v_3$) is 4.00 μm, and even at the moderately low spectral resolution of the data in Figure 1 the gas band would appear significantly different from the laboratory frost sample. The fact that the ($v_1 + v_3$) band is variable in strength at different locations on Io is further evidence that it originates on the surface in an irregular distribution. While there may be a component of the overall absorption in the 4-μm spectral region that is contributed by molecules of SO_2 adsorbed on the grains of a very porous soil or microregolith, considerations of the band strength and its central wavelength of the primary absorption indicate that it originates in sulfur dioxide ice or frost on the surface of Io. Details of the identification of the state of the SO_2 on Io are given by Howell et al. (11) and in the review by Nash et al. (3).

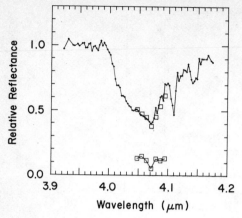

Figure 2. A higher resolution spectrum of Io (solid line) than
that shown in Figure 1. The broad absorption feature is SO_2,
while the sharp bands at 4.11 and 4.14 μm are the ^{34}S and ^{18}O
isotopic bands in the sulfur dioxide on Io's surface. The lower
squares are the data obtained during eclipse, while the upper
squares are those obtained immediately after the satellite
emerged from Jupiter's shadow. From Howell et al. (13).

The laboratory transmission spectrum of SO_2 ice reproduced
here in Figure 1 is from the data of Barbe et al. (10), who
identify several isotopes of oxygen and sulfur. Howell,
Cruikshank, and Geballe (Figure 2, and in a paper in preparation)
have obtained new spectra of Io with substantially higher
spectral resolution than that shown in Figure 1. The new data
show individual spectral bands of SO_2 in various combinations of
the ^{32}S, ^{34}S, ^{16}O, and ^{18}O isotopes.

DISTRIBUTION OF SO_2 ON THE SURFACE

In the earliest spectroscopic studies by Cruikshank et al.
(4) it appeared that the 4-μm absorption band strength is
variable with position on Io. The satellite is in locked
synchronous rotation with its revolution around Jupiter so that
the same hemisphere is directed toward Jupiter at all times.
From Earth most of the satellite's surface can be seen as it
revolves about Jupiter in a cycle synchronized with its orbital
period of 1.769 days. Nelson et al. (12) found that the
ultraviolet absorption is also dependent upon longitude on Io.

In a more thorough study of the infrared spectral data,
Howell et al. (11) have shown a clear pattern in the distribution of the sulfur dioxide from the variable strength of the 4-μm
absorption. In Figure 3, from the Howell et al. work, both the

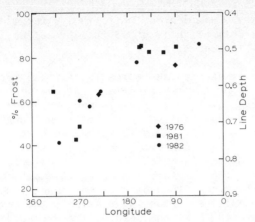

Figure 3. Plotted against the longitude on Io's surface is
the depth of the 4.07-μm SO_2 frost band from telescopic observations. Also shown on the ordinate is the fractional coverage of
SO_2 frost on the satellite's surface according to detail given by
Howell et al. (11), from which this figure is reproduced.

depth of the 4-μm band and the implied (modeled) fractional surface coverage are shown versus longitude on Io. The central
meridian (0°) of Io is the center of the hemisphere directed
toward Jupiter. The greatest abundance of sulfur dioxide frost
is found in a band of longitudes centered at approximately 90°,
consistent with the conclusions of Nelson et al. (12). W. M.
Sinton (in preparation) has refined this study of the SO_2 distribution using a filter observation technique and finds maximum
frost absorption centered near 70° with a minimum near 300°.

     To a first approximation, the distribution in the strength
of the SO_2 frost band is consistent with the distribution of the
"white plains unit" observed in the Voyager pictures of Io (see
Smith et al. (14, 15) ; Schaber (16) ; Howell et al. (11)).
Howell et al. (11) have shown, however, that there must be a
significant amount of the frost mixed with the darker soil outside the white plains units mapped by the geologists in order to
account for the strength of the 4-μm absorption band. This is
reflected in the band depth shown as an ordinate in Figure 2,
where even in the zones of minimum SO_2 absorption, some 50% of
the surface must have exposed frost.

     While the infrared observations presented here indicate a
surface dominated by SO_2, the UV observations discussed in Nash
et al. (17) suggest that at most 20% of the surface is covered
with optically thick frost. The Voyager UV observations also
show only minor amounts of SO_2 in units other than the white
plains. These estimates are based on the fact that the frost is

bright in the 0.32- to 0.42-μm region, while the sulfur is dark.
The observed albedo is only slightly above that of sulfur,
suggesting only a small amount of frost.

    This apparent inconsistency can be resolved if one notes
that the relative albedo of the two materials is reversed in the
two cases. In the ultraviolet, the frost is much brighter than
the sulfur, while in the 4-μm band, the sulfur is much brighter
than the frost. The Nash et al. analysis was based on the stated
assumption that all spatial domains with an SO_2 spectral component were covered with optically thick SO_2 frost. However, if
the two components are mixed on a scale smaller than the optical
pathlength of a typical photon, then at each wavelength, the
darker component will dominate, producing precisely the results
seen. The more intimate the mix, the less of the darker component is necessary. Conversely, if the mix can be assumed to be
intimate and the albedo is significantly above that of the dark
component, then its concentration must be relatively small. It
would seem that this is the case on Io. Unfortunately, the
temperatures and the natures of the materials involved make it
extremely difficult to perform the laboratory modeling that has
been done in the analogous case of silicate/water ice mixture.

    If both ultraviolet and infrared data were available for the
individual color units, then it would be possible to apply the
above arguments to them. This would allow us to discern the
extent of the mixing and estimate the relative amounts of sulfur
and SO_2 in each unit. However, spatially resolved information is
available only in the ultraviolet; we can only assume that the
intimate mixing, present in most units, is present in all. In
that case, the white plains, previously estimated to consist of
50% SO_2, actually contain a higher percentage. Soderblom et al.
(18) suggest that small amounts of SO_2 may be mixed through all
of Io's surface types. If the mix is intimate, then the percentages of SO_2 must again be larger.

    Howell et al. (11) found no evidence for changes in the
SO_2 abundance at a given longitude over a period of a year.
Comparing their 1981/1982 data with the earlier Cruikshank et al.
(4) data from 1976, they found no evidence for changes in a
six-year period, estimating that no more than 10% change in the
strength of the main spectral band could have occurred.

SO_2 FROST AND POST-ECLIPSE PHENOMENA

    On each of Io's revolutions about Jupiter, the satellite
traverses the shadow of the planet. During these eclipses, when
there is no direct sunlight falling on the satellite, the surface
temperature falls by several tens of degrees. Binder and

Cruikshank (19) found from telescopic measurements that Io
emerges from Jupiter's shadow excessively bright by about 10%,
and that after 10-15 minutes it fades to its normal brightness.
This phenomenon has been confirmed by some other observers, but
it appears to be intermittent. Fanale et al. (20) showed that
condensation and sublimation of SO_2 frost on some regions of Io's
surface could account for the observed effect, but a complete
understanding of the phenomenon is still lacking. We note it
here in the context of SO_2 frost or ice on Io because of the possible role that this volatile substance may play in causing the
post-eclipse brightening effect.

SO_2 FROST IN IO'S VOLCANIC PLUMES

The geometry of Jupiter, Io, Earth, and the Sun allows Io to
be visible from Earth while still in Jupiter's shadow at some
times. During eclipses the thermal radiation of Io's surface can
be detected in the absence of reflected sunlight. Sinton (21)
and Johnson et al. (22) have shown that there is substantial
thermal radiation at 4 μm wavelength, and that during the geometric configurations of the eclipses, most of the 4-μm radiation
emanates from one major volcanic area on the satellite (Johnson
et al. (22)).

Howell et al. (13) observed the spectrum of Io in the
region of the 4.07-μm SO_2 band in 1983 and found the band in
absorption during eclipse. Because the bulk of the 4-μm background radiation emanates from one major thermal region, and
because the implied surface temperature of that region is far
above the sublimation temperature of SO_2 ice, Howell et al.
(13) conclude that the band they observed is formed by ice
particles, or snowflakes, suspended above the satellite's surface. The thermal region visible from Earth during the
eclipses is the Loki region, found during the Voyager missions to
be the site of a major eruptive plume (Strom and Schneider (23)).
In this same plume, Pearl et al. (9) found SO_2 gas absorption
with the infrared spectrometer on board Voyager 1.

Figure 2 shows the Howell et al. (13) spectrum obtained
during eclipse, with the normal spectrum (diffusely reflected
sunlight) shown for comparison. The coincidence in wavelength of
the band observed in and out of eclipse indicates that the
eclipse band is indeed SO_2 ice rather than gas; the central wavelength of the gas band is 4.0 μm. Howell et al. have modeled the
strength of the band using laboratory data on SO_2 ice (G. T.
Sill, private communicaton) and derive a pathlength of ~4 x 10^{-4}
cm through the ice. If the density of the ice is 1 g cm^{-3}, the
line-of-sight abundance is 4 x 10^{-4} g cm^{-2}.

We can estimate the rate at which solid SO_2 is deposited on the surface of Io by making a plausible assumption about the dimensions of the background thermal area, and then use the mean dimensions of the Loki plume complex and the size of the outfall halo described by Strom and Schneider (23). The Loki plume complex emanates from an area of dimensions 40 x 180 km, according to Strom and Schneider (23). If the total area of the hot zone contributing the 4-μm radiation observed from Earth is about twice the area of the plume source, the total area is about 1.5 x 10^{14} cm^2. The abundance of SO_2 ice in the plume is then 6 x 10^{10} g. The outer diameter of the diffuse outfall halo around the Loki plume complex is 850 km. If the majority of the mass from the plume falls in a ring with this outer dimension and an inner diameter of 500 km, the total area covered is 3.7 x 10^{15} cm^2. Thus, 1.6 x 10^{-5} g cm^{-2} of SO_2 ice will be deposited in one transit time defined by the time that an individual particle spends in the space above Io after it is ejected from the volcanic vent. Ignoring all effects but gravity, if the plume height is on the order of 100 km, the transit time is about 900 sec. The annual deposition rate of solid SO_2 on the surface in the halo is then 0.56 cm per year, if the plume system is continuously active and if the Howell et al. (13) observations refer to a mean solid SO_2 abundance. This value lies toward the upper end of the wide range (10^{-3}-10 cm per year) suggested by Johnson and Soderblom (24) from consideration of a number of factors independent of the above argument. The Johnson and Soderblom analysis for Loki in particular includes particles in the range 0.01-0.1 μm and gives 10^{-2}-1 cm per year. The presence of large particles would increase the deposition rate, of course.

MONITORING IO'S VOLCANOES

The confluence of numerous discoveries in the last few years offers the exciting prospect of monitoring in a quantitative way the activity of at least one major volcano on another planetary body indefinitely into the future. The highlights of these discoveries are (1) active volcanism and its morphological characteristics as seen from the Voyager spacecraft, (2) recognition that spectral features seen from Earth-based telescopes are caused by the SO_2 ice in those volcanic eruptive plumes, and (3) recognition that at certain critical geometries when Io can be observed in eclipse from Earth, one major volcano on the satellite is visible. Because this geometry is repeated at every eclipse, Loki can be monitored in terms of the evolution of the thermal region (from infrared photometry) and the SO_2 content of the eruption plume. With improved spectral resolution, it may be possible to distinguish in the ground-based telescopic data the contributions of SO_2 ice and gas to the eruption process.

SUMMARY

The presence of sulfur dioxide ice on Io has been established, though questions remain as to its areal surface coverage
and the nature of its mixing with the satellite's surface microregolith (soil). The volcanic plumes are clearly the dominant
source of the SO_2 ice, while sublimation and sputtering are at
work to remove the ice from the surface. The problems of removal
and redistribution of SO_2 on the surface of Io are complex and
not fully resolved. Fanale et al. (25) have reviewed the problem recently, and reference is made to their paper for a detailed
treatment of the known and unknown elements in the discussion.

REFERENCES

1. Morrison, D., Ed., 1982, "Satellites of Jupiter", Univ. of
   Arizona Press, Tucson.

2. Burns, J. A. and Morrison, D., Eds., 1985, "Natural Satellites", Univ. of Arizona Press, Tucson, in press.

3. Nash, D. B., Carr, M. H., Gradie, J., Hunten, D. M. and Yoder,
   C. F., 1985, Io. In "Natural Satellites", eds. J. A. Burns
   and D. Morrison, Univ. of Arizona Press, Tucson, in press.

4. Cruikshank, D. P., Jones, T. J. and Pilcher, C. B., 1978,
   Absorption bands in the spectrum of Io. Astrophys. J. 225,
   pp. L89-L92.

5. Pollack, J. B., Witteborn, F. C., Erickson, E. F., Strecker,
   D. W., Baldwin, B. J. and Bunch, T. E., 1978, Near-infrared
   spectra of the Galilean satellites : Observations and compositional implications, Icarus 36, pp. 271-303.

6. Fanale, F. P., Brown, R. H., Cruikshank, D. P. and Clark, R.
   N., 1979, Significance of absorption features in Io's IR
   reflectance spectrum, Nature 280, pp. 761-763.

7. Smythe, W. D., Nelson, R. M. and Nash, D. B., 1979, Spectral
   evidence for SO_2 frost or adsorbate on Io's surface, Nature
   280, p. 766.

8. Hapke, B., 1979, Io's surface and environs : A magmatic-
   volatile model, Geophys. Res. Lett. 6, pp. 799-802.

9. Pearl, J., Hanel, R., Kunde, V., Maguire, W., Fox, K., Gupta,
   S., Ponnamperuma, C. and Raulin, F., 1979, Identification of
   gaseous SO_2 and new upper limits for other gases on Io, Nature 280, pp. 755-758.

10. Barbe, A., Delahaigue, A. and Jouve, P., 1971, Spectre infra-
     rouge à l'état solid des molécules isotopiques de SO_2. Spec-
     trochimica Acta 27 A, pp. 1439-1446.

11. Howell, R. R., Cruikshank, D. P. and Fanale, F. P., 1984, Sul-
     fur dioxide on Io : Spatial distribution and physical state,
     Icarus, 57, pp. 83-92.

12. Nelson, R. M., Lane, A. L., Matson, D. L., Fanale, F. P.,
     Nash, D. B. and Johnson, T. V., 1980, Io : Longitudinal dis-
     tribution of sulfur dioxide frost, Science 210, pp. 784-786.

13. Howell, R. R., Cruikshank, D. P. and Geballe, T. R., 1984, A
     ground-based detection of solid sulfur dioxide in the plumes
     of Io, Science (submitted).

14. Smith, B. A. and the Voyager Imaging Team, 1979, The Jupiter
     system through the eyes of Voyager 1, Science 204, pp. 951-
     972.

15. Smith, B. A. and the Voyager Imaging Team, 1979, The Galilean
     satellites and Jupiter : Voyager 2 imaging science results,
     Science 206, pp. 927-950.

16. Schaber, G. G., 1982, The geology of Io. In "Satellites of
     Jupiter", ed. D. Morrison, Univ. of Arizona Press, Tucson,
     pp. 556-597.

17. Nash, D. B., Fanale, F. P. and Nelson, R. M., 1980, SO_2
     frost : UV-visible reflectivity and Io surface coverage,
     Geophys. Res. Lett. 7, pp. 665-668.

18. Soderblom, L., Johnson, T., Morrison, D., Danielson, E.,
     Smith, B., Veverka, J., Cook, A., Sagan, C., Kupferman, P.,
     Pieri, D., Mosher, J., Avis, C., Gradie, J. and Clancy, T.,
     1980, Spectrophotometry of Io : Preliminary Voyager 1 re-
     sults, Geophys. Res. Lett. 7, pp. 963-966.

19. Binder, A. and Cruikshank, D. P., 1964, Evidence for an at-
     mosphere on Io, Icarus 3, pp. 299-305.

20. Fanale, F. P., Banerdt, W. B. and Cruikshank, D. P., 1981,
     Io : Could SO_2 condensation/sublimation cause the sometimes
     reported post-eclipse brightening ? Geophys. Res. Lett. 8,
     pp. 625-628.

21. Sinton, W. M., 1981, The thermal emission spectrum of Io and a
     determination of the heat flux from its hot spots, J. Geophys.
     Res. 86, pp. 3122-3128.

22. Johnson, T. V., Morrison, D., Matson, D. L., Veeder, G. J.,
    Brown, R. H. and Nelson, R. M., 1984, Io volcanic hot spots :
    Stability and longitudinal distribution, Science 226, pp.
    134-137.

23. Strom, R. G. and Schneider, N. M., 1982, Volcanic eruption
    plumes on Io. In "Satellites of Jupiter", ed. D. Morrison,
    Univ. of Arizona Press, Tucson, pp. 598-633.

24. Johnson, T. V. and Soderblom, L. A., 1982, Volcanic eruptions
    on Io : Implications for surface evolution and mass loss.
    In "Satellites of Jupiter", ed. D. Morrison, Univ. of Arizo-
    na Press, Tucson, pp. 634-646.

25. Fanale, F. P., Banerdt , W. B., Elson, L. S., Johnson, T. V.
    and Zurek, R. W., 1982, Io'surface : Its phase composition
    and influence on Io's atmosphere and Jupiter's magneto-
    sphere. In "Satellites of Jupiter", ed. D. Morrison, Univ.
    of Arizona Press, Tucson, pp. 756-781.

METHANE ICE ON TRITON AND PLUTO

Dale P. Cruikshank

Institute for Astronomy
University of Hawaii
Honolulu, Hawaii USA 96822

R. H. Brown

Planetary Geosciences Division
Hawaii Institute of Geophysics
University of Hawaii
Honolulu, Hawaii USA 96822

R. N. Clark

U. S. Geological Survey
Branch of Geophysics and Remote Sensing
Box 25046
Denver Federal Center
Denver, Colorado 80225 USA

ABSTRACT

    Near-infrared spectra (0.8-2.5 μm) of Triton and Pluto
reveal six absorption bands attributed to the methane molecule.
Additional considerations of temperature, spatial variability on
the surfaces of these bodies, and the character of the absorption
bands indicates that frozen methane is the principal contributor
to the spectral signatures observed. Triton shows an additional
absorption band which is tentatively identified as molecular
nitrogen, probably in the liquid state. The red color observed
for both of these objects in the photovisual spectral region
(0.3-1.0 μm) may result from photochemical derivatives of methane
and nitrogen, known from laboratory work to have such coloration.

817

J. Klinger et al. (eds.), Ices in the Solar System, 817–827.
© *1985 by D. Reidel Publishing Company.*

INTRODUCTION

The presence of water ice on the surfaces of some bodies in the outer solar system has been known or strongly suspected since the first infrared studies by Kuiper in the 1950s and reported by him first in an extraordinarily brief but pregnant abstract (Kuiper (1)). While other ices are predicted in modern cosmochemical schemes of the origin of the planetary system (Lewis (2) , (3)), the first evidence for methane in a solid state in the outer solar system was obtained by Cruikshank et al. (4) from near-infrared observations of Pluto in which the 1.7-μm band was found. Methane ice on Pluto was confirmed tentatively by Lebofsky et al. (5) and positively by Soifer et al. (6) and Cruikshank and Silvaggio (7).

Spectrophotometric observations of Triton in the region 1.4-2.5 μm, made in 1978, showed the strong combination band of methane at 2.3 μm (Cruikshank and Silvaggio (8)), and while it was originally thought to be primarily absorption in gaseous methane, it is now known to be due largely to the ice. In addition to the six methane bands now observable in the region 0.8-2.6 μm on Triton (Cruikshank and Apt (9)), there is evidence for another spectrally active component tentatively identified as molecular nitrogen, probably in the liquid state (Cruikshank et al. (10)).

The study of solid methane-bearing objects in the outer solar system is just beginning, and to date the required laboratory spectroscopic support work is somewhat lacking. Still, the identification of solid methane on Pluto and Triton, and (so far) on no other object closer to the Sun than 30 AU, provides useful boundary conditions to models of the chemistry of the outer solar system as well as to our first exploratory efforts to understand the bodies most distant from the Sun.

This paper is a brief review of the state of our knowledge of the surface chemistry of Triton and Pluto during the time when near-infrared spectrophotometric studies are advancing to the point where faint objects (visual magnitude 14) can be studied with large telescopes and the most sensitive infrared detectors.

TRITON

The spectrum of Triton has been explored at relatively low spectral resolution from 0.3 to 2.5 μm. From 0.3 to 0.6 μm, the spectrum increases in reflectance without any distinct absorption features. Longward of 0.8 μm, absorption bands appear, as shown in the composite spectrum of Triton in Figure 1, in which the 0.8- to 2.5-μm data are from Cruikshank and Apt (9) and Cruikshank

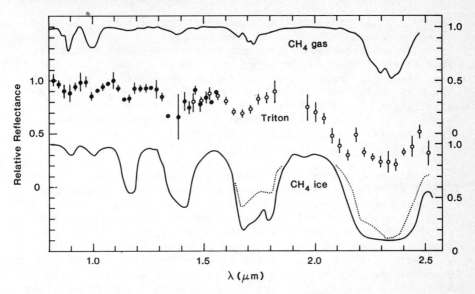

Figure 1. Composite spectrum of Triton, 0.8–2.5 μm. Solid dots
represent data from 18.6 May 1982; open circles are data from 2.4
June 1980. The synthetic methane gas spectrum at the top of the
figure was calculated for T = 55 K and 50 m-Am density at 10^{-3}
bar by J. Apt. The bottom traces are laboratory spectra of methane ice in diffuse reflection; the dashed line is a sample prepared with a small amount of carbon black mixed with the ice.
From Cruikshank and Apt (9).

et al. (10). The majority of these bands are identified as due
to methane on the basis of comparisons with synthetic laboratory
spectra of gaseous methane calculated for the low-temperature
conditions on Triton (Cruikshank and Apt [9]) and with laboratory spectra of methane ice. In high-resolution spectra, Apt et
al. (11) have found the 0.89-μm band of methane and have noted
that it is weaker than the same band in the spectrum of Pluto.

    In addition to the six bands in the Triton spectrum that
match the methane spectrum, there is an additional feature centered at 2.16 μm located on the steep slope of the 2.3-μm methane
band that cannot itself be attributed to methane either in the
gaseous or solid state. Cruikshank et al. (10) have tentatively identified this band as the density-induced (2-0) absorption
band of molecular nitrogen. The identification is regarded as
tentative because it is based upon the presence of a single band,
and the apparent coincidence between the central wavelengths of
N_2 observed in the laboratory and the band on Triton. From a
consideration of the expected temperature of Triton's surface and
the phase equilibrium of nitrogen, Cruikshank et al. (12), (10)

showed that in order for the observed spectral band to be produced in gaseous nitrogen, the surface pressure would exceed that
at which condensation would occur at the relevant temperature.
Thus, the nitrogen should exist as a liquid or solid, depending
upon the exact temperature. Laboratory observations show that
the 2.16-μm feature also occurs in liquid nitrogen. From a determination of the absorption coefficient in the band, Cruikshank
et al. (10) calculated that the spectral feature on Triton
could be produced by absorption through at least a few tens of
centimeters of liquid nitrogen on the surface of the satellite.

This result raises interesting possibilities about the atmosphere of Triton, its interaction with the liquid nitrogen and
the solid methane, and the nature of the red color of the satellite. Estimates of the temperature of Triton are complicated
because of the unknown effects of the atmosphere. It is probable
that the atmosphere itself has not yet been observed because it
is mostly nitrogen at a surface pressure regulated by the temperature of the liquid; the equilibrium vapor pressure of nitrogen
at $T = 64°K$ is 0.13 atm. The vapor pressure of methane is less
by about 10^{-3}, so that this gas is likely a minor constituent of
the satellite's atmosphere.

Additional observational results show that the strengths of
the methane bands on Triton vary with its orbital position
(Cruikshank and Apt (9)). Because Triton is in locked synchronous rotation in its orbit around Neptune, the orbital variability reflects a nonuniform distribution of methane on the surface
of the satellite. There are presently no observations about the
possible variability of the strength of the nitrogen band.
Triton does not show a pronounced photometric variability with
its orbital position. Franz (13) found an amplitude of about
0.06 mag (at $\lambda = 0.56$ μm), with the maximum at western elongation
(leading hemisphere). The variability of the methane band
strength appears to be in the sense that the strongest bands are
also found near western elongation, though this requires confirmation.

Methane is highly soluble in liquid nitrogen and has the
effect of raising the freezing point of the mixture. The implications of this for Triton have been explored in a very preliminary way in the paper by Cruikshank et al. (10), from which it
appears that the colder portions of the satellite, particularly
those near the pole in extended darkness at the present season,
are sufficiently cold to permit the nitrogen sea to freeze.
Whether or not there is a diurnal freeze-thaw cycle depends on
the heat capacity of the sea and its global extent, as well as
the possibility that materials which may lower the freezing point
are dissolved in it.

In attempts to model the infrared spectrum of Triton with laboratory observations of methane ice and liquid nitrogen, Cruikshank et al. (12), (10) found that an additional component was necessary in order to match the shape of the continuum at various wavelengths. The absorption spectrum of water ice provides the additional component needed to fit the Triton spectrum to a precision commensurate with the quality of the telescopic data for the satellite. The best-fit model for Triton's

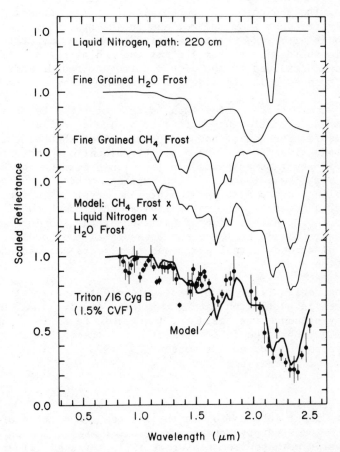

Figure 2. Model of the Triton spectrum with resolution ($\Delta\lambda/\lambda$) of 1.5%, together with the individual components used in the calculation. Emphasis was placed on the spectral region 1.8-2.5 μm in trying to achieve a fit. Error bars on the Triton points are ±1 σ calculated from the mean of several individual runs through the spectrum, each corrected for extinction and ratioed to a solar-type star. From Cruikshank et al. (10).

spectrum, together with the individual components shown separately, is given in Figure 2. In order for photons to penetrate the large distance in the nitrogen, the other components can be only fractions of a percent of the nitrogen.

While a detailed study of the short wavelength end of the spectrum where Triton shows a distinct reddish color has not yet been accomplished, interesting possibilities arise from a consideration of the fact that other methane-bearing bodies in the outer solar system also have reddish hues. In the case of Saturn's Titan, the red color is probably that of the aerosol photochemically produced from the methane in the upper atmosphere. Pluto is also red. Photochemistry on Triton may produce reddish organic solid matter from the methane and nitrogen. If there is methane dissolved in the liquid nitrogen, red organic matter may be suspended in the liquid.

In summary, the surface of Triton is characterized by solid methane, either as a continuous surface or as icebergs floating in a sea of liquid nitrogen of unknown depth. Reddish photochemical products may give the surface a slight coloration, and water ice may occur as crystals suspended in the liquid nitrogen or as a solid mixed with the methane frost on expanses of solid surface (spectral modeling favors a suspension of fine crystals in the liquid). In this scenario, the satellite has an atmosphere of nitrogen with other possible minor constituents; the surface pressure may be regulated by the vapor pressure at the local temperature, but strong diurnal and seasonal effects are probable (see also Trafton (14)).

No methane derivatives, such as acetylene, ethane, or ethylene, can be seen in our Triton data (Cruikshank et al. (10)), nor can we find any evidence for solid CO, thought to be an important clathrate-forming substance in the formation of the bodies in the outer solar system (Lunine and Stevenson (15)).

PLUTO

The first direct information on the surface composition of Pluto came from the study by Cruikshank et al. (4), who found evidence for solid methane from measurements in infrared filter bands selected to distinguish among the ices of water, methane, and ammonia; a spectrally bland rocky surface would have also been evident from the data obtained. Spectrophotometric observations (Cruikshank and Silvaggio (7) ; Soifer et al. (6)) subsequently confirmed the strong bands of methane in the near-infrared. In Figure 3 we show a spectrum of Pluto from 0.8 to 2.5 μm in which six distinct methane bands are apparent. Comparison with synthetic methane spectra computed by J. Apt for the low-

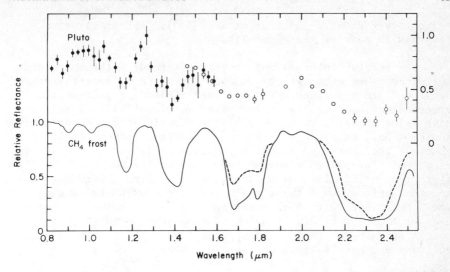

Figure 3. Composite spectrum of Pluto, 0.8-2.5 μm. Open circles
are the data from Soifer et al. (6); solid dots with ±1 σ
error bars are data obtained by the present authors at the NASA
Infrared Telescope Facility, Mauna Kea, Hawaii. The methane
frost spectrum is the same as that reproduced in Figure 1.

temperature conditions on Pluto, and with laboratory spectra of
methane ice, suggest that the Pluto spectrum is best matched by
the ice spectrum. In further support of the ice interpretation
is the observation that the strength of the infrared bands is
variable with the planet's rotation, indicating a nonuniform
distribution of methane ice across the surface (Cruikshank and
Brown, in preparation).

    Observations of the spectrum of Pluto in the region 0.6-1.0
μm at which higher spectral resolution than the near-infrared
spectrophotometry show additional methane bands (Fink et al.
(16); Apt et al. (11)). Structure in the bands, particularly
that at 0.89 μm, indicates that some of the absorption is due to
methane gas. Buie and Fink (17) find that the strengths of the
bands in this spectral region vary in phase with Pluto's lightcurve, indicating that a component of the absorption is attribu-
table to methane ice having bands coincident with the gas phase
bands.

    Pluto has a distinct reddish color in the photovisual spec-
tral region (0.3-1.0 μm), as seen in the data from several observers (Harris (18); Fix et al. (19); Lane et al. (20); Bell et al.

(21); Barker et al. (22); Apt et al. ⟨11⟩). Apt et al. show a
figure with many of these results.

The reddish color is not characteristic of methane ice and
therefore represents direct evidence for a second spectral component on Pluto's surface. The earliest five-color photometric
observations made in the early 1950s (Harris (18)) show the reddish color clearly; when the early observations are compared with
those of Bell et al. (21) there is no clear indication of
change. Therefore, we cannot say with certainty that the surface
component progressively coming into view at the current epoch is
alone responsible for the reddish overall color. There may be a
colored surface constituent mixed intimately with the methane
ice. Photometry with high precision may help resolve the question of the surface distribution of components of different
colors and albedos during the approaching epoch of mutual
eclipses and occultations of Pluto and Charon.

As we have already noted, Triton is also reddish in the
photovisual spectral range, as is the atmosphere of Titan. The
fact that Pluto and the other two objects are rich in methane, in
one phase or another, plus the laboratory results that show that
tholins, hydrocarbon derivatives of methane made under planetary
conditions (Khare et al. (23)), are red, suggest some common
causal connection. The reddish colors of Pluto and Triton may
indicate the presence of surface deposits of complex hydrocarbons
of the general type that cause the photochemical smog in the
atmosphere of Titan.

It is important to consider the atmosphere of Pluto in the
context of possible atmospheres on other bodies of comparable
size. Fink et al. (16) found the first direct spectroscopic
evidence for a methane atmosphere on Pluto, while Cruikshank and
Silvaggio (7) speculated that there must be a tenuous atmosphere because of the high vapor pressure of CH_4. Fink et al.
(16) found structure in the 0.89-μm methane band and, by comparison with synthetic spectra, inferred an atmosphere with 27 ±
7 m-Am of methane, giving a surface pressure of 0.15 mbar. This
derived abundance does not take into account coincident absorption by methane ice, so the derived gas abundance should be less
than 27 m-Am.

Trafton (24) showed that for an isothermal atmosphere in
contact with the surface, hydrodynamic escape of a quantity of
methane amounts to a mass loss equivalent to that of the entire
mass of Pluto over the age of the solar system. He proposed that
the loss of methane would be slowed by diffusion through a heavier and undetected atmospheric constituent. Hunten and Watson
(25) have pointed out, however, that the adiabatic expansion of
an escaping atmosphere will produce a cold trap unless there is

sufficient heat input in the high atmosphere. Plausible sources
of heat are quantitatively insufficient in the case of Pluto (and
perhaps also for Triton), and the escape rate should be slowed by
the cold trap to a value consistent with the presence of solid
methane on the surface of the planet. The escape parameter
defined by Hunten and Watson (25) differs only by a factor of 3
for Charon (Pluto's satellite), thus indicating that only a modest amount of methane will be lost from the satellite as well as
from the planet.

OTHER CONSIDERATIONS

        Methane may occur as an ice on other planetary bodies in the
outer solar system where it has not yet been identified. The
presence of water ice may mask small amounts of methane either as
an adsorbed species or as a clathrate. Water ice is found on
many planetary satellite systems from the Jupiter system through
the Uranus system, and some of these icy surfaces may contain
methane that has not yet been detected because of the low spectral resolution thus far obtainable in the telescopic observations. Work is continuing to improve both spectral resolution
and signal precision in the hope of detecting (or establishing
limits on the presence of) methane, ammonia, and other possible
molecules entrained in the ices of the satellites of Jupiter,
Saturn, and Uranus.

REFERENCES

1. Kuiper, G. P., 1957, IR observations of planets and satelli-
    tes, Astron. J. 62, p. 245 (abstract).

2. Lewis, J. S., 1972, Low temperature condensation from the so-
    lar nebula, Icarus 16, pp. 241-252.

3. Lewis, J. S., 1974, The temperature gradient in the solar ne-
    bula, Science 186, pp. 440-443.

4. Cruikshank, D. P., Pilcher, C. B. and Morrison, D., 1976,
    Pluto : Evidence for methane frost, Science 194, pp. 835-
    837.

5. Lebofsky, L. A., Rieke, G. H. and Lebofsky, M. J., 1979, Sur-
    face composition of Pluto, Icarus 37, pp. 554-558.

6. Soifer, B. T., Neugebauer, G. and Matthews, K., 1980, The
    1.5-2.5 μm spectrum of Pluto, Astron. J. 85, pp. 166-167.

7. Cruikshank, D. P. and Silvaggio, P. M., 1980, The surface
    and atmosphere of Pluto, Icarus 41, pp. 96-102.

8. Cruikshank, D. P. and Silvaggio, P. M., 1979, Triton : A sa-
    tellite with an atmosphere, Astrophys. J. 233, pp. 1016-
    1020.

9. Cruikshank, D. P. and Apt, J., 1984, Methane on Triton : Phy-
    sical state and distribution, Icarus 58, pp. 306-311.

10. Cruikshank, D. P., Brown, R. H. and Clark, R. N., 1984, Ni-
    trogen on Triton, Icarus 58, pp. 293-305.

11. Apt, J., Carleton, N. P. and Mackay, C. D., 1983, Methane on
    Triton and Pluto : New CCD spectra, Astrophys. J. 270, pp.
    342-350.

12. Cruikshank, D. P., Brown, R. H. and Clark, R. N., 1983, Ni-
    trogen on Triton, Bull. Amer. Astron. Soc. 15, p. 857
    (abstract).

13. Franz, O. G., 1981, UBV photometry of Triton, Icarus 45,
    pp. 602-606.

14. Trafton, L., 1984, Seasonal variations in Triton's atmosphe-
    ric mass and composition. In "Uranus and Neptune", ed. J. T.
    Bergstralh, NASA Conf. Pub. 2330, pp. 481-493.

15. Lunine, J. I. and Stevenson, D. J., 1983, The role of clath-
    rates in the formation and evolution of icy satellites.
    Paper presented at IAU Colloquium 77, Natural Satellites,
    Ithaca, NY, July.

16. Fink, U., Smith, B. A., Benner, D. C., Johnson, J. R., Reit-
    sema, H. J. and Westphal, J. A., 1980, Detection of a CH_4
    atmosphere on Pluto, Icarus 44, pp. 62-71.

17. Buie, M. W. and Fink, U., 1983, Lightcurve spectrophotome-
    try of methane absorptions on Pluto, Bull. Amer. Astron.
    Soc. 15, p. 860 (abstract).

18. Harris, L., 1961, Photometry and colorimetry of planets and
    satellites. In "Planets and Satellites", eds. G. P. Kuiper
    and B. M. Middlehurst, Univ. of Chicago Press, Chicago, pp.
    272-342.

19. Fix, J. D., Neff, J. S. and Kelsey, L. A., 1970, Spectropho-
    tometry of Pluto, Astron. J. 75, pp. 895-896.

20. Lane, W. A., Neff, J. S. and Fix, J. D., 1976, A measurement of the relative reflectance of Pluto at 0.86 micron. Pub. Astron. Soc. Pacific 88, pp. 77-79.

21. Bell, J. F., Clark, R. N., McCord, T. B. and Cruikshank, D. P., 1979, Reflection spectra of Pluto and three distant satellites, Bull. Amer. Astron. Soc. 11, p. 570 (abstract).

22. Barker, E. S., Cochran, W. D. and Cochran, A. L., 1980, Spectrophotometry of Pluto from 3500 to 7350 Å, Icarus 44, pp. 43-52.

23. Khare, B. N., Sagan, C., Arakawa, E. T., Suits, F., Callicott, T. A. and Williams, M. W., 1983, Optical constants of Titan tholin aerosols, Bull. Amer. Astron. Soc. 15, pp. 842-843.

24. Trafton, L., 1980, Does Pluto have a substantial atmosphere ? Icarus 44, pp. 53-61.

25. Hunten, D. M. and Watson, A. J., 1982, Stability of Pluto's atmosphere, Icarus 51, 665-667.

GEOLOGY OF ICY SATELLITES

William B. McKinnon

Department of Earth and Planetary Sciences and
McDonnell Center for the Space Sciences
Washington University, Saint Louis, MO 63130

ABSTRACT. The geology of the major icy satellites of Jupiter, Saturn, Uranus, and Neptune is discussed and related to the four major processes that shape icy satellite surfaces: impact cratering, volcanism, tectonism, and interactions with planetary magnetospheres and solar radiation.

1. INTRODUCTION

To describe the geology of icy satellites may at first seem difficult due to the bewildering variety of icy objects in the solar system, but it is useful to remember that geology is not planet-specific, but constitutes a suite of processes that operate on each planet or satellite. Given the complex interactions possible among these processes, the number of different appearances a surface may present is effectively transcendental. This paper is restricted to the major satellites of the outer planets (Jupiter, Saturn, Uranus, and Neptune), which are somewhat arbitrarily defined to include only those objects whose physical shape is mainly determined by self-gravity. The icy Galilean satellites of Jupiter, for which the most data exist, are emphasized.

Four major geologic processes shape the surfaces of icy satellites: impact cratering, volcanism, tectonism, and interactions with planetary magnetospheres and solar radiation. Mass wasting, in a restrictive sense, is a minor process. The two satellites with significant atmospheres, Titan and Triton, constitute a special case in that aeolian, fluvial, pluvial, and chemical weathering effects (ultimately insolation driven) are possible.

J. Klinger et al. (eds.), Ices in the Solar System, 829–856.
© *1985 by D. Reidel Publishing Company.*

1.1 Impact Cratering

Structures caused by hypervelocity collisions (velocities >2 km/s for water-ice targets [1]) are found in abundance on satellites in the outer solar system. They differ remarkably little from more familiar terrestrial examples and attest to the fact that the entire solar system has been subjected to significant bombardment over its ~4.5 billion year history (see [2] for a comprehensive discussion). Impact can lead to novel effects on these worlds, however, mostly due to the rheology of ices and the small dimensions of some satellites. These will be brought out in the detailed discussion below.

1.2 Volcanism

The eruption to the surface of fluids is a major way that planets and satellites transport internal heat. Volcanism on icy satellites may involve melting, as it does on the terrestrial planets, and can act as a powerful agent of differentiation as well as resurfacing. The possible volcanic fluids are numerous and are contained within the quinary system $H_2O-NH_3-CH_4-CO_2-N_2$, the boundaries of which define the petrology of the outer solar system. In addition, water ice near its melting point is also a potential volcanic fluid, due to its relatively low effective viscosity.

1.3 Tectonism

Satellite surfaces deform in characteristic ways when subject to regional stress. The sources of stress that drive this tectonic deformation are varied: some come from within the satellite (convective motions, volume changes, etc.); some are directly or indirectly due to tidal interactions with the primary (tidal flexing, shape changes due to despinning and orbital recession); others are due to the gravitational potential energy stored in topography (viscous relaxation and isostatic rebound); and some are ultimately derived from large impacts (multiple rings, for example). The pattern of deformation may be due to brittle failure, leading to observed faults and joints, or ductile failure.

1.4 Interactions with Planetary Magnetospheres and Solar Radiation

Jupiter and Saturn are known to have magnetospheres, Uranus is inferred to have one (e.g., [3]), and Neptune probably has one as well. The atmosphereless icy satellites are thus subject to significant charged particle bombardment. Surface grains are sputtered by these charged particles, their sizes are reduced, and the satellites suffer net erosion (e.g., [4]). Depending on solar distance and ice type, insolation-driven sublimation may also be important. Sublimating molecules are generally retained on icy satellites, however. The net effect is the growth of larger grains at the expense of smaller ones (the surface/volume ratio decreases). The interaction of charged particles and photons is a major factor in determining a satellite's albedo. In addition, magnetospheric ions may be implanted and both ions and solar radiation can catalyze surface chemical reactions.

Below, I briefly describe the geology of the icy satellites, with an emphasis on these four major processes.

2. THE ICY GALILEAN SATELLITES

The Galilean satellites, Io, Europa, Ganymede, and Callisto, constitute a suite of planetary-scale objects ranging in scale from slightly sub-lunar (Europa) to larger than Mercury (Ganymede). Io is an apparently water-free silicate object. (Here I use the term silicate in a broad sense that incorporates Fe-Ni metal as well). Europa is mostly silicate with some water ice ($\lesssim$10% by mass) and Ganymede and Callisto are approximately half silicate and half ice by mass. They accreted in a nebula of gas and dust, which according to best current understanding, formed when the protosolar nebula in the Jovian region collapsed hydrodynamically onto a massive protojovian core (5,6). Temperatures in at least the outer portion of this protojovian nebula were low enough to permit the condensation of water ice, which along with silicate, was then available for planetesimal accretion and satellite growth.

Satellite formation times are extremely short, $<10^6$ yr (7,8), principally due to the short circulation times of material about Jupiter. For this reason it is unlikely that gas-free Safronov-type accretion models apply to the Galilean satellites. Most likely, the presence of gas during accretion profoundly influenced the structure and composition of these bodies, and reduced accretion times even further (7,9). To this extent the protojovian nebula diverged from the protosolar nebula as a planetary nursery.

In the context of this model, the lack of water on innermost Io and the relatively small amount on Europa (the next-most distant from Jupiter) are due to higher nebula temperatures close to Jupiter. These elevated temperatures result from gravitational energy released by the contraction of proto-Jupiter (10). Europa's ice complement would be due to the dehydration of incorporated hydrous and hydrated silicates in an equilibrium condensation scenario, or due to icy planetesimals scattered inward from the outer fringes of its "feeding zone" in a disequilibrium version (11). The details of accretion in a relatively dense, planetary nebula are complex, however, and the possibility of substantial radial drift of satellite semimajor axes has been noted (9). The presently observed density gradient, monotonically decreasing outwards, may not be simply due to a nebular temperature gradient. Shoemaker suggests that large impacts have preferentially stripped away primordial, icy mantles from Io and Europa (12). Io and Europa are most affected because of the gravitational focusing of heliocentric projectiles towards Jupiter (13). It is also useful to remark at this point that the

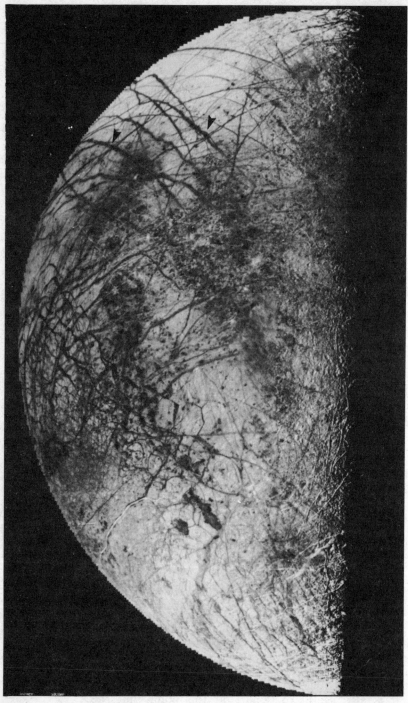

Fig. 1. Computer-processed Voyager 2 mosaic of Europa. North is up.

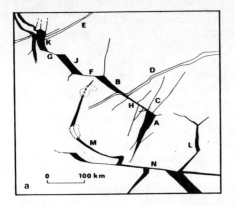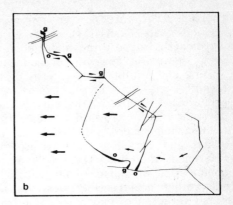

Fig. 2.   Reconstruction of wedge-shaped bands on Europa.   Configuration (a) at present and (b) after bands (A) and (B) are closed.
Right-lateral strike-slip motion along lineaments (F), (G), and (H) is
indicated by offsets of crosscutting lineaments.   From (55).

approximately <50/50 ice/silicate mass ratio for Ganymede and Callisto
is substantially less than the solar value of ~60/40 (derived from
[14]).   This may again be related to Shoemaker's suggestion, or some
other mechanism may be involved (e.g., [15]).

    As the geology of the icy Galilean satellites provides strong
constraints on these most profound questions of planetary and
satellite accretion, they should be kept in mind during the following
discussion.   This is especially important insofar as no accepted
paradigm of satellite formation has yet emerged.

2.1   Europa

Europa is generally perceived as a relatively smooth, ice-covered
silicate sphere, one whose surface abounds with enigmatic darker
markings and lineaments (Fig. 1).   Indeed, internal models allow for a
surface ice shell of ~100-km thickness (16).   More to the point,
however, is the observation that no silicate mountains of either
constructional or impact origin are seen;   a minimum of several
kilometers of ice is required.

    Visible and near-infrared refectance spectra are unequivocal in
determining that Europa's optical surface is essentially pure water
ice, with only a few weight percent impurities or contaminants (17,18,
and references therein).   Radar observations imply that this ice
extends to a depth of at least several meters, if not several tens of
meters (19,20).   Both Europa's high geometric albedo at 10-100 cm
radar wavelengths and its anomalous circular polarization ratio
require multiple scattering from regolith interfaces;   such
scattering is possible because of the extremely low extinction

coefficient of water ice at temperatures relevant to satellite surfaces (~50-150 K) (21). VLA observations at somewhat shorter wavelengths support the high geometric radar albedo (22,23).

There are, however, pronounced spectral differences between the leading and trailing hemispheres of orbital motion. (All the satellites discussed in this paper are tidally locked to their primaries, keeping, as the earth's moon does, one hemisphere constantly facing toward the direction of orbital motion). On the trailing hemisphere, infrared band depths are deeper (17,24), and visible and ultraviolet albedos especially are lower (25-29, and references therein). These asymmetries are plausibly ascribed to a combination of enhanced magnetospheric sputtering on the trailing side, which preferentially destroys small ice grains and thus lowers visual refectance and increases infrared band depth (24), and enhanced implantation of magnetospheric sulfur on the trailing side, which could account for the lower UV spectral reflectance (26,30). The effects of these mechanisms are probably modified by preferential micrometeorite gardening of the leading hemisphere (24,31). The UV asymmetry is global in nature and not sensitive to the underlying geology (28); presumably the infrared asymmetry behaves similarly, but this won't be tested until the Galileo orbiter reaches Europa.

In contrast to the growing understanding of Europa's global spectral properties, explanations for the albedos and colors of the various dark lineaments and mottled surface units are presently lacking. It is certainly possible that some combination of slight contamination, radiation darkening, and larger grain size is responsible, but the bulk composition of darker materials on Europa's surface may still be close to pure water ice (18). It should be noted that darkness is relative on Europa; dark material albedos are actually on the order of 0.5 in the visible, about 25% less than those of bright materials (27).

The highest resolution images of Europa were obtained by Voyager 2. The mosaic in Fig. 1 has been photometrically processed to bring out detail near the limb (32). The surface is generally divided into bright plains and slightly darker mottled terrains. Numerous dark lineaments are seen (32-37); major ones in the north (towards top) are globally arcuate and connect to apparently similar lineaments in lower resolution Voyager 1 images (33). In many instances a bright central stripe can be discerned (arrows mark examples); these lineaments are termed triple bands (35,36) and tend to be somewhat irregular in outline. In the south-central region the dark lineaments appear more curvilinear and in certain instances decidedly wedged-shaped (38). Linear and cycloidal ridges abound near the terminator; their overall relief does not exceed a few hundred meters (32).

Impact craters are scarce on the Europan surface, at least down to a limit of ~10 km on Voyager 2 high resolution images (only five have been identified with confidence) (36). The first order interpretation is that the Europan surface is youthful. Estimates based on a cratering flux model are 30-200 m.y., with the lower value preferred (13). It has been recognized, however, that many craters

may have viscously relaxed away over geologic time (13,26,39-41); the
rheology of water ice is sufficiently soft at temperatures appropriate
to Europa's icy crust (~100 K) for this to occur (42-44). Hence the
present Europan surface may have existed for far longer than the above
estimates of the crater retention age indicate. On the other hand,
numerous crater albedo scars, or palimpsests (45), are not seen
either. Such palimpsests may not be obviously differentiable from
bright plains material, but they would certainly interrupt if not
obliterate any preexisting lineaments (36). Thus the surface of
Europa (that is, its ice crust and albedo patterns) cannot be as old
as the heavily cratered terrain of Ganymede, and probably cannot be as
old as the grooved terrains (see below), which have been modeled as
3.0-3.9 b.y. old (13).

The mottled terrains are believed to be younger than the bright
plains because lineaments extending from the bright plains into the
mottled terrains are disrupted (35,36). The cause of the disruption
is obscure, however. The fact that the leading/trailing hemisphere
visual albedo asymmetry derives from the mottled terrains and not from
the bright plains, which are more globally uniform (29), suggests some
rejuvenation mechanism may be involved (41,46); possibly, the bright
plains are the younger.

Even though most lineaments are defined solely on the basis of
albedo patterns, it seems clear that the majority represent tectonic
structures; they are probably the result of faulting and fracturing
and subsequent modification. The great lengths and simple geometries
of many of the lineaments and bands argue for two conditions: a
reasonably "uniform" lithosphere and simple patterns of driving
stress. The condition of uniformity refers to mechanical strength and
coherence. One way to satisfy this condition is for the lithosphere
to be truly young, say, the result of freezing of a global ocean.
Alternately, since the Europan lithosphere is at a sufficiently high
fraction of its melting temperature, it can be thought of as a
high-grade metamorphic terrain. Fractures will tend to anneal and its
tectonic "memory" will be more limited than the lithospheres of the
moon or Mars, for example. Hence, uniformity is a natural consequence
of the rheology of an ice crust under the conditions of reasonably
high heat flow (see below). This does not preclude fault reactivation
(35,36), but the lifetimes of faults as mechanical discontinuities are
limited.

The requirement of simple stress patterns immediately suggests
those due to volume changes, tidal despinning, and orbital evolution
and eccentricity variation (47,48). Indeed, the fracture patterns
predicted from such stress systems are a reasonably good match to the
observed geometries (49,50). Arcuate dark bands and triple bands away
from the anti-Jove point are consistent with strike-slip faulting due
to orbital recession, and dark lineaments and bands near the anti-Jove
point are consistent with tension fracturing due to orbital
eccentricity variations. The latter either occurred before orbital
recession stresses built up, or as is more likely, after they
substantially relaxed. Tension fracturing near the anti-Jove point is

also supported by polygon analysis (34). The ridges seen at the
terminator are believed to be compressional, and are probably due to
some combination of tidal stress and planetary volume decrease.
Despinning, per se, is considered to have occurred far too early to
contribute to present tectonics. It is interesting to note that these
inferences are consistent with models of Europa's orbital evolution
(51,52).

　　　Much remains to be understood about the tectonics of Europa.
The inferences above are based on analyses of only 25% of the surface,
and most of the lineaments used to test various stress models are seen
only as albedo features. Little morphological or structural
information is available, except for the ridges (or flexi) and
wedge-shaped bands. As an illustration, structural models for triple
bands have been proposed that have little to do with strike-slip
faults (53,54). In addition, analysis of the wedge-shaped bands shows
that they formed by tension fracturing and the rotation of crustal
blocks (Fig. 2) (55). Significant E-W and NE-SW transcurrent offsets
are observed throughout the region of sub-parallel curvilinear and
wedge-shaped dark bands in Fig. 1. These bands mark a major NW-SE
fracture belt or zone involving significant crustal extension.
Although the fracture pattern may have been initiated by a tidal or
global expansion stress field, such simple models do not directly give
the amount of strain implied here or the sense of shear. Possibly the
extension is due to differential accumulation of strain from the rest
of the ice shell. Equally plausibly, the extension may be due to a
different stress source such as convection in the warm ice that may
underlie the colder lithosphere.

　　　This evidence for crustal rotation and the filling of tension
fractures is in itself strong evidence for mechanical decoupling of
the Europan ice shell from the silicate interior, at least in this
region. This mechanical configuration is supported by thermal models
(16,41), although disagreements persist (56). The most extreme
tidally heated models raise the possibility that the ice shell is
decoupled by liquid water, and that this water may erupt to the
surface (41). Rotation of a decoupled ice shell with respect to the
tidal axes may also enter into the fracture history (57,58).
Mechanically, the rift zone is consistent with decoupling by either
warm ice or liquid water (55).

　　　The criterion for uniformity of the lithosphere argues against
stresses and fault patterns in the outer portion of the silicate
interior (the silicate lithosphere) being transmitted directly to a
thin, brittle ice shell. It is implausible that any real silicate
lithosphere could behave in such a uniform manner unless restrictive
rheological assumptions are met. An example of simple, global fault
geometries in a silicate lithosphere might be the ridges, transforms,
and trenches of the terrestrial sea floor, but the point is that the
terrestrial oceanic lithosphere is very young.

　　　Substantive progress in understanding Europan geology will have
to await the arrival of the Galileo orbiter at Jupiter. This
understanding is vital, however, in deciphering Europa's petrological.

thermal, and orbital evolution, and hence, conditions in the
protojovian nebula.

2.2   Ganymede and Callisto

Ganymede and Callisto differ markedly in appearance compared to
Europa.   Ganymede (Fig. 3) is divided into two general terrain types,
dark and bright, has polar frost deposits, and is covered with
abundant impact craters.   Callisto (Fig. 4) is even more divergent;
it appears nearly uniformly dark and very heavily cratered.   Both the
dark terrain on Ganymede and the dark surface of Callisto are brighter
than the lunar highlands (normal visual albedos are ~0.3, ~0.2, and
0.11, respectively [27,59]), an indication that icy compositions may
be involved;   such are anticipated given the mean densities of both
objects (1.93 and 1.83 g/cm^3 for Ganymede and Callisto, respectively
[60]).   Furthermore, most of the compositional inferences for the
surface of Europa based on near-infrared reflectance spectra and radar
and passive microwave observations hold true (to a degree) for
Ganymede and Callisto.   Specifically, Clark found the optical surfaces
of Ganymede and Callisto to be ~90 wt% and 30-90 wt% water ice,
respectively (17,18).   More recent work limits the amount of dark
grains in Ganymede's surface to 3 wt% and that in Callisto's surface
to an astounding 7-8 wt% (61).   In addition, both Ganymede and
Callisto have high geometric radar albedos (compared to the
terrestrial planets) and anomalous circular polarization ratios
(19,20,22,23).   Perhaps the surfaces and regoliths of these two
satellites are, despite appearances, close to pure ice in composition,
with darker units being more contaminated.

A number of considerations temper the robustness of these
conclusions.   The modeling of the infrared band depths assumes an
intimate mixture of ice and contaminating particles;   a degree of
areal segregation of ice and dark material could modify the results.
There is obvious evidence for some segregation at scales resolved by
Voyager, i.e., the terrain dichotomy on Ganymede, ice patches on
north-facing slopes on Callisto (62), and crater deposits on both
objects;   evidence also exists for segregation below the limit of
Voyager resolution for both satellites (63,64).   Furthermore, with
respect to the radar and microwave observations, both the geometric
albedos and anomalous polarizations decrease from Europa to Ganymede
to Callisto.   None have been modeled in detail in terms of an
ice-silicate regolith, and conclusions about Callisto in particular
should be considered preliminary (20).

The hemispheric spectral asymmetries observed on Europa are seen
on Ganymede as well, but in a weaker form (24-28).   Presumably the
interaction of the Jovian magnetosphere with Ganymede is reduced
compared to that at Europa.   Callisto exhibits little hemispheric
asymmetry in infrared band depth at all (24).   It is not exactly clear
why this should be, but one possibility is an extremely tenuous
atmosphere, derived from sublimation or sputtering, that stands off
most magnetospheric particles (65).   Callisto does have a

Fig. 3.   Global and regional views of Ganymede taken by Voyager 2 at
ranges of 1,200,000 and 86,000 km, respectively;   north is up.   The
basic division of Ganymede into dark, ancient, heavily cratered
terrains, and brighter, younger, grooved terrains is seen.   Polar
frost deposits, and young, bright rim and ray craters are also
visible.   Circular patches and parallel lineaments within the large
region of dark terrain are old crater palimpsests and elements of a
rimmed furrow system, respectively.   The lane of grooved terrain in
the foreground image is Uruk Sulcus.   (Frames FDS 20608.11
[background] and 20638.32 [foreground]).

leading/trailing hemisphere asymmetry in UV and visual reflectance,
but in a sense opposite to that observed on Europa and Ganymede
(25,26,28); this is possibly related to enhanced micrometeorite
gardening on the leading hemisphere as well as a lack of
magnetospheric interaction (65,66). The same process may also be
responsible for the well-known hemispheric asymmetry in visual
polarization properties (67-69).

The general abundance of craters on all terrains on Ganymede and
Callisto clearly indicates great geologic age compared to Europa.
Furthermore, there appears to be a correlation of increasing crater
density with decreasing surface albedo. The dark, heavily cratered
surface of Callisto is modeled to be generally older than 4.0
b.y. (13,45); the somewhat less heavily cratered and dark terrains of
Ganymede are still ancient, possibly ~3.8-4.0 b.y. old (45,70). The
brighter grooved and smooth terrains on Ganymede have lower crater
densities, although the most heavily cratered grooved terrain, near
the south pole, has a comparable density to the least cratered dark
terrain (70). One salient feature of all the cratered surfaces of
Ganymede and Callisto is the dearth of craters $\gtrsim 60$ km in diameter
compared to the preserved record on the terrestrial planets (71,72).
This fact, combined with spatial distribution statistics (73,74) and
scaling arguments (2), implies that the projectile population in the
Jovian region was different from that in the terrestrial planet zone.
Because the model crater retention ages assume a lunar-like variation
of the projectile flux with time, the absolute values are subject to
revision. Specifically, it is plausible that all of the terrains are
older than given above.

The correlation of crater density with albedo suggests that the
darkness of the ancient terrains on both satellites is due to
meteoritic contamination (75-77). As certain units of both dark and
bright terrain on Ganymede have comparable crater densities, however,
such contamination must predate the observed crater population in
large measure, a point argued on other grounds by (27). It is not
implausible that the dark ancient surfaces of these objects
accumulated material for a period before their crusts were stable
enough to retain craters. The contaminant also appears to be
spectrally red (78), which may help constrain its origin. The
alternate possibility that the dark terrains are primordial mixtures
of the ice and rock that formed each satellite is not favored here
because of the remote sensing arguments above, but it cannot be
conclusively rejected.

The craters themselves have been the subject of detailed study
(45,79). Morphologically, they are to first order similar to craters
on the terrestrial planets. The most significant differences are a
low simple-to-complex crater transition diameter (<5 km) compared to
the moon, the prevalence of the central pit form above ~20-km
diameter, highly flattened or unusual crater forms on the ancient
terrains, and the nature of the multiple ring and furrow systems.
These can be understood (to a degree) in terms of the mechanics of
impact into icy surfaces and the long-term effects of viscous

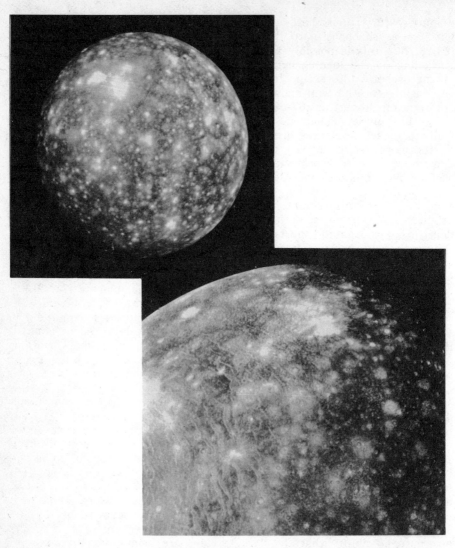

Fig. 4. Global and regional views of Callisto taken by Voyagers 2 and 1 at ranges of 1,100,000 and 320,000 km, respectively; north is ~up. Callisto is heavily cratered; the bright region in the background image is the Asgard double multiringed system, and a portion of the great Valhalla multiringed system is seen in the foreground image. (Frames FDS 20583.17 [background] and 16419.00 [foreground]).

relaxation of topography (2,45,77,79). The highly flattened craters are particularly noteworthy; in the most extreme examples topography is reduced to the point that, at the limit of resolution, the "crater" is only an albedo scar, or palimpsest.

All units on Ganymede are heavily tectonized, although the younger grooved terrains are the most spectacularly affected. The older dark terrains contain, upon close inspection, evidence for multiple episodes of furrow formation (80,81). The most prominent are rimmed and form arcuate parallel and sub-parallel sets of great length (>1000 km in some cases) (32). (The comments above regarding young and uniform, icy lithospheres are applicable here as well). These arcuate rimmed furrows are best interpreted as graben formed in an extensional tectonic regime and then subjected to viscous relaxation (70,77). They are analogous to portions of the great impact ring structures on Callisto (32,33), and this remains the most observationally and theoretically supportable interpretation (82,83). The various ring and furrow systems can be used as probes of the rheological (and hence thermal) structures of the upper mantles of both satellites --- the lithospheres of both satellites are inferred to have cooled and thickened with age with Callisto being more thermally "evolved" at any given time; furthermore, the upper mantles of both bodies are seen to have been predominantly solid at the time of impact. Differences exist, however, between the furrow systems on Ganymede and the ring structures on Callisto (81,84), potentially indicating differences in the type of mantle rheologic structure for each body (2,85).

Only about 40% of Ganymede's surface is dark, ancient cratered terrain. The rest is brighter and younger and characterized by abundant parallel groove sets or, in some cases, a smooth surface at Voyager resolutions (32,33). The grooved and smooth terrains form regions, swaths, and lanes (sulci) which divide the dark terrain into assorted polygons. The polygons give the impression of being separated and translated from one another, but there is actually little unequivocal evidence for large-scale transcurrent motion on Ganymede (86,87). The boundaries between grooved and cratered terrain are sharp, with the grooved terrain being slightly lower topographically (70,87,88). These structural relationships suggest that grooved terrain formation is initiated by the filling of rifts, wide grabens, or troughs with bright terrain material (87). Direct evidence for this interpretation comes from analysis of dark halo craters in the grooved terrain (Fig. 5). These craters apparently excavate cratered terrain that underlies grooved terrain, and at least in one well-characterized area (Uruk Sulcus) the overall thickness of the grooved terrain material is 1.0-1.5 km (89).

Bright terrain material must be close to pure water ice in composition, i.e., if the amount of dark grains in Ganymede's optical surface, globally averaged, is 3 wt%, then the amount in the optical surface of the bright terrains must be smaller still. The physical state of bright terrain material during emplacement, however, is not clear. Liquid water or slush is favored by freeboard arguments (70) but encounters hydraulic difficulties if it is to rise as a fluid-filled crack through a less-dense ice lithosphere. Similarly,

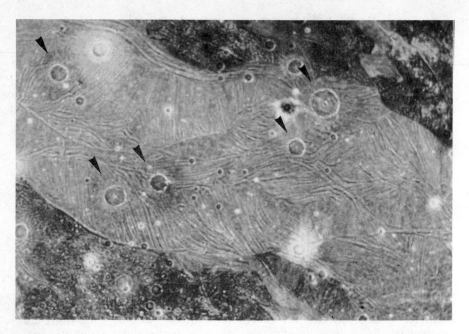

Fig. 5. High resolution image of a lane of grooved terrain, Uruk Sulcus (see Figure 3), containing dark halo craters (arrows) and multiple groove sets. Portions of Voyager 2 frames FDS 20637.17 and 20637.20 were mosaiced and contrast stretched. Scene is approximately 900 km across.

the ascent of warm, mobile ice is not contraindicated by density arguments but might more naturally lead to diapirism and not to structurally confined resurfacing. Detailed mechanical modeling has not been attempted, though, and Voyager resolution is insufficient to rule out the morphologies that would result from the flow of what would be essentially "glacial" ice. If liquid water or slush erupted to the surface, vigorous boiling would ensue; the polar shrouds have been interpreted as ultimately resulting from this process (90), although a sputtering (46) or impact origin is also possible.

The grooves themselves are linear to curvilinear troughs. They are generally hundreds of kilometers long, and often form regularly spaced parallel to sub-parallel sets (Fig. 5) (32,33); spacings range between approximately 5 and 13 km, and crest-to-trough topographic amplitudes are on the order of a few 100 m (91,92). Groove morphology is strongly suggestive of extensional tectonics (70,86,87,93). Small grooves and groove pairs probably form initially as tension fractures or graben; by analogy, groove sets may be akin to complex horst and graben terrains, although taken by themselves they could conceivably be due to compression. Complicating such identifications are the effects of viscous relaxation and mass wasting (mainly impact degradation) (94). Stages in groove terrain development have been identified, beginning with the formation of master grooves, separation of the terrain into polygons by secondary grooves, and finally the formation of groove sets within the polygons (88). Extensional driving stresses are inferred. Global expansion is often invoked as the cause (94,95), but it is now recognized that the expression of this stress field was highly variable over the satellite, that additional local and regional stresses (such as due to convective upwelling or thermal contraction) were important, or both (92,80,96).

It is important to stress that no endogenic tectonic patterns have been observed on Callisto to date, and this fundamental dichotomy between Ganymede and Callisto, two worlds of similar size and bulk composition and sharing nearby, adjacent orbits, remains a fundamental issue in comparative planetology.

A popular, but not necessarily correct, notion is that this difference can be traced back to a lack of differentiation for Callisto while at least partial differentiation affected Ganymede (97). By itself, however, this concept avoids the question of how a large, differentiated ice-rock satellite produces grooved terrain, nor for that matter does it seriously consider that such a satellite might evolve to a geologically inactive state before its cratering record is established. A better understanding of grooved terrain formation, or at least better constraints on its origin, is necessary to firmly link internal structure and evolution with surface expression. A few points seem clear: (1) extensional tectonics dominate; (2) stresses concentrate in regions of groove formation; (3) resurfaced material is differentiated (i.e., clean ice); and (4) groove formation and resurfacing are causally connected. This last point is not trivial, for it suggests that the same internal process is responsible for the tectonics and the volcanism. Beyond this, however, lie a number of mechanisms or models proposed to rationalize surface activity on Ganymede and its absence on Callisto (9,16,70,80,95,96,98-104).

Possible differences in the internal structures of Ganymede and Callisto ultimately lead back to the accretional state, so the question of grooved terrain bears directly on conditions in the protojovian nebula. It should be noted that the most detailed accretion model actually predicts partially differentiated structures for both satellites (9), making the above discussion more pointed. Constraints on internal structures provided by Galileo, via measurement of satellite second order gravity fields, will be welcome indeed.

2.3 Summary

Finally, the icy Galilean satellites are remarkable in that their geological properties are arranged in an orderly progression away from Jupiter. Bulk ice content increases from Europa outward to Callisto, yet surface ice content displays the opposite trend. Magnetospheric interaction decreases away from Jupiter while the influence of solar radiation increases. The amount and longevity of tectonic and volcanic activity declines from Europa to Callisto, and quite naturally surface age and the influence of cratering increase.

Many mysteries remain to be solved regarding these three "planets," but a number of well-posed questions and issues exist. Many of these will hopefully be answered or resolved when the Galileo orbiter reaches the Jupiter system. The next generation of questions should be even more exciting.

3. SATELLITES OF SATURN

The Saturnian satellite system differs from its Jovian counterpart in that only one Galilean-scale satellite, Titan, is present. The other "classical" satellites are intermediate in size, ~500-1500 km in diameter. Numerous other smaller satellites were discovered by Voyager (105,106), and of course, the rings represent untold numbers of extremely small satellites. The mean densities of all the major satellites of Saturn save Titan range between ~1.2 and ~1.4 g/cm^3 (with considerable uncertainties for some), indicating ice-rich compositions whose ice/rock ratios bracket the solar value (105,107). Titan's density (1.88 g/cm^3) is intermediate to that of Ganymede and Callisto, and as Titan is nearly as large as Ganymede, it is ice-rich as well.

The intermediate-sized satellites are arranged outward from Saturn in three size pairs: Mimas-Enceladus (~500 km in diameter), Tethys-Dione (~1000 km in diameter), and Rhea-Iapetus (~1500 km in diameter). Titan and the smaller, irregularly shaped Hyperion lie between Rhea and Iapetus. The presence of a single large satellite and numerous smaller ones (mostly orbiting closer to Saturn) suggests a less massive protosaturnian nebula, multiple epochs of satellite formation, or both (108,109). Although a formation scenario similar to that outlined above for the Galilean satellites seems likely, predicting the mass distribution of this satellite system is well beyond the present state of the art. Even the monotonic density decrease away from the primary exhibited by the Galilean satellites is apparently missing at Saturn. The protosaturnian nebula concept does, however, predict the condensation of ices more volatile than water ice.

Most geological knowledge of the Saturnian satellites stems from the Voyager encounters (105,106). The discussion below attempts to put what has been learned into the perspective of this paper; it does not aim to be comprehensive. A more complete geologic overview can be found in (110), a summary of ground-based studies in (111), and a discussion of Titan in (112).

3.1 Mimas, Tethys, Dione, and Rhea

The Cassinian satellites Tethys, Dione, and Rhea, along with Mimas, are worlds whose geology is dominated by impact cratering. Large craters exist on the leading hemispheres of Mimas (Herschel, 130-km-diameter), Tethys (Odysseus, 400-km-diameter), and Rhea (unnamed, 390-km-diameter). Amata (~250-km-diameter), on the poorly-imaged trailing hemisphere of Dione, is probably an impact structure as well. The kinetic energies of the impactors that created Herschel and Odysseus are within a factor of ~3 and ~10 of the gravitational binding energies of Mimas and Tethys, respectively (2). Whether such impacts are close to the threshold for catastrophic disruption and dispersal is uncertain, but surely cratering events of this sort catastrophically reorganize satellite surface geology, first by spallation and seismic disruption, and then by global ejecta deposition and cratering by secondary fragments ejected into Saturnian orbit (2). A primary bombardment so fierce as to cataclysmically disrupt and reassemble these satellites has been postulated (106), but this relies on suitably large impactors (109). If the projectile population at Saturn is similar to that at Jupiter, which is plausible and compatible with crater size-frequency statistics (2), then the likelihood of disruption is greatly reduced.

Evidence for impact-induced global-scale fracturing on these bodies includes simple fissures or grooves on Mimas (perhaps similar to those induced on the martian moon Phobos by the crater Stickney [113]), abundant lineaments on Tethys (114), and the "wispy" linea on Dione's trailing hemisphere that appear geometrically related to Amata (115,116). The globe-girdling rift system on Tethys, Ithaca Chasma, appears related to Odysseus, but in a more complicated manner. It follows a rough great circle centered on Odysseus for at least 270°. (The region where the rift may connect to a full 360° was not imaged). The rift appears to be several kilometers deep and to have developed as a multi-stage graben. A possible stress source is the viscous flow induced by the relaxation of Odysseus (106).

Endogenic tectonic activity on these satellites appears limited. Possible effects of global expansion and compression have been noted (114-116).

Despite the dominance of cratered terrains, all four satellites have resurfaced regions (117,118). Dione and Tethys, in particular, have lightly cratered plains units, regions where relatively smooth-lying material has been emplaced over older terrain. Rhea apparently has similar units, but of greater age (117). Some form of icy volcanism is implied, possibly involving ammonia-rich fluids (see Enceladus, below). Plains on Tethys and Dione are the best characterized; both contain linear to broadly sinuous troughs (termed chasmata on Dione) that morphologically resemble collapsed lava tubes on the earth and moon. In contrast, resurfaced regions on Mimas are identified solely on the basis of lower crater density; they may only reflect some previous cratering event(s).

The compositions of the optical surfaces of these satellites are essentially pure water ice (119,120, and references therein). Albedos are also generally very high ($\gtrsim 0.5$) and spectrally flat in the visible (121,122), attesting to a low level of contamination, radiation darkening, etc. Unfortunately, these worlds are too distant, too

small, and too cold to be detected by radio or radar techniques, so
direct measurement of their regolith compositions is not yet possible.
Still, the extreme iciness of their optical surfaces may imply
differentiation. Such small satellites are generally not expected to
accretionally melt (107), but repeated volcanism could have created a
silicate-free crust. Creating an ammonia-free crust is another matter
(see below). Note that there is no direct evidence for reassembly of
these satellites from compositionally distinct (i.e., silicate-rich
and silicate-free) sections, as might be expected in the "mega-impact"
scenario above.

The trailing hemispheres of Tethys, Dione, and Rhea are darker
than their leading hemispheres. A darkening mechanism acting in
conjunction with enhanced meteorite bombardment (and erosion, due to
low satellite gravity) on the leading sides appears to be responsible
(105). The darkening mechanism is unknown, but is probably
magnetospheric in origin (120). Brighter, apparently younger, "wispy"
materials stand out against the relatively darker backgrounds on Dione
and Rhea. The wisps are related to fault patterns; they may merely
be fresh talus, or more intriguingly, evidence for relatively recent
volcanic activity. Some form of pyroclastic deposit has been
postulated (105), and this would imply the incorporation of methane
during accretion.

3.2 Enceladus

Enceladus is remarkable in that, despite its small size
(500-km-diameter), it is the geologically youngest satellite of Saturn
(106). Most of its terrains are plains units; some have no
detectable craters at Voyager resolution. There are abundant tectonic
landforms (fissures and ridges) as well as evidence for the viscous
relaxation of even relatively small craters (down to 8 km in diameter
[123]). Furthermore, Enceladus is the most reflective object in the
solar system, with a geometric albedo of ~1.0 (121).

The key to this vigorous endogenic activity is probably tidal
heating (124-126). The geophysically preferred plains-forming "magma"
(on all the Saturnian satellites) is $2H_2O \cdot NH_3$ eutectic or peritectic
melt (107,127,128). Whether the melt is peritectic or eutectic
depends on the overall petrologic character of the satellite's ices.
(Silicates are inert at the relevant temperatures). Basically, the
petrology of icy satellites can be considered within the $H_2O-NH_3-CH_4$
ternary system. These are the major condensable species in reduced
form. Extending the dimensions of the system to include CO_2 and N_2
as end-members is simple; qualitatively, they behave similarly to CH_4
in that they form a clathrate with water ice (128). By themselves,
water and ammonia ice form three solid compounds: the two pure end
members, and a hemi-, mono-, and dihydrate. An equilibrium solar
mixture (~15 mole% NH_3) is part water ice, part dihydrate. Melting
occurs incongruently at ~175K, yielding a peritectic liquid of
approximately dihydrate composition, which is buoyant and fluid enough
to act as a resurfacing agent. Considerably less internal heating is

necessary to form this melt than to melt pure water ice, mainly because solid-state convection is inhibited at 175K. Sufficient methane partial pressure during condensation causes the water ice to be clathrated ($\sim 6H_2O \cdot CH_4$), but otherwise does not affect the melting relationship above because the melt is more ammonia-rich than the dihydrate. Sufficient CH_4 partial pressure causes the dihydrate to become unstable in favor of the monohydrate, however (129). Upon heating an assemblage of clathrate and monohydrate to $\sim 175K$, a eutectic melt forms with nearly the same composition as the peritectic melt. The melt is less ammonia-rich than the monohydrate, so clathrate must be consumed to make it. The melt thus becomes methane charged, and the potential for explosive volcanism is obvious. The only way to incorporate methane into the peritectic melt is by high degrees of partial melting or by assimilation of methane at shallow crustal levels (128).

Despite all these arguments, there is no evidence for ammonia in the visible and near-IR spectra taken to date of Enceladus or any other Saturnian satellite (120). Possibly, ammonia is preferentially removed from the optical surfaces of the satellites by UV photolysis (127) or sputtering (130).

The key to this puzzle may lie in the E ring, a tenuous ring whose peak particle concentration lies at Enceladus (131). These particles are now known to be spherical and small (2.0-2.5 μm in diameter) (132), implying derivation from liquid droplets (132,133) and a short ring lifetime in the absence of resupply (see 134). The particles are probably linked to Enceladus' bright, frosty surface, and the implication of recent ($< 10^4$ yr ago) activity of some sort is strong. Different scenarios or models have been proposed: impact- or tide-triggered liquid-water geysering (135), melt or condensed vapor products from a cometary impact (133), and condensation products from the sublimation of $2H_2O \cdot NH_3$ flood eruptions (136). Each predicts different compositions for the resulting ice spheres: water ice for the first, either water or $2H_2O \cdot NH_3$ ice for the second (depending on target composition), and ammonia ice for the third. Present photometric evidence admits H_2O ice, $2H_2O \cdot NH_3$ ice, but not NH_3 ice (132). The case of the E ring is far from closed, however. The short lifetimes of E-ring particles imply that the E ring represents the best chance for confirming the presence of ammonia in the satellite system. Further ground-based observations are clearly important for Enceladus (and Saturnian satellite) geophysics.

3.3 Iapetus

Iapetus is quite ice-rich (considering its density of ~ 1.16 g/cm^3) and heavily cratered (118). It is best-known for its pronounced hemispheric albedo asymmetry --- a very dark leading side and and a bright trailing side. The trailing side now appears to be mostly water ice, and the dark material, which neither completely covers nor is exclusively contained on the leading side, is of very low albedo (<0.08) and spectrally red (137). Early hypotheses postulated either preferential leading-hemisphere water-ice erosion or trailing-hemisphere water-ice deposition, with the presumption that Iapetus was fundamentally a dark, silicate body (138,139). These ideas are no longer considered relevant, given Iapetus' icy nature. Current hypotheses range from the implantation of dark dust from Phoebe, Saturn's outer retrograde satellite (140), to "carbonaceous" volcanism (105), UV photochemistry in impact-exposed methane clathrate (141), or even more exotic possibilities (142). The spatial distribution of dark material clearly implicates Phoebe dust, but some mechanism or mechanisms must account for the fine details of its (or some related material's) distribution as well as the red color (Phoebe itself is spectrally neutral) (131). Add to this the possibility that Iapetus has exchanged its leading and trailing hemispheres, due to large (>100-km-diameter) impacts, over geologic time (2), and the mystery surrounding this satellite deepens. Those hypotheses that deal with organic matter on the surface of Iapetus, in particular, have potentially profound ramifications (143).

3.4 Titan

Titan's geology is unknown, as its hazy atmosphere is opaque in the visible. Whether it even has a solid surface is under discussion (144,145). Nevertheless, it should possess an icy crust of some sort, one which has been bombarded and subjected to tectonism and volcanism. These last two are especially expected as Titan is a Ganymede-sized planet that has incorporated more volatile ices such as NH_3 and CH_4 (this inferred from its atmospheric constitution, mainly N_2 and CH_4) (112). Titan may have an (ethane-methane) ocean, and thus the potential exists for fluvial (stream and river) or even pluvial (rainfall) processes. Obviously, interesting chemical cycles exist, many of them organic (143).

4. SATELLITES OF URANUS

Relatively little is known about the satellites of Uranus, but we are poised, as of this writing, on the eve of the Voyager 2 encounter with the Uranian system, in January, 1986. In addition, remarkable new insights are continually being provided by ground-based observation and interpretation.

The five satellites, Miranda, Ariel, Umbriel, Titania, and Oberon form a compact, regular system. Miranda is the smallest (~500 km in diameter), Ariel and Umbriel are a larger pair (~1200 km in diameter), and Titania and Oberon are a still larger pair (~1800 km in diameter) (146). This step-wise increase in scale is reminiscent of the intermediate-sized satellites of Saturn, although the Uranian satellites are generally larger. This system, like those of Jupiter and Saturn above, is believed to have formed in a circumplanetary accretion disk. Uranus is sufficiently different from Jupiter and Saturn in terms of composition and structure, however, that the formation of a protouranian nebula is not necessarily obvious, so its mere existence is significant.

The surfaces of the Uranian satellites are characterized by low albedos (except for Miranda, $\lesssim 0.3$), large opposition effects, and water-ice compositions (146-148, and references therein). The low albedos are consistent with a small admixture of dark, spectrally neutral material; this material may be carbon-rich and related to the low albedo of the Uranian rings (120,147). The opposition effect (a brightness surge as zero phase angle is approached) may be due to a unique porosity structure in the optical surface. As sublimation of water ice is not an important process at ~80 K, the approximate surface temperature of these satellites (24), grain sizes are likely to be small unless sputtering by magnetospheric particles (149) is important.

There is some correlation between albedo and orbital distance from Uranus; the outer three satellites are the darkest (albedos are ~0.2), Ariel is brighter (albedo ~0.3), and Miranda may be brighter still (albedo ~0.5). This pattern may reflect degrees of internal activity (150) or intrinsic compositional differences. In particular, Miranda may have a basically pure water-ice surface similar to that of the bright Saturnian satellites (148). The albedo pattern may also be related to the inverted density trend among the satellites. The most recent astrometry-based determination (151) implies that the densities of Oberon and Titania are ~2.5 g/cm^3, while those of the inner satellites, including Miranda (152), are 1-2 g/cm^3. The error bars on these estimates are large enough to make this trend uncertain, but provocative explanations can be suggested. One involves the CO/CH$_4$ kinetics in the protouranian nebula (153). As discussed in (11,154), the conversion of CO to CH$_4$ in the protosolar nebula may be kinetically inhibited at low temperatures, while in a warmer, denser planetary nebula it would proceed to equilibrium (i.e., nearly complete reduction). The protouranian nebula could have been cold enough and at low enough density in its outer portions, however, that CO could have been the dominant carbon-bearing species there. CO locks up a great deal of oxygen that would normally form water ice, so an ice-rock satellite could have its ice/rock mass ratio changed from 60/40 to ~30/70. (This calculation ignores NH$_3$ ice, but as nitrogen is a much less abundant ice-former than carbon or oxygen [14], its presence or absence cannot substantially alter these ratios). Naturally, for this scenario to work, there must be little convective communication between the H$_2$O- and CH$_4$-rich inner nebula and the H$_2$O-poor and CO-rich outer portion. An alternate scenario involves creating the protouranian nebula by giant impact, with possible shock effects on composition (see 155).

These inferences are tantalizing, but the true study of Uranian satellite geology will begin with the arrival of Voyager. We expect to see unveiled the effects of cratering, volcanism, tectonism, and interaction with magnetospheric particles; possibly, interaction of solar radiation with ices more volatile than water ice will be apparent as well. This is especially true for Miranda and Ariel, which are presently targeted for the closest encounters.

5. TRITON

Neptune does not possess a regular satellite system to our knowledge,
but it does possess Triton, a large satellite following an inclined,
retrograde orbit. Precious little is known about Triton; even its
size and mass are uncertain (156-158). The best radiometric
determination gives a radius of ~1750 km (159), and a plausible
density for an ice-rock satellite of this size is ~1.6 g/cm^3. Other
significant data concern surface composition; methane ice, water ice,
and nitrogen in the solid state (probably liquid) have been detected
(158). The identification of nitrogen, either in large lakes, oceans,
or fields of extremely large (tens of centimeters across!) crystals,
implies a substantial N_2 atmosphere, possibly up to 0.3 bars worth
(158), and hence the possibility exists for weather, seasons, climate,
chemical cycles, etc. Unlike Titan with its opaque atmosphere, Triton
is more amenable to observation; it is more quickly available as
well, for an encounter with Voyager 2 is planned for August, 1989. If
Triton's unique orbit is taken as evidence of a capture origin, then
it probably once had a very active tectonic and volcanic regime driven
by capture-derived tidal energy (157). Such an episode of vigorous
tidal heating would have led to catastrophic melting of Triton's ices,
forming a primordial atmosphere more massive than at present.
Triton's unique surface composition may be due to a condensed fraction
of that atmosphere. Furthermore, pluvial and fluvial activity may
have once existed on Triton.
        The remarks above for Titan and the satellites of Uranus,
concerning the general geologic processes expected, also apply to
Triton. At both Uranus and Neptune, new icy satellites may be
discovered by Voyager as well.

6. PROSPECTS

On one level, the prospects before us are the geological introduction
to six new satellites by Voyager 2, a detailed examination of the icy
Galilean satellites by the Galileo orbiter, and the first true
constraints on the global and surface properties of Pluto's presumably
icy satellite, Charon, as mutual occultations in that system occur for
the next few years. Downstream lies a potential return to the
Saturnian system, the Cassini mission, and perhaps orbiters and probes
of Uranus and Neptune as well. More fundamental, however, are the
challenges to understanding that the geological processes of icy
worlds present when compared with the relatively familiar processes of
the terrestrial planets. For example, could multiringed systems such
as found on Callisto have formed on the early earth? Why don't we
see evidence for plate tectonics on Ganymede? Can an ice-mantle
planet have plate tectonics? If small icy satellites differentiated,

what was the energy source? Was it operating during accretion of the
inner solar system? How do giant planets and satellites really
accrete? Is there a role for very large planetesimals? What were the
relative timescales for bombardment of the inner and outer solar
system? The answers to these and other questions will take years to
work out, but is it too bold to suppose that a reasonable paradigm of
solar system origin and evolution will exist by the end of the
century? And could this not form a basis for a new century of
detailed (and manned) exploration of the solar system?

7. ACKNOWLEDGMENTS

I sincerely thank A. Dollfus, R. Smoluchowski, T.J. Ahrens, and
D. Morrison, for making it possible for me to attend this unique
meeting, and especially J. Klinger for extreme patience in awaiting
this manuscript. This research supported by NASA Grant NAGW-432.

8. REFERENCES

(1)   Gaffney, E.S. 1984, Geophys. Res. Lett. 11, pp. 121-123.
(2)   Chapman, C.R., and McKinnon, W.B. 1985, in Natural Satellites,
       J.A. Burns, ed., University of Arizona Press, Tucson.
(3)   Voight, G.-H., Hill, T.W., and Dessler, A.J. 1983,
       Astrophys. J. 266, pp. 390-401.
(4)   Lanzerotti, L.J., Maclennan, C.G., Brown, W.L., Johnson, R.E.,
       Barton, L.A., Reimann, C.T., Garrett, J.W., and Boring, J.W.
       1983, J. Geophys. Res. 88, pp. 8765-8770.
(5)   Mizuno, H. 1980, Progr. Theor. Phys. 64, pp. 544-557.
(6)   Stevenson, D.J. 1982, Planet. Space Sci. 30, pp. 755-764.
(7)   Weidenschilling, S.K. 1981, in The Comparative Study of the
       Planets, A. Coradini and M. Fulchignoni, eds., D. Reidel,
       pp. 49-59.
(8)   Shoberg, T.G. 1982, M.A. thesis, Washington University,
       Saint Louis, 132 pp.
(9)   Lunine, J.I., and Stevenson, D.J. 1982, Icarus 52, pp. 14-39.
(10)   Bodenheimer, P., Grossman, A.S., DeCampli, W.M., Marcy, G.,
       and Pollack, J.B. 1980, Icarus 41, pp. 293-308.
(11)   Prinn, R.G., and Fegley, B., Jr. 1981, Astrophys. J. 249,
       pp. 308-317.
(12)   Shoemaker, E.M. 1984, Urey Lecture, 16th DPS Meeting, Kona,
       Hawaii.
(13)   Shoemaker, E.M., and Wolfe, R.M. 1982, in Satellites of
       Jupiter, D. Morrison, ed., University of Arizona Press,
       Tucson, pp. 277-339.
(14)   Anders, E., and Ebihara, M. 1982, Geochim. Cosmochim. Acta 46,
       pp. 2363-2380.
(15)   Ahrens, T.J., and O'Keefe, J.D. 1984, Lunar Planet. Sci. XV,
       pp. 3-4.

(16)   Cassen, P.M., Peale, S.J., and Reynolds, R.T. 1982, in
        Satellites of Jupiter, D. Morrison, ed., University of Arizona
        Press, Tucson, pp. 93-128.
(17)   Clark, R.N. 1980, Icarus 44, pp. 388-409.
(18)   Clark, R.N. 1981, Icarus 49, pp. 244-257.
(19)   Goldstein, R.M., and Green, R.R. 1980, Science 207,
        pp. 179-180.
(20)   Ostro, S.J. 1982, in Satellites of Jupiter, D. Morrison, ed.,
        University of Arizona Press, Tucson, pp. 213-236.
(21)   Warren, S.J. 1984, Appl. Opt. 23, pp. 1206-1225.
(22)   De Pater, I., Brown, R.A., and Dickel, J.R. 1984, Icarus 57,
        pp. 93-101.
(23)   Muhleman, D., Berge, G., and Rudy, D. 1984, Bull. Am. Astron.
        Soc. 16, p. 686.
(24)   Clark, R.N., Fanale, F.P., and Zent, A.P. 1983, Icarus 56,
        pp. 233-245.
(25)   Johnson, T.V., and Pilcher, C.B. 1977, in Planetary Satellites,
        J.A. Burns, ed., University of Arizona Press, Tucson,
        pp. 232-268.
(26)   Morrison, D., and Morrison, N.D. 1977, in Planetary Satellites,
        J.A. Burns, ed., University of Arizona Press, Tucson,
        pp. 363-378.
(27)   Johnson, T.V., Soderblom, L.A., Mosher, J.A., Danielson, G.E.,
        Cook, A.F., and Kupferman, P. 1982, J. Geophys. Res. 88,
        pp. 5789-5805.
(28)   Nelson, M.L., McCord, T.B., Clark, R.N., Johnson, T.V., and
        Matson, D. 1983, Lunar Planet. Sci. XIV, pp. 554-555.
(29)   Buratti, B., and Veverka, J. 1983, Icarus 55, pp. 93-110.
(30)   Lane, A.L., Nelson, R.M., and Matson, D.L. 1981, Nature 292,
        pp. 33-39.
(31)   McEwen, A.S. 1984, Lunar Planet. Sci. XV, pp. 525-526.
(32)   Smith, B.A., et al. 1979, Science 206, pp. 927-950.
(33)   Smith, B.A., et al. 1979, Science 204, pp. 951-972.
(34)   Pieri, D.C. 1981, Nature 289, pp. 17-21.
(35)   Lucchitta, B.K., Soderblom, L.A., and Ferguson, H.M. 1981,
        Proc. Lunar Planet. Sci., 12B, pp. 1555-1567.
(36)   Lucchitta, B.K., and Soderblom, L.A. 1982, in Satellites of
        Jupiter, D. Morrison, ed., pp. 521-555.
(37)   Schenk, P.M. 1984, Adv. Planet. Geol., NASA TM-86247,
        pp. 3-111.
(38)   Schenk, P.M., and Seyfert, C.K. 1980, Eos Trans. AGU 61,
        p. 286.
(39)   Johnson, T.V., and McGetchin, T.R. 1973, Icarus 18,
        pp. 612-620.
(40)   Parmentier, E.M., and Head, J.W. 1981, Icarus 47, pp. 100-111.
(41)   Squyres, S.W., Reynolds, R.T., Cassen, P.M., and Peale,
        S.J. 1983, Nature 301, pp. 225-226.
(42)   Goodman, D.J., Frost, H.J., and Ashby, M.F. 1981, Phil. Mag.
        A43, pp. 665-695.
(43)   Weertman, J. 1983, Ann. Rev. Earth Planet. Sci. 11,

pp. 215-240.

(44) Durham, W.B., Heard, H.C., and Kirby, S.H. 1983, Proc. Lunar Planet. Sci. Conf. 14th, J. Geophys. Res. 88, pp. B377-B392.

(45) Passey, Q.R., and Shoemaker, E.M. 1982, in Satellites of Jupiter, D. Morrison, ed., pp. 379-434.

(46) Sieveka, E.M., and Johnson, R.E. 1982, Icarus 51, pp. 528-548.

(47) Melosh, H.J. 1977, Icarus 31, pp. 221-243.

(48) Melosh, H.J. 1980, Icarus 43, pp. 334-337.

(49) Helfenstein, P., and Parmentier, E.M. 1980, Proc. Lunar Planet. Sci. Conf. 11th, pp. 1987-1998.

(50) Helfenstein, P., and Parmentier, E.M. 1983, Icarus 53, pp. 415-430.

(51) Yoder, C.F. 1979, Nature 279, pp. 767-770.

(52) Yoder, C.F., and Peale, S.J. 1981, Icarus 47, pp. 1-35.

(53) Finnerty, A.A., Ransford, G.A., Pieri, D.C., and Collerson, K.D. 1981, Nature 289, pp. 24-27.

(54) Golombek, M., and Bruckenthal, E. 1983, Lunar Planet. Sci. XIV, pp. 251-252.

(55) Schenk, P.M. 1985, preprint.

(56) Ransford, G.A., Finnerty, A.A., and Collerson, K.D. 1981, Nature 289, pp. 21-24.

(57) Greenberg, R., and Weidenschilling, S.J. 1984, Icarus, pp. 186-196.

(58) Parmentier, E.M., and Helfenstein, P. 1984, Bull. Am. Astron. Soc. 16, p. 686.

(59) Squyres, S.W., and Veverka, J. 1981, Icarus 46, pp. 137-155.

(60) Morrison, D. 1982, in Satellites of Jupiter, D. Morrison, ed., pp. 3-43.

(61) Lucey, P. 1984, paper presented at Ices in the Solar System, NATO Adv. Res. Workshop, Nice, France.

(62) Spencer, J.R., and Maloney, P.R. 1984, Geophys. Res. Lett. 11, pp. 1223-1226.

(63) Spencer, J.R. 1983, abs. Natural Satellites, IAU Colloquium 77, p. 25.

(64) Spencer, J.R. 1984, Bull. Am. Astron. Soc. 16, p. 685.

(65) Wolff, R.S., and Mendis, D.A. 1983, J. Geophys. Res. 88, pp. 4749-4769.

(66) Pang, K., Ajello, J.M., Lumme, K., and Bowell, E. 1983, Proc. Lunar Planet. Sci. Conf. 13th, J. Geophys. Res. 88, pp. A569-A576.

(67) Dollfus, A. 1975, Icarus 25, pp. 416-431.

(68) Veverka, J. 1977, in Planetary Satellites, J.A. Burns, ed., University of Arizona Press, Tucson, pp. 210-231.

(69) Mandeville, J.-C., Geake, J.E., and Dollfus, A. 1980, Icarus 41, pp. 343-355.

(70) Shoemaker, E.M., Lucchitta, B.K., Wilhelms, D.E., Plescia, J.B., and Squyres, S.W. 1982, in Satellites of Jupiter, D. Morrison, ed., pp. 435-520.

(71) Strom, R.G., Woronow, A., and Gurnis, M. 1981, J. Geophys. Res. 86, pp. 8659-8674.

(72)   Woronow, A., Strom, R.G., and Gurnis, M. 1982, in Satellites of
       Jupiter, D. Morrison, ed., pp. 237-276.
(73)   Woronow, A., and Strom, R.G. 1982, Geophys. Res. Lett. 8,
       pp. 891-894.
(74)   Gurnis, M. 1985, preprint.
(75)   Pollack, J.B., Witteborn, F.C., Erickson, E.F., Strecker, D.W.,
       Baldwin, B.J., and Bunch, T.E. 1978, Icarus 36, pp. 271-303.
(76)   Hartmann, W.K. 1980, Icarus 44, pp. 441-453.
(77)   McKinnon, W.B., and Melosh, H.J. 1980, Icarus 44, pp. 454-471.
(78)   Schenk, P.M., and McKinnon, W.B. 1984, Bull. Am. Astron.
       Soc. 16, p. 683.
(79)   Greeley, R., Fink, J.F., Gault, D.E., and Guest, J.E. 1982,
       in Satellites of Jupiter, D. Morrison, ed., pp. 340-378.
(80)   McKinnon, W.B. 1981, Proc. Lunar Planet. Sci., 12B,
       pp. 1585-1597.
(81)   Casacchia, R., and Strom, R.G. 1984, Proc. Lunar Planet. Sci.
       Conf. 14th, J. Geophys. Res. 89, pp. B419-B428.
(82)   McKinnon, W.B. 1981, Multi-ring Basins, Proc. Lunar Planet.
       Sci., 12A, pp. 259-273.
(83)   Melosh, H.J. 1982, J. Geophys. Res. 87, pp. 1880-1890.
(84)   Zuber, M.T., and Parmentier, E.M. 1984, Icarus 60, pp. 200-210.
(85)   McKinnon, W.B. 1982, Lunar Planet. Sci. XIII, pp. 499-500.
(86)   Lucchitta, B.K. 1980, Icarus 44, pp. 481-501.
(87)   Parmentier, E.M., Squyres, S.W., Head, J.W.,
       and Allison, M.L. 1982, Nature 295, pp. 290-293.
(88)   Golombek, M.P., and Allison, M.L. 1981, Geophys. Res. Lett. 8,
       pp. 1139-1142.
(89)   Schenk, P.M., and McKinnon, W.B. 1985, Proc. Lunar Planet. Sci.
       Conf. 15th, J. Geophys. Res. 90.
(90)   Shaya, E.J., and Pilcher, C.B. 1984, Icarus 58, pp. 74-80.
(91)   Squyres, S.W. 1981, Icarus 46, pp. 156-168.
(92)   Grimm, R.E., and Squyres, S.W. 1985, J. Geophys. Res. 90,
       pp. 2013-2021.
(93)   Golombek, M.P. 1982, Proc. Lunar Planet. Sci. Conf. 13th,
       J. Geophys. Res. 87, pp. A77-A83.
(94)   Squyres, S.W. 1982, Icarus 52, pp. 545-559.
(95)   Squyres, S.W. 1981, Geophys. Res. Lett. 7, pp. 593-596.
(96)   Zuber, M.T., and Parmentier, E.M. 1984, Proc. Lunar Planet.
       Sci. Conf. 14th, J. Geophys. Res. 89, pp. B429-B437.
(97)   Schubert, G., Stevenson, D.J., and Ellsworth, K. 1981,
       Icarus 47, pp. 46-59.
(98)   Thurber, C.H., Hsui, A.T., and Toksoz, M.N. 1980, Proc.
       Lunar Planet. Sci. Conf. 11th, pp. 1957-1977.
(99)   Cassen, P., Peale, S.J., and Reynolds, R.T. 1980, Icarus 41,
       pp. 232-239.
(100) Kawakami, S., and Mizutani, H. 1980, Proc. 13th ISAS Lunar
       Planet. Symp., Univ. Tokyo, pp. 330-345.
(101) Coradini, A., Federico, C., and Lanciano, P. 1981, in
       The Comparative Study of the Planets, A. Coradini and
       M. Fulchignoni, eds., D. Reidel, pp. 61-70.

(102) Greenberg, R. 1982, in Satellites of Jupiter, D. Morrison,
       ed., University of Arizona Press, Tucson, pp. 65-89.
(103) Kirk, R.L., and Stevenson, D.J. 1983, Lunar Planet. Sci. XIV,
       pp. 373-374.
(104) Friedson, A.J., and Stevenson, D.J. 1983, Icarus 56, pp. 1-14.
(105) Smith, B.A., et al. 1981, Science 212, pp. 163-191.
(106) Smith, B.A., et al. 1982, Science 215, pp. 504-537.
(107) Ellsworth, K., and Schubert, G. 1983, Icarus 54, pp. 490-510.
(108) Stevenson, D.J. 1982, Ann. Rev. Earth Planet. Sci. 10,
       pp. 257-295.
(109) Pollack, J.B., and Consolmagno, G. 1984, in Saturn, T. Gehrels
       and M.S. Matthews, eds., University of Arizona Press, Tucson,
       pp. 811-866.
(110) Morrison, D., Johnson, T.V., Shoemaker, E.M., Soderblom, L.A.,
       Thomas, P., Veverka, J., and Smith, B.A. 1984, in Saturn,
       T. Gehrels and M.S. Matthews, eds., University of Arizona
       Press, Tucson, pp. 811-866.
(111) Cruikshank, D.P., Veverka, J., and Lebofsky, L.A. 1984,
       in Saturn, T. Gehrels and M.S. Matthews, eds., University of
       Arizona Press, Tucson, pp. 811-866.
(112) Hunten, D.M., Tomasko, M.G., Flasar, F.M., Samuelson, R.E.,
       Strobel, D.F., and Stevenson, D.J. 1984, in Saturn, T. Gehrels
       and M.S. Matthews, eds., University of Arizona Press, Tucson,
       pp. 811-866.
(113) Thomas, P., and Veverka, J. 1979, Icarus 40, pp. 394-405.
(114) Moore, J.M., and Ahern, J.L. 1982, Proc. Lunar Planet. Sci.
       Conf. 12th, J. Geophys. Res. 88, pp. A577-A584.
(115) Plescia, J.B. 1983, Icarus 56, pp. 255-277.
(116) Moore, J.M. 1984, Icarus 59, pp. 205-220.
(117) Plescia, J.B., and Boyce, J.M. 1982, Nature 295, pp. 285-290.
(118) Plescia, J.B., and Boyce, J.M. 1983, Nature 301, pp. 666-670.
(119) Clark, R.N., and Owensby, P.D. 1981, Icarus 46, pp. 354-360.
(120) Clark, R.N., Brown, R.H., Owensby, P.D., and Steele, A. 1984,
       Icarus 58, pp. 265-281.
(121) Buratti, B., and Veverka, J. 1984, Icarus 58, pp. 254-264.
(122) Buratti, B. 1984, Icarus 59, pp. 392-405.
(123) Passey, Q.R. 1983, Icarus 53, pp. 105-120.
(124) Yoder, C.F. 1979, Nature 279, pp. 767-770.
(125) Squyres, S.W., Reynolds, R.T., Cassen, P.M., and Peale, S.J.
       1983, Icarus 53, pp. 319-331.
(126) Lissauer, J.J., Peale, S.J., and Cuzzi, J.N. 1984, Icarus 58,
       pp. 159-168.
(127) Consolmagno, G.J., and Lewis, J.S. 1978, Icarus 34,
       pp. 280-293.
(128) Stevenson, D.J. 1982, Nature 298, pp. 142-144.
(129) Lunine, J.I. 1984, Ph.D Thesis, Caltech, 326 pp.
(130) Lanzerotti, L.J., Brown, W.L., Marcantonio, K.J., and Johnson,
       R.E. 1984, Nature 312, pp. 139-140.
(131) Baum, W.A., Kreidl, T., Westphal, J.A., Danielson, G.E.,
       Seidelmann, P.K., Pascu, D., and Currie, D.G. 1981, Icarus 47,

pp. 84-96.

(132) Pang, K.D., Voge, C.C., Rhoads, J.W., and Ajello, J.M. 1984,
       J. Geophys. Res. 89, pp. 9459-9470.

(133) McKinnon, W.B. 1983, Lunar Planet. Sci. XIV, pp. 487-488.

(134) Haff, P.K., Eviatar, A., and Siscoe, G.L. 1983, Icarus 56,
       pp. 426-438.

(135) Terrile, R.J., and Cook, A.F. 1981, Lunar Planet. Sci. XII,
       suppl. A, pp. 10-11.

(136) Herkenhoff, K.E., and Stevenson, D.J. 1984, Lunar Planet.
       Sci. XV, pp. 361-362.

(137) Cruikshank, D.P., Bell, J.F., Gaffey, M.J., Brown, R.H.,
       Howell, R., Beerman, C., and Rognstad, M. 1983, Icarus 53,
       pp. 90-104.

(138) Cook, A.F., and Franklin, F.A. 1970, Icarus 13, pp. 282-291.

(139) Peterson, C.F. 1975, Icarus 24, pp. 499-503.

(140) Soter, S. 1974, paper presented at "Planetary Satellites",
       Ithaca, NY.

(141) Squyres, S.W., and Sagan, C. 1983, Nature 303, pp. 782-785.

(142) Clarke, A.C. 1968, 2001: A Space Odyssesy, New American
       Library, Inc., 221 pp.

(143) Sagan, C., Khare, B.N., and Lewis, J.S. 1984, in Saturn,
       T. Gehrels and M.S. Matthews, eds., University of Arizona
       Press, Tucson, pp. 788-807.

(144) Dermott, S.F., and Sagan, C. 1982, Nature 300, pp. 731-733.

(145) Lunine, J.I., Stevenson, D.J., and Yung, Y.L. 1983,
       Science 222, p. 1229-1230.

(146) Brown, R.H., Cruikshank, D.P., and Morrison, D. 1982,
       Nature 300, pp. 423-425.

(147) Brown, R.H., and Cruikshank, D.P. 1983, Icarus 55, pp. 83-92.

(148) Brown, R.H., and Clark, R.N. 1984, Icarus 58, pp. 288-292.

(149) Cheng, A.F., and Lanzerotti, L.J. 1978, J. Geophys. Res. 83,
       pp. 2597-2602.

(150) Squyres, S.W., and Reynolds, R.T. 1983, Eos Trans. AGU 64,
       p. 746.

(151) Veillet, Ch. 1983, Ph.D Thesis, Université de Paris, 171 pp.

(152) Greenberg, R. 1984, Uranus and Neptune, NASA CP-2330,
       pp. 463-480.

(153) McKinnon, W.B. 1984, Abs. NATO Workshop, Ices in the Solar
       System, p. 80.

(154) Lewis, J.S., and Prinn, R.G. 1980, Astrophys. J. 238,
       pp. 357-364.

(155) Stevenson, D.J. 1984, Uranus and Neptune, NASA CP-2330,
       pp. 405-421.

(156) Cruikshank, D.P., Stockton, A., Dyck, H.M., Becklin, E.,
       and Macy, W., Jr. 1979, Icarus 40, pp. 104-114.

(157) McKinnon, W.B. 1984, Nature 311, pp. 355-358.

(158) Cruikshank, D.P., Brown, R.H., and Clark, R.N. 1984,
       Icarus 58, pp. 293-305.

(159) Lebofsky, L.A., Rieke, G.H., and Lebofsky, M.J. 1982,
       Bull. Am. Astron. Soc. 14, p. 766.

Part VI

Summary of the Highlights
of the Conference

SUMMARY OF THE HIGHLIGHTS OF THE CONFERENCE

R. Smoluchowski

Department of Astronomy
University of Texas at Austin
Austin, Texas 78712   USA

Attempts at being unbiased in summarizing the results of a
scientific conference such as ICES IN THE SOLAR SYSTEM usually
fail because of unavoidable personal preferences for particular
topics. Nevertheless, I shall do my best to be impartial in
presenting the highlights of the subjects discussed during this
very fruitful and interesting meeting.

1. PROPERTIES OF ICES

The analysis of many problems involving ices in the solar
system often has been limited to the simplest properties of water-
ice in normal conditions because of lack of good other data. The
new results reported at the conference improve greatly this
situation. We have now a better Hugoniot for water-ice, which
includes the effects of yielding and of phase changes. Also
available are new data on melting under pressure and a detailed
diagram of the H_2O-ice including a new easy transformation of
hexagonal ice at 77 K. Also microwave data are available which
permit remote determination of the thickness of ice layers.
There appear to be two kinds of amorphous H_2O-ice whose structure
(presence of water clusters?) depends on the rate of cooling and
on surface conditions; they devitrify at 160–177 K. Investigations
of the important binary NH_3-H_2O system at various temperatures in
the water-rich region from 1 to 4 GPa have indicated that the
highest pressure at which a homogeneous liquid can exist occurs
at 15 percent NH_3. The results are of importance for the
question of the presence of liquid water on certain satellites.

Much progress can be noted in measuring and understanding

859

J. Klinger et al. (eds.), Ices in the Solar System, 859–864.
© *1985 by D. Reidel Publishing Company.*

mechanical properties of water-ice which are so important for the analysis of cratering and of the motion and rheology of ices on satellites. Detailed experimental and theoretical studies have been made of impacts on ice and of the associated scaling laws for craters. New general scaling laws have been derived and applied to various features in the solar system. These laws are now based on firmer ground so that statisical studies have better validity. The numerical coefficients in the constitutive equation of creep (i.e. the viscosity) have been determined for water-ice VI at pressures from 0.9 to 1.7 GPa and 250 to 300 K. This ice, as well as ice VII, may exist in the larger satellites of the giant planets and permit convection. The response of icy bodies to impacts and the transition between ductile and brittle deformation of hexagonal ice, which occurs near 190 K, were studied in considerable quantitative detail. As expected, brittle behavior is enhanced by lower pressure and temperature and by higher strain rate. It is concluded that layers of hexagonal ice on icy bodies such as comets are much weaker than it has been usually assumed. The relaxation of craters made in ice can be estimated and the transition between craters controlled by the strength of ice and those controlled by gravity evaluated.

Further progress is to be noted in our knowledge of the rather elusive clathrate hydrates. On the Earth methane hydrate occurs in Siberia and Alaska and it should be present on Uranus and Neptune. The polar ice caps on Mars and the comets should contain CO_2 hydrate. The order-disorder and other phase transitions in CH_4 ice up to 60 kbar and 350 K have been further investigated. Detailed studies of infrared absorption and scattering by ice frost and of frost on ice (10 to 100 μm in size) have been made with particular attention to the important shift of the position of the maximum of the 3 μm line with increasing path length in the solid. This shift, towards shorter wavelength, is accompanied by increasing linewidth. In view of these observations a positive identification of amorphous water-ice based on its 3-μm band width and shape signatures will be unfortunately very difficult. There appear to exist important linear relationships between purity, albedo and the V-J and J-K color indices of ices on Trojan asteroids and on satellites of the outer planets. All these results are of great value for remote sensing of ices and for understanding their behavior on bodies in the solar system. Also the effect of charged particle bombardment on the albedo of ices and the associated sputtering and erosion phenomena have been carefully investigated.

2.   ICES ON PLANETS AND SATELLITES

A number of observations and deductions concerning ices on

planets and their satellites have been reported. There are
indications that Mars has a layer of permafrost which, on the
equator, is about 3.5 km thick and reaches about 8 km near the
poles. Its salinity may be high enough to prevent from freezing
liquid water located below the lithosphere. If true, this could
open interesting biological possibilities. There is also
evidence, based on surface polygonal structure, of large climatic
changes in the past. Jupiter's Io is covered with SO_2-frost
(10^{-3} g cm^{-2}) which may either be free or be entrapped in pores
in the surface layer. Although there are several hot spots on
this satellite one of them, the Loki crater, accounts for nearly
all Io's heat emission. This hot spot is observable even when
the satellite is in the planet's shadow. The heat emission of
Io seems to have been overestimated in the past by a factor of
two. The surfaces of the other Galilean satellites are, roughly,
a 50-50 mixture of rocks and water-ice. Europa shows fairly
uniform faulting and tidal flexing of ice implying its decoupling
from the underlying rock. These observations could be accounted
for by an underlying layer of liquid water. There is also the
possibility of the presence of NH_3 which lowers drastically the
melting point of water-ice. Spectral asymmetries can be
explained by sputtering and implantation of magnetospheric ions.
Although on Ganymede there is some evidence for relative
rotation of parts of the crust and extensional tectonics, so
far, there is no suggestion for terrestrial-type plate tectonics.
The grooved terrain has been carefully studied. Ganymede's polar
caps are probably remnants of thicker layers formed at a time
when the satellite was geologically active; later they were
eroded by impacts and thermal migration. Very likely Callisto
never had such thick layers and thus has no polar caps. It
shows, however, evidence of tidal flexing and despinning in the
Valhalla region. As compared to inner planets both these
satellites show very few craters larger than about 60 km in
diameter and those craters which are there have undergone
viscous relaxation.

On Saturnian satellites the ices are probably water ice and
are at least 3.5 billion years old. Their cratering history and
resurfacing, which has eliminated many large craters, has been
studied in detail. Certain areas on Enceladus are 3.75 billion
years old, other areas have been resurfaced recently, presumably
by a $2H_2O.NH_3$ peritectic although there is no direct evidence
for NH_3. These processes appear to be driven by tidal heating.
It has been suggested that the ten-fold brightness contrast on
Iapetus may be explained by an early plasma irradiation and an
impact which altered the satellite's orientation. Other
possibilities are dust from Phoebe and carbonaceous ejecta. The
shape of Saturn's icy satellites agrees with the expected maximum
size of non-spherical, irregular bodies. The albedos of these
satellites range from 0.4 to 0.8, the darker ones being presumably

covered with less pure ice. The surface of Titan (and Triton) is
dominated by solids much more volatile than water ice. There may
be liquid methane or methane-ethane-nitrogen oceans with bedrock
made of solid acetylene resting on a layer of H_2O- or H_2O-NH_3
ices including perhaps hydrocarbon clathrate hydrates. The
various gases entrapped in these clathrates presumably accreted
on the early planetesimals from which this, and other, satellites
were made. Later they were partly released into the atmosphere.
The satellites of Uranus are very likely icy, show strong
opposition brightening and have low albedos. Presence of carbon
and of carbonaceous chondrites is suspected. It is tentatively
suggested that Neptune's Triton may be similar to Titan, may have
a nitrogen-rich atmosphere and liquid nitrogen oceans on which
solid methane would float. IR studies indicate that Pluto is
almost certain to contain a large proportion of CH_4-ice but the
poorly known density of this planet does not permit making better
quantitative estimates.

3. COMETARY ICES

There is various evidence suggesting that at least some icy
cometary bodies did not form directly in and from the solar
nebula but that they are agglomerates of smaller icy grains which
have been irradiated for a time longer than that normally
associated with the formation of large icy bodies in the primeval
nebula. One argument is based on the presence in comets of S_2
rather than the expected S_8; the former requiring intensive UV
irradiation. The presence of sulfur might alter the surface
OH^--sites into SH^--sites with various interesting consequences.
Similarly the presence of HCO, formed from H_2CO, in several comets
suggests that they originated not in the solar nebula but in an
interstellar cloud associated with it. The oxygen content of
comets is better explained as a product of dissociation of H_2O
rather than of CO_2. Assuming that cometary bodies contain H_2O
and CO_2 but no CO it is possible to conclude that these bodies
were formed around 78 K which, depending on the model of the
nebula, occurred somewhere between the orbits of Saturn and
Uranus. Larger distances would be indicated for N_2 containing
cometary nuclei. Further information about the origin of comets
could be obtained if the ratio of deuterated to normal water-ice
in their nuclei were known. Formation of H_2CO in cometary nuclei
by cosmic rays can lead to an estimate of the flux of the latter
in the outer reaches of the solar nebula. Certain cometary
bodies and meteorites may have formed from the same icy
interstellar dust but gradually lost many volatiles. The
"missing carbon" in comets is presumably accounted for by complex
organic compounds. A detailed study has been made of the
distribution of volatile and refractory grains in comae and of
their life times at various heliocentric distances. In

particular, a model for the formation of the relatively high
density Brownlee particles from much less dense aggregates has
been proposed. Whipple's dirty ice model of cometary nuclei
appears to be as fruitful as ever. The presence of H_2O-ice on
P/Bowell and perhaps also of CO_2-ice on P/Kohutek has been
established. The UV albedo of the cometary grains has been
measured. In general the dynamics, the parent molecules, the
reaction rates and the life times of various molecular species in
the comae are still poorly understood and much work in this area
is needed. Cometary brightness of various comets appears to fit
the behavior of H_2O- as well as CO_2-ices. No comet is known
whose brightness would be controlled by vaporization of CH_4 or
CO ices or of condensed N_2.

The heat transport, phase changes and densification of ices
in cometary nuclei seems to be rather well understood. The
results agree with the observed brightness asymmetry of the comae
and tails, and with the sudden explosions of comets such as
P/Holmes and cometary splittings. One can define a thermal time
constant of an icy cometary nucleus and deduce the internal
temperature. This result permits taking into account rotation
of the nucleus and making predictions about the aging processes
such as fading. The feedback of heat between dusty coma and
nucleus has been investigated in detail indicating that the
nightside temperature is strongly affected by the dust while the
daytime side is not. Applications of these concepts to P/Halley
have been made. The lag between subsolar direction and the
direction of maximum activity appears to be correlated with
evaporation through a dusty layer. The P/SW1 may be an example
of this effect.

4. ICES IN PLANETARY RINGS AND INTERPLANETARY GRAINS

Many important aspects of icy particles in planetary rings
have been investigated. In particular, microwave studies show
that the thickness of the ice layer on the large particles in
Saturn's ring is about 30±10 cm. The size distribution of these
particles follows a r^{-3} law which has an upper cut-off so that
the visible ring surface is dominated by particles in the cm
range while the ring's total mass by those in the 5-10 m range.
The rings are 10-100 m thick, the relative velocities of particles
being of the order of cm s^{-1}. The larger particles are, however,
fairly loosely bound aggregates of smaller units so that even
weak collisions break them up; the smaller particles, in turn,
can stick together and accrete to larger sizes or clusters.
There is thus a continuous flux of matter between the particles
and across the rings including deposition of freshly sputtered
water molecules so that the particles, which are probably made of
amorphous ice, are continuously restructured and renewed. The

intriguing polarization anomalies of the rings may be related to these processes. The mechanics of these processes, the restitution coefficients, the erosion and sputtering by ions and electrons, the change of albedo, etc. have been analyzed in considerable detail. Spectral studies suggest a 50 μm water-ice frost containing some deeply imbedded carbon and silicate impurities. The presence or absence of halos around the Sun, when seen through the rings, may indicate whether these rings are dominated by crystalline or by amorphous water-ice. Laboratory studies have shown that Mie theory accounts well for the scattering of IR by amorphous ice particles.

The rings of Jupiter are not icy because of their flat, slightly reddish, spectrum, lack of absorption lines and low albedo. The same is true of Uranian rings which have a low albedo and no characteristic IR features. They are, however, very likely made of methane-ice from which the hydrogen atoms in the surface layer have been driven off leaving a dark carbon deposit behind. The icy particles of the E-ring, which encloses the orbit of Enceladus and which presumably originate in meteoroid impacts on the satellite, are strongly bombarded and have a life time of 10^2 to 10^5 years.

A great deal of theoretical and experimental work has been done to elucidate the early stages and further evolution of icy grains. Investigations of the 3 μm band on ice-covered, i.e. layered, grains permits a study of their evolution. Computer models indicate that the mantles of these grains may contain other ices besides water-ice and be enriched in molecules such as CO_2, CO, etc. Laboratory studies of photoprocesses and of other irradiations, such as by solar wind or by solar flare protons, clarify the mechanisms of chemical changes, of the formation of free radicals and of the escape of volatiles. The explosive recombination reactions involving free radicals may lead to the scattering of the processed matter in space and to a recycling time of the order of 10^9 years. The effect of irradiations and entrapped gases affects also the sticking probability of grains and thus the rate of formation of larger bodies.

5. CONCLUSION

The above summary of the highlights of the conference indicates the importance of ices for the understanding of many aspects of the origin and evolution of the solar system. The opportunity to hear and to discuss freely the various topics played an essential role in furthering the creative exchange of ideas.

Indexes

REMARKS :

  If a page number is underlined, it indicates that the reference is to an author of a paper (Author names index) or to an object or a subject listed in the title of a paper (Object names index or Subject index).

  When a word appears several times in a paper, the index indicates only the first occurence of this word in the paper.

AUTHOR INDEX

A

Abebe, M. : 37.
Ackerson, K. L. : 335.
Adams, N. G. : 244.
Adamson, A. W. : 88, 417.
A'Hearn, M. F. : 195, 200, 243, 259, 272, 312, 354, 364,
                371, 383, 393, 394, 417, 444, 452, 453
                463, 469, 473, 483, 507, 515, 668
Ahern, J. L. : 855
Ahrens, T. J. : 127, 628, 631, 708.
Aiello, S. : 198.
Aikin, A. C. : 756.
Ajello, J. M. : 675, 740, 856.
Ajello, M. M. : 332.
Alben, R. : 88.
Alfven, H. : 241.
Allamandola, L. J. : 198, 271, 394, 470.
Allen, C. C. : 603.
Allen, K. W. : 77.
Allen, M. A. : 756.
Aller, L. H. : 654.
Allison, M. L. : 766, 854.
Altenhoff, W. J. : 219, 474, 484.
Al'Tshuler, L. V. : 145.
Alvarez, J. M. : 257.
Anderman, I. : 106.
Anders, E. : 217, 234, 240, 587, 602, 851.
Andersland, O. B. : 580.
Anderson, D. L. : 653.
Anderson, D. M. : 565, 602.
Anderson, G. D. : 127, 636.
Anderson, H. H. : 293.
Anderson, J. D. : 709.
Angell, C. A. : 34.
Anicich, V. G. : 271.
Apt, J. : 628, 818.
Arakawa, E. T. : 827.
Arduini, M. : 440.
Armstrong, T. P. : 297, 312, 331, 689.
Arpigny, C. : 346, 483.
Arrhenius, G. : 241.
Arvidson, R. E. : 563.
Asay, J. R. : 146.

867

OBJECT INDEX

Names of artificial and natural objects in the
Solar System with structures on them, and other astronomical objects. Names of asteroids, comets, meteor streams
and meteorites can be found under the corresponding general headings ; structures on (and rings around) celestial bodies can be found under the names of these bodies.

A

Adrastea (J 14) : 670.
Amalthea : 632, 672, 699.
Apollo spacecrafts : 792.
Ariel : 623, 711, 848.
Asteroids :
  - Ceres (1) : 173, 703.
  - Chiron (2060) : 172, 323.
  - Hektor (624) : 169.
  - 1983 TB : 350.
Atlas (S 17) : 659, 694, 707.

B

Bechlin-Neugebauer object : 392.

C

Callisto : 89, 109, 174, 449, 622, 632, 715, 733, 760,
           767, 781, 831, 861.
Callisto (Structures on -) :
  - Asgard : 781, 840.
  - Valhalla : 781, 840, 861.
Calypso (S 14) : 694, 707.
Charon : 328, 824, 850.
Comets :
  - Arend-Roland (1957 III) : 506.
  - Austin (1982 VI) : 458.
  - Bennett (1970 II) : 372, 506.
  - P/Biela : 353.
  - P/Borrelly : 454.
  - Borelly (1903 IV) : 377.
  - Borrelly-Brooks (1900 II) : 377.
  - Bowell (1982 I) : 39, 257, 350, 443, 453, 507, 863.
  - P/Bowell-Skiff : 178.

903

A